Coding the Matrix

Linear Algebra through Applications to Computer Science

Edition 1

PHILIP N. KLEIN
Brown University

Newtonian Press, 2013

The companion website is at `codingthematrix.com`. There you will find, in digital form, the data, examples, and support code you need to solve the problems given in the book. Auto-grading will be provided for many of the problems.

Contents

Introduction

> Tourist on Fifty-Seventh Street, Manhattan: "Pardon me—could you tell me how to get to Carnegie Hall?"
>
> Native New Yorker: "Practice, practice!"

There's a scene in the movie *The Matrix* in which Neo is strapped in a chair and Morpheus inserts into a machine what looks like a seventies-era videotape cartridge. As the tape plays, knowledge of how to fight streams into Neo's brain. After a very short time, he has become an expert.

I would be delighted if I could strap my students into chairs and quickly stream knowledge of linear algebra into their brain, but brains don't learn that way. The input device is rarely the bottleneck. Students need lots of practice—but what kind of practice?

No doubt students need to practice the basic numerical calculations, such as matrix-matrix multiplication, that underlie elementary linear algebra and that seem to fill all their time in traditional cookbook-style courses on linear algebra. No doubt students need to find proofs and counterexamples to exercise their understanding of the abstract concepts of which linear algebra is constructed.

However, they also need practice in using linear algebra to think about problems in other domains, and in actually *using* linear-algebra computations to address these problems. These are the skills they most need from a linear-algebra class when they go on to study other topics such as graphics and machine learning. This book is aimed at students of computer science; such students are best served by seeing applications from their field because these are the applications that will be most meaningful for them.

Moreover, a linear-algebra instructor whose pupils are students of computer science has a special advantage: her students are computationally sophisticated. They have a learning modality that most students don't—they can learn through reading, writing, debugging, and using computer programs.

For example, there are several ways of writing a program for matrix-vector or matrix-matrix multiplication, each providing its own kernel of insight into the meaning of the operation—and the experience of writing such programs is more effective in conveying this meaning and cementing the relationships between the operations than spending the equivalent time carrying out hand calculations.

Computational sophistication also helps students in the more abstract, mathematical aspects of linear algebra. Acquaintance with object-oriented programming helps a student grasp the notion of a *field*—a set of values together with operations on them. Acquaintance with subtyping prepares a student to understand that some vector spaces are inner product spaces. Familiarity with loops or recursion helps a student understand procedural proofs, e.g. of the existence of a basis or an orthogonal basis.

Computational thinking is the term suggested by Jeannette Wing, former head of the National Science Foundation's directorate on Computer and Information Science and Engineering, to refer to the skills and concepts that a student of computer science can bring to bear. For this book, computational thinking is the road to mastering elementary linear algebra.

Companion website

The companion website is at `codingthematrix.com`. There you will find, in digital form, the data, examples, and support code you need to solve the problems given in the book.

Intended audience

This book is accessible to a student who is an experienced programmer. Most students who take my course have had at least two semesters of introductory computer science, or have previously learned programming on their own. In addition, it is desirable that the student has has some exposure (in prior semesters or concurrently) in proof techniques such as are studied in a Discrete Math course.

The student's prior programming experience can be in pretty much any programming language; this book uses Python, and the first two labs are devoted to bringing the student up to speed in Python programming. Moreover, the programs we write in this book are not particularly sophisticated. For example, we provide stencil code that obviates the need for the student to have studied object-oriented programming.

Some sections of the text, marked with *, provide supplementary mathematical material but are not crucial for the reader's understanding.

Labs

An important part of the book is the labs. For each chapter, there is a lab assignment in which the student is expected to write small programs and use some modules we provide, generally to carry out a task or series of tasks related to an application of the concepts recently covered or about to be covered. Doing the labs "keeps it real", grounding the student's study of linear algebra in getting something done, something meaningful in its own right but also illustrative of the concepts.

In my course, there is a lab section each week, a two-hour period in which the students carry out the lab assignment. Course staff are available during this period, not to supervise but to assist when necessary. The goal is to help the students move through the lab assignment efficiently, and to get them unstuck when they encounter obstacles. The students are expected to have prepared by reviewing the previous week's course material and reading through the lab assignment.

Most students experience the labs as the most fun part of the course—it is where they discover the power of the knowledge they are acquiring to help them accomplish something that has meaning in the world of computer science.

Programming language

The book uses Python, not a programming language with built-in support for vectors and matrices. This gives students the opportunity to build vectors and matrices out of the data structures that Python does provide. Using their own implementations of vector and matrix provides transparency. Python does provide complex numbers, sets, lists (sequences), and dictionaries (which we use for representing functions). In addition, Python provides *comprehensions*, expressions that create sets, lists, or dictionaries using a simple and powerful syntax that resembles the mathematical notation for defining sets. Using this syntax, many of the procedures we write require only a single line of code.

Students are not expected to know Python at the beginning; the first two labs form an introduction to Python, and the examples throughout the text reinforce the ideas.

Vector and Matrix representations

The traditional concrete representation for a vector is as a sequence of field elements. This book uses that representation but also uses another, especially in Python programs: a vector as a function mapping a finite set D to a field. Similarly, the traditional representation for a matrix is as a two-dimensional array or grid of field elements. We use this representation but also use another: a matrix as a function from the Cartesian product $R \times C$ of two finite sets to a field.

These more general representations allow the vectors and matrices to be more directly connected to the application. For example, it is traditional in information retrieval to represent a document as a vector in which, for each word, the vector specifies the number of occurences of the word in the document. In this book, we define such a vector as a function from the domain D of English words to the set of real numbers. Another example: when representing, say, a 1024×768 black-and-white image as a vector, we define the vector as a function from the domain $D = \{1, \ldots, 1024\} \times \{1, .., 768\}$ to the real numbers. The function specifies, for each pixel (i, j), the image intensity of that pixel.

From the programmer's perspective, it is certainly more convenient to directly index vectors by strings (in the case of words) or tuples (in the case of pixels). However, a more important advantage is this: having to choose a domain D for vectors gets us thinking about the application from the vector perspective.

Another advantage is analogous to that of type-checking in programs or unit-checking in physical calculations. For an $R \times C$ matrix A, the matrix-vector product $A\boldsymbol{x}$ is only legal if $\boldsymbol{x}$ is a C-vector; the matrix-matrix product AB is only legal if C is the set of row-labels of B. These constraints further reinforce the meanings of the operations.

Finally, allowing arbitrary finite sets (not just sequences of consecutive integers) to label the elements helps make it clear that the order of elements in a vector or matrix is not always (or even often) significant.

Fundamental Questions

The book is driven not just by applications but also by fundamental questions and computational problems that arise in studying these applications. Here are some of the fundamental questions:

- How can we tell whether a solution to a linear system is unique?
- How can we find the number of solutions to a linear system over $GF(2)$?
- How can we tell if a set $\mathcal{V}$ of vectors is equal to the span of vectors $\boldsymbol{v}_1, \ldots, \boldsymbol{v}_n$?
- For a system of linear equations, what other linear equations are implied?
- How can we tell if a matrix is invertible?
- Can every vector space be represented as the solution set of a homogeneous linear system?

Fundamental Computational Problems

There are a few computational problems that are central to linear algebra. In the book, these arise in a variety of forms as we examine various applications, and we explore the connections between them. Here are examples:

- Find the solution to a matrix equation $M\boldsymbol{x} = \boldsymbol{b}$.
- Find the vector $\boldsymbol{x}$ minimizing the distance between $M\boldsymbol{x}$ and $\boldsymbol{b}$.
- Given vector $\boldsymbol{b}$, find the closest vector to $\boldsymbol{b}$ whose representation in a given basis is k-sparse.
- Find the solution to a matrix inequality $M\boldsymbol{x} \leq \boldsymbol{b}$.
- Given a matrix M, find the closest matrix to M whose rank is at most k.

Multiple representations

The most important theme of this book is the idea of multiple different representations for the same object. This theme should be familiar to computer scientists. In linear algebra, it arises again and again:

- Representing a vector space by generators or by homogeneous linear equations.
- Different bases for the same vector space.

- Different data structures used to represent a vector or a matrix.
- Different decompositions of a matrix.

Multiple fields

In order to illustrate the generality of the ideas of linear algebra and in order to address a broader range of applications, the book deals with three different fields: the real numbers, the complex numbers, and the finite field $GF(2)$. Most examples are over the real numbers because they are most familiar to the reader. The complex numbers serve as a warm-up for vectors since they can be used to represent points in the plane and transformations on these points. The complex numbers also come up in the discussion of the finite Fourier transform and in eigenvalues. The finite field $GF(2)$ comes up in many applications involving information, such as encryption, authentication, checksums, network coding, secret-sharing, and error-correcting codes.

The multiple codes help to illustrate the idea of an inner-product space. There is a very simple inner product for vectors over the reals; there is a slightly more complicated inner product for vectors over the complex numbers; and there is no inner-product for vectors over a finite field.

Acknowledgements

Thanks to my Brown University colleague John F. Hughes, computer scientist and recovering mathematician. I have learned a great deal from our conversations, and this book owes much to him.

Thanks to my Brown University colleague Dan Abramovich, mathematician, who has shared his insights on the tradeoffs between abstraction and simplicity of presentation.

Thanks to the students who have served as teaching assistants for the Brown University course on which this book is based and who have helped prepare some of the problems and labs, especially Sarah Meikeljohn, Shay Mozes, Olga Ohrimenko, Matthew Malin, Alexandra Berke, Anson Rosenthal, and Eli Fox-Epstein.

Thanks to Rosemary Simpson for her work on the index for this book.

Thanks to the creator of xkcd, Randall Munroe, for giving permission to include some of his work in this book.

Thanks, finally, to my family for their support and understanding.

Chapter 0

The Function (and other mathematical and computational preliminaries)

> Later generations will regard
> Mengenlehre [set theory] as a disease
> from which one has recovered.
> attributed to Poincáre

The basic mathematical concepts that inform our study of vectors and matrices are sets, sequences (lists), functions, and probability theory.

This chapter also includes an introduction to Python, the programming language we use to (i) model the mathematical objects of interest, (ii) write computational procedures, and (iii) carry out data analyses.

0.1 Set terminology and notation

The reader is likely to be familiar with the idea of a *set*, a collection of mathematical objects in which each object is considered to occur at most once. The objects belonging to a set are its *elements*. We use curly braces to indicate a set specified by explicitly enumerating its elements. For example, $\{\heartsuit, \spadesuit, \clubsuit, \diamondsuit\}$ is the set of suits in a traditional deck of cards. The order in which elements are listed is not significant; a set imposes no order among its elements.

The symbol $\in$ is used to indicate that an object belongs to a set (equivalently, that the set *contains* the object). For example, $\heartsuit \in \{\heartsuit, \spadesuit, \clubsuit, \diamondsuit\}$.

One set S_1 is *contained in* another set S_2 (written $S_1 \subseteq S_2$) if every element of S_1 belongs to S_2. Two sets are equal if they contain exactly the same elements. A convenient way to prove that two sets are equal consists of two steps: (1) prove the first set is contained in the second, and (2) prove the second is contained in the first.

A set can be infinite. In Chapter 1, we discuss the set $\mathbb{R}$, which consists of all real numbers, and the set $\mathbb{C}$, which consists of all complex numbers.

If a set S is not infinite, we use $|S|$ to denote its *cardinality*, the number of elements it contains. For example, the set of suits has cardinality 4.

0.2 Cartesian product

One from column A, one from column B.

The *Cartesian product* of two sets A and B is the set of all pairs (a, b) where $a \in A$ and $b \in B$.

Example 0.2.1: For the sets $A = \{1, 2, 3\}$ and $B = \{\heartsuit, \spadesuit, \clubsuit, \diamondsuit\}$, the Cartesian product is

$$\{(1, \heartsuit), (2, \heartsuit), (3, \heartsuit), (1, \spadesuit), (2, \spadesuit), (3, \spadesuit), (1, \clubsuit), (2, \clubsuit), (3, \clubsuit), (1, \diamondsuit), (2, \diamondsuit), (3, \diamondsuit)\}$$

Quiz 0.2.2: What is the cardinality of $A \times B$ in Example 0.2.1 (Page 2)?

Answer

$|A \times B| = 12$.

Proposition 0.2.3: For finite sets A and B, $|A \times B| = |A| \cdot |B|$.

Quiz 0.2.4: What is the cardinality of $\{1, 2, 3, \ldots, 10, J, Q, K\} \times \{\heartsuit, \spadesuit, \clubsuit, \diamondsuit\}$?

Answer

We use Proposition 0.2.3. The cardinality of the first set is 13, and the cardinality of the second set is 4, so the cardinality of the Cartesian product is $13 \cdot 4$, which is 52.

The Cartesian product is named for René Descartes, whom we shall discuss in Chapter 6.

0.3 The function

Mathematicians never die—they just lose function.

Loosely speaking, a function is a rule that, for each element in some set D of possible inputs, assigns a possible output. The output is said to be the *image* of the input under the function and the input is a *pre-image* of the output. The set D of possible inputs is called the *domain* of the function.

Formally, a *function* is a (possibly infinite) set of pairs (a, b) no two of which share the same first entry.

Example 0.3.1: The doubling function with domain $\{1, 2, 3, \ldots\}$ is

$$\{(1, 2), (2, 4), (3, 6), (4, 8), \ldots\}$$

The domain can itself consist of pairs of numbers.

Example 0.3.2: The multiplication function with domain $\{1, 2, 3, \ldots\} \times \{1, 2, 3, \ldots\}$ looks something like this:

$$\{((1, 1), 1), ((1, 2), 2), \ldots, ((2, 1), 2), ((2, 2), 4), ((2, 3), 6), \ldots\}$$

For a function named f, the image of q under f is denoted by $f(q)$. If $r = f(q)$, we say that q *maps to* r *under* f. The notation for "q maps to r" is $q \mapsto r$. (This notation omits specifying the function; it is useful when there is no ambiguity about which function is intended.)

It is convenient when specifying a function to specify a *co-domain* for the function. The co-domain is a set from which the function's output values are chosen. Note that one has some leeway in choosing the co-domain since not all of its members need be outputs.

The notation

$$f : D \longrightarrow F$$

means that f is a function whose domain is the set D and whose *co-domain* (the set of possible outputs) is the set F. (More briefly: "a function from D to F", or "a function that maps D to F.")

Example 0.3.3: Caesar was said to have used a cryptosystem in which each letter was replaced with the one three steps forward in the alphabet (wrapping around for X,Y, and Z).[a] Thus the plaintext MATRIX would be encrypted as the cyphertext PDWULA. The function that maps each plaintext letter to its cyphertext replacement could be written as

$$A \mapsto D, B \mapsto E, C \mapsto F, D \mapsto G, W \mapsto Z, X \mapsto A, Y \mapsto B, Z \mapsto C$$

This function's domain and co-domain are both the alphabet $\{A, B, \ldots, Z\}$.

[a] Some imaginary historians have conjectured that Caesar's assasination can be attributed to his use of such a weak cryptosystem.

Example 0.3.4: The cosine function, *cos*, maps from the set of real numbers (indicated by $\mathbb{R}$) to the set of real numbers. We would therefore write

$$\cos : \mathbb{R} \longrightarrow \mathbb{R}$$

Of course, the outputs of the cos function do not include all real numbers, only those between -1 and 1.

The *image* of a function f is the set of images of all domain elements. That is, the image of f is the set of elements of the co-domain that actually occur as outputs. For example, the image of Caesar's encryption function is the entire alphabet, and the image of the cosine function is the set of numbers between -1 and 1.

Example 0.3.5: Consider the function *prod* that takes as input a pair of integers greater than 1 and outputs their product. The domain (set of inputs) is the set of pairs of integers greater than 1. We choose to define the co-domain to be the set of all integers greater than 1. The image of the function, however, is the set of *composite* integers since no domain element maps to a prime number.

0.3.1 Functions versus procedures, versus computational problems

There are two other concepts that are closely related to functions and that enter into our story, and we must take some care to distinguish them.

- A *procedure* is a precise description of a computation; it accepts *inputs* (called *arguments*) and produces an output (called the *return value*).

  Example 0.3.6: This example illustrates the Python syntax for defining procedures:

  ```
  def mul(p,q): return p*q
  ```

  In the hope of avoiding confusion, we diverge from the common practice of referring to procedures as "functions".

- A *computational problem* is an input-output specification that a procedure might be required to satisfy.

Example 0.3.7: – *input:* a pair (p, q) of integers greater than 1
– *output:* the product pq

Example 0.3.8:

– *input:* an integer m greater than 1
– *output:* a pair (p, q) of integers whose product is m

How do these concepts differ from one another?

- Unlike a procedure, a function or computational problem does not give us any idea how to compute the output from the input. There are often many different procedures that satisfy the same input-output specification or that implement the same function. For integer multiplication, there is ordinary long multiplication (you learned this in elementary school), the Karatsuba algorithm (used by Python for long-integer multiplication), the faster Schönhage-Strassen algorithm (which uses the Fast Fourier Transform, discussed in Chapter 10), and the even faster Fürer algorithm, which was discovered in 2007.
- Sometimes the same procedure can be used for different functions. For example, the Python procedure `mul` can be used for multiplying negative integers and numbers that are not integers.
- Unlike a function, a computational problem need not specify a unique output for every input; for Example 0.3.8 (Page 4), if the input is 12, the output could be $(2, 6)$ or $(3, 4)$ or $(4, 3)$ or $(6, 2)$.

0.3.2 The two computational problems related to a function

All the king's horses and all the king's men
Couldn't put Humpty together again.

Although *function* and *computational problem* are defined differently, they are clearly related. For each function f, there is a corresponding computational problem:

The forward problem: Given an element a of f's domain, compute $f(a)$, the image of a under f.

Example 0.3.7 (Page 4) is the computational problem that corresponds in this sense to the function defined in Example 0.3.2 (Page 2).

However, there is another computational problem associated with a function:

The backward problem: Given an element r of the co-domain of the function, compute any pre-image (or report that none exists).

How very different are these two computational problems? Suppose there is a procedure $P(x)$ for computing the image under f of any element of the domain. An obvious procedure for computing the pre-image of r is to iterate through each of the domain elements q, and, one by one, apply the procedure $P(x)$ on q to see if the output matches r.

This approach seems ridiculously profligate—even if the domain is finite, it might be so large that the time required for solving the pre-image problem would be much more than that for $P(x)$—and yet there is no better approach that works for all functions.

Indeed, consider Example 0.3.7 (Page 4) (integer multiplication) and Example 0.3.8 (Page 4) (integer factoring). The fact that integer multiplication is computationally easy while integer factoring is computationally difficult is in fact the basis for the security of the RSA cryptosystem, which is at the heart of secure commerce over the world-wide web.

And yet, as we will see in this book, finding pre-images can be quite useful. What is one to do?

In this context, the generality of the concept of function is also a weakness. To misquote *Spiderman*,

With great generality comes great computational difficulty.

This principle suggests that we consider the pre-image problem not for arbitrary functions but for specific families of functions. Yet here too there is a risk. If the family of functions is too restrictive, the existence of fast procedures for solving the pre-image problem will have no relevance to real-world problems. We must navigate between the Scylla of computational intractability and the Charybdis of inapplicability.

In linear algebra, we will discover a sweet spot. The family of *linear functions*, which are introduced in Chapter 4, manage to model enough of the world to be immensely useful. At the same time, the pre-image problem can be solved for such functions.

0.3.3 Notation for the set of functions with given domain and co-domain

For sets D and F, we use the notation F^D to denote all functions from D to F. For example, the set of functions from the set W of words to the set $\mathbb{R}$ of real numbers is denoted $\mathbb{R}^W$.

This notation derives from a mathematical "pun":

Fact 0.3.9: For any finite sets D and F, $|D^F| = |D|^{|F|}$.

0.3.4 Identity function

For any domain D, there is a function $\text{id}_D : D \longrightarrow D$ called the *identity function* for D, defined by

$$\text{id}_D(d) = d$$

for every $d \in D$.

0.3.5 Composition of functions

The operation *functional composition* combines two functions to get a new function. We will later define matrix multiplication in terms of functional composition. Given two functions $f : A \longrightarrow B$ and $g : B \longrightarrow C$, the function $g \circ f$, called the composition of g and f, is a function whose domain is A and its co-domain is C. It is defined by the rule

$$(g \circ f)(x) = g(f(x))$$

for every $x \in A$.

If the image of f is not contained in the domain of g then $g \circ f$ is not a legal expression.

Example 0.3.10: Say the domain and co-domains of f and g are $\mathbb{R}$, and $f(x) = x + 1$ and $g(y) = y^2$. Then $g \circ f(x) = (x + 1)^2$.

Example 0.3.11: Define the function

$$f : \{A, B, C, \ldots, Z\} \longrightarrow \{0, 1, 2, \ldots, 25\}$$

by

$$A \mapsto 0, B \mapsto 1, C \mapsto 2, \cdots, Z \mapsto 25$$

Define the function g as follows. The domain and co-domain of g are both the set $\{0, 1, 2, \ldots, 25\}$, and $g(x) = (x + 3) \bmod 26$. For a third function h, the domain is $\{0, ...25\}$ and the co-domain is $\{A, ..., Z\}$, and $0 \mapsto A$, $1 \mapsto B$, etc. Then $h \circ (g \circ f)$ is a function that implements the Caesar cypher as described in Example 0.3.3 (Page 3).

For building intuition, we can use a diagram to represent composition of functions with finite domains and co-domains. Figure 1 depicts the three functions of Example 0.3.11 (Page 5) being composed.

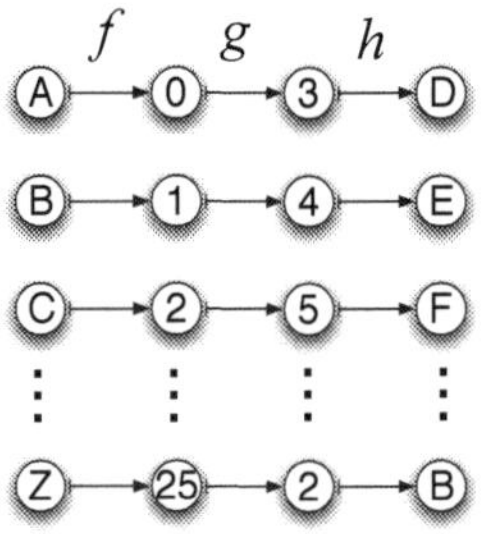

Figure 1: This figure represents the composition of the functions f, g, h. Each function is represented by arrows from circles representing its domain to circles representing its co-domain. The composition of the three functions is represented by following three arrows.

0.3.6 Associativity of function composition

Next we show that composition of functions is *associative*:

Proposition 0.3.12 (Associativity of composition): For functions f, g, h,

$$h \circ (g \circ f) = (h \circ g) \circ f$$

if the compositions are legal.

Proof

Let x be any member of the domain of f.

$$\begin{aligned} (h \circ (g \circ f))(x) &= h((g \circ f)(x)) \text{ by definition of } h \circ (g \circ f)) \\ &= h(g(f(x))) \text{ by definition of } g \circ f \\ &= (h \circ g)(f(x)) \text{ by definition of } h \circ g \\ &= ((h \circ g) \circ f)(x) \text{ by definition of } (h \circ g) \circ f \end{aligned}$$

□

Associativity means that parentheses are unnecessary in composition expression: since $h \circ (g \circ f)$ is the same as $(h \circ g) \circ f$, we can write either of them as simply $h \circ g \circ f$.

0.3.7 Functional inverse

Let us take the perspective of a lieutenant of Caesar who has received a cyphertext: PDWULA. To obtain the plaintext, the lieutenant must find for each letter in the cyphertext the letter that maps to it under the encryption function (the function of Example 0.3.3 (Page 3)). That is, he must find the letter that maps to P (namely M), the letter that maps to D (namely A), and so on. In doing so, he can be seen to be applying *another* function to each of the letters of the cyphertext, specifically the function that reverses the effect of the encryption function. This function is said to be the *functional inverse* of the encryption function.

For another example, consider the functions f and h in Example 0.3.11 (Page 5): f is a function from $\{A, \ldots, Z\}$ to $\{0, \ldots, 25\}$ and h is a function from $\{0, \ldots, 25\}$ to $\{A, \ldots, Z\}$. Each one reverses the effect of the other. That is, $h \circ f$ is the identity function on $\{A, \ldots, Z\}$, and $f \circ h$ is the identity function on $\{0, \ldots, 25\}$. We say that h is the functional inverse of f. There is no reason for privileging f, however; f is the functional inverse of h as well.

In general,

Definition 0.3.13: We say that functions f and g are *functional inverses* of each other if

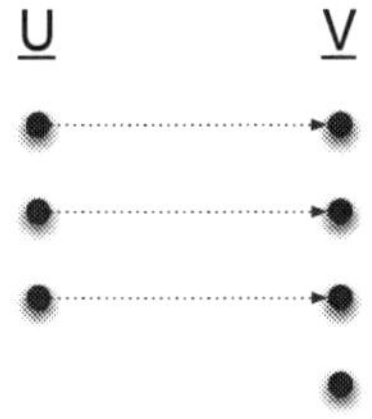

Figure 2: A function $f : U \to V$ is depicted that is not onto, because the fourth element of the co-domain is not the image under f of any element

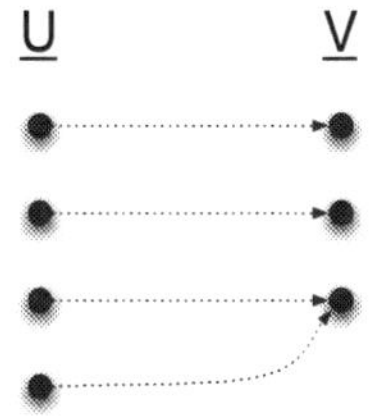

Figure 3: A function $f : U \to V$ is depicted that is not one-to-one, because the third element of the co-domain is the image under f of more than one element.

- $f \circ g$ is defined and is the identity function on the domain of g, and
- $g \circ f$ is defined and is the identity function on the domain of f.

Not every function has an inverse. A function that has an inverse is said to be *invertible*. Examples of noninvertible functions are shown in Figures 2 and 3

Definition 0.3.14: Consider a function $f : D \longrightarrow F$. We say that f is *one-to-one* if for every $x, y \in D$, $f(x) = f(y)$ implies $x = y$. We say that f is *onto* if, for every $z \in F$, there exists $x \in D$ such that $f(x) = z$.

Example 0.3.15: Consider the function *prod* defined in Example 0.3.5 (Page 3). Since a prime number has no pre-image, this function is not onto. Since there are multiple pairs of integers, e.g. $(2, 3)$ and $(3, 2)$, that map to the same integer, the function is also not one-to-one.

Lemma 0.3.16: An invertible function is one-to-one.

Proof

Suppose f is not one-to-one, and let x_1 and x_2 be distinct elements of the domain such that $f(x_1) = f(x_2)$. Let $y = f(x_1)$. Assume for a contradiction that f is invertible. The definition of inverse implies that $f^{-1}(y) = x_1$ and also $f^{-1}(y) = x_2$, but both cannot be true. □

Lemma 0.3.17: An invertible function is onto.

Proof

Suppose f is not onto, and let $\hat{y}$ be an element of the co-domain such that $\hat{y}$ is not the image of any domain element. Assume for a contradiction that f is invertible. Then $\hat{y}$ has an image $\hat{x}$ under f^{-1}. The definition of inverse implies that $f(\hat{x}) = \hat{y}$, a contradiction. □

Theorem 0.3.18 (Function Invertibility Theorem): A function is invertible iff it is one-to-one and onto.

Proof

Lemmas 0.3.16 and 0.3.17 show that an invertible function is one-to-one and onto. Suppose conversely that f is a function that is one-to-one and onto. We define a function g whose domain is the co-domain of f as follows:

> For each element $\hat{y}$ of the co-domain of f, since f is onto, f's domain contains some element $\hat{x}$ for which $f(\hat{x}) = \hat{y}$; we define $g(\hat{y}) = \hat{x}$.

We claim that $g \circ f$ is the identity function on f's domain. Let $\hat{x}$ be any element of f's domain, and let $\hat{y} = f(\hat{x})$. Because f is one-to-one, $\hat{x}$ is the only element of f's domain whose image under f is $\hat{y}$, so $g(\hat{y}) = \hat{x}$. This shows $g \circ f$ is the identity function.

We also claim that $f \circ g$ is the identity function on g's domain. Let $\hat{y}$ be any element of g's domain. By the definition of g, $f(g(\hat{y})) = \hat{y}$. □

Lemma 0.3.19: Every function has at most one functional inverse.

Proof

Let $f : U \to V$ be an invertible function. Suppose that g_1 and g_2 are inverses of f. We show that, for every element $v \in V$, $g_1(v) = g_2(v)$, so g_1 and g_2 are the same function.

Let $v \in V$ be any element of the co-domain of f. Since f is onto (by Lemma 0.3.17), there is some element $u \in U$ such that $v = f(u)$. By definition of inverse, $g_1(v) = u$ and $g_2(v) = u$. Thus $g_1(v) = g_2(v)$. □

0.3.8 Invertibility of the composition of invertible functions

In Example 0.3.11 (Page 5), we saw that the composition of three functions is a function that implements the Caesar cypher. The three functions being composed are all invertible, and the result of composition is also invertible. This is not a coincidence:

Lemma 0.3.20: If f and g are invertible functions and $f \circ g$ exists then $f \circ g$ is invertible and $(f \circ g)^{-1} = g^{-1} \circ f^{-1}$.

Problem 0.3.21: Prove Lemma 0.3.20.

Problem 0.3.22: Use diagrams like those of Figures 1, 2, and 3 to specify functions g and f that are a counterexample to the following:

False Assertion 0.3.23: Suppose that f and g are functions and $f \circ g$ is invertible. Then f and g are invertible.

□

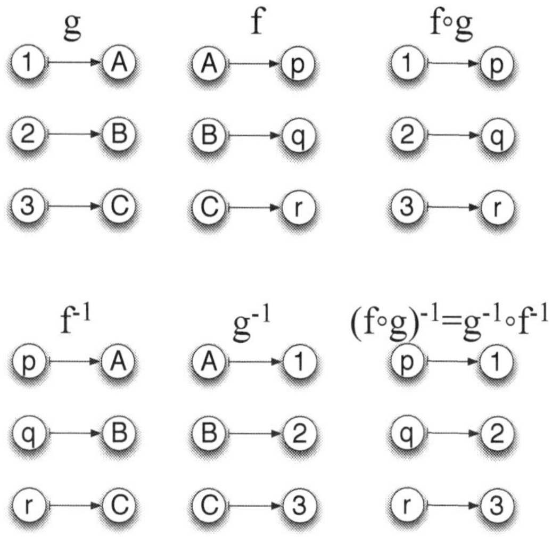

Figure 4: The top part of this figure shows two invertible functions f and g, and their composition $f \circ g$. Note that the composition $f \circ g$ is invertible. This illustrates Lemma 0.3.20. The bottom part of this figure shows g^{-1}, f^{-1} and $(f \circ g)^{-1}$. Note that $(f \circ g)^{-1} = g^{-1} \circ f^{-1}$. This illustrates Lemma 0.3.20.

0.4 Probability

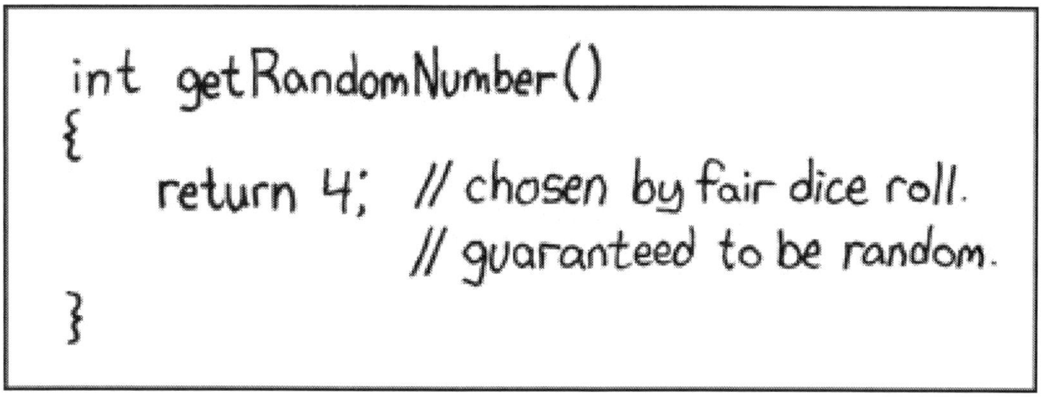

Random Number (`http://xkcd.com/221/`)

One important use of vectors and matrices arises in probability. For example, this is how they arise in Google's PageRank method. We will therefore study very rudimentary probability theory in this course.

In probability theory, nothing ever happens—probability theory is just about what *could* happen, and how likely it is to happen. Probability theory is a calculus of probabilities. It is used to make predictions about a hypothetical experiment. (Once something actually happens, you use *statistics* to figure out what it means.)

0.4.1 Probability distributions

A function $\Pr(\cdot)$ from a finite domain Ω to the set $\mathbb{R}^+$ of nonnegative reals is a *(discrete) probability distribution* if $\sum_{\omega \in \Omega} \Pr(\omega) = 1$. We refer to the elements of the domain as *outcomes*. The image of an outcome under $\Pr(\cdot)$ is called the *probability* of the outcome. The probabilities are supposed to be proportional to the *relative likelihoods* of outcomes. Here I use the term *likelihood* to mean the common-sense notion, and *probability* to mean the mathematical abstraction of it.

Psychic, `http://xkcd.com/628/`

Uniform distributions

For the simplest examples, all the outcomes are equally likely, so they are all assigned the same probabilities. In such a case, we say that the probability distribution is *uniform*.

Example 0.4.1: To model the flipping of a single coin, $\Omega = \{\text{heads}, \text{tails}\}$. We assume that the two outcomes are equally likely, so we assign them the same probability: $\Pr(\text{heads}) = \Pr(\text{tails})$. Since we require the sum to be 1, $\Pr(\text{heads}) = 1/2$ and $\Pr(\text{tails}) = 1/2$. In Python, we would write the probability distribution as

```
>>> Pr = {'heads':1/2, 'tails':1/2}
```

Example 0.4.2: To model the roll of a single die, $\Omega = \{1, 2, 3, 4, 5, 6\}$, and $\Pr(1) = \Pr(2) = \cdots = \Pr(6)$. Since the probabilities of the six outcomes must sum to 1, each of these probabilities must be $1/6$. In Python,

```
>>> Pr = {1:1/6, 2:1/6, 3:1/6, 4:1/6, 5:1/6, 6:1/6}
```

Example 0.4.3: To model the flipping of two coins, a penny and a nickel, $\Omega = \{HH, HT, TH, TT\}$, and each of the outcomes has the same probability, $1/4$. In Python,

```
>>> Pr = {('H', 'H'):1/4, ('H', 'T'):1/4, ('T','H'):1/4, ('T','T'):1/4}
```

Nonuniform distributions

In more complicated situations, different outcomes have different probabilities.

Example 0.4.4: Let $\Omega = \{A, B, C, \ldots, Z\}$, and let's assign probabilities according to how likely you are to draw each letter at the beginning of a Scrabble game. Here is the number of tiles with each letter in Scrabble:

A	9	B	2	C	2	D	4
E	12	F	2	G	3	H	2
I	9	J	1	K	1	L	1
M	2	N	6	O	8	P	2
Q	1	R	6	S	4	T	6
U	4	V	2	W	2	X	1
Y	2	Z	1				

The likelihood of drawing an R is twice that of drawing a G, thrice that of drawing a C, and six times that of drawing a Z. We need to assign probabilities that are proportional to these likelihoods. We must have some number c such that, for each letter, the probability of drawing that letter should be c times the number of copies of that letter.

$$\Pr[\text{drawing letter X}] = c \cdot \text{number of copies of letter X}$$

Summing over all letters, we get

$$1 = c \cdot \text{total number of tiles}$$

Since the total number of tiles is 95, we define $c = 1/95$. The probability of drawing an E is therefore 12/95, which is about .126. The probability of drawing an A is 9/95, and so on. In Python, the probability distribution is

```
{'A':9/95, 'B':2/95, 'C':2/95, 'D':4/95, 'E':12/95, 'F':2/95,
           'G':3/95, 'H':2/95, 'I':9/95, 'J':1/95, 'K':1/95,  'L':1/95,
           'M':2/95, 'N':6/95, 'O':8/95, 'P':2/95, 'Q':1/95,  'R':6/95,
           'S':4/95, 'T':6/95, 'U':4/95, 'V':2/95, 'W':2/95,  'X':1/95,
           'Y':2/95, 'Z':1/95}
```

0.4.2 Events, and adding probabilities

In Example 0.4.4 (Page 10), what is the probability of drawing a *vowel* from the bag?

A set of outcomes is called an *event*. For example, the event of drawing a vowel is represented by the set $\{A, E, I, O, U\}$.

Principle 0.4.5 (Fundamental Principle of Probability Theory): The probability of an event is the sum of probabilities of the outcomes making up the event.

According to this principle, the probability of a vowel is

$$9/95 + 12/95 + 9/95 + 8/95 + 4/95$$

which is 42/95.

0.4.3 Applying a function to a random input

Now we think about applying a function to a random input. Since the input to the function is random, the output should also be considered random. Given the probability distribution of the input and a specification of the function, we can use probability theory to derive the probability distribution of the output.

Example 0.4.6: Define the function $f : \{1,2,3,4,5,6\} \longrightarrow \{0,1\}$ by

$$f(x) = \begin{cases} 0 & \text{if } x \text{ is even} \\ 1 & \text{if } x \text{ is odd} \end{cases}$$

Consider the experiment in which we roll a single die (as in Example 0.4.2 (Page 10)), yielding one of the numbers in $\{1,2,3,4,5,6\}$, and then we apply $f(\cdot)$ to that number, yielding either a 0 or a 1. What is the probability function for the outcome of this experiment?

The outcome of the experiment is 0 if the rolled die shows 2, 4, or 6. As discussed in Example 0.4.2 (Page 10), each of these possibilies has probability 1/6. By the Fundamental Principle of Probability Theory, therefore, the output of the function is 0 with probability $1/6 + 1/6 + 1/6$, which is $1/2$. Similarly, the output of the function is 1 with probability $1/2$. Thus the probability distribution of the output of the function is `{0: 1/2., 1:1/2.}`.

Quiz 0.4.7: Consider the flipping of a penny and a nickel, described in Example 0.4.3 (Page 10). The outcome is a pair (x, y) where each of x and y is `'H'` or `'T'` (heads or tails). Define the function

$$f : \{(\text{'H'}, \text{'H'})\ (\text{'H'}, \text{'T'}),\ (\text{'T'},\text{'H'}),\ (\text{'T'},\text{'T'})\}$$

by

$$f((x, y)) = \text{the number of H's represented}$$

Give the probability distribution for the output of the function.

Answer

`{0: 1/4., 1:1/2., 2:1/4.}`

Example 0.4.8 (Caesar plays Scrabble): Recall that the function f defined in Example 0.3.11 (Page 5) maps A to 0, B to 1, and so on. Consider the experiment in which f is applied to a letter selected randomly according to the probability distribution described in Example 0.4.4 (Page 10). What is the probability distribution of the output?

Because f is an invertible function, there is one and only one input for which the output is 0, namely A. Thus the probability of the output being 0 is exactly the same as the probability of the input being A, namely $9/95.$. Similarly, for each of the integers 0 through 25 comprising the co-domain of f, there is exactly one letter that maps to that integer, so the probability of that integer equals the probability of that letter. The probability distribution is thus

```
{0:9/95., 1:2/95., 2:2/95., 3:4/95., 4:12/95., 5:2/95.,
          6:3/95., 7:2/95., 8:9/95., 9:1/95., 10:1/95.,  11:1/95.,
          12:2/95., 13:6/95., 14:8/95., 15:2/95., 16:1/95.,  17:6/95.,
          18:4/95., 19:6/95., 20:4/95., 21:2/95., 22:2/95.,  23:1/95.,
          24:2/95., 25:1/95.}
```

The previous example illustrates that, if the function is invertible, the probabilities are preserved: the probabilities of the various outputs match the probabilities of the inputs. It follows that, if the input is chosen according to a uniform distribution, the distribution of the output is also uniform.

Example 0.4.9: In Caesar's Cyphersystem, one encrypts a letter by advancing it three positions. Of course, the number k of positions by which to advance need not be three; it can be any integer from 0 to 25. We refer to k as the *key*. Suppose we select the key k according to the uniform distribution on $\{0, 1, \ldots, 25\}$, and use it to encrypt the letter P. Let $w : \{0, 1, \ldots, 25\} \longrightarrow \{A, B, \ldots, Z\}$ be the the function mapping the key to the cyphertext:

$$\begin{aligned} w(k) &= h(f(P) + k \bmod 26) \\ &= h(15 + k \bmod 26) \end{aligned}$$

The function $w(\cdot)$ is invertible. The input is chosen according to the uniform distribution, so the distribution of the output is also uniform. Thus when the key is chosen randomly, the cyphertext is equally likely to be any of the twenty-six letters.

0.4.4 Perfect secrecy

Cryptography (`http://xkcd.com/153/`)

We apply the idea of Example 0.4.9 (Page 12) to some even simpler cryptosystems. A cryptosystem must satisfy two obvious requirements:

- the intended recipient of an encrypted message must be able to decrypt it, and

- someone for whom the message was not intended should *not* be able to decrypt it.

The first requirement is straightforward. As for the second, we must dispense with a misconception about security of cryptosystems. The idea that one can keep information secure by not revealing the *method* by which it was secured is often called, disparagingly, *security through obscurity*. This approach was critiqued in 1881 by a professor of German, Jean-Guillame-Hubert-Victor-François-Alexandre-August Kerckhoffs von Niewenhof, known as August Kerckhoffs. The *Kerckhoffs Doctrine* is that *the security of a cryptosystem should depend only on the secrecy of the key used, not on the secrecy of the system itself.*

There is an encryption method that meets Kerchoffs' stringent requirement. It is utterly unbreakable if used correctly.[1] Suppose Alice and Bob work for the British military. Bob is the commander of some troops stationed in Boston harbor. Alice is the admiral, stationed several miles away. At a certain moment, Alice must convey a one-bit message p (the plaintext) to Bob: whether to attack by land or by sea (0=land, 1=sea). Their plan, agreed upon in advance, is that Alice will encrypt the message, obtaining a one-bit *cyphertext* c, and send the cyphertext c to Bob by hanging one or two lanterns (say, one lantern = 0, two lanterns = 1). They are aware that the fate of a colony might depend on the secrecy of their communication. (As it happens, a rebel, Eve, knows of the plan and will be observing.)

Let's go back in time. Alice and Bob are consulting with their cryptography expert, who suggests the following scheme:

[1]For an historically significant occurence of the former Soviet Union failing to use it correctly, look up VENONA.

Bad Scheme: Alice and Bob randomly choose k from $\{\clubsuit, \heartsuit, \spadesuit\}$ according to the uniform probability function ($\text{pr}(\clubsuit) = 1/3, \text{pr}(\heartsuit) = 1/3, \text{pr}(\spadesuit) = 1/3$). Alice and Bob must both know k but must keep it secret. It is the *key*.
When it is time for Alice to use the key to encrypt her plaintext message p, obtaining the cyphertext c, she refers to the following table:

p	k	c
0	$\clubsuit$	0
0	$\heartsuit$	1
0	$\spadesuit$	1
1	$\clubsuit$	1
1	$\heartsuit$	0
1	$\spadesuit$	0

The good news is that this cryptosystem satisfies the first requirement of cryptosystems: it will enable Bob, who knows the key k and receives the cyphertext c, to determine the plaintext p. No two rows of the table have the same k-value and c-value.

The bad news is that this scheme leaks information to Eve. Suppose the message turns out to be 0. In this case, $c = 0$ if $k = \clubsuit$ (which happens with probability 1/3), and $c = 1$ if $k = \heartsuit$ or $k = \spadesuit$ (which, by the Fundamental Principle of Probability Theory, happens with probability 2/3). Thus in this case $c = 1$ is twice as likely as $c = 0$. Now suppose the message turns out to be 1. In this case, a similar analysis shows that $c = 0$ is twice as likely as $c = 1$.

Therefore, when Eve sees the cyphertext c, she learns something about the plaintext p. Learning c doesn't allow Eve to determine the value of p with certainty, but she can revise her estimate of the chance that $p = 0$. For example, suppose that, before seeing c, Eve believed $p = 0$ and $p = 1$ were equally likely. If she sees $c = 1$ then she can infer that $p = 0$ is twice as likely as $p = 1$. The exact calculation depends on Bayes' Rule, which is beyond the scope of this analysis but is quite simple.

Confronted with this argument, the cryptographer changes the scheme simply by removing $\spadesuit$ as a possible value for p.

Good Scheme: Alice and Bob randomly choose k from $\{\clubsuit, \heartsuit\}$ according to the uniform probability function ($\text{pr}(\clubsuit) = 1/2, \text{pr}(\heartsuit) = 1/2$
When it is time for Alice to encrypt her plaintext message p, obtaining the cyphertext c, she uses the following table:

p	k	c
0	$\clubsuit$	0
0	$\heartsuit$	1
1	$\clubsuit$	1
1	$\heartsuit$	0

0.4.5 Perfect secrecy and invertible functions

Consider the functions

$$f_0 : \{\clubsuit, \heartsuit\} \longrightarrow \{0, 1\}$$

and

$$f_1 : \{\clubsuit, \heartsuit\} \longrightarrow \{0, 1\}$$

defined by

$$f_0(x) = \text{encryption of 0 when the key is } x$$

$$f_1(x) = \text{encryption of 1 when the key is } x$$

Each of these functions is invertible. Consequently, for each function, if the input x is chosen uniformly at random, the output will also be distributed according to the uniform distribution. This in turn means that the probability distribution of the output does not depend on whether 0 or 1 is being encrypted, so knowing the output gives Eve no information about which is being encrypted. We say the scheme achieves *perfect secrecy*.

0.5 *Lab: Introduction to Python—sets, lists, dictionaries, and comprehensions*

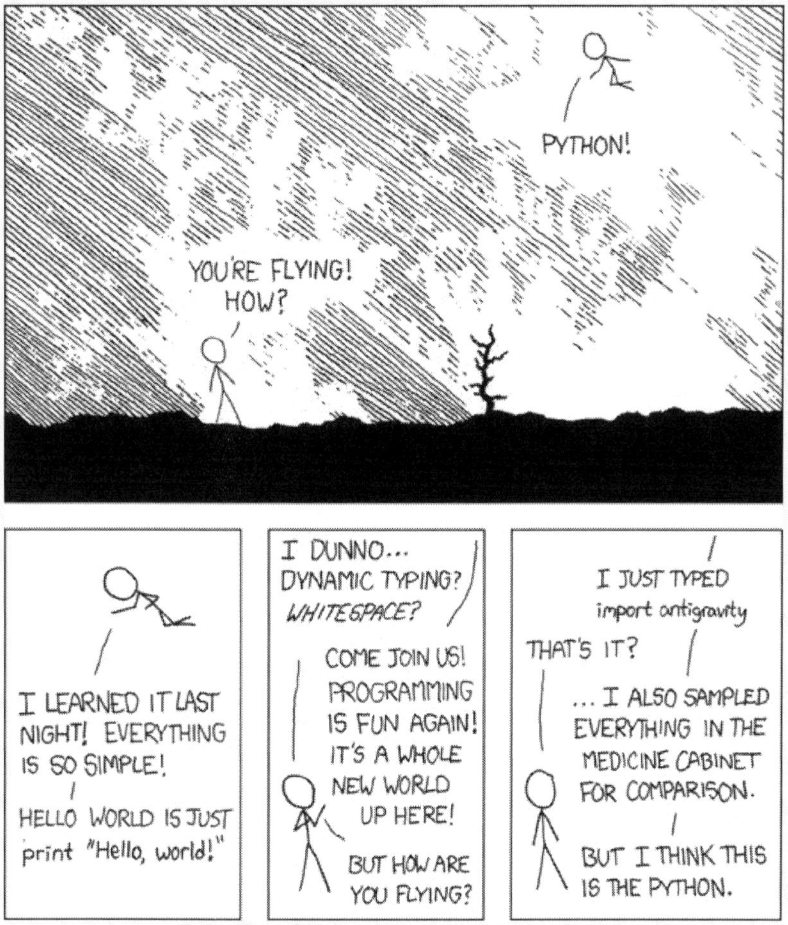

Python `http://xkcd.com/353/`

We will be writing all our code in Python (Version 3.x). In writing Python code, we emphasize the use of *comprehensions*, which allow one to express computations over the elements of a set, list, or dictionary without a traditional for-loop. Use of comprehensions leads to more compact and more readable code, code that more clearly expresses the mathematical idea behind the computation being expressed. Comprehensions might be new to even some readers who are familiar with Python, and we encourage those readers to at least skim the material on this topic.

To start Python, simply open a console (also called a shell or a terminal or, under Windows, a "Command Prompt" or "MS-DOS Prompt"), and type `python3` (or perhaps just `python`) to the console (or shell or terminal or Command Prompt) and hit the `Enter` key. After a few lines telling you what version you are using (e.g., Python 3.3.3), you should see `>>>` followed by a space. This is the *prompt*; it indicates that Python is waiting for you to type something. When you type an expression and hit the `Enter` key, Python evaluates the expression and prints the result, and then prints another prompt. To get out of this environment, type `quit()` and `Enter`, or `Control-D`. To interrupt Python when it is running too long, type `Control-C`.

This environment is sometimes called a *REPL*, an acronym for "read-eval-print loop." It reads what you type, evaluates it, and prints the result if any. In this assignment, you will interact with Python primarily through the REPL. In each task, you are asked to come up with an expression of a certain form.

There are two other ways to run Python code. You can import a *module* from within the REPL, and you can run a Python script from the command line (outside the REPL). We will discuss modules and importing in the next lab assignment. This will be an important part of your interaction with Python.

0.5.1 *Simple expressions*

Arithmetic and numbers

You can use Python as a calculator for carrying out arithmetic computations. The binary operators +, *, -, / work as you would expect. To take the negative of a number, use - as a unary operator (as in -9). Exponentiation is represented by the binary operator **, and truncating integer division is //. Finding the remainder when one integer is divided by another (modulo) is done using the % operator. As usual, ** has precedence over * and / and //, which have precedence over + and -, and parentheses can be used for grouping.

To get Python to carry out a calculation, type the expression and press the Enter/Return key:

```
>>> 44+11*4-6/11.
87.454545454545454
>>>
```

Python prints the answer and then prints the prompt again.

Task 0.5.1: Use Python to find the number of minutes in a week.

Task 0.5.2: Use Python to find the remainder of 2304811 divided by 47 without using the modulo operator %. (Hint: Use //.)

Python uses a traditional programming notation for scientific notation. The notation `6.022e23` denotes the value 6.02×10^{23}, and `6.626e-34` denotes the value 6.626×10^{-34}. As we will discover, since Python uses limited-precision arithmetic, there are round-off errors:

```
>>> 1e16 + 1
1e16
```

Strings

A string is a series of characters that starts and ends with a single-quote mark. Enter a string, and Python will repeat it back to you:

```
>>> 'This sentence is false.'
'This sentence is false.'
```

You can also use double-quote marks; this is useful if your string itself contains a single-quote mark:

```
>>> "So's this one."
"So's this one."
```

Python is doing what it usually does: it *evaluates* (finds the value of) the expression it is given and prints the value. The value of a string is just the string itself.

Comparisons and conditions and Booleans

You can compare values (strings and numbers, for example) using the operators ==, < , >, <=, >=, and !=. (The operator != is inequality.)

```
>>> 5 == 4
False
>>> 4 == 4
True
```

The value of such a comparison is a Boolean value (True or False). An expression whose value is a boolean is called a Boolean expression.

Boolean operators such as `and` and `or` and `not` can be used to form more complicated Boolean expressions.

```
>> True and False
False
>>> True and not (5 == 4)
True
```

Task 0.5.3: Enter a Boolean expression to test whether the sum of 673 and 909 is divisible by 3.

0.5.2 *Assignment statements*

The following is a *statement*, not an expression. Python executes it but produces neither an error message nor a value.

```
>>> mynum = 4+1
```

The result is that henceforth the variable `mynum` is bound to the value 5. Consequently, when Python evaluates the expression consisting solely of `mynum`, the resulting value is 5. We say therefore that the value of `mynum` is 5.

A bit of terminology: the variable being assigned to is called the *left-hand side* of an assignment, and the expression whose value is assigned is called the *right-hand side*.

A variable name must start with a letter and must exclude certain special symbols such as the dot (period). The underscore _ is allowed in a variable name. A variable can be bound to a value of any type. You can rebind `mynum` to a string:

```
>>> mynum = 'Brown'
```

This binding lasts until you assign some other value to `mynum` or until you end your Python session. It is called a *top-level* binding. We will encounter cases of binding variables to values where the bindings are temporary.

It is important to remember (and second nature to most experienced programmers) that an assignment statement binds a variable to the *value* of an expression, not to the expression itself. Python first evaluates the right-hand side and only then assigns the resulting value to the left-hand side. This is the behavior of most programming languages.

Consider the following assignments.

```
>>> x = 5+4
>>> y = 2 * x
>>> y
18
>>> x = 12
>>> y
18
```

In the second assignment, `y` is assigned the value of the expression `2 * x`. The value of that expression is 9, so `y` is bound to 18. In the third assignment, `x` is bound to 12. This does not change the fact that `y` is bound to 18.

0.5.3 *Conditional expressions*

There is a syntax for conditional expressions:

⟨*expression*⟩ `if` ⟨*condition*⟩ `else` ⟨*expression*⟩

The *condition* should be a Boolean expression. Python evaluates the condition; depending on whether it is `True` or `False`, Python then evaluates either the first or second *expression*, and uses the result as the result of the entire conditional expression.

For example, the value of the expression `x if x>0 else -x` is the absolute value of `x`.

Task 0.5.4: Assign the value -9 to `x` and 1/2 to `y`. Predict the value of the following expression, then enter it to check your prediction:

```
2**(y+1/2) if x+10<0 else 2**(y-1/2)
```

0.5.4 *Sets*

Python provides some simple data structures for grouping together multiple values, and integrates them with the rest of the language. These data structures are called *collections*. We start with sets.

A set is an unordered collection in which each value occurs at most once. You can use curly braces to give an expression whose value is a set. Python prints sets using curly braces.

```
>>> {1+2, 3, "a"}
{'a', 3}
>>> {2, 1, 3}
{1, 2, 3}
```

Note that duplicates are eliminated and that the order in which the elements of the output are printed does not necessarily match the order of the input elements.

The *cardinality* of a set S is the number of elements in the set. In Mathese we write $|S|$ for the cardinality of set S. In Python, the cardinality of a set is obtained using the procedure `len(·)`.

```
>>> len({'a', 'b', 'c', 'a', 'a'})
3
```

Summing

The sum of elements of collection of values is obtained using the procedure `sum(·)`.

```
>>> sum({1,2,3})
6
```

If for some reason (we'll see one later) you want to start the sum not at zero but at some other value, supply that value as a second argument to `sum(·)`:

```
>>> sum({1,2,3}, 10)
16
```

Testing set membership

Membership in a set can be tested using the `in` operator and the `not in` operator. If S is a set, x `in` S is a Boolean expression that evaluates to `True` if the value of x is a member of the set S, and `False` otherwise. The value of a `not in` expression is just the opposite

```
>>> S={1,2,3}
>>> 2 in S
True
>>> 4 in S
False
>>> 4 not in S
True
```

Set union and intersection

The *union* of two sets S and T is a new set that contains every value that is a member of S or a member of T (or both). Python uses the vertical bar | as the *union* operator:

```
>>> {1,2,3} | {2,3,4}
{1, 2, 3, 4}
```

The *intersection* of S and T is a new set that contains every value that is a member of both S and T. Python uses the ampersand & as the *intersection* operator:

```
>>> {1,2,3} & {2,3,4}
{2, 3}
```

Mutating a set

A value that can be altered is a *mutable* value. Sets are mutable; elements can be added and removed using the add and remove methods:

```
>>> S={1,2,3}
>>> S.add(4)
>>> S.remove(2)
>>> S
{1, 3, 4}
```

The syntax using the dot should be familiar to students of object-oriented programming languages such as Java and C++. The operations add(·) and remove(·) are *methods.* You can think of a method as a procedure that takes an extra argument, the value of the expression to the left of the dot.

Python provides a method update(...) to add to a set all the elements of another collection (e.g. a set or a list):

```
>>> S.update({4, 5, 6})
>>> S
{1, 3, 4, 5, 6}
```

Similarly, one can intersect a set with another collection, removing from the set all elements not in the other collection:

```
>>> S.intersection_update({5,6,7,8,9})
>>> S
{5, 6}
```

Suppose two variables are bound to the same value. A mutation to the value made through one variable is seen by the other variable.

```
>>> T=S
>>> T.remove(5)
>>> S
{6}
```

This behavior reflects the fact that Python stores only one copy of the underlying data structure. After Python executes the assignment statement `T=S`, both `T` and `S` point to the same data structure. This aspect of Python will be important to us: many different variables can point to the same huge set without causing a blow-up of storage requirements.

Python provides a method for copying a collection such as a set:

```
>>> U=S.copy()
>>> U.add(5)
>>> S
{1, 3}
```

The assignment statement binds `U` not to the value of `S` but to a copy of that value, so mutations to the value of `U` don't affect the value of `S`.

Set comprehensions

Python provides for expressions called *comprehensions* that let you build collections out of other collections. We will be using comprehensions a lot because they are useful in constructing an expression whose value is a collection, and they mimic traditional mathematical notation. Here's an example:

```
>>> {2*x for x in {1,2,3} }
{2, 4, 6}
```

This is said to be a set comprehension *over* the set `{1,2,3}`. It is called a set comprehension because its value is a set. The notation is similar to the traditional mathematical notation for expressing sets in terms of other sets, in this case $\{2x \, : \, x \in \{1,2,3\}\}$. To compute the value, Python iterates over the elements of the set `{1,2,3}`, temporarily binding the control variable `x` to each element in turn and evaluating the expression `2*x` in the context of that binding. Each of the values obtained is an element of the final set. (The bindings of `x` during the evaluation of the comprehension do not persist after the evaluation completes.)

Task 0.5.5: Write a comprehension over $\{1,2,3,4,5\}$ whose value is the set consisting of the squares of the first five positive integers.

Task 0.5.6: Write a comprehension over $\{0,1,2,3,4\}$ whose value is the set consisting of the first five powers of two, starting with 2^0.

Using the union operator `|` or the intersection operator `&`, you can write set expressions for the union or intersection of two sets, and use such expressions in a comprehension:

```
>>> {x*x for x in S | {5, 7}}
{1, 25, 49, 9}
```

By adding the phrase `if` ⟨*condition*⟩ at the end of the comprehension (before the closing brace "}"), you can skip some of the values in the set being iterated over:

```
>>> {x*x for x in S | {5, 7}  if x > 2}
{9, 49, 25}
```

I call the conditional clause a *filter*.

You can write a comprehension that iterates over the Cartesian product of two sets:

```
>>>{x*y for x in {1,2,3} for y in {2,3,4}}
{2, 3, 4, 6, 8, 9, 12}
```

This comprehension constructs the set of the products of every combination of `x` and `y`. I call this a *double comprehension*.

Task 0.5.7: The value of the previous comprehension,

`{x*y for x in {1,2,3} for y in {2,3,4}}`

is a seven-element set. Replace `{1,2,3}` and `{2,3,4}` with two other three-element sets so that the value becomes a nine-element set.

Here is an example of a double comprehension with a filter:

```
>>> {x*y for x in {1,2,3} for y in {2,3,4} if x != y}
{2, 3, 4, 6, 8, 12}
```

Task 0.5.8: Replace `{1,2,3}` and `{2,3,4}` in the previous comprehension with two disjoint (i.e. non-overlapping) three-element sets so that the value becomes a five-element set.

Task 0.5.9: Assume that `S` and `T` are assigned sets. Without using the *intersection operator* `&`, write a comprehension over `S` whose value is the intersection of `S` and `T`. Hint: Use a membership test in a filter at the end of the comprehension.

Try out your comprehension with `S = {1,2,3,4}` and `T = {3,4,5,6}`.

Remarks

The empty set is represented by `set()`. You would think that `{}` would work but, as we will see, that notation is used for something else.

You cannot make a set that has a set as element. This has nothing to do with Cantor's Paradox—-Python imposes the restriction that the elements of a set must not be mutable, and sets are mutable. The reason for this restriction will be clear to a student of data structures from the error message in the following example:

```
>>> {{1,2},3}
Traceback (most recent call last):
  File "<stdin>", line 1, in <module>
TypeError: unhashable type: 'set'
```

There is a nonmutable version of set called *frozenset*. Frozensets can be elements of sets. However, we won't be using them.

0.5.5 *Lists*

Python represents sequences of values using *lists*. In a list, order is significant and repeated elements are allowed. The notation for lists uses square brackets instead of curly braces. The empy list is represented by `[]`.

```
>>> [1,1+1,3,2,3]
[1, 2, 3, 2, 3]
```

There are no restrictions on the elements of lists. A list can contain a set or another list.

```
>>> [[1,1+1,4-1],{2*2,5,6}, "yo"]
[[1, 2, 3], {4, 5, 6}, 'yo']
```

However, a set cannot contain a list since lists are mutable.

The *length* of a list, obtained using the procedure `len(·)`, is the number of elements in the list, even though some of those elements may themselves be lists, and even though some elements might have the same value:

```
>>> len([[1,1+1,4-1],{2*2,5,6}, "yo", "yo"])
4
```

As we saw in the section on sets, the sum of elements of a collection can be computed using `sum(·)`

```
>>> sum([1,1,0,1,0,1,0])
4
>>> sum([1,1,0,1,0,1,0], -9)
-5
```

In the second example, the second argument to `sum(·)` is the value to start with.

Task 0.5.10: Write an expression whose value is the average of the elements of the list `[20, 10, 15, 75]`.

List concatenation

You can combine the elements in one list with the elements in another list to form a new list (without changing the original lists) using the + operator.

```
>>> [1,2,3]+["my", "word"]
[1, 2, 3, 'my', 'word']
>>> mylist = [4,8,12]
>>> mylist + ["my", "word"]
[4, 8, 12, 'my', 'word']
>>> mylist
[4, 8, 12]
```

You can use `sum(·)` on a collection of lists, obtaining the concatenation of all the lists, by providing `[]` as the second argument.

```
>>> sum([ [1,2,3], [4,5,6], [7,8,9] ])
Traceback (most recent call last):
  File "<stdin>", line 1, in <module>
TypeError: unsupported operand type(s) for +: 'int' and 'list'
>>> sum([ [1,2,3], [4,5,6], [7,8,9] ], [])
[1, 2, 3, 4, 5, 6, 7, 8, 9]
```

List comprehensions

Next we discuss how to write a list comprehension (a comprehension whose value is a list). In the following example, a list is constructed by iterating over the elements in a set.

```
>>> [2*x for x in {2,1,3,4,5} ]
[2, 4, 6, 8, 10]
```

Note that the order of elements in the resulting list might not correspond to the order of elements in the set since the latter order is not significant.

You can also use a comprehension that constructs a list by iterating over the elements in a list:

```
>>> [ 2*x for x in [2,1,3,4,5] ]
[4, 2, 6, 8, 10]
```

Note that the list `[2,1,3,4,5]` specifies the order among its elements. In evaluating the comprehension Python iterates through them in that order. Therefore the order of elements in the resulting list corresponds to the order in the list iterated over.

You can also write list comprehensions that iterate over multiple collections using two control variables. As I mentioned in the context of sets, I call these "double comprehensions". Here is an example of a list comprehension over two lists.

```
>>> [ x*y for x in [1,2,3] for y in [10,20,30] ]
[10, 20, 30, 20, 40, 60, 30, 60, 90]
```

The resulting list has an element for every combination of an element of `[1,2,3]` with an element of `[10,20,30]`.

We can use a comprehension over two sets to form the Cartesian product.

Task 0.5.11: Write a double list comprehension over the lists `['A','B','C']` and `[1,2,3]` whose value is the list of all possible two-element lists [letter, number]. That is, the value is

```
[['A', 1], ['A', 2], ['A', 3], ['B', 1], ['B', 2],['B', 3],
['C', 1], ['C', 2], ['C', 3]]
```

Task 0.5.12: Suppose `LofL` has been assigned a list whose elements are themselves lists of numbers. Write an expression that evaluates to the sum of all the numbers in all the lists. The expression has the form

```
sum([sum(...
```

and includes one comprehension. Test your expression after assigning `[[.25, .75, .1], [-1, 0], [4, 4, 4, 4]]` to `LofL`. Note that your expression should work for a list of any length.

Obtaining elements of a list by indexing

Donald Knuth `http://xkcd.com/163/`

There are two ways to obtain an individual element of a list. The first is by indexing. As in some other languages (Java and c++, for example) indexing is done using square brackets around the index. Here is an example. Note that the first element of the list has index 0.

```
>>> mylist[0]
4
>>> ['in','the','CIT'][1]
'the'
```

Slices: A *slice* of a list is a new list consisting of a consecutive subsequence of elements of the old list, namely those indexed by a range of integers. The range is specified by a colon-separated pair $i : j$ consisting of the index i as the first element and j as one past the index of the last element. Thus `mylist[1:3]` is the list consisting of elements 1 and 2 of `mylist`.

Prefixes: If the first element i of the pair is 0, it can be omitted, so `mylist[:2]` consists of the first 2 elements of `mylist`. This notation is useful for obtaining a prefix of a list.

Suffixes: If the second element j of the pair is the length of the list, it can be omitted, so `mylist[1:]` consists of all elements of `mylist` except element 0.

```
>>> L = [0,10,20,30,40,50,60,70,80,90]
>>> L[:5]
[0, 10, 20, 30, 40]
>>> L[5:]
[50, 60, 70, 80, 90]
```

Slices that skip You can use a colon-separated *triple* a:b:c if you want the slice to include every c^{th} element. For example, here is how you can extract from `L` the list consisting of even-indexed elements and the list consisting of odd-indexed elements:

```
>>> L[::2]
[0, 20, 40, 60, 80]
>>> L[1::2]
[10, 30, 50, 70, 90]
```

Obtaining elements of a list by unpacking

The second way to obtain individual elements is by *unpacking*. Instead of assigning a list to a single variable as in `mylist =[4,8,12]`, one can assign to a list of variables:

```
>>> [x,y,z] = [4*1, 4*2, 4*3]
>>> x
4
>>> y
8
```

I called the left-hand side of the assignment a "list of variables," but beware: this is a notational fiction. Python does not allow you to create a value that is a list of variables. The assignment is simply a convenient way to assign to each of the variables appearing in the left-hand side.

Task 0.5.13: Find out what happens if the length of the left-hand side list does not match the length of the right-hand side list.

Unpacking can similarly be used in comprehensions:

```
>>> listoflists = [[1,1],[2,4],[3, 9]]
>>> [y for [x,y] in listoflists]
[1, 4, 9]
```

Here the two-element list `[x,y]` iterates over all elements of `listoflists`. This would result in an error message if some element of `listoflists` were not a two-element list.

Mutating a list: indexing on the left-hand side of =

You can mutate a list, replacing its ith element, using indexing on the left-hand side of the =, analogous to an assignment statement:

```
>>> mylist = [30, 20, 10]
>>> mylist[1] = 0
>>> mylist
[30, 0, 10]
```

Slices can also be used on the left-hand side but we will not use this.

0.5.6 *Tuples*

Like a list, a tuple is an ordered sequence of elements. However, tuples are immutable so they can be elements of sets. The notation for tuples is the same as that for lists except that ordinary parentheses are used instead of square brackets.

```
>>> (1,1+1,3)
(1, 2, 3)
>>> {0, (1,2)} | {(3,4,5)}
{(1, 2), 0, (3, 4, 5)}
```

Obtaining elements of a tuple by indexing and unpacking

You can use indexing to obtain an element of a tuple.

```
>>> mytuple = ("all", "my", "books")
>>> mytuple[1]
'my'
>>> (1, {"A", "B"}, 3.14)[2]
3.14
```

You can also use unpacking with tuples. Here is an example of top-level variable assignment:

```
>>> (a,b) = (1,5-3)
>>> a
1
```

In some contexts, you can get away without the parentheses, e.g.

```
>>> a,b = (1,5-3)
```

or even

```
>>> a,b = 1,5-3
```

You can use unpacking in a comprehension:

```
>>> [y for (x,y) in [(1,'A'),(2,'B'),(3,'C')] ]
['A', 'B', 'C']
```

Task 0.5.14: Suppose S is a set of integers, e.g. $\{-4, -2, 1, 2, 5, 0\}$. Write a triple comprehension whose value is a list of all three-element tuples (i, j, k) such that i, j, k are elements of S whose sum is zero.

Task 0.5.15: Modify the comprehension of the previous task so that the resulting list does not include $(0, 0, 0)$. Hint: add a filter.

Task 0.5.16: Further modify the expression so that its value is not the list of all such tuples but is the first such tuple.

The previous task provided a way to compute three elements i, j, k of S whose sum is zero—if there exist three such elements. Suppose you wanted to determine if there were a hundred elements of S whose sum is zero. What would go wrong if you used the approach used in the previous task? Can you think of a clever way to quickly and reliably solve the problem, even if the integers making up S are very large? (If so, see me immediately to collect your Ph.D.)

Obtaining a list or set from another collection

Python can compute a set from another collection (e.g. a list) using the constructor `set(·)`. Similarly, the constructor `list(·)` computes a list, and the constructor `tuple(·)` computes a tuple

```
>>> set(range(10))
{0, 1, 2, 3, 4, 5, 6, 7, 8, 9}
>>> set([1,2,3])
{1, 2, 3}
>>> list({1,2,3})
[1, 2, 3]
>>>
>>> set((1,2,3))
{1, 2, 3}
```

Task 0.5.17: Find an example of a list L such that `len(L)` and `len(list(set(L)))` are different.

0.5.7 Other things to iterate over

Tuple comprehensions—not! Generators

One would expect to be able to create a tuple using the usual comprehension syntax, e.g. `(i for i in range(10))` but the value of this expression is not a tuple. It is a *generator*. Generators are a very powerful feature of Python but we don't study them here. Note, however, that one can write a comprehension over a generator instead of over a list or set or tuple. Alternatively, one can use `set(·)` or `list(·)` or `tuple(·)` to transform a generator into a set or list or tuple.

Ranges

A range plays the role of a list consisting of the elements of an arithmetic progression. For any integer n, `range(`n`)` represents the sequence of integers from 0 through $n-1$. For example, `range(10)` represents the integers from 0 through 9. Therefore, the value of the following comprehension is the sum of the squares of these integers: `sum({i*i for i in range(10)})`.

Even though a range represents a sequence, it is not a list. Generally we will either iterate through the elements of the range or use `set(·)` or `list(·)` to turn the range into a set or list.

```
>>> list(range(10))
[0, 1, 2, 3, 4, 5, 6, 7, 8, 9]
```

Task 0.5.18: Write a comprehension over a range of the form `range(`n`)` such that the value of the comprehension is the set of odd numbers from 1 to 99.

You can form a range with one, two, or three arguments. The expression `range(`a`,`b`)` represents the sequence of integers $a, a+1, a+2, \ldots, b-1$. The expression `range(`a`,`b`,`c`)` represents $a, a+c, a+2c, \ldots$ (stopping just before b).

Zip

Another collection that can be iterated over is a *zip*. A zip is constructed from other collections all of the same length. Each element of the zip is a tuple consisting of one element from each of the input collections.

```
>>> list(zip([1,3,5],[2,4,6]))
[(1, 2), (3, 4), (5, 6)]
>>> characters = ['Neo', 'Morpheus', 'Trinity']
>>> actors = ['Keanu', 'Laurence', 'Carrie-Anne']
>>> set(zip(characters, actors))
{('Trinity', 'Carrie-Anne'), ('Neo', 'Keanu'), ('Morpheus', 'Laurence')}
>>> [character+' is played by '+actor
...      for (character,actor) in zip(characters,actors)]
['Neo is played by Keanu', 'Morpheus is played by Laurence',
 'Trinity is played by Carrie-Anne']
```

Task 0.5.19: Assign to L the list consisting of the first five letters `['A','B','C','D','E']`. Next, use L in an expression whose value is

`[(0, 'A'), (1, 'B'), (2, 'C'), (3, 'D'), (4, 'E')]`

Your expression should use a range and a zip, but should not use a comprehension.

Task 0.5.20: Starting from the lists `[10, 25, 40]` and `[1, 15, 20]`, write a comprehension whose value is the three-element list in which the first element is the sum of 10 and 1, the second is the sum of 25 and 15, and the third is the sum of 40 and 20. Your expression should use `zip` but not `list`.

`reversed`

To iterate through the elements of a list L in reverse order, use `reversed(L)`, which does not change the list L:

```
>>> [x*x for x in reversed([4, 5, 10])]
[100, 25, 16]
```

0.5.8 *Dictionaries*

We will often have occasion to use functions with finite domains. Python provides collections, called *dictionaries*, that are suitable for representing such functions. Conceptually, a dictionary is a set of key-value pairs. The syntax for specifying a dictionary in terms of its key-value pairs therefore resembles the syntax for sets—it uses curly braces—except that instead of listing the elements of the set, one lists the key-value pairs. In this syntax, each key-value pair is written using *colon* notation: an expression for the key, followed by the colon, followed by an expression for the value:

$$key : value$$

The function f that maps each letter in the alphabet to its rank in the alphabet could be written as

```
{'A':0, 'B':1, 'C':2, 'D':3, 'E':4, 'F':5, 'G':6, 'H':7, 'I':8,
 'J':9, 'K':10, 'L':11, 'M':12, 'N':13, 'O':14, 'P':15, 'Q':16,
 'R':17, 'S':18, 'T':19, 'U':20, 'V':21, 'W':22, 'X':23, 'Y':24,
 'Z':25}
```

As in sets, the order of the key-value pairs is irrelevant, and the keys must be immutable (no sets or lists or dictionaries). For us, the keys will mostly be integers, strings, or tuples of integers and strings.

The keys and values can be specified with expressions.

```
>>> {2+1:'thr'+'ee', 2*2:'fo'+'ur'}
{3: 'three', 4: 'four'}
```

To each key in a dictionary there corresponds only one value. If a dictionary is given multiple values for the same key, only one value will be associated with that key.

```
>>> {0:'zero', 0:'nothing'}
{0: 'nothing'}
```

Indexing into a dictionary

Obtaining the value corresponding to a particular key uses the same syntax as indexing a list or tuple: right after the dictionary expression, use square brackets around the key:

```
>>> {4:"four", 3:'three'}[4]
'four'
>>> mydict = {'Neo':'Keanu', 'Morpheus':'Laurence',
 'Trinity':'Carrie-Anne'}
>>> mydict['Neo']
'Keanu'
```

If the key is not represented in the dictionary, Python considers it an error:

```
>>> mydict['Oracle']
Traceback (most recent call last):
  File "<stdin>", line 1, in <module>
KeyError: 'Oracle'
```

Testing dictionary membership

You can check whether a key is in a dictionary using the `in` operator we earlier used for testing membership in a set:

```
>>> 'Oracle' in mydict
False
>>> mydict['Oracle'] if 'Oracle' in mydict else 'NOT PRESENT'
'NOT PRESENT'
>>> mydict['Neo'] if 'Neo' in mydict else 'NOT PRESENT'
'Keanu'
```

Lists of dictionaries

Task 0.5.21: Suppose `dlist` is a list of dictionaries and `k` is a key that appears in all the dictionaries in `dlist`. Write a comprehension that evaluates to the list whose i^{th} element is the value corresponding to key `k` in the ith dictionary in `dlist`.

Test your comprehension with some data. Here are some example data.

```
dlist = [{'James':'Sean', 'director':'Terence'}, {'James':'Roger',
'director':'Lewis'}, {'James':'Pierce', 'director':'Roger'}]
k = 'James'
```

Task 0.5.22: Modify the comprehension in Task 0.5.21 to handle the case in which `k` might not appear in all the dictionaries. The comprehension evaluates to the list whose i^{th} element is the value corresponding to key `k` in the ith dictionary in `dlist` if that dictionary contains that key, and `'NOT PRESENT'` otherwise.

Test your comprehension with `k = 'Bilbo'` and `k = 'Frodo'` and with the following list of dictionaries:

```
dlist = [{'Bilbo':'Ian','Frodo':'Elijah'},
         {'Bilbo':'Martin','Thorin':'Richard'}]
```

Mutating a dictionary: indexing on the left-hand side of =

You can mutate a dictionary, mapping a (new or old) key to a given value, using the syntax used for assigning a list element, namely using the index syntax on the left-hand side of an assignment:

```
>>> mydict['Agent Smith'] = 'Hugo'
>>> mydict['Neo'] = 'Philip'
>>> mydict
{'Neo': 'Philip', 'Agent Smith': 'Hugo', 'Trinity': 'Carrie-Anne',
 'Morpheus': 'Laurence'}
```

Dictionary comprehensions

You can construct a dictionary using a comprehension.

```
>>> { k:v for (k,v) in [(3,2),(4,0),(100,1)] }
{3: 2, 4: 0, 100: 1}
>>> { (x,y):x*y for x in [1,2,3] for y in [1,2,3] }
{(1, 2): 2, (3, 2): 6, (1, 3): 3, (3, 3): 9, (3, 1): 3,
 (2, 1): 2, (2, 3): 6, (2, 2): 4, (1, 1): 1}
```

Task 0.5.23: Using `range`, write a comprehension whose value is a dictionary. The keys should be the integers from 0 to 99 and the value corresponding to a key should be the square of the key.

Task 0.5.24: Assign some set to the variable `D`, e.g. `D ={'red','white','blue'}`. Now write a comprehension that evaluates to a dictionary that represents the identity function on `D`.

Task 0.5.25: Using the variables base=10 and digits=set(range(base)), write a dictionary comprehension that maps each integer between zero and nine hundred ninety nine to the list of three digits that represents that integer in base 10. That is, the value should be

```
{0: [0, 0, 0], 1: [0, 0,  1], 2: [0, 0, 2], 3: [0, 0, 3], ...,
 10: [0, 1, 0], 11: [0, 1, 1], 12: [0, 1, 2], ...,
 999: [9, 9, 9]}
```

Your expression should work for any base. For example, if you instead assign 2 to base and assign {0,1} to digits, the value should be

```
{0: [0, 0, 0], 1: [0, 0, 1], 2: [0, 1, 0], 3: [0, 1, 1],
 ..., 7: [1, 1, 1]}
```

Comprehensions that iterate over dictionaries

You can write list comprehensions that iterate over the keys or the values of a dictionary, using keys() or values():

```
>>> [2*x for x in {4:'a',3:'b'}.keys() ]
[6, 8]
>>> [x for x in {4:'a', 3:'b'}.values()]
['b', 'a']
```

Given two dictionaries A and B, you can write comprehensions that iterate over the union or intersection of the keys, using the *union* operator | and intersection operator & we learned about in Section 0.5.4.

```
>>> [k for k in {'a':1, 'b':2}.keys() | {'b':3, 'c':4}.keys()]
['a', 'c', 'b']
>>> [k for k in {'a':1, 'b':2}.keys() & {'b':3, 'c':4}.keys()]
['b']
```

Often you'll want a comprehension that iterates over the (key, value) pairs of a dictionary, using items(). Each pair is a tuple.

```
>>> [myitem for myitem in mydict.items()]
[('Neo', 'Philip'), ('Morpheus', 'Laurence'),
 ('Trinity', 'Carrie-Anne'), ('Agent Smith', 'Hugo')]
```

Since the items are tuples, you can access the key and value separately using unpacking:

```
>>> [k + " is played by " + v for (k,v) in mydict.items()]
['Neo is played by Philip, 'Agent Smith is played by Hugo',
'Trinity is played by Carrie-Anne', 'Morpheus is played by Laurence']
>>> [2*k+v for (k,v) in {4:0,3:2, 100:1}.items() ]
[8, 8, 201]
```

Task 0.5.26: Suppose d is a dictionary that maps some employee IDs (a subset of the integers from 0 to $n-1$) to salaries. Suppose L is an n-element list whose i^{th} element is the name of employee number i. Your goal is to write a comprehension whose value is a dictionary mapping employee names to salaries. You can assume that employee names are distinct. However, not every employee ID is represented in d.

Test your comprehension with the following data:

```
id2salary = {0:1000.0, 3:990, 1:1200.50}
names = ['Larry', 'Curly', '', 'Moe']
```

0.5.9 *Defining one-line procedures*

The procedure *twice* : $\mathbb{R} \longrightarrow \mathbb{R}$ that returns twice its input can be written in Python as follows:

```
def twice(z): return 2*z
```

The word `def` introduces a procedure definition. The name of the function being defined is `twice`. The variable `z` is called the *formal argument* to the procedure. Once this procedure is defined, you can invoke it using the usual notation: the name of the procedure followed by an expression in parenthesis, e.g. `twice(1+2)`

The value 3 of the expression `1+2` is the *actual argument* to the procedure. When the procedure is invoked, the formal argument (the variable) is temporarily bound to the actual argument, and the body of the procedure is executed. At the end, the binding of the actual argument is removed. (The binding was temporary.)

Task 0.5.27: Try entering the definition of `twice(z)`. After you enter the definition, you will see the ellipsis. Just press enter. Next, try invoking the procedure on some actual arguments. Just for fun, try strings or lists. Finally, verify that the variable `z` is now not bound to any value by asking Python to evaluate the expression consisting of `z`.

Task 0.5.28: Define a one-line procedure `nextInts(`L`)` specified as follows:

- *input:* list L of integers
- *output:* list of integers whose i^{th} element is one more than the ith element of L
- *example:* input $[1, 5, 7]$, output $[2, 6, 8]$.

Task 0.5.29: Define a one-line procedure `cubes(`L`)` specified as follows:

- *input:* list L of numbers
- *output:* list of numbers whose i^{th} element is the cube of the ith element of L
- *example:* input $[1, 2, 3]$, output $[1, 8, 27]$.

Task 0.5.30: Define a one-line procedure `dict2list(`*dct*`,`*keylist*`)` with this spec:

- *input:* dictionary *dct*, list *keylist* consisting of the keys of *dct*
- *output:* list L such that $L[i] = dct[\text{keylist}[i]]$ for $i = 0, 1, 2, \ldots, \text{len}(keylist) - 1$
- *example:* input *dct*=`{'a':'A', 'b':'B', 'c':'C'}` and *keylist*=`['b','c','a']`, output `['B', 'C', 'A']`

Task 0.5.31: Define a one-line procedure `list2dict(`L`, `*keylist*`)` specified as follows:

- *input:* list L, list *keylist* of immutable items
- *output:* dictionary that maps keylist$[i]$ to $L[i]$ for $i = 0, 1, 2, \ldots, \text{len}(L) - 1$
- *example:* input L=`['A','B','C']` and *keylist*=`['a','b','c']`, output `{'a':'A', 'b':'B', 'c':'C'}`

Hint: Use a comprehension that iterates over a zip or a range.

Task 0.5.32: Write a procedure `all_3_digit_numbers(base, digits)` with the following spec:

- *input:* a positive integer *base* and the set *digits* which should be $\{0, 1, 2, \ldots, base-1\}$.
- *output:* the set of all three-digit numbers where the base is *base*

For example,

```
>>> all_3_digit_numbers(2, {0,1})
{0, 1, 2, 3, 4, 5, 6, 7}
>>> all_3_digit_numbers(3, {0,1,2})
{0, 1, 2, 3, 4, 5, 6, 7, 8, 9, 10, 11, 12, 13, 14, 15, 16, 17, 18,
 19, 20, 21, 22, 23, 24, 25, 26}
>>> all_3_digit_numbers(10, {0,1,2,3,4,5,6,7,8,9})
{0, 1, 2, 3, 4, 5, 6, 7, 8, 9, 10, 11, 12, 13, 14, 15, 16, 17, 18,
 19, 20, 21, 22, 23, 24, 25, 26, 27, 28, 29, 30, 31, 32, 33, 34, 35,
 ...
985, 986, 987, 988, 989, 990, 991, 992, 993, 994, 995, 996, 997, 998, 999}
```

0.6 *Lab: Python—modules and control structures—and inverse index*

In this lab, you will create a simple search engine. One procedure will be responsible for reading in a large collection of documents and indexing them to facilitate quick responses to subsequent search queries. Other procedures will use the index to answer the search queries.

The main purpose of this lab is to give you more Python programming practice.

0.6.1 *Using existing modules*

Python comes with an extensive library, consisting of components called *modules*. In order to use the definitions defined in a module, you must either import the module itself or import the specific definitions you want to use from the module. If you import the module, you must refer to a procedure or variable defined therein by using its *qualified name*, i.e. the name of the module followed by a dot followed by the short name.

For example, the library `math` includes many mathematical procedures such as square-root, cosine, and natural logarithm, and mathematical constants such as π and e.

Task 0.6.1: Import the math module using the command

```
>>> import math
```

Call the built-in procedure `help(modulename)` on the module you have just imported:

```
>>> help(math)
```

This will cause the console to show documentation on the module. You can move forward by typing f and backward by typing b, and you can quit looking at the documentation by typing q.

Use procedures defined by the math module to compute the square root of 3, and raise it to the power of 2. The result might not be what you expect. Keep in mind that Python represents nonintegral real numbers with limited precision, so the answers it gives are only approximate.

Next compute the square root of -1, the cosine of π, and the natural logarithm of e.

The short name of the square-root function is `sqrt` so its qualified name is `math.sqrt`. The short names of the cosine and the natural logarithm are `cos` and `log`, and the short names of π and e are `pi` and `e`.

The second way to bring a procedure or variable from a module into your Python environment is to specifically import the item itself from the module, using the syntax

`from` ⟨*module name*⟩ `import` ⟨*short name*⟩

after which you can refer to it using its short name.

Task 0.6.2: The module `random` defines a procedure `randint(a,b)` that returns an integer chosen uniformly at random from among $\{a, a+1, \ldots, b\}$. Import this procedure using the command

```
>>> from random import randint
```

Try calling `randint` a few times. Then write a one-line procedure `movie_review(name)` that takes as argument a string naming a movie, and returns a string review selected uniformly at random from among two or more alternatives (Suggestions: "See it!", "A gem!", "Ideological claptrap!")

0.6.2 *Creating your own modules*

You can create your own modules simply by entering the text of your procedure definitions and variable assignments in a file whose name consists of the module name you choose, followed by `.py`. Use a text editor such as kate or vim or, my personal favorite, emacs.

The file can itself contain import statements, enabling the code in the file to make use of definitions from other modules.

If the file is in the current working directory when you start up Python, you can import the module.[a]

Task 0.6.3: In Tasks 0.5.30 and 0.5.31 of Lab 0.5, you wrote procedures `dict2list(dct, keylist)` and `list2dict(L, keylist)`. Download the file `dictutil.py` from `http://resources.codingthematrix.com`. (That site hosts support code and sample data for the problems in this book.) Edit the provided file `dictutil.py` and edit it, replacing each occurence of pass with the appropriate statement. Import this module, and test the procedures. We will have occasion to use this module in the future.

[a] There is an environment variable, `PYTHONPATH`, that governs the sequence of directories in which Python searches for modules.

Reloading

You will probably find it useful when debugging your own module to be able to edit it and load the edited version into your current Python session. Python provides the procedure `reload(module)` in the module `imp`. To import this procedure, use the command

```
>>> from imp import reload
```

Note that if you import a specific definition using the `from ... import ...` syntax then you cannot reload it.

Task 0.6.4: Edit `dictutil.py`. Define a procedure `listrange2dict(L)` with this spec:

- *input:* a list L
- *output:* a dictionary that, for $i = 0, 1, 2, \ldots, \text{len}(L) - 1$, maps i to $L[i]$

You can write this procedure from scratch or write it in terms of `list2dict(L, keylist)`.
Use the statement

```
>>> reload(dictutil)
```

to reload your module, and then test `listrange2dict` on the list `['A','B','C']`.

0.6.3 *Loops and conditional statements*

Comprehensions are not the only way to loop over elements of a set, list, dictionary, tuple, range, or zip. For the traditionalist programmer, there are *for-loops*: `for x in {1,2,3}: print(x)`. In this statement, the variable `x` is bound to each of the elements of the set in turn, and the statement `print(x)` is executed in the context of that binding.

There are also *while-loops*: `while v[i] == 0: i = i+1`.

There are also conditional statements (as opposed to conditional expressions):
`if x > 0: print("positive")`

0.6.4 *Grouping in Python using indentation*

You will sometimes need to define loops or conditional statements in which the body consists of more than one statement. Most programming languages have a way of grouping a series of statements into a block. For example, c and Java use curly braces around the sequence of statements.

Python uses *indentation* to indicate grouping of statements. **All the statements forming a block should be indented the same number of spaces.** Python is very picky about this. Python files we provide will use **four** spaces to indent. Also, don't mix tabs with spaces in the same block. In fact, I recommend you avoid using tabs for indentation with Python.

Statements at the top level should have no indentation. The group of statements forming the body of a control statement should be indented more than the control statement. Here's an example:

```
for x in [1,2,3]:
    y = x*x
    print(y)
```

This prints 1, 4, and 9. (After the loop is executed, `y` remains bound to 9 and `x` remains bound to 3.)

Task 0.6.5: Type the above for-loop into Python. You will see that, after you enter the first line, Python prints an ellipsis (...) to indicate that it is expecting an indented block of statements. Type a space or two before entering the next line. Python will again print the ellipsis. Type a space or two (same number of spaces as before) and enter the next line. Once again Python will print an ellipsis. Press *enter*, and Python should execute the loop.

The same use of indentation can be used used in conditional statements and in procedure definitions.

```
def quadratic(a,b,c):
   discriminant = math.sqrt(b*b - 4*a*c)
   return ((-b + discriminant)/(2*a), (-b - discriminant)/(2*a))
```

You can nest as deeply as you like:

```
def print_greater_quadratic(L):
  for a, b, c in L:
    plus, minus = quadratic(a, b, c)
    if plus > minus:
      print(plus)
    else:
      print(minus)
```

Many text editors help you handle indentation when you write Python code. For example, if you are using emacs to edit a file with a .py suffix, after you type a line ending with a colon and hit return, emacs will automatically indent the next line the proper amount, making it easy for you to start entering lines belonging to a block. After you enter each line and hit Return, emacs will again indent the next line. However, emacs doesn't know when you have written the last line of a block; when you need to write the first line outside of that block, you should hit Delete to unindent.

0.6.5 *Breaking out of a loop*

As in many other programming languages, when Python executes the `break` statement, the loop execution is terminated, and execution continues immediately after the innermost nested loop containing the statement.

```
>>> s = "There is no spoon."
>>> for i in range(len(s)):
...   if s[i] == 'n':
...     break
...
>>> i
9
```

0.6.6 *Reading from a file*

In Python, a *file* object is used to refer to and access a file. The expression `open('stories_small.txt')` returns a file object that allows access to the file with the name given. You can use a comprehension or for-loop to loop over the lines in the file

```
>>> f = open('stories_big.txt')
>>> for line in f:
...  print(line)
```

or, if the file is not too big, use `list(·)` to directly obtain a list of the lines in the file, e.g.

```
>>> f = open('stories_small.txt')
>>> stories = list(f)
>>> len(stories)
50
```

In order to read from the file again, one way is to first create a new file object by calling `open` again.

0.6.7 *Mini-search engine*

Now, for the core of the lab, you will be writing a program that acts as a sort of search engine.

Given a file of "documents" where each document occupies a line of the file, you are to build a data structure (called an *inverse index*) that allows you to identify those documents containing a given word. We will identify the documents by *document number*: the document represented by the first line of the file is document number 0, that represented by the second line is document number 1, and so on.

You can use a method defined for strings, `split()`, which splits the string at spaces into substrings, and returns a list of these substrings:

```
>>> mystr = 'Ask not what you can do for your country.'
>>> mystr.split()
['Ask', 'not', 'what', 'you', 'can', 'do', 'for', 'your', 'country.']
```

Note that the period is considered part of a substring. To make this lab easier, we have prepared a file of documents in which punctuation are separated from words by spaces.

Often one wants to iterate through the elements of a list while keeping track of the indices of the elements. Python provides `enumerate(L)` for this purpose.

```
>>> list(enumerate(['A','B','C']))
[(0, 'A'), (1, 'B'), (2, 'C')]
>>> [i*x for (i,x) in enumerate([10,20,30,40,50])]
[0, 20, 60, 120, 200]
>>> [i*s for (i,s) in enumerate(['A','B','C','D','E'])]
['', 'B', 'CC', 'DDD', 'EEEE']
```

Task 0.6.6: Write a procedure `makeInverseIndex(strlist)` that, given a list of strings (documents), returns a dictionary that maps each word to the set consisting of the document numbers of documents in which that word appears. This dictionary is called an *inverse index*. (Hint: use `enumerate`.)

Task 0.6.7: Write a procedure `orSearch(inverseIndex, query)` which takes an inverse index and a list of words `query`, and returns the set of document numbers specifying all documents that conain *any* of the words in `query`.

Task 0.6.8: Write a procedure `andSearch(inverseIndex, query)` which takes an inverse index and a list of words `query`, and returns the set of document numbers specifying all documents that contain *all* of the words in `query`.

Try out your procedures on these two provided files:

- `stories_small.txt`
- `stories_big.txt`

0.7 Review questions

- What does the notation $f : A \longrightarrow B$ mean?
- What are the criteria for f to be an invertible function?
- What is associativity of functional composition?
- What are the criteria for a function to be a probability function?
- What is the Fundamental Principle of Probability Theory?
- If the input to an invertible function is chosen randomly according to the uniform distribution, what is the distribution of the output?

0.8 Problems

Python comprehension problems

For each of the following problems, write the one-line procedure using a comprehension.

Problem 0.8.1: `increments(L)`
input: list L of numbers
output: list of numbers in which the ith element is one plus the ith element of L.
Example: `increments([1,5,7])` should return `[2,6,8]`.

Problem 0.8.2: `cubes(L)`
input: list L of numbers
output: list of numbers in which the ith element is the cube of the ith element of L.
Example: given $[1, 2, 3]$ return $[1, 8, 27]$.

Problem 0.8.3: `tuple_sum(A, B)`
input: lists A and B of the same length, where each element in each list is a pair (x, y) of numbers
output: list of pairs (x, y) in which the first element of the ith pair is the sum of the first element of the ith pair in A and the first element of the ith pair in B
example: given lists $[(1, 2), (10, 20)]$ and $[(3, 4), (30, 40)]$, return $[(4, 6), (40, 60)]$.

Problem 0.8.4: `inv_dict(d)`
input: dictionary d representing an invertible function f
output: dictionary representing the inverse of f, the returned dictionary's keys are the values of d and its values are the keys of d
example: given an English-French dictionary
`{'thank you': 'merci', 'goodbye':  'au revoir'}`
return a French-English dictionary
`{'merci':'thank you', 'au revoir':'goodbye'}`

Problem 0.8.5: First write a procedure `row(p, n)` with the following spec:

- *input:* integer p, integer n
- *output:* n-element list such that element i is $p + i$
- *example:* given $p = 10$ and $n = 4$, return $[10, 11, 12, 13]$

Next write a comprehension whose value is a 15-element list of 20-element lists such that the jth element of the ith list is $i + j$. You can use `row(p)` in your comprehension.

Finally, write the same comprehension but without using `row(p)`. Hint: replace the call to `row(p, n)` with the comprehension that forms the body of `row(p, n)`.

Functional inverse

Problem 0.8.6: Is the following function invertible? If yes, explain why. If not, can you change domain and/or codomain of the function to make it invertible? Provide the drawing.

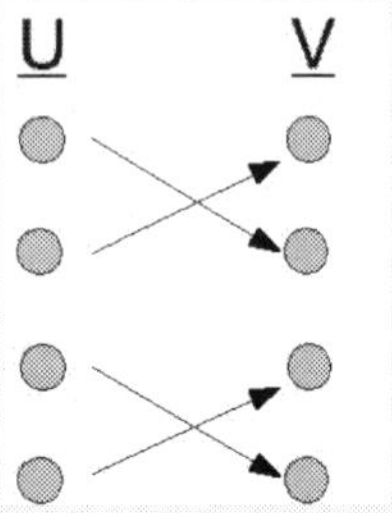

Problem 0.8.7: Is the following function invertible? If yes, explain why. If not, can you change domain and/or codomain of the function to make it invertible? Provide the drawing.

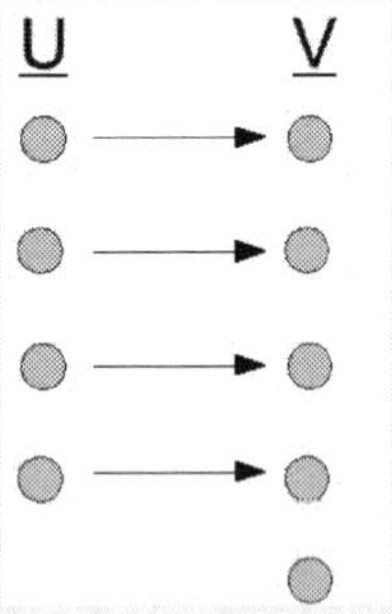

Functional composition

Problem 0.8.8: Let $f : \mathbb{R} \to \mathbb{R}$ where $f(x) = abs(x)$. Is there a choice of domain and co-domain for the function $g(x)$ with rule $g(x) = \sqrt{x}$ such that $g \circ f$ is defined? If so, specify it. If not, explain why not. Could you change domain and/or codomain of f or g so that $g \circ f$ will be defined?

Problem 0.8.9: Consider functions f and g in the following figure:

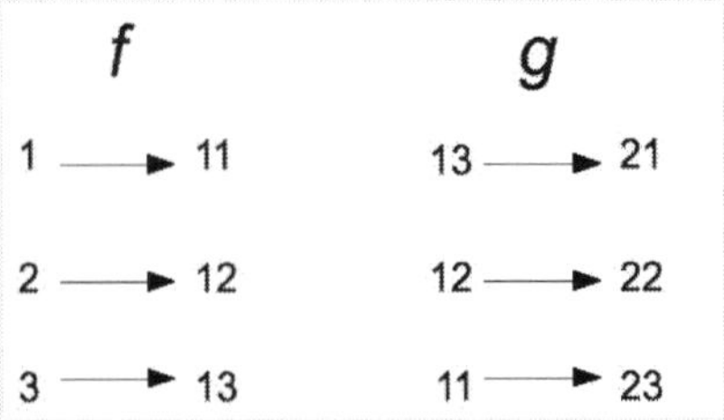

Is $f \circ g$ defined? If so, draw it, otherwise explain why not.

Probability

Problem 0.8.10: A function $f(x) = x+1$ with domain $\{1, 2, 3, 5, 6\}$ and codomain $\{2, 3, 4, 6, 7\}$ has the following probability function on its domain: $\Pr(1) = 0.5$, $\Pr(2) = 0.2$ and $\Pr(3) = \Pr(5) = \Pr(6) = 0.1$. What is the probability of getting an even number as an output of $f(x)$? An odd number?

Problem 0.8.11: A function $g(x) = x \; mod \; 3$ with domain $\{1, 2, 3, 4, 5, 6, 7\}$ and codomain $\{0, 1, 2\}$ has the following probability function on its domain: $\Pr(1) = \Pr(2) = \Pr(3) = 0.2$ and $\Pr(4) = \Pr(5) = \Pr(6) = \Pr(7) = 0.1$. What is the probability of getting 1 as an output of $g(x)$? What is the probability of getting 0 or 2?

Chapter 1

The Field

...the different branches of
Arithmetic—Ambition, Distraction,
Uglification, and Derision.
Lewis Carroll, *Alice in Wonderland*

We introduce the notion of a field, a collection of values with a *plus* operation and a *times* operation. The reader is familiar with the field of *real numbers* but perhaps not with the field of *complex numbers* or the field consisting just of zero and one. We discuss these fields and give examples of applications.

1.1 Introduction to complex numbers

If you stick to real numbers, there are no solutions to the equation $x^2 = -1$. To fill this void, mathematicians invented **i**. That's a bold letter i, and it's pronounced "i", but it is usually defined as *the square root of minus 1.*

Guest Week: Bill Amend (excerpt, `http://xkcd.com/824`)

By definition,

$$\mathbf{i}^2 = -1$$

Multiplying both sides by 9, we get

$$9\mathbf{i}^2 = -9$$

which can be transformed to

$$(3\mathbf{i})^2 = -9$$

Thus $3\mathbf{i}$ is the solution to the equation $x^2 = -9$. Similarly, for any positive number b, the solution to $x^2 = -b$ is $\sqrt{b}$ times $\mathbf{i}$. The product of a real number and $\mathbf{i}$ is called an *imaginary* number.

What about the equation $(x-1)^2 = -9$? We can solve this by setting $x - 1 = 3\mathbf{i}$, which yields $x = 1 + 3\mathbf{i}$. The sum of a real number and an imaginary number is called a *complex number.* A complex number has a *real part* and an *imaginary part.*

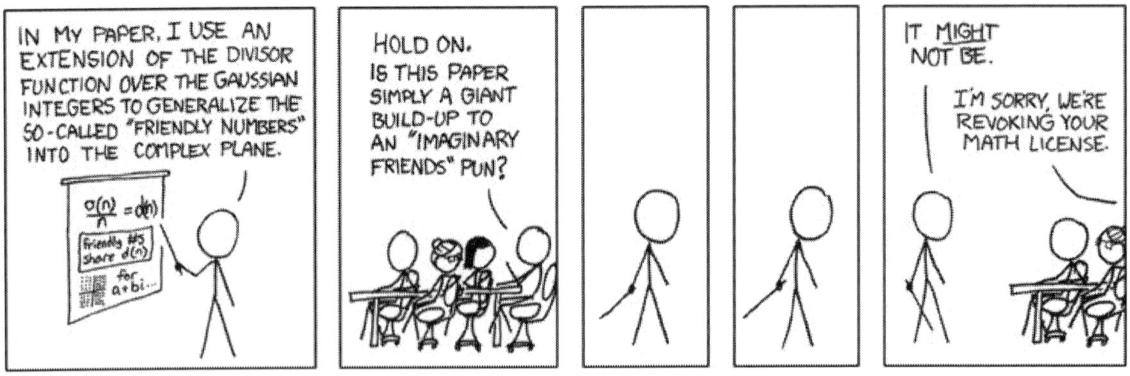

Math Paper (`http://xkcd.com/410`)

1.2 Complex numbers in Python

Python supports complex numbers. The square root of -9, the imaginary number 3**i**, is written `3j`.

```
>>> 3j
3j
```

Thus `j` plays the role of **i**. (In electrical engineering, i means "current")

The square root of -1, the imaginary number **i**, is written `1j` so as to avoid confusion with the variable `j`.

```
>>> j
Traceback (most recent call last):
  File "<stdin>", line 1, in <module>
NameError: name 'j' is not defined
>>> 1j
1j
```

Since Python allows the use of + to add a real number and an imaginary number, you can write the complex solution to $(x-1)^2 = -9$ as `1+3j`:

```
>>> 1+3j
(1+3j)
```

In fact, the operators +, -, *, /, and ** all work with complex numbers. When you add two complex numbers, the real parts are added and the imaginary parts are added.

```
>>> (1+3j) + (10+20j)
(11+23j)
>>> x=1+3j
>>> (x-1)**2
(-9+0j)
```

Python considers the value `(-9+0j)` to be a complex number even though its imaginary part is zero.

As in ordinary arithmetic, multiplication has precedence over addition; exponentiation has precedence over multiplication. These precedence rules are illustrated in the following evaluations.

```
>>> 1+2j*3
(1+6j)
>>> 4*3j**2
(-36+0j)
```

You can obtain the real and imaginary parts of a complex number using a dot notation.

```
>>> x.real
1.0
>>> x.imag
3.0
```

It is not an accident that the notation is that used in object-oriented programming languages to access instance variables (a.k.a. member variables). The complex numbers form a class in Python.

```
>>> type(1+2j)
<class 'complex'>
```

This class defines the procedures (a.k.a. methods, a.k.a. member functions) used in arithmetic operations on complex numbers.

1.3 Abstracting over *fields*

In programming languages, use of the same name (e.g. $+$) for different procedures operating on values of different datatypes is called *overloading*. Here's an example of why it's useful in the present context. Let us write a procedure `solve1(a,b, c)` to solve an equation of the form $ax + b = c$ where a is nonzero:

```
>>> def solve1(a,b,c): return (c-b)/a
```

It's a pretty simple procedure. It's the procedure you would write even if you had never heard of complex numbers. Let us use it to solve the equation $10x + 5 = 30$:

```
>>> solve1(10, 5, 30)
2.5
```

The remarkable thing, however, is that the same procedure can be used to solve equations involving complex numbers. Let's use it to solve the equation $(10 + 5\mathbf{i})x + 5 = 20$:

```
>>> solve1(10+5j, 5, 20)
(1.2-0.6j)
```

The procedure works even with complex arguments because the correctness of the procedure does not depend on what kind of numbers are supplied to it; it depends only on the fact that the *divide* operator is the inverse of the *multiply* operator and the *subtract* operator is the inverse of the *add* operator.

The power of this idea goes well beyond this simple procedure. Much of linear algebra—concepts, theorems, and, yes, procedures—works not just for the real numbers but also for the complex numbers and for other kinds of numbers as well. The strategy for achieving this is simple:

- The concepts, theorems, and procedures are stated in terms of the arithmetic operators +, -, *, and /.
- They assume only that these operators satisfy certain basic laws, such as commutativity ($a + b = b + a$) and distributivity ($a(b + c) = ab + ac$).

Because the concepts, theorems, and procedures rely only on these basic laws, we can "plug in" any system of numbers, called a *field*.[1] Different fields arise in different applications.

In this book, we illustrate the generality of linear algebra with three different fields.

- $\mathbb{R}$, the field of real numbers,
- $\mathbb{C}$, the field of complex numbers, and
- $GF(2)$, a field that consists of 0 and 1.

In object-oriented programming, one can use the name of a class to refer to the set of instances of that class, e.g. we refer to instances of the class `Rectangle` as Rectangles. In mathematics, one uses the name of the field, e.g. $\mathbb{R}$ or $GF(2)$, to refer also to the set of values.

[1] For the reader who knows about object-oriented programming, a field is analogous to a class satisfying an interface that requires it to possess methods for the arithmetic operators.

1.4 Playing with $\mathbb{C}$

Because each complex number `z` consists of two ordinary numbers, `z.real` and `z.imag`, it is traditional to think of `z` as specifying a *point*, a location, in the plane (the *complex plane*).

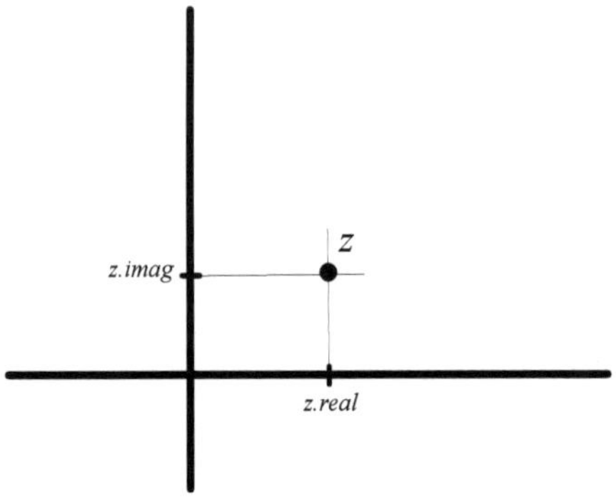

To build intuition, let us use a set S of complex numbers to represent a black-and-white image. For each location in the complex plane where we want a dot, we include the corresponding complex number in S.

The following figure shows $S = \{2+2\mathbf{i}, 3+2\mathbf{i}, 1.75+1\mathbf{i}, 2+1\mathbf{i}, 2.25+1\mathbf{i}, 2.5+1\mathbf{i}, 2.75+1\mathbf{i}, 3+1\mathbf{i}, 3.25+1\mathbf{i}\}$:

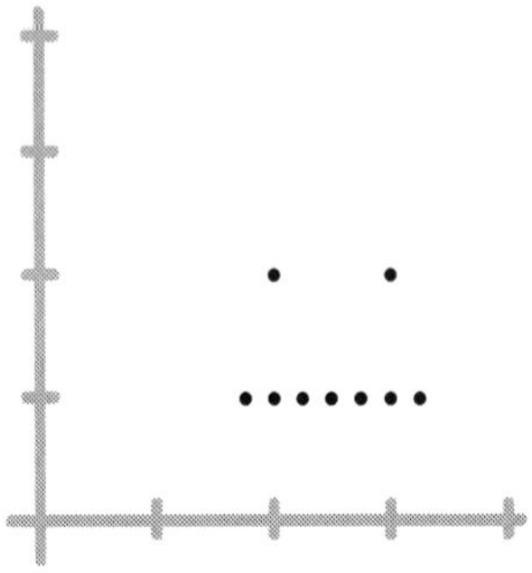

Task 1.4.1: First, assign to the variable S a list or set consisting of the complex numbers listed above.

We have provided a module `plotting` for showing points in the complex plane. The module defines a procedure `plot`. Import this class from the module as follows:

```
>>> from plotting import plot
```

Next, plot the points in S as follows:

```
>>>> plot(S, 4)
```

Python should create a window displaying the points of `S` in the complex plane. The first argument to `plot` is a collection of complex numbers (or 2-tuples). The second argument sets the scale of the plot; in this case, the window can show complex numbers whose real and imaginary parts have absolute value less than 4. The scale argument is optional and defaults to 1, and there is another optional argument that sets the size of the dots.

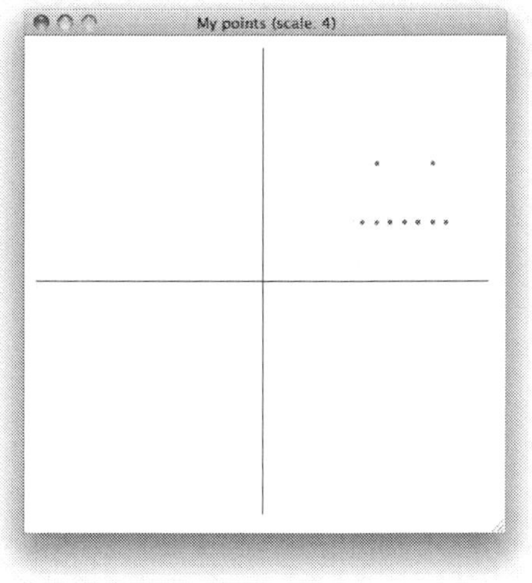

1.4.1 The absolute value of a complex number

The *absolute value* of a complex number z, written $|z|$ (and, in Python, `abs(z)`) is the distance from the origin to the corresponding point in the complex plane.

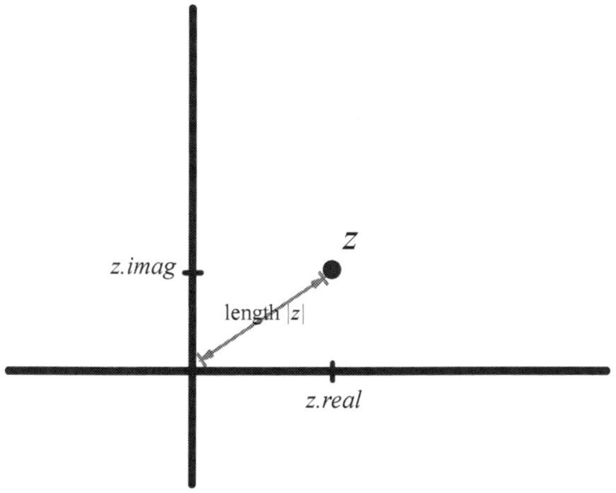

By the Pythagorean Theorem, $|z|^2 = (z.\text{real})^2 + (z.\text{imag})^2$.

```
>>> abs(3+4j)
5.0
>>> abs(1+1j)
1.4142135623730951
```

Definition 1.4.2: The *conjugate* of a complex number z, written $\bar{z}$, is defined as z.real $-$ z.imag.

In Python, we write `z.conjugate()`.

```
>>> (3+4j).conjugate()
(3-4j)
```

Using the fact that $\mathbf{i}^2 = -1$, we can get a formula for $|z|^2$ in terms of z and $\bar{z}$:

$$|z|^2 = z \cdot \bar{z} \tag{1.1}$$

Proof

$$\begin{aligned}
z \cdot \bar{z} &= (z.\text{real} + z.\text{imag}\,\mathbf{i}) \cdot (z.\text{real} - z.\text{imag}\,\mathbf{i}) \\
&= z.\text{real} \cdot z.\text{real} - z.\text{real} \cdot z.\text{imag}\,\mathbf{i} + z.\text{imag}\,\mathbf{i} \cdot z.\text{real} - z.\text{imag}\,\mathbf{i} \cdot z.\text{imag}\,\mathbf{i} \\
&= z.\text{real}^2 - z.\text{imag}\,\mathbf{i} \cdot z.\text{imag}\,\mathbf{i} \\
&= z.\text{real}^2 - z.\text{imag} \cdot z.\text{imag}\,\mathbf{i}^2 \\
&= z.\text{real}^2 + z.\text{imag} \cdot z.\text{imag}
\end{aligned}$$

where the last equality uses the fact that $\mathbf{i}^2 = -1$. □

1.4.2 Adding complex numbers

Suppose we add a complex number, say $1 + 2\mathbf{i}$, to each complex number z in S. That is, we derive a new set by applying the following function to each element of S:

$$f(z) = 1 + 2\mathbf{i} + z$$

This function increases each real coordinate (the x coordinate) by 1 and increases each imaginary coordinate (the y coordinate) by 2. The effect is to shift the picture one unit to the right and two units up:

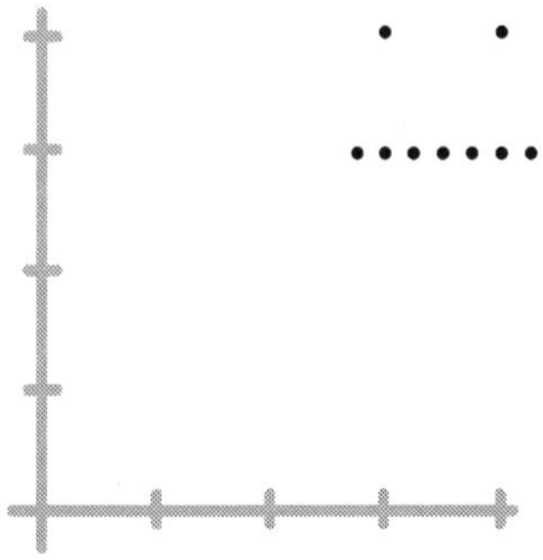

This transformation of the numbers in S is called a *translation*. A translation has the form

$$f(z) = z_0 + z \tag{1.2}$$

where z_0 is a complex number. Translations can take the picture anywhere in the complex plane. For example, adding a number z_0 whose real coordinate is negative would have the effect of translating the picture to the left.

Task 1.4.3: Create a new plot using a comprehension to provide a set of points derived from S by adding $1 + 2\mathbf{i}$ to each:

```
>>> plot({1+2j+z for z in S}, 4)
```

Quiz 1.4.4: The "left eye" of the set S of complex numbers is located at $2 + 2\mathbf{i}$. For what value of z_0 does the translation $f(z) = z_0 + z$ move the left eye to the origin?

Answer

$z_0 = -2 - 2\mathbf{i}$. That is, the translation is $f(z) = -2 - 2\mathbf{i} + z$.

Problem 1.4.5: Show that, for any two distinct points z_1 and z_2,

- there is a translation that maps z_1 to z_2,
- there is a translation that maps z_2 to z_1, and

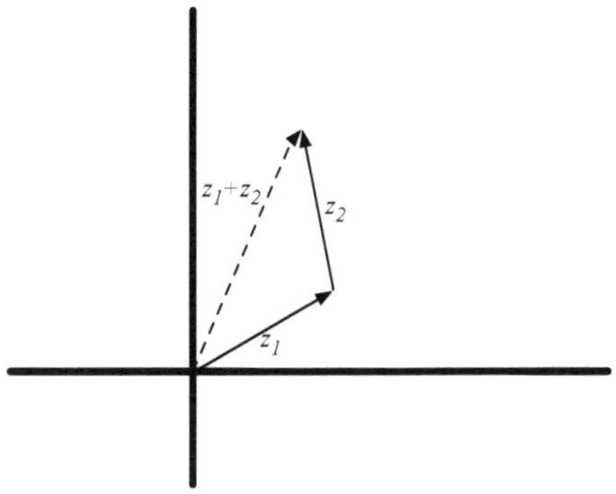

Figure 1.1: This figure illustrates the geometric interpretation of complex-number addition.

- there is *no* translation that both maps z_1 to z_2 and z_2 to z_1.

Complex numbers as arrows It is helpful to visualize a translation $f(z)$ by an arrow. The tail of the arrow is located at any point z in the complex plane; the head of the arrow is then located at the point $f(z)$, the translation of z. Of course, this representation is not unique.

Since a translation has the form $f(z) = z_0 + z$, we represent the translation by the complex number z_0. It is therefore appropriate to represent the complex number z_0 by an arrow.

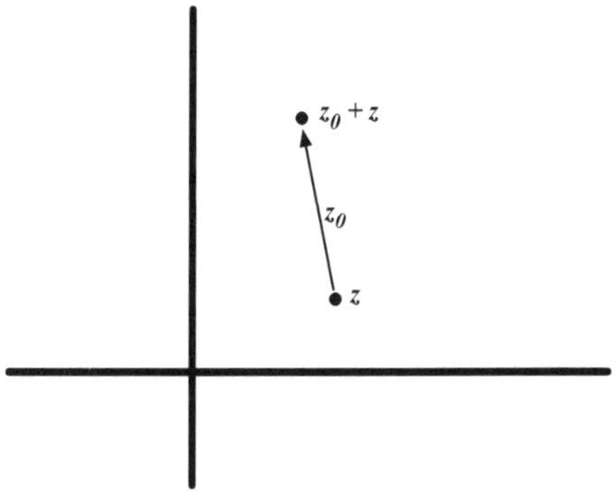

Again, the representation is not unique. For example, the vector $z_0 = 5 - 2\mathbf{i}$ can be represented by an arrow whose tail is at $0 + 0\mathbf{i}$ and whose head is at $5 - 2\mathbf{i}$, or one whose tail is at $1 + 1\mathbf{i}$ and whose head is at $6 - 1\mathbf{i}$, or....

Problem 1.4.6: Draw a diagram representing the complex number $z_0 = -3 + 3\mathbf{i}$ using two arrows with their tails located at different points.

Composing translations, adding arrows Let $f_1(z) = z_1 + z$ and $f_2(z) = z_2 + z$ be two translations. Then their composition is also a translation:

$$\begin{aligned}(f_2 \circ f_1)(z) &= f_2(f_1(z)) \\ &= f_2(z_1 + z) \\ &= z_2 + z_1 + z\end{aligned}$$

and is defined by $z \mapsto (z_2 + z_1) + z$. The idea that two translations can be collapsed into one is illustrated by Figure 1.1, in which each translation is represented by an arrow.

The translation arrow labeled by z_1 takes a point (in this case, the origin) to another point, which in turn is mapped by z_2 to a third point. The arrow mapping the origin to the third point is the composition of the two other translations, so, by the reasoning above, is $z_1 + z_2$.

1.4.3 Multiplying complex numbers by a positive real number

Now suppose we halve each complex number in S:

$$g(z) = \frac{1}{2}z$$

This operation simply halves the real coordinate and the imaginary coordinate of each complex number. The effect on the picture is to move all the points closer from the origin but also closer to each other:

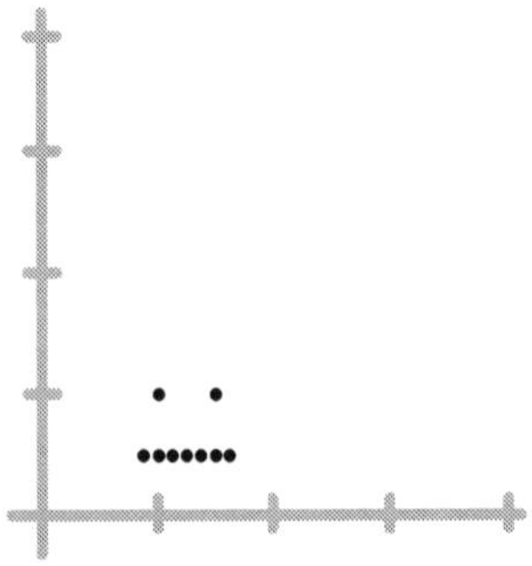

This operation is called *scaling*. The scale of the picture has changed. Similarly, doubling each complex number moves the points farther from the origin and from each other.

Task 1.4.7: Create a new plot titled "My scaled points" using a comprehension as in Task 1.4.3. The points in the new plot should be halves of the points in S.

1.4.4 Multiplying complex numbers by a negative number: rotation by 180 degrees

Here is the result of multiplying each complex number by -1:

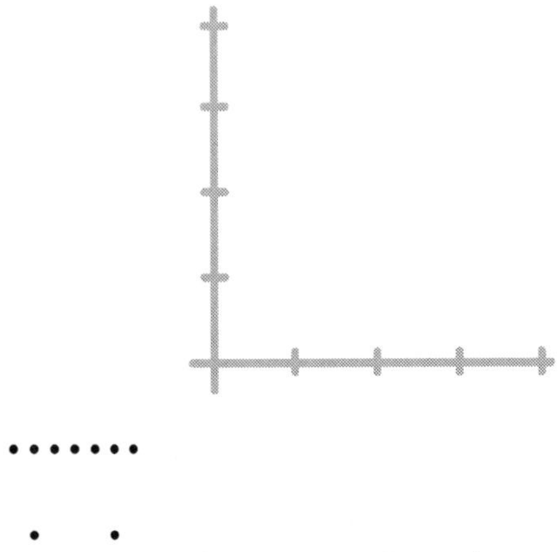

Think of the points as drawn on a shape that rotates about the origin; this picture is the result of rotating the shape by 180 degrees.

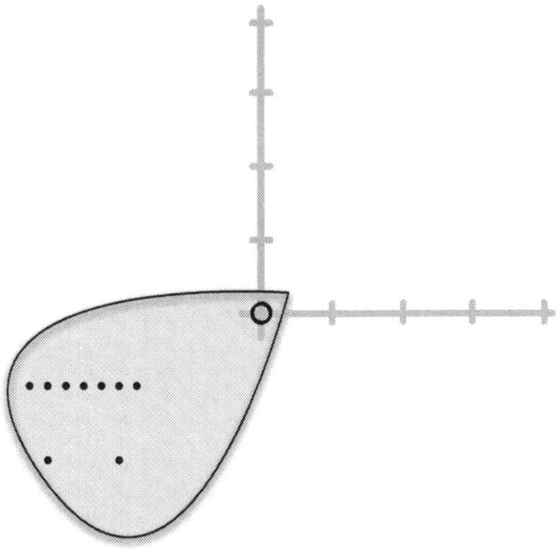

1.4.5 Multiplying by i: rotation by 90 degrees

"The number you have dialed is imaginary. Please rotate your phone by ninety degrees and try again."

How can we rotate the shape by only 90 degrees?

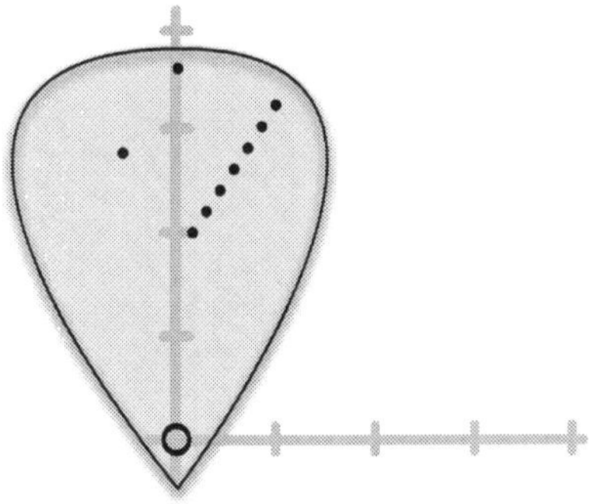

For this effect, a point located at (x, y) must be moved to $(-y, x)$. The complex number located at (x, y) is $x + \mathbf{i}y$. Now is our chance to use the fact that $\mathbf{i}^2 = -1$. We use the function

$$h(z) = \mathbf{i} \cdot z$$

Multiplying $x + \mathbf{i}y$ by $\mathbf{i}$ yields $\mathbf{i}x + \mathbf{i}^2 y$, which is $\mathbf{i}x - y$, which is the complex number represented by the point $(-y, x)$.

Task 1.4.8: Create a new plot "Rotated and scaled" in which the points of S are rotated by 90 degrees and scaled by 1/2. Use a comprehension in which the points of S are multiplied by a single complex number.

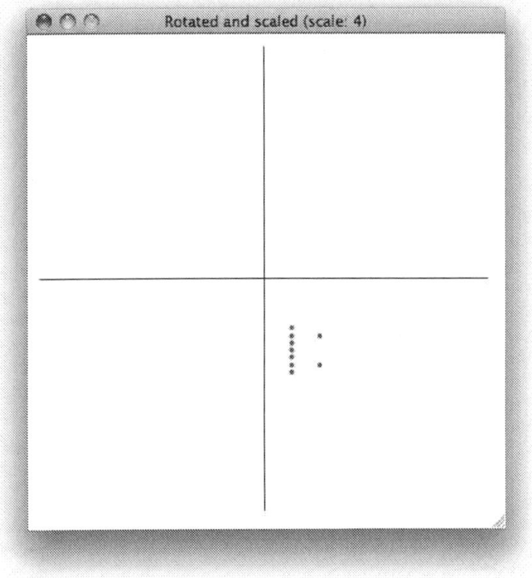

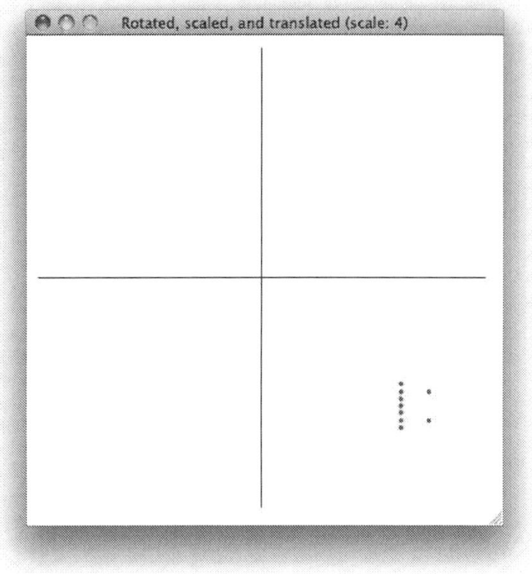

Task 1.4.9: Using a comprehension, create a new plot in which the points of S are rotated by 90 degrees, scaled by 1/2, and then shifted down by one unit and to the right two units. Use a comprehension in which the points of S are multiplied by one complex number and added to another.

Task 1.4.10: We have provided a module `image` with a procedure `file2image(filename)` that reads in an image stored in a file in the `.png` format. Import this procedure and invoke it, providing as argument the name of a file containing an image in this format, assigning the returned value to variable `data`. An example grayscale image, `img01.png`, is available for download.

The value of `data` is a list of lists, and `data[y][x]` is the intensity of pixel `(x,y)`. Pixel `(0,0)` is at the bottom-left of the image, and pixel `(width-1, height-1)` is at the top-right. The intensity of a pixel is a number between 0 (black) and 255 (white).

Use a comprehension to assign to a list `pts` the set of complex numbers $x + y\mathbf{i}$ such that the image intensity of pixel (x, y) is less than 120, and plot the list `pts`.

Task 1.4.11: Write a Python procedure `f(z)` that, applied to the points in `pts`, would translate those points to result in a set of points centered at the origin. Write a comprehension in terms of `pts` and `f` whose value is the set of translated points, and plot the value.

Task 1.4.12: Repeat Task 1.4.8 with the points in pts instead of the points in S.

1.4.6 The unit circle in the complex plane: *argument* and angle

We shall see that it is not a coincidence that rotation by 180 or 90 degrees can be represented by complex multiplication: any rotation can be so represented. However, it is convenient to use *radians* instead of degrees to measure the angle of rotation.

The *argument* of a complex number on the unit circle

Consider the unit circle—the circle of radius one, centered at the origin of the complex plane.

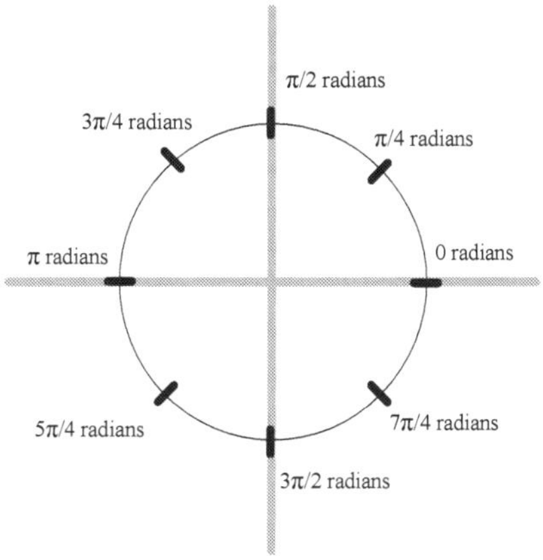

A point z on the circle is represented by the distance an ant would have to travel counterclockwise along the circle to get to z if the ant started at $1 + 0\mathbf{i}$, the rightmost point of the circle. We call this number the *argument* of z.

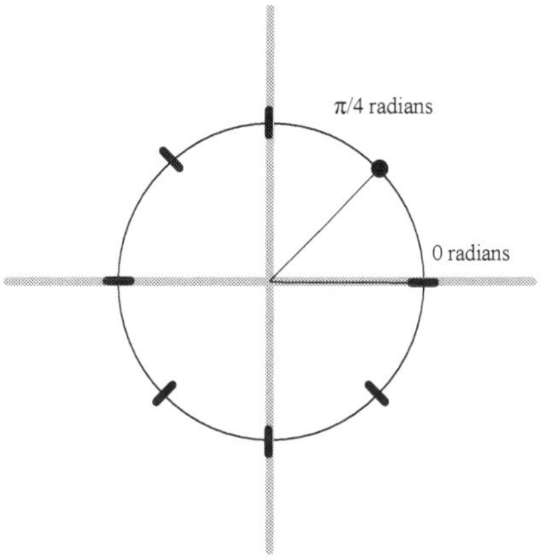

Example 1.4.13: Since the circumference of the circle is 2π, the point halfway around the circle has an argument of π, and the point one-eighth of the way around has an argument of $\pi/4$.

The angle formed by two complex numbers on the unit circle

We have seen how to label points on the unit circle by distances. We can similarly assign a number to the *angle* formed by the line segments from the origin to two points z_1 and z_2 on the circle. The angle, measured in radians, is the distance along the circle traversed by an ant walking counter-

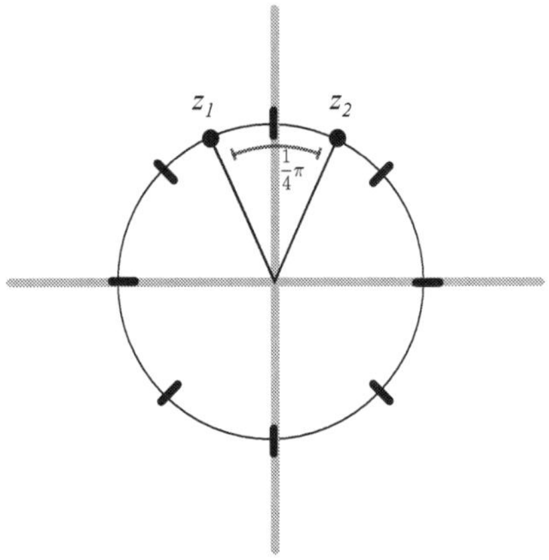

clockwise from z_2 to z_1.

Example 1.4.14: Let z_1 be the point on the circle that has argument $\frac{5}{16}\pi$, and let z_2 be the point on the circle that has argument $\frac{3}{16}\pi$. An ant starting at z_2 and traveling to z_1 would travel a distance of $\frac{1}{8}\pi$ counterclockwise along the circle, so $\frac{1}{8}\pi$ is the angle between the origin-to-z_1 line segment and the origin-to-z_2 line segment.

Remark 1.4.15: The argument of z is the angle formed by z with $1 + 0\mathbf{i}$.

1.4.7 Euler's formula

He calculated just as men breathe, as eagles sustain themselves in the air.
Said of Leonhard Euler

We turn to a formula due to Leonhard Euler, a remarkable mathematician who contributed to the foundation for many subfields of mathematics: number theory and algebra, complex analysis, calculus, differential geometry, fluid mechanics, topology, graph theory, and even music theory and cartography. Euler's formula states that, for any real number θ, $e^{\mathbf{i}\cdot\theta}$ is the point z on the unit circle with argument θ. Here e is the famous transcendental number 2.718281828....

Example 1.4.16: The point $-1 + 0\mathbf{i}$ has argument π. Plugging π into Euler's formula yields the surprising equation $e^{\mathbf{i}\pi} + 1 = 0$.

e to the π times $\mathbf{i}$ (`http://xkcd.com/179/`)

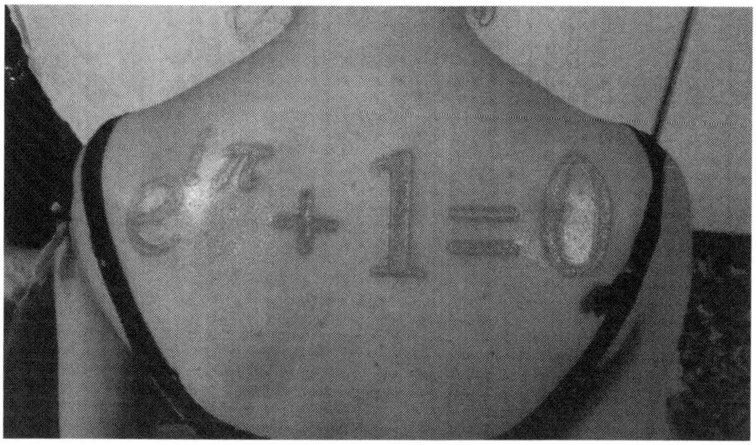

photo taken at 3PiCon in Springfield, MA, by Cory Doctorow

Task 1.4.17: From the module `math`, import the definitions `e` and `pi`. Let `n` be the integer 20. Let `w` be the complex number $e^{2\pi \mathrm{i}/n}$. Write a comprehension yielding the list consisting of $w^0, w^1, w^2, \ldots, w^{n-1}$. Plot these complex numbers.

1.4.8 Polar representation for complex numbers

Euler's formula gives us a convenient representation for complex numbers that lie on the unit circle. Now consider any complex number z. Let L be the line segment in the complex plane from the origin to z, and let z' be the point at which this line segment intersects the unit circle.

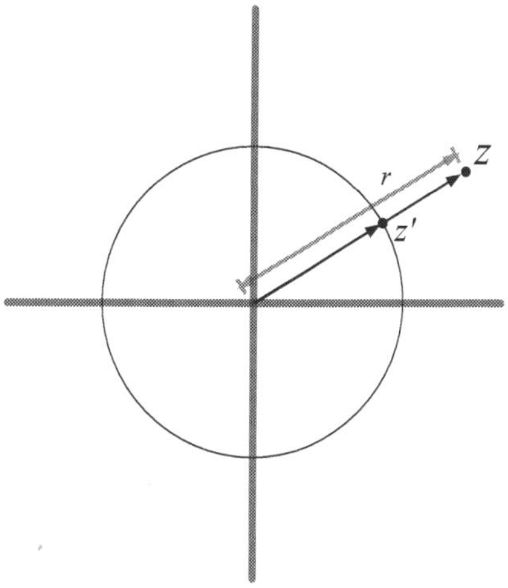

Let r be the length of the line segment to z. Viewing z' as the result of scaling down z, we have

$$z' = \frac{1}{r} z$$

Let θ be the argument of z'. Euler's formula tells us that $z' = e^{\theta \mathrm{i}}$. We therefore obtain

$$z = re^{\theta \mathrm{i}}$$

The astute student might recognize that r and θ are the polar coordinates of z. In the context of complex numbers, we define the *argument* of z to be θ, and we define the *absolute value* of z (written $|z|$) to be r.

1.4.9 The First Law of Exponentiation

When powers multiply, their exponents add:

$$e^u e^v = e^{u+v}$$

We can use this rule to help us understand how to rotate a complex number z. We can write

$$z = re^{\theta \mathrm{i}}$$

where $r = |z|$ and $\theta = \arg z$.

1.4.10 Rotation by τ radians

Let τ be a number of radians. The rotation of z by τ should have the same absolute value as z but its argument should be τ more than that of z, i.e. it should be $re^{(\theta+\tau)\mathbf{i}}$. How do we obtain this number from z?

$$\begin{aligned} re^{(\theta+\tau)\mathbf{i}} &= re^{\theta\mathbf{i}}e^{\tau\mathbf{i}} \\ &= ze^{\tau\mathbf{i}} \end{aligned}$$

Thus the function that rotates by τ is simply

$$f(z) = ze^{\tau\mathbf{i}}$$

Task 1.4.18: Recall from Task 1.4.1 the set S of complex numbers. Write a comprehension whose value is the set consisting of rotations by $\pi/4$ of the elements of S. Plot the value of this comprehension.

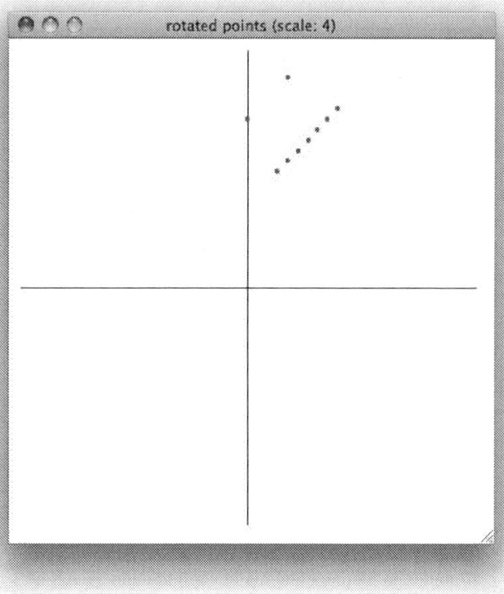

Task 1.4.19: Similarly, recall from Task 1.4.10 the list pts of points derived from an image. Plot the rotation by $\pi/4$ of the complex numbers comprising pts.

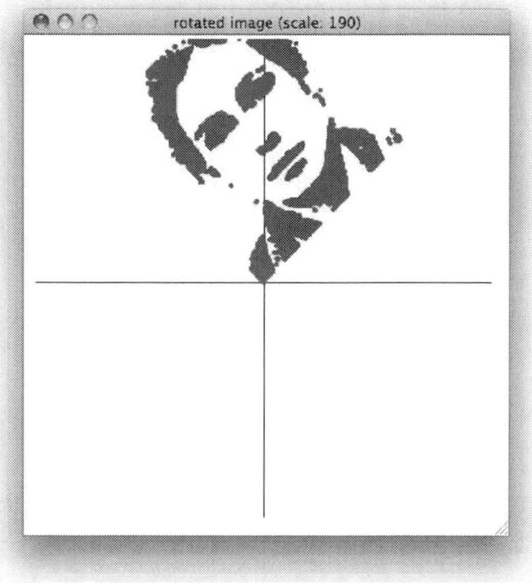

1.4.11 Combining operations

Task 1.4.20: Write a comprehension that transforms the set pts by translating it so the image is centered, then rotating it by $\pi/4$, then scaling it by half. Plot the result.

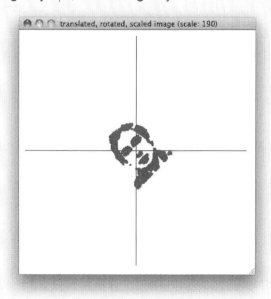

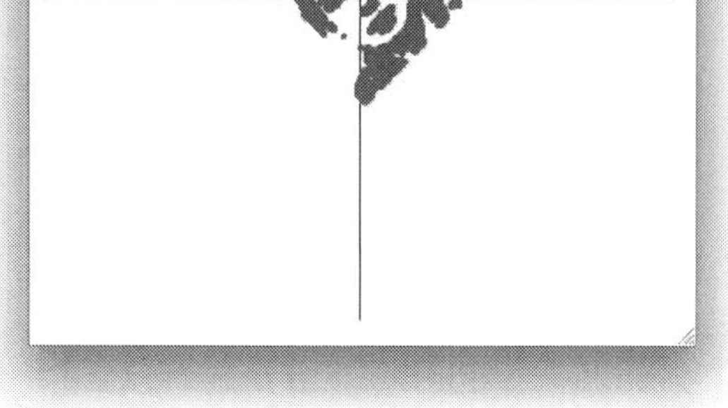

Because the complex numbers form a field, familiar algebraic rules can be used. For example, $a \cdot (b \cdot z) = (a \cdot b) \cdot z$. Using this rule, two scaling operations can be combined into one; scaling by 2 and then by 3 is equivalent to scaling by 6.

Similarly, since rotation is carried out by multiplication, two rotations can be combined into one; rotating by $\frac{\pi}{4}$ (multiplying by $e^{\frac{\pi}{4}\mathbf{i}}$) and and then rotating by $\frac{\pi}{3}$ (multiplying by $e^{\frac{\pi}{3}\mathbf{i}}$) is equivalent to multiplying by $e^{\frac{\pi}{4}\mathbf{i}} \cdot e^{\frac{\pi}{3}\mathbf{i}}$, which is equal to $e^{\frac{\pi}{4}\mathbf{i}+\frac{\pi}{3}\mathbf{i}}$, i.e. rotating by $\frac{\pi}{4} + \frac{\pi}{3}$.

Since scaling and rotation both consist in multiplication, a rotation and a scaling can be combined: rotating by $\frac{\pi}{4}$ (multiplying by $e^{\frac{\pi}{4}\mathbf{i}}$) and then scaling by $\frac{1}{2}$ is equivalent to multiplying by $\frac{1}{2}e^{\frac{\pi}{4}\mathbf{i}}$.

1.4.12 Beyond two dimensions

The complex numbers are so convenient for transforming images—and, more generally, sets of points in the plane—one might ask whether there is a similar approach to operating on points in three dimensions. We discuss this in the next chapter.

1.5 Playing with $GF(2)$

$GF(2)$ is short for *Galois Field 2*. Galois was a mathematician, born in 1811, who while in his teens essentially founded the field of abstract algebra. He died in a duel at age twenty.

The field $GF(2)$ is very easy to describe. It has two elements, 0 and 1. Arithmetic over $GF(2)$ can be summarized in two small tables:

×	0	1
0	0	0
1	0	1

+	0	1
0	0	1
1	1	0

Addition is modulo 2. It is equivalent to exclusive-or. In particular, $1 + 1 = 0$.

Subtraction is identical to addition. The negative of 1 is again 1, and the negative of 0 is again 0.

Multiplication in $GF(2)$ is just like ordinary multiplication of 0 and 1: multiplication by 0 yields 0, and 1 times 1 is 1. You can divide by 1 (as usual, you get the number you started with) but dividing by zero is illegal (as usual).

We provide a module, `GF2`, with a very simple implementation of $GF(2)$. It defines a value, `one`, that acts as the element 1 of $GF(2)$. Ordinary zero plays the role of the element 0 of $GF(2)$. (For visual consistency, the module defines `zero` to be the value 0.)

```
>>> from GF2 import one
>>> one*one
one
>>> one*0
0
>>> one + 0
one
>>> one+one
0
>>> -one
one
```

1.5.1 Perfect secrecy revisited

In Chapter 0, we described a cryptosystem that achieves perfect secrecy (in transmitting a single bit). Alice and Bob randomly choose the key k uniformly from $\{\clubsuit, \heartsuit\}$. Subsequently, Alice uses the following encryption function to transform the plaintext bit p to a cyphertext bit c:

p	k	c
0	$\clubsuit$	0
0	$\heartsuit$	1
1	$\clubsuit$	1
1	$\heartsuit$	0

The encryption method is just $GF(2)$ addition in disguise! When we replace $\clubsuit$ with 0 and $\heartsuit$ with 1, the encryption table becomes the addition table for $GF(2)$:

p	k	c
0	0	0
0	1	1
1	0	1
1	1	0

For each plaintext $p \in GF(2)$, the function $k \mapsto k + p$ (mapping $GF(2)$ to $GF(2)$) is invertible (hence one-to-one and onto). Therefore, when the key k is chosen uniformly at random, the cyphertext is also distributed uniformly. This shows that the scheme achieves perfect secrecy.

Using integers instead of $GF(2)$

Why couldn't Alice and Bob use, say, ordinary integers instead of $GF(2)$? After all, for each $x \in \mathbb{Z}$, the function $y \mapsto x + y$ mapping $\mathbb{Z}$ to $\mathbb{Z}$ is also invertible. The reason this cannot work as a cryptosystem is that there is no uniform distribution over $\mathbb{Z}$, so the first step—choosing a key—is impossible.

Encrypting long messages

How, then, are we to encrypt a long message? Students of computer science know that a long message can be represented by a long string of bits. Suppose the message to be encrypted will consist of n bits. Alice and Bob should select an equally long sequence of key bits $k_1 \ldots k_n$. Now, once Alice has selected the plaintext $p_1 \ldots p_n$, she obtains the cyphertext $c_1 \ldots c_n$ one bit at a time:

$$\begin{aligned} c_1 &= k_1 + p_1 \\ c_2 &= k_2 + p_2 \\ &\vdots \\ c_n &= k_n + p_n \end{aligned}$$

We argue informally that this system has perfect secrecy. The earlier argument shows that each bit c_i of cyphertext tells Eve nothing about the corresponding bit p_i of plaintext; certainly the bit c_i tells Eve nothing about any of the other bits of plaintext. From this we infer that the system has perfect secrecy.

Our description of the multi-bit system is a bit cumbersome, and the argument for perfect secrecy is rather sketchy. In Chapter 2, we show that using vectors over $GF(2)$ simplify the presentation.

The one-time pad

The cryptosystem we have described is called the *one-time pad.* As suggested by the name, it is crucial that each bit of key be used only once, i.e. that each bit of plaintext be encrypted with its bit of key. This can be a burden for two parties that are separated for long periods of time because the two parties must agree before separating on many bits of key.

Starting in 1930, the Soviet Union used the one-time pad for communication. During World War II, however, they ran out of bits of key and began to re-use some of the bits. The US and Great Britain happen to discover this; they exploited it (in a top-secret project codenamed VENONA) to partially decrypt some 1% of the encrypted messages, revealing, for example, the involvement of Julius Rosenberg and Alger Hiss in espionage.

Problem 1.5.1: An 11-symbol message has been encrypted as follows. Each symbol is represented by a number between 0 and 26 ($A \mapsto 0, B \mapsto 1, \ldots, Z \mapsto 25, space \mapsto 26$). Each number is represented by a five-bit binary sequence ($0 \mapsto 00000, 1 \mapsto 00001, ..., 26 \mapsto 11010$). Finally, the resulting sequence of 55 bits is encrypted using a flawed version of the one-time pad: the key is not 55 random bits but 11 copies of the same sequence of 5 random bits. The cyphertext is

10101 00100 10101 01011 11001 00011 01011 10101 00100 11001 11010

Try to find the plaintext.

1.5.2 Network coding

Consider the problem of streaming video through a network. Here is a simple example network:

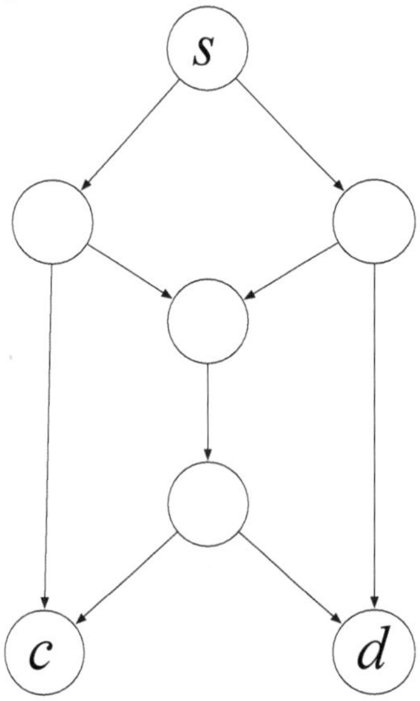

The node at the top labeled s needs to stream a video to each of the two customer nodes, labeled c and d, at the bottom. Each link in the network has a capacity of 1 megabit per second. The video stream, however, requires 2 megabits per second. If there were only one customer, this would be no problem; as shown below, the network can handle two simultaneous 1-megabit-per-second streams from s to c:

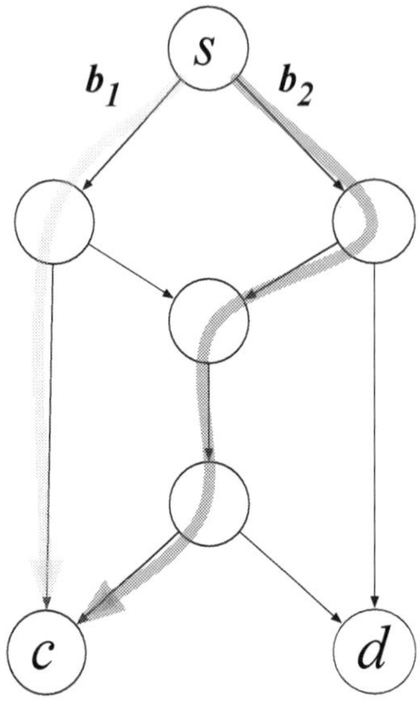

A million times a second, one bit b_1 is sent along one path and another bit b_2 is sent along another path. Thus the total rate of bits delivered to the customer is 2 megabits per second.

However, as shown below, we can't use the same scheme to deliver two bitstreams to each of two customers because the streams contend for bandwidth on one of the network links.

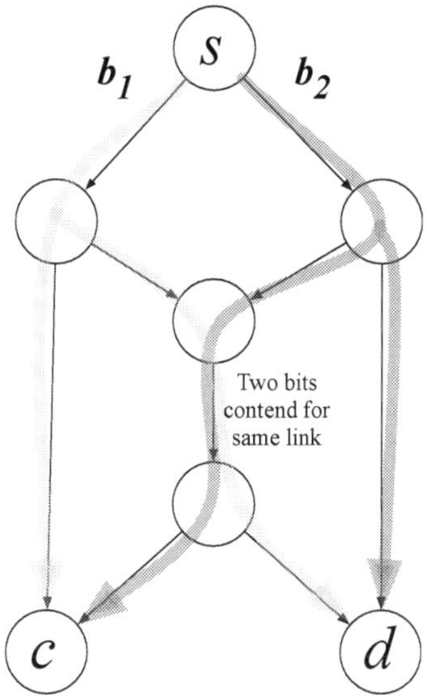

$GF(2)$ to the rescue! We can use the fact that network nodes can do a tiny bit (!) of computation. The scheme is depicted here:

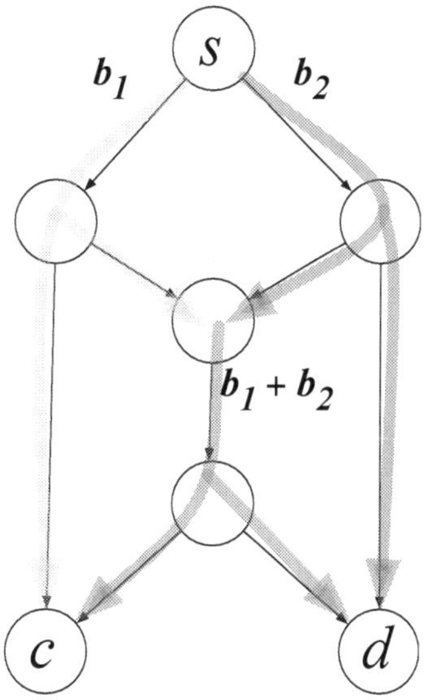

At the centermost node, the bits b_1 and b_2 arrive and are combined by $GF(2)$ addition to obtain a single bit. That single bit is transmitted as shown to the two customers c and d. Customer c receives bit b_1 and the sum $b_1 + b_2$, so can also compute the bit b_2. Customer d receives bit b_2 and the sum $b_1 + b_2$, so can also compute the bit b_1.

We have shown that a network that appears to support streaming only one megabit per second to a pair of customers actually supports streaming two megabits per second. This approach to routing can of course be generalized to larger networks and more customers; the idea is called *network coding*.

1.6 Review questions

- Name three fields.
- What is the conjugate of a complex number? What does it have to do with the absolute value of a complex number?
- How does complex-number addition work?
- How does complex-number multiplication work?
- How can translation be defined in terms of complex numbers?
- How can scaling be defined in terms of complex numbers?

- How can rotation by 180 degrees be defined in terms of complex numbers?
- How can rotation by 90 degrees be defined in terms of complex numbers?
- How does addition of $GF(2)$ values work?
- How does multiplication of $GF(2)$ values work?

1.7 Problems

Python comprehension problems

Write each of the following three procedures using a comprehension:

Problem 1.7.1: `my_filter(L, num)`
input: list of numbers and a number.
output: list of numbers not containing a multiple of `num`.
example: given `list = [1,2,4,5,7]` and `num = 2`, return `[1,5,7]`.

Problem 1.7.2: `my_lists(L)`
input: list `L` of non-negative integers.
output: a list of lists: for every element x in `L` create a list containing $1, 2, \ldots, x$.
example: given `[1,2,4]` return `[[1],[1,2],[1,2,3,4]]`. *example:* given `[0]` return `[[]]`.

Problem 1.7.3: `my_function_composition(f,g)`
input: two functions f and g, represented as dictionaries, such that $g \circ f$ exists.
output: dictionary that represents the function $g \circ f$.
example: given $f = \{0{:}\text{'a'}, 1{:}\text{'b'}\}$ and $g = \{\text{'a'}{:}\text{'apple'}, \text{'b'}{:}\text{'banana'}\}$, return $\{0{:}\text{'apple'}, 1{:}\text{'banana'}\}$.

Python loop problems

For procedures in the following problems, use the following format:

```
def <ProcedureName>(L):
    current = ...
    for x in L:
        current = ...
    return current
```

The value your procedure initially assigns to `current` turns out to be the return value in the case when the input list `L` is empty. This provides us insight into how the answer should be defined in that case. Note: You are not allowed to use Python built-in procedures `sum(·)` and `min(·)`.

Problem 1.7.4: `mySum(L)`
Input: list of numbers
Output: sum of numbers in the list

Problem 1.7.5: `myProduct(L)`
input: list of numbers
output: product of numbers in the list

Problem 1.7.6: `myMin(L)`
input: list of numbers
output: minimum number in the list

Problem 1.7.7: `myConcat(L)`
input: list of strings
output: concatenation of all the strings in L

Problem 1.7.8: `myUnion(L)`
input: list of sets
output: the union of all sets in L.

In each of the above problems, the value of `current` is combined with an element of `myList` using some operation $\diamond$. In order that the procedure return the correct result, `current` should be initialized with the *identity element* for the operation $\diamond$, i.e. the value i such that $i \diamond x = x$ for any value x.

It is a consequence of the structure of the procedure that, when the input list is empty, the output value is the initial value of `current` (since in this case the body of the loop is never executed). It is convenient to define this to be the correct output!

Problem 1.7.9: Keeping in mind the comments above, what should be the answer for each of the following?

1. the sum of the numbers in an empty set;
2. the product of the numbers in an empty set;
3. the minimum of the numbers in an empty set;
4. the concatenation of an empty list of strings;
5. the union of an empty list of sets

What goes wrong when we try to apply this reasoning to define the intersection of an empty list of sets?

Complex addition practice

Problem 1.7.10: Each of the following problems asks for the sum of two complex numbers. For each, write the solution and illustrate it with a diagram like that of Figure 1.1. The arrows you draw should (roughly) correspond to the vectors being added.

a. $(3 + 1\mathbf{i}) + (2 + 2\mathbf{i})$

b. $(-1 + 2\mathbf{i}) + (1 - 1\mathbf{i})$

c. $(2 + 0\mathbf{i}) + (-3 + .001\mathbf{i})$

d. $4(0 + 2\mathbf{i}) + (.001 + 1\mathbf{i})$

Multiplication of exponentials

Problem 1.7.11: Use the First Rule of Exponentiation (Section 1.4.9) to express the product of two exponentials as a single exponential. For example, $e^{(\pi/4)\mathbf{i}}e^{(\pi/4)\mathbf{i}} = e^{(\pi/2)\mathbf{i}}$.

a. $e^{1\mathbf{i}}e^{2\mathbf{i}}$

b. $e^{(\pi/4)\mathbf{i}}e^{(2\pi/3)\mathbf{i}}$

c. $e^{-(\pi/4)\mathbf{i}}e^{(2\pi/3)\mathbf{i}}$

Combining operations on complex numbers

Problem 1.7.12: Write a procedure `transform(a,b, L)` with the following spec:

- *input:* complex numbers a and b, and a list L of complex numbers
- *output:* the list of complex numbers obtained by applying $f(z) = az + b$ to each complex number in L

Next, for each of the following problems, explain which value to choose for `a` and `b` in order to achieve the specified transformation. If there is no way to achieve the transformation, explain.

a. Translate z one unit up and one unit to the right, then rotate ninety degrees clockwise, then scale by two.

b. Scale the real part by two and the imaginary part by three, then rotate by forty-five degrees counterclockwise, and then translate down two units and left three units.

$GF(2)$ arithmetic

Problem 1.7.13: For each of the following problems, calculate the answer over $GF(2)$.

a. $1 + 1 + 1 + 0$

b. $1 \cdot 1 + 0 \cdot 1 + 0 \cdot 0 + 1 \cdot 1$

c. $(1 + 1 + 1) \cdot (1 + 1 + 1 + 1)$

Network coding

Problem 1.7.14: Copy the example network used in Section 1.5.2. Suppose the bits that need to be transmitted in a given moment are $b_1 = 1$ and $b_2 = 1$. Label each link of the network with the bit transmitted across it according to the network-coding scheme. Show how the customer nodes c and d can recover b_1 and b_2.

Chapter 2

The Vector

> One of the principal objects of theoretical research in my department of knowledge is to find the point of view from which the subject appears in its greatest simplicity.
>
> Josiah Willard Gibbs

Josiah Gibbs, the inventor of modern vector analysis, was up against stiff competition. The dominant system of analysis, *quaternions*, had been invented by Sir William Rowan Hamilton. Hamilton had been a bona fide prodigy. By age five, he was reported to have learned Latin, Greek, and Hebrew. By age ten, he had learned twelve languages, including Persian, Arabic, Hindustani and Sanskrit.

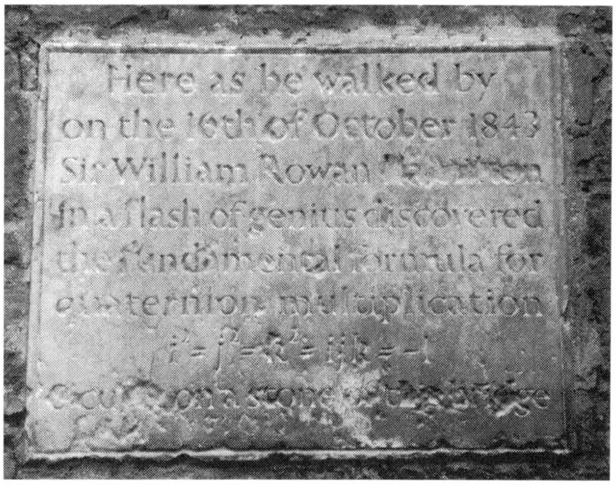

Figure 2.1: William Rowan Hamilton, the inventor of the theory of quaternions, and the plaque on Brougham Bridge, Dublin, commemorating Hamilton's act of vandalism

Hamilton was a Trinity man. His uncle (who raised him) had gone to Trinity College in Dublin, and Hamilton matriculated there. He was first in every subject. However, he did not complete college; while still an undergraduate, he was appointed Professor of Astronomy.

Among Hamilton's contributions to mathematics is his elegant theory of quaternions. We saw in Chapter 1 that the field of complex numbers makes it simple to describe transformations on points in the plane, such as translations, rotations, and scalings. Hamilton struggled to find a similar approach. When the solution came to him while he was walking along Dublin's Royal Canal with his wife, he committed a particularly egregious form of vandalism, carving the defining equations in the stone of Brougham Bridge.

Hamilton described his epipheny in a letter to a friend:

> And here there dawned on me the notion that we must admit, in some sense, a fourth dimension of space for the purpose of calculating with triples ... An electric circuit seemed to close, and a spark flashed forth.

Quaternions occupied much of Hamilton's subsequent life.

Josiah Willard Gibbs, on the other hand, was a Yale man. His father, Josiah Willard Gibbs, was a professor at Yale, and the son matriculated there at age fifteen. He got his Ph.D. at Yale, tutored at Yale, and spent three years studying in Europe, after which he returned to become a professor at Yale and remained there for the rest of his life. He developed *vector analysis* as an alternative to quaternions.

Figure 2.2: Josiah Willard Gibbs, the inventor of vector analysis

For twenty years vector analysis did not appear in published form (the primary source was unpublished notes) until Gibbs finally agreed to publish a book on the topic. It began to displace the theory of quaternions because it was more convenient to use.

However, it had the drawback of having been invented by an American. The eminent British physicist Peter Guthrie Tait, a former student of Hamilton and a partisan of quaternions, attacked mercilessly, writing, for example,

> Professor Willard Gibbs must be ranked as one of the retarders of ... progress in virtue of his pamphlet on *Vector Analysis*; a sort of hermaphrodite monster."
>
> Tait, *Elementary Treatise on Quaternions*

Today, quaternions are still used, especially in representing rotations in three dimensions. It has its advocates in computer graphics and computer vision. However, it is safe to say that, in the end, vector analysis won out. It is used in nearly every field of science and engineering, in economics, in mathematics, and, of course, in computer science.

2.1 What is a vector?

No, not that kind of vector!

The word *vector* comes from the Latin for "carrier". We don't plan to study pests; the term comes from a vector's propensity to move something from one location to another.

In some traditional math courses on linear algebra, we are taught to think of a vector as a list of numbers:

$$[3.14159, 2.718281828, -1.0, 2.0]$$

You need to know this way of writing a vector because it is commonly used.[1] Indeed, we will sometimes represent vectors using Python's lists.

Definition 2.1.1: A vector with four entries, each of which is a real number, is called a *4-vector over* $\mathbb{R}$.

The entries of a vector must all be drawn from a single field. As discussed in the previous chapter, three examples of fields are $\mathbb{R}$, $\mathbb{C}$, and $GF(2)$. Therefore we can have vectors over each of these fields.

Definition 2.1.2: For a field $\mathbb{F}$ and a positive integer n, a vector with n entries, each belonging to $\mathbb{F}$, is called an *n-vector over* $\mathbb{F}$. The set of n-vectors over $\mathbb{F}$ is denoted $\mathbb{F}^n$.

For example, the set of 4-vectors over $\mathbb{R}$ is written $\mathbb{R}^4$.

This notation might remind you of the notation $\mathbb{F}^D$ for the set of *functions* from D to $\mathbb{F}$. Indeed, I suggest you interpret $\mathbb{F}^d$ as shorthand for $\mathbb{F}^{\{0,1,2,3,\ldots,d-1\}}$ According to this interpretation, $\mathbb{F}^d$ is the set of functions from $\{0, 1, \ldots, d-1\}$ to $\mathbb{F}$.

For example, the 4-vector we started with, $[3.14159, 2.718281828, -1.0, 2.0]$, is in fact the function

$$\begin{array}{rcl} 0 & \mapsto & 3.14159 \\ 1 & \mapsto & 2.718281828 \\ 2 & \mapsto & -1.0 \\ 3 & \mapsto & 2.0 \end{array}$$

[1] Often parentheses are used instead of brackets.

2.2 Vectors are functions

excerpt from *Matrix Revisited* (http://xkcd.com/566/)

Once we embrace this interpretation—once we accept that vectors are functions—a world of applications opens to us.

Example 2.2.1: Documents as vectors: Here's an example from a discipline called *information retrieval* that addresses the problem of finding information you want from a corpus of documents.

Much work in information retrieval has been based on an extremely simple model that disregards grammar entirely: the *word-bag* model of documents. A document is considered just a *multiset* (also called a *bag*) of words. (A multiset is like a set but can contain more than one copy of an element. The number of copies is called the *multiplicity* of the element.)

We can represent a bag of words by a function f whose domain is the set of words and whose co-domain is the set of real numbers. The image of a word is its multiplicity. Let WORDS be the set of words (e.g. English words). We write

$$f : \text{WORDS} \longrightarrow \mathbb{R}$$

to indicate that f maps from WORDS to $\mathbb{R}$.

Such a function can be interpreted as representing a vector. We would call it a *WORDS-vector over* $\mathbb{R}$.

Definition 2.2.2: For a finite set D and a field $\mathbb{F}$, a D*-vector over* $\mathbb{F}$ is a function from D to $\mathbb{F}$.

This is a computer scientist's definition; it lends itself to representation in a data structure. It differs in two important ways from a mathematician's definition.

- I require the domain D to be finite. This has important mathematical consequences: we will state theorems that would not be true if D were allowed to be infinite. There are important mathematical questions that are best modeled using functions with infinite domains, and you will encounter them if you continue in mathematics.
- The traditional, abstract approach to linear algebra does not directly define vectors at all. Just as a field is defined as a set of values with some operations (+, -, *, /) that satisfy certain algebraic laws, a vector space is defined as a set with some operations that satisfy certain algebraic laws; then vectors are the things in that set. This approach is more general but it is more abstract, hence harder for some people to grasp. If you continue in mathematics, you will become very familiar with the abstract approach.

Returning to the more concrete approach we take in this book, According to the notation from Section 0.3.3, we use $\mathbb{F}^D$ to denote the set of functions with domain D and co-domain $\mathbb{F}$, i.e. the set of all D-vectors over $\mathbb{F}$.

Example 2.2.3: To illustrate this notation for vectors as functions, consider the following:

(a.) $\mathbb{R}^{WORDS}$: The set of all $WORDS$-vectors over $\mathbb{R}$, seen in Example 2.2.1 (Page 66).

(b.) $GF(2)^{\{0,1,\ldots,n-1\}}$: The set of all n-vectors over $GF(2)$

2.2.1 Representation of vectors using Python dictionaries

We will sometimes use Python's lists to represent vectors. However, we have decreed that a vector is a function with finite domain, and Python's dictionaries are a convenient representation of functions with finite domains. Therefore we often use dictionaries in representing vectors.

For example, the 4-vector of Section 2.1 could be represented as `{0:3.14159, 1:2.718281828, 2:-1.0, 3:2.0}`.

In Example 2.2.1 (Page 66) we discussed the word-bag model of documents, in which a document is represented by a WORDS-vector over $\mathbb{R}$. We could represent such a vector as a dictionary but the dictionary would consist of perhaps two hundred thousand key-value pairs. Since a typical document uses a small subset of the words in WORDS, most of the values would be equal to zero. In information-retrieval, one typically has many documents; representing each of them by a two-hundred-thousand-element dictionary would be profligate. Instead, we adopt the convention of allowing the omission of key-value pairs whose values are zero. This is called a *sparse representation*. For example, the document "The rain in Spain falls mainly on the plain" would be represented by the dictionary

```
{'on': 1, 'Spain': 1, 'in': 1, 'plain': 1, 'the': 2, 'mainly': 1,
                                        'rain': 1, 'falls': 1}
```

There is no need to explicitly represent the fact that this vector assigns zero to `'snow'`, `'France'`, `'primarily'`, `'savannah'`, and the other elements of WORDS.

2.2.2 Sparsity

A vector most of whose values are zero is called a *sparse* vector. If no more than k of the entries are nonzero, we say the vector is k-*sparse*. A k-sparse vector can be represented using space proportional to k. Therefore, for example, when we represent a corpus of documents by WORD-vectors, the storage required is proportional to the total number of words comprising all the documents.

Vectors that represent data acquired via physical sensors (e.g. images or sound) are not likely to be sparse. In a future chapter, we will consider a computational problem in which the goal, given a vector and a parameter k, is to find the "closest" k-sparse vector. After we learn what it means for vectors to be close, it will be straightforward to solve this computational problem.

A solution to this computational problem would seem to be the key to *compressing* images and audio segments, i.e. representing them compactly so more can be stored in the same amount of computer memory. This is correct, but there is a hitch: unfortunately, the vectors representing images or sound are not even close to sparse vectors. In Section 5.2, we indicate the way around this obstacle. In Chapter 10, we explore some compression schemes based on the idea.

In Chapter 4, we introduce matrices and their representation. Because matrices are often sparse, in order to save on storage and computational time we will again use a dictionary representation in which zero values need not be represented.

However, many matrices arising in real-world problems are not sparse in the obvious sense. In Chapter 11, we investigate another form of sparsity for matrices, *low rank*. Low-rank matrices arise in analyzing data to discover factors that explain the data. We consider a computational probem in which the goal, given a matrix and a parameter k, is to find the closest matrix whose rank is at most k. We show that linear algebra provides a solution for this computational problem. It is at the heart of a widely used method called *principal component analysis*, and we will explore some of its applications.

2.3 What can we represent with vectors?

We've seen two examples of what we can represent with vectors: multisets and sets. Now I want to give some more examples.

binary string An n-bit binary string 10111011, e.g. the secret key to a cryptosystem, can be represented by an n-vector over $GF(2)$, $[1, 0, 1, 1, 1, 0, 1, 1]$. We will see how some simple cryptographic schemes can be specified and analyzed using linear algebra.

attributes In learning theory, we will consider data sets in which each item is represented by a collection of *attribute names* and *attribute values*. This collection is in turn represented by a function that maps attribute names to the corresponding values.

For example, perhaps the items are congresspeople. Each congressperson is represented by his or her votes on a set of bills. A single vote is represented by +1, -1, or 0 (*aye*, *nay*, or *abstain*). We will see in Lab 2.12 a method for measuring the difference between two congresspersons' voting policies.

Perhaps the items are consumers. Each consumer is represented by his or her *age*, *education level*, and *income*, e.g.

```
>>> Jane = {'age':30, 'education level':16, 'income':85000}
```

Given data on which consumers liked a particular product, one might want to come up with a function that predicted, for a new consumer vector, whether the consumer would like the product. This is an example of *machine learning*. In Lab 8.4, we will consider vectors that describe tissue samples, and use a rudimentary machine-learning technique to try to predict whether a cancer is benign or malignant.

State of a system We will also use functions/vectors to represent different states of an evolving system. The state the world might be represented, for example, by specifying the the population of each of the five most populous countries:

```
{'China':1341670000, 'India':1192570000, 'US':308745538,
 'Indonesia':237556363, 'Brazil':190732694}
```

We will see in Chapter 12 that linear algebra provides a way to analyze a system that evolves over time according to simple known rules.

Probability distribution Since a finite probability distribution is a function from a finite domain to the real numbers, e.g.

```
{1:1/6, 2:1/6, 3:1/6, 4:1/6, 5:1/6, 6:1/6}
```

it can be considered a vector. We will see in Chapter 12 that linear algebra provides a way to analyze a random process that evolves over time according to simple probabilistic rules. One such random process underlies the original definition of PageRank, the method by which Google ranks pages.

Image A black-and-white 1024×768 image can be viewed as a function from the set of pairs $\{(i, j) : 0 \le i < 1024, 0 \le j < 768\}$ to the real numbers, and hence as a vector. The pixel-coordinate pair (i, j) maps to a number, called the *intensity* of pixel (i, j). We will study several applications of representing images by vectors, e.g. subsampling, blurring, searching for a specified subimage, and face detection.

Example 2.3.1: As an example of a black and white image, consider an 4x8 gradient, represented as a vector in dictionary form (and as an image), where 0 is black and 255 is white:

```
{(0,0): 0,    (0,1): 0,    (0,2): 0,    (0,3): 0,
 (1,0): 32,   (1,1): 32,   (1,2): 32,   (1,3): 32,
 (2,0): 64,   (2,1): 64,   (2,2): 64,   (2,3): 64,
 (3,0): 96,   (3,1): 96,   (3,2): 96,   (3,3): 96,
 (4,0): 128,  (4,1): 128,  (4,2): 128,  (4,3): 128,
 (5,0): 160,  (5,1): 160,  (5,2): 160,  (5,3): 160,
 (6,0): 192,  (6,1): 192,  (6,2): 192,  (6,3): 192,
 (7,0): 224,  (7,1): 224,  (7,2): 224,  (7,3): 224}
```

Point in space We saw in Chapter 1 that points in the plane could be represented by complex numbers. Here and henceforth, we use *vectors* to represent points in the plane, in three dimensions, and in higher-dimensional spaces.

Task 2.3.2: In this task, we will represent a vector using a Python list.
In Python, assign to the variable L a list of 2-element lists:

```
>>> L = [[2, 2], [3, 2], [1.75, 1], [2, 1], [2.25, 1], [2.5, 1], [2.75,
  1], [3, 1], [3.25, 1]]
```

Use the plot module described in Task 1.4.1 to plot these 2-vectors.

```
>>> plot(L, 4)
```

Unlike complex numbers, vectors can represent points in a higher-dimensional space, e.g. a three-dimensional space:

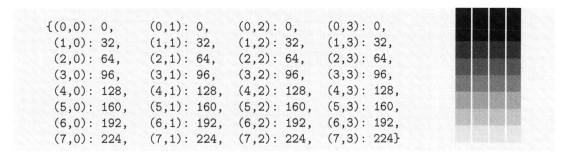

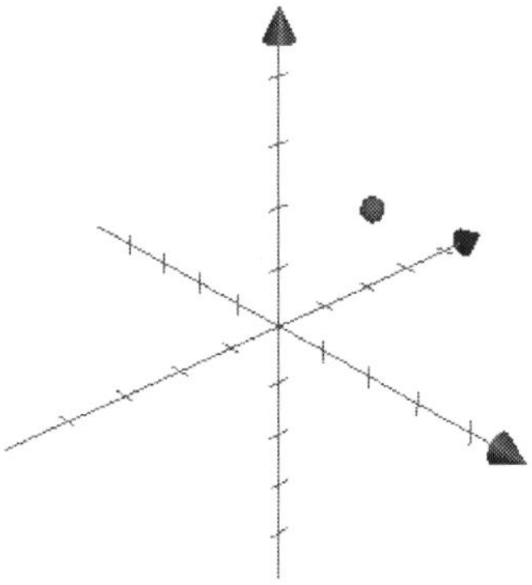

2.4 Vector addition

We have seen examples of what vectors can represent. Now we study the operations performed with vectors. We have seen that vectors are useful for representing geometric points. The concept of a vector originated in geometry, and it is in the context of geometry that the basic vector operations are most easily motivated. We start with *vector addition*.

2.4.1 Translation and vector addition

We saw in Chapter 1 that *translation* was achieved in the complex plane by a function $f(z) = z_0 + z$ that adds a complex number z_0 to its input complex number; here we similarly achieve translation by a function $f(\boldsymbol{v}) = \boldsymbol{v}_0 + \boldsymbol{v}$ that adds a vector to its input vector.

Definition 2.4.1: Addition of n-vectors is defined in terms of addition of corresponding entries:

$$[u_1, u_2, \ldots, u_n] + [v_1, v_2, \ldots, v_n] = [u_1 + v_1, u_2 + v_2, \ldots, u_n + v_n]$$

For 2-vectors represented in Python as 2-element lists, the addition procedure is as follows:

```
def add2(v,w):
    return [v[0]+w[0], v[1]+w[1]]
```

Quiz 2.4.2: Write the translation "go east one mile and north two miles" as a function from 2-vectors to 2-vectors, using vector addition. Next, show the result of applying this function to the vectors $[4, 4]$ and $[-4, -4]$.

Answer

$f(\boldsymbol{v}) = [1, 2] + \boldsymbol{v}$

$$\begin{aligned} f([4,4]) &= [5,6] \\ f([-4,-4]) &= [-3,-2] \end{aligned}$$

Since a vector such as $[1, 2]$ corresponds to a translation, we can think of the vector as "carrying" something from one point to another, e.g. from $[4, 4]$ to $[5, 6]$ or from $[-4, -4]$ to $[-3, -2]$. This is the sense in which a vector is a carrier.

Task 2.4.3: Recall the list L defined in Task 2.3.2. Enter the procedure definition for 2-vector addition, and use a comprehension to plot the points obtained from L by adding $[1, 2]$ to each:

```
>>> plot([add2(v, [1,2]) for v in L], 4)
```

Quiz 2.4.4: Suppose we represent n-vectors by n-element lists. Write a procedure addn to compute the sum of two vectors so represented.

Answer

```
def addn(v, w): return [x+y for (x,y) in zip(v,w)]
```

or

```
def addn(v, w): return [v[i]+w[i] for i in range(len(v))]
```

Every field $\mathbb{F}$ has a zero element, so the set $\mathbb{F}^D$ of D-vectors over $\mathbb{F}$ necessarily has a *zero vector*, a vector all of whose entries have value zero. I denote this vector by $\mathbf{0}_D$, or merely by $\mathbf{0}$ if it is not necessary to specify D.

The function $f(\boldsymbol{v}) = \boldsymbol{v} + \mathbf{0}$ is a translation that leaves its input unchanged.

2.4.2 Vector addition is associative and commutative

Two properties of addition in a field are *associativity*

$$(x + y) + z = x + (y + z)$$

and *commutativity*

$$x + y = y + x$$

Since vector addition is defined in terms of an associative and commutative operation, it too is associative and commutative:

Proposition 2.4.5 (Associativity and Commutativity of Vector Addition): For any vectors $\boldsymbol{u}, \boldsymbol{v}, \boldsymbol{w}$,

$$(\boldsymbol{u} + \boldsymbol{v}) + \boldsymbol{w} = \boldsymbol{u} + (\boldsymbol{v} + \boldsymbol{w})$$

and

$$u + v = v + u$$

2.4.3 Vectors as arrows

Like complex numbers in the plane, n-vectors over $\mathbb{R}$ can be visualized as arrows in $\mathbb{R}^n$. The 2-vector $[3, 1.5]$ can be represented by an arrow with its tail at the origin and its head at $(3, 1.5)$

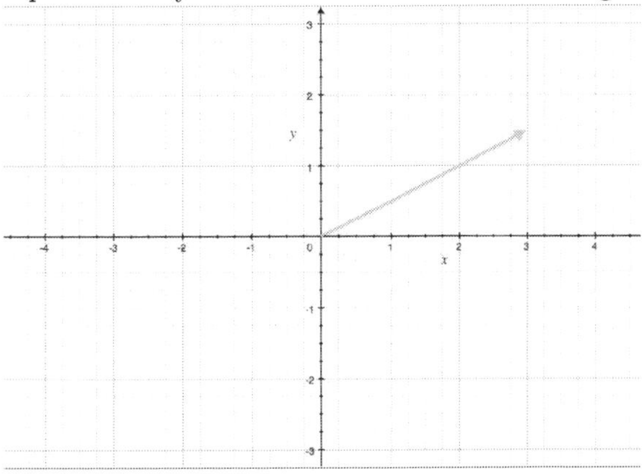

or, equivalently, by an arrow whose tail is at $(-2, -1)$ and whose head is at $(1, 0.5)$:

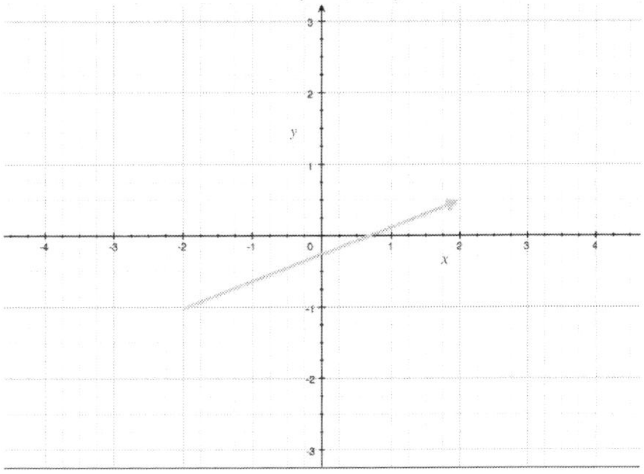

Exercise 2.4.6: Draw a diagram representing the vector $[-2, 4]$ using two different arrows.

In three dimensions, for example, the vector $[1, 2, 3]$ can be represented by an arrow whose tail is at the origin and whose head is at $[1, 2, 3]$

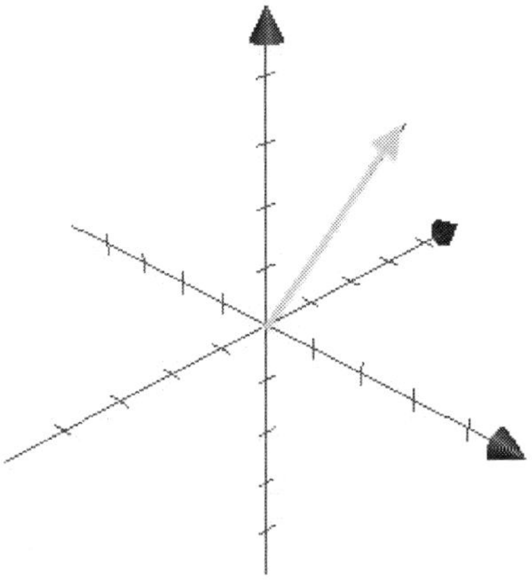

or by an arrow whose tail is at $[0, 1, 0]$ and whose head is at $[1, 3, 3]$:

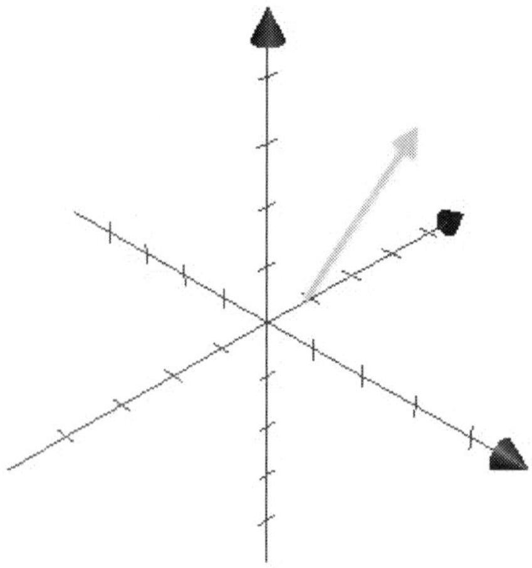

Like complex numbers, addition of vectors over $\mathbb{R}$ can be visualized using arrows. To add $\boldsymbol{u}$ and $\boldsymbol{v}$, place the tail of $\boldsymbol{v}$'s arrow on the head of $\boldsymbol{u}$'s arrow, and draw a new arrow (to represent the sum) from the tail of $\boldsymbol{u}$ to the head of $\boldsymbol{v}$.

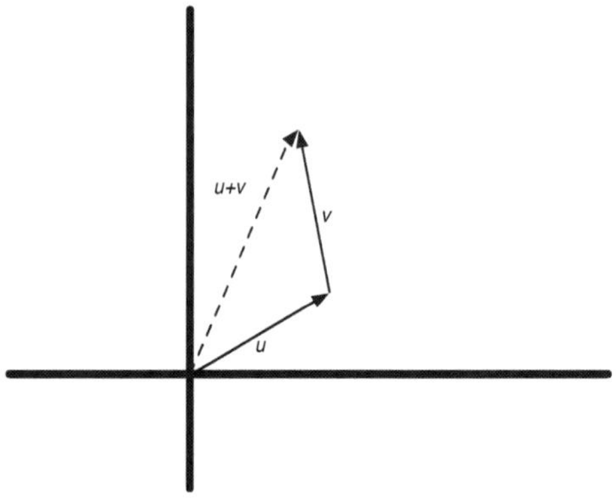

We can interpret this diagram as follows: the translation corresponding to $\boldsymbol{u}$ can be composed with the translation corresponding to $\boldsymbol{v}$ to obtain the translation corresponding to $\boldsymbol{u}+\boldsymbol{v}$.

Exercise 2.4.7: Draw a diagram illustrating $[-2,4]+[1,2]$.

2.5 Scalar-vector multiplication

We saw in Chapter 1 that *scaling* could be represented in the complex plane by a function $f(z) = r\,z$ that multiplies its complex-number input by a positive real number r, and that multiplying by a negative number achieves a simultaneous scaling and rotation by 180 degrees. The analogous operation for vectors is called *scalar-vector multiplication*. In the context of vectors, a field element (e.g. a number) is called a *scalar* because it can be be used to scale a vector via multiplication. In this book, we typically use Greek letters (e.g. α, β, γ) to denote scalars.

Definition 2.5.1: Multiplying a vector $\boldsymbol{v}$ by a scalar α is defined as multiplying each entry of $\boldsymbol{v}$ by α:

$$\alpha\,[v_1, v_2, \ldots, v_n] = [\alpha\,v_1, \alpha\,v_2, \ldots, \alpha\,v_n]$$

Example 2.5.2: $2\,[5,4,10] = [2\cdot 5, 2\cdot 4, 2\cdot 10] = [10,8,20]$

Quiz 2.5.3: Suppose we represent n-vectors by n-element lists. Write a procedure `scalar_vector_mult(alpha, v)` that multiplies the vector `v` by the scalar `alpha`.

Answer

```
def scalar_vector_mult(alpha, v):
    return [alpha*v[i] for i in range(len(v))]
```

Task 2.5.4: Plot the result of scaling the vectors in L by 0.5, then plot the result of scaling them by -0.5.

How shall we interpret an expression such as $2\,[1,2,3] + [10,20,30]$? Do we carry out the scalar-vector multiplication first or the vector addition? Just as multiplication has precedence over addition in ordinary arithmetic, scalar-vector multiplication has precedence over vector addition. Thus (unless parentheses indicate otherwise) scalar-vector multiplication happens first, and the result of the expression above is $[2,4,6] + [10,20,30]$, which is $[12,24,36]$.

2.5.1 Scaling arrows

Scaling a vector over $\mathbb{R}$ by a positive real number changes the length of the corresponding arrow without changing its direction. For example, an arrow representing the vector $[3, 1.5]$ is this:

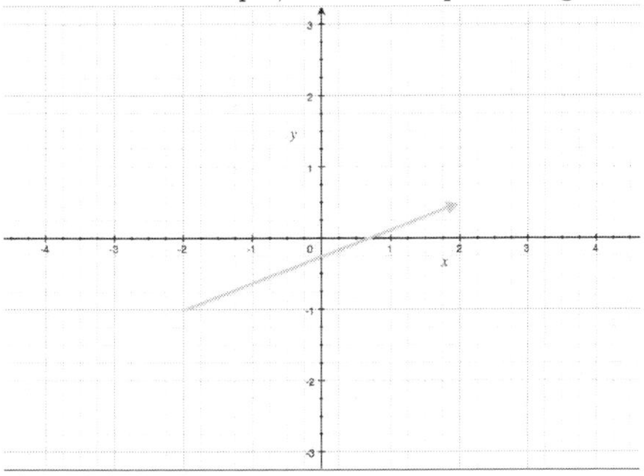

and an arrow representing two times this vector is this:

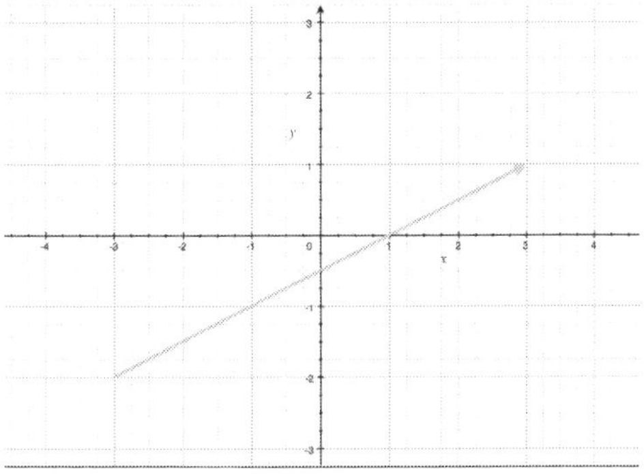

The vector $[3, 1.5]$ corresponds to the translation $f(\boldsymbol{v}) = [3, 1.5] + \boldsymbol{v}$, and two times this vector ($[6, 3]$) corresponds to a translation in the same direction but twice as far.

Multiplying a vector by a negative number negates all the entries. As we have seen in connection with complex numbers, this reverses the direction of the corresponding arrow. For example, negative two times $[3, 1.5]$ is $[-6, -3]$, which is represented by the arrow

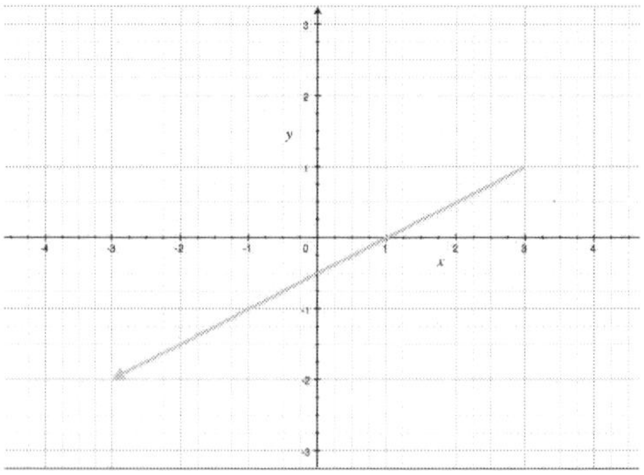

2.5.2 Associativity of scalar-vector multiplication

Multiplying a vector by a scalar and then multiplying the result by another scalar can be simplified:

Proposition 2.5.5 (Associativity of scalar-vector multiplication): $\alpha(\beta \boldsymbol{v}) = (\alpha\beta)\boldsymbol{v}$

Proof

To show that the left-hand side equals the right-hand side, we show that each entry of the left-hand side equals the corresponding enty of the right-hand side. For each element k of the domain D, entry k of $\beta\boldsymbol{v}$ is $\beta\boldsymbol{v}[k]$, so entry k of $\alpha(\beta\boldsymbol{v})$ is $\alpha(\beta\boldsymbol{v}[k])$. Entry k of $(\alpha\beta)\boldsymbol{v}$ is $(\alpha\beta)\boldsymbol{v}[k]$. By the field's associative law, $\alpha(\beta\boldsymbol{v}[k])$ and $(\alpha\beta)\boldsymbol{v}[k]$ are equal. □

2.5.3 Line segments through the origin

Let $\boldsymbol{v}$ be the 2-vector $[3, 2]$ over $\mathbb{R}$. Consider the set of scalars $\{0, 0.1, 0.2, 0.3, \ldots, 0.9, 1.0\}$. For each scalar α in this set, $\alpha\,\boldsymbol{v}$ is a vector that is somewhat shorter than $\boldsymbol{v}$ but points in the same direction:

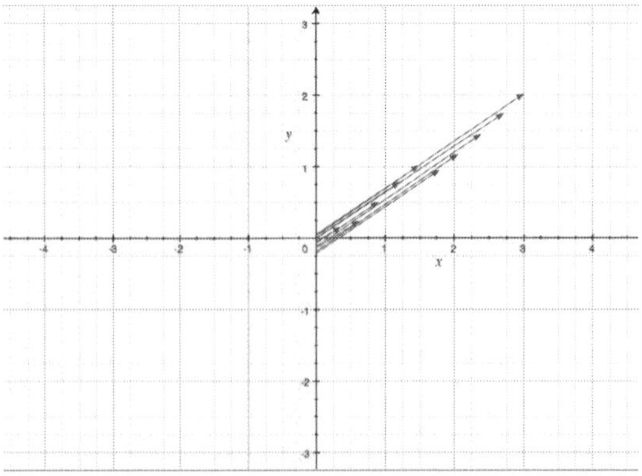

The following plot shows the points obtained by multiplying each of the scalars by $\boldsymbol{v}$:

```
plot([scalar_vector_mult(i/10, v) for i in range(11)], 5)
```

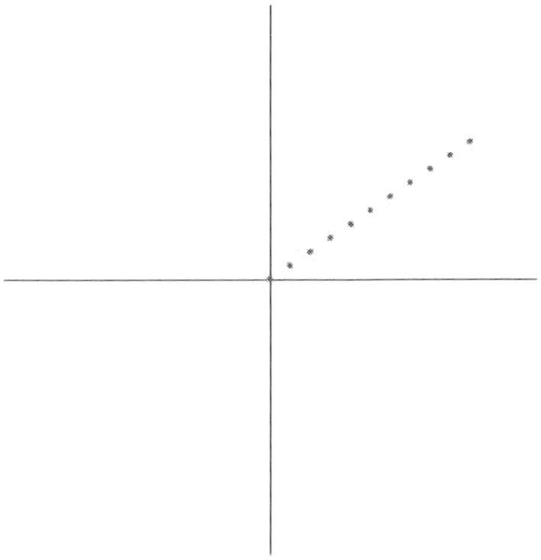

Hmm, seems to be tracing out the line segment from the origin to the point $(3, 2)$. What if we include as scalar multipliers *all* the real numbers between 0 and 1? The set of points

$$\{\alpha \boldsymbol{v} \; : \; \alpha \in \mathbb{R}, 0 \leq \alpha \leq 1\}$$

forms the line segment between the origin and $\boldsymbol{v}$. We can visualize this by plotting not all such points (even Python lacks the power to process an uncountably infinite set of points) but a sufficiently dense sample, say a hundred points:

```
plot([scalar_vector_mult(i/100, v) for i in range(101)], 5)
```

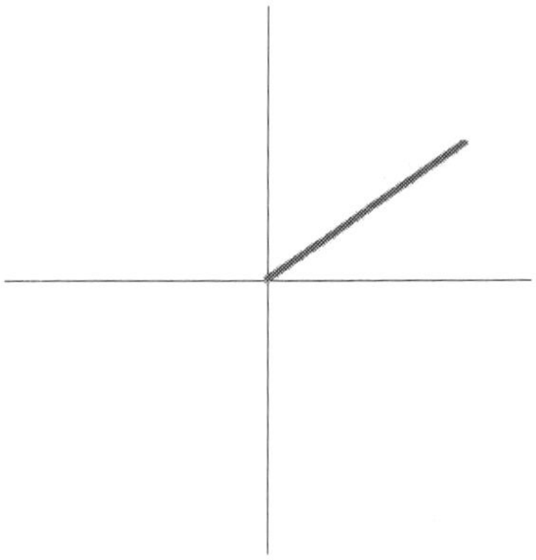

2.5.4 Lines through the origin

As long as we have permitted an infinite set of scales, let's go all out. What shape do we obtain when α ranges over all real numbers? The scalars bigger than 1 give rise to somewhat larger copies of $\boldsymbol{v}$. The negative scalars give rise to vectors pointing in the opposite direction. Putting these together,

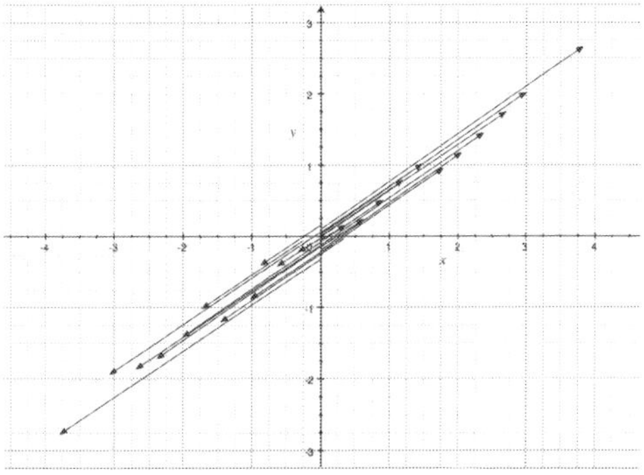

we see that the points of

$$\{\alpha \boldsymbol{v} \; : \; \alpha \in \mathbb{R}\}$$

forms the (infinite) line through the origin and through $\boldsymbol{v}$:

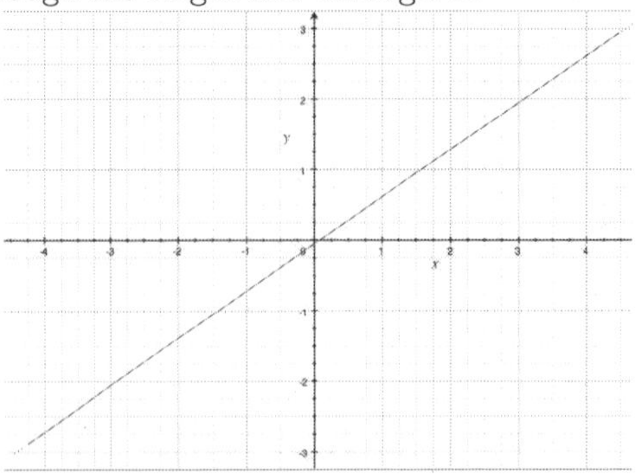

Review question: *Express the line segment between the origin and another point through the origin as a set of scalar multiples of a single vector.*

Review question: *Express a line through the origin as the set of scalar multiples of a single vector.*

2.6 Combining vector addition and scalar multiplication

2.6.1 Line segments and lines that don't go through the origin

Great—we can describe the set of points forming a line or line segment through the origin. It would be a lot more useful if we could describe the set of points forming an arbitrary line or line segment; we could then, for example, plot street maps:

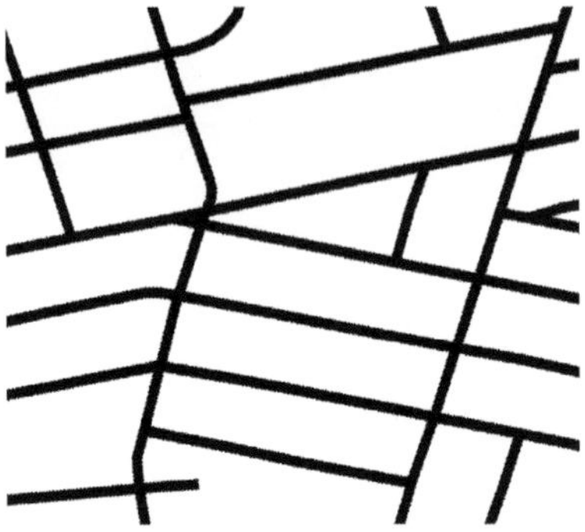

We already know that the points forming the segment from $[0, 0]$ to $[3, 2]$ are $\{\alpha\,[3, 2] \;:\; \alpha \in \mathbb{R}, 0 \leq \alpha \leq 1\}$. By applying the translation $[x, y] \mapsto [x + 0.5, y + 1]$ to these points,

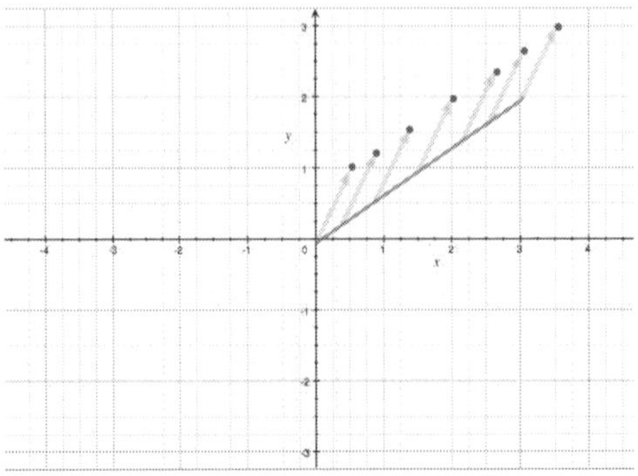

we obtain the line segment from $[0.5, 1]$ to $[3.5, 3]$:

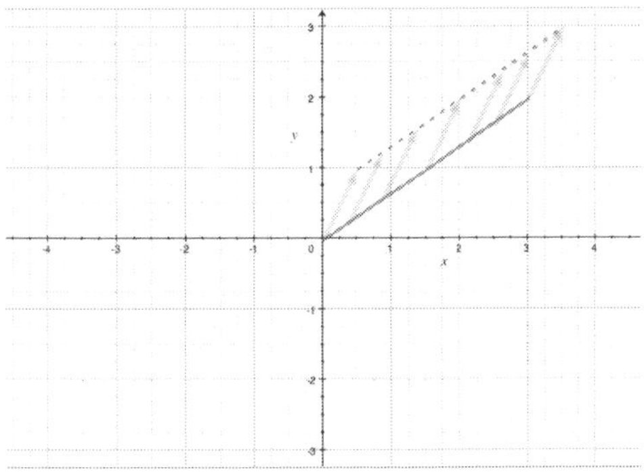

Thus the set of points making up this line segment is:

$$\{\alpha\,[3,2] + [0.5,1] \; : \; \alpha \in \mathbb{R}, 0 \leq \alpha \leq 1\}$$

Accordingly, we can plot the line segment using the following statement:

```
plot([add2(scalar_vector_mult(i/100., [3,2]), [0.5,1]) for i in range(101)], 4)
```

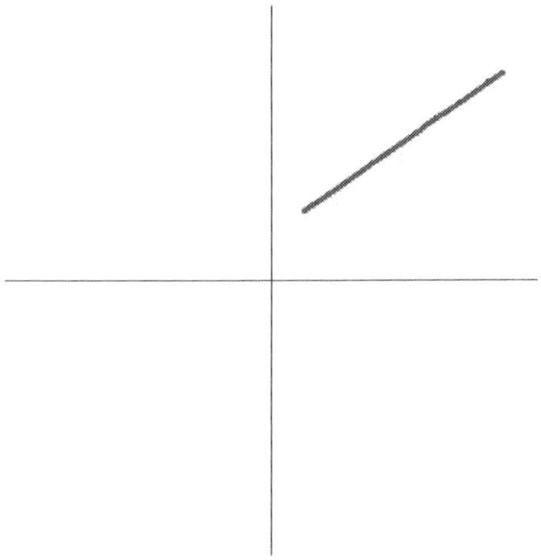

We can similarly represent the entire line through two given points. For example, we know that the line through $[0,0]$ and $[3,2]$ is $\{\alpha\,[3,2] \; : \; \alpha \in \mathbb{R}\}$. Adding $[0.5,1]$ to each point in this set gives us the line through $[0.5,1]$ and $[3.5,3]$: $\{[0.5,1] + \alpha\,[3,2] \; : \; \alpha \in \mathbb{R}\}$.

Exercise 2.6.1: Given points $\boldsymbol{u} = [2,3]$ and $\boldsymbol{v} = [5,7]$ in $\mathbb{R}^2$, what is the point $\boldsymbol{w}$ such that the origin-to-$\boldsymbol{w}$ line segment can be translated to yield the $\boldsymbol{u}$-to-$\boldsymbol{v}$ line segment? And what is the translation vector that is applied to both endpoints?

Exercise 2.6.2: Given a pair of points, $\boldsymbol{u} = [1,4]$, $\boldsymbol{v} = [6,3]$ in $\mathbb{R}^2$, write a mathematical expressing giving the set of points making up the line segment between the points.

2.6.2 Distributive laws for scalar-vector multiplication and vector addition

To get a better understanding of this formulation of line segments and lines, we make use of two properties that arise in combining scalar-vector multiplication and vector adddition. Both arise from the distributive law for fields, $x(y+z) = xy + xz$.

Proposition 2.6.3 (***Scalar-vector multiplication distributes over vector addition):***

$$\alpha(\boldsymbol{u} + \boldsymbol{v}) = \alpha\boldsymbol{u} + \alpha\boldsymbol{v} \tag{2.1}$$

Example 2.6.4: As an example, consider the multiplication:

$$2\,([1,2,3]+[3,4,4]) = 2\,[4,6,7] = [8,12,14]$$

which is the same as:

$$2\,([1,2,3]+[3,4,4]) = 2\,[1,2,3]+2\,[3,4,4] = [2,4,6]+[6,8,8] = [8,12,14]$$

Proof

We use the same approach as used in the proof of Proposition 2.5.5. To show that the left-hand side of Equation 2.1 equals the right-hand side, we show that each entry of the left-hand side equals the corresponding entry of the right-hand side.

For each element k of the domain D, entry k of $(\boldsymbol{u}+\boldsymbol{v})$ is $\boldsymbol{u}[k]+\boldsymbol{v}[k]$, so entry k of $\alpha\,(\boldsymbol{u}+\boldsymbol{v})$ is $\alpha\,(\boldsymbol{u}[k]+\boldsymbol{v}[k])$.

Entry k of $\alpha\,\boldsymbol{u}$ is $\alpha\boldsymbol{u}[k]$ and entry k of $\alpha\,\boldsymbol{v}$ is $\alpha\,\boldsymbol{v}[k]$, so entry k of $\alpha\,\boldsymbol{u}+\alpha\,\boldsymbol{v}$ is $\alpha\,\boldsymbol{u}[k]+\alpha\,\boldsymbol{v}[k]$.

Finally, by the distributive law for fields, $\alpha(\boldsymbol{u}[k]+\boldsymbol{v}[k]) = \alpha\,\boldsymbol{u}[k]+\alpha\,\boldsymbol{v}[k]$. □

Proposition 2.6.5 (scalar-vector multiplication distributes over scalar addition):

$$(\alpha+\beta)\boldsymbol{u} = \alpha\boldsymbol{u}+\beta\boldsymbol{u}$$

Problem 2.6.6: Prove Proposition 2.6.5.

2.6.3 First look at convex combinations

It might seem odd that the form of the expression for the set of points making up the $[0.5,1]$-to-$[3.5,3]$ segment, $\{\alpha\,[3,2]+[0.5,1]\ :\ \alpha\in\mathbb{R}, 0\le\alpha\le 1\}$, mentions one endpoint but not the other. This asymmetry is infelicitous. Using a bit of vector algebra, we can obtain a nicer expression:

$$\begin{aligned}
\alpha\,[3,2]+[0.5,1] &= \alpha\,([3.5,3]-[0.5,1])+[0.5,1] \\
&= \alpha\,[3.5,3]-\alpha\,[0.5,1]+[0.5,1] \text{ by Proposition 2.6.3} \\
&= \alpha\,[3.5,3]+(1-\alpha)\,[0.5,1] \text{ by Proposition 2.6.5} \\
&= \alpha\,[3.5,3]+\beta\,[0.5,1]
\end{aligned}$$

where $\beta = 1-\alpha$. We now can write an expression for the $[0.5,1]$-to-$[3.5,3]$ segment

$$\{\alpha\,[3.5,3]+\beta\,[0.5,1]\ :\ \alpha,\beta\in\mathbb{R}, \alpha,\beta\ge 0, \alpha+\beta=1\}$$

that is symmetric in the two endpoints.

An expression of the form $\alpha\,\boldsymbol{u}+\beta\,\boldsymbol{v}$ where $\alpha,\beta\ge 0$ and $\alpha+\beta=1$ is called a *convex combination* of $\boldsymbol{u}$ and $\boldsymbol{v}$. Based on the example, we are led to the following assertion, which is true for any pair $\boldsymbol{u},\boldsymbol{v}$ of distinct n-vectors over $\mathbb{R}$:

Proposition 2.6.7: The $\boldsymbol{u}$-to-$\boldsymbol{v}$ line segment consists of the set of convex combinations of $\boldsymbol{u}$ and $\boldsymbol{v}$.

Example 2.6.8: The table below shows some convex combinations of pairs of 1- and 2-vectors over $\mathbb{R}$:

1. $\mathbf{u_1} = [2]$, $\mathbf{v_1} = [12]$

2. $\mathbf{u_2} = \begin{bmatrix} 5 \\ 2 \end{bmatrix}$, $\mathbf{v_2} = \begin{bmatrix} 10 \\ -6 \end{bmatrix}$

	$\alpha = 1$ $\beta = 0$	$\alpha = .75$ $\beta = .25$	$\alpha = .5$ $\beta = .5$	$\alpha = .25$ $\beta = .75$	$\alpha = 0$ $\beta = 1$
$\alpha\mathbf{u_1} + \beta\mathbf{v_1}$	$[2]$	$[4.5]$	$[7]$	$[9.5]$	$[12]$
$\alpha\mathbf{u_2} + \beta\mathbf{v_2}$	$\begin{bmatrix} 5 \\ 2 \end{bmatrix}$	$\begin{bmatrix} 6.25 \\ -2 \end{bmatrix}$	$\begin{bmatrix} 7.5 \\ -2 \end{bmatrix}$	$\begin{bmatrix} 8.75 \\ -4 \end{bmatrix}$	$\begin{bmatrix} 10 \\ -6 \end{bmatrix}$

Task 2.6.9: Write a python procedure `segment(pt1, pt2)` that, given points represented as 2-element lists, returns a list of a hundred points spaced evenly along the line segment whose endpoints are the two points

Plot the hundred points resulting when `pt1` $= [3.5, 3]$ and `pt2` $= [0.5, 1]$

Example 2.6.10: Let's consider the convex combinations of a pair of vectors that represent images,

$\boldsymbol{u} =$  and $\boldsymbol{v} =$ .

For example, with scalars $\frac{1}{2}$ and $\frac{1}{2}$, the convex combination, which is the average, looks like this:

$\frac{1}{2}$  $+ \frac{1}{2}$  $=$ 

To represent the "line segment" between the two face images, we can take a number of convex combinations:

$1\boldsymbol{u} + 0\boldsymbol{v}$ $\frac{7}{8}\boldsymbol{u} + \frac{1}{8}\boldsymbol{v}$ $\frac{6}{8}\boldsymbol{u} + \frac{2}{8}\boldsymbol{v}$ $\frac{5}{8}\boldsymbol{u} + \frac{3}{8}\boldsymbol{v}$ $\frac{4}{8}\boldsymbol{u} + \frac{4}{8}\boldsymbol{v}$ $\frac{3}{8}\boldsymbol{u} + \frac{5}{8}\boldsymbol{v}$ $\frac{2}{8}\boldsymbol{u} + \frac{6}{8}\boldsymbol{v}$ $\frac{1}{8}\boldsymbol{u} + \frac{7}{8}\boldsymbol{v}$ $0\boldsymbol{u} + 1\boldsymbol{v}$

By using these images as frames in a video, we get the effect of a crossfade.

2.6.4 First look at affine combinations

What about the infinite line through $[0.5, 1]$ and $[3.5, 3]$? We saw that this line consists of the points of $\{[0.5, 1] + \alpha\,[3, 2] \;:\; \alpha \in \mathbb{R}\}$. Using a similar argument, we can rewrite this set as

$$\{\alpha\,[3.5, 3] + \beta\,[0.5, 1] \;:\; \alpha \in \mathbb{R}, \beta \in R, \alpha + \beta = 1\}$$

An expression of the form $\alpha\,\boldsymbol{u} + \beta\,\boldsymbol{v}$ where $\alpha + \beta = 1$ is called an *affine* combination of $\boldsymbol{u}$ and $\boldsymbol{v}$. Based on the example, we are led to the following assertion:

Hypothesis 2.6.11: The line through $\boldsymbol{u}$ and $\boldsymbol{v}$ consists of the set of affine combinations of $\boldsymbol{u}$ and $\boldsymbol{v}$.

In Chapter 3, we will explore affine and convex combinations of more than two vectors.

2.7 Dictionary-based representations of vectors

In Section 2.2, I proposed that a vector is a function from some domain D to a field. In Section 2.2.1, I proposed to represent such a function using a Python dictionary. It is convenient to define a Python class `Vec` so that an instance has two fields (also known as *instance variables*, also known as *attributes*):

- `f`, the function, represented by a Python dictionary, and
- `D`, the domain of the function, represented by a Python set.

We adopt the convention described in Section 2.2.1 in which entries with value zero may be omitted from the dictionary f. This enables sparse vectors to be represented compactly.

It might seem a bad idea to require that each instance of `Vec` keep track of the domain. For example, as I pointed out in Section 2.2.1, in information retrieval one typically has many documents, each including only a very small subset of words; it would waste memory to duplicate the entire list of allowed words with each document. Fortunately, as we saw in Section 0.5.4, Python allows many variables (or instance variables) to point to the same set in memory. Thus, if we are careful, we can ensure that all the vectors representing documents point to the same domain.

The Python code required to define the class `Vec` is

```
class Vec:
    def __init__(self, labels, function):
        self.D = labels
        self.f = function
```

Once Python has processed this definition, you can create an instance of `Vec` like so:

```
>>> Vec({'A','B','C'}, {'A':1})
```

The first argument is assigned to the new instance's `D` field, and the second is assigned to the `f` field. The value of this expression will be the new instance. You can assign the value to a variable

```
>>> v = Vec({'A','B','C'}, {'A':1})
```

and subsequently access the two fields of `v`, e.g.:

```
>>> for d in v.D:
...   if d in v.f:
...    print(v.f[d])
...
1.0
```

Quiz 2.7.1: Write a procedure `zero_vec(D)` with the following spec:

- *input:* a set `D`
- *output:* an instance of `Vec` representing a `D`-vector all of whose entries have value zero

Answer

Exploiting the sparse-representation convention, we can write the procedure like this:

```
def zero_vec(D): return Vec(D, {})
```

Without the convention, one could write it like this:

```
def zero_vec(D): return Vec(D, {d:0 for d in D})
```

The procedure `zero_vec(D)` is defined in the provided file `vecutil.py`.

2.7.1 Setter and getter

In the following quizzes, you will write procedures that work with the class-based representation of vectors. In a later problem, you will incorporate some of these procedures into a module that defines the class Vec.

The following procedure can be used to assign a value to a specified entry of a Vec v:

```
def setitem(v, d, val): v.f[d] = val
```

The second argument d should be a member of the domain v.D. The procedure can be used, for example, as follows:

```
>>> setitem(v, 'B', 2.)
```

Quiz 2.7.2: Write a procedure getitem(v, d) with the following spec:

- *input:* an instance v of Vec, and an element d of the set v.D
- *output:* the value of entry d of v

Write your procedure in a way that takes into account the sparse-representation convention. Hint: the procedure can be written in one-line using a conditional expression (Section 0.5.3). You can use your procedure to obtain the 'A' entry of the vector v we defined earlier:

```
>>> getitem(v, 'A')
1
```

Answer

The following solution uses a conditional expression:

```
def getitem(v,d): return v.f[d] if d in v.f else 0
```

Using an if-statement, you could write it like this:

```
def getitem(v,d):
    if d in v.f:
        return v.f[d]
    else:
        return 0
```

2.7.2 Scalar-vector multiplication

Quiz 2.7.3: Write a procedure scalar_mul(v, alpha) with the following spec:

- *input:* an instance of Vec and a scalar alpha
- *output:* a new instance of Vec that represents the scalar-vector product alpha times v.

There is a nice way to ensure that the output vector is as sparse as the input vector, but you are not required to ensure this. You can use getitem(v, d) in your procedure but are not required to. Be careful to ensure that your procedure does not modify the vector it is passed as argument; it creates a new instance of Vec. However, the new instance should point to the same set D as the old instance.

Try it out on the vector v:

```
>>> scalar_mul(v, 2)
<__main__.Vec object at 0x10058cd10>
```

Okay, that's not so enlightening. Let's look at the dictionary of the resulting Vec:

```
>>> scalar_mul(v, 2).f
{'A': 2.0, 'C': 0, 'B': 4.0}
```

Answer

The following procedure does not preserve sparsity.

```
def scalar_mul(v, alpha):
 return Vec(v.D, {d:alpha*getitem(v,d) for d in v.D})
```

To preserve sparsity, you can instead write

```
def scalar_mul(v, alpha):
 return Vec(v.D, {d:alpha*value for d,value in v.f.items()})
```

2.7.3 Addition

Quiz 2.7.4: Write a procedure add(u, v) with the following spec:

- *input:* instances u and v of Vec
- *output:* an instance of Vec that is the vector sum of u and v

Here's an example of the procedure being used:

```
>>> u = Vec(v.D, {'A':5., 'C':10.})
>>> add(u,v)
<__main__.Vec object at 0x10058cd10>
>>> add(u,v).f
{'A': 6.0, 'C': 10.0, 'B': 2.0}
```

You are encouraged to use getitem(v, d) in order to tolerate sparse representations. You are encouraged *not* to try to make the output vector sparse. Finally, you are encouraged to use a dictionary comprehension to define the dictionary for the new instance of Vec.

Answer

```
def add(u, v):
 return Vec(u.D,{d:getitem(u,d)+getitem(v,d) for d in u.D})
```

2.7.4 Vector negative, invertibility of vector addition, and vector subtraction

The *negative* of a vector $\boldsymbol{v}$ is the vector $-\boldsymbol{v}$ obtained by negating each element of $\boldsymbol{v}$. If we interpret $\boldsymbol{v}$ as an arrow, its negative $-\boldsymbol{v}$ is the arrow of the same length pointed in the exactly opposite direction.

If we interpret $\boldsymbol{v}$ as a translation (e.g. "Go east two miles and north three miles"), its negative (e.g., "Go east negative two miles and north negative three miles") is the inverse translation. Applying one translation and then another leaves you back where you started.

Vector subtraction is defined in terms of vector addition and negative: $\boldsymbol{u} - \boldsymbol{v}$ is defined as $\boldsymbol{u} + (-\boldsymbol{v})$. This definition is equivalent to the obvious definition of vector subtraction: subtract corresponding elements.

Vector subtraction is the inverse of vector addition. For some vector $\boldsymbol{w}$, consider the function

$$f(\boldsymbol{v}) = \boldsymbol{v} + \boldsymbol{w}$$

that adds $\boldsymbol{w}$ to its input and the function

$$g(\boldsymbol{v}) = \boldsymbol{v} - \boldsymbol{w}$$

that subtracts $\boldsymbol{w}$ from its input. One function translates its input by $\boldsymbol{w}$ and the other translates *its* input by $-\boldsymbol{w}$. These functions are inverses of each other. Indeed,

$$\begin{aligned}(g \circ f)(\boldsymbol{v}) &= g(f(\boldsymbol{v})) \\ &= g(\boldsymbol{v}+\boldsymbol{w}) \\ &= \boldsymbol{v}+\boldsymbol{w}-\boldsymbol{w} \\ &= \boldsymbol{v}\end{aligned}$$

Quiz 2.7.5: Write a Python procedure `neg(v)` with the following spec:

- *input:* an instance v of Vec
- *output:* a dictionary representing the negative of v

Here's an example of the procedure being used:

```
>>> neg(v).f
{'A': -1.0, 'C': 0, 'B': -2.0}
```

There are two ways to write the procedure. One is by explicitly computing the `.f` field of the output vector using a comprehension. The other way is by using an appropriate call to the procedure `scalar_mul` you defined in Quiz 2.7.3.

Answer

```
def neg(v):
    return Vec(v.D, {d:-getitem(v, d) for d in v.D})
```
or

```
def neg(v):
    return Vec(v.D, {key:-value for key, value in v.f.items()})
```

or

```
def neg(v): return scalar_mul(v, -1)
```

2.8 Vectors over $GF(2)$

So far we have studied only vectors over $\mathbb{R}$. In this section, we consider vectors over $GF(2)$, and give some example applications. Remember from Section 1.5 that $GF(2)$ is a field in which the only values are 0 and 1, and adding 1 and 1 gives 0, and subtracting is the same as adding.

For the sake of brevity, we will sometimes write specific n-vectors over $GF(2)$ as n-bit binary strings. For example, we write 1101 for the 4-vector whose only zero is in its third entry.

Quiz 2.8.1: $GF(2)$ *vector addition practice:* What is $1101 + 0111$? (Note: it is the same as $1101 - 0111$.)

Answer

1010

2.8.1 Perfect secrecy re-revisited

Recall Alice and Bob and their need for perfect secrecy. We saw In Section 1.5.1 that encrypting a single-bit plaintext consisted of adding that bit to a single-bit key, using $GF(2)$ addition. We saw also that, to encrypt a sequence of bits of plaintext, it sufficed to just encrypt each bit with its own bit of key. That process can be expressed more compactly using addition of vectors over $GF(2)$.

Suppose Alice needs to send a ten-bit plaintext $\boldsymbol{p}$ to Bob.

Vernam's cypher: Alice and Bob randomly choose a 10-vector $\boldsymbol{k}$.
Alice computes the cyphertext $\boldsymbol{c}$ according to the formula

$$\boldsymbol{c} = \boldsymbol{p} + \boldsymbol{k}$$

where the sum is a *vector* sum.

The first thing to check is that this cryptosystem is decryptable—that Bob, who knows $\boldsymbol{k}$ and $\boldsymbol{c}$, can recover $\boldsymbol{p}$. He does so using the equation

$$\boldsymbol{p} = \boldsymbol{c} - \boldsymbol{k} \tag{2.2}$$

Example 2.8.2: For example, Alice and Bob agree on the following 10-vector as a key:

$$\boldsymbol{k} = [0, 1, 1, 0, 1, 0, 0, 0, 0, 1]$$

Alice wants to send this message to Bob:

$$\boldsymbol{p} = [0, 0, 0, 1, 1, 1, 0, 1, 0, 1]$$

She encrypts it by adding $\boldsymbol{p}$ to $\boldsymbol{k}$:

$$\boldsymbol{c} = \boldsymbol{k} + \boldsymbol{p} = [0, 1, 1, 0, 1, 0, 0, 0, 0, 1] + [0, 0, 0, 1, 1, 1, 0, 1, 0, 1]$$

$$\boldsymbol{c} = [0, 1, 1, 1, 0, 1, 0, 1, 0, 0]$$

When Bob receives $\boldsymbol{c}$, he decrypts it by adding $\boldsymbol{k}$:

$$\boldsymbol{c} + \boldsymbol{k} = [0, 1, 1, 1, 0, 1, 0, 1, 0, 0] + [0, 1, 1, 0, 1, 0, 0, 0, 0, 1] = [0, 0, 0, 1, 1, 1, 0, 1, 0, 1]$$

which is the original message.

Next we check that the system satisfies perfect secrecy. The argument should be familiar. For each plaintext $\boldsymbol{p}$, the function $\boldsymbol{k} \mapsto \boldsymbol{k} + \boldsymbol{p}$ is one-to-one and onto, hence invertible. Since the key $\boldsymbol{k}$ is chosen uniformly at random, therefore, the cyphertext $\boldsymbol{c}$ is also distributed uniformly.

2.8.2 All-or-nothing secret-sharing using $GF(2)$

I have a secret: the midterm exam. I've represented it as an n-vector $\boldsymbol{v}$ over $GF(2)$. I want to provide it to my two teaching assistants, Alice and Bob (A and B) so they can administer the midterm while I'm taking a vacation. However, I don't completely trust them. One TA might be bribed by a student into giving out the exam ahead of time, so I don't want to simply provide each TA with the exam.

I therefore want to take precautions. I provide pieces to the TAs in such a way that the two TAs can jointly reconstruct the secret but that neither of the TAs all alone gains any information whatsoever.

Here's how I do that. I choose a n-vector $\boldsymbol{v}_A$ over $GF(2)$ randomly according to the uniform distribution. (That involves a lot of coin-flips.) I then compute another n-vector $\boldsymbol{v}_B$ by

$$\boldsymbol{v}_B := \boldsymbol{v} - \boldsymbol{v}_A$$

Finally, I provide Alice with $\boldsymbol{v}_A$ and Bob with $\boldsymbol{v}_B$, and I leave for vacation.

When the time comes to administer the exam, the two TAs convene to reconstruct the exam from the vectors they have been given. They simply add together their vectors

$$\boldsymbol{v}_A + \boldsymbol{v}_B$$

The definition of $\boldsymbol{v}_B$ ensures that this sum is in fact the secret vector $\boldsymbol{v}$.

How secure is this scheme against a single devious TA? Assume that Alice is corrupt and wants to sell information about the exam. Assume that Bob is honest, so Alice cannot get his help in her evil plan. What does Alice learn from her piece, $\boldsymbol{v}_A$, about the exam $\boldsymbol{v}$? Since Alice's piece was chosen uniformly at random, she learns nothing from it.

Now suppose instead that Bob is the corrupt TA. What does he learn from his piece, $\boldsymbol{v}_B$, about the exam? Define the function $f : GF(2)^n \longrightarrow GF(2)^n$ by

$$f(\boldsymbol{x}) = \boldsymbol{v} - \boldsymbol{x}$$

The function $g(\boldsymbol{y}) = \boldsymbol{v} + \boldsymbol{x}$ is the inverse of f, so f is an invertible function.[2] Therefore, since $\boldsymbol{v}_B = f(\boldsymbol{v}_A)$ and $\boldsymbol{v}_A$ is chosen according to the uniform distribution, the distribution of $\boldsymbol{v}_B$ is also uniform. This shows that Bob learns nothing about the secret from his piece. The secret-sharing scheme is secure.

The company RSA recently introduced a product based on this idea:

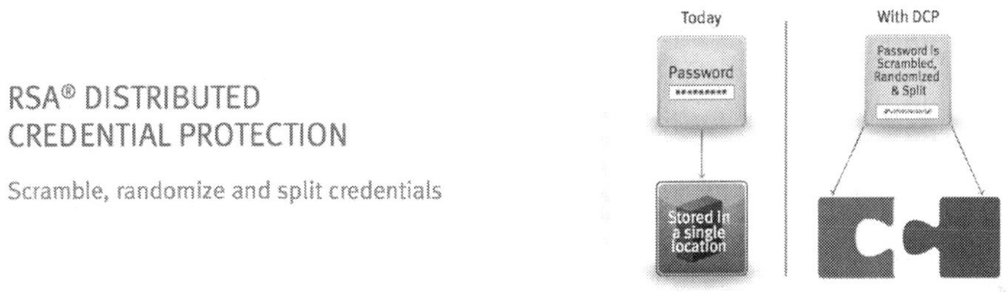

The main idea is to split each password into two parts, and store the two parts on two different servers. An attack on only one server does not compromise the security of the passwords.

Problem 2.8.3: Explain how to share an n-bit secret among *three* TAs so that a cabal consisting of any *two* of them learns nothing about the secret.

2.8.3 *Lights Out*

Vectors over $GF(2)$ can be used to analyze a puzzle called *Lights Out*. It is a five-by-five grid of lighted buttons.

[2]In fact, because we are using $GF(2)$, it turns out that g is the same function as f, but that is not important here.

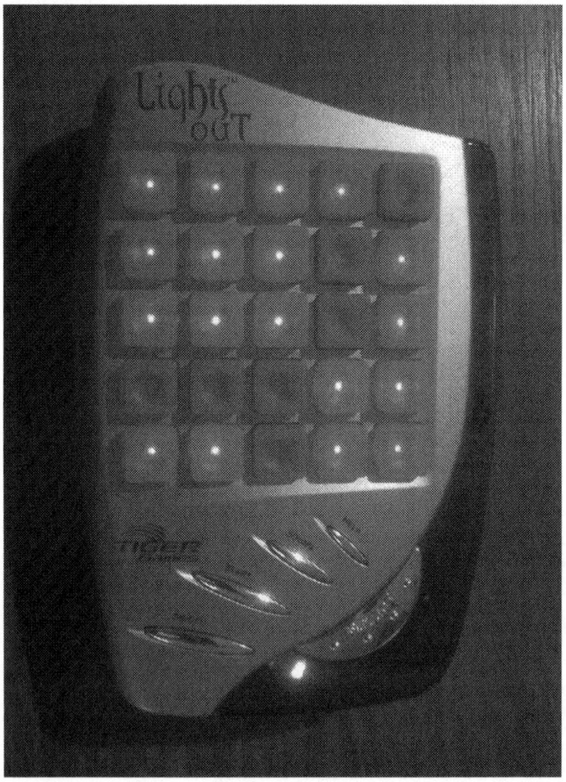

Initially some lights are on and some are off. When you push a button, you switch the corresponding light (from on to off or vice versa) but you also switch the lights of the button's four neighbors. The goal is to turn out all the lights.

Solving this problem is a computational problem:

> **Computational Problem 2.8.4:** *Solving Lights Out:*
> Given an initial configuration of lights, find a sequence of button-pushes that turns out all the lights, or report that none exists.

This Computational Problem raises a question:

> **Question 2.8.5:** Is there a way to solve the puzzle for every possible starting configuration?

Of course, the Question and the Computational Problem can be studied not just for the traditional five-by-five version of Lights Out, but for a version of any dimensions.

What do vectors have to do with Lights Out? The state of the puzzle can be represented by a vector over $GF(2)$ with one entry for each of the grid positions. A convenient domain is the set of tuples (0,0), (0,1), ..., (4,3), (4,4). We adopt the convention of representing a light that is on by a *one* and a light that is off by a *zero*. Thus the state of the puzzle in the picture is

```
{(0,0):one, (0,1):one, (0,2):one, (0,3):one, (0,4):0,
 (1,0):one, (1,1):one, (1,2):one, (1,3):0,   (1,4):one,
 (2,0):one, (2,1):one, (2,2):one, (2,3):0,   (2,4):one,
 (3,0):0,   (3,1):0,   (3,2):0,   (3,3):one, (3,4):one,
 (4,0):one, (4,1):one, (4,2):0,   (4,3):one, (4,4):one}
```

Let $\boldsymbol{s}$ denote this vector.

A *move* consists of pushing a button, which changes the state of the puzzle. For example, pushing the top-left button (0,0) flips the light at (0,0), at (0,1), and at (1,0). Therefore this change can be represented by the "button vector"

```
{(0,0):one, (0,1):one, (1,0):one}
```

Let $\boldsymbol{v}_{0,0}$ denote this vector.

The new state resulting when you start $\boldsymbol{s}$ and then push button $(0,0)$ is represented by the vector $\boldsymbol{s} + \boldsymbol{v}_{0,0}$. Why?

- For each entry (i,j) for which $\boldsymbol{v}_{0,0}$ is zero, entries (i,j) in $\boldsymbol{s}$ and $\boldsymbol{s} + \boldsymbol{v}_{0,0}$ are the same.

- For each entry (i,j) for which $\boldsymbol{v}_{0,0}$ is one, entries (i,j) in $\boldsymbol{s}$ and $\boldsymbol{s} + \boldsymbol{v}_{0,0}$ differ.

We chose the vector $\boldsymbol{v}_{0.0}$ to have one's in exactly the positions that change when you push button $(0,0)$.

The state of the puzzle is now

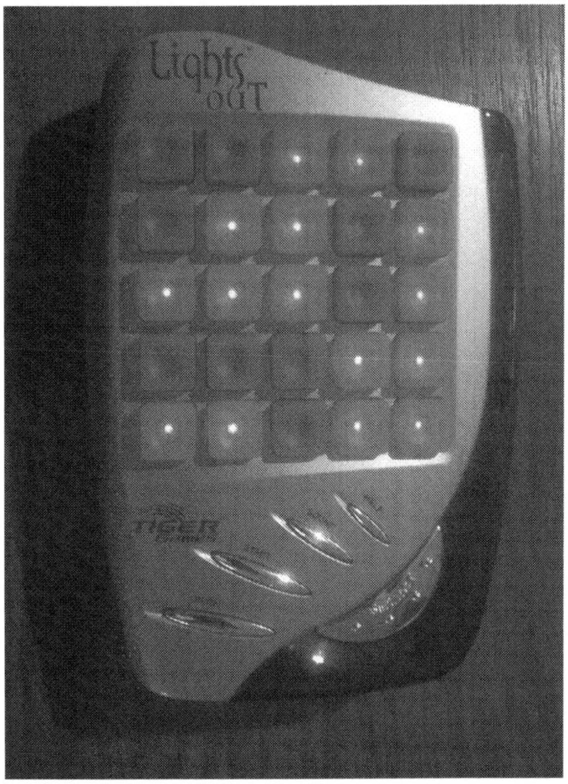

Next we will push the button at (1, 1), which flips the lights at (1,1), (0,1), (1,0), (2,1), and (1,2). This button corresponds to the vector

```
{(1,1):one, (0,1):one, (1,0):one, (2,1):one, (1,2):one}
```

which we denote by $\boldsymbol{v}_{1,1}$. Before the button is pushed, the state is represented by the vector $\boldsymbol{s} + \boldsymbol{v}_{0,0}$. After the button is pushed, the state is represented by the vector $\boldsymbol{s} + \boldsymbol{v}_{0,0} + \boldsymbol{v}_{1,1}$.

Summarizing, executing a move—pushing a button—means updating the state of the puzzle as:

new state := old state + button vector

where by addition we mean addition of vectors over $GF(2)$. Thus a button vector can be viewed as a translation.

Here is an example of solving an instance of the 3×3 puzzle:

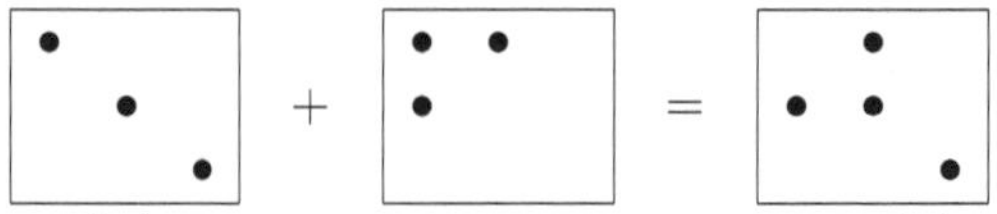

state move new state

Returning to the 5×5 case, there are twenty-five buttons, and for each button there is a corresponding button vector. We can use these vectors to help us solve the puzzle. Given an initial state of the lights, the goal is to find a sequence of button-pushes that turn off all the lights. Translated into the language of vectors, the problem is: given a vector $\boldsymbol{s}$ representing the initial state, select a sequence of button vectors $\boldsymbol{v}_1, \ldots, \boldsymbol{v}_m$ such that

$$(\cdots((\boldsymbol{s} + \boldsymbol{v}_1) + \boldsymbol{v}_2) \cdots) + \boldsymbol{v}_m = \text{the zero vector}$$

By the associativity of vector addition, the parenthesization on the left-hand side is irrelevant, so we can instead write

$$\boldsymbol{s} + \boldsymbol{v}_1 + \boldsymbol{v}_2 + \cdots + \boldsymbol{v}_m = \text{the all-zeroes vector}$$

Let us add $\boldsymbol{s}$ to both sides. Since a vector plus itself is the all-zeroes vector, and adding the all-zeroes vector does nothing, we obtain

$$\boldsymbol{v}_1 + \boldsymbol{v}_2 + \cdots + \boldsymbol{v}_m = \boldsymbol{s}$$

If a particular button's vector appears twice on the left-hand side, the two occurences cancel each other out. Thus we can restrict attention to solutions in which each button vector appears at most once.

By the commutativity of vector addition, the order of addition is irrelevant. Thus finding out how to get from a given initial state of the lights to the completely dark state is equivalent to selecting a subset of the button vectors whose sum is the vector $\boldsymbol{s}$ corresponding to the given initial state. If we could solve this puzzle, just think of the energy savings!

For practice, let's try the 2×2 version of the puzzle. The button vectors for the 2×2 puzzle are:

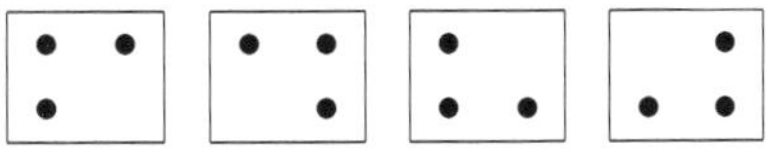

where the black dots represent ones.

Quiz 2.8.6: Find the subset of the button vectors whose sum is 

Answer

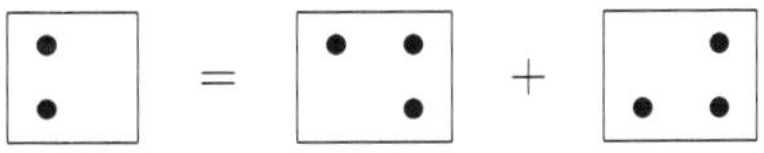

Now that we know how to model *Lights Out* in terms of vectors, we can see Computational Problem 2.8.4 (*Solving Lights Out*) as a special case of a more general problem:

Computational Problem 2.8.7: *Representing a given vector as a sum of a subset of other given vectors over* $GF(2)$

- *input:* a vector $\boldsymbol{s}$ and a list L of vectors over $GF(2)$
- *output:* A subset of the vectors in L whose sum is $\boldsymbol{s}$, or a report that there is no such subset.

There is a brute-force way to compute a solution to this problem: try each possible subset of vectors in L. The number of possibilities is $2^{|L|}$, 2 to the power of the cardinality of L. For example, in the *Lights Out* problem, L consists of twenty-five vectors, one for each of the buttons. The number of possible subsets is 2^{25}, which is 33,554,432.

However, there is a much slicker way to compute the solution. In Chapter 3, we introduce a still more general problem, Computational Problem 3.1.8. In Chapter 7, we will describe an algorithm to solve it. The algorithm is relevant not only to *Lights Out* problems but to other, perhaps more serious problems such as factoring integers.

2.9 Dot-product

For two D-vectors $\boldsymbol{u}$ and $\boldsymbol{v}$, the *dot-product* is the sum of the product of corresponding entries:

$$\boldsymbol{u}\cdot\boldsymbol{v} = \sum_{k\in D} \boldsymbol{u}[k]\ \boldsymbol{v}[k]$$

For example, for traditional vectors $\boldsymbol{u} = [u_1,\ldots,u_n]$ and $\boldsymbol{v} = [v_1,\ldots,v_n]$,

$$\boldsymbol{u}\cdot\boldsymbol{v} = u_1v_1 + u_2v_2 + \cdots + u_nv_n$$

Note that the output is a scalar, not a vector. For this reason, the dot-product is sometimes called the *scalar product* of vectors.

Example 2.9.1: Consider the dot-product of $[1,1,1,1,1]$ with $[10,20,0,40,-100]$. To find the dot-product, we can write the two vectors so that corresponding entries are aligned, multiply the pairs of corresponding entries, and sum the resulting products.

	1		1		1		1		1	
•	10		20		0		40		-100	
	10	+	20	+	0	+	40	+	(-100)	= -30

In general, the dot-product of an all-ones vector with a second vector equals the sum of entries of the second vector:

$$\begin{aligned}[1,1,\ldots,1]\cdot[v_1,v_2,\ldots,v_n] &= 1\cdot v_1 + 1\cdot v_2 + \cdots + 1\cdot v_n\\ &= v_1+v_2+\cdots+v_n\end{aligned}$$

Example 2.9.2: Consider the dot-product of $[0,0,0,0,1]$ with $[10,20,0,40,-100]$.

	0		0		0		0		1	
•	10		20		0		40		-100	
	0	+	0	+	0	+	0	+	(-100)	= -100

In general, if only one entry of $\boldsymbol{u}$, say the ith entry, is 1, and all other entries of $\boldsymbol{u}$ are zero, $\boldsymbol{u}\cdot\boldsymbol{v}$ is the ith entry of $\boldsymbol{v}$:

$$\begin{aligned}&[0,0,\cdots,0,1,0,\cdots,0,0]\cdot[v_1,v_2,\cdots,v_{i-1},v_i,v_{i+1},\ldots,v_n]\\ =\ & 0\cdot v_1 + 0\cdot v_2 + \cdots + 0\cdot v_{i-1} + 1\cdot v_i + 0\cdot v_{i+1} + \cdots + 0\cdot v_n\\ =\ & 1\cdot v_i\\ =\ & v_i\end{aligned}$$

Quiz 2.9.3: Express the average of the entries of an n-vector $\boldsymbol{v}$ as a dot-product.

Answer

Let $\boldsymbol{u}$ be the vector in which every entry is $1/n$. Then $\boldsymbol{u}\cdot\boldsymbol{v}$ is the average of the entries of $\boldsymbol{v}$.

Quiz 2.9.4: Write a procedure `list_dot(u, v)` with the following spec:

- *input:* equal-length lists u and v of field elements
- *output:* the dot-product of u and v interpreted as vectors

Use the `sum(·)` procedure together with a list comprehension.

Answer

```
def list_dot(u, v): return sum([u[i]*v[i] for i in range(len(u))])
```

or

```
def list_dot(u, v): return sum([a*b for (a,b) in zip(u,v)])
```

2.9.1 Total cost or benefit

Example 2.9.5: Suppose D is a set of foods, e.g. four ingredients of beer:

$$D = \{\text{hops, malt, water, yeast}\}$$

A *cost* vector maps each food to a price per unit amount:

$$cost = \text{Vec}(D, \{\text{hops}: \$2.50/ounce, \text{malt}: \$1.50/pound, \text{water}: \$0.006, \text{yeast}: \$0.45/gram\})$$

A *quantity* vector maps each food to an amount (e.g. measured in pounds). For example, here is the amount of each of the four ingredients going into about six gallons of stout:
quantity = Vec({hops:6 ounces, malt:14 pounds, water:7 gallons, yeast:11 grams})
The total cost is the dot-product of *cost* with *quantity*:

$$\textit{cost} \cdot \textit{quantity} = \$2.50 \cdot 6 + \$1.50 \cdot 14 + \$0.006 \cdot 7 + \$0.45 \cdot 11 = \$40.992$$

A *value* vector maps each food to its caloric content per pound:

$$\textit{value} = \{\text{hops} : 0, \text{malt} : 960, \text{water} : 0, \text{yeast} : 3.25\}$$

The total calories represented by six gallons of stout is the dot-product of *value* with *quantity*:

$$\textit{value} \cdot \textit{quantity} = 0 \cdot 6 + 960 \cdot 14 + 7 \cdot 0 + 3.25 \cdot 11 = 13475.75$$

2.9.2 Linear equations

Definition 2.9.6: A *linear equation* is an equation of the form $\boldsymbol{a} \cdot \boldsymbol{x} = \beta$, where $\boldsymbol{a}$ is a vector, β is a scalar, and $\boldsymbol{x}$ is a vector variable.

The scalar β is called the *right-hand side* of the linear equation because it is conventionally written on the right of the equals sign.

Example 2.9.7: *Sensor node energy utilization:* Sensor networks are made up of small, cheap sensor nodes. Each sensor node consists of some hardware components (e.g., radio, temperature sensor, memory, CPU). Often a sensor node is battery-driven and located in a remote place, so designers care about how each component's power consumption. Define

$$D = \{\textit{radio}, \textit{sensor}, \textit{memory}, \textit{CPU}\}$$

The function mapping each hardware component to its power consumption is a vector that we will call $\boldsymbol{rate}$:

$$\boldsymbol{rate} = \text{Vec}(D, \{\textit{memory} : 0.06\text{W}, \textit{radio} : 0.1\text{W}, \textit{sensor} : 0.004\text{W}, \textit{CPU} : 0.0025\text{W}\})$$

The function mapping each component to the amount of time it is on during a test period is a vector that we will call *duration*:

$$\boldsymbol{duration} = \text{Vec}(D, \{\textit{memory} : 1.0\text{s}, \textit{radio} : 0.2\text{s}, \textit{sensor} : 0.5\text{s}, \textit{CPU} : 1.0\text{s}\})$$

The total energy consumed by the sensor node during the test period is the dot-product of $\boldsymbol{rate}$ and $\boldsymbol{duration}$:

$$\boldsymbol{duration} \cdot \boldsymbol{rate} = 625J$$

measured in Joules (equivalently, Watt-seconds)

```
>>> D = {'memory', 'radio', 'sensor', 'CPU'}
>>> rate = Vec(D, {'memory':0.06, 'radio':0.1, 'sensor':0.004, 'CPU':0.0025})
>>> duration = Vec(D, {'memory':1.0, 'radio':0.2, 'sensor':0.5, 'CPU':1.0})
>>> rate*duration
0.0845
```

Now suppose that in reality we don't know the power consumption of each hardware component; the values of the entries of *rate* are unknowns. Perhaps we can calculate these values by testing the total power consumed during each of several test periods. Suppose that there are three test periods. For $i = 1, 2, 3$, we have a vector $\boldsymbol{duration}_i$ giving the amount of time each hardware component is on during test period i, and a scalar β_i giving the total power used during the test period. We consider $\boldsymbol{rate}$ a vector-valued variable, and we write down what we know in terms of five linear equations involving that variable:

$$\begin{aligned} \boldsymbol{duration}_1 \cdot \boldsymbol{rate} &= \beta_1 \\ \boldsymbol{duration}_2 \cdot \boldsymbol{rate} &= \beta_2 \\ \boldsymbol{duration}_3 \cdot \boldsymbol{rate} &= \beta_3 \end{aligned}$$

Can we compute the entries of ***rate*** from these equations? This amounts to two questions:

1. Is there an algorithm to find a vector that satisfies these linear equations?
2. Is there exactly one solution, one vector that satisfies the linear equations?

Even if there is an algorithm to compute *some* vector that satisfies the linear equations, we cannot be sure that the solution we compute is in fact the vector we are seeking unless there is only one vector that satisfies the equations.

Example 2.9.8: Here are some duration vectors:

```
>>> duration1 = Vec(D, {'memory':1.0, 'radio':0.2, 'sensor':0.5, 'CPU':1.0})
>>> duration2 = Vec(D, {'sensor':0.2, 'CPU':0.4})
>>> duration3 = Vec(D, {'memory':0.3, 'CPU':0.1})
```

Can we find a vector `rate` such that `duration1*rate = 0.11195`, `duration2*rate = 0.00158`, and `duration3*rate = 0.02422`? And is there only one such vector?

Quiz 2.9.9: Using the data in the following table, calculate the rate of energy consumption of each of the hardware components. The table specifies for each of four test periods how long each hardware component operates and how much charge is transferred through the sensor node.

	radio	sensor	memory	CPU	TOTAL ENERGY CONSUMED
test 0	1.0 sec	1.0 sec	0 sec	0 sec	1.5 J
test 1	2.0 sec	1.0 sec	0	0	2.5 J
test 2	0	0	1.0 sec	1.0 sec	1.5 J
test 3	0	0	0	1.0 sec	1 W

Answer

radio	sensor	memory	CPU
1 W	0.5 W	0.5 W	1 W

Definition 2.9.10: In general, a *system of linear equations* (often abbreviated *linear system*) is a collection of equations:

$$\begin{aligned} \boldsymbol{a}_1 \cdot \boldsymbol{x} &= \beta_1 \\ \boldsymbol{a}_2 \cdot \boldsymbol{x} &= \beta_2 \\ &\vdots \\ \boldsymbol{a}_m \cdot \boldsymbol{x} &= \beta_m \end{aligned} \tag{2.3}$$

where $\boldsymbol{x}$ is a vector variable. A *solution* is a vector $\hat{\boldsymbol{x}}$ that satisfies all the equations.

With these definitions in hand, we return to the two questions raised in connection with estimating the energy consumption of sensor-node components. First, the question of uniqueness:

> **Question 2.9.11:** *Uniqueness of solution to a linear system*
> For a given linear system (such as 2.3), how can we tell if there is only one solution?

Second, the question of computing a solution:

> **Computational Problem 2.9.12:** *Solving a linear system*
>
> - *input:* a list of vectors $\boldsymbol{a}_1, \ldots, \boldsymbol{a}_m$, and corresponding scalars $\beta_1, \ldots, \beta_m$ (the right-hand sides)
> - *output:* a vector $\hat{\boldsymbol{x}}$ satisfying the linear system 2.3 or a report that none exists.

Computational Problem 2.8.7, *Representing a given vector as a sum of a subset of other given vectors over* $GF(2)$, will turn out to be a special case of this problem. We will explore the connections in the next couple of chapters. In later chapters, we will describe algorithms to solve the computational problems.

2.9.3 Measuring similarity

Dot-product can be used to measure the similarity between vectors over $\mathbb{R}$.

Comparing voting records

In Lab 2.12, you will compare the voting records of senators using dot-product. The domain D is a set of bills the Senate voted on. Each senator is represented by a vector that maps a bill to $\{+1, -1, 0\}$, corresponding to *Aye*, *Nay*, or *Abstain*. Consider the dot-product of two Senators, e.g. (then) Senator Obama and Senator McCain. For each bill, if the two senators agreed (if they both voted in favor or both voted against), the product of the corresponding entries is 1. If the senators disagreed (if one voted in favor and one voted against), the product is -1. If one or both abstained, the product is zero. Adding up these products gives us a measure of how much they agree: the higher the sum, the greater the agreement. A positive sum indicates general agreement, and a negative sum indicates general disagreement.

Comparing audio segments

Suppose you have a short audio clip and want to search for occurences of it in a longer audio segment. How would you go about searching for the needle (the short audio clip) in the long segment (the haystack)?

We pursue this idea in the next example. In preparation, we consider a simpler problem: measuring the similarity between two audio segments of the same length.

Mathematically, an audio segment is a waveform, a continuous function of time:

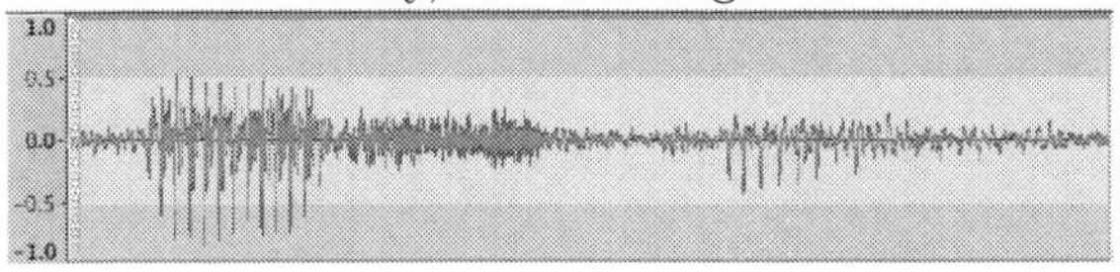

The value of the function is *amplitude*. The amplitude oscillates between being a positive number and being a negative number. How positive and how negative depends on the volume of the audio.

On a digital computer, the audio segment is represented by a sequence of numbers, values of the continuous function sampled at regular time intervals, e.g. 44,100 times a second:

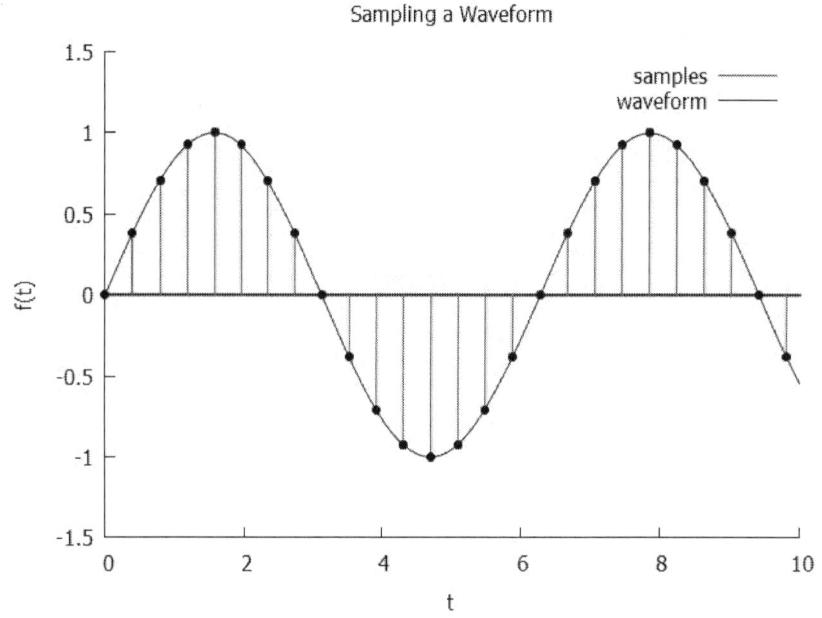

Let's first consider the task of *comparing* two equally long audio segments. Suppose we have two segments, each consisting of n samples, represented as n-vectors $\boldsymbol{u}$ and $\boldsymbol{v}$.

5	-6	9	-9	-5	-9	-5	5	-8	-5	-9	9	8	-5	-9	6	-2	-4	-9	-1	-1	-9	-3
5	-3	-9	0	-1	3	0	-2	-1	6	0	0	-4	5	-7	1	-9	0	-1	0	9	5	-3

One simple way to compare them is using dot-product $\sum_{i=1}^{n} \boldsymbol{u}[i]\ \boldsymbol{v}[i]$. Term i in this sum is positive if $\boldsymbol{u}[i]$ and $\boldsymbol{v}[i]$ have the same sign, and negative if they have opposite signs. Thus, once again, the greater the agreement, the greater the value of the dot-product.

Nearly identical audio segments (even if they differ in loudness) will produce a higher value than different segments. However, the bad news is that if the two segments are even slightly off in tempo or pitch, the dot-product will be small, probably close to zero. (There are other techniques to address differences of this kind.)

Finding an audio clip

Back to the problem of finding a needle (a short audio clip) in a haystack (a long audio segment). Suppose, for example, that the haystack consists of 23 samples and the needle consists of 11 samples.

Suppose we suspect that samples 10 through 22 of the long segment match up with the samples comprising the short clip. To verify that suspicion, we can form the vector consisting of samples 10 through 22 of the long segment, and compute the dot-product of that vector with the short clip:

5	-6	9	-9	-5	-9	-5	5	-8	-5	-9	9	8	-5	-9	6	-2	-4	-9	-1	-1	-9	-3
										2	7	4	-3	0	-1	-6	4	5	-8	-9		

Of course, we ordinarily have no idea where in the long segment we might find the short clip. It might start at position 0, or position 1, or ... or 12. There are $23 - 11 + 1$ possible starting positions (not counting those positions too close to the end for the short clip to appear there). We can evaluate each of these possibilities by computing an appropriate dot-product.

5	-6	9	-9	-5	-9	-5	5	-8	-5	-9	9	8	-5	-9	6	-2	-4	-9	-1	-1	-9	-3
2	7	4	-3	0	-1	-6	4	5	-8	-9												

5	-6	9	-9	-5	-9	-5	5	-8	-5	-9	9	8	-5	-9	6	-2	-4	-9	-1	-1	-9	-3
	2	7	4	-3	0	-1	-6	4	5	-8	-9											

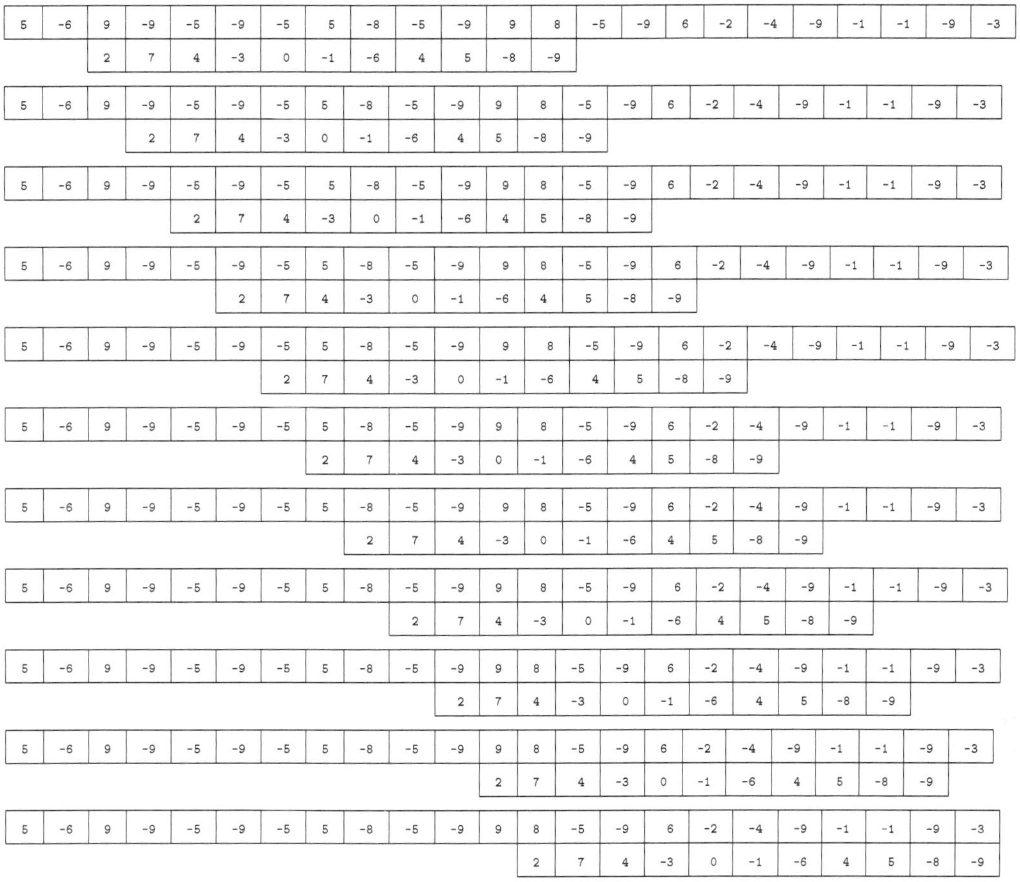

For this example, the long segment consists of 23 numbers and the short clip consists of 11 numbers, so we end up with 23 − 11 dot-products. We put these twelve numbers in an output vector.

Quiz 2.9.13: Suppose the haystack is $[1, -1, 1, 1, 1, -1, 1, 1, 1]$ and the needle is $[1, -1, 1, 1, -1, 1]$. Compute the dot-products and indicate which position achieves the best match.

Answer

The dot-products are $[2, 2, 0, 0]$, so the best matches start at position 0 and 1 of the haystack.

Quiz 2.9.14: This method of searching is not universally applicable. Say we wanted to locate the short clip $[1, 2, 3]$ in the longer segment $[1, 2, 3, 4, 5, 6]$. What would the dot-product method select as the best match?

Answer

There are 4 possible starts to our vector, and taking the dot product at each yields the following vector:

$$[1 + 4 + 9, 2 + 6 + 12, 3 + 8 + 15, 4 + 10 + 18] = [14, 20, 26, 32]$$

By that measure, the best match is to start at 4, which is obviously not right.

Now you will write a program to carry out these dot-products.

Quiz 2.9.15: Write a procedure `dot_product_list(needle,haystack)` with the following spec:

- *input:* a short list `needle` and a long list `haystack`, both containing numbers
- *output:* a list of length `len(haystack)-len(needle)` such that entry i of the output list equals the dot-product of the needle with the equal-length sublist of `haystack` starting at position i

Your procedure should use a comprehension and use the procedure `list_dot(u,v)` from Quiz 2.9.4. Hint: you can use slices as described in Section 0.5.5.

Answer

```
def dot_product_list(needle, haystack):
    s = len(needle)
    return [dot(needle, haystack[i:i+s]) for i in range(len(haystack)-s)]
```

First look at linear filters

In Section 2.9.3, we compared a short needle to a slice of a long haystack by turning the slice into a vector and taking the dot-product of the slice with the needle. Here is another way to compute the same number: we turn the needle into a longer vector by padding it with zeroes, and then calculate the dot-product of the padded vector with the haystack:

5	-6	9	-9	-5	-9	-5	5	-8	-5	-9	9	8	-5	-9	6	-2	-4	-9	-1	-1	-9	-3
0	0	0	0	0	0	0	0	0	0	2	7	4	-3	0	-1	-6	4	5	-8	-9	0	0

We can similarly compute the dot-products corresponding to other alignments of the needle vector with the haystack vector. This process is an example of applying a *linear filter*. The short clip plays the role of the *kernel* of the filter. In a more realistic example, both the needle vector and the haystack vector would be much longer. Imagine if the haystack were of length 5,000,000 and the needle were of length 50,000. We would have to compute almost 5,000,000 dot-products, each involving about 50,000 nonzero numbers. This would take quite a while.

Fortunately, there is a computational shortcut. In Chapter 4, we observe that matrix-vector multiplication is a convenient notation for computing the output vector $\boldsymbol{w}$ from the input vector $\boldsymbol{u}$ and the kernel. In Chapter 10, we give an algorithm for quickly computing all these dot-products. The algorithm draws on an idea we study further in Chapter 12.

2.9.4 Dot-product of vectors over $GF(2)$

We have seen some applications of dot-products of vectors over $\mathbb{R}$. Now we consider dot-products of vectors over $GF(2)$.

Example 2.9.16: Consider the dot-product of 11111 and 10101:

$$\begin{array}{cccccccccccc} & 1 & & 1 & & 1 & & 1 & & 1 & & \\ \bullet & 1 & & 0 & & 1 & & 0 & & 1 & & \\ \hline & 1 & + & 0 & + & 1 & + & 0 & + & 1 & = & 1 \end{array}$$

Next, consider the dot-product of 11111 and 00101:

$$\begin{array}{cccccccccccc} & 1 & & 1 & & 1 & & 1 & & 1 & & \\ \bullet & 1 & & 0 & & 1 & & 0 & & 1 & & \\ \hline & 0 & + & 0 & + & 1 & + & 0 & + & 1 & = & 0 \end{array}$$

In general, when you take the dot-product of an all-ones vector with a second vector, the value is the parity of the second vector: 0 if the number of ones is even, 1 if the number of ones is odd.

2.9.5 Parity bit

When data are stored or transmitted, errors can occur. Often a system is designed to detect such errors if they occur infrequently. The most basic method of error detection is a *parity check bit.* To reliably transmit an n-bit sequence, one computes one additional bit, the parity bit, as the parity of the n-bit sequence, and sends that along with the n-bit sequence.

For example, the PCI (peripheral component interconnect) bus in a computer

has a PAR line that transmits the parity bit. A mismatch generally causes a processor interrupt.

Parity check has its weaknesses:

- If there are exactly two bit errors (more generally, an even number of errors), the parity check will not detect the problem. In Section 3.6.4, I discuss *checksum functions*, which do a better job of catching errors.
- In case there is a single error, parity check doesn't tell you in which bit position the error has occurred. In Section 4.7.3, I discuss *error-correcting codes*, which can locate the error.

2.9.6 Simple authentication scheme

We consider schemes that enable a human to log onto a computer over an insecure network. Such a scheme is called an *authentication scheme* since it provides a way for the human to give evidence that he is who he says he is. The most familiar such scheme is based on *passwords*: Harry, the human, sends his password to Carole, the computer, and the computer verifies that it is the correct password.

This scheme is a disaster if there is an eavesdropper, Eve, who can read the bits going over the network. Eve need only observe one log-on before she learns the password and can subsequently log on as Harry.

A scheme that is more secure against eavesdroppers is a *challenge-response* scheme: A human tries to log on to Carole. In a series of *trials*, Carole repeatedly asks the human questions that someone not possessing the password would be unlikely to answer correctly. If the human answers each of several questions correctly, Carole concludes that the human knows the password.

Here is a simple challenge-response scheme. Suppose the password is a n-bit string, i.e. an n-vector $\hat{\boldsymbol{x}}$ over $GF(2)$, chosen uniformly at random. In the ith trial, Carole selects a nonzero n-vector $\boldsymbol{a}_i$, a *challenge vector*, and sends it to the human. The human sends back a single bit β_i, which is supposed to be the dot-product of $\boldsymbol{a}_i$ and the password $\hat{\boldsymbol{x}}$, and Carole checks whether $\beta_i = \boldsymbol{a}_i \cdot \hat{\boldsymbol{x}}$. If the human passes enough trials, Carole concludes that the human knows the password, and allows the human to log in.

Example 2.9.17: The password is $\hat{\boldsymbol{x}} = 10111$. Harry initiates log-in. In response, Carole selects the challenge vector $\boldsymbol{a}_1 = 01011$ and sends it to Harry. Harry computes the dot-product $\boldsymbol{a}_1 \cdot \hat{\boldsymbol{x}}$:

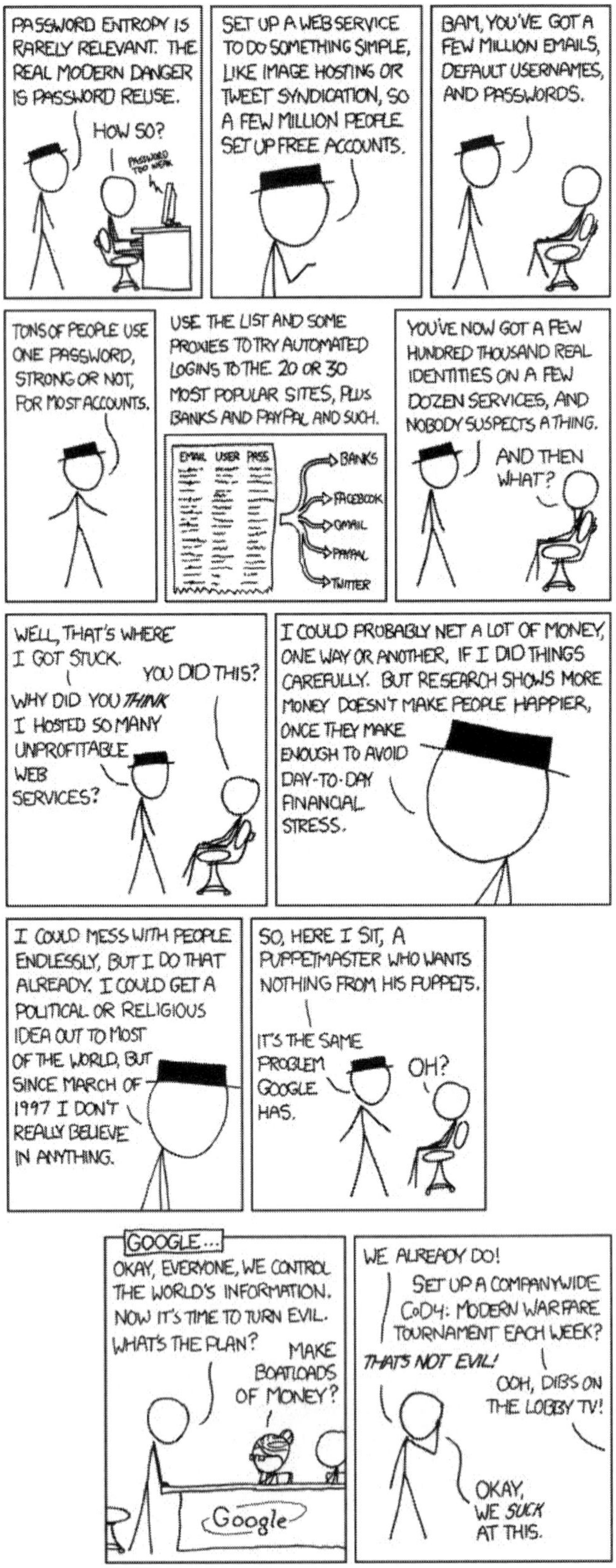

Figure 2.3: *Password Reuse* (`http://xkcd.com/792/`)

	0		1		0		1		1		
•	1		0		1		1		1		
	0	+	0	+	0	+	1	+	1	=	0

and responds by sending the resulting bit $\beta_1 = 0$ back to Carole.

Next, Carole sends the challenge vector $\boldsymbol{a}_2 = 11110$ to Harry. Harry computes the dot-product $\boldsymbol{a}_2 \cdot \hat{\boldsymbol{x}}$:

	1		1		1		1		0		
•	1		0		1		1		1		
	1	+	0	+	1	+	1	+	0	=	1

and responds by sending the resulting bit $\beta_2 = 1$ back to Carole.

This continues for a certain number k of trials. Carole lets Harry log in if $\beta_1 = \boldsymbol{a}_1 \cdot \hat{\boldsymbol{x}}, \beta_2 = \boldsymbol{a}_2 \cdot \hat{\boldsymbol{x}}, \ldots, \beta_k = \boldsymbol{a}_k \cdot \hat{\boldsymbol{x}}$.

2.9.7 Attacking the simple authentication scheme

We consider how Eve might attack this scheme. Suppose she eavesdrops on m trials in which Harry correctly responds. She learns a sequence of challenge vectors $\boldsymbol{a}_1, \boldsymbol{a}_2, \ldots, \boldsymbol{a}_m$ and the corresponding response bits $\beta_1, \beta_2, \ldots, \beta_m$. *What do these tell Eve about the password?*

Since the password is unknown to Eve, she represents it by a vector-valued variable $\boldsymbol{x}$. Since Eve knows that Harry correctly computed the response bits, she knows that the following linear equations are true:

$$\begin{aligned} \boldsymbol{a}_1 \cdot \boldsymbol{x} &= \beta_1 \\ \boldsymbol{a}_2 \cdot \boldsymbol{x} &= \beta_2 \\ &\vdots \\ \boldsymbol{a}_m \cdot \boldsymbol{x} &= \beta_m \end{aligned} \tag{2.4}$$

Perhaps Eve can compute the password by using an algorithm for Computational Problem 2.9.12, solving a linear system! Well, perhaps she can find *some* solution to the system of equations but is it the correct one? We need to consider Question 2.9.11: does the linear system have a unique solution?

Perhaps uniqueness is too much to hope for. Eve would likely be satisfied if the number of solutions were not too large, as long as she could compute them all and then try them out one by one. Thus we are interested in the following Question and Computational Problem:

Question 2.9.18: *Number of solutions to a linear system over $GF(2)$*
How many solutions are there to a given linear system over $GF(2)$?

Computational Problem 2.9.19: *Computing all solutions to a linear system over $GF(2)$*
Find all solutions to a given linear system over $GF(2)$.

However, Eve has another avenue of attack. Perhaps even without precisely identifying the password, she can use her knowledge of Harry's response bits to derive the answers to future challenges! For which future challenge vectors $\boldsymbol{a}$ can the dot-products with $\boldsymbol{x}$ be computed from the m equations? Stated more generally:

Question 2.9.20: Does a system of linear equations imply any other linear equations? If so, what other linear equations?

We next study properties of dot-product, one of which helps address this Question.

2.9.8 Algebraic properties of the dot-product

In this section we introduce some simple but powerful algebraic properties of the dot-product. These hold regardless of the choice of field (e.g. $\mathbb{R}$ or $GF(2)$).

Commutativity When you take a dot-product of two vectors, the order of the two does not matter:

Proposition 2.9.21 (Commutativity of dot-product): $\boldsymbol{u} \cdot \boldsymbol{v} = \boldsymbol{v} \cdot \boldsymbol{u}$

Commutativity of the dot-product follows from the fact that scalar-scalar multiplication is commutative:

Proof

$$\begin{aligned}[u_1, u_2, \ldots, u_n] \cdot [v_1, v_2, \ldots, v_n] &= u_1 v_1 + u_2 v_2 + \cdots + u_n v_n \\ &= v_1 u_1 + v_2 u_2 + \cdots + v_n u_n \\ &= [v_1, v_2, \ldots, v_n] \cdot [u_1, u_2, \ldots, u_n]\end{aligned}$$

□

Homogeneity The next property relates dot-product to scalar-vector multiplication: multiplying one of the vectors in the dot-product is equivalent to multiplying the value of the dot-product.

Proposition 2.9.22 (Homogeneity of dot-product): $(\alpha\, \boldsymbol{u}) \cdot \boldsymbol{v} = \alpha\, (\boldsymbol{u} \cdot \boldsymbol{v})$

Problem 2.9.23: Prove Proposition 2.9.22.

Problem 2.9.24: Show that $(\alpha\, \boldsymbol{u}) \cdot (\alpha\, \boldsymbol{v}) = \alpha\, (\boldsymbol{u} \cdot \boldsymbol{v})$ is *not* always true by giving a counterexample.

Distributivity The final property relates dot-product to vector addition.

Proposition 2.9.25 (Dot-product distributes over vector addition): $(\boldsymbol{u} + \boldsymbol{v}) \cdot \boldsymbol{w} = \boldsymbol{u} \cdot \boldsymbol{w} + \boldsymbol{v} \cdot \boldsymbol{w}$

Proof

Write $\boldsymbol{u} = [u_1, \ldots, u_n]$, $\boldsymbol{v} = [v_1, \ldots, v_n]$ and $\boldsymbol{w} = [w_1, \ldots, w_n]$.

$$\begin{aligned}
(\boldsymbol{u}+\boldsymbol{v})\cdot\boldsymbol{w} &= ([u_1,\ldots,u_n] + [v_1,\ldots,v_n])\cdot[w_1,\ldots,w_n] \\
&= [u_1+v_1,\ldots,u_n+v_n]\cdot[w_1,\ldots,w_n] \\
&= (u_1+v_1)w_1 + \cdots + (u_n+v_n)w_n \\
&= u_1w_1 + v_1w_1 + \cdots + u_nw_n + v_nw_n \\
&= (u_1w_1 + \cdots + u_nw_n) + (v_1w_1 + \cdots + v_nw_n) \\
&= [u_1,\ldots,u_n]\cdot[w_1,\ldots,w_n] + [v_1,\ldots,v_n]\cdot[w_1,\ldots,w_n]
\end{aligned}$$

□

Problem 2.9.26: Show by giving a counterexample that $(\boldsymbol{u}+\boldsymbol{v})\cdot(\boldsymbol{w}+\boldsymbol{x}) = \boldsymbol{u}\cdot\boldsymbol{w} + \boldsymbol{v}\cdot\boldsymbol{x}$ is *not* true.

Example 2.9.27: We first give an example of the distributive property for vectors over the reals: $[27, 37, 47]\cdot[2,1,1] = [20,30,40]\cdot[2,1,1] + [7,7,7]\cdot[2,1,1]$:

$$\begin{array}{rccccccc}
 & 20 & & 30 & & 40 & & \\
\bullet & 2 & & 1 & & 1 & & \\
\hline
 & 20\cdot 2 & + & 30\cdot 1 & + & 40\cdot 1 & = & 110
\end{array}$$

$$\begin{array}{rccccccc}
 & 7 & & 7 & & 7 & & \\
\bullet & 2 & & 1 & & 1 & & \\
\hline
 & 7\cdot 2 & + & 7\cdot 1 & + & 7\cdot 1 & = & 28
\end{array}$$

$$\begin{array}{rccccccc}
 & 27 & & 37 & & 47 & & \\
\bullet & 2 & & 1 & & 1 & & \\
\hline
 & 27\cdot 2 & + & 37\cdot 1 & + & 47\cdot 1 & = & 138
\end{array}$$

2.9.9 Attacking the simple authentication scheme, revisited

I asked in Section 2.9.7 whether Eve can use her knowledge of Harry's responses to some challenges to derive the answers to others. We address that question by using the distributive property for vectors over $GF(2)$.

Example 2.9.28: This example builds on Example 2.9.17 (Page 97). Carole had previously sent Harry the challenge vectors 01011 and 11110, and Eve had observed that the response bits were 0 and 1. Suppose Eve subsequently tries to log in as Harry, and Carole happens to send her as a challenge vector the sum of 01011 and 11110. Eve can use the distributive property to compute the dot-product of this sum with the password $\boldsymbol{x}$ even though she does not know the password:

$$\begin{array}{rcccc}
(01011 + 11110)\cdot\boldsymbol{x} & = & 01011\cdot\boldsymbol{x} & + & 11110\cdot\boldsymbol{x} \\
 & = & 0 & + & 1 \\
 & = & & 1 &
\end{array}$$

Since you know the password, you can verify that this is indeed the correct response to the challenge vector.

This idea can be taken further. For example, suppose Carole sends a challenge vector that is the sum of three previously observed challenge vectors. Eve can compute the response bit (the dot-product with the password) as the sum of the responses to the three previous challenge vectors.

Indeed, the following math shows that Eve can compute the right response to the sum of any number of previous challenges for which she has the right response:

$$\begin{array}{ll} \text{if} & \boldsymbol{a}_1 \cdot \boldsymbol{x} = \beta_1 \\ \text{and} & \boldsymbol{a}_2 \cdot \boldsymbol{x} = \beta_2 \\ \vdots & \vdots \\ \text{and} & \boldsymbol{a}_k \cdot \boldsymbol{x} = \beta_k \\ \text{then} & (\boldsymbol{a}_1 + \boldsymbol{a}_2 + \cdots + \boldsymbol{a}_k) \cdot \boldsymbol{x} = (\beta_1 + \beta_2 + \cdots + \beta_k) \end{array}$$

Problem 2.9.29: Eve knows the following challenges and responses:

challenge	response
110011	0
101010	0
111011	1
001100	1

Show how she can derive the right responses to the challenges 011101 and 000100.

Imagine that Eve has observed hundreds of challenges $\boldsymbol{a}_1, \ldots, \boldsymbol{a}_n$ and responses $\beta_1, \ldots, \beta_n$, and that she now wants to respond to the challenge $\boldsymbol{a}$. She must try to find a subset of $\boldsymbol{a}_1, \ldots, \boldsymbol{a}_n$ whose sum equals $\boldsymbol{a}$.

Question 2.9.20 asks: Does a system of linear equations imply any other linear equations? The example suggests a partial answer:

$$\begin{array}{ll} \text{if} & \boldsymbol{a}_1 \cdot \boldsymbol{x} = \beta_1 \\ \text{and} & \boldsymbol{a}_2 \cdot \boldsymbol{x} = \beta_2 \\ \vdots & \vdots \\ \text{and} & \boldsymbol{a}_k \cdot \boldsymbol{x} = \beta_k \\ \text{then} & (\boldsymbol{a}_1 + \boldsymbol{a}_2 + \cdots + \boldsymbol{a}_k) \cdot \boldsymbol{x} = (\beta_1 + \beta_2 + \cdots + \beta_k) \end{array}$$

Therefore, from observing challenge vectors and the response bits, Eve can derive the response to any challenge vector that is the sum of *any subset* of previously observed challenge vectors.

That presumes, of course, that she can *recognize* that the new challenge vector can be expressed as such a sum, and determine which sum! This is precisely Computational Problem 2.8.7. We are starting to see the power of computational problems in linear algebra; the same computational problem arises in addressing solving a puzzle and attacking an authentication scheme! Of course, there are many other settings in which this problem arises.

2.10 Our implementation of `Vec`

In Section 2.7, we gave the definition of a rudimentary Python class for representing vectors, and we developed some procedures for manipulating this representation.

2.10.1 Syntax for manipulating `Vec`s

We will expand our class definition of `Vec` to provide some notational conveniences:

operation	syntax
vector addition	`u+v`
vector negation	`-v`
vector subtraction	`u-v`
scalar-vector multiplication	`alpha*v`
division of a vector by a scalar	`v/alpha`
dot-product	`u*v`
getting value of an entry	`v[d]`
setting value of an entry	`v[d] = ...`
testing vector equality	`u == v`
pretty-printing a vector	`print(v)`
copying a vector	`v.copy()`

In addition, if an expression has as a result a `Vec` instance, the value of the expression will be presented not as an obscure Python incantation

```
>>> v
<__main__.Vec object at 0x10058cad0>
```

but as an expression whose value is a vector:

```
>>> v
Vec({'A', 'B', 'C'},{'A': 1.0})
```

2.10.2 The implementation

In Problem 2.14.10, you will implement `Vec`. However, since this book is not about the intricacies of defining classes in Python, you need not write the class definition; it will be provided for you. All you need to do is fill in the missing bodies of some procedures, most of which you wrote in Section 2.7.

2.10.3 Using Vecs

You will write the bodies of named procedures such as `setitem(v, d, val)` and `add(u,v)` and `scalar_mul(v, alpha)`. However, in actually using Vecs in other code, you must use operators instead of named procedures, e.g.

```
>>> v['a'] = 1.0
```

instead of

```
>>> setitem(v, 'a', 1.0)
```

and

```
>>> b = b - (b*v)*v
```

instead of

```
>>> b = add(b, neg(scalar_mul(v, dot(b,v))))
```

In fact, in code outside the `vec` module that uses `Vec`, you will import just `Vec` from the `vec` module:

```
from vec import Vec
```

so the named procedures will not be imported into the namespace. Those named procedures in the `vec` module are intended to be used *only* inside the `vec` module itself.

2.10.4 Printing Vecs

The class `Vec` defines a procedure that turns an instance into a string for the purpose of printing:

```
>>> print(v)

 A B C
------
 1 0 0
```

The procedure for pretty-printing a vector `v` must select some order on the domain `v.D`. Ours uses `sorted(v.D, key=hash)`, which agrees with numerical order on numbers and with alphabetical order on strings, and which does something reasonable on tuples.

2.10.5 Copying Vecs

The `Vec` class defines a `.copy()` method. This method, called on an instance of `Vec`, returns a new instance that is equal to the old instance. It shares the domain `.D` with the old instance. but has a new function `.f` that is initially equal to that of the old instance.

Ordinarily you won't need to copy `Vecs`. The scalar-vector multiplication and vector addition operations return new instances of `Vec` and do not mutate their inputs.

2.10.6 From list to Vec

The Vec class is a useful way of representing vectors, but it is not the only such representation. As mentioned in Section 2.1, we will sometimes represent vectors by lists. A list L can be viewed as a function from $\{0, 1, 2, \ldots, \text{len}(L) - 1\}$, so it is possible to convert from a list-based representation to a dictionary-based representation.

Quiz 2.10.1: Write a procedure list2vec(L) with the following spec:

- *input:* a list L of field elements
- *output:* an instance $\boldsymbol{v}$ of Vec with domain $\{0, 1, 2, \ldots, \text{len}(L) - 1\}$ such that $\boldsymbol{v}[i] = L[i]$ for each integer i in the domain

Answer

```
def list2vec(L):
  return Vec(set(range(len(L))), {k:x for k,x in enumerate(L)})
```

or

```
def list2vec(L):
  return Vec(set(range(len(L))), {k:L[k] for k in range(len(L))})
```

This procedure facilitates quickly creating small Vec examples. The procedure definition is included in the provided file vecutil.py.

2.11 Solving a triangular system of linear equations

As a step towards Computational Problem 2.9.12 (Solving a linear system), we describe an algorithm for solving a system if the system has a special form.

2.11.1 Upper-triangular systems

A *upper-triangular system of linear equations* has the form

$$\begin{array}{ccccccccccc}
[\; a_{11}, & a_{12}, & a_{13}, & a_{14}, & \cdots & a_{1,n-1}, & a_{1,n} \;] & \cdot & \boldsymbol{x} & = & \beta_1 \\
[\; 0, & a_{22}, & a_{23}, & a_{24}, & \cdots & a_{2,n-1}, & a_{2,n} \;] & \cdot & \boldsymbol{x} & = & \beta_2 \\
[\; 0, & 0, & a_{33}, & a_{34}, & \cdots & a_{3,n-1}, & a_{3,n} \;] & \cdot & \boldsymbol{x} & = & \beta_3 \\
 & & & & \vdots & & & & & & \\
[\; 0, & 0, & 0, & 0, & \cdots & a_{n-1,n-1}, & a_{n-1,n} \;] & \cdot & \boldsymbol{x} & = & \beta_{n-1} \\
[\; 0, & 0, & 0, & 0, & \cdots & 0, & a_{n,n} \;] & \cdot & \boldsymbol{x} & = & \beta_n
\end{array}$$

That is,

- the first vector need not have any zeroes,
- the second vector has a zero in the first position,
- the third vector has zeroes in the first and second positions,
- the fourth vector has zeroes in the first, second, and third positions,

  $\vdots$

- the $n - 1^{st}$ vector is all zeroes except possibly for the $n - 1^{st}$ and n^{th} entries, and

- the nth vector is all zeroes except possibly for the nth entry.

Example 2.11.1: Here's an example using 4-vectors:

$$\begin{array}{rcl} [\ 1,\ 0.5,\ -2,\ 4\ ] \cdot \boldsymbol{x} & = & -8 \\ [\ 0,\ 3,\ 3,\ 2\ ] \cdot \boldsymbol{x} & = & 3 \\ [\ 0,\ 0,\ 1,\ 5\ ] \cdot \boldsymbol{x} & = & -4 \\ [\ 0,\ 0,\ 0,\ 2\ ] \cdot \boldsymbol{x} & = & 6 \end{array}$$

The right-hand sides are -8, 3, -4, and 6.

The origin of the term *upper-triangular system* should be apparent by considering the positions of the nonzero entries: they form a triangle:

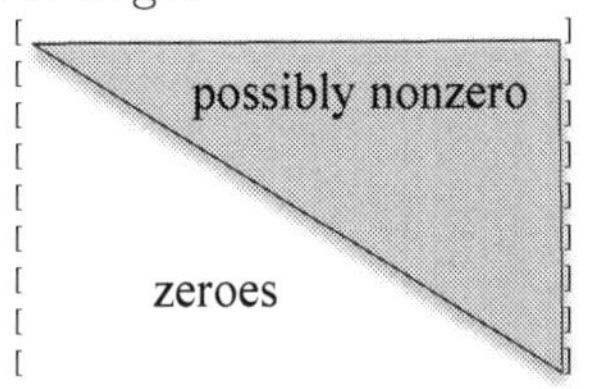

Writing $\boldsymbol{x} = [x_1, x_2, x_3, x_4]$ and using the definition of dot-product, we can rewrite this system as four ordinary equations in the (scalar) unknowns x_1, x_2, x_3, x_4:

$$\begin{array}{rcrcrcrcr} 1x_1 & + & 0.5x_2 & - & 2x_3 & + & 4x_4 & = & -8 \\ & & 3x_2 & + & 3x_3 & + & 2x_4 & = & 3 \\ & & & & 1x_3 & + & 5x_4 & = & -4 \\ & & & & & & 2x_4 & = & 6 \end{array}$$

2.11.2 Backward substitution

This suggests a solution strategy. First, solve for x_4 using the fourth equation. Plug the resulting value for x_4 into the third equation, and solve for x_3. Plug the values for x_3 and x_4 into the second equation and solve for x_2. Plug the values for x_2, x_3, and x_4 into the first equation and solve for x_1. In each iteration, only one variable needs to be solved for.

Thus the above system is solved as follows:

$$\begin{array}{llcccccc} & 2x_4 & = & 6 & & & & \\ \text{so} & x_4 & & & = & 6/2 & = & 3 \\ & & & & & & & \\ & 1x_3 & = & -4 - 5x_4 & = & -4 - 5(3) & = & -19 \\ \text{so} & x_3 & & & = & -19/1 & = & -19 \\ & & & & & & & \\ & 3x_2 & = & 3 - 3x_3 - 2x_4 & = & 3 - 2(3) - 3(-19) & = & 54 \\ \text{so} & x_2 & & & = & 54/3 & = & 18 \\ & & & & & & & \\ & 1x_1 & = & -8 - 0.5x_2 + 2x_3 - 4x_4 & = & -8 - 4(3) + 2(-19) - 0.5(18) & = & -67 \\ \text{so} & x_1 & & & = & -67/1 & = & -67 \end{array}$$

The algorithm I have illustrated is called *backward substitution* ("backward" because it starts with the last equation and works its way towards the first).

Quiz 2.11.2: Using the above technique, solve the following system by hand:

$$\begin{array}{rcrcrcr} 2x_1 & + & 3x_2 & - & 4x_3 & = & 10 \\ & & 1x_2 & + & 2x_3 & = & 3 \\ & & & & 5x_3 & = & 15 \end{array}$$

Answer

$$x_3 = 15/5 = 3$$

$$x_2 = 3 - 2x_3 = -3$$

$$x_1 = (10 + 4x_3 - 3x_2)/2 = (10 + 12 + 9)/2 = 31/2$$

Exercise 2.11.3: Solve the following system:

$$\begin{array}{rcrcrcr} 1x_1 & - & 3x_2 & - & 2x_3 & = & 7 \\ & & 2x_2 & + & 4x_3 & = & 4 \\ & & & & -10x_3 & = & 12 \end{array}$$

2.11.3 First implementation of backward substitution

There is a convenient way to express this algorithm in terms of vectors and dot-products. The procedure initializes the solution vector x to the all-zeroes vector. The procedure will populate x entry by entry, starting at the last entry. By the beginning of the entry in which x_i will be populated, entries $x_{i+1}, x_{i+2}, \ldots, x_n$ will have already been populated and the other entries are zero, so the procedure can use a dot-product to calculate the part of the expression that involves variables whose values are already known:

$$\text{entry } a_{ii} \cdot \text{ value of } x_i = \beta_i - \text{(expression involving known variables)}$$

so

$$\text{value of } x_i = \frac{\beta_i - \text{(expression involving known variables)}}{a_{ii}}$$

Using this idea, let's write a procedure `triangular_solve_n(rowlist, b)` with the following spec:

- *input:* for some integer n, a triangular system consisting of a list `rowlist` of n-vectors, and a length-n list $\boldsymbol{b}$ of numbers
- *output:* a vector $\hat{\boldsymbol{x}}$ such that, for $i = 0, 1, \ldots, n-1$, the dot-product of `rowlist`$[i]$ with $\hat{\boldsymbol{x}}$ equals $\boldsymbol{b}[i]$

The `n` in the name indicates that this procedure requires each of the vectors in `rowlist` to have domain $\{0, 1, 2, \ldots, n-1\}$. (We will later write a procedure without this requirement.)

Here is the code:

```
def triangular_solve_n(rowlist, b):
    D = rowlist[0].D
    n = len(D)
    assert D == set(range(n))
    x = zero_vec(D)
    for i in reversed(range(n)):
        x[i] = (b[i] - rowlist[i] * x)/rowlist[i][i]
    return x
```

Exercise 2.11.4: Enter `triangular_solve_n` into Python and try it out on the example system above.

2.11.4 When does the algorithm work?

The *backward substitution* algorithm does not work on all upper triangular systems of equations. If `rowlist[i][i]` is zero for some `i`, the algorithm will fail. We must therefore require when using this algorithm that these entries are not zero. Thus the spec given above is incomplete.

If these entries are nonzero so the algorithm *does* succeed, it will have found the *only* solution to the system of linear equations. The proof is by induction; it is based on the observation that the value assigned to a variable in each iteration is the *only* possible value for that variable that is consistent with the values assigned to variables in previous iterations.

Proposition 2.11.5: For a triangular system specified by a length-n list `rowlist` of n-vectors and an n-vector $\boldsymbol{b}$, if `rowlist`$[i][i] \neq 0$ for $i = 0, 1, \ldots, n-1$ then the solution found by `triangular_solve_n(rowlist, b)` is the *only* solution to the system.

On the other hand,

Proposition 2.11.6: For a length-n list `rowlist` of n-vector, if `rowlist`$[i][i] = 0$ for some integer i then there is a vector $\boldsymbol{b}$ for which the triangular system has no solution.

Proof

Let k be the largest integer less than n such that `rowlist`$[k][k] = 0$. Define $\boldsymbol{b}$ to be a vector whose entries are all zero except for entry k which is nonzero. The algorithm iterates $i = n-1, n-2, \ldots, k+1$. In each of these iterations, the value of `x` before the iteration is the zero vector, and `b[i]` is zero, so `x[i]` is assigned zero. In each of these iterations, the value assigned is is the only possible value consistent with the values assigned to variables in previous iterations.

Finally, the algorithm gets to $i = k$. The equation considered at this point is

`rowlist[k][k]*x[k]+rowlist[k][k+1]*x[k+1]+ ⋯ +rowlist[k][n-1]*x[n-1] =` *nonzero*

but the variables `x[k+1]`, `x[k+2]`, `x[n-1]` have all been forced to be zero, and `rowlist[k][k]` is zero, so the left-hand side of the equation is zero, so the equation cannot be satisfied. □

2.11.5 Backward substitution with arbitrary-domain vectors

Next we write a procedure `triangular_solve(rowlist, label_list, b)` to solve a triangular system in which the domain of the vectors in `rowlist` need not be $\{0, 1, 2, \ldots, n-1\}$. What does it mean for a system to be triangular? The argument `label_list` is a list that specifies an ordering of the domain. For the system to be triangular,

- the first vector in `rowlist` need not have any zeroes,
- the second vector has a zero in the entry labeled by the first element of `label_list`,
- the third vector has zeroes in the entries labeled by the first two elements of `label_list`,

and so on.

The spec of the procedure is:

- *input:* for some positive integer n, a list *rowlist* of n Vecs all having the same n-element domain D, a list *label_list* consisting of the elements of D, and a list b consisting of n numbers
  such that, for $i = 0, 1, \ldots, n-1$,
  - *rowlist*$[i]$[label_list$[j]$] is zero for $j = 0, 1, 2, \ldots, i-1$ and is nonzero for $j = i$
- *output:* the Vec $\boldsymbol{x}$ such that, for $i = 0, 1, \ldots, n-1$, the dot-product of rowlist$[i]$ and $\boldsymbol{x}$ equals $b[i]$.

The procedure involves making small changes to the procedure given in Section 2.11.3.

Here I illustrate how the procedure is used.

```
>>> label_list = ['a','b','c','d']
>>> D = set(label_list)
>>> rowlist=[Vec(D,{'a':4, 'b':-2,'c':0.5,'d':1}), Vec(D,{'b':2,'c':3,'d':3}),
                Vec(D,{'c':5, 'd':1}), Vec(D,{'d':2.})]
>>> b = [6, -4, 3, -8]
>>> triangular_solve(rowlist, label_list, b)
Vec({'d', 'b', 'c', 'a'},{'d': -4.0, 'b': 1.9, 'c': 1.4, 'a': 3.275})
```

Here is the code for `triangular_solve`. Note that it uses the procedure `zero_vec(D)`.

```
def triangular_solve(rowlist, label_list, b):
    D = rowlist[0].D
    x = zero_vec(D)
    for j in reversed(range(len(D))):
        c = label_list[j]
        row = rowlist[j]
        x[c] = (b[j] - x*row)/row[c]
    return x
```

The procedures `triangular_solve(rowlist, label_list, b)` and `triangular_solve_n(rowlist, b)` are provided in the module `triangular`.

2.12 *Lab: Comparing voting records using dot-product*

In this lab, we will represent a US senator's voting record as a vector over $\mathbb{R}$, and will use dot-products to compare voting records. For this lab, we will just use a list to represent a vector.

2.12.1 *Motivation*

These are troubled times. You might not have noticed from atop the ivory tower, but take our word for it that the current sociopolitical landscape is in a state of abject turmoil. Now is the time for a hero. Now is the time for someone to take up the mantle of protector, of the people's shepherd. Now is the time for linear algebra.

In this lab, we will use vectors to evaluate objectively the political mindset of the senators who represent us. Each senator's voting record can be represented as a vector, where each element of that vector represents how that senator voted on a given piece of legislation. By looking at the difference between the "voting vectors" of two senators, we can dispel the fog of politics and see just where our representatives stand.

Or, rather, stood. Our data are a bit dated. On the bright side, you get to see how Obama did as a senator. In case you want to try out your code on data from more recent years, we will post more data files on `resources.codingthematrix.com`.

2.12.2 *Reading in the file*

As in the last lab, the information you need to work with is stored in a whitespace-delimited text file. The senatorial voting records for the 109th Congress can be found in `voting_record_dump109.txt`.

Each line of the file represents the voting record of a different senator. In case you've forgotten how to read in the file, you can do it like this:

```
>>> f = open('voting_record_dump109.txt')
>>> mylist = list(f)
```

You can use the `split(·)` procedure to split each line of the file into a list; the first element of the list will be the senator's name, the second will be his/her party affiliation (R or D), the third will be his/her home state, and the remaining elements of the list will be that senator's voting record on a collection of bills. A "1" represents a 'yea' vote, a "-1" a 'nay', and a "0" an abstention.

Task 2.12.1: Write a procedure `create_voting_dict(strlist)` that, given a list of strings (voting records from the source file), returns a dictionary that maps the last name of a senator to a list of numbers representing that senator's voting record. You will need to use the built-in procedure `int(·)` to convert a string representation of an integer (e.g. '1') to the actual integer (e.g. 1).

2.12.3 *Two ways to use dot-product to compare vectors*

Suppose $\boldsymbol{u}$ and $\boldsymbol{v}$ are two vectors. Let's take the simple case (relevant to the current lab) in which the entries are all 1, 0, or -1. Recall that the dot-product of $\boldsymbol{u}$ and $\boldsymbol{v}$ is defined as

$$\boldsymbol{u} \cdot \boldsymbol{v} = \sum_k \boldsymbol{u}[k]\boldsymbol{v}[k]$$

Consider the kth entry. If both $\boldsymbol{u}[k]$ and $\boldsymbol{v}[k]$ are 1, the corresponding term in the sum is 1. If both $\boldsymbol{u}[k]$ and $\boldsymbol{v}[k]$ are -1, the corresponding term in the sum is also 1. Thus a term in the sum that is 1 indicates agreement. If, on the other hand, $\boldsymbol{u}[k]$ and $\boldsymbol{v}[k]$ have different signs, the corresponding term is -1. Thus a term in the sum that is -1 indicates disagreement. (If one or both of $\boldsymbol{u}[k]$ and $\boldsymbol{v}[k]$ are zero then the term is zero, reflecting the fact that those entries provide no evidence of either agreement or disagreement.) The dot-product of $\boldsymbol{u}$ and $\boldsymbol{v}$ therefore is a measure of how much $\boldsymbol{u}$ and $\boldsymbol{v}$ are in agreement.

2.12.4 *Policy comparison*

We would like to determine just how like-minded two given senators are. We will use the dot-product of vectors $\boldsymbol{u}$ and $\boldsymbol{v}$ to judge how often two senators are in agreement.

Task 2.12.2: Write a procedure `policy_compare(sen_a, sen_b, voting_dict)` that, given two names of senators and a dictionary mapping senator names to lists representing voting records, returns the dot-product representing the degree of similarity between two senators' voting policies.

Task 2.12.3: Write a procedure `most_similar(sen, voting_dict)` that, given the name of a senator and a dictionary mapping senator names to lists representing voting records, returns the name of the senator whose political mindset is most like the input senator (excluding, of course, the input senator him/herself).

Task 2.12.4: Write a very similar procedure `least_similar(sen, voting_dict)` that returns the name of the senator whose voting record agrees the least with the senator whose name is `sen`.

Task 2.12.5: Use these procedures to figure out which senator is most like Rhode Island legend Lincoln Chafee. Then use these procedures to see who disagrees most with Pennsylvania's Rick Santorum. Give their names.

Task 2.12.6: How similar are the voting records of the two senators from your favorite state?

2.12.5 *Not your average Democrat*

Task 2.12.7: Write a procedure `find_average_similarity(sen, sen_set, voting_dict)` that, given the name `sen` of a senator, compares that senator's voting record to the voting records of all senators whose names are in `sen_set`, computing a dot-product for each, and then returns the average dot-product.

Use your procedure to compute which senator has the greatest average similarity with the set of Democrats (you can extract this set from the input file).

In the last task, you had to compare each senator's record to the voting record of each Democrat senator. If you were doing the same computation with, say, the movie preferences of all Netflix subscribers, it would take far too long to be practical.

Next we see that there is a computational shortcut, based on an algebraic property of the dot-product: the *distributive* property:

$$(\boldsymbol{v}_1 + \boldsymbol{v}_2) \cdot \boldsymbol{x} = \boldsymbol{v}_1 \cdot \boldsymbol{x} + \boldsymbol{v}_2 \cdot \boldsymbol{x}$$

Task 2.12.8: Write a procedure `find_average_record(sen_set, voting_dict)` that, given a set of names of senators, finds the average voting record. That is, perform vector addition on the lists representing their voting records, and then divide the sum by the number of vectors. The result should be a vector.

Use this procedure to compute the average voting record for the set of Democrats, and assign the result to the variable `average_Democrat_record`. Next find which senator's voting record is most similar to the average Democrat voting record. Did you get the same result as in Task 2.12.7? Can you explain?

2.12.6 *Bitter Rivals*

Task 2.12.9: Write a procedure `bitter_rivals(voting_dict)` to find which two senators disagree the most.

This task again requires comparing each pair of voting records. Can this be done faster than the obvious way? There is a slightly more efficient algorithm, using *fast matrix multiplication*. We will study matrix multiplication later, although we won't cover the theoretically fast algorithms.

2.12.7 *Open-ended study*

You have just coded a set of simple yet powerful tools for sifting the truth from the sordid flour of contemporary politics. Use your new abilities to answer at least one of the following questions (or make up one of your own):

- Who/which is the most Republican/Democratic senator/state?
- Is John McCain really a maverick?
- Is Barack Obama really an extremist?
- Which two senators are the most bitter rivals?
- Which senator has the most political opponents? (Assume two senators are opponents if their dot-product is very negative, i.e. is less than some negative threshold.)

2.13 Review Questions

- What is vector addition?
- What is the geometric interpretation of vector addition?

- What is scalar-vector multiplication?
- What is the distributive property that involves scalar-vector multiplication but not vector addition?
- What is the distributive property that involves both scalar-vector multiplication and vector addition?
- How is scalar-vector multiplication used to represent the line through the origin and a given point?
- How are scalar-vector multiplication and vector addition used to represent the line through a pair of given points?
- What is dot-product?
- What is the *homogeneity* property that relates dot-product to scalar-vector multiplication?
- What is the distributive property property that relates dot-product to vector addition?
- What is a linear equation (expressed using dot-product)?
- What is a linear system?
- What is an upper-triangular linear system?
- How can one solve an upper-triangular linear system?

2.14 Problems

Vector addition practice

Problem 2.14.1: For vectors $\boldsymbol{v} = [-1, 3]$ and $\boldsymbol{u} = [0, 4]$, find the vectors $\boldsymbol{v} + \boldsymbol{u}$, $\boldsymbol{v} - \boldsymbol{u}$, and $3\boldsymbol{v} - 2\boldsymbol{u}$. Draw these vectors as arrows on the same graph..

Problem 2.14.2: Given the vectors $\boldsymbol{v} = [2, -1, 5]$ and $\boldsymbol{u} = [-1, 1, 1]$, find the vectors $\boldsymbol{v} + \boldsymbol{u}$, $\boldsymbol{v} - \boldsymbol{u}$, $2\boldsymbol{v} - \boldsymbol{u}$, and $\boldsymbol{v} + 2\boldsymbol{u}$.

Problem 2.14.3: For the vectors $\boldsymbol{v} = [0, one, one]$ and $\boldsymbol{u} = [one, one, one]$ over $GF(2)$, find $\boldsymbol{v} + \boldsymbol{u}$ and $\boldsymbol{v} + \boldsymbol{u} + \boldsymbol{u}$.

Expressing one $GF(2)$ vector as a sum of others

Problem 2.14.4: Here are six 7-vectors over $GF(2)$:

a =	1100000	d =	0001100
b =	0110000	e =	0000110
c =	0011000	f =	0000011

For each of the following vectors $\boldsymbol{u}$, find a subset of the above vectors whose sum is $\boldsymbol{u}$, or report that no such subset exists.

1. $\boldsymbol{u} = 0010010$
2. $\boldsymbol{u} = 0100010$

Problem 2.14.5: Here are six 7-vectors over $GF(2)$:

a =	1110000	d =	0001110
b =	0111000	e =	0000111
c =	0011100	f =	0000011

For each of the following vectors $\boldsymbol{u}$, find a subset of the above vectors whose sum is $\boldsymbol{u}$, or report that no such subset exists.

1. $\boldsymbol{u} = 0010010$
2. $\boldsymbol{u} = 0100010$

Finding a solution to linear equations over $GF(2)$

Problem 2.14.6: Find a vector $\boldsymbol{x} = [x_1, x_2, x_3, x_4]$ over $GF(2)$ satisfying the following linear equations:

$$\begin{aligned} 1100 \cdot \boldsymbol{x} &= 1 \\ 1010 \cdot \boldsymbol{x} &= 1 \\ 1111 \cdot \boldsymbol{x} &= 1 \end{aligned}$$

Show that $\boldsymbol{x} + 1111$ also satisfies the equations.

Formulating equations using dot-product

Problem 2.14.7: Consider the equations

$$\begin{array}{ccccccccc} 2x_0 & + & 3x_1 & - & 4x_2 & + & x_3 & = & 10 \\ x_0 & - & 5x_1 & + & 2x_2 & + & 0x_3 & = & 35 \\ 4x_0 & + & x_1 & - & x_2 & - & x_3 & = & 8 \end{array}$$

Your job is not to solve these equations but to formulate them using dot-product. In particular, come up with three vectors `v1`, `v2`, and `v3` represented as lists so that the above equations are equivalent to

$$\begin{aligned} \texttt{v1} \cdot \boldsymbol{x} &= 10 \\ \texttt{v2} \cdot \boldsymbol{x} &= 35 \\ \texttt{v3} \cdot \boldsymbol{x} &= 8 \end{aligned}$$

where x is a 4-vector over $\mathbb{R}$.

Plotting lines and line segments

Problem 2.14.8: Use the `plot` module to plot

(a) a substantial portion of the line through [-1.5,2] and [3,0], and

(b) the line segment between [2,1] and [-2,2].

For each, provide the Python statements you used and the plot obtained.

Practice with dot-product

Problem 2.14.9: For each of the following pairs of vectors $\boldsymbol{u}$ and $\boldsymbol{v}$ over $\mathbb{R}$, evaluate the expression $\boldsymbol{u} \cdot \boldsymbol{v}$:

(a) $\boldsymbol{u} = [1, 0], \boldsymbol{v} = [5, 4321]$

(b) $\boldsymbol{u} = [0, 1], \boldsymbol{v} = [12345, 6]$

(c) $\boldsymbol{u} = [-1, 3], \boldsymbol{v} = [5, 7]$

(d) $\boldsymbol{u} = [-\frac{\sqrt{2}}{2}, \frac{\sqrt{2}}{2}], \boldsymbol{v} = [\frac{\sqrt{2}}{2}, -\frac{\sqrt{2}}{2}]$

Writing procedures for the Vec class

Problem 2.14.10: Download the file vec.py to your computer, and edit it. The file defines procedures using the Python statement pass, which does nothing. You can import the vec module and create instancs of Vec but the operations such as * and + currently do nothing. Your job is to replace each occurence of the pass statement with appropriate code. Your code for a procedure can include calls to others of the seven. You should make no changes to the class definition.

Docstrings At the beginning of each procedure body is a multi-line string (deliminated by triple quote marks). This is called a documentation string (*docstring*). It specifies what the procedure should do.

Doctests The documentation string we provide for a procedure also includes examples of the functionality that procedure is supposed to provide to Vecs. The examples show an interaction with Python: statements and expressions are evaluated by Python, and Python's responses are shown. These examples are provided to you as tests (called *doctests*). You should make sure that your procedure is written in such a way that the behavior of your Vec implementation matches that in the examples. If not, your implementation is incorrect.[a]

Python provides convenient ways to test whether a module such as vec passes all its doctests. You don't even need to be in a Python session. From a console, make sure your current working directory is the one containing vec.py, and type

```
python3 -m doctest vec.py
```

to the console, where python3 is the name of your Python executable. If your implementation passes all the tests, this command will print nothing. Otherwise, the command prints information on which tests were failed.

You can also test a module's doctest from within a Python session:

```
>>> import doctest
>>> doctest.testfile("vec.py")
```

Assertions For most of the procedures to be written, the first statement after the docstring is an *assertion*. Executing an assertion verifies that the condition is true, and raises an error if not. The assertions are there to detect errors in the use of the procedures. Take a look at the assertions to make sure you understand them. You can take them out, but you do so at your own risk.

Arbitrary set as domain: Our vector implementation allows the domain to be, for example, a set of strings. Do not make the mistake of assuming that the domain consists of integers. If your code includes len or range, you're doing it wrong.

Sparse representation: Your procedures should be able to cope with our sparse representation, i.e. an element in the domain v.D that is not a key of the dictionary v.f. For example, getitem(v, k) should return a value for every domain element even if k is not a key of v.f. However, your procedures need *not* make any effort to retain sparsity when adding two vectors. That is, for two instances u and v of Vec, it is okay if every element of u.D is represented explicitly in the dictionary of the instance u+v.

Several other procedures need to be written with the sparsity convention in mind. For example, two vectors can be equal even if their .f fields are not equal: one vector's .f field can contain a key-value pair in which the value is zero, and the other vector's `.f` field can omit this particular key. For this reason, the `equal(u, v)` procedure needs to be written with care.

[a]The examples provided for each procedure are supposed to test that procedure; however, note that, since equality is used in tests for procedures other than `equal(u,v)`, a bug in your definition for `equal(u,v)` could cause another procedure's test to fail.

Chapter 3

The Vector Space

> [Geometry of the ancients] ... is so exclusively restricted to the consideration of figures that it can exercise the understanding only on condition of greatly fatiguing the imagination....
> *René Descartes*, **Discourse on Method**

In the course of discussing applications of vectors in the previous chapter, we encountered four Questions. We will soon encounter two more. However, we won't answer any of the Questions in this chapter; instead, we will turn them into new and deeper Questions. The answers will come in Chapters 5 and 6. In this chapter, we will encounter the concept of vector spaces, a concept that underlies the answers and everything else we do in this book.

3.1 Linear combination

3.1.1 Definition of linear combination

Definition 3.1.1: Suppose $\boldsymbol{v}_1, \ldots, \boldsymbol{v}_n$ are vectors. We define a *linear combination* of $\boldsymbol{v}_1, \ldots, \boldsymbol{v}_n$ to be a sum

$$\alpha_1 \boldsymbol{v}_1 + \cdots + \alpha_n \boldsymbol{v}_n$$

where $\alpha_1, \ldots, \alpha_n$ are scalars. In this context, we refer to $\alpha_1, \ldots, \alpha_n$ as the *coefficients* in this linear combination. In particular, α_1 is the coefficient of $\boldsymbol{v}_1$ in the linear combination, α_2 is the coefficient of $\boldsymbol{v}_2$, and so on.

Example 3.1.2: One linear combination of $[2, 3.5]$ and $[4, 10]$ is

$$-5\,[2, 3.5] + 2\,[4, 10]$$

which is equal to $[-5 \cdot 2, -5 \cdot 3.5] + [2 \cdot 4, 2 \cdot 10]$, which is equal to $[-10, -17.5] + [8, 20]$, which is $[-2, 2.5]$.

Another linear combination of the same vectors is

$$0\,[2, 3.5] + 0\,[4, 10]$$

which is equal to the zero vector $[0, 0]$.

If all the coefficients in a linear combination are zero, we say that it is a *trivial* linear combination

3.1.2 Uses of linear combinations

Example 3.1.3: *Stock portfolios:* Let D be the set of stocks. A D-vector over $\mathbb{R}$ represents a portfolio, i.e. it maps each stock to a number of shares owned.

Suppose that there are n mutual funds. For $i = 1, \ldots, n$, each share of mutual fund i represents ownership of a certain amount of each stock, and can therefore be represented by a D-vector $\boldsymbol{v}_i$. Let α_i be the number of shares of stock fund i that you own. Then your total implied ownership of stocks is represented by the linear combination

$$\alpha_1 \boldsymbol{v}_1 + \cdots + \alpha_n \boldsymbol{v}_n$$

Example 3.1.4: *Diet design:* In the 1930's and 1940's the US military wanted to find the minimum-cost diet that would satisfy a soldier's nutritional requirements. An economist, George Stigler, considered seventy-seven different foods (wheat flour, evaporated milk, cabbage ...) and nine nutritional requirements (calories, Vitamin A, riboflavin...). For each food, he calculated how much a unit of that food satisfied each of nine nutritional requirements. The results can be represented by seventy-seven 9-vectors $\boldsymbol{v}_i$, one for each food.

A possible diet is represented by an amount of each food: one pound wheat flour, half a pound of cabbage, etc. For $i = 1, \ldots, 77$, let α_i be the amount of food i specified by the diet. Then the linear combination

$$\alpha_1 \boldsymbol{v}_1 + \cdots + \alpha_{77} \boldsymbol{v}_{77}$$

represents the total nutritional value provided by that diet.

In Chapter 13, we will study how to find the minimum-cost diet achieving specified nutritional goals.

Example 3.1.5: *Average face:* As mentioned in Section 2.3, black-and-white images, e.g. of faces, can be stored as vectors. A linear combination of three such vectors, with coefficients 1/3 and 1/3 and 1/3, yields an *average* of the three faces.

The idea of average faces arises later in the book, when we describe a method for face detection.

Example 3.1.6: *Products and resources:* The JunkCo factory makes things using five resources: metal, concrete, plastic, water, and electricity. Let D be this set of resources. The factory has the ability to make five different products.

Here is a fabricated table that shows how much of each resource is used in making each product, on a per-item basis:

	metal	concrete	plastic	water	electricity
garden gnome	0	1.3	.2	.8	.4
hula hoop	0	0	1.5	.4	.3
slinky	.25	0	0	.2	.7
silly putty	0	0	.3	.7	.5
salad shooter	.15	0	.5	.4	.8

The ith product's resource utilization is stored in a D-vector $\boldsymbol{v}_i$ over $\mathbb{R}$. For example, a gnome is represented by

$$\boldsymbol{v}_{\text{gnome}} = \texttt{Vec(D,\{'concrete':1.3,'plastic':.2,'water':.8,'electricity':.4\})}$$

Suppose the factory plans to make α_{gnome} garden gnomes, α_{hoop} hula hoops, α_{slinky} slinkies, α_{putty} silly putties, and α_{shooter} salad shooters. The total resource utilization is expressed as a linear combination

$$\alpha_{\text{gnome}}\, \boldsymbol{v}_{\text{gnome}} + \alpha_{\text{hoop}}\, \boldsymbol{v}_{\text{hoop}} + \alpha_{\text{slinky}}\, \boldsymbol{v}_{\text{slinky}} + \alpha_{\text{putty}}\, \boldsymbol{v}_{\text{putty}} + \alpha_{\text{shooter}}\, \boldsymbol{v}_{\text{shooter}}$$

For example, suppose JunkCo decides to make 240 gnomes, 55 hoops, 150 slinkies, 133 putties, and 90 shooters. Here's how the linear combination can be written in Python using our Vec class:

```
>>> D = {'metal','concrete','plastic','water','electricity'}
>>> v_gnome=Vec(D,{'concrete':1.3,'plastic':.2,'water':.8,'electricity':.4})
>>> v_hoop =Vec(D, {'plastic':1.5, 'water':.4, 'electricity':.3})
>>> v_slinky = Vec(D, {'metal':.25, 'water':.2, 'electricity':.7})
>>> v_putty = Vec(D, {'plastic':.3, 'water':.7, 'electricity':.5})
>>> v_shooter = Vec(D, {'metal':.15, 'plastic':.5, 'water':.4,'electricity':.8})

>>> print(240*v_gnome + 55*v_hoop + 150*v_slinky + 133*v_putty + 90*v_shooter)

 plastic metal concrete water electricity
------------------------------------------
     215    51      312   373         356
```

We build on this example in the next section.

3.1.3 From coefficients to linear combination

For a length-n list $[\boldsymbol{v}_1, \ldots, \boldsymbol{v}_n]$ of vectors, there is a function f that maps each length-n list $[\alpha_1, \ldots, \alpha_n]$ of coefficients to the corresponding linear combination $\alpha_1\, \boldsymbol{v}_1 + \cdots + \alpha_n\, \boldsymbol{v}_n$. As discussed in Section 0.3.2, there are two related computational problems, the *forward* problem (given an element of the domain, find the image under the function) and the *backward* problem (given an element of the co-domain, find any pre-image if there is one).

Solving the *forward* problem is easy.

Quiz 3.1.7: Define a procedure `lin_comb(vlist, clist)` with the following spec:

- *input:* a list `vlist` of vectors, a list `clist` of the same length consisting of scalars
- *output:* the vector that is the linear combination of the vectors in `vlist` with corresponding coefficients `clist`

Answer

```
def lin_comb(vlist,clist):
  return sum([coeff*v for (coeff,v) in zip(clist, vlist)])
```

or

```
def lin_comb(vlist,clist):
  return sum([clist[i]*vlist[i] for i in range(len(vlist))])
```

For example, the JunkCo factory can use this procedure for the forward problem: given an amount of each product, the factory can compute how much of each resource will be required.

3.1.4 From linear combination to coefficients

Suppose, however, you are an industrial spy. Your goal is to figure out how many garden gnomes the JunkCo factory is manufacturing. To do this, you can sneakily observe how much of each resource the factory is consuming. That is, you can acquire the vector $\boldsymbol{b}$ that is the output of the function f.

The first question is: can you solve the *backward* problem? That is, can you obtain a pre-image of $\boldsymbol{b}$ under f? The second question is: how can we tell whether there is a single solution? If there are multiple pre-images of $\boldsymbol{b}$, we cannot be confident that we have calculated the true number of garden gnomes.

The first question is a computational problem:

Computational Problem 3.1.8: *Expressing a given vector as a linear combination of other given vectors*

- *input:* a vector $\boldsymbol{b}$ and a list $[\boldsymbol{v}_1, \ldots, \boldsymbol{v}_n]$ of n vectors
- *output:* a list $[\alpha_1, \ldots, \alpha_n]$ of coefficients such that

$$\boldsymbol{b} = \alpha_1\, \boldsymbol{v}_1 + \cdots + \alpha_n\, \boldsymbol{v}_n$$

or a report that none exists.

In Chapter 4, we will see that finding a linear combination of given vectors $\boldsymbol{v}_1, \ldots, \boldsymbol{v}_n$ that equals a given vector $\boldsymbol{b}$ is equivalent to solving a linear system. Therefore the above Computational Problem is equivalent to Computational Problem 2.9.12, *Solving a system of linear equations*, and the question of whether there is at most a single solution is equivalent to Question 2.9.11, *Uniqueness of solutions to systems of linear equations*.

Example 3.1.9: *(Lights Out)* We saw in Section 2.8.3 that the state of the *Lights Out* puzzle could be represented by a vector over $GF(2)$, and that each button corresponds to a "button" vector over $GF(2)$.

Let $\boldsymbol{s}$ denote the initial state of the puzzle. We saw that finding a solution to the puzzle (which buttons to push in order to turn off all the lights) is equivalent to finding a subset of the button vectors whose sum is $\boldsymbol{s}$.

We can in turn formulate this problem using the notion of linear combinations. Over $GF(2)$, the only coefficients are zero and one. A linear combination of the twenty-five button vectors

$$\alpha_{0,0}\boldsymbol{v}_{0,0} + \alpha_{0,1}\boldsymbol{v}_{0,1} + \cdots + \alpha_{4,4}\boldsymbol{v}_{4,4}$$

is the sum of some subset of button vectors, namely those whose corresponding coefficents are one.

Our goal, then, is to find a linear combination of the twenty-five button vectors whose value is $\boldsymbol{s}$:

$$\boldsymbol{s} = \alpha_{0,0}\boldsymbol{v}_{0,0} + \alpha_{0,1}\boldsymbol{v}_{0,1} + \cdots + \alpha_{4,4}\boldsymbol{v}_{4,4} \tag{3.1}$$

That is, once again we must solve Computational Problem 3.1.8.

Quiz 3.1.10: For practice, we use the 2×2 version of *Lights Out*. Show how to express

$s =$ 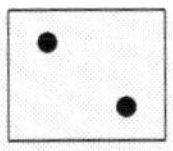 as a linear combination of the button vectors

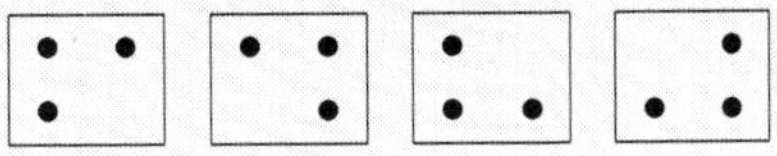

Answer

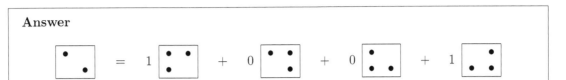

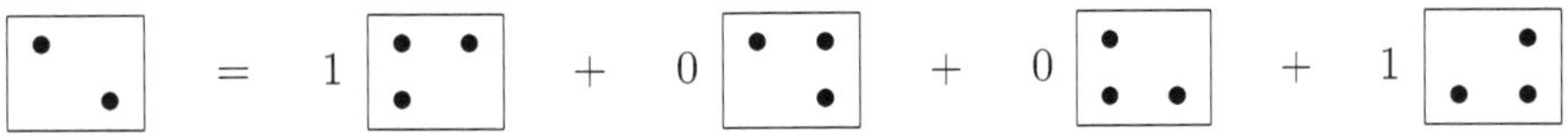

3.2 Span

3.2.1 Definition of span

Definition 3.2.1: The set of all linear combinations of vectors $v_1, \ldots, v_n$ is called the *span* of these vectors, and is written Span $\{v_1, \ldots, v_n\}$.

For vectors over infinite fields such as $\mathbb{R}$ or over $\mathbb{C}$, the span is usually an infinite set. In the next section, we discuss the geometry of such a set. For vectors over $GF(2)$, a finite field, the span is finite.

Quiz 3.2.2: How many vectors are in Span $\{[1, 1], [0, 1]\}$ over the field $GF(2)$?

Answer

The linear combinations are

$$\begin{aligned} 0\,[1,1] + 0\,[0,1] &= [0,0] \\ 0\,[1,1] + 1\,[0,1] &= [0,1] \\ 1\,[1,1] + 0\,[0,1] &= [1,1] \\ 1\,[1,1] + 1\,[0,1] &= [1,0] \end{aligned}$$

Thus there are four vectors in the span.

Quiz 3.2.3: How many vectors are in Span $\{[1, 1]\}$ over the field $GF(2)$?

Answer

The linear combinations are

$$\begin{aligned} 0\,[1,1] &= [0,0] \\ 1\,[1,1] &= [1,1] \end{aligned}$$

Thus there are two vectors in the span.

Quiz 3.2.4: How many vectors are in the span of an empty set of 2-vectors?

Answer

Don't make the mistake of thinking that there are *no* linear combinations, i.e. *no* assignments of numbers to coefficients. There is one such assignment: the empty assignment. Taking the sum of this empty set of vectors (and thinking back to Problem 1.7.9), we obtain $[0, 0]$.

Quiz 3.2.5: How many vectors are in the span of the 2-vector $[2, 3]$ over $\mathbb{R}$?

Answer

There are an infinite number. The span is $\{\alpha\,[2, 3] \;:\; \alpha \in \mathbb{R}\}$, which, as we saw in Section 2.5.3, consists of the points on the line through the origin and $[2, 3]$.

Quiz 3.2.6: For which 2-vector $\boldsymbol{v}$ over $\mathbb{R}$ does Span $\{\boldsymbol{v}\}$ consists of a finite number of vectors?

Answer

The zero vector $[0, 0]$.

3.2.2 A system of linear equations implies other equations

Example 3.2.7: Recall the simple authentication scheme from Section 2.9.6. The secret password is a vector $\hat{\boldsymbol{x}}$ over $GF(2)$. The computer tests the human's knowledge of the password by sending a *challenge* vector $\boldsymbol{a}$; the human must respond with the dot-product $\boldsymbol{a} \cdot \hat{\boldsymbol{x}}$.

Meanwhile, the eavesdropper, Eve, is observing all their communication. Suppose Eve has observed the challenges $\boldsymbol{a}_1 = [1, 1, 1, 0, 0]$, $\boldsymbol{a}_2 = [0, 1, 1, 1, 0]$, $\boldsymbol{a}_3 = [0, 0, 1, 1, 1]$ and the corresponding responses $\beta_1 = 1$, $\beta_2 = 0$, $\beta_3 = 1$. For what possible challenge vectors can Eve derive the right response?

We consider all linear combinations of $\boldsymbol{a}_1, \boldsymbol{a}_2, \boldsymbol{a}_3$. Since there are three vectors, there are three coefficients $\alpha_1, \alpha_2, \alpha_3$ to choose. For each coefficient α_i, there are two choices, 0 and 1. Therefore there are eight vectors in the span. Here is a table of them:

$$\begin{aligned}
0\,[1,1,1,0,0] + 0\,[0,1,1,1,0] + 0\,[0,0,1,1,1] &= [0,0,0,0,0]\\
1\,[1,1,1,0,0] + 0\,[0,1,1,1,0] + 0\,[0,0,1,1,1] &= [1,1,1,0,0]\\
0\,[1,1,1,0,0] + 1\,[0,1,1,1,0] + 0\,[0,0,1,1,1] &= [0,1,1,1,0]\\
1\,[1,1,1,0,0] + 1\,[0,1,1,1,0] + 0\,[0,0,1,1,1] &= [1,0,0,1,0]\\
0\,[1,1,1,0,0] + 0\,[0,1,1,1,0] + 1\,[0,0,1,1,1] &= [0,0,1,1,1]\\
1\,[1,1,1,0,0] + 0\,[0,1,1,1,0] + 1\,[0,0,1,1,1] &= [1,1,0,1,1]\\
0\,[1,1,1,0,0] + 1\,[0,1,1,1,0] + 1\,[0,0,1,1,1] &= [0,1,0,0,1]\\
1\,[1,1,1,0,0] + 1\,[0,1,1,1,0] + 1\,[0,0,1,1,1] &= [1,0,1,0,1]
\end{aligned}$$

If the challenge is in the span, Eve can calculate the right response to it. For example, suppose the challenge is $[1, 0, 1, 0, 1]$, the last vector in the table. We see from the table that

$$[1,0,1,0,1] = 1\,[1,1,1,0,0] + 1\,[0,1,1,1,0] + 1\,[0,0,1,1,1]$$

Therefore

$$\begin{aligned}[1,0,1,0,1]\cdot\hat{x} &= (1\,[1,1,1,0,0]+1\,[0,1,1,1,0]+1\,[0,0,1,1,1])\cdot\hat{x} \\ &= 1\,[1,1,1,0,0]\cdot\hat{x}+1\,[0,1,1,1,0]\cdot\hat{x}+1\,[0,0,1,1,1]\cdot\hat{x} && \text{by distributivity} \\ &= 1\,([1,1,1,0,0]\cdot\hat{x})+1\,([0,1,1,1,0]\cdot\hat{x})+1\,([0,0,1,1,1]\cdot\hat{x}) && \text{by homogeneity} \\ &= 1\beta_1+1\beta_2+1\beta_3 \\ &= 1\cdot 1+1\cdot 0+1\cdot 1 \\ &= 0\end{aligned}$$

More generally, if you know that a vector $\hat{\boldsymbol{x}}$ satisfies linear equations

$$\begin{aligned}\boldsymbol{a}_1\cdot\boldsymbol{x} &= \beta_1 \\ &\vdots \\ \boldsymbol{a}_m\cdot\boldsymbol{x} &= \beta_m\end{aligned}$$

over any field then you can calculate the dot-product with $\hat{\boldsymbol{x}}$ of any vector $\boldsymbol{a}$ that is in the span of $\boldsymbol{a}_1,\ldots,\boldsymbol{a}_m$.

Suppose $\boldsymbol{a}=\alpha_1\,\boldsymbol{a}_1+\cdots+\alpha_m\,\boldsymbol{a}_m$. Then

$$\begin{aligned}\boldsymbol{a}\cdot\boldsymbol{x} &= (\alpha_1\,\boldsymbol{a}_1+\cdots+\alpha_m\,\boldsymbol{a}_m)\cdot\boldsymbol{x} \\ &= \alpha_1\,\boldsymbol{a}_1\cdot\boldsymbol{x}+\cdots+\alpha_m\,\boldsymbol{a}_m\cdot\boldsymbol{x} && \text{by distributivity} \\ &= \alpha_1\,(\boldsymbol{a}_1\cdot\boldsymbol{x})+\cdots+\alpha_m\,(\boldsymbol{a}_m\cdot\boldsymbol{x}) && \text{by homogeneity} \\ &= \alpha_1\,\beta_1+\cdots+\alpha_m\,\beta_m\end{aligned}$$

This math addresses Question 2.9.20: *Does a system of linear equations imply any other linear equations? If so, what other linear equations?* The system of linear equations implies a linear equation of the form $\boldsymbol{a}\cdot\boldsymbol{x}=\beta$ for every vector $\boldsymbol{a}$ in the span of $\boldsymbol{a}_1,\ldots,\boldsymbol{a}_m$.

But we have only partially answered the Question, for we have not yet shown that these are the *only* linear equations implied by the system. We will show this in a later chapter.

Example 3.2.8: (*Attacking the simple authentication scheme:*) Suppose Eve has already seen a collection of challenge vectors $\boldsymbol{a}_1,\ldots,\boldsymbol{a}_m$ for which she knows the responses. She can answer any challenge in Span $\{\boldsymbol{a}_1,\ldots,\boldsymbol{a}_m\}$. Does that include all possible challenges? This is equivalent to asking if $GF(2)^n$ equals Span $\{\boldsymbol{a}_1,\ldots,\boldsymbol{a}_m\}$.

3.2.3 Generators

Definition 3.2.9: Let $\mathcal{V}$ be a set of vectors. If $\boldsymbol{v}_1,\ldots,\boldsymbol{v}_n$ are vectors such that $\mathcal{V}=$ Span $\{\boldsymbol{v}_1,\ldots,\boldsymbol{v}_n\}$ then we say $\{\boldsymbol{v}_1,\ldots,\boldsymbol{v}_n\}$ is a *generating set* for $\mathcal{V}$, and we refer to the vectors $\boldsymbol{v}_1,\ldots,\boldsymbol{v}_n$ as *generators* for $\mathcal{V}$.

Example 3.2.10: Let $\mathcal{V}$ be the set $\{00000, 11100, 01110, 10010, 00111, 11011, 01001, 10101\}$ of 5-vectors over $GF(2)$. We saw in Example 3.2.7 (Page 120) that these eight vectors are exactly the span of 11100, 01110, and 00111. Therefore 11100, 01110, and 00111 form a generating set for $\mathcal{V}$.

Example 3.2.11: I claim that $\{[3,0,0],[0,2,0],[0,0,1]\}$ is a generating set for $\mathbb{R}^3$. To prove that claim, I must show that the set of linear combinations of these three vectors is equal to $\mathbb{R}^3$. That means I must show two things:

1. Every linear combination is a vector in $\mathbb{R}^3$.
2. Every vector in $\mathbb{R}^3$ is a linear combination.

The first statement is pretty obvious since $\mathbb{R}^3$ includes all 3-vectors over $\mathbb{R}$. To prove the second statement, let $[x, y, z]$ be any vector in $\mathbb{R}^3$. I must demonstrate that $[x, y, z]$ can be written as a linear combination, i.e. I must specify the coefficients in terms of x, y, and z. Here goes:

$$[x, y, z] = (x/3)\,[3, 0, 0] + (y/2)\,[0, 2, 0] + z\,[0, 0, 1]$$

3.2.4 Linear combinations of linear combinations

I claim that another generating set for $\mathbb{R}^3$ is $\{[1, 0, 0], [1, 1, 0], [1, 1, 1]\}$. This time, I prove that their span includes all of $\mathbb{R}^3$ by writing each of the three vectors in Example 3.2.11 (Page 121) as a linear combination:

$$\begin{aligned}
[3, 0, 0] &= 3\,[1, 0, 0] \\
[0, 2, 0] &= -2\,[1, 0, 0] + 2\,[1, 1, 0] \\
[0, 0, 1] &= -1\,[1, 0, 0] - 1\,[1, 1, 0] + 1\,[1, 1, 1]
\end{aligned}$$

Why is that sufficient? Because each of the old vectors can in turn be written as a linear combination of the new vectors, I can convert any linear combination of the old vectors into a linear combination of the new vectors. We saw in Example 3.2.11 (Page 121) that any 3-vector $[x, y, z]$ can be written as a linear combination of the old vectors, hence it can be written as a linear combination of the new vectors.

Let's go through that explicitly. First we write $[x, y, z]$ as a linear combination of the old vectors:

$$[x, y, z] = (x/3)\,[3, 0, 0] + (y/2)\,[0, 2, 0] + z\,[0, 0, 1]$$

Next, we replace each old vector with an equivalent linear combination of the new vectors:

$$[x, y, z] = (x/3)\left(3\,[1, 0, 0]\right) + (y/2)\left(-2\,[1, 0, 0] + 2\,[1, 1, 0]\right) + z\left(-1\,[1, 0, 0] - 1\,[1, 1, 0] + 1\,[1, 1, 1]\right)$$

Next, we multiply through, using associativity of scalar-vector multiplication (Proposition 2.5.5) and the fact that scalar multiplication distributes over vector addition (Proposition 2.6.3):

$$[x, y, z] = x\,[1, 0, 0] - y\,[1, 0, 0] + y\,[1, 1, 0] - z\,[1, 0, 0] - z\,[1, 1, 0] + z\,[1, 1, 1]$$

Finally, we collect like terms, using the fact that scalar-vector multiplication distributes over scalar addition (Proposition 2.6.5):

$$[x, y, z] = (x - y - z)\,[1, 0, 0] + (y - z)\,[1, 1, 0] + z\,[1, 1, 1]$$

We have shown that an arbitrary vector in $\mathbb{R}^3$ can be written as a linear combination of $[1, 0, 0]$, $[1, 1, 0]$, and $[1, 1, 1]$. This shows that $\mathbb{R}^3$ is a subset of Span $\{[1, 0, 0], [1, 1, 0], [1, 1, 1]\}$.

Of course, every linear combination of these vectors belongs to $\mathbb{R}^3$, which means Span $\{[1, 0, 0], [1, 1, 0], [1, 1, 1]\}$ is a subset of $\mathbb{R}^3$. Since each of these two sets is a subset of the other, they are equal.

Quiz 3.2.12: Write each of the old vectors $[3, 0, 0]$, $[0, 2, 0]$, and $[0, 0, 1]$ as a linear combination of new vectors $[2, 0, 1]$, $[1, 0, 2]$, $[2, 2, 2]$, and $[0, 1, 0]$.

Answer

$$[3,0,0] = 2\,[2,0,1] - 1\,[1,0,2] + 0\,[2,2,2]$$
$$[0,2,0] = -\frac{2}{3}\,[2,0,1] - \frac{2}{3}\,[1,0,2] + 1\,[2,2,2]$$
$$[0,0,1] = -\frac{1}{3}\,[2,0,1] + \frac{2}{3}\,[1,0,2] + 0\,[2,2,2]$$

3.2.5 Standard generators

We saw a formula expressing $[x, y, z]$ as a linear combination of the vectors $[3,0,0]$, $[0,2,0]$, and $[0,0,1]$. The formula was particularly simple because of the special form of those three vectors. It gets even simpler if instead we use $[1,0,0]$, $[0,1,0]$, and $[0,0,1]$:

$$[x, y, z] = x\,[1,0,0] + y\,[0,1,0] + z\,[0,0,1]$$

The simplicity of this formula suggests that these vectors are the most "natural" generators for $\mathbb{R}^3$. Indeed, the coordinate representation of $[x, y, z]$ in terms of these generators is $[x, y, z]$.

We call these three vectors the *standard* generators for $\mathbb{R}^3$. We denote them by $\boldsymbol{e}_0, \boldsymbol{e}_1, \boldsymbol{e}_2$ (when it is understood we are working with vectors in $\mathbb{R}^3$).

When we are working with, for example, $\mathbb{R}^4$, we use $\boldsymbol{e}_0, \boldsymbol{e}_1, \boldsymbol{e}_2, \boldsymbol{e}_3$ to refer to $[1,0,0,0]$, $[0,1,0,0]$, $[0,0,1,0]$,$[0,0,0,1]$.

For any positive integer n, the standard generators for $\mathbb{R}^n$ are:

$$\begin{array}{rcl} \boldsymbol{e}_0 & = & [1,0,0,0,\ldots,0] \\ \boldsymbol{e}_1 & = & [0,1,0,0,\ldots,0] \\ \boldsymbol{e}_2 & = & [0,0,1,0,\ldots,0] \\ & \vdots & \\ \boldsymbol{e}_{n-1} & = & [0,0,0,0,\ldots,1] \end{array}$$

where e_i is all zeroes except for a 1 in position i.

Naturally, for any finite domain D and field $\mathbb{F}$, there are standard generators for $\mathbb{F}^D$. We define them as follows. For each $k \in D$, $\boldsymbol{e}_k$ is the function $\{k : 1\}$. That is, $\boldsymbol{e}_k$ maps k to 1 and maps all other domain elements to zero.

It is easy to prove that what we call "standard generators" for $\mathbb{F}^D$ are indeed generators for $\mathbb{F}^D$. We omit the proof since it is not very illuminating.

Quiz 3.2.13: Write a procedure `standard(D, one)` that, given a domain `D` and given the number one for the field, returns the list of standard generators for $\mathbb{R}^D$. (The number one is provided as an argument so that the procedure can support use of $GF(2)$.)

Answer

```
>>> def standard(D, one): return [Vec(D, {k:one}) for k in D]
```

Example 3.2.14: (Solvability of 2×2 *Lights Out*:) Can 2×2 *Lights Out* be solved from every starting configuration? This is equivalent to asking whether the 2×2 button vectors

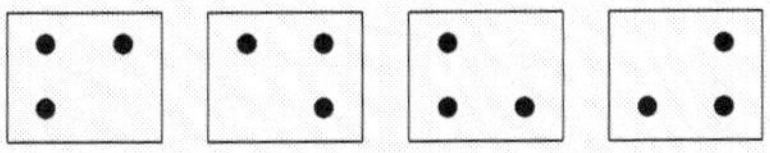

are generators for $GF(2)^D$, where $D = \{(0,0), (0,1), (1,0), (1,1)\}$.

To prove that the answer is yes, it suffices to show that each of the standard generators can be written as a linear combination of the button vectors:

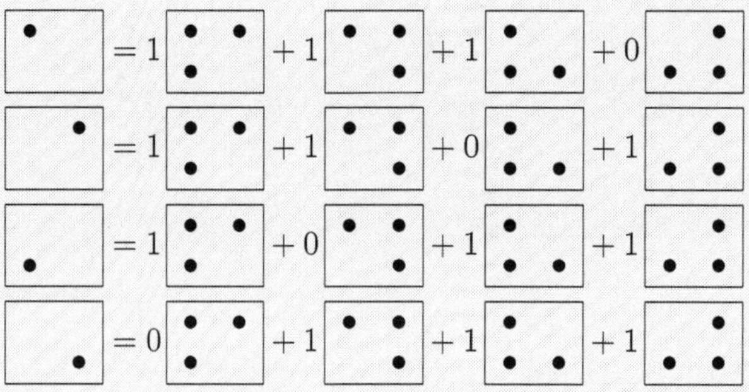

Exercise 3.2.15: For each of the subproblems, you are to investigate whether the given vectors span $\mathbb{R}^2$. If possible, write each of the standard generators for $\mathbb{R}^2$ as a linear combination of the given vectors. If doing this is impossible for one of the subproblems, you should first add one additional vector and then do it.

1. $[1, 2], [3, 4]$
2. $[1, 1], [2, 2], [3, 3]$
3. $[1, 1], [1, -1], [0, 1]$

Exercise 3.2.16: You are given the vectors $[1, 1, 1]$, $[0.4, 1.3, -2.2]$. Add one additional vector and express each of the standard generators for $\mathbb{R}^3$ as a linear combination of the three vectors.

3.3 The geometry of sets of vectors

In Chapter 2, we saw how to write lines and line segments in terms of vectors. In a physical simulation or graphics application, we might need to manipulate higher-dimensional geometrical objects such as planes—perhaps we need to represent a wall or the surface of a table, or perhaps we are representing the surface of a complicated three-dimensional object by many flat polygons glued together. In this section, we informally investigate the geometry of the span of vectors over $\mathbb{R}$, and, as a bonus, the geometry of other kinds of sets of vectors.

3.3.1 The geometry of the span of vectors over $\mathbb{R}$

Consider the set of all linear combinations of a single nonzero vector $\boldsymbol{v}$:

$$\text{Span } \{\boldsymbol{v}\} = \{\alpha \boldsymbol{v} \ : \ \alpha \in \mathbb{R}\}$$

We saw in Section 2.5.3 that this set forms the line through the origin and the point $\boldsymbol{v}$. A line is a one-dimensional geometrical object.

An even simpler case is the span of an empty set of vectors. We saw in Quiz 3.2.4 that the span consists of exactly one vector, the zero vector. Thus in this case the span consists of a point, which we consider a zero-dimensional geometrical object.

What about the span of two vectors? Perhaps it is a two-dimensional geometric object, i.e. a plane?

Example 3.3.1: What is Span $\{[1, 0], [0, 1]\}$? These vectors are the standard generators for $\mathbb{R}^2$, so every 2-vector is in the span. Thus Span $\{[1, 0], [0, 1]\}$ includes all points in the Euclidean plane.

Example 3.3.2: What is Span $\{[1,2],[3,4]\}$? You might have shown in Exercise 3.2.15 that the standard generators for $\mathbb{R}^2$ can be written as linear combinations of these vectors, so again we see that the set of linear combinations of the two vectors includes all points in the plane.

Example 3.3.3: What about the span of two 3-vectors? The linear combinations of $[1,0,1.65]$ and $[0,1,1]$ form a plane through the origin; part of this plane is shown below:

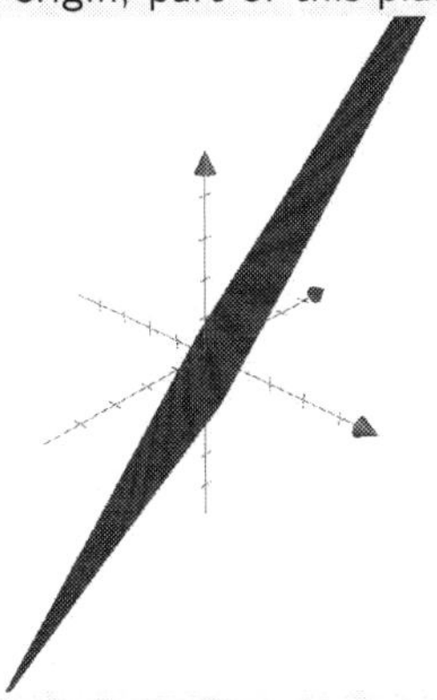

We can use these two vectors in plotting the plane. Here is a plot of the points in the set $\{\alpha\,[1,0.1.65]+\beta\,[0,1,1]\ :\alpha\in\{-5,-4,\ldots,3,4\},$
$\beta\in\{-5,-4,\ldots,3,4\}\}$:

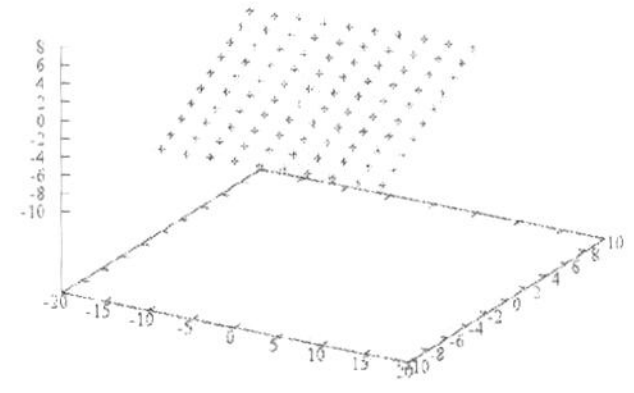

Example 3.3.4: Do every two distinct vectors span a plane? What about Span $\{[1,2],[2,4]\}$? For any pair of coefficients α_1 and α_2,

$$\begin{aligned}\alpha_1[1,2]+\alpha_2[2,4] &= \alpha_1[1,2]+\alpha_2(2\,[1,2])\\ &= \alpha_1[1,2]+(\alpha_2\cdot 2)[1,2]\\ &= (\alpha_1+2\alpha_2)[1,2]\end{aligned}$$

This shows that Span $\{[1,2],[2,4]\}$ = Span $\{[1,2]\}$, and we know from Section 2.5.3 that Span $\{[1,2]\}$ forms a line, not a plane.

These examples lead us to believe that the span of two vectors over $\mathbb{R}$ forms a plane *or* a lower-dimensional object (a line or a point). Note that the span of *any* collection of vectors must include the origin, because the trivial linear combination (all coefficients equal to zero) is included in the set.

The pattern begins to become clear:

- The span of zero vectors forms a point—a zero-dimensional object—which must be the origin.
- The span of one vector forms a line through the origin—a one-dimensional object—or a point, the origin.
- The span of two vectors forms a plane through the origin—a two-dimensional object—or a line through the origin or a point, the origin.

A geometric object such as a point, a line, or a plane is called a *flat*. There are higher-dimensional flats too. All of $\mathbb{R}^3$ is a three-dimensional flat. Although it is hard to envision, one can define a three-dimensional flat within four-dimensional space, $\mathbb{R}^4$, and so on.

Generalizing from our observations, we are led to hypothesize:

Hypothesis 3.3.5: The span of k vectors over $\mathbb{R}$ forms a k-dimensional flat containing the origin or a flat of lower dimension containing the origin.

Observing this pattern raises the following Question:

Question 3.3.6: How can we tell if the span of a given collection of k vectors forms a k-dimensional object? More generally, given a collection of vectors, how can we predict the dimensionality of the span?

The question will be answered starting in Chapter 6.

3.3.2 The geometry of solution sets of homogeneous linear systems

Perhaps a more familiar way to specify a plane is with an equation, e.g. $\{(x, y, z) \in \mathbb{R}^3 : ax + by + cz = d\}$. For now, we want to focus on planes that contain the origin $(0, 0, 0)$. For $(0, 0, 0)$ to satisfy the equation $ax + by + cz = d$, it must be that d equals zero.

Example 3.3.7: The plane depicted earlier, Span $\{[1, 0, 1.65], [0, 1, 1]\}$, can be represented as

$$\{(x, y, z) \in \mathbb{R}^3 \;:\; 1.65x + 1y - 1z = 0\}$$

We can rewrite the equation using dot-product, obtaining

$$\{[x, y, z] \in \mathbb{R}^3 \;:\; [1.65, 1, -1] \cdot [x, y, z] = 0\}$$

Thus the plane is the solution set of a linear equation with right-hand side zero.

Definition 3.3.8: A linear equation with right-hand side zero is a *homogeneous* linear equation.

Example 3.3.9: The line

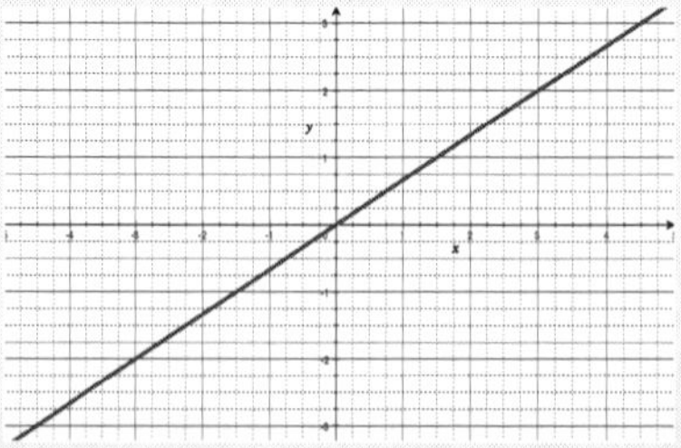

can be represented as Span $\{[3, 2]\}$ but it can also be represented as

$$\{[x, y] \in \mathbb{R}^2 \;:\; 2x - 3y = 0\}$$

That is, the line is the solution set of a homogeneous linear equation.

Example 3.3.10: This line

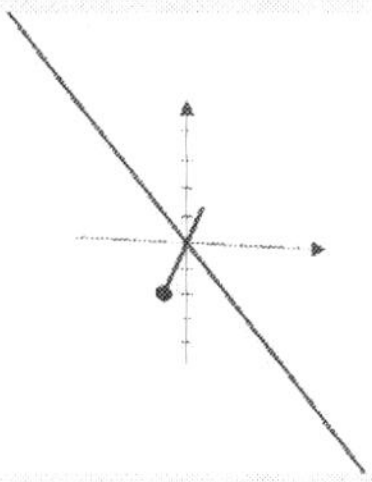

can be represented as Span $\{[1, -2, -2]\}$. It can also be represented as the solution set of a pair of homogeneous linear equations

$$\{[x, y, z] \in \mathbb{R}^3 \ : [4, -1, 1] \cdot [x, y, z] = 0, [0, 1, 1] \cdot [x, y, z] = 0\}$$

That is, the line consists of the set of triples $[x, y, z]$ that satisfy both of these two homogeneous linear equations.

Definition 3.3.11: A linear system (collection of linear equations) with all right-hand sides zero is called a *homogeneous* linear system.

Generalizing from our two examples, we are led to hypothesize:

Hypothesis 3.3.12: A flat containing the origin is the solution set of a homogeneous linear system.

We are not yet in a position to formally justify our hypotheses or even to formally define *flat*. We are working towards developing the notions that underlie that definition.

3.3.3 The two representations of flats containing the origin

A well-established theme in computer science is the usefulness of multiple representations for the same data. We have seen two ways to represent a flat containing the origin:

- as the span of some vectors, and
- as the solution set to a homogeneous linear system.

Each of these representations has its uses. Or, to misquote Hat Guy,

> Different tasks call for different representations.

Suppose you want to find the plane containing two given lines, the line Span $\{[4, -1, 1]\}$ and the line Span $\{[0, 1, 1]\}$.

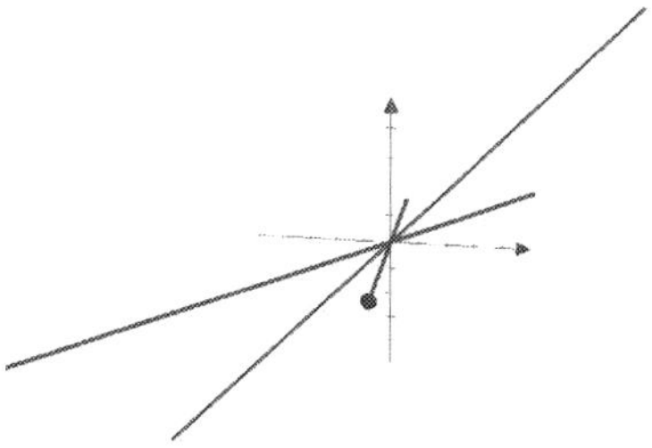

Since the lines are represented as spans, it is easy to obtain the solution: The plane containing these two lines is Span $\{[4, -1, 1], [0, 1, 1]\}$:

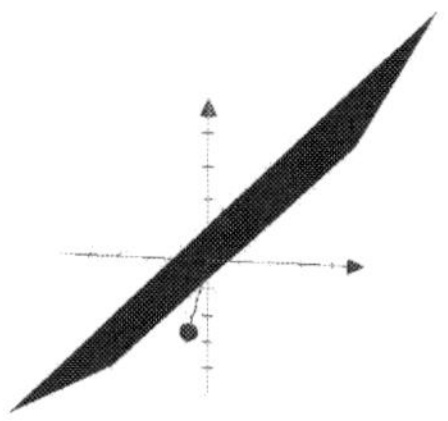

On the other hand, suppose you want to find the intersection of two given planes, the plane $\{[x, y, z] \,:\, [4, -1, 1] \cdot [x, y, z] = 0\}$ and the plane $\{[x, y, z] \,:\, [0, 1, 1] \cdot [x, y, z] = 0\}$:

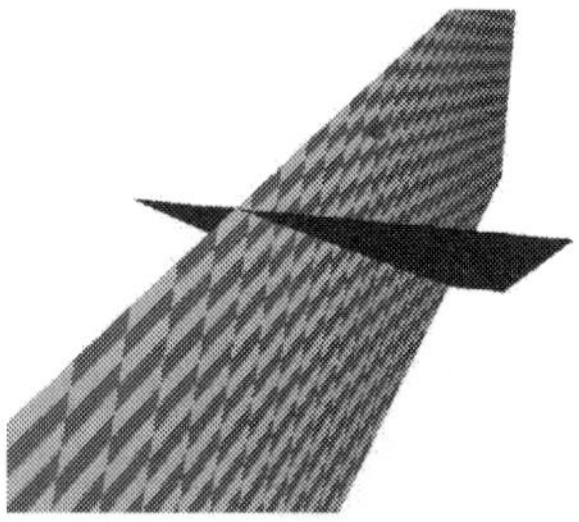

Since each plane is represented as the solution set of a homogeneous linear equation, it is easy to obtain the solution. The set of points that belong to both planes is the set of vectors satisfying both equations: $\{[x, y, z] \,:\, [4, -1, 1] \cdot [x, y, z] = 0, [0, 1, 1] \cdot [x, y, z] = 0\}$.

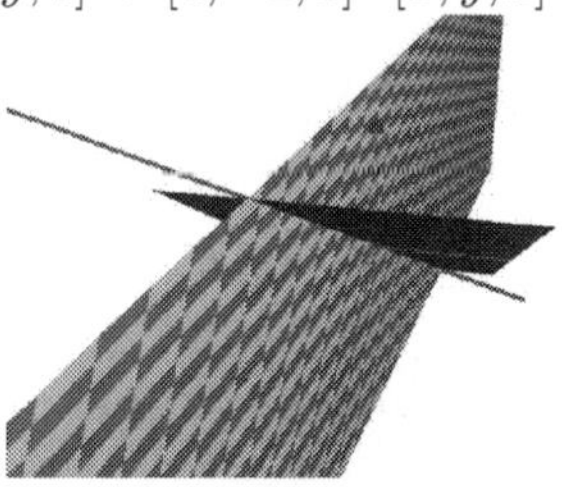

Since each representation is useful, we would like to be able to transform from one representation to another. Is this possible? Can any set represented as the span of vectors also be represented as the solution set of a homogeneous linear system? What about the other way round? We further discuss these conversion problems in Section 6.5. We first need to better understand the underlying mathematics.

3.4 Vector spaces

3.4.1 What's common to the two representations?

Our goal is understanding the connection between these two representations. We will see that a subset $\mathcal{V}$ of $\mathbb{F}^D$, whether $\mathcal{V}$ is the span of some D-vectors over $\mathbb{F}$ or the solution set of a linear system, has three properties:

Property V1: $\mathcal{V}$ contains the zero vector,

Property V2: For every vector $\boldsymbol{v}$, if $\mathcal{V}$ contains $\boldsymbol{v}$ then it contains $\alpha\,\boldsymbol{v}$ for every scalar α, is closed under scalar-vector multiplication, and

Property V3: For every pair $\boldsymbol{u}$ and $\boldsymbol{v}$ of vectors, if $\mathcal{V}$ contains $\boldsymbol{u}$ and $\boldsymbol{v}$ then it contains $\boldsymbol{u} + \boldsymbol{v}$.

First suppose $\mathcal{V} = \text{Span}\ \{\boldsymbol{v}_1, \ldots, \boldsymbol{v}_n\}$. Then $\mathcal{V}$ satisfies

- Property V1 because

$$0\,\boldsymbol{v}_1 + \cdots + 0\,\boldsymbol{v}_n$$

- Property V2 because

$$\text{if } \boldsymbol{v} = \beta_1\,\boldsymbol{v}_1 + \cdots + \beta_n\,\boldsymbol{v}_n \text{ then } \alpha\,\boldsymbol{v} = \alpha\,\beta_1\boldsymbol{v}_1 + \cdots + \alpha\,\beta_n\,\boldsymbol{v}_n$$

- Property V3 because

$$\begin{array}{lrcl} \text{if} & \boldsymbol{u} & = & \alpha_1\,\boldsymbol{v}_1 + \cdots + \alpha_n\,\boldsymbol{v}_1 \\ \text{and} & \boldsymbol{v} & = & \beta_1\,\boldsymbol{v}_1 + \cdots + \beta_n\,\boldsymbol{v}_n \\ \text{then} & \boldsymbol{u}+\boldsymbol{v} & = & (\alpha_1+\beta_1)\boldsymbol{v}_1 + \cdots + (\alpha_n+\beta_n)\boldsymbol{v}_n \end{array}$$

Now suppose $\mathcal{V}$ is the solution set $\{\boldsymbol{x} : \boldsymbol{a}_1 \cdot \boldsymbol{x} = 0, \quad \ldots, \quad \boldsymbol{a}_m \cdot \boldsymbol{x} = 0\}$ of a linear system. Then $\mathcal{V}$ satisfies

- Property V1 because

$$\boldsymbol{a}_1 \cdot \mathbf{0} = 0, \quad \ldots, \quad \boldsymbol{a}_m \cdot \mathbf{0} = 0$$

- Property V2 because

$$\begin{array}{lrcllrcl} \text{if} & \boldsymbol{a}_1 \cdot \boldsymbol{v} & = & 0, & \ldots, & \boldsymbol{a}_m \cdot \boldsymbol{v} & = & 0 \\ \text{then} & \alpha\,(\boldsymbol{a}_1 \cdot \boldsymbol{v}) & = & 0, & \cdots, & \alpha\,(\boldsymbol{a}_m \cdot \boldsymbol{v}) & = & 0 \\ \text{so} & \boldsymbol{a}_1 \cdot (\alpha\,\boldsymbol{v}) & = & 0, & \cdots, & \boldsymbol{a}_m \cdot (\alpha\,\boldsymbol{v}) & = & 0 \end{array}$$

- Property V3 because

$$\begin{array}{lrcllrcl} \text{if} & \boldsymbol{a}_1 \cdot \boldsymbol{u} & = & 0, & \ldots, & \boldsymbol{a}_m \cdot \boldsymbol{u} & = & 0 \\ \text{and} & \boldsymbol{a}_1 \cdot \boldsymbol{v} & = & 0, & \ldots, & \boldsymbol{a}_m \cdot \boldsymbol{v} & = & 0 \\ \text{then} & \boldsymbol{a}_1 \cdot \boldsymbol{u} + \boldsymbol{a}_1 \cdot \boldsymbol{v} & = & 0, & \ldots, & \boldsymbol{a}_m \cdot \boldsymbol{u} + \boldsymbol{a}_m \cdot \boldsymbol{v} & = & 0 \\ \text{so} & \boldsymbol{a}_1 \cdot (\boldsymbol{u}+\boldsymbol{v}) & = & 0, & \ldots, & \boldsymbol{a}_m \cdot (\boldsymbol{u}+\boldsymbol{v}) & = & 0 \end{array}$$

3.4.2 Definition and examples of vector space

We use Properties V1, V2, and V3 to define a notion that encompasses both kinds of representations: spans of vectors, and solution sets of homogeneous linear systems.

Definition 3.4.1: A set $\mathcal{V}$ of vectors is called a *vector space* if it satisfies Properties V1, V2, and V3.

Example 3.4.2: We have seen that the span of some vectors is a vector space.

Example 3.4.3: We have seen that the solution set of a homogeneous linear system is a vector space.

Example 3.4.4: A flat (such as a line or a plane) that contains the origin can be written as the span of some vectors or as the solution set of a homogeneous linear system, and therefore such a flat is a vector space.

The statement "If $\mathcal{V}$ contains $\boldsymbol{v}$ then it contains $\alpha\,\boldsymbol{v}$ for every scalar α" is expressed in Mathese as

"$\mathcal{V}$ is *closed under* scalar-vector multiplication."

The statement "if $\mathcal{V}$ contains $\boldsymbol{u}$ and $\boldsymbol{v}$ then it contains $\boldsymbol{u}+\boldsymbol{v}$" is expressed in Mathese as

"$\mathcal{V}$ is *closed under* vector addition."

(In general, we say a set is *closed* under an operation if the set contains any object produced by that operation using inputs from the set.)

What about $\mathbb{F}^D$ itself?

Example 3.4.5: For any field $\mathbb{F}$ and any finite domain D, the set $\mathbb{F}^D$ of D-vectors over F is a vector space. Why? Well, $\mathbb{F}^D$ contains a zero vector and is closed under scalar-vector multiplication and vector addition. For example, $\mathbb{R}^2$ and $\mathbb{R}^3$ and $GF(2)^4$ are all vector spaces.

What is the smallest subset of $\mathbb{F}^D$ that is a vector space?

Proposition 3.4.6: For any field $\mathbb{F}$ and any finite domain D, the singleton set consisting of the zero vector $\mathbf{0}_D$ is a vector space.

Proof

The set $\{\mathbf{0}_D\}$ certainly contains the zero vector, so Property V1 holds. For any scalar α, $\alpha\,\mathbf{0}_D = \mathbf{0}_D$, so Property V2 holds: $\{\mathbf{0}_D\}$ is closed under scalar-vector multiplication. Finally, $\mathbf{0}_D + \mathbf{0}_D = \mathbf{0}_D$, so Property V3 holds: $\{\mathbf{0}_D\}$ is closed under vector addition. □

Definition 3.4.7: A vector space consisting only of a zero vector is a *trivial* vector space.

Quiz 3.4.8: What is the minimum number of vectors whose span is $\{\mathbf{0}_D\}$?

Answer

The answer is zero. As we discussed in the answer to Quiz 3.2.4, $\{\mathbf{0}_D\}$ equals the span of the empty set of D-vectors. It is true, as discussed in the answer to Quiz 3.2.6, that $\{\mathbf{0}_D\}$ is the span of $\{\mathbf{0}_D\}$, but this just illustrates that there are different sets with the same span. We are often interested in the set with the smallest size.

3.4.3 Subspaces

Definition 3.4.9: If $\mathcal{V}$ and $\mathcal{W}$ are vector spaces and $\mathcal{V}$ is a subset of $\mathcal{W}$, we say $\mathcal{V}$ is a *subspace* of $\mathcal{W}$.

Remember that a set is considered a subset of itself, so one subspace of $\mathcal{W}$ is $\mathcal{W}$ itself.

Example 3.4.10: The only subspace of $\{[0, 0]\}$ is itself.

Example 3.4.11: The set $\{[0, 0]\}$ is a subspace of $\{\alpha\,[2, 1] \;:\; \alpha \in \mathbb{R}\}$, which is in turn a subspace of $\mathbb{R}^2$.

Example 3.4.12: The set $\mathbb{R}^2$ is *not* a subspace of $\mathbb{R}^3$ since $\mathbb{R}^2$ is not contained in $\mathbb{R}^3$; indeed, $\mathbb{R}^2$ consists of 2-vectors and $\mathbb{R}^3$ contains no 2-vectors.

Example 3.4.13: What vector spaces are contained in $\mathbb{R}^2$?

- The smallest is $\{[0, 0]\}$.
- The largest is $\mathbb{R}^2$ itself.
- For any nonzero vector $[a, b]$, the line through the origin and $[a, b]$, Span $\{[a, b]\}$, is a

vector space.

Does $\mathbb{R}^2$ have any other subspaces? Suppose $\mathcal{V}$ is a subspace of $\mathbb{R}^2$. Assume it has some nonzero vector $[a, b]$, and assume it also has some other vector $[c, d]$ such that $[c, d]$ is not in Span $\{[a, b]\}$. We prove that in this case $\mathcal{V} = \mathbb{R}^2$.

Lemma 3.4.14: *$ad \neq bc$*

Proof

Since $[a, b] \neq [0, 0]$, either $a \neq 0$ or $b \neq 0$ (or both).
Case 1: $a \neq 0$. In this case, define $\alpha = c/a$. Since $[c, d]$ is not in Span $\{[a, b]\}$, it must be that $[c, d] \neq \alpha\,[a, b]$. Because $c = \alpha\, a$, it must be that $d \neq \alpha\, b$. Substituting c/a for α, we infer that $d \neq \frac{c}{a} b$. Multiplying through by a, we infer that $ad \neq cb$.

Case 2: $b \neq 0$. In this case, define $\alpha = d/b$. Since $[c, d] \neq \alpha\,[a, b]$, we infer that $c \neq \alpha\, a$. Substituting for α and multiplying through by b, we infer that $ad \neq cb$. □

Now we show that $\mathcal{V} = \mathbb{R}^2$. To show that, we show that every vector in $\mathbb{R}^2$ can be written as a linear combination of just two vectors in $\mathcal{V}$, namely $[a, b]$ and $[c, d]$.

Let $[p, q]$ be any vector in $\mathbb{R}^2$. Define $\alpha = \frac{dp-cq}{ad-bc}$ and $\beta = \frac{aq-bp}{ad-bc}$. Then

$$\begin{aligned}
&\alpha\,[a, b] + \beta\,[c, d] \\
&= \frac{1}{ad-bc}[(pd-qc)a + (aq-bp)c, (pd-qc)b + (aq-bp)d] \\
&= \frac{1}{ad-bc}[adp - bcp, adq - bcq] \\
&= [p, q]
\end{aligned}$$

We have shown that $[p, q]$ is equal to a linear combination of $[a, b]$ and $[c, d]$. Since $[p, q]$ is an arbitrary element of $\mathbb{R}^2$, we have shown that $\mathbb{R}^2 = $ Span $\{[a, b], [c, d]\}$.

Since $\mathcal{V}$ contains $[a, b]$ and $[c, d]$ and is closed under scalar-vector multiplication and vector addition, it contains all of Span $\{[a, b], [c, d]\}$. This proves that $\mathcal{V}$ contains all of $\mathbb{R}^2$. Since every vector in $\mathcal{V}$ belongs to $\mathbb{R}^2$, $\mathcal{V}$ is also a subset of $\mathbb{R}^2$. Since each of $\mathcal{V}$ and $\mathbb{R}^2$ is a subset of the other, they must be equal.

We came to the concept of vector space by considering two ways of forming a set:

- as the span of some vectors, and
- as the solution set of a homogeneous linear system.

Each of these is a vector space. In particular, each is a subspace of $\mathbb{F}^D$ for some field $\mathbb{F}$ and some domain D.

What about the converse?

Question 3.4.15: Can any subspace of $\mathbb{F}^D$ be expressed as the span of a finite set of vectors?

Question 3.4.16: Can any subspace of $\mathbb{F}^D$ be expressed as the solution set of a homogeneous linear system?

We will see in Chapter 6 that the answers are yes, and yes. Establishing this, however, requires we learn some more mathematics.

3.4.4 *Abstract vector spaces

I am tempted to state more simply that any *vector space* can be expressed as the span of a finite number of vectors and as the solution set of a homogeneous linear system. However, that is not true according to the formal definitions of Mathematics.

In this book, I have defined a vector as a function from a finite domain D to a field $\mathbb{F}$. However, modern mathematics tends to define things in terms of the axioms they satisfy rather than in terms of their internal structure. (I informally raised this idea in discussing the notion of a field.)

Following this more abstract approach, one does not define the notion of a vector; instead, one defines a *vector space* over a field $\mathbb{F}$ to be any set $\mathcal{V}$ that is equipped with a *addition* operation and a *scalar-multiplication* operation (satisfying certain axioms) and that satisfies Properties V1, V2, and V3. The elements of $\mathcal{V}$, whatever they happen to be, play the role of vectors.

This definition avoids committing to a specific internal structure for vectors and consequently allows for a much broader class of mathematical objects to be considered vectors. For example, the set of all functions from $\mathbb{R}$ to $\mathbb{R}$ is a vector space according to the abstract definition. The question of whether a subspace of this space is the span of a finite set of vectors is a deeper mathematical question than we can address in this book.

I avoid the abstract approach in this book because I find that the more concrete notion of vector is helpful in developing intuition. However, if you go deeper into mathematics, you should expect to encounter this approach.

3.5 Affine spaces

What about points, lines, planes, etc. that do not include the origin?

3.5.1 Flats that don't go through the origin

In Section 2.6.1, we observed that a line segment not through the origin could be obtained from a line segment through the origin by translation, i.e. by applying a function such as $f([x, y]) = [x, y] + [0.5, 1]$.

How can we represent a line not through the origin? Two approaches were outlined in Section 2.6.4. First, we start with a line that does go through the origin

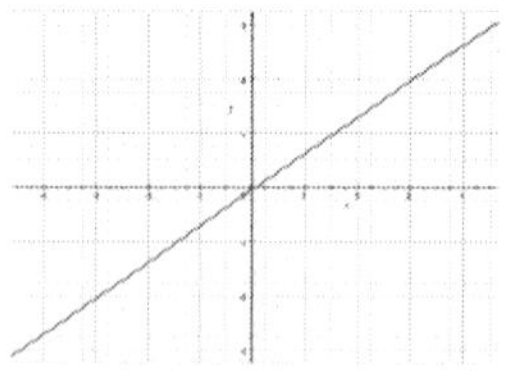

We now know that the points of this line form a vector space $\mathcal{V}$.

We can choose a vector $\boldsymbol{a}$ and add it to every vector in $\mathcal{V}$:

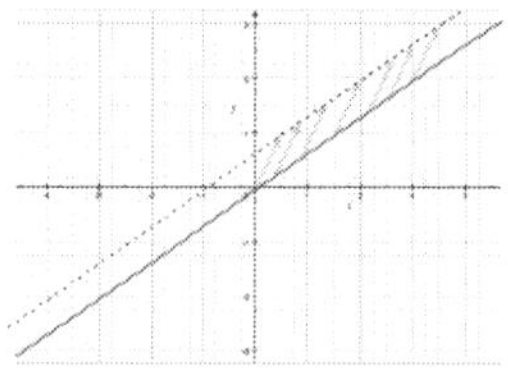

In Mathese, we would write the resulting set as

$$\{\boldsymbol{a} + \boldsymbol{v} \; : \; \boldsymbol{v} \in \mathcal{V}\}$$

We will abbreviate this set expression as $\boldsymbol{a} + \mathcal{V}$.

The resulting set is a line that goes through $\boldsymbol{a}$ (and not through the origin):

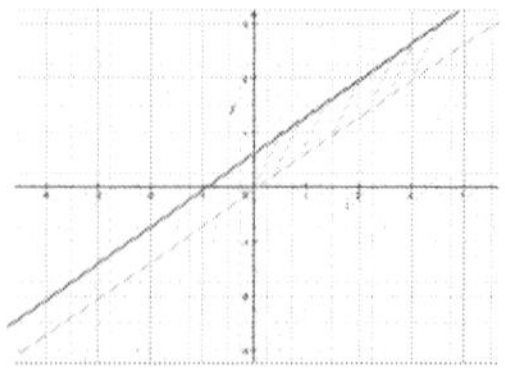

Now let's carry out the same process on a plane.

Example 3.5.1: There is one plane through the points $\boldsymbol{u}_1 = [1, 0, 4.4]$, $\boldsymbol{u}_2 = [0, 1, 4]$, and $\boldsymbol{u}_3 = [0, 0, 3]$:

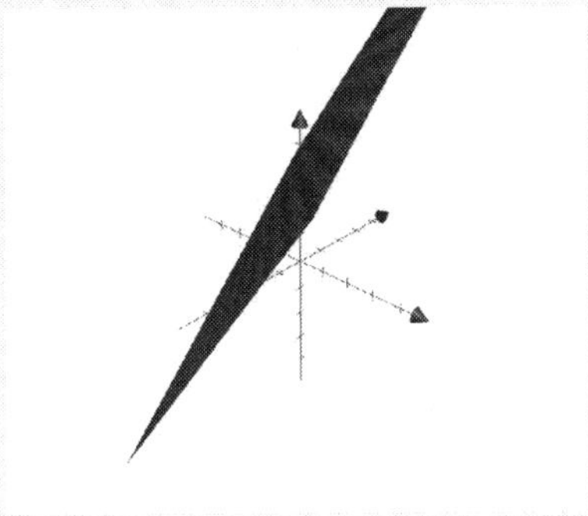

How can we write the set of points in the plane as a translation of a vector space?

Define $\boldsymbol{a} = \boldsymbol{u}_2 - \boldsymbol{u}_1$ and $\boldsymbol{b} = \boldsymbol{u}_3 - \boldsymbol{u}_1$, and let $\mathcal{V}$ be the vector space Span $\{\boldsymbol{a}, \boldsymbol{b}\}$. Then the points of $\mathcal{V}$ form a plane:

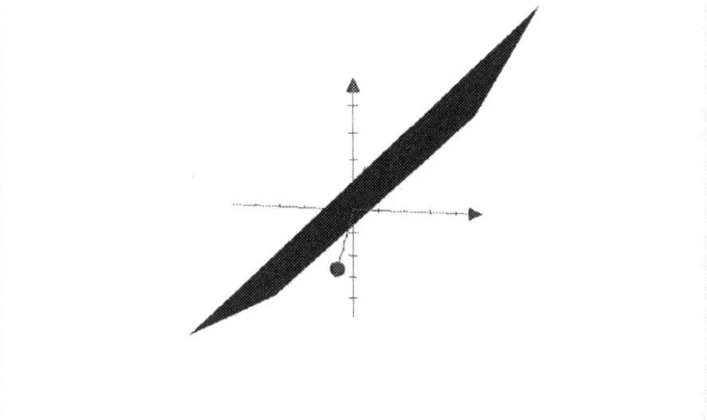

Now consider the set

$$\boldsymbol{u}_1 + \mathcal{V}$$

Intuitively, the translation of a plane remains a plane. Note in addition that $\boldsymbol{u}_1 + \mathcal{V}$ contains

- the point $\boldsymbol{u}_1$ since $\mathcal{V}$ contains the zero vector,
- the point $\boldsymbol{u}_2$ since $\mathcal{V}$ contains $\boldsymbol{u}_2 - \boldsymbol{u}_1$, and
- the point $\boldsymbol{u}_3$ since $\mathcal{V}$ contains $\boldsymbol{u}_3 - \boldsymbol{u}_1$.

Since the plane $\boldsymbol{u}_1 + \mathcal{V}$ contains $\boldsymbol{u}_1$, $\boldsymbol{u}_2$, and $\boldsymbol{u}_3$, it must be the unique plane through those points.

3.5.2 Affine combinations

In Section 2.6.4, we saw another way to write the line through points $\boldsymbol{u}$ and $\boldsymbol{v}$: as the set of affine combinations of $\boldsymbol{u}$ and $\boldsymbol{v}$. Here we generalize that notion as well.

Definition 3.5.2: A linear combination $\alpha_1 \boldsymbol{u}_1 + \cdots + \alpha_n \boldsymbol{u}_n$ is called an *affine* combination if the coefficients sum to one.

Example 3.5.3: The linear combination $2\,[10., 20.] + 3\,[0, 10.] + (-4)\,[30., 40.]$ is a an *affine* combination of the vectors because $2 + 3 + (-4) = 1$.

Example 3.5.4: In Example 3.5.1 (Page 133), we wrote the plane through $\boldsymbol{u}_1$, $\boldsymbol{u}_2$, and $\boldsymbol{u}_3$ as

$$\boldsymbol{u}_1 + \mathcal{V}$$

where $\mathcal{V} = \text{Span}\ \{\boldsymbol{u}_2 - \boldsymbol{u}_1, \boldsymbol{u}_3 - \boldsymbol{u}_1\}$.

The vectors in $\mathcal{V}$ are the vectors that can be written as linear combinations

$$\alpha\,(\boldsymbol{u}_2 - \boldsymbol{u}_1) + \beta\,(\boldsymbol{u}_3 - \boldsymbol{u}_1)$$

so the vectors in $\boldsymbol{u}_1 + \mathcal{V}$ are the vectors that can be written as

$$\boldsymbol{u}_1 + \alpha\,(\boldsymbol{u}_2 - \boldsymbol{u}_1) + \beta\,(\boldsymbol{u}_3 - \boldsymbol{u}_1)$$

which can be rewritten as

$$(1 - \alpha - \beta)\,\boldsymbol{u}_1 + \alpha, \boldsymbol{u}_2 + \beta\,\boldsymbol{u}_3$$

Let $\gamma = 1 - \alpha - \beta$. Then the above expression can be rewritten as the affine combination

$$\gamma\,\boldsymbol{u}_1 + \alpha\,\boldsymbol{u}_2 + \beta\,\boldsymbol{u}_3$$

That is, the vectors in $\boldsymbol{u}_1 + \mathcal{V}$ are exactly the set of all affine combinations of $\boldsymbol{u}_1$, $\boldsymbol{u}_2$, and $\boldsymbol{u}_3$.

The set of all affine combinations of a collection of vectors is called the *affine hull* of that collection.

Example 3.5.5: What is the affine hull of $\{[0.5, 1], [3.5, 3]\}$? We saw in Section 2.6.4 that the set of affine combinations,

$$\{\alpha\,[3.5, 3] + \beta\,[0.5, 1]\ :\ \alpha \in \mathbb{R}, \beta \in R, \alpha + \beta = 1\}$$

is the line through $[0.5, 1]$ and $[3.5, 3]$.

Example 3.5.6: What is the affine hull of $\{[1, 2, 3]\}$? It is the set of linear combinations $\alpha\,[1, 2, 3]$ where the coefficients sum to one—but there is only one coefficient, α, so we require $\alpha = 1$. Thus the affine hull consists of a single vector, $[1, 2, 3]$.

In the examples, we have seen,

- the affine hull of a one-vector collection is a single point (the one vector in the collection), i.e. a 0-dimensional object;
- the affine hull of a two-vector collection is a line (the line through the two vectors), i.e. a 1-dimensional object;
- the affine hull of a three-vector collection is a plane (the plane through the three vectors), i.e. a 2-dimensional object.

However, let's not jump to conclusions.

Example 3.5.7: What is the affine hull of $\{[2, 3], [3, 4], [4, 5]\}$?

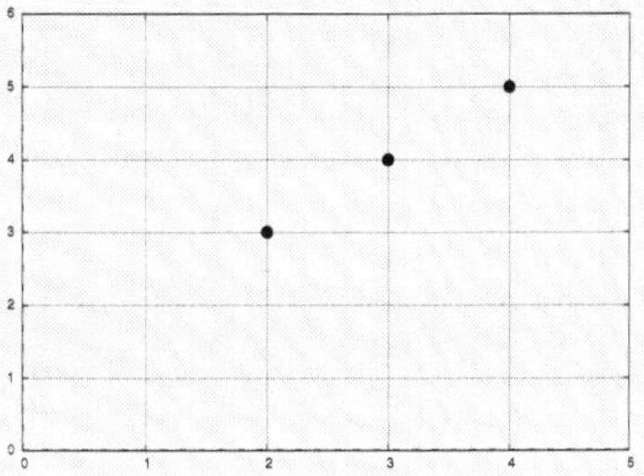

These points all lie on a line. The affine hull is therefore that line, rather than a plane.

Like the span of vectors, the affine hull of vectors can end up being a lower-dimensional object than you would predict just from the number of vectors. Just as we asked in Question 3.3.6 about spans, we might ask a new question: how can we predict the dimensionality of an affine hull? In Example 3.5.1 (Page 133), the affine hull of $\boldsymbol{u}_1, \boldsymbol{u}_2, \boldsymbol{u}_3$ is the translation of Span $\{\boldsymbol{u}_2 - \boldsymbol{u}_1, \boldsymbol{u}_3 - \boldsymbol{u}_1\}$, so, our intuition tell us, the dimensionality of the affine hull is the same as that of Span $\{\boldsymbol{u}_2 - \boldsymbol{u}_1, \boldsymbol{u}_3 - \boldsymbol{u}_1\}$. Thus, in this case, the question about affine hull is not really a new question.

More generally, we will see in in Section 3.5.3 every affine hull of some vectors is the translation of the span of some other vectors, so questions about the dimensionality of the former can be replaced with questions about the dimensionality of the latter.

3.5.3 Affine spaces

Definition 3.5.8: An *affine space* is the result of translating a vector space. That is, a set $\mathcal{A}$ is an affine space if there is a vector $\boldsymbol{a}$ and a vector space $\mathcal{V}$ such that

$$\mathcal{A} = \{\boldsymbol{a} + \boldsymbol{v} \; : \; \boldsymbol{v} \in \mathcal{V}\}$$

i.e. $\mathcal{A} = \boldsymbol{a} + \mathcal{V}$.

A *flat*, it can now be told, is just an affine space that is a subset of $\mathbb{R}^n$ for some n.

Example 3.5.9: We saw in Example 3.5.1 (Page 133) that the plane through the points $\boldsymbol{u}_1 = [1, 0, 4.4]$, $\boldsymbol{u}_2 = [0, 1, 4]$, and $\boldsymbol{u}_3 = [0, 0, 3]$ can be written as the result of adding $\boldsymbol{u}_1$ to each point in the span of $\boldsymbol{u}_2 - \boldsymbol{u}_1$ and $\boldsymbol{u}_3 - \boldsymbol{u}_1$. Since Span $\{\boldsymbol{u}_2 - \boldsymbol{u}_1, \boldsymbol{u}_3 - \boldsymbol{u}_1\}$ is a vector space, it follows that the plane through $\boldsymbol{u}_1$, $\boldsymbol{u}_2$, and $\boldsymbol{u}_3$ is an affine space.

We also saw in Section 3.5.2 that the plane is the set of *affine* combinations of $\boldsymbol{u}_1$, $\boldsymbol{u}_2$, and $\boldsymbol{u}_3$. Thus, in this case at least, the affine combination of the vectors is an affine space. Is this true generally?

Lemma 3.5.10: For any vectors $\boldsymbol{u}_1, \ldots, \boldsymbol{u}_n$,

$$\{\alpha_1 \boldsymbol{u}_1 + \cdots + \alpha_n \boldsymbol{u}_n \; : \sum_{i=1}^{n} \alpha_i = 1\} = \{\boldsymbol{u}_1 + \boldsymbol{v} \; : \; \boldsymbol{v} \in \text{Span } \{\boldsymbol{u}_2 - \boldsymbol{u}_1, \ldots, \boldsymbol{u}_n - \boldsymbol{u}_1\}\} \quad (3.2)$$

In words, the affine hull of $\boldsymbol{u}_1, \ldots, \boldsymbol{u}_n$ equals the set obtained by adding $\boldsymbol{u}_1$ to each vector in the span of $\boldsymbol{u}_2 - \boldsymbol{u}_1, \ldots, \boldsymbol{u}_n - \boldsymbol{u}_1$.

The lemma shows that the affine hull of vectors is an affine space. Knowing this will help us, for example, learn how to find the intersection of a plane with a line.

The proof follows the calculations in Example 3.5.4 (Page 134).

Proof

Every vector in Span $\{\boldsymbol{u}_2 - \boldsymbol{u}_1, \ldots, \boldsymbol{u}_n - \boldsymbol{u}_1\}$ can be written in the form

$$\alpha_2 (\boldsymbol{u}_2 - \boldsymbol{u}_1) + \cdots + \alpha_n (\boldsymbol{u}_n - \boldsymbol{u}_1)$$

so every vector in the right-hand side of Equation 3.2 can be written in the form

$$\boldsymbol{u}_1 + \alpha_2 (\boldsymbol{u}_2 - \boldsymbol{u}_1) + \cdots + \alpha_n (\boldsymbol{u}_n - \boldsymbol{u}_1)$$

which can be rewritten (using homogeneity and distributivity as in Example 3.5.1 (Page 133)) as

$$(1 - \alpha_2 - \cdots - \alpha_n)\, \boldsymbol{u}_1 + \alpha_2\, \boldsymbol{u}_2 + \cdots + \alpha_n\, \boldsymbol{u}_n \tag{3.3}$$

which is an affine combination of $\boldsymbol{u}_1, \boldsymbol{u}_2, \ldots, \boldsymbol{u}_n$ since the coefficients sum to one. Thus every vector in the right-hand side of Equation 3.2 is in the left-hand side.

Conversely, for every vector $\alpha_1\, \boldsymbol{u}_1 + \alpha_2\, \boldsymbol{u}_2 + \cdots + \alpha_n\, \boldsymbol{u}_n$ in the left-hand side, since $\sum_{i=1}^{n} \alpha_i = 1$, we infer $\alpha_1 = 1 - \alpha_2 - \cdots - \alpha_n$, so the vector can be written as in Line 3.3, which shows that the vector is in the right-hand side. □

We now have two representations of an affine space:

- as $\boldsymbol{a} + \mathcal{V}$ where $\mathcal{V}$ is the span of some vectors, and
- as the affine hull of some vectors.

These representations are not fundamentally different; as we have seen, it is easy to convert between one representation and the other. Next, we discuss a representation that is quite different.

3.5.4 Representing an affine space as the solution set of a linear system

In Section 3.3.2, we saw examples in which a flat containing the origin could be represented as the solution set of a homogeneous linear system. Here we represent a flat not containing the origin as the solution set of a linear system that is *not* homogeneous.

Example 3.5.11: We saw in Example 3.5.1 (Page 133) that the plane through the points $[1, 0, 4.4]$, $[0, 1, 4]$, and $[0, 0, 3]$ is the affine hull of those points. However, the plane is also the solution set of the equation $1.4x + y - z = -3$, i.e. the plane is

$$\{[x, y, z] \in \mathbb{R}^3 \ : \ [1.4, 1, -1] \cdot [x, y, z] = -3\}$$

Example 3.5.12: We saw in Section 2.6.4 (see also Example 3.5.5 (Page 134)) that the line

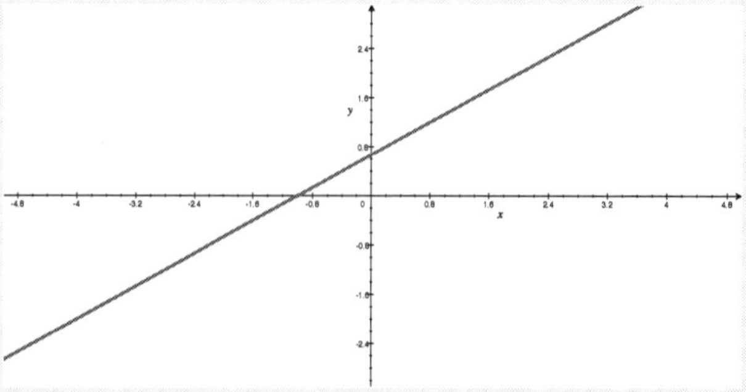

through $[0.5, 1]$ and $[3.5, 3]$ consists of the set of all affine combinations of $[0.5, 1]$ and $[3.5, 3]$.

This line is also the solution set of the equation $2x - 3y = -2$, i.e. the set

$$\{[x,y] \in \mathbb{R}^2 \;:\; [2,-3] \cdot [x,y] = -2\}$$

Example 3.5.13: The line

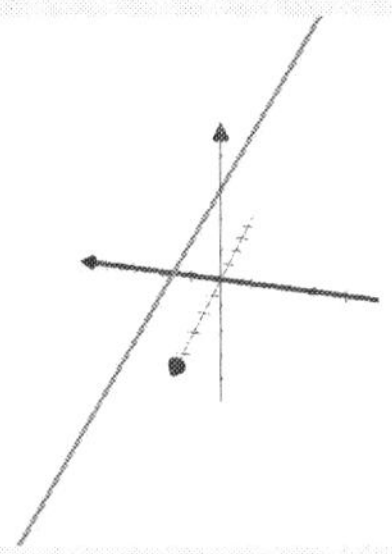

can be represented as the set of all affine combinations of $[1,2,1]$ and $[1,2,-2]$. The line is also the solution set of the linear system consisting of the equations $4x - y + z = 3$ and $y + z = 3$, i.e. the set

$$\{[x,y,z] \in \mathbb{R}^3 \;:\; [4,-1,1] \cdot [x,y,z] = 3, [0,1,1] \cdot [x,y,z] = 3\}$$

3.5.5 The two representations, revisited

As we saw in Section 3.3.3 in the context of flats containing the origin, having two representations can be useful.

Example 3.5.14: Suppose you are given two lines

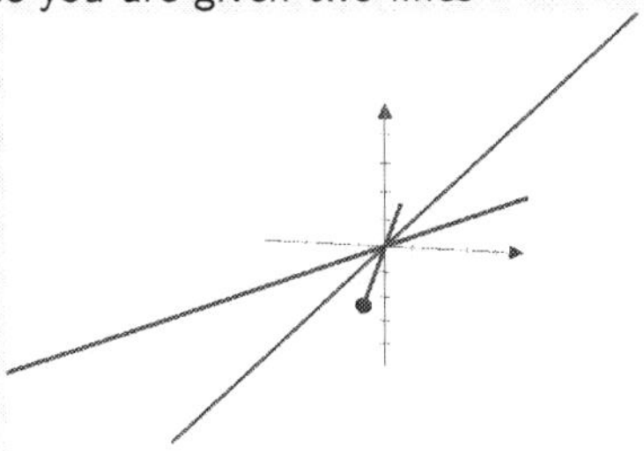

and want to find the plane containing the two lines.

The first line is Span $\{[4,-1,1]\}$. The second line is Span $\{[0,1,1]\}$. Therefore the plane containing these two lines is Span $\{[4,-1,1],[0,1,1]\}$:

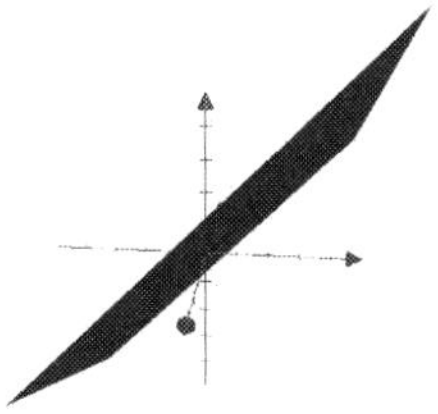

Next we give an example using the second kind of representation.

Example 3.5.15: Now you are given two planes through the origin:

and your goal is to find the the intersection.

The first plane is $\{[x, y, z] \; : \; [4, -1, 1] \cdot [x, y, z] = 0\}$. The second plane is $\{[x, y, z] \; : \; [0, 1, 1] \cdot [x, y, z] = 0\}$. We are representing each plane as the solution set of a linear system with right-hand sides zero. The set of points comprising the intersection is exactly the set of points that satisfy both equations,

$$\{[x, y, z] \; : \; [4, -1, 1] \cdot [x, y, z] = 0, [0, 1, 1] \cdot [x, y, z] = 0\}$$

This set of points forms a line, but to draw the line it is helpful to find its representation as the span of a vector. We will learn later how to go from a linear system with zero right-hand sides to a set of generators for the solution set. It turns out the solution set is Span $\{[1, 2, -2]\}$.

Because different representations facilitate different operations, it is useful to be able to convert between different representations of the same geometric object. We'll illustrate this using an example that arises in computer graphics. A scene is often constructed of thousands of triangles. How can we test whether a beam of light strikes a particular triangle, and, if so, where on that triangle?

Let's say the triangle's corners are located at the vectors $\boldsymbol{v}_0$, $\boldsymbol{v}_1$, and $\boldsymbol{v}_2$. Then the plane containing the triangle is the affine hull of these vectors.

Next, suppose a beam of light originates at a point $\boldsymbol{b}$, and heads in the direction of the arrow representing the vector $\boldsymbol{d}$. The beam of light forms the ray consisting of the set of points

$$\{\boldsymbol{b} + \alpha \, \boldsymbol{d} \; : \; \alpha \in \mathbb{R}, \alpha \geq 0\}$$

which in turn forms part of the line

$$\{\boldsymbol{b} + \alpha \, \boldsymbol{d} \; : \; \alpha \in \mathbb{R}\}$$

So far we are using the first kind of representation for the triangle, for the plane containing the triangle, for the ray and for the line containing the ray. To find out whether the beam of light strikes the triangle, we find the intersection of

- the plane containing the triangle, and
- the line containing the ray of light.

Usually, the intersection will consist of a single point. We can then test whether that point lies in the triangle and whether it belongs to the ray.

But how can we find the intersection of the plane and the line? We use the *second* kind of representation:

- we find the representation of the plane as the solution set of one linear system, and
- we find the representation of the line as the solution set of another linear system.

The set of points belonging to both the plane and the line is exactly the set of points in the solution sets of both linear systems, which is the set of points in the solution set of the new linear system consisting of all the equations in both of the two original linear systems.

Example 3.5.16: Suppose the vertices of the triangle are the points $[1, 1, 1]$, $[2, 2, 3]$, and $[-1, 3, 0]$.

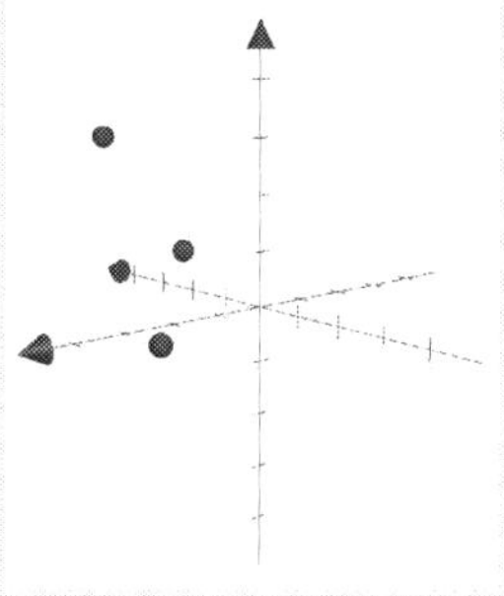

The triangle looks like this:

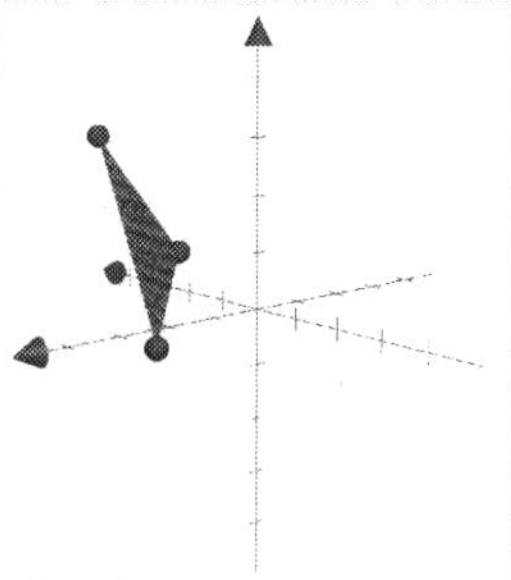

The ray of light originates at $\boldsymbol{p} = [-2.2, 0.8, 3.1]$ and moves in the direction $\boldsymbol{d} = [1.55, 0.65, -0.7]$.

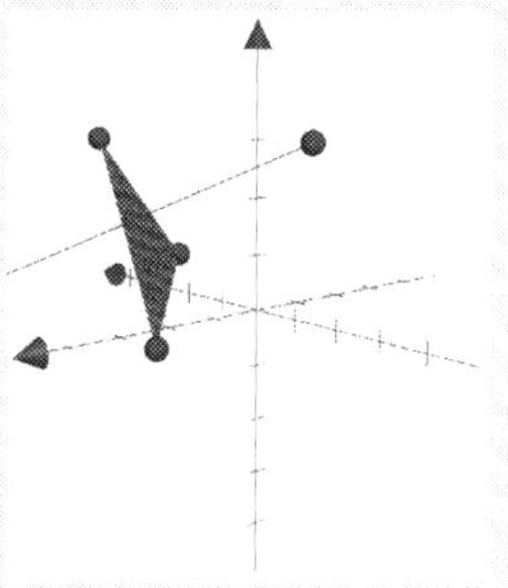

We can see from the picture that the ray does indeed intersect the triangle, but how can a computer discover that?

Here we show the plane containing the triangle:

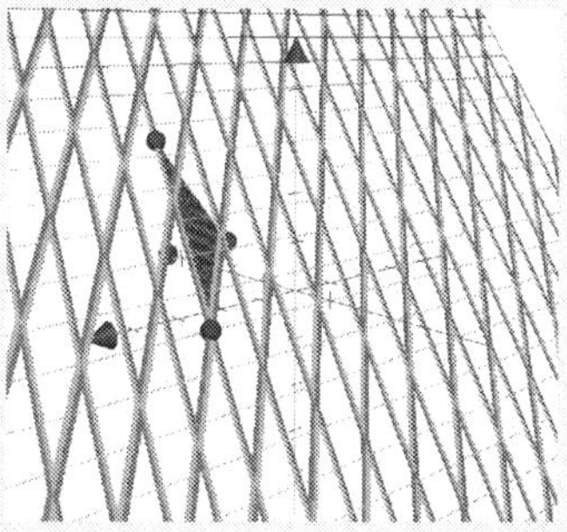

We will later learn to find a linear equation whose solution space is the plane. One such equation turns out to be $[5, 3, -4] \cdot [x, y, z] = 4$.

We will later learn to find a linear system whose solution space is the line containing the ray of light. One such system, it turns out, is

$$\begin{aligned}[0.275..., -0.303..., 0.327...] \cdot [x, y, z] &= 0.1659...\\ [0, 0.536..., 0.498...] \cdot [x, y, z] &= 1.975...\end{aligned}$$

To find the intersection of the plane and the line, we put all these linear equations together, obtaining

$$\begin{aligned}[5, 3, -4] \cdot [x, y, z] &= 4\\ [0.275..., -0.303..., 0.327...] \cdot [x, y, z] &= 0.1659...\\ [0, 0.536..., 0.498...] \cdot [x, y, z] &= 1.975...\end{aligned}$$

The solution set of this combined linear system consists of the points belonging to both the plane and the line. We are using the second kind of representation. In this case, the solution set consists of just one point. To find that point, we convert back to the first kind of representation.

We will later learn algorithms to solve a linear system. The solution turns out to be $\boldsymbol{w} = [0.9, 2.1, 1.7]$. The point $\boldsymbol{w}$, therefore is the intersection of the plane and the line.

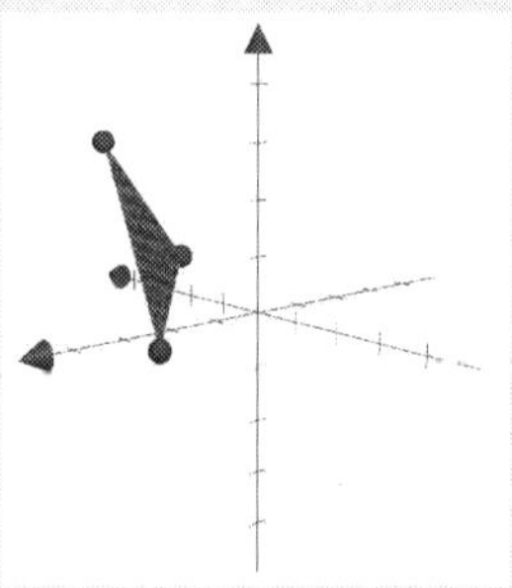

Once we have found the point of intersection of the line and the plane, how do we find out whether the intersection belongs to the triangle and the ray? For this, we return to the first kind of representation.

Example 3.5.17: The point of intersection lies in the plane and is therefore an affine combination of the vertices:

$$\boldsymbol{w} = \alpha_0 [1, 1, 1] + \alpha_1 [2, 2, 3] + \alpha_2 [-1, 3, 0]$$

which means that the coefficients sum up to one. The point is in the triangle if it is a *convex* combination of the vertices. We will learn that in this case there is only one way to represent the point as an affine combination, and we will learn how to find the coefficients:

$$\boldsymbol{w} = 0.2 [1, 1, 1] + 0.5 [2, 2, 3] + 0.3 [-1, 3, 0]$$

Since the coefficients are nonnegative, we know that the point of intersection is indeed in the triangle.

There is one more thing to check. We should check that the intersection point lies in the 'half' of the line that comprises the ray. The ray is the set of points $\{\boldsymbol{p} + \alpha\, \boldsymbol{d} \; : \; \alpha \in \mathbb{R}, \alpha \geq 0\}$. The line is the set of points $\{\boldsymbol{p} + \alpha\, \boldsymbol{d} \; : \; \alpha \in \mathbb{R}, \alpha \geq 0\}$. To see if the intersection point $\boldsymbol{w}$ is in the ray, we find the unique value of α such that $\boldsymbol{w} = \boldsymbol{p} + \alpha\, \boldsymbol{d}$, and we check that this value is nonnegative.

The vector equation $\boldsymbol{w} = \boldsymbol{p} + \alpha\, \boldsymbol{d}$ is equivalent to three scalar equations, one for each of the entries of the vector. To find the value of α, let's just consider the first entry. The first entry of $\boldsymbol{w}$ is 0.9, the first entry of $\boldsymbol{p}$ is -2.2, and the first entry of $\boldsymbol{d}$ is 3.1, so α must satisfy the equation

$$0.9 = -2.2 + \alpha\; 1.55$$

which we can solve, obtaining $\alpha = 2$. Since the value of α is nonnegative, the intersection point does indeed belong to the ray.

In this example, we needed to convert between the two kinds of representations of a flat, (1) as a set of of linear combinations and (2) as the solution set of a linear system.

3.6 Linear systems, homogeneous and otherwise

In Section 3.4, we saw that the solution set of a homogeneous linear system is a vector space. What about the solution set of an arbitrary linear system? Is that an affine space? Yes, with an exception: the case in which the solution set is empty.

3.6.1 The homogeneous linear system corresponding to a general linear system

In Section 2.9.2, we considered the problem of calculating the rate of power consumption for hardware components of a sensor node. We formulated this as the problem of finding a solution to a linear system over $\mathbb{R}$, and we asked (Question 2.9.11): *how can we tell if there is only one solution?*

In Section 2.9.7, we considered an attack on a simple authentication scheme. We found a way in which Eve, an eavesdropper, might calculate the password from observing authentication trials. We formulated this as the problem of finding a solution to a system of linear equations over $GF(2)$, and we asked (Question 2.9.18): *how many solutions are there to a given linear system over $GF(2)$?*

We shall see that, in each of these applications, the first question can be addressed by studying the corresponding system of *homogeneous* linear equations, i.e. where each right-hand side is replaced by a zero.

Lemma 3.6.1: Let $\boldsymbol{u}_1$ be a solution to the system of linear equations

$$\begin{aligned} \boldsymbol{a}_1 \cdot \boldsymbol{x} &= \beta_1 \\ &\vdots \\ \boldsymbol{a}_m \cdot \boldsymbol{x} &= \beta_m \end{aligned} \tag{3.4}$$

Then another vector $\boldsymbol{u}_2$ is also a solution if and only if the difference $\boldsymbol{u}_2 - \boldsymbol{u}_1$ is a solution to the system of corresponding *homogeneous* equations

$$\begin{aligned} \boldsymbol{a}_1 \cdot \boldsymbol{x} &= 0 \\ &\vdots \\ \boldsymbol{a}_m \cdot \boldsymbol{x} &= 0 \end{aligned} \tag{3.5}$$

Proof

For $i = 1, \ldots, m$, we have $\boldsymbol{a}_i \cdot \boldsymbol{u}_1 = \beta_i$, so $\boldsymbol{a}_i \cdot \boldsymbol{u}_2 = \beta_i$ iff $\boldsymbol{a}_i \cdot \boldsymbol{u}_2 - \boldsymbol{a}_i \cdot \boldsymbol{u}_1 = 0$ iff $\boldsymbol{a}_i \cdot (\boldsymbol{u}_2 - \boldsymbol{u}_1) = 0$.
□

The set of solutions to a homogeneous linear system is a vector space $\mathcal{V}$. We can restate the assertion of Lemma 3.6.1:

> $\boldsymbol{u}_2$ is a solution to the original linear system (3.4) if and only if $\boldsymbol{u}_2 - \boldsymbol{u}_1$ is in $\mathcal{V}$ where $\mathcal{V}$ is the solution set of the homogeneous linear system (3.5.

Substituting $\boldsymbol{v}$ for $\boldsymbol{u}_2 - \boldsymbol{u}_1$ (which implies $\boldsymbol{u}_2 = \boldsymbol{u}_1 + \boldsymbol{v}$), we reformulate it as:

$\boldsymbol{u}_1 + \boldsymbol{v}$ is a solution to the original linear system if and only if $\boldsymbol{v}$ is in $\mathcal{V}$

which can be reworded as:

$$\{\text{solutions to original linear system}\} = \{\boldsymbol{u}_1 + \boldsymbol{v} \;:\; \boldsymbol{v} \in \mathcal{V}\} \tag{3.6}$$

The set on the right-hand side is an affine space!

Theorem 3.6.2: For any linear system, the set of solutions either is empty or is an affine space.

Proof

If the linear system has no solution, the solution set is empty. If it has at least one solution $\boldsymbol{u}_1$ then the solution set is $\{\boldsymbol{u}_1 + \boldsymbol{v} \;:\; \boldsymbol{v} \in \mathcal{V}\}$. □

We asked in Question 3.4.16 whether every vector space is the solution space of a homogeneous system (and indicated that the answer is yes). An analogous question is *Is every affine space the solution set of a linear system?* That the answer is yes follows from the fact that the answer to the previous question is yes.

Example 3.6.3: The solution set of the linear system

$$\begin{bmatrix} 0 & 0 \end{bmatrix} \cdot \boldsymbol{x} = 1$$

is the empty set.

The solution set of the linear system

$$\begin{bmatrix} 1 & 0 \end{bmatrix} \cdot \boldsymbol{x} = 2$$
$$\begin{bmatrix} 0 & 1 \end{bmatrix} \cdot \boldsymbol{x} = 5$$

is the singleton set $\left\{\begin{bmatrix} 2 \\ 5 \end{bmatrix}\right\}$, which can be written as

$$\left\{\begin{bmatrix} 2 \\ 5 \end{bmatrix} + \boldsymbol{v} \;:\; \boldsymbol{v} \in \left\{\begin{bmatrix} 0 \\ 0 \end{bmatrix}\right\}\right\}$$

The solution set of the linear system

$$\begin{bmatrix} 2 & -5 \end{bmatrix} \cdot \boldsymbol{x} = 1$$
$$\begin{bmatrix} 4 & -10 \end{bmatrix} \cdot \boldsymbol{x} = 2$$

is the set $\left\{\begin{bmatrix} -2 \\ -1 \end{bmatrix} + \alpha \begin{bmatrix} 1 \\ 2.5 \end{bmatrix} \;:\; \alpha \in \mathbb{R}\right\}$, which can be written as

$$\left\{\begin{bmatrix} -2 \\ -1 \end{bmatrix} + \boldsymbol{v} \;:\; \boldsymbol{v} \in \text{Span}\left\{\begin{bmatrix} 1 \\ 2.5 \end{bmatrix}\right\}\right\}$$

3.6.2 Number of solutions revisited

We can now give a partial answer to Question 2.9.11 (*How can we tell if a linear system has only one solution?*):

Corollary 3.6.4: Suppose a linear system has a solution. The solution is unique if and only if the only solution to the corresponding homogeneous linear system is the zero vector.

The question about uniqueness of solution is therefore replaced with

Question 3.6.5: How can we tell if a homogeneous linear system has only a trivial solution?

Moreover, Question 2.9.18 (*How many solutions are there to a given system of linear equations over* $GF(2)$*?*) is partially addressed by Equation 3.6, which tells us that the number of solutions equals $|\mathcal{V}|$, the cardinality of the vector space consisting of solutions to the corresponding homogeneous system.

The question about counting solutions to a linear system over $GF(2)$ thus becomes

Question 3.6.6: How can we find the number of solutions to a homogeneous linear system over $GF(2)$?

In addressing these questions, we will make use of the fact that the solution set for a homogeneous linear system is a vector space.

3.6.3 Towards intersecting a plane and a line

Here's an example of how Theorem 3.6.2 could help us: an approach to compute the intersection of a plane and a line:

Step 1: Since the plane is an affine space, we hope to represent it as the solution set of a linear system.

Step 2: Since the line is an affine space, we hope to represent it as the solution set of a second linear system.

Step 3: Combine the two linear systems to form a single linear system consisting of all the linear equations from the two. The solutions to the combined linear system are the points that are on both the plane and the line.

The solution set of the combined linear system might consist of many vectors (if the line lies within the plane) or just one (the point at which the line intersects the plane).

This approach sounds promising—but so far we don't know how to carry it out. Stay tuned.

3.6.4 Checksum functions

This section gives another application of homogeneous linear equations.

A *checksum* for a big chunk of data or program is a small chunk of data used to verify that the big chunk has not been altered. For example, here is a fragment of the download page for Python:

MD5 checksums and sizes of the released files:

```
3c63a6d97333f4da35976b6a0755eb67  12732276  Python-3.2.2.tgz
9d763097a13a59ff53428c9e4d098a05  10743647  Python-3.2.2.tar.bz2
3720ce9460597e49264bbb63b48b946d   8923224  Python-3.2.2.tar.xz
f6001a9b2be57ecfbefa865e50698cdf  19519332  python-3.2.2-macosx10.3.dmg
8fe82d14dbb2e96a84fd6fa1985b6f73  16226426  python-3.2.2-macosx10.6.dmg
cccb03e14146f7ef82907cf12bf5883c  18241506  python-3.2.2-pdb.zip
72d11475c986182bcb0e5c91acec45bc  19940424  python-3.2.2.amd64-pdb.zip
ddeb3e3fb93ab5a900adb6f04edab21e  18542592  python-3.2.2.amd64.msi
8afb1b01e8fab738e7b234eb4fe3955c  18034688  python-3.2.2.msi
```

With each downloadable Python release is listed the checksum and the size.

A *checksum function* is a function that maps a large file of data to a small chunk of data, the checksum. Since the number of possible checksums is much smaller than the number of possible files, there is no one-to-one checksum function: there will always be pairs of distinct files that map to the same checksum. The goal of using a checksum function is to detect accidental corruption of a file during transmission or storage.

Here we seek is a function such that a *random* corruption is likely detectable: for any file F, a random change to the file probably leads to a change in the checksum.

We describe an impractical but instructive checksum function. The input is a "file" represented as an n-bit vector over $GF(2)$. The output is a 64-vector. The function is specified by sixty-four n-vectors $\boldsymbol{a}_1, \ldots, \boldsymbol{a}_{64}$. The function is then defined as follows:

$$\boldsymbol{x} \mapsto [\boldsymbol{a}_1 \cdot \boldsymbol{x}, \ldots, \boldsymbol{a}_{64} \cdot \boldsymbol{x}]$$

Suppose $\boldsymbol{p}$ is a "file". We model corruption as the addition of a random n-vector $\boldsymbol{e}$ (the *error*), so the corrupted version of the file is $\boldsymbol{p}+\boldsymbol{e}$. We want to find a formula for the probability that the corrupted file has the same checksum as the original file.

The checksum for the original file is $[\beta_1, \ldots, \beta_m]$, where $\beta_i = \boldsymbol{a}_i \cdot \boldsymbol{p}$ for $i = 1, \ldots, m$. For $i = 1, \ldots, m$, bit i of the checksum of the corrupted file is $\boldsymbol{a}_i \cdot (\boldsymbol{p}+\boldsymbol{e})$. Since dot-product distributes over vector addition (Proposition 2.9.25), this is equal to $\boldsymbol{a}_i \cdot \boldsymbol{p} + \boldsymbol{a} \cdot \boldsymbol{e}$. Thus bit i of the checksum of the corrupted file equals that of the original file if and only if $\boldsymbol{a}_i \cdot \boldsymbol{p} + \boldsymbol{a}_i \cdot \boldsymbol{e} = \boldsymbol{a}_i \cdot \boldsymbol{p}$—that is, if and only if $\boldsymbol{a}_i \cdot \boldsymbol{e} = 0$.

Thus the entire checksum of the corrupted file is the same as that of the original if and only if $\boldsymbol{a}_i \cdot \boldsymbol{e} = 0$ for $i = 1, \ldots$ and m, if and only if $\boldsymbol{e}$ belongs to the solution set for the homogeneous linear system

$$\begin{aligned} \boldsymbol{a}_1 \cdot \boldsymbol{x} &= 0 \\ &\vdots \\ \boldsymbol{a}_m \cdot \boldsymbol{x} &= 0 \end{aligned}$$

The probability that a random n-vector $\boldsymbol{e}$ belongs to the solution set is

$$\frac{\text{number of vectors in solution set}}{\text{number of } n\text{-vectors over } GF(2)}$$

We know that the number of n-vectors over $GF(2)$ is 2^n. To calculate this probability, therefore, we once again need an answer to Question 3.6.6: *How can we find the number of solutions to a homogeneous linear system over $GF(2)$?*

3.7 Review questions

- What is a linear combination?
- What are coefficients?
- What is the span of vectors?
- What are standard generators?
- What are examples of flats?
- What is a homogeneous linear equation?
- What is a homogeneous linear system?
- What are the two kinds of representations of flats containing the origin?
- What is a vector space?
- What is a subspace?
- What is an affine combination?
- What is the affine hull of vectors?
- What is an affine space?
- What are the two kinds of representations of flats not containing the origin?
- Is the solution set of a linear system always an affine space?

3.8 Problems

Vec review

Vectors in containers

Problem 3.8.1:

1. Write and test a procedure `vec_select` using a comprehension for the following computational problem:

   - *input:* a list `veclist` of vectors over the same domain, and an element k of the domain
   - *output:* the sublist of `veclist` consisting of the vectors `v` in `veclist` where `v[k]` is zero

2. Write and test a procedure `vec_sum` using the built-in procedure `sum(·)` for the following:

   - *input:* a list `veclist` of vectors, and a set `D` that is the common domain of these vectors
   - *output:* the vector sum of the vectors in `veclist`.

   Your procedure must work even if `veclist` has length 0.

   Hint: Recall from the Python Lab that `sum(·)` optionally takes a second argument, which is the element to start the sum with. This can be a vector.

   Disclaimer: The `Vec` class is defined in such a way that, for a vector `v`, the expression `0 + v` evaluates to `v`. This was done precisely so that `sum([v1,v2,... vk])` will correctly evaluate to the sum of the vectors when the number of vectors is nonzero. However, this won't work when the number of vectors is zero.

3. Put your procedures together to obtain a procedure `vec_select_sum` for the following:

   - *input:* a set `D`, a list `veclist` of vectors with domain `D`, and an element k of the domain
   - *output:* the sum of all vectors `v` in `veclist` where `v[k]` is zero

Problem 3.8.2: Write and test a procedure `scale_vecs(vecdict)` for the following:

- *input:* A dictionary `vecdict` mapping positive numbers to vectors (instances of `Vec`)
- *output:* a list of vectors, one for each item in `vecdict`. If `vecdict` contains a key k mapping to a vector $\boldsymbol{v}$, the output should contain the vector $(1/k)\boldsymbol{v}$

Linear combinations

Constructing the span of given vectors over $GF(2)$

Problem 3.8.3: Write a procedure `GF2_span` with the following spec:

- *input:* a set D of labels and a list L of vectors over $GF(2)$ with label-set D
- *output:* the list of all linear combinations of the vectors in L

(Hint: use a loop (or recursion) and a comprehension. Be sure to test your procedure on examples where L is an empty list.)

Problem 3.8.4: Let a, b be real numbers. Consider the equation $z = ax + by$. Prove that there are two 3-vectors $\boldsymbol{v}_1, \boldsymbol{v}_2$ such that the set of points $[x, y, z]$ satisfying the equation is exactly the set of linear combinations of $\boldsymbol{v}_1$ and $\boldsymbol{v}_2$. (Hint: Specify the vectors using formulas involving a, b.)

Problem 3.8.5: Let a, b, c be real numbers. Consider the equation $z = ax+by+c$. Prove that there are three 3-vectors $\boldsymbol{v}_0, \boldsymbol{v}_1, \boldsymbol{v}_2$ such that the set of points $[x, y, z]$ satisfying the equation is exactly

$$\{\boldsymbol{v}_0 + \alpha_1\, \boldsymbol{v}_1 + \alpha_2\, \boldsymbol{v}_2 \;:\; \alpha_1 \in \mathbb{R}, \alpha_2 \in \mathbb{R}\}$$

(Hint: Specify the vectors using formulas involving a, b, c.)

Sets of linear combinations and geometry

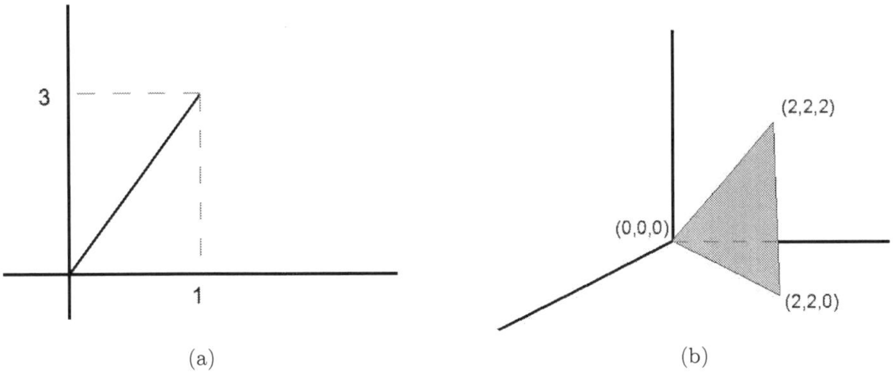

Figure 3.1: Figures for Problem 3.8.6.

Problem 3.8.6: Express the line segment in Figure 3.1(a) using a set of linear combinations. Do the same for the plane containing the triangle in Figure 3.1(b).

Vector spaces

Problem 3.8.7: Prove or give a counterexample: "$\{[x, y, z] : x, y, z \in \mathbb{R}, x + y + z = 1\}$ forms a vector space."

Problem 3.8.8: Prove or give a counterexample: $\{[x, y, z] : x, y, z \in \mathbb{R}$ and $x + y + z = 0\}$ is a vector space.

Problem 3.8.9: Prove or give a counterexample: $\{[x_1, x_2, x_3, x_4, x_5] : x_1, x_2, x_3, x_4, x_5 \in \mathbb{R}, x_2 = 0$ or $x_5 = 0\}$ is a vector space.

Problem 3.8.10: Explain your answers.

1. Let $\mathcal{V}$ be the set of 5-vectors over $GF(2)$ that have an even number of 1's. Is $\mathcal{V}$ a vector space?
2. Let $\mathcal{V}$ be the set of 5-vectors over $GF(2)$ that have an odd number of 1's. Is $\mathcal{V}$ a vector space?

Chapter 4

The Matrix

> **Neo:** What is the Matrix?
> **Trinity:** The answer is out there, Neo, and it's looking for you, and it will find you if you want it to.
>
> *The Matrix*, 1999

4.1 What is a matrix?

4.1.1 Traditional matrices

Traditionally, a matrix over $\mathbb{F}$ is a two-dimensional array whose entries are elements of $\mathbb{F}$. Here is a matrix over $\mathbb{R}$:

$$\begin{bmatrix} 1 & 2 & 3 \\ 10 & 20 & 30 \end{bmatrix}$$

This matrix has two rows and three columns, so we call it a 2×3 matrix. It is traditional to refer to the rows and columns by numbers. Row 1 is $\begin{bmatrix} 1 & 2 & 3 \end{bmatrix}$ and row 2 is $\begin{bmatrix} 10 & 20 & 30 \end{bmatrix}$; column 1 is $\begin{bmatrix} 1 \\ 10 \end{bmatrix}$, column 2 is $\begin{bmatrix} 2 \\ 20 \end{bmatrix}$, and column 3 is $\begin{bmatrix} 3 \\ 30 \end{bmatrix}$.

In general, a matrix with m rows and n columns is called an $m \times n$ matrix. . For a matrix A, the i, j *element* is defined to be the element in the ith row and the jth column, and is traditionally written $A_{i,j}$ or A_{ij}. We will often use the Pythonese notation, $A[i, j]$.

Row i is the vector

$$\begin{bmatrix} A[i,0], & A[i,1], & A[i,2], & \cdots & A[i,m-1] \end{bmatrix}$$

and column j is the vector

$$\begin{bmatrix} A[0,j], & A[1,j], & A[2,j], & \cdots & A[n-1,j] \end{bmatrix}$$

Representing a traditional matrix by a list of row-lists

How can we represent a matrix? Perhaps the first representation that comes to mind is a *list of row-lists*: —each row of the matrix A is represented by a list of numbers, and the matrix is represented by a list L of these lists. That is, a list L such that

$$A[i, j] = L[i][j] \text{ for every } 0 \le i < m \text{ and } 0 \le j < n$$

For example, the matrix $\begin{bmatrix} 1 & 2 & 3 \\ 10 & 20 & 30 \end{bmatrix}$ would be represented by `[[1,2,3],[10,20,30]]`.

Quiz 4.1.1: Write a nested comprehension whose value is list-of-*row*-list representation of a

Figure 4.1: The Matrix Revisited (excerpt) `http://xkcd.com/566/`

3×4 matrix all of whose elements are zero:

$$\begin{bmatrix} 0 & 0 & 0 & 0 \\ 0 & 0 & 0 & 0 \\ 0 & 0 & 0 & 0 \end{bmatrix}$$

Hint: first write a comprehension for a typical row, then use that expression in a comprehension for the list of lists.

Answer

```
>>> [[0 for j in range(4)] for i in range(3)]
[[0, 0, 0, 0], [0, 0, 0, 0], [0, 0, 0, 0]]
```

Representing a traditional matrix by a list of column-lists

As you will see, one aspect of matrices that makes them so convenient and beautiful is the duality between rows and columns. Anything you can do with columns, you can do with rows. Thus we can represent a matrix A by a *list of column-lists*; that is, a list L such that

$$A[i,j] = L[j][i] \text{ for every } 0 \leq i < m \text{ and } 0 \leq j < n$$

For example, the matrix $\begin{bmatrix} 1 & 2 & 3 \\ 10 & 20 & 30 \end{bmatrix}$ would be represented by `[[1,10],[2,20],[3,30]]`.

Quiz 4.1.2: Write a nested comprehension whose value is list-of-*column*-lists representation of a 3×4 matrix whose i,j element is $i - j$:

$$\begin{bmatrix} 0 & -1 & -2 & -3 \\ 1 & 0 & -1 & -2 \\ 2 & 1 & 0 & -1 \end{bmatrix}$$

Hint: First write a compresion for column j, assuming j is bound to an integer. Then use that expression in a comprehension in which j is the control variable.

Answer

```
>>> [[i-j for i in range(3)] for j in range(4)]
[[0, 1, 2], [-1, 0, 1], [-2, -1, 0], [-3, -2, -1]]
```

4.1.2 The matrix revealed

We will often use the traditional notation in examples. However, just as we find it helpful to define vectors whose entries are identified by elements of an arbitrary finite set, we would like to be able to refer to a matrix's rows and columns using arbitrary finite sets,

As we have defined a D-vector over $\mathbb{F}$ to be a function from a set D to $\mathbb{F}$, so we define a $R \times C$ *matrix over* $\mathbb{F}$ to be a function from the Cartesian product $R \times C$. We refer to the elements of R as *row labels* and we refer to the elements of C as *column labels*.

Example 4.1.3: Here is an example in which $R =$ `{'a', 'b'}` and $C =$ `{'#', '@', '?'}`:

	@	#	?
a	1	2	3
b	10	20	30

The column labels are given atop the columns, and the row labels are listed to the left of the rows.

Formally, this matrix is a function from $R \times C$ to $\mathbb{R}$. We can represent the function using Python's dictionary notation:

```
{('a','@'):1, ('a','#'):2, ('a', '?'):3, ('b', '@'):10, ('b', '#'):20,
 ('b','?'):30}
```

4.1.3 Rows, columns, and entries

Much of the power of matrices comes from our ability to interpret the rows and columns of a matrix as vectors. For the matrix of Example 4.1.3 (Page 151):

- row `'a'` is the vector `Vec({'@', '#', '?'}, {'@':1, '#':2, '?':3})`
- row `'b'` is the vector `Vec({'@', '#', '?'}, {'@':10, '#':20, '?':30})`
- column `'#'` is the vector `Vec({'a','b'}, {'a':2, 'b':20})`
- column `'@'` is the vector `Vec({'a','b'}, {'a':1, 'b':10})`

Quiz 4.1.4: Give a Python expression using Vec for column `'?'`.

Answer

```
Vec({'a','b'}, {'a':3, 'b':30})
```

For an $R \times C$ matrix M, and for $r \in R$ and $c \in C$, the r, c *element of* M is defined to be whatever the pair (r, c) maps to, and is written $M_{r,c}$ or $M[r, c]$. The rows and columns are defined as follows:

- For $r \in R$, row r is the C-vector such that, for each element $c \in C$, entry c is $M[r, c]$, and
- for $c \in C$, column c is the R-vector such that, for each element $r \in R$, entry r is $M[r, c]$.

We denote row r of M by $M[r, :]$ or $M_{r,:}$ and we denote column c of M by $M[:, c]$ or $M_{:,c}$.

Dict-of-rows representation

Since I have said that each row of a matrix is a vector, we can represent each row by an instance of `Vec`. To map row-labels to the rows, we use a dictionary. I call this representation a *rowdict*. For example, the rowdict representation of the matrix of Example 4.1.3 (Page 151) is:

```
{'a': Vec({'#', '@', '?'}, {'@':1, '#':2, '?':3}),
 'b': Vec({'#', '@', '?'}, {'@':10, '#':20, '?':30})}
```

Dict-of-columns representation

The duality of rows and colums suggests a representation consisting of a dictionary mapping column-labels to the columns represented as instances of Vec. I call this representation a *coldict*.

Quiz 4.1.5: Give a Python expression whose value is the coldict representation of the matrix of Example 4.1.3 (Page 151).

Answer

```
{'#': Vec({'a','b'}, {'a':2, 'b':20}),
 '@': Vec({'a','b'}, {'a':1, 'b':10}),
 '?': Vec({'a','b'}, {'a':3, 'b':30})}
```

4.1.4 Our Python implementation of matrices

We have defined several different representations of matrices, and will later define still more. It is convenient, however, to define a class Mat, analogous to our vector class Vec, for representing matrices. An instance of Mat will have two fields:

- D, which will be bound to a *pair* (R, C) of sets (unlike Vec, in which D is a single set);
- f, which will be bound to a dictionary representing the function that maps pairs $(r, c) \in R \times C$ to field elements.

We will follow the sparsity convention we used in representing vectors: entries of the matrix whose values are zero need not be represented in the dictionary. Sparsity for matrices is more important than for vectors since matrices tend to be much bigger: a C-vector has $|C|$ entries but an $R \times C$ matrix has $|R| \cdot |C|$ entries.

One key difference between our representations of vectors and matrices is the use of the D field. In a vector, the value of D is a set, and the keys of the dictionary are elements of this set. In a matrix, the value of D is a pair (R, C) of sets, and the keys of the dictionary are elements of the Cartesian product $R \times C$. The reason for this choice is that storing the entire set $R \times C$ would require too much space for large sparse matrices.

The Python code required to define the class Mat is

```
class Mat:
    def __init__(self, labels, function):
        self.D = labels
        self.f = function
```

Once Python has processed this definition, you can create an instance of Mat like so:

```
>>> M=Mat(({'a','b'}, {'@', '#', '?'}), {('a','@'):1, ('a','#'):2,
            ('a','?'):3, ('b','@'):10, ('b','#'):20, ('b','?'):30})
```

As with Vec, the first argument is assigned to the new instance's D field, and the second is assigned to the f field.

As with Vec, we will write procedures to manipulate instances of Mat, and eventually give a more elaborate class definition for Mat, one that allows use of operators such as * and that includes pretty printing as in

```
>>> print(M)
        #  @  ?
      ---------
 a  |   2  1  3
 b  |  20 10 30
```

4.1.5 Identity matrix

Definition 4.1.6: For a finite set D, the $D \times D$ *identity matrix* is the matrix whose row-label set and column-label set are both D, and in which entry (d, d) is a 1 for every $d \in D$ (and all other entries are zero). We denote it by $\mathbb{1}_D$. Usually the set D is clear from the context, and the identity matrix is written $\mathbb{1}$, without the subscript.

For example, here is the `{'a','b','c'}`×`{'a','b','c'}` identity matrix:

```
      a b c
     -------
a  |  1 0 0
b  |  0 1 0
c  |  0 0 1
```

Quiz 4.1.7: Write an expression for the `{'a','b','c'}`×`{'a','b','c'}` identity matrix represented as an instance of `Mat`.

Answer

```
Mat(({'a','b','c'},{'a','b','c'}),{('a','a'):1,('b','b'):1,('c','c'):1})
```

Quiz 4.1.8: Write a one-line procedure `identity(`D`)` that, given a finite set D, returns the $D \times D$ identity matrix represented as an instance of `Mat`.

Answer

```
def identity(D): return Mat((D,D), {(d,d):1 for d in D})
```

4.1.6 Converting between matrix representations

Since we will be using different matrix representations, it is convenient to be able to convert between them.

Quiz 4.1.9: Write a one-line procedure `mat2rowdict(A)` that, given an instance of `Mat`, returns the rowdict representation of the same matrix. Use dictionary comprehensions.

```
>>> mat2rowdict(M)
{'a': Vec({'@', '#', '?'},{'@': 1, '#': 2, '?': 3}),
 'b': Vec({'@', '#', '?'},{'@': 10, '#': 20, '?': 30})}
```

Hint: First write the expression whose value is the row `r` `Vec`; the `F` field's value is defined by a dictionary comprehension. Second, use that expression in a dictionary comprehension in which r is the control variable.

Answer

Assuming `r` is bound to one of `M`'s row-labels, row `r` is the value of the expression

```
Vec(A.D[1],{c:A[r,c] for c in A.D[1]})
```

We want to use this expression as the value corresponding to key `r` in a dictionary comprehension:

```
{r:... for r in A.D[0]}
```

Putting these two expressions together, we define the procedure as follows:

```
def mat2rowdict(A):
 return {r:Vec(A.D[1],{c:A[r,c] for c in A.D[1]}) for r in A.D[0]}
```

Quiz 4.1.10: Write a one-line procedure `mat2coldict(A)` that, given an instance of `Mat`, returns the coldict representation of the same matrix. Use dictionary comprehensions.

```
>>> mat2coldict(M)
{'@': Vec({'a', 'b'},{'a': 1, 'b': 10}),
 '#': Vec({'a', 'b'},{'a': 2, 'b': 20}),
 '?': Vec({'a', 'b'},{'a': 3, 'b': 30})}
```

Answer

```
def mat2coldict(A):
 return {c:Vec(A.D[0],{r:A[r,c] for r in A.D[0]}) for c in A.D[1]}
```

4.1.7 `matutil.py`

The file `matutil.py` is provided. We will be using this module in the future. It contains the procedure `identity(D)` from Quiz 4.1.8 and the conversion procedures from Section 4.1.6. It also contains the procedures `rowdict2mat(rowdict)` and `coldict2mat(coldict)`, which are the inverses, respectively, of `mat2rowdict(A)` and `mat2coldict(A)`. [1] It also contains the procedure `listlist2mat(L)` that, given a list `L` of lists of field elements, returns an instance of `Mat` whose rows correspond to the lists that are elements of `L`. This procedure is convenient for easily creating small example matrices:

```
>>> A=listlist2mat([[10,20,30,40],[50,60,70,80]])
>>> print(A)

        0  1  2  3
      -------------
 0  |  10 20 30 40
 1  |  50 60 70 80
```

4.2 Column space and row space

Matrices serve many roles, but one is a way of packing together vectors. There are two ways of interpreting a matrix as a bunch of vectors: a bunch of columns and a bunch of rows.

Correspondingly, there are two vector spaces associated with a matrix:

Definition 4.2.1: For a matrix M,

- the *column space* of M, written Col M, is the vector space spanned by the columns of M, and
- *row space* of M, written Row M, is the vector space spanned by the rows of M.

Example 4.2.2: The column space of $\begin{bmatrix} 1 & 2 & 3 \\ 10 & 20 & 30 \end{bmatrix}$ is Span $\{[1,10],[2,20],[3,30]\}$. In this case, the column space is equal to Span $\{[1,10]\}$ since $[2,20]$ and $[3,30]$ are scalar multiples of

[1] For each of the procedures `rowdict2mat(rowdict)`, the argument can be either a dictionary of vectors or a list of vectors.

[1, 10].

The row space of the same matrix is Span {[1, 2, 3], [10, 20, 30]}. In this case, the span is equal to Span {[1, 2, 3]} since [10, 20, 30] is a scalar multiple of [1, 2, 3].

We will get a deeper understanding of the significance of the column space and row space in Sections 4.5.1, 4.5.2 and 4.10.6. In Section 4.7, we will learn about one more important vector space associated with a matrix.

4.3 Matrices as vectors

Presently we will describe the operations that make matrices useful. First, we observe that a matrix can be interpreted as a vector. In particular, an $R \times S$ matrix over $\mathbb{F}$ is a function from $R \times S$ to $\mathbb{F}$, so it can be interpreted as an $R \times S$-vector over $\mathbb{F}$. Using this interpretation, we can perform the usual vector operations on matrices, *scalar-vector multiplication* and *vector addition*. Our full implementation of the Mat class will include these operations. (We won't be using dot-product with matrices).

Quiz 4.3.1: Write the procedure mat2vec(M) that, given an instance of Mat, returns the corresponding instance of Vec. As an example, we show the result of applying this procedure to the matrix M given in Example 4.1.3 (Page 151):

```
>>> print(mat2vec(M))
 ('a', '#') ('a', '?') ('a', '@') ('b', '#') ('b', '?') ('b', '@')
--------------------------------------------------------------------
          2          3          1         20         30         10
```

Answer

```
def mat2vec(M):
  return Vec({(r,s) for r in M.D[0] for s in M.D[1]}, M.f)
```

We won't need mat2vec(M) since Mat will include vector operations.

4.4 Transpose

Transposing a matrix means swapping its rows and columns.

Definition 4.4.1: The transpose of an $P \times Q$ matrix, written M^T, is a $Q \times P$ matrix such that $(M^T)_{j,i} = M_{i,j}$ for every $i \in P, j \in Q$.

Quiz 4.4.2: Write the procedure transpose(M) that, given an instance of Mat representing a matrix, returns the representation of the transpose of that matrix.

```
>>> print(transpose(M))
        a  b
      ------
 #  |   2 20
 @  |   1 10
 ?  |   3 30
```

Answer

```
def transpose(M):
    return Mat((M.D[1], M.D[0]), {(q,p):v for (p,q),v in M.F.items()})
```

We say a matrix M is a *symmetric matrix* if $M^T = M$.

Example 4.4.3: The matrix $\begin{bmatrix} 1 & 2 \\ 3 & 4 \end{bmatrix}$ is not symmetric but the matrix $\begin{bmatrix} 1 & 2 \\ 2 & 4 \end{bmatrix}$ is symmetric.

4.5 Matrix-vector and vector-matrix multiplication in terms of linear combinations

What do we do with matrices? Mostly we multiply them by vectors. There are two ways to multiply a matrix by a vector: *matrix-vector* multiplication and *vector-matrix* multiplication. For each, I will give two equivalent definitions of multiplication: one in terms of linear combinations and one in terms of dot-products. The reader needs to absorb all these definitions because different contexts call for different interpretations.

4.5.1 Matrix-vector multiplication in terms of linear combinations

Definition 4.5.1 (*Linear-combinations* definition of matrix-vector multiplication): Let M be an $R \times C$ matrix over $\mathbb{F}$. Let $\boldsymbol{v}$ be a C-vector over $\mathbb{F}$. Then $M * \boldsymbol{v}$ is the linear combination

$$\sum_{c \in C} \boldsymbol{v}[c] \text{ (column } c \text{ of } M)$$

If M is an $R \times C$ matrix but $\boldsymbol{v}$ is not a C-vector then the product $M * \boldsymbol{v}$ is illegal.

In the traditional-matrix case, if M is an $m \times n$ matrix over $\mathbb{F}$ then $M * \boldsymbol{v}$ is legal only if $\boldsymbol{v}$ is an n-vector over $\mathbb{F}$. That is, the number of columns of the matrix must match the number of entries of the vector.

Example 4.5.2: Let's consider an example using a traditional matrix:

$$\begin{bmatrix} 1 & 2 & 3 \\ 10 & 20 & 30 \end{bmatrix} * [7,0,4] = 7\,[1,10] + 0\,[2,20] + 4\,[3,30] = [7,70] + [0,0] + [12,120] = [19,190]$$

Example 4.5.3: What about $\begin{bmatrix} 1 & 2 & 3 \\ 10 & 20 & 30 \end{bmatrix}$ times the vector $[7,0]$? This is illegal: you can't multiply a 2×3 matrix times a 2-vector. The matrix has three columns but the vector has two entries.

Example 4.5.4: Now we do an example with a matrix with more interesting row and column labels:

	@	#	?
a	2	1	3
b	20	10	30

*

@	#	?
0.5	5	-1

=

a	b
3.0	30.0

Example 4.5.5: *Lights Out:* In Example 3.1.9 (Page 118), we saw that a solution to a *Lights Out* puzzle (which buttons to press to turn out the lights) is a linear combination of "button vectors." Now we can write such a linear combination as a matrix-vector product where the columns of the matrix are button vectors.

For example, the linear combination

$$1\ \Box\ +\ 0\ \Box\ +\ 0\ \Box\ +\ 1\ \Box$$

can be written as

$$\left[\ \Box\ \Big|\ \Box\ \Big|\ \Box\ \Big|\ \Box\ \right] * [1,0,0,1]$$

4.5.2   Vector-matrix multiplication in terms of linear combinations

We have seen a definition of matrix-vector multiplication in terms of linear combinations of *columns* of a matrix. We now define vector-matrix multiplication in terms of linear combinations of the *rows* of a matrix.

Definition 4.5.6 (*Linear-combinations* definition of vector-matrix multiplication):
Let M be an $R \times C$ matrix. Let w be an R-vector. Then $w * M$ is the linear combination

$$\sum_{r \in R} w[r]\ (\text{row } r \text{ of } M)$$

If M is an $R \times C$ matrix but w is not an R-vector then the product $w * M$ is illegal.

This is a good moment to point out that matrix-vector multiplication is different from vector-matrix multiplication; in fact, often $M * v$ is a legal product but $v * M$ is not or vice versa. Because we are used to assuming commutativity when we multiply numbers, the noncommutativity of multiplication between matrices and vectors can take some getting used to.

Example 4.5.7:

$$[3,4]\ *\ \begin{bmatrix} 1 & 2 & 3 \\ 10 & 20 & 30 \end{bmatrix} = 3\,[1,2,3]\ +\ 4\,[10,20,30]$$
$$= [3,6,9]\ +\ [40,80,120] = [43,86,129]$$

Example 4.5.8: What about $[3,4,5] * \begin{bmatrix} 1 & 2 & 3 \\ 10 & 20 & 30 \end{bmatrix}$? This is illegal: you can't multiply a 3-vector and a 2×3 matrix. The number of entries of the vector must match the number of rows of the matrix.

Remark 4.5.9: Transpose swaps rows and columns. The rows of M are the columns of M^T. We could therefore define $w * M$ as $M^T * w$. However, implementing it that way would be a mistake—transpose creates a completely new matrix, and, if the matrix is big, it is inefficient to do that just for the sake of computing a vector-matrix product.

Example 4.5.10: In Section 3.1.2, we gave examples of applications of linear combinations. Recall the JunkCo factory data table from Example 3.1.6 (Page 116):

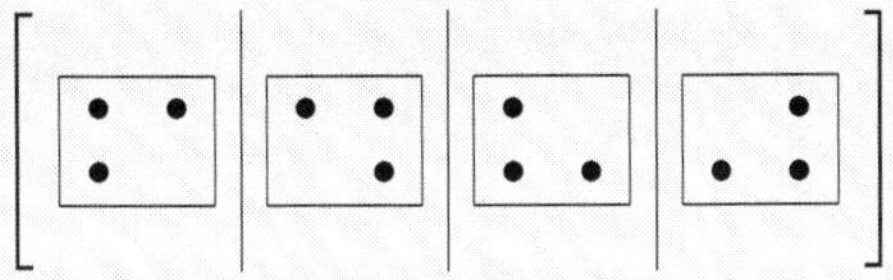

	metal	concrete	plastic	water	electricity
garden gnome	0	1.3	.2	.8	.4
hula hoop	0	0	1.5	.4	.3
slinky	.25	0	0	.2	.7
silly putty	0	0	.3	.7	.5
salad shooter	.15	0	.5	.4	.8

Corresponding to each product is a vector. In Example 3.1.6 (Page 116), we defined the vectors

v_gnome, v_hoop, v_slinky, v_putty, and v_shooter,

each with domain

{'metal','concrete','plastic','water','electricity'}

We can construct a matrix M whose rows are these vectors:

```
>>> rowdict = {'gnome':v_gnome, 'hoop':v_hoop, 'slinky':v_slinky,
               'putty':v_putty, 'shooter':v_shooter}
>>> M = rowdict2mat(rowdict)
>>> print(M)

          plastic metal concrete water electricity
         ------------------------------------------
   putty |     0.3     0        0   0.7         0.5
   gnome |     0.2     0      1.3   0.8         0.4
  slinky |       0  0.25        0   0.2         0.7
    hoop |     1.5     0        0   0.4         0.3
 shooter |     0.5  0.15        0   0.4         0.8
```

In that example, JunkCo decided on quantities $\alpha_{\text{gnome}}, \alpha_{\text{hoop}}, \alpha_{\text{slinky}}, \alpha_{\text{putty}}, \alpha_{\text{shooter}}$ for the products. We saw that the the vector giving the total utilization of each resource, a vector whose domain is {*metal, concrete, plastic, water, electricity*}, is a linear combination of the rows of the table where the coefficient for product p is α_p.

We can obtain the total-utilization vector as a vector-matrix product

$$[\alpha_{\text{gnome}}, \alpha_{\text{hoop}}, \alpha_{\text{slinky}}, \alpha_{\text{putty}}, \alpha_{\text{shooter}}] * M \tag{4.1}$$

Here's how we can compute the total utilization in Python using vector-matrix multiplication. Note the use of the asterisk * as the multiplication operator.

```
>>> R = {'gnome', 'hoop', 'slinky', 'putty', 'shooter'}
>>> u = Vec(R, {'putty':133, 'gnome':240, 'slinky':150, 'hoop':55,
                'shooter':90})
>>> print(u*M)

 plastic metal concrete water electricity
-----------------------------------------
     215    51      312   373         356
```

4.5.3 Formulating *expressing a given vector as a linear-combination* as a matrix-vector equation

We have learned that a linear combination can be expressed as a matrix-vector or vector-matrix product. We now use that idea to reformulate the problem of *expressing a given vector as a linear-combination*.

Example 4.5.11: Recall the *industrial espionage* problem of Section 3.1.4: given the JunkCo factory data table, and given the amount of resources consumed, compute the quantity of the products produced. Let $\boldsymbol{b}$ be the vector of resources consumed. Define $\boldsymbol{x}$ to be a vector variable.

In view of 4.1, we obtain a matrix-vector equation:

$$\boldsymbol{x} * M = \boldsymbol{b}$$

Solving the industrial espionage problem amounts to solving this equation.

Example 4.5.12: In Example 3.1.9 (Page 118), we said that, for a given initial state $\boldsymbol{s}$ of *Lights Out*, the problem of figuring out which buttons to push to turn all lights out could be expressed as the problem of expressing $\boldsymbol{s}$ as a linear combination (over $GF(2)$) of the button vectors. In Example 4.5.5 (Page 156), we further pointed out that the linear combination of button vectors could be written as a matrix-vector product $B * \boldsymbol{x}$ where B is a matrix whose columns are the the button vectors. Thus the problem of finding the correct coefficients can be expressed as the problem of finding a vector $\boldsymbol{x}$ such that $B * \boldsymbol{x} = \boldsymbol{s}$.

Here we give a Python procedure to create a dictionary of button-vectors for $n \times n$ *Lights Out*. Note that we use the value one defined in the module GF2.

```
def button_vectors(n):
 D = {(i,j) for i in range(n) for j in range(n)}
 vecdict={(i,j):Vec(D,dict([((x,j),one) for x in range(max(i-1,0), min(i+2,n))]
                           +[((i,y),one) for y in range(max(j-1,0), min(j+2,n))]))
                           for (i,j) in D}
 return vecdict
```

Entry (i, j) of the returned dictionary is the button-vector corresponding to button (i, j).

Now we can construct the matrix B whose columns are button-vectors for 5×5 *Lights Out*:

```
>>> B = coldict2mat(button_vectors(5))
```

Suppose we want to find out which button vectors to press when the puzzle starts from a particular configuration, e.g. when only the middle light is on. We create a vector $\boldsymbol{s}$ representing that configuration:

```
>>> s = Vec(b.D, {(2,2):one})
```

Now we need to solve the equation $B * \boldsymbol{x} = \boldsymbol{s}$.

4.5.4 Solving a matrix-vector equation

In each of the above examples—and in many more applications—we face the following computational problem.

Computational Problem 4.5.13: *Solving a matrix-vector equation*

- *input:* an $R \times C$ matrix A and an R-vector $\boldsymbol{b}$
- *output:* the C-vector $\hat{\boldsymbol{x}}$ such that $A * \hat{\boldsymbol{x}} = \boldsymbol{b}$

Though we have specified the computational problem as solving an equation of the form $A * \boldsymbol{x} = \boldsymbol{b}$, an algorithm for this problem would also suffice to solve a matrix-vector equation of the form $\boldsymbol{x} * A = \boldsymbol{b}$ since we could apply the algorithm to the transpose A^T of A.

Example 4.5.14: In Example 3.4.13 (Page 130), we considered Span $\{[a, b], [c, d]\}$ where $a, b, c, d \in \mathbb{R}$.

1. We showed that, if $[c, d]$ is not in Span $\{[a, b]\}$ then $ad \neq bc$.
2. If that is the case, we showed that, for every vector $[p, q]$ in $\mathbb{R}^2$, there are coefficients α

and β such that

$$[p, q] = \alpha\, [a, b] + \beta\, [c, d] \tag{4.2}$$

In Part 2, we actually gave formulas for α and β in terms of p, q, a, b, c, d: $\alpha = \frac{dp-cq}{ad-bc}$ and $\beta = \frac{aq-bp}{ad-bc}$.

Note that Equation 4.2 can be rewritten as a matrix vector equation:

$$\begin{bmatrix} a & c \\ b & d \end{bmatrix} * [\alpha, \beta] = [p, q]$$

Thus the formulas for α and β give an algorithm for solving a matrix-vector equation in which the matrix is 2×2 and the second column is not in the span of the first.

For example, to solve the matrix equation $\begin{bmatrix} 1 & 2 \\ 3 & 4 \end{bmatrix} * [\alpha, \beta] = [-1, 1]$, we set $\alpha = \frac{4 \cdot -1 - 2 \cdot 1}{1 \cdot 4 - 2 \cdot 3} = \frac{-6}{-2} = 3$ and $\beta = \frac{1 \cdot 1 - 3 \cdot -1}{1 \cdot 4 - 2 \cdot 3} = \frac{4}{-2} = -2$

In later chapters, we will study algorithms for this computational problem. For now, I have provided a module `solver` that implements these algorithms. It contains a procedure `solve(A, b)` with the following spec:

- *input:* an instance A of `Mat`, and an instance $\boldsymbol{v}$ of `Vec`
- *output:* a vector $\boldsymbol{u}$ such that $A\boldsymbol{u} = \boldsymbol{v}$ (to within some error tolerance) if there is *any* such vector $\boldsymbol{u}$

Note that the output vector might *not* be a solution to the matrix-vector equation. In particular, if there is *no* solution to the matrix-vector equation, the vector returned by `solve(A,b)` is not a solution. You should therefore check each answer `u` you get from `solver(A,b)` by comparing `A*u` to `b`.

Moreover, if the matrix and vector are over $\mathbb{R}$, the calculations use Python's limited-precision arithmetic operations. Even if the equation $A*\boldsymbol{x} = \boldsymbol{b}$ has a solution, the vector `u` returned might not be an exact solution.

Example 4.5.15: We use `solve(A,b)` to solve the industrial espionage problem. Suppose we observe that JunkCo uses 51 units of metal, 312 units of concrete, 215 units of plastic, 373.1 units of water, and 356 units of electricity. We represent these observations by a vector $\boldsymbol{b}$:

```
>>> C = {'metal','concrete','plastic','water','electricity'}
>>> b = Vec(C, {'water':373.1,'concrete':312.0,'plastic':215.4,
                'metal':51.0,'electricity':356.0})
```

We want to solve the vector-matrix equation $\boldsymbol{x} * M = \boldsymbol{b}$ where M is the matrix defined in Example 4.5.10 (Page 157). Since `solve(A,b)` solves a matrix-vector equation, we supply the transpose of M as the first argument A:

```
>>> solution = solve(M.transpose(), b)
>>> print(solution)

 putty gnome slinky hoop shooter
---------------------------------
   133   240    150   55      90
```

Does this vector solve the equation? We can test it by computing the *residual vector* (often called the *residual*):

```
>>> residual = b - solution*M
```

If the solution were exact, the residual would be the zero vector. An easy way to see if the residual is *almost* the zero vector is to calculate the sum of squares of its entries, which is just its dot-product with itself:

```
>>> residual * residual
1.819555009546577e-25
```

About 10^{-25}, so zero for our purposes!

However, we cannot yet truly be confident we have penetrated the secrets of JunkCo. Perhaps the solution we have computed is not the only solution to the equation! More on this topic later.

Example 4.5.16: Continuing with Example 4.5.12 (Page 159), we use `solve(A,b)` to solve 5×5 *Lights Out* starting from a state in which only the middle light is on:

```
>>> s = Vec(b.D, {(2,2):one})
>>> sol = solve(B, s)
```

You can check that this is indeed a solution:

```
>>> B*sol == s
True
```

Here there is no issue of accuracy since elements of $GF(2)$ are represented precisely. Moreover, for this problem we don't care if there are multiple solutions to the equation. This solution tells us one collection of buttons to press:

```
>>> [(i,j) for (i,j) in sol.D if sol[i,j] == one]
[(4,0),(2,2),(4,1),(3,2),(0,4),(1,4),(2,3),(1,0),(0,1),(2,0),(0,2)]
```

4.6 Matrix-vector multiplication in terms of dot-products

We will also define matrix-vector product in terms of dot-products.

4.6.1 Definitions

Definition 4.6.1 (*Dot-Product* Definition of Matrix-Vector Multiplication): If M is an $R \times C$ matrix and $\boldsymbol{u}$ is a C-vector then $M * \boldsymbol{u}$ is the R-vector $\boldsymbol{v}$ such that $\boldsymbol{v}[r]$ is the dot-product of row r of M with $\boldsymbol{u}$.

Example 4.6.2: Consider the matrix-vector product

$$\begin{bmatrix} 1 & 2 \\ 3 & 4 \\ 10 & 0 \end{bmatrix} * [3, -1]$$

The product is a 3-vector. The first entry is the dot-product of the first row, $[1, 2]$, with $[3, -1]$, which is $1 \cdot 3 + 2 \cdot (-1) = 1$. The second entry is the dot-product of the second row, $[3, 4]$, with $[3, -1]$, which is $3 \cdot 3 + 4 \cdot (-1) = 5$. The third entry is $10 \cdot 3 + 0 \cdot (-1) = 30$. Thus the product is $[1, 5, 30]$.

$$\begin{bmatrix} 1 & 2 \\ 3 & 4 \\ 10 & 0 \end{bmatrix} * [3, -1] \quad = \quad [\quad [1,2] \cdot [3,-1], \quad [3,4] \cdot [3,-1], \quad [10,0] \cdot [3,-1] \quad] \quad = \quad [1, 5, 30]$$

Vector-matrix multiplication is defined in terms of dot-products with the columns.

Definition 4.6.3 (*Dot-Product* Definition of Vector-Matrix Multiplication): If M is an $R \times C$ matrix and $\boldsymbol{u}$ is a R-vector then $\boldsymbol{u} * M$ is the C-vector $\boldsymbol{v}$ such that $\boldsymbol{v}[c]$ is the dot-product of $\boldsymbol{u}$ with column c of M.

4.6.2 Example applications

Example 4.6.4: You are given a high-resolution image. You would like a lower-resolution version to put on your web page so the page will load more quickly.

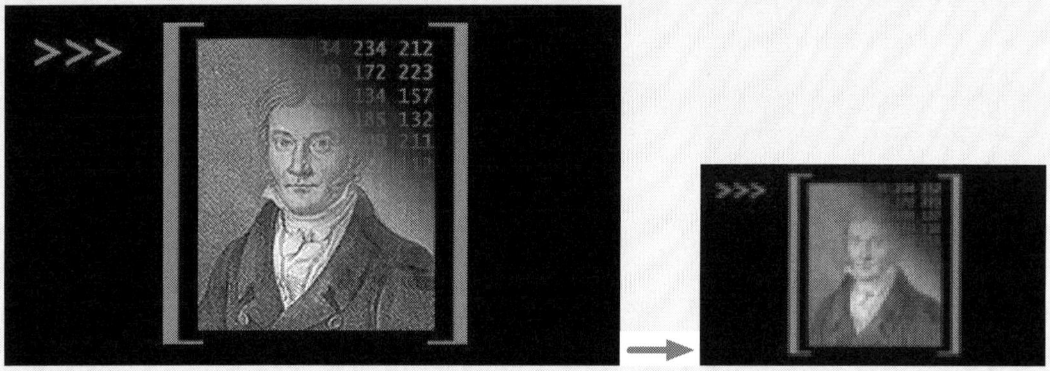

You therefore seek to *downsample* the image.

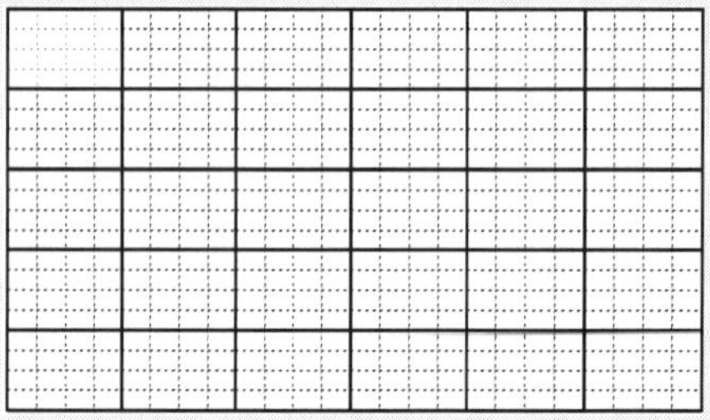

Each pixel of the low-res image (represented as a solid rectangle) corresponds to a little grid of pixels of the high-res image (represented as dotted rectangles). The intensity value of a pixel of the low-res image is the *average* of the intensity values of the corresponding pixels of the high-res image.

Let's represent the high-res image as a vector $\boldsymbol{u}$. We saw in Quiz 2.9.3 that averaging can be expressed as a dot-product. In downsampling, for each pixel of the low-res image to be created, the intensity is computed as the average of a subset of the entries of $\boldsymbol{u}$; this, too, can be expressed as a dot-product. Computing the low-res image thus requires one dot-product for each pixel of that image.

Employing the dot-product definition of matrix-vector multiplication, we can construct a matrix M whose rows are the vectors that must be dotted with $\boldsymbol{u}$. The column-labels of M are the pixel coordinates of the high-res image. The row-labels of M are the pixel coordinates of the low-res image. We write $\boldsymbol{v} = M * \boldsymbol{u}$ where $\boldsymbol{v}$ is a vector representing the low-res image.

Suppose the high-res image has dimensions 3000×2000 and our goal is to create a low-res image with dimensions 750×500. The high-res image is represented by a vector $\boldsymbol{u}$ whose domain is $\{0, 1, \ldots, 2999\} \times \{0, 1, \ldots, 1999\}$ and the low-res image is represented by a vector $\boldsymbol{v}$ whose domain is $\{0, 1, \ldots, 749\} \times \{0, 1, \ldots, 499\}$.

The matrix M has column-label set $\{0, 1, \ldots, 2999\} \times \{0, 1, \ldots, 1999\}$ and row-label set $\{0, 1, \ldots, 749\} \times \{0, 1, \ldots, 499\}$. For each low-res pixel coordinate pair (i, j), the corresponding row of M is the vector that is all zeroes except for the 4×4 grid of high-res pixel coordinates

$$(4i, 4j), (4i, 4j+1), (4i, 4j+2), (4i, 4j+3), (4i+1, 4j), (4i+1, 4j+1), \ldots, (4i+3, 4j+3)$$

where the values are $\frac{1}{16}$.

Here is the Python code to construct the matrix M.

```
D_high = {(i,j) for i in range(3000) for j in range(2000)}
D_low ={(i,j) for i in range(750) for j in range(500)}
M = Mat((D_low, D_high),
    {((i,j), (4*i+m, 4*j+n)):1./16 for m in range(4) for n in range(4)
```

```
for i in range(750) for j in range(500)})
```

However, you would never actually want to create this matrix! I provide the code just for illustration.

Example 4.6.5: You are given an image and a set of pixel-coordinate pairs forming regions in the image, and you wish to produce version of the image in which the regions are blurry.

Perhaps the regions are faces, and you want to blur them to protect the subjects' privacy. Once again, the transformation can be formulated as matrix-vector multiplication $M * v$. (Once again, there is no reason you would actually want to construct the matrix explicitly, but the existence of such a matrix is useful in quickly computing the transformation, as we discuss in Chapter 10.)

This time, the input image and output image have the same dimensions. For each pixel that needs to be blurred, the intensity is computed as an average of the intensities of many nearby pixels. Once again, we use the fact that average can be computed as a dot-product and that matrix-vector multiplication can be interpreted as carrying out many dot-products, one for each row of the matrix.

Averaging treats all nearby pixels equally. This tends to produce undesirable visual artifacts and is not a faithful analogue of the kind of blur we see with our eyes. A *Gaussian* blur more heavily weights very nearby pixels; the weights go down (according to a specific formula) with distance from the center.

Whether blurring is done using simple averaging or weighted averaging, the transformation is an example of a *linear filter*, as mentioned in Section 2.9.3.

Example 4.6.6: As in Section 2.9.3, searching for an audio clip within an audio segment can be formulated as finding many dot-products, one for each of the possible locations of the audio clip or subimage. It is convenient to formulate finding these dot-products as a matrix-vector product.

Supppose we are trying to find the sequence $[0, 1, -1]$ in the longer sequence

$$[0, 0, -1, 2, 3, -1, 0, 1, -1, -1]$$

We need to compute one dot-product for each of the possible positions of the short sequence within the long sequence. The long sequence has ten entries, so there are ten possible positions for the short sequence, hence ten dot-products to compute.

You might think a couple of these positions are not allowed since these positions do not leave enough room for matching all the entries of the short sequence. However, we adopt a *wrap-around* convention: we look for the short sequence starting at the end of the long sequence, and wrapping around to the beginning. It is exactly as if the long sequence were written on a circular strip.

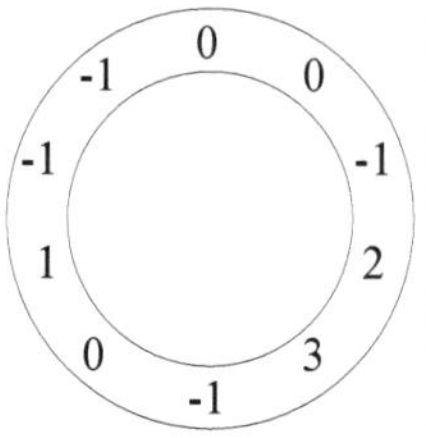

We formulate computing the ten dot-products as a product of a ten-row matrix with the ten-element long sequence:

$$\begin{bmatrix} 0 & 1 & -1 & 0 & 0 & 0 & 0 & 0 & 0 & 0 \\ 0 & 0 & 1 & -1 & 0 & 0 & 0 & 0 & 0 & 0 \\ 0 & 0 & 0 & 1 & -1 & 0 & 0 & 0 & 0 & 0 \\ 0 & 0 & 0 & 0 & 1 & -1 & 0 & 0 & 0 & 0 \\ 0 & 0 & 0 & 0 & 0 & 1 & -1 & 0 & 0 & 0 \\ 0 & 0 & 0 & 0 & 0 & 0 & 1 & -1 & 0 & 0 \\ 0 & 0 & 0 & 0 & 0 & 0 & 0 & 1 & -1 & 0 \\ 0 & 0 & 0 & 0 & 0 & 0 & 0 & 0 & 1 & -1 \\ -1 & 0 & 0 & 0 & 0 & 0 & 0 & 0 & 0 & 1 \\ 1 & -1 & 0 & 0 & 0 & 0 & 0 & 0 & 0 & 0 \end{bmatrix} * [0, 0, -1, 2, 3, -1, 0, 1, -1, -1]$$

The product is the vector $[1, -3, -1, 4, -1, -1, 2, 0, -1, 0]$. The second-biggest dot-product, 2, indeed occurs at the best-matching position, though the biggest dot-product, 5, occurs at a not-so-great match.

Why adopt the wrap-around convention? It allows us to use a remarkable algorithm to compute the matrix-vector product much more quickly than would seem possible. The *Fast Fourier Transform* (FFT) algorithm, described in Chapter 10, makes use of the fact that the matrix has a special form.

4.6.3 Formulating a system of linear equations as a matrix-vector equation

In Section 2.9.2, we defined a linear equation as an equation of the form $\boldsymbol{a} \cdot \boldsymbol{x} = \beta$, and we defined a system of linear equations as a collection of such equations:

$$\begin{aligned} \boldsymbol{a}_1 \cdot \boldsymbol{x} &= \beta_1 \\ \boldsymbol{a}_2 \cdot \boldsymbol{x} &= \beta_2 \\ &\vdots \\ \boldsymbol{a}_m \cdot \boldsymbol{x} &= \beta_m \end{aligned}$$

Using the dot-product definition of matrix-vector multiplication, we can rewrite this system of equations as a single matrix-vector equation. Let A be the matrix whose rows are $\boldsymbol{a}_1, \boldsymbol{a}_2, \ldots, \boldsymbol{a}_m$. Let $\boldsymbol{b}$ be the vector $[\beta_1, \beta_2, \ldots, \beta_m]$. Then the system of linear equations is equivalent to the matrix-vector equation $A * \boldsymbol{x} = \boldsymbol{b}$.

Example 4.6.7: Recall that in Example 2.9.7 (Page 91) we studied current consumption of hardware components in sensor nodes. Define D = {'radio', 'sensor', 'memory', 'CPU'}. Our goal was to compute a D-vector that, for each hardware component, gives the current drawn by that component.

We have five test periods. For $i = 0, 1, 2, 3, 4$, there is a vector $\boldsymbol{duration}_i$ giving the amount of time each hardware component is on during test period i.

```
>>> D = {'radio', 'sensor', 'memory', 'CPU'}
>>> v0 = Vec(D, {'radio':.1, 'CPU':.3})
>>> v1 = Vec(D, {'sensor':.2, 'CPU':.4})
>>> v2 = Vec(D, {'memory':.3, 'CPU':.1})
>>> v3 = Vec(D, {'memory':.5, 'CPU':.4})
```

```
>>> v4 = Vec(D, {'radio':.2, 'CPU':.5})
```

We are trying to compute a D-vector rate such that
v0*rate = 140, v1*rate = 170, v2*rate = 60, v3*rate = 170, and v4*rate = 250
We can formulate this system of equations as a matrix-vector equation:

$$\begin{bmatrix} \text{v0} \\ \hline \text{v1} \\ \hline \text{v2} \\ \hline \text{v3} \\ \hline \text{v4} \end{bmatrix} * [x_0, x_1, x_2, x_3, x_4] = [140, 170, 60, 170, 250]$$

To carry out the computation in Python, we construct the vector

```
>>> b = Vec({0, 1, 2, 3, 4},{0: 140.0, 1: 170.0, 2: 60.0, 3: 170.0, 4: 250.0})
```

and construct a matrix A whose rows are v0, v1, v2, v3, and v4:

```
>>> A = rowdict2mat([v0,v1,v2,v3,v4])
```

Next we solve the matrix-vector equation A*x=b:

```
>>> rate = solve(A, b)
```

obtaining the vector

```
Vec(D, {'radio':500, 'sensor':250, 'memory':100, 'CPU':300})
```

Now that we recognize that systems of linear equations can be formulated as matrix-vector equations, we can reformulate problems and questions involving linear equations as problems involving matrix-vector equations:

- *Solving a linear system* (Computational Problem 2.9.12) becomes *solving a matrix equation* (Computational Problem 4.5.13).
- The question *how many solutions are there to a linear system over* $GF(2)$ (Question 2.9.18), which came up in connection with attacking the authentication scheme (Section 2.9.7), becomes the question *how many solutions are there to a matrix-vector equation over* $GF(2)$.
- Computational Problem 2.9.19, *computing all solutions to a linear system over* $GF(2)$, becomes *computing all solutions to a matrix-vector equation over* $GF(2)$.

4.6.4 Triangular systems and triangular matrices

In Section 2.11, we described an algorithm to solve a triangular system of linear equations. We have just seen that a system of linear equations can be formulated as a matrix-vector equation. Let's see what happens when we start with a triangular system.

Example 4.6.8: Reformulating the triangular system of Example 2.11.1 (Page 105) as a matrix-vector equation, we obtain

$$\begin{bmatrix} 1 & 0.5 & -2 & 4 \\ 0 & 3 & 3 & 2 \\ 0 & 0 & 1 & 5 \\ 0 & 0 & 0 & 2 \end{bmatrix} * \boldsymbol{x} = [-8, 3, -4, 6]$$

Because we started with a triangular system, the resulting matrix has a special form: the first entry of the second row is zero, the first and second entries of the third row are zero, and the first and second and third entries of the fourth row are zero. Since the nonzero entries form a triangle, the matrix itself is called a *triangular matrix*.

Definition 4.6.9: An $n \times n$ *upper-triangular* matrix A is a matrix with the property that $A_{ij} = 0$ for $i > j$.

Note that the entries forming the triangle can be be zero or nonzero.

The definition applies to traditional matrices. To generalize to our matrices with arbitrary row- and column-label sets, we specify orderings of the label-sets.

Definition 4.6.10: Let R and C be finite sets. Let L_R be a list of the elements of R, and let L_C be a list of the elements of C. An $R \times C$ matrix A is *triangular* with respect to L_R and L_C if

$$A[L_R[i], L_C[j]] = 0$$

for $j > i$.

Example 4.6.11: The {a, b c}× {@, #, ?} matrix

	@	#	?
a	0	2	3
b	10	20	30
c	0	35	0

is triangular with respect to [a, b c] and [@, ?, #]. We can see this by reordering the rows and columns according to the list orders:

	@	?	#
b	10	30	20
a	0	3	2
c	0	0	35

To facilitate viewing a matrix with reordered rows and columns, the class Mat will provide a pretty-printing method that takes two arguments, the lists L_R and L_C:

```
>>> A = Mat(({'a','b','c'}, {'#',  '@', '?'}),
...          {('a','#'):2, ('a','?'):3,
...          ('b','@'):10, ('b','#'):20, ('b','?'):30,
...          ('c','#'):35})
>>>
>>> print(A)

        #  ?  @
      ----------
 a  |   2  3  0
 b  |  20 30 10
 c  |  35  0  0

>>> A.pp(['b','a','c'], ['@','?','#'])

        @  ?  #
      ----------
 b  |  10 30 20
 a  |   0  3  2
 c  |   0  0 35
```

Problem 4.6.12: (For the student with knowledge of graph algorithms) Design an algorithm that, for a given matrix, finds a list of a row-labels and a list of column-labels with respect to which the matrix is triangular (or report that no such lists exist).

4.6.5 Algebraic properties of matrix-vector multiplication

We use the dot-product interpretation of matrix-vector multiplication to derive two crucial properties. We will use the first property in the next section, in characterizing the solutions to a matrix-vector equation and in error-correcting codes.

Proposition 4.6.13: Let M be an $R \times C$ matrix.

- For any C-vector $\boldsymbol{v}$ and any scalar α,

$$M * (\alpha\, \boldsymbol{v}) = \alpha\, (M * \boldsymbol{v}) \tag{4.3}$$

- For any C-vectors $\boldsymbol{u}$ and $\boldsymbol{v}$,

$$M * (\boldsymbol{u} + \boldsymbol{v}) = M * \boldsymbol{u} + M * \boldsymbol{v} \tag{4.4}$$

Proof

To show Equation 4.3 holds, we need only show that, for each $r \in R$, entry r of the left-hand side equals entry r of the right-hand side. By the dot-product interpretation of matrix-vector multiplication,

- entry r of the left-hand side equals the dot-product of row r of M with $\alpha\boldsymbol{v}$, and
- entry r of the right-hand side equals α times the dot-product of row r of M with $\boldsymbol{v}$.

These two quantities are equal by the homogeneity of dot-product, Proposition 2.9.22.

The proof of Equation 4.4 is similar; we leave it as an exercise. □

Problem 4.6.14: Prove Equation 4.4.

4.7 Null space

4.7.1 Homogeneous linear systems and matrix equations

In Section 3.6, we introduced homogeneous linear systems, i.e. systems of linear equations in which all right-hand side values were zero. Such a system can of course be formulated as a matrix-vector equation $A * \boldsymbol{x} = \boldsymbol{0}$ where the right-hand side is a zero vector.

Definition 4.7.1: The *null space* of a matrix A is the set $\{\boldsymbol{v} \ : \ A * \boldsymbol{v} = \boldsymbol{0}\}$. It is written Null A.

Since Null A is the solution set of a homogeneous linear system, it is a vector space (Section 3.4.1).

Example 4.7.2: Let $A = \left[\begin{array}{c|c|c} 1 & 4 & 5 \\ 2 & 5 & 7 \\ 3 & 6 & 9 \end{array}\right]$. Since the sum of the first two columns equals the third column, $A * [1, 1, -1]$ is the zero vector. Thus $[1, 1, -1]$ is in Null A. By Equation 4.3, for any scalar α, $A * (\alpha\, [1, 1, -1])$ is also the zero vector, so $\alpha\, [1, 1, -1]$ is also in Null A. For example, $[2, 2, -2]$ is in Null A.

Problem 4.7.3: For each of the given matrices, find a nonzero vector in the null space of the matrix.

1. $\begin{bmatrix} 1 & 0 & 1 \end{bmatrix}$

2. $\begin{bmatrix} 2 & 0 & 0 \\ 0 & 1 & 1 \end{bmatrix}$

3. $\begin{bmatrix} 1 & 0 & 0 \\ 0 & 0 & 0 \\ 0 & 0 & 1 \end{bmatrix}$

Here we make use of Equation 4.4:

Lemma 4.7.4: For any $R \times C$ matrix A and C-vector $\boldsymbol{v}$, a vector $\boldsymbol{z}$ is in the null space of A if and only if $A * (\boldsymbol{v} + \boldsymbol{z}) = A * \boldsymbol{v}$.

Proof

The statement is equivalent to the following statements:

1. if the vector $\boldsymbol{z}$ is in the null space of A then $A * (\boldsymbol{v} + \boldsymbol{z}) = A * \boldsymbol{v}$;
2. if $A * (\boldsymbol{v} + \boldsymbol{z}) = A * \boldsymbol{v}$ then $\boldsymbol{z}$ is in the null space of A.

For simplicity, we prove these two statements separately.

1. Suppose $\boldsymbol{z}$ is in the null space of A. Then

$$A * (\boldsymbol{v} + \boldsymbol{z}) = A * \boldsymbol{v} + A * \boldsymbol{z} = A * \boldsymbol{v} + \mathbf{0} = A * \boldsymbol{v}$$

2. Suppose $A * (\boldsymbol{v} + \boldsymbol{z}) = A * \boldsymbol{v}$. Then

$$\begin{aligned} A * (\boldsymbol{v} + \boldsymbol{z}) &= A * \boldsymbol{v} \\ A * \boldsymbol{v} + A * \boldsymbol{z} &= A * \boldsymbol{v} \\ A * \boldsymbol{z} &= \mathbf{0} \end{aligned}$$

□

4.7.2 The solution space of a matrix-vector equation

In Lemma 3.6.1 (Section 3.6.1), we saw that two solutions to a system of linear equations differ by a vector that solves the corresponding system of homogeneous equations. We restate and reprove the result in terms of matrix-vector equations:

Corollary 4.7.5: Suppose $\boldsymbol{u}_1$ is a solution to the matrix equation $A * \boldsymbol{x} = \boldsymbol{b}$. Then $\boldsymbol{u}_2$ is also a solution if and only if $\boldsymbol{u}_1 - \boldsymbol{u}_2$ belongs to the null space of A.

Proof

Since $A * \boldsymbol{u}_1 = \boldsymbol{b}$, we know that

$A * \boldsymbol{u}_2 = \boldsymbol{b}$ if and only if $A * \boldsymbol{u}_2 = A * \boldsymbol{u}_1$.

Applying Lemma 4.7.4 with $\boldsymbol{v} = \boldsymbol{u}_2$ and $\boldsymbol{z} = \boldsymbol{u}_1 - \boldsymbol{u}_2$, we infer:

$A * \boldsymbol{u}_2 = A * \boldsymbol{u}_1$ if and only if $\boldsymbol{u}_1 - \boldsymbol{u}_2$ is in the null space of A.

Combining these two statements proves the corollary. □

While studying a method for calculating the rate of power consumption for hardware components (Section 2.9.2), we asked about uniqueness of a solution to a system of linear equations. We saw in Corollary 3.6.4 that uniqueness depended on whether the corresponding homogeneous system have only the trivial solution. Here is the same corollary, stated in matrix terminology:

Corollary 4.7.6: Suppose a matrix-vector equation $A\boldsymbol{x} = \boldsymbol{b}$ has a solution. The solution is unique if and only if the null space of A consists solely of the zero vector.

Thus uniqueness of a solution comes down to the following question:

Question 4.7.7: How can we tell if the null space of a matrix consist solely of the zero vector?

This is just a restatement using matrix terminology, of Question 3.6.5, *How can we tell if a homogeneous linear system has only the trivial solution?*

While studying an attack on an authentication scheme in Section 2.9.7, we became interested in counting the solutions to a system of linear equations over $GF(2)$ (Question 2.9.18). In Section 3.6.1 we saw that this was equivalent to counting the solutions to a homogeneous system (Question 3.6.6). Here we restate this problem in terms of matrices:

Question 4.7.8: How can we find the cardinality of the null space of a matrix over $GF(2)$?

4.7.3 Introduction to error-correcting codes

Richard Hamming was getting, he later recalled, "very annoyed." He worked for Bell Laboratories in New Jersey but needed to use a computer located in New York. This was a very early computer, built using electromechanical relays, and it was somewhat unreliable. However, the computer could detect when an error occured, and when it did, it would restart the current computation. After three tries, however, it would go on to the next computation.

Hamming was, in his words, "low man on the totem pole", so he didn't get much use of the computer during the work week. However, nobody else was using it during the weekend. Hamming was allowed to submit a bunch of computations on Friday afternoon; the computer would run them during the weekend, and Hamming would be able to collect the results.

However, he came in one Monday to collect his results, and found that something went wrong, and all the computations failed. He tried again the following weekend—the same thing happened. Peeved, he asked himself: if the computer can detect that its input has an error, why can't it tell me where the error is?

Hamming had long known one solution to this problem: replication. if you are worried about occasional bit errors, write your bit string three times: for each bit position, if the three bit strings differ in that position, choose the bit that occurs twice. However, this solution uses more bits than necessary.

As a result of this experience, Hamming invented *error-correcting codes.* The first code he invented is now called the *Hamming code* and is still used, e.g. in flash memory. He and other researchers subsequently discovered many other error-correcting codes. Error-correcting codes are ubiquitious today; they are used in many kinds of transmission (including WIFI, cell phones, communication with satellites and spacecraft, and digital television) and storage (RAM, disk drives, flash memory, CDs, and DVDs).

The Hamming code is what we now call a *linear binary block code*:

- *linear* because it is based on linear algebra,

- *binary* because the input and output are assumed to be in binary, and
- *block* because the code involves a fixed-length sequence of bits.

The transmission or storage of data is modeled by a *noisy channel*, a tube through which you can push vectors but which sometimes flips bits. A block of bits is represented by a vector over $GF(2)$. A binary block code defines a function $f : GF(2)^m \longrightarrow GF(2)^n$. (In the Hamming code, m is 4 and n is 7.)

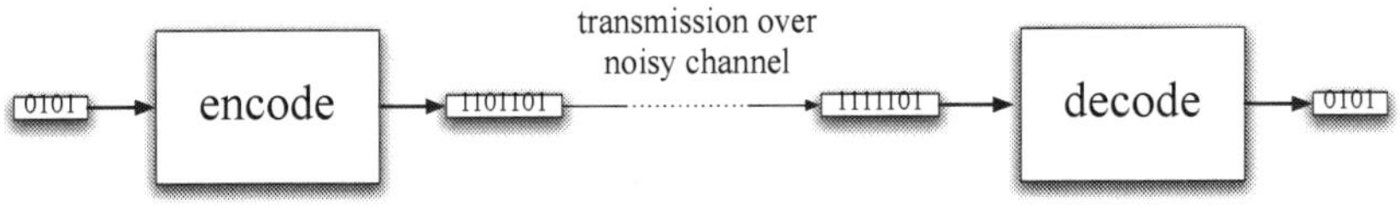

When you have a block of m bits you want to be reliably received at the other end, you first use f to transform it to an n-vector, which you then push through the noisy channel. At the other end of the noisy channel, the recipient gets an n-vector that might differ from the original in some bit positions; the recipient must somehow figure out which bits were changed as the vector passed through the noisy channel.

We denote by $\mathcal{C}$ the set of encodings, the image of f—the set of n-vectors that can be injected into the noisy channel. The vectors of $\mathcal{C}$ are called *codewords*.

4.7.4 Linear codes

Let $\boldsymbol{c}$ denote the codeword injected into the noisy channel, and let $\tilde{\boldsymbol{c}}$ denote the vector (not necessarily a codeword) that comes out the other end. Ordinarily, $\tilde{\boldsymbol{c}}$ differs from $\boldsymbol{c}$ only in a small number of bit positions, the positions in which the noisy channel introduced errors. We write

$$\tilde{\boldsymbol{c}} = \boldsymbol{c} + \boldsymbol{e}$$

where $\boldsymbol{e}$ is the vector with 1's in the error positions. We refer to $\boldsymbol{e}$ as the *error vector*.

The recipient gets $\tilde{\boldsymbol{c}}$ and needs to figure out $\boldsymbol{e}$ in order to figure out $\boldsymbol{c}$. How?

In a linear code, the set $\mathcal{C}$ of codewords is the null space of a matrix H. This simplifies the job of the recipient. Using Equation 4.4, we see

$$H * \tilde{\boldsymbol{c}} = H * (\boldsymbol{c} + \boldsymbol{e}) = H * \boldsymbol{c} + H * \boldsymbol{e} = \boldsymbol{0} + H * \boldsymbol{e} = H * \boldsymbol{e}$$

because $\boldsymbol{c}$ is in the null space of H.

Thus the recipient knows something useful about $\boldsymbol{e}$: she knows $H * \boldsymbol{e}$ (because it is the same as $H * \tilde{\boldsymbol{c}}$, which she can compute). The vector $H * \boldsymbol{e}$ is called the *error syndrome*. If the error syndrome is the zero vector then the recipient assumes that $\boldsymbol{e}$ is all zeroes, i.e. that no error has been introduced. If the error syndrome is a nonzero vector then the recipient knows that an error has occured, i.e. that $\boldsymbol{e}$ is not all zeroes. The recipient needs to figure out $\boldsymbol{e}$ from the vector $H * \boldsymbol{e}$. The method for doing this depends on the particular code being used.

4.7.5 The Hamming Code

In the Hamming code, the codewords are 7-vectors, and

$$H = \begin{bmatrix} 0 & 0 & 0 & 1 & 1 & 1 & 1 \\ 0 & 1 & 1 & 0 & 0 & 1 & 1 \\ 1 & 0 & 1 & 0 & 1 & 0 & 1 \end{bmatrix}$$

Notice anything special about the columns and their order?

Now suppose that the noisy channel introduces at most one bit error. Then $\boldsymbol{e}$ has only one 1. Can you determine the position of the bit error from the matrix-vector product $H * \boldsymbol{e}$?

Example 4.7.9: Suppose $\boldsymbol{e}$ has a 1 in its third position, $\boldsymbol{e} = [0, 0, 1, 0, 0, 0, 0]$. Then $H * \boldsymbol{e}$ is the third column of H, which is $[0, 1, 1]$.

As long as $\boldsymbol{e}$ has at most one bit error, the position of the bit can be determined from $H * \boldsymbol{e}$. This shows that the Hamming code allows the recipient to correct one-bit errors.

Quiz 4.7.10: Suppose $H * e$ is $[1, 1, 0]$. What is e?

Answer

$[0, 0, 0, 0, 0, 1, 0]$.

Quiz 4.7.11: Show that the Hamming code does not allow the recipient to correct two-bit errors: give two different error vectors, e_1 and e_2, each with at most two 1's, such that $H * e_1 = H * e_2$.

Answer

There are many acceptable answers, e.g. $\boldsymbol{e}_1 = [1, 1, 0, 0, 0, 0, 0]$ and $\boldsymbol{e}_2 = [0, 0, 1, 0, 0, 0, 0]$ or $\boldsymbol{e}_1 = [0, 0, 1, 0, 0, 1, 0]$ and $\boldsymbol{e}_2 = [0, 1, 0, 0, 0, 0, 1]$.

Next we show that the Hamming code allows *detection* of errors as long as the number of errors is no more than two. Remember that the recipient assumes that no error has occured if $H * \boldsymbol{e}$ is the zero vector. Is there a way to set exactly two 1's in $\boldsymbol{e}$ so as to achieve $H * \boldsymbol{e} = \mathbf{0}$?

When $\boldsymbol{e}$ has two 1's, $H * \boldsymbol{e}$ is the sum of the two corresponding columns of H. If the sum of two columns is $\mathbf{0}$ then (by $GF(2)$ arithmetic) the two columns must be equal.

Example 4.7.12: Suppose $\boldsymbol{e} = [0, 0, 1, 0, 0, 0, 1]$. Then $H * \boldsymbol{e} = [0, 1, 1] + [1, 1, 1] = [1, 0, 0]$

Note, however, that a two-bit error can get misinterpreted as a one-bit error. In the example, if the recipient assumes at most one error, she will conclude that the error vector is $\boldsymbol{e} = [0, 0, 0, 1, 0, 0, 0]$.

In Lab 4.14, we will implement the Hamming code and try it out.

4.8 Computing sparse matrix-vector product

For computing products of matrices with vectors, we could use the linear-combinations or dot-products definitions but they are not very convenient for exploiting sparsity.

By combining the definition of dot-product with the dot-product definition of matrix-vector multiplication, we obtain the following equivalent definition.

Definition 4.8.1 (*Ordinary* Definition of Matrix-Vector Multiplication:): If M is an $R \times C$ matrix and $\boldsymbol{u}$ is a C-vector then $M * \boldsymbol{u}$ is the R-vector $\boldsymbol{v}$ such that, for each $r \in R$,

$$v[r] = \sum_{c \in C} M[r, c]\boldsymbol{u}[c] \tag{4.5}$$

The most straightforward way to implement matrix-vector multiplication based on this definition is:

1 for i in R:
2 $v[i] := \sum_{j \in C} M[i, j]u[j]$

However, this doesn't take advantage of the fact that many entries of M are zero and do not even appear in our sparse representation of M. We could try implementing the sum in Line 2 in a clever way, omitting those terms corresponding to entries of M that do not appear in our sparse representation. However, our representation does not support doing this efficiently. The more general idea is sound, however: iterate over the entries of M that are actually represented.

The trick is to initialize the output vector $\boldsymbol{v}$ to the zero vector, and then iterate over the nonzero entries of M, adding terms as specified by Equation 4.5.

```
initialize v to zero vector
for each pair (i, j) such that the sparse representation specifies M[i, j],
    v[i] = v[i] + M[i, j]u[j]
```

A similar algorithm can be used to compute a vector-matrix product.

Remark 4.8.2: This algorithm makes no effort to exploit sparsity in the vector. When doing matrix-vector or vector-matrix multiplication, it is not generally worthwhile to try to exploit sparsity in the vector.

Remark 4.8.3: There could be zeroes in the output vector but such zeroes are considered "accidental" and are so rare that it is not worth trying to notice their occurence.

4.9 The matrix meets the function

4.9.1 From matrix to function

For every matrix M, we can use matrix-vector multiplication to define a function $\boldsymbol{x} \mapsto M * \boldsymbol{x}$. The study of the matrix M is in part the study of this function, and vice versa. It is convenient to have a name by which we can refer to this function. There is no traditional name for this function; just in this section, we will refer to it by f_M. Formally, we define it as follows: if M is an $R \times C$ matrix over a field $\mathbb{F}$ then the function $f_M : \mathbb{F}^C \longrightarrow \mathbb{F}^R$ is defined by $f_M(\boldsymbol{x}) = M * \boldsymbol{x}$

This is not a traditional definition in linear algebra; I introduce it here for pedagogical purposes.

Example 4.9.1: Let M be the matrix

	#	@	?
a	1	2	3
b	10	20	30

Then the domain of the function f_M is $\mathbb{R}^{\{\#, @, ?\}}$ and the co-domain is $\mathbb{R}^{\{a, b\}}$. The image, for example, of the vector

#	@	?
2	2	-2

is the vector

a	b
0	0

Problem 4.9.2: Recall that M^T is the transpose of M. The function corresponding to M^T is f_{M^T}

1. What is the domain of f_{M^T}?
2. What is the co-domain?
3. Give a vector in the domain of f_{M^T} whose image is the all-zeroes vector.

4.9.2 From function to matrix

Suppose we have a function $f_M : \mathbb{F}^A \longrightarrow \mathbb{F}^B$ corresponding to some matrix M but we don't happen to know the matrix M. We want to compute the matrix M such that $f_M(\boldsymbol{x}) = M * \boldsymbol{x}$.

Let's first figure out the column-label set for M. Since the domain of f_M is $\mathbb{F}^A$, we know that $\boldsymbol{x}$ is an A-vector. For the product $M * \boldsymbol{x}$ to even be legal, we need the column-label set of M to be A.

Since the co-domain of f_M is $\mathbb{F}^B$, we know that the result of multiplying M by $\boldsymbol{x}$ must be a B-vector. In order for that to be the case, we need the row-label set of M to be B.

So far, so good. We know M must be a $B \times A$ matrix. But what should its entries be? To find them, we use the linear-combinations definition of matrix-vector product.

Remember the standard generators for $\mathbb{F}^A$: for each element $a \in A$, there is a generator $\boldsymbol{e}_a$ that maps a to one and maps every other element of A to zero. By the linear-combinations definition, $M * \boldsymbol{e}_a$ is column a of M. This shows that column a of M must equal $f_M(\boldsymbol{e}_a)$.

4.9.3 Examples of deriving the matrix

In this section, we give some examples illustrating how one derives the matrix from a function, *assuming* that there is some matrix M such that the function is $\boldsymbol{x} \mapsto M * \boldsymbol{x}$. *Warning:* In at least one of these examples, that assumption is not true.

Example 4.9.3: Let $s(\cdot)$ be the function from $\mathbb{R}^2$ to $\mathbb{R}^2$ that scales the x-coordinate by 2.

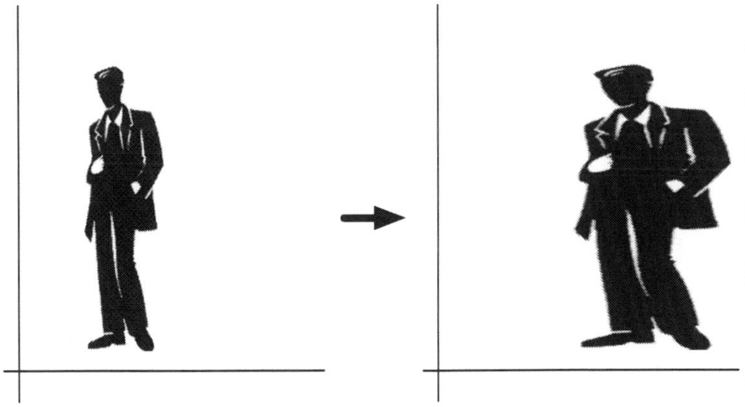

Assume that $s([x, y]) = M * [x, y]$ for some matrix M. The image of $[1, 0]$ is $[2, 0]$ and the image of $[0, 1]$ is $[0, 1]$, so $M = \begin{bmatrix} 2 & 0 \\ 0 & 1 \end{bmatrix}$.

Example 4.9.4: Let $r_{90}(\cdot)$ be the function from $\mathbb{R}^2$ to $\mathbb{R}^2$ that rotates points in 2D by ninety degrees counterclockwise around the origin.

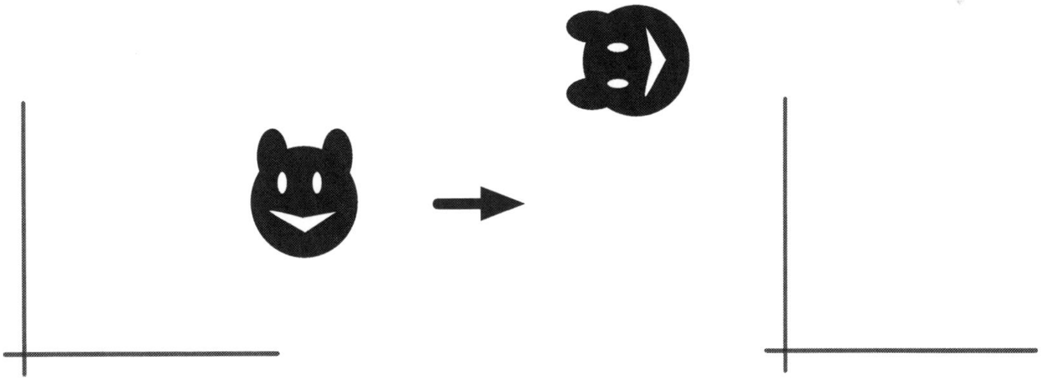

Let's assume for now that $r_{90}([x, y]) = M * [x, y]$ for some matrix M. To find M, we find the image under this function of the two standard generators $[1, 0]$ and $[0, 1]$.

Rotating the point $[1, 0]$ by ninety degrees about the origin yields $[0, 1]$, so this must be the first column of M.

Rotating the point $[0, 1]$ by ninety degrees yields $[-1, 0]$, so this must be the second column of M. Therefore $M = \begin{bmatrix} 0 & -1 \\ 1 & 0 \end{bmatrix}$.

Example 4.9.5: For an angle θ, let $r_\theta(\cdot)$ be the function from $\mathbb{R}^2$ to $\mathbb{R}^2$ that rotates points around the origin counterclockwise by θ. Assume $r_\theta([x, y]) = M * [x, y]$ for some matrix M.

Rotating the point $[1, 0]$ by θ gives us the point $[\cos\theta, \sin\theta]$, which must therefore be the first column of M.

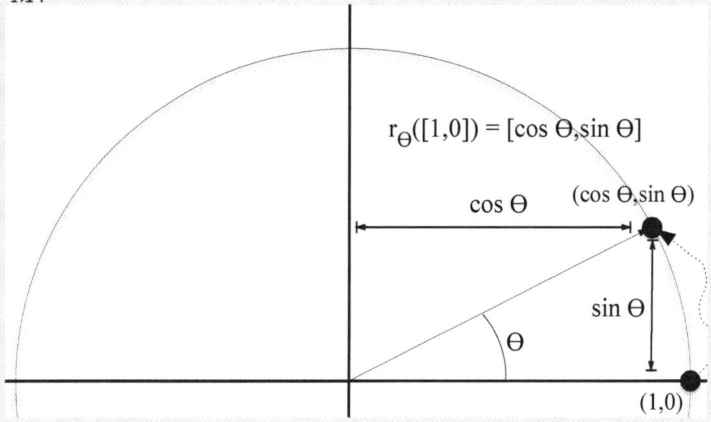

Rotating the point $[0,1]$ by θ gives us the point $[-\sin\theta, \cos\theta]$, so this must be the second column of M.

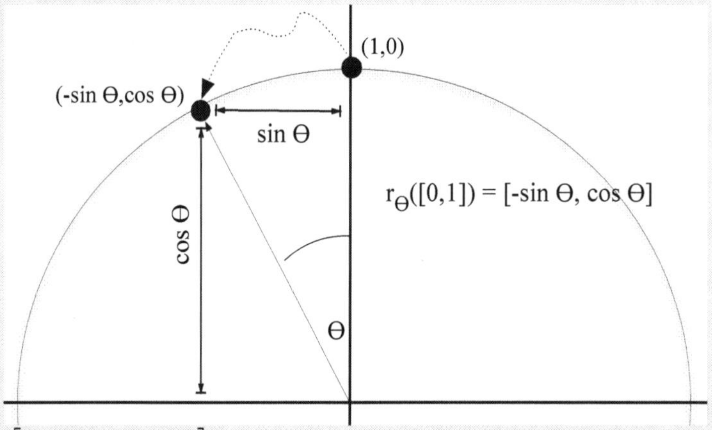

Thus $M = \begin{bmatrix} \cos\theta & -\sin\theta \\ \sin\theta & \cos\theta \end{bmatrix}$.

For example, for rotating by thirty degrees, the matrix is $\begin{bmatrix} \frac{\sqrt{3}}{2} & -\frac{1}{2} \\ \frac{1}{2} & \frac{\sqrt{3}}{2} \end{bmatrix}$. Finally, we have caught up with complex numbers, for which rotation by a given angle is simply multiplication (Section 1.4.10).

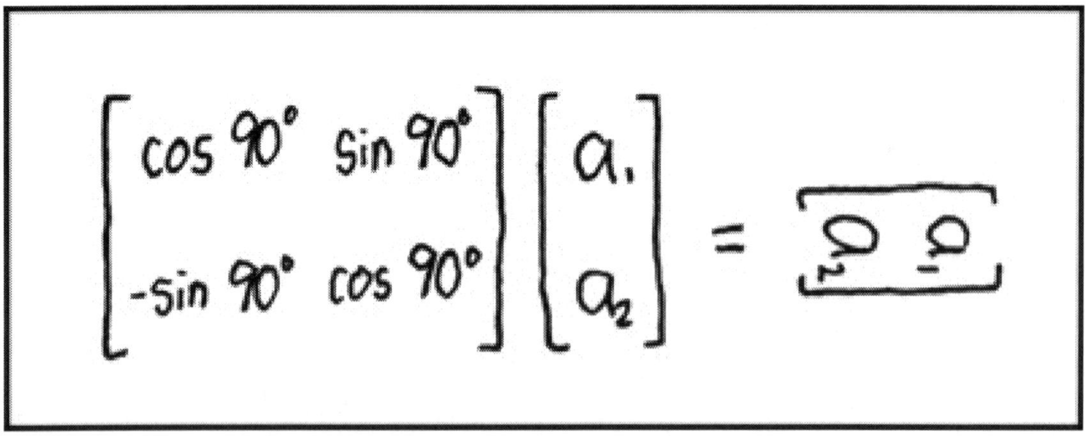

Matrix Transform (http://xkcd.com/824)

Example 4.9.6: Let $t(\cdot)$ be the function from $\mathbb{R}^2$ to $\mathbb{R}^2$ that translates a point one unit to the

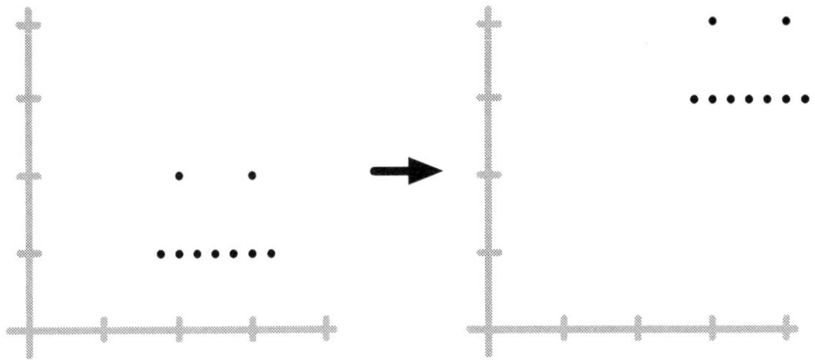

right and two units up.
Assume that $t([x,y]) = M * [x,y]$ for some matrix M. The image of $[1,0]$ is $[2,2]$ and the image of $[0,1]$ is $[1,3]$, so $M = \begin{bmatrix} 2 & 1 \\ 2 & 3 \end{bmatrix}$.

4.10 Linear functions

In each of the examples, we *assumed* that the function could be expressed in terms of matrix-vector multiplication, but this assumption turns out not to be valid in all these examples. How can we tell whether a function can be so expressed?

4.10.1 Which functions can be expressed as a matrix-vector product

In Section 3.4, we identified three properties, Property V1, Property V2, and Property V3, that hold of

- the span of some vectors, and
- the solution set of a homogeneous linear system.

We called any set of vectors satisfying Properties V1, V2, and V3 a *vector space.*

Here we take a similar approach. In Section 4.6.5, we proved two algebraic properties of matrix-vector multiplication. We now use those algebraic properties to define a special kind of function, *linear functions.*

4.10.2 Definition and simple examples

Definition 4.10.1: Let $\mathcal{U}$ and $\mathcal{V}$ be vector spaces over a field $\mathbb{F}$. A function $f : \mathcal{U} \longrightarrow \mathcal{V}$ is called a *linear* function if it satisfies the following two properties:

Property L1: For any vector $\boldsymbol{u}$ in the domain of f and any scalar α in $\mathbb{F}$,

$$f(\alpha\, \boldsymbol{u}) = \alpha\, f(\boldsymbol{u})$$

Property L2: For any two vectors $\boldsymbol{u}$ and $\boldsymbol{v}$ in the domain of f,

$$f(\boldsymbol{u} + \boldsymbol{v}) = f(\boldsymbol{u}) + f(\boldsymbol{v})$$

(A synonym for *linear function* is *linear transformation.*)
Let M be an $R \times C$ matrix over a field $\mathbb{F}$, and define

$$f : \mathbb{F}^C \longrightarrow \mathbb{F}^R$$

by $f(\boldsymbol{x}) = M * \boldsymbol{x}$. The domain and co-domain are vector spaces. By Proposition 4.6.13, the function f satisfies Properties L1 and L2. Thus f is a linear function. We have proved:

Proposition 4.10.2: For any matrix M, the function $\boldsymbol{x} \mapsto M * \boldsymbol{x}$ is a linear function.

Here is a special case.

Lemma 4.10.3: For any C-vector $\boldsymbol{a}$ over $\mathbb{F}$, the function $f : \mathbb{F}^C \longrightarrow \mathbb{F}$ defined by $f(\boldsymbol{x}) = \boldsymbol{a} \cdot \boldsymbol{x}$ is a linear function.

Proof

Let A be the $\{0\} \times C$ matrix whose only row is $\boldsymbol{a}$. Then $f(\boldsymbol{x}) = A * \boldsymbol{x}$, so the lemma follows from Proposition 4.10.2. □

Bilinearity of dot-product Lemma 4.10.3 states that, for any vector $\boldsymbol{w}$, the function $\boldsymbol{x} \mapsto \boldsymbol{w} \cdot \boldsymbol{x}$ is a linear function of $\boldsymbol{x}$. Thus the dot-product function $f(\boldsymbol{x}, \boldsymbol{y}) = \boldsymbol{x} \cdot \boldsymbol{y}$ is linear in its first argument (i.e. if we plug in a vector for the second argument). By the symmetry of the dot-product (Proposition 2.9.21), the dot-product function is also linear in its second argument. We say that the dot-product function is *bilinear* to mean that it is linear in each of its arguments.

Example 4.10.4: Let $\mathbb{F}$ be any field. The function from $\mathbb{F}^2$ to $\mathbb{F}$ defined by $(x, y) \mapsto x + y$ is a linear function. You can prove this using bilinearity of dot-product.

Quiz 4.10.5: Show that the function with domain $\mathbb{R}^2$ defined by $[x, y] \mapsto xy$ is *not* a linear function by giving inputs for which the function violates either Property L1 or Property L2.

Answer

$$f([1,1] + [1,1]) = f([2,2]) = 4$$
$$f([1,1]) + f([1,1]) = 1 + 1$$

Quiz 4.10.6: Show that rotation by ninety degrees, $r_{90}(\cdot)$, is a linear function.

Answer

The scalar-multiplication property, Property L1, is proved as follows:

$$\begin{aligned} \alpha\, f([x, y]) &= \alpha\, [-y, x] \\ &= [-\alpha y, \alpha x] \\ &= f([\alpha x, \alpha y]) \\ &= f(\alpha\, [x, y]) \end{aligned}$$

The vector-addition property, Property L2, is proved similiarly:

$$\begin{aligned} f([x_1, y_1]) + f([x_2, y_2]) &= [-y_1, x_1] + [-y_2, x_2] \\ &= [-(y_1 + y_2), x_1 + x_2] \\ &= f([x_1 + x_2, y_1 + y_2]) \end{aligned}$$

Exercise 4.10.7: Define $g : \mathbb{R}^2 \longrightarrow \mathbb{R}^3$ by $g([x, y]) = [x, y, 1]$. Is g a linear function? If so, prove it. If not, give a counterexample.

Exercise 4.10.8: Define $h : \mathbb{R}^2 \longrightarrow \mathbb{R}^2$ to be the function that takes a point $[x, y]$ to its reflection about the y-axis. Give an explicit (i.e. algebraic) definition of h. Is it a linear function? Explain your answer.

Problem 4.10.9: In at least one of the examples in Section 4.9.3, the function cannot be written as $f(\boldsymbol{x}) = M * \boldsymbol{x}$. Which one? Demonstrate using a numerical example that the function does not satisfy the Properties L1 and L2 that define linear functions.

4.10.3 Linear functions and zero vectors

Lemma 4.10.10: If $f : \mathcal{U} \longrightarrow \mathcal{V}$ is a linear function then f maps the zero vector of $\mathcal{U}$ to the zero vector of $\mathcal{V}$.

Proof

Let $\mathbf{0}$ denote the zero vector of $\mathcal{U}$, and let $\mathbf{0}_{\mathcal{V}}$ denote the zero vector of $\mathcal{V}$.

$$f(\mathbf{0}) = f(\mathbf{0} + \mathbf{0}) = f(\mathbf{0}) + f(\mathbf{0})$$

Subtracting $f(\mathbf{0})$ from both sides, we obtain

$$\mathbf{0}_{\mathcal{V}} = f(\mathbf{0})$$

□

Definition 4.10.11: Analogous to the null space of a matrix (Definition 4.7.1), we define the *kernel* of a linear function f to be $\{\boldsymbol{v} \ : \ f(\boldsymbol{v}) = \mathbf{0}\}$. We denote the kernel of f by Ker f.

Lemma 4.10.12: The kernel of a linear function is a vector space.

Problem 4.10.13: Prove Lemma 4.10.12 by showing that Ker f satisfies Properties V1, V2, and V3 of vector spaces (Section 3.4).

4.10.4 What do linear functions have to do with lines?

Suppose $f : \mathcal{U} \longrightarrow \mathcal{V}$ is a linear function. Let $\boldsymbol{u}_1$ and $\boldsymbol{u}_2$ be two vectors in $\mathcal{U}$, and consider a linear combination $\alpha_1 \boldsymbol{u}_1 + \alpha_2 \boldsymbol{u}_2$ and its image under f.

$$\begin{aligned} f(\alpha_1 \boldsymbol{v}_1 + \alpha_2 \boldsymbol{v}_2) &= f(\alpha_1 \boldsymbol{v}_1) + f(\alpha_2 \boldsymbol{v}_2) && \text{by Property L2} \\ &= \alpha_1 f(\boldsymbol{v}_1) + \alpha_2 f(\boldsymbol{v}_2) && \text{by Property L1} \end{aligned}$$

We interpret this as follows: the image of a linear combination of $\boldsymbol{u}_1$ and $\boldsymbol{u}_2$ is the corresponding linear combination of $f(\boldsymbol{u}_1)$ and $f(\boldsymbol{u}_2)$.

What are the geometric implications?

Let's focus on the case where the domain $\mathcal{U}$ is $\mathbb{R}^n$. The line through the points $\boldsymbol{u}_1$ and $\boldsymbol{u}_2$ is the affine hull of $\boldsymbol{u}_1$ and $\boldsymbol{u}_2$, i.e. the set of all affine combinations:

$$\{\alpha_1 \boldsymbol{u}_1 + \alpha_2 \boldsymbol{u}_2 \ : \ \alpha_1, \alpha_2 \in \mathbb{R}, \alpha_1 + \alpha_2 = 1\}$$

What is the set of images under f of all these affine combinations? It is

$$\{f(\alpha_1 \boldsymbol{u}_1 + \alpha_2 \boldsymbol{u}_2) \ : \ \alpha_1, \alpha_2 \in \mathbb{R}, \alpha_1 + \alpha_2 = 1\}$$

which is equal to

$$\{\alpha_1 f(\boldsymbol{u}_1) + \alpha_2 f(\boldsymbol{u}_2) \;:\; \alpha_1, \alpha_2 \in \mathbb{R}, \alpha_1 + \alpha_2 = 1\}$$

which is the set of all affine combinations of $f(\boldsymbol{u}_1)$ and $f(\boldsymbol{u}_2)$.

This shows:

The image under f of the line through $\boldsymbol{u}_1$ and $\boldsymbol{u}_2$ is the "line" through $f(\boldsymbol{u}_1)$ and $f(\boldsymbol{u}_2)$.

The reason for the scare-quotes is that f might map $\boldsymbol{u}_1$ and $\boldsymbol{u}_2$ to the same point! The set of affine combinations of two identical points is the set consisting just of that one point.

The argument we have given about the image of a linear combination can of course be extended to handle a linear combination of more than two vectors.

Proposition 4.10.14: For a linear function f, for any vectors $\boldsymbol{u}_1, \ldots, \boldsymbol{u}_n$ in the domain of f and any scalars $\alpha_1, \ldots, \alpha_n$,

$$f(\alpha_1 \boldsymbol{u}_1 + \cdots + \alpha_n \boldsymbol{u}_n) = \alpha_1 f(\boldsymbol{u}_1) + \cdots + \alpha_n f(\boldsymbol{u}_n)$$

Therefore the image under a linear function of any flat is another flat.

4.10.5 Linear functions that are one-to-one

Using the notion of kernel, we can give a nice criterion for whether a linear function is one-to-one.

Lemma 4.10.15 (One-to-One Lemma): A linear function is one-to-one if and only if its kernel is a trivial vector space.

Proof

Let $f : \mathcal{V} \longrightarrow \mathcal{W}$ be a linear function. We prove two directions.

Suppose Ker f contains some nonzero vector $\boldsymbol{v}$, so $f(\boldsymbol{v}) = \boldsymbol{0}_{\mathcal{V}}$. By Lemma 4.10.10, $f(\boldsymbol{0}) = \boldsymbol{0}_{\mathcal{V}}$ as well, so f is not one-to-one.

Suppose Ker $f = \{\boldsymbol{0}\}$. Let $\boldsymbol{v}_1, \boldsymbol{v}_2$ be any vectors such that $f(\boldsymbol{v}_1) = f(\boldsymbol{v}_2)$. Then $f(\boldsymbol{v}_1) - f(\boldsymbol{v}_2) = \boldsymbol{0}_{\mathcal{V}}$ so, by linearity, $f(\boldsymbol{v}_1 - \boldsymbol{v}_2) = \boldsymbol{0}_{\mathcal{V}}$, so $\boldsymbol{v}_1 - \boldsymbol{v}_2 \in$ Ker f. Since Ker f consists solely of $\boldsymbol{0}$, it follows that $\boldsymbol{v}_1 - \boldsymbol{v}_2 = \boldsymbol{0}$, so $\boldsymbol{v}_1 = \boldsymbol{v}_2$. □

This simple lemma gives us a fresh perspective on the question of uniqueness of solution to a linear system. Consider the function $f(\boldsymbol{x}) = A * \boldsymbol{x}$. Solving a linear system $A * \boldsymbol{x} = \boldsymbol{b}$ can be interpreted as finding a pre-image of $\boldsymbol{b}$ under f. If a pre-image exists, it is guaranteed to be unique if f is one-to-one.

4.10.6 Linear functions that are onto?

The One-to-One Lemma gives us a nice criterion for determining whether a linear function is one-to-one. What about onto?

Recall that the *image* of a function f with domain $\mathcal{V}$ is the set $\{f(v) \;:\; v \in \mathcal{V}\}$. Recall that a function f being onto means that the image of the function equals the co-domain.

Question 4.10.16: How can we tell if a linear function is onto?

When $f : \mathcal{V} \longrightarrow \mathcal{W}$ is a linear function, we denote the image of f by Im f. Thus asking whether f is onto is asking whether Im $f = \mathcal{W}$.

Example 4.10.17: (Solvability of *Lights Out*) Can 3×3 *Lights Out* be solved from any initial configuration? (Question 2.8.5).

As we saw in Example 4.5.5 (Page 156), we can use a matrix to address *Lights Out*. We construct a matrix M whose columns are the button vectors:

$M =$ 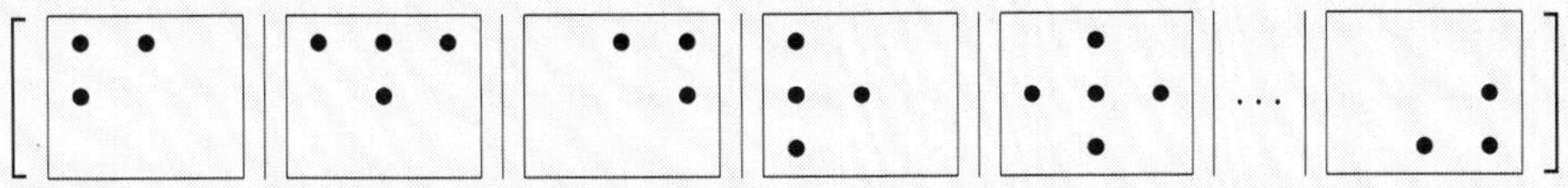

The set of solvable initial configurations (those from which it is possible to turn out all lights) is the set of all linear combinations of these columns, the column space of the matrix. We saw in Example 3.2.14 (Page 123) that, in the case of 2×2 *Lights Out*, every initial configuration is solvable in this case. What about 3×3 *Lights Out*?

Let $D = \{(0,0), \ldots, (2,2)\}$. Let $f : GF(2)^D \longrightarrow GF(2)^D$ be defined by $f(\boldsymbol{x}) = M * \boldsymbol{x}$. The set of solvable initial configurations is the image of f. The set of all initial configurations is the co-domain of f. Therefore, the question of whether every position is solvable is exactly the question of whether f is onto.

We can make one step towards answering Question 4.10.16.

Lemma 4.10.18: The image of a linear function is a subspace of the function's co-domain.

Proof

Let $f : \mathcal{V} \longrightarrow \mathcal{W}$ be a linear function. Clearly Im f is a subset of $\mathcal{W}$. To show that Im f is a subspace of $\mathcal{W}$, we must show that Im f satisfies Properties V1, V2, and V3 of a vector space.

- *V1:* We saw in Lemma 4.10.10 that f maps the zero vector of $\mathcal{V}$ to the zero vector of $\mathcal{W}$, so the zero vector of $\mathcal{W}$ belongs to Im f.
- *V2:* Let $\boldsymbol{w}$ be a vector in Im f. By definition of Im f, there must be a vector $\boldsymbol{v}$ in $\mathcal{V}$ such that $f(\boldsymbol{v}) = \boldsymbol{w}$. For any scalar α,

  $$\alpha\, \boldsymbol{w} = \alpha\, f(\boldsymbol{v}) = f(\alpha\, \boldsymbol{v})$$

  so $\alpha\, \boldsymbol{w}$ is in Im f.
- *V3:* Let $\boldsymbol{w}_1$ and $\boldsymbol{w}_2$ be vectors in Im f. By definition of Im f, there must be vectors $\boldsymbol{v}_1$ and $\boldsymbol{v}_2$ in $\mathcal{V}$ such that $f(\boldsymbol{v}_1) = \boldsymbol{w}_1$ and $f(\boldsymbol{v}_2) = \boldsymbol{w}_2$. By Property L1 of linear functions, $\boldsymbol{w}_1 + \boldsymbol{w}_2 = f(\boldsymbol{v}_1) + f(\boldsymbol{v}_2) = f(\boldsymbol{v}_1 + \boldsymbol{v}_2)$, so $\boldsymbol{w}_1 + \boldsymbol{w}_2$ is in Im f.

□

The complete answer to Question 4.10.16 must wait until Chapter 6.

4.10.7 A linear function from $\mathbb{F}^C$ to $\mathbb{F}^R$ can be represented by a matrix

Suppose $f : \mathbb{F}^C \longrightarrow \mathbb{F}^R$ is a linear function. We can use the method of Section 4.9.2 to obtain a matrix M: for each $c \in C$, column c of M is the image under f of the standard generator $\boldsymbol{e}_c$.

How do we know the resulting matrix M satisfies $f(\boldsymbol{x}) = M * \boldsymbol{x}$? Linearity! For any vector $\boldsymbol{x} \in \mathbb{F}^C$, for each $c \in C$, let α_c be the value of entry c of $\boldsymbol{x}$. Then $\boldsymbol{x} = \sum_{c \in C} \alpha_c\, \boldsymbol{e}_c$. Because f is linear, $f(\boldsymbol{x}) = \sum_{c \in C} \alpha_c f(\boldsymbol{e}_c)$.

On the other hand, by the linear-combinations definition of matrix-vector multiplication, $M * \boldsymbol{x}$ is the linear combination of M's columns where the coefficients are the scalars α_c (for $c \in C$). We defined M to be the matrix whose column c is $f(\boldsymbol{e}_c)$ for each $c \in C$, so $M\boldsymbol{x}$ also equals $\sum_{c \in C} \alpha_c f(\boldsymbol{e}_c)$. This shows that $f(\boldsymbol{x}) = M * \boldsymbol{x}$ for every vector $\boldsymbol{x} \in \mathbb{F}^C$.

We summarize this result in a lemma.

Lemma 4.10.19: If $f : \mathbb{F}^C \longrightarrow \mathbb{F}^R$ is a linear function then there is an $R \times C$ matrix M over $\mathbb{F}$ such that $f(\boldsymbol{x}) = M * \boldsymbol{x}$ for every vector $\boldsymbol{x} \in \mathbb{F}^C$.

4.10.8 Diagonal matrices

Let $d_1, \ldots, d_n$ be real numbers. Let $f : \mathbb{R}^n \longrightarrow \mathbb{R}^n$ be the function such that $f([x_1, \ldots, x_n]) = [d_1 x_1, \ldots, d_n x_n]$. The matrix corresponding to this function is

$$\begin{bmatrix} d_1 & & \\ & \ddots & \\ & & d_n \end{bmatrix}$$

Such a matrix is called a *diagonal* matrix because the only entries allowed to be nonzero form a diagonal.

Definition 4.10.20: For a domain D, a $D \times D$ matrix M is a *diagonal* matrix if $M[r, c] = 0$ for every pair $r, c \in D$ such that $r \neq c$.

Diagonal matrices are very important in Chapters 11 and 12.

Quiz 4.10.21: Write a procedure `diag(D, entries)` with the following spec:

- *input:* a set `D` and a dictionary `entries` mapping `D` to elements of a field
- *output:* a diagonal matrix such that entry `(d,d)` is `entries[d]`

Answer

```
def diag(D, entries):
 return Mat((D,D), {(d,d):entries[d] for d in D})
```

A particularly simple and useful diagonal matrix is the *identity* matrix, defined in Section 4.1.5. For example, here is the $\{\text{a}, \text{b}, \text{c}\} \times \{\text{a}, \text{b}, \text{c}\}$ identity matrix:

```
       a b c
      -------
 a  |  1 0 0
 b  |  0 1 0
 c  |  0 0 1
```

Recall that we refer to it as $\mathbb{1}_D$ or just $\mathbb{1}$.

Why is it called the identity matrix? Consider the function $f : \mathbb{F}^D \longrightarrow \mathbb{F}^D$ defined by $f(\boldsymbol{x}) = \mathbb{1} * \boldsymbol{x}$. Since $\mathbb{1} * \boldsymbol{x} = \boldsymbol{x}$, the function f is the identity function on $\mathbb{F}^D$.

4.11 Matrix-matrix multiplication

We can also multiply a pair of matrices. Suppose A is an $R \times S$ matrix and B is an $S \times T$ matrix. Then it is legal to multiply A times B, and the result is a $R \times T$ matrix. The traditional way of writing "A times B" is simply AB, with no operator in between the matrices.

(In our `Mat` class implementing matrices, however, we will use the $*$ operator to signify matrix-matrix multiplication.)

Note that the product AB is different from the product BA, and in fact one product might be legal while the other is legal. Matrix multiplication is *not* commutative.

4.11.1 Matrix-matrix multiplication in terms of matrix-vector and vector-matrix multiplication

We give two equivalent definitions of matrix-matrix multiplication, one in terms of matrix-vector multiplication and one in terms of vector-matrix multiplication.

Definition 4.11.1 (*Vector-matrix* definition of matrix-matrix multiplication): For each row-label r of A,

$$\text{row } r \text{ of } AB = (\text{row } r \text{ of } A) * B \tag{4.6}$$

Example 4.11.2: Here is a matrix A that differs only slightly from the 3×3 identity matrix:

$$A = \begin{bmatrix} 1 & 0 & 0 \\ 2 & 1 & 0 \\ 0 & 0 & 1 \end{bmatrix}$$

Consider the product AB where B is a $3 \times n$ matrix. In order to use the vector-matrix definition of matrix-matrix multiplication, we think of A as consisting of three rows:

$$A = \begin{bmatrix} 1 & 0 & 0 \\ \hline 2 & 1 & 0 \\ \hline 0 & 0 & 1 \end{bmatrix}$$

Row i of the matrix-matrix product AB is the vector-matrix product

$$(\text{row } i \text{ of } A) * B$$

This product, according to the linear-combinations definition of vector-matrix multiplication, is the linear combination of the rows of B in which the coefficients are the entries of row i of A.

Writing B in terms of its rows,

$$B = \begin{bmatrix} \boldsymbol{b}_1 \\ \hline \boldsymbol{b}_2 \\ \hline \boldsymbol{b}_3 \end{bmatrix}$$

Then

$$\begin{aligned} \text{row 1 of } AB &= 1\,\boldsymbol{b}_1 + 0\,\boldsymbol{b}_2 + 0\,\boldsymbol{b}_3 &= \boldsymbol{b}_1 \\ \text{row 2 of } AB &= 2\,\boldsymbol{b}_1 + 1\,\boldsymbol{b}_2 + 0\,\boldsymbol{b}_3 &= 2\boldsymbol{b}_1 + \boldsymbol{b}_2 \\ \text{row 3 of } AB &= 0\,\boldsymbol{b}_1 + 0\,\boldsymbol{b}_2 + 1\,\boldsymbol{b}_3 &= \boldsymbol{b}_3 \end{aligned}$$

The effect of left-multiplication by A is adding twice row 1 to row 2.

Matrix A in Example 4.11.2 (Page 181) is an *elementary row-addition matrix*, a matrix that is an identity matrix plus at most one off-diagonal nonzero entry. Multiplying on the left by an elementary row-addition matrix adds a multiple of one row to another. We make use of these matrices in an algorithm in Chapter 7.

Definition 4.11.3 (*Matrix-vector* definition of matrix-matrix multiplication): For each column-label s of B,

$$\text{column } s \text{ of } AB = A * (\text{column } s \text{ of } B) \tag{4.7}$$

Example 4.11.4: Let $A = \begin{bmatrix} 1 & 2 \\ -1 & 1 \end{bmatrix}$ and let B be the matrix whose columns are $[4, 3]$,

$[2,1]$, and $[0,-1]$, i.e.

$$B = \left[\begin{array}{c|c|c} 4 & 2 & 0 \\ 3 & 1 & -1 \end{array}\right]$$

Now AB is the matrix whose column i is the result of multiplying A by column i of B. Since $A * [4,3] = [10,-1]$, $A * [2,1] = [4,-1]$, and $A * [0,-1] = [-2,-1]$,

$$AB = \left[\begin{array}{c|c|c} 10 & 4 & -2 \\ -1 & -1 & -1 \end{array}\right]$$

Example 4.11.5: The matrix from Example 4.9.5 (Page 173) that rotates points in $\mathbb{R}^2$ by thirty degrees is

$$A = \begin{bmatrix} \cos\theta & -\sin\theta \\ \sin\theta & \cos\theta \end{bmatrix} = \begin{bmatrix} \frac{\sqrt{3}}{2} & -\frac{1}{2} \\ \frac{1}{2} & \frac{\sqrt{3}}{2} \end{bmatrix}$$

We form a matrix B whose columns are the points in $\mathbb{R}^2$ belonging to the list L in Task 2.3.2:

$$B = \left[\begin{array}{c|c|c|c|c|c|c|c|c} 2 & 3 & 1.75 & 2 & 2.25 & 2.5 & 2.75 & 3 & 3.25 \\ 2 & 2 & 1 & 1 & 1 & 1 & 1 & 1 & 1 \end{array}\right]$$

Now AB is the matrix whose column i is the result of multiplying column i of B on the left by A, i.e. rotating the ith point in L by thirty degrees.

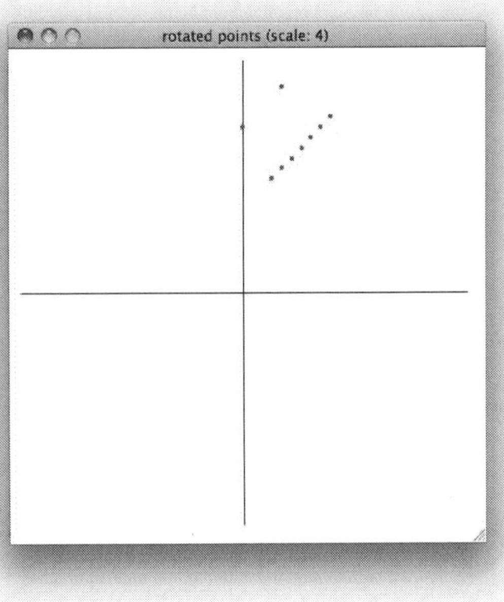

Example 4.11.6: In Example 3.2.11 (Page 121), I state equations showing that the "old" vectors $[3,0,0]$, $[0,2,0]$, and $[0,0,1]$ can each be written as a linear combination of "new" vectors $[1,0,0]$, $[1,1,0]$, and $[1,1,1]$:

$$[3,0,0] = 3\,[1,0,0] + 0,\ [1,1,0] + 0\,[1,1,1]$$
$$[0,2,0] = -2\,[1,0,0] + 2\,[1,1,0] + 0\,[1,1,1]$$
$$[0,0,1] = 0\,[1,0,0] - 1\,[1,1,0] + 1\,[1,1,1]$$

We rewrite these equations using the the linear-combinations definition of matrix-vector multiplication:

$$[3,0,0] = \begin{bmatrix} 1 & 1 & 1 \\ 0 & 1 & 1 \\ 0 & 0 & 1 \end{bmatrix} * [3,0,0]$$

$$[0,2,0] = \begin{bmatrix} 1 & 1 & 1 \\ 0 & 1 & 1 \\ 0 & 0 & 1 \end{bmatrix} * [-2,2,0]$$

$$[0,0,1] = \begin{bmatrix} 1 & 1 & 1 \\ 0 & 1 & 1 \\ 0 & 0 & 1 \end{bmatrix} * [0,-1,1]$$

We combine these three equations to form one equation, using the matrix-vector definition of matrix-matrix multiplication:

$$\begin{bmatrix} 3 & 0 & 0 \\ 0 & 2 & 0 \\ 0 & 0 & 1 \end{bmatrix} = \begin{bmatrix} 1 & 1 & 1 \\ 0 & 1 & 1 \\ 0 & 0 & 1 \end{bmatrix} \begin{bmatrix} 3 & -2 & 0 \\ 0 & 2 & -1 \\ 0 & 0 & 1 \end{bmatrix}$$

The matrix-vector and vector-matrix definitions suggest that matrix-matrix multiplication exists simply as a convenient notation for a collection of matrix-vector products or vector-matrix products. However, matrix-matrix multiplication has a deeper meaning, which we discuss in Section 4.11.3.

Meanwhile, note that by combining a definition of matrix-matrix multiplication with a definition of matrix-vector or vector-matrix multiplication, you can get finer-grained definitions of matrix-matrix multiplication. For example, by combining the matrix-vector definition of matrix-matrix multiplication with the dot-product definition of matrix-vector multiplication, we get:

Definition 4.11.7 (*Dot-product* definition of matrix-matrix multiplication): Entry rc of AB is the dot-product of row r of A with column c of B.

In Problem 4.17.19, you will work with United Nations (UN) voting data. You will build a matrix A each row of which is the voting record of a different country in the UN. You will use matrix-matrix multiplication to find the dot-product of the voting records of every pair of countries. Using these data, you can find out which pairs of countries are in greatest disagreement.

You will find that matrix-matrix multiplication can take quite a long time! Researchers have discovered faster algorithms for matrix-matrix multiplication that are especially helpful when the matrices are very large and dense and roughly square. (Strassen's algorithm was the first and is still the most practical.)

There is an even faster algorithm to compute all the dot-products *approximately* (but accurately enough to find the top few pairs of countries in greatest disagreement).

4.11.2 Graphs, incidence matrices, and counting paths

In the movie *Good Will Hunting*, at the end of class the professor announces,

"I also put an advanced Fourier system on the main hallway chalkboard. I'm hoping that one of you might prove it by the end of the semester. Now the person to do so will not only be in my good graces but will also go on to fame and fortune, having their accomplishment recorded and their name printed in the auspicious MIT Tech. Former winners include Nobel laureates, Fields Medal winners, renowned astrophysicists, and lowly MIT professors."

Must be a tough problem, eh?

The hero, Will Hunting, who works as a janitor at MIT, sees the problem and surreptitiously writes down the solution.

The class is abuzz over the weekend—who is the mysterious student who cracked this problem?

The problem has nothing to do with Fourier systems. It has to do with representing and manipulating a *graph* using a matrix.

Graphs

Informally, a *graph* has points, called *vertices* or *nodes*, and links, called *edges*. Here is a diagram of the graph appearing in Will's problem, but keep in mind that the graph doesn't specify the geometric positions of the nodes or edges—just which edges connect which nodes.

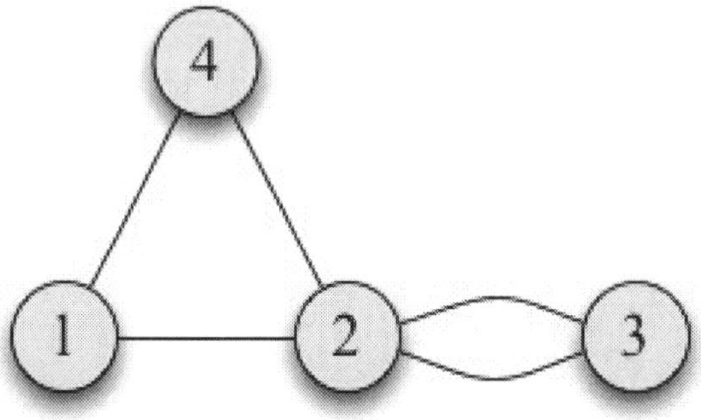

The nodes of this graph are labeled 1, 2, 3, and 4. There are two edges with endpoints 2 and 3, one edge with endpoints 1 and 2, and so on.

Adjacency matrix

The first part of Will's problem is to find the adjacency matrix of the graph. The adjacency matrix A of a graph G is the $D \times D$ matrix where D is the set of node labels. In Will's graph, $D = \{1, 2, 3, 4\}$. For any pair i, j of nodes, $A[i, j]$ is the number of edges with endpoints i and j. Therefore the adjacency matrix of Will's graph is

	1	2	3	4
1	0	1	0	1
2	1	0	2	1
3	0	2	0	0
4	1	1	0	0

Note that this matrix is symmetric. This reflects the fact that if an edge has endpoints i and j then it has endpoints j and i. Much later we discuss *directed* graphs, for which things are a bit more complicated.

Note also that the diagonal elements of the matrix are zeroes. This reflects the fact that Will's graph has no *self-loops*. A self-loop is an edge whose two endpoints are the same node.

Walks

The second part of the problem addresses *walks* in the graph. A walk is a sequence of alternating nodes and edges

$$v_0 \; e_0 \; v_1 \; e_1 \; \cdots \; e_{k-1} \; v_k$$

in which each edge is immediately between its two endpoints.

Here is a diagram of Will's graph with the edges labeled:

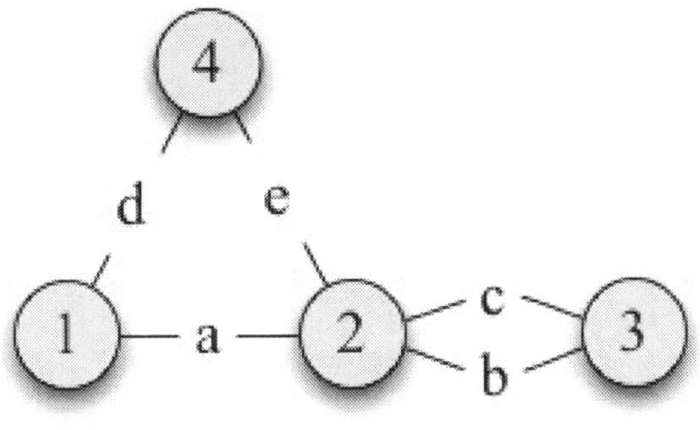

and here is the same diagram with the walk 3 c 2 e 4 e 2 shown:

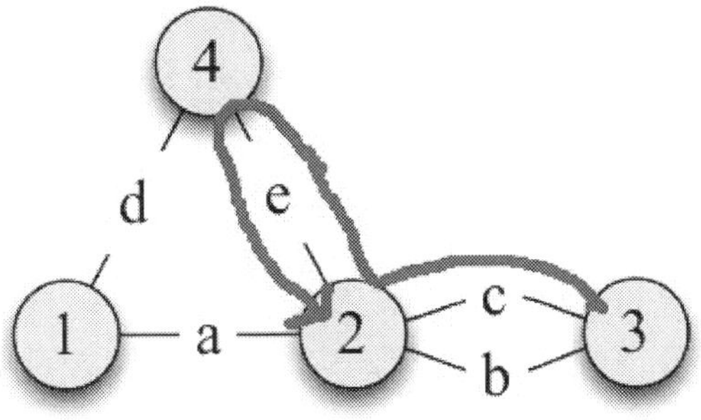

Note that a walk can use an edge multiple times. Also, a walk need not visit all the nodes of a graph. (A walk that visits all nodes is a traveling-salesperson tour; finding the shortest such tour is a famous example of a computationally difficult problem.)

Will's problem concerns three-step walks, by which is meant walks consisting of three edges. For example, here are all the three-step walks from node 3 to node 2:

3 c 2 e 4 e 2, 3 b 2 e 4 e 2, 3 c 2 c 3 c 2, 3 c 2 c 3 b 2, 3 c 2 b 3 c 2,
3 c 2 b 3 b 2, 3 b 2 c 3 c 2, 3 b 2 c 3 b 2, 3 b 2 b 3 c 2, 3 b 2 b 3 b 2

for a total of ten. Or, wait.... Did I miss any?

Computing the number of walks

Matrix-matrix multiplication can be used to compute, for every pair i, j of nodes, the number of two-step walks from i to j, the number of three-steps from i to j, and so on.

First, note that the adjacency matrix A itself encodes the number of one-step walks. For each pair i, j of nodes, $A[i, j]$ is the number of edges with endpoints i and j, and therefore the number of one-step walks from i-to-j.

What about two-step walks? A two-step walk from i to j consists of a one-step walk from i to some node k, followed by a one-step walk from k to j. Thus the number of two-step walks from i to j equals

	number of one-step walks from i to 1	$\times$	number of one-step walks from 1 to j
+	number of one-step walks from i to 2	$\times$	number of one-step walks from 2 to j
+	number of one-step walks from i to 3	$\times$	number of one-step walks from 3 to j
+	number of one-step walks from i to 4	$\times$	number of one-step walks from 4 to j

This has the form of a dot-product! Row i of A is a vector $\boldsymbol{u}$ such that $\boldsymbol{u}[k]$ is the number of one-step walks from i to k. Column j of A is a vector $\boldsymbol{v}$ such that $\boldsymbol{v}[k]$ is the number of one-step walks from k to j. Therefore the dot-product of row i with column j is the number of two-step walks from i to j. By the dot-product definition of matrix-matrix multiplication, therefore, the product AA encodes the number of two-step walks.

```
>>> D = {1,2,3,4}
```

```
>>> A = Mat((D,D), {(1,2):1, (1,4):1, (2,1):1, (2,3):2, (2,4):1, (3,2):2,
                    (4,1):1, (4,2):1})
>>> print(A*A)

       1 2 3 4
     ---------
 1  |  2 1 2 1
 2  |  1 6 0 1
 3  |  2 0 4 2
 4  |  1 1 2 2
```

Now we consider three-step walks. A three-step walk from i to j consists of a two-step walk from i to some node k, followed by a one-step walk from k to j. Thus the number of three-step walks from i to j equals

	number of two-step walks from i to 1	$\times$	number of one-step walks from 1 to j
$+$	number of two-step walks from i to 2	$\times$	number of one-step walks from 2 to j
$+$	number of two-step walks from i to 3	$\times$	number of one-step walks from 3 to j
$+$	number of two-step walks from i to 4	$\times$	number of one-step walks from 4 to j

We already know that AA gives the number of two-step walks. Again using the dot-product definition of matrix-matrix multiplication, the product $(AA)A$ gives the number of three-step walks:

```
>>> print((A*A)*A)

       1  2  3 4
     -----------
 1  |  2  7  2 3
 2  |  7  2 12 7
 3  |  2 12  0 2
 4  |  3  7  2 2
```

Oops, there are not ten but twelve three-step walks from 3 to 2. I missed the walks 3 c 2 a 1 a 2 and 3 b 2 a 1 a 2. Anyway, we're half the way towards solving Will's problem. The problems about generating functions are not much harder; they make use of polynomials (Chapter 10) and determinants (Chapter 12). We will not be able to cover generating functions in this book, but rest assured that they are not beyond your ability and are quite elegant.

Why (apart from undying fame) would you want to compute the number of k-step walks between pairs of nodes in a graph? These numbers can serve as a crude way to measure how closely coupled a pair of nodes are in a graph modeling a social network, although there are assuredly better and faster ways to do so.

4.11.3 Matrix-matrix multiplication and function composition

The matrices A and B define functions via matrix-vector multiplication: $f_A(\boldsymbol{y}) = A * \boldsymbol{y}$ and $f_B(\boldsymbol{x}) = B * \boldsymbol{x}$. Naturally, the matrix AB resulting from multiplying the two also defines a function $f_{AB}(\boldsymbol{x}) = (AB) * \boldsymbol{x}$. There's something interesting about this function:

Lemma 4.11.8 (Matrix-Multiplication Lemma): $f_{AB} = f_A \circ f_B$

Proof

For notational convenience, we assume traditional row- and column-labels.

Write B in terms of columns.

$$B = \begin{bmatrix} \Bigg| & & \Bigg| \\ \boldsymbol{b}_1 & \cdots & \boldsymbol{b}_n \\ \Bigg| & & \Bigg| \end{bmatrix}$$

By the matrix-vector definition of matrix-matrix multiplication, column j of AB is $A *$ (column j of B).

For any n-vector $\boldsymbol{x} = [x_1, \ldots, x_n]$,

$$\begin{aligned} f_B(\boldsymbol{x}) &= B * \boldsymbol{x} && \text{by definition of } f_B \\ &= x_1\boldsymbol{b}_1 + \cdots + x_n\boldsymbol{b}_n && \text{by the linear combinations def. of matrix-vector multiplication} \end{aligned}$$

Therefore

$$\begin{aligned} f_A(f_B(\boldsymbol{x})) &= f_A(x_1\boldsymbol{b}_1 + \cdots x_n\boldsymbol{b}_n) \\ &= x_1(f_A(\boldsymbol{b}_1)) + \cdots + x_n(f_A(\boldsymbol{b}_n)) && \text{by linearity of } f_A \\ &= x_1(A\boldsymbol{b}_1) + \cdots + x_n(A\boldsymbol{b}_n) && \text{by definition of } f_A \\ &= x_1(\text{column 1 of } AB) + \cdots + x_n(\text{column } n \text{ of } AB) && \text{by matrix-vector def. of} AB \\ &= (AB) * \boldsymbol{x} && \text{by linear-comb. def. of} \\ & && \text{matrix-vector multiplication} \\ &= f_{AB}(\boldsymbol{x}) && \text{by definition of } f_{AB} \end{aligned}$$

□

Example 4.11.9: Because function composition is not commutative, it should not be surprising that matrix-matrix multiplication is also not commutative. Consider the functions $f([x_1, x_2]) = [x_1 + x_2, x_2]$ and $g([x_1, x_2]) = [x_1, x_1 + x_2]$. These correspond to the matrices

$$A = \begin{bmatrix} 1 & 1 \\ 0 & 1 \end{bmatrix} \text{ and } B = \begin{bmatrix} 1 & 0 \\ 1 & 1 \end{bmatrix}$$

which are both elementary row-addition matrices. Let's look at the composition of the functions

$$\begin{aligned} f \circ g([x_1, x_2]) = f([x_1, x_1 + x_2]) &= [2x_1 + x_2, x_1 + x_2] \\ g \circ f([x_1, x_2]) = g([x_1 + x_2, x_2]) &= [x_1 + x_2, x_1 + 2x_2] \end{aligned}$$

The corresponding matrix-matrix products are:

$$\begin{aligned} AB &= \begin{bmatrix} 1 & 1 \\ 0 & 1 \end{bmatrix} \begin{bmatrix} 1 & 0 \\ 1 & 1 \end{bmatrix} = \begin{bmatrix} 2 & 1 \\ 1 & 1 \end{bmatrix} \\ BA &= \begin{bmatrix} 1 & 0 \\ 1 & 1 \end{bmatrix} \begin{bmatrix} 1 & 1 \\ 0 & 1 \end{bmatrix} = \begin{bmatrix} 1 & 1 \\ 1 & 2 \end{bmatrix} \end{aligned}$$

illustrating that matrix-matrix multiplication is not commutative.

Example 4.11.10: However, often the product of specific matrices does not depend on their order. The matrices A and B of Example 4.11.9 (Page 187) are both elementary row-addition matrices, but one adds row 1 to row 2 and the other adds row 2 to row 1

Consider instead three elementary row-addition matrices, each of which adds a multiple of row 1 to another row.

- The matrix $B = \begin{bmatrix} 1 & 0 & 0 & 0 \\ 2 & 1 & 0 & 0 \\ 0 & 0 & 1 & 0 \\ 0 & 0 & 0 & 1 \end{bmatrix}$ adds two times row 1 to row 2.
- The matrix $C = \begin{bmatrix} 1 & 0 & 0 & 0 \\ 0 & 1 & 0 & 0 \\ 3 & 0 & 1 & 0 \\ 0 & 0 & 0 & 1 \end{bmatrix}$ adds three times row 1 to row 3.
- The matrix $D = \begin{bmatrix} 1 & 0 & 0 & 0 \\ 2 & 1 & 0 & 0 \\ 0 & 0 & 1 & 0 \\ 4 & 0 & 0 & 1 \end{bmatrix}$ adds four times row 1 to row 4.

Consider the matrix $A = \begin{bmatrix} 1 & 0 & 0 & 0 \\ 2 & 1 & 0 & 0 \\ 3 & 0 & 1 & 0 \\ 4 & 0 & 0 & 1 \end{bmatrix}$ that performs all these additions at once. The additions can be performed one at a time in any order, suggesting that the product of B, C, and D is always A, regardless of order.

Because function composition is associative, the Matrix-Multiplication Lemma (Lemma 4.11.8) implies the following: corollary.

Corollary 4.11.11: Matrix-matrix multiplication is associative.

Example 4.11.12:

$$\begin{bmatrix} 1 & 0 \\ 1 & 1 \end{bmatrix}\left(\begin{bmatrix} 1 & 1 \\ 0 & 1 \end{bmatrix}\begin{bmatrix} -1 & 3 \\ 1 & 2 \end{bmatrix}\right) = \begin{bmatrix} 1 & 0 \\ 1 & 1 \end{bmatrix}\begin{bmatrix} 0 & 5 \\ 1 & 2 \end{bmatrix} = \begin{bmatrix} 0 & 5 \\ 1 & 7 \end{bmatrix}$$

$$\left(\begin{bmatrix} 1 & 0 \\ 1 & 1 \end{bmatrix}\begin{bmatrix} 1 & 1 \\ 0 & 1 \end{bmatrix}\right)\begin{bmatrix} -1 & 3 \\ 1 & 2 \end{bmatrix} = \begin{bmatrix} 1 & 1 \\ 1 & 2 \end{bmatrix}\begin{bmatrix} -1 & 3 \\ 1 & 2 \end{bmatrix} = \begin{bmatrix} 0 & 5 \\ 1 & 7 \end{bmatrix}$$

Example 4.11.13: Recall from Section 4.11.2 that, for a graph G and its adjacency matrix A, the product $(AA)A$ gives, for each pair i, j of nodes, the number of three-step walks from i to j. The same reasoning shows that$((AA)A)A$ gives the number of four-step walks, and so on. Because matrix-matrix multiplication is associative, the parentheses don't matter, and we can write this product as $AAAA$.

The k-fold product of a matrix A with itself,

$$\underbrace{AA\cdots A}_{k \text{ times}}$$

is written A^k, and spoken "A to the power of k".

4.11.4 Transpose of matrix-matrix product

Transpose interacts in a predictable way with matrix-matrix multiplication.

Proposition 4.11.14: For matrices A and B,

$$(AB)^T = B^T A^T$$

Example 4.11.15:

$$\begin{bmatrix} 1 & 2 \\ 3 & 4 \end{bmatrix} \begin{bmatrix} 5 & 0 \\ 1 & 2 \end{bmatrix} = \begin{bmatrix} 7 & 4 \\ 19 & 8 \end{bmatrix}$$

$$\begin{aligned} \begin{bmatrix} 5 & 0 \\ 1 & 2 \end{bmatrix}^T \begin{bmatrix} 1 & 2 \\ 3 & 4 \end{bmatrix}^T &= \begin{bmatrix} 5 & 1 \\ 0 & 2 \end{bmatrix} \begin{bmatrix} 1 & 3 \\ 2 & 4 \end{bmatrix} \\ &= \begin{bmatrix} 7 & 19 \\ 4 & 8 \end{bmatrix} \end{aligned}$$

Proof

Suppose A is an $R \times S$ matrix and B is an $S \times T$ matrix. Then, by the dot-product definition of matrix-matrix multiplication, for every $r \in R$ and $t \in T$,

$$\begin{aligned} \text{entry } t, r \text{ of } (AB)^T &= \text{entry } r, t \text{ of } AB &&= (\text{row } r \text{ of } A) \cdot (\text{column } t \text{ of } B) \\ \text{entry } t, r \text{ of } B^T A^T & &&= (\text{row } t \text{ of } B^T) \cdot (\text{column } r \text{ of } A^T) \\ & &&= (\text{column } t \text{ of } B) \cdot (\text{row } r \text{ of } A) \end{aligned}$$

Finally, unlike matrix-matrix multiplication, vector dot-product *is* commutative, so

$$(\text{row } r \text{ of } A) \cdot (\text{column } t \text{ of } B) = (\text{column } t \text{ of } B) \cdot (\text{row } r \text{ of } A)$$

□

Note that the order of multiplication is reversed. You might expect the equation "$(AB)^T = A^T B^T$" to be true instead but row- and column-labels can show that this equation doesn't make sense. Suppose A is an $R \times S$ matrix and B is an $S \times T$ matrix:

- Since the column-label set of A matches the row-label set of B, the product AB is defined.
- However, A^T is an $S \times R$ matrix and B is a $T \times S$ matrix, so the column-label set of A^T does *not* match the row-label set of B^T, so $A^T B^T$ is *not* defined.
- On the other hand, the column-label set of B^T *does* match the row-label set of A^T, so $B^T A^T$ *is* defined.

Example 4.11.16: This example shows that, even if both $B^T A^T$ and $A^T B^T$ are legal, only the former equals $(AB)^T$.

$$\begin{aligned} \begin{bmatrix} 1 & 2 \\ 3 & 4 \end{bmatrix}^T \begin{bmatrix} 5 & 0 \\ 1 & 2 \end{bmatrix}^T &= \begin{bmatrix} 1 & 3 \\ 2 & 4 \end{bmatrix} \begin{bmatrix} 5 & 1 \\ 0 & 2 \end{bmatrix} \\ &= \begin{bmatrix} 5 & 7 \\ 10 & 10 \end{bmatrix} \end{aligned}$$

Compare this result to Example 4.11.15 (Page 189).

4.11.5 Column vector and row vector

Column Vector An $m \times 1$ matrix is called a *column vector* because it acts like a vector when multiplied from its left. Consider the matrix-matrix product

$$\begin{bmatrix} & & \\ & M & \\ & & \end{bmatrix} \begin{bmatrix} u_1 \\ \vdots \\ u_n \end{bmatrix}$$

By the matrix-vector definition of matrix-matrix multiplication, the result of multiplying M by a matrix with only one column $\boldsymbol{u}$ is in turn a matrix with only one column:

$$\left[\begin{array}{c} \\ M \\ \\ \end{array}\right]\left[\begin{array}{c} u_1 \\ \vdots \\ u_n \end{array}\right] = \left[\begin{array}{c} v_1 \\ \vdots \\ v_m \end{array}\right] \tag{4.8}$$

By interpreting $\left[\begin{array}{c} u_1 \\ \vdots \\ u_n \end{array}\right]$ as a vector $\boldsymbol{u}$ and interpreting $\left[\begin{array}{c} v_1 \\ \vdots \\ v_m \end{array}\right]$ as a vector $\boldsymbol{v}$, we reinterpret Equation (4.8) as the matrix-vector equation $M * \boldsymbol{u} = \boldsymbol{v}$.

Row vector There is an alternative way to interpret a vector as a matrix: a matrix with only one row. Such a matrix is called a *row vector*. Multiplying a row vector on its right by a matrix M acts like vector-matrix multiplication.

$$\left[\begin{array}{ccc} v_1 & \dots & v_m \end{array}\right]\left[\begin{array}{c} \\ M \\ \\ \end{array}\right] = \left[\begin{array}{ccc} u_1 & \cdots & u_n \end{array}\right]$$

4.11.6 Every vector is interpreted as a column vector

The convention in linear algebra when writing expressions involving matrices and vectors is to interpret every vector as a column vector. Thus a matrix-vector product is traditionally written $M\boldsymbol{v}$.

The corresponding notation for a vector-matrix product is $\boldsymbol{v}^T M$. This might seem odd—it doesn't make sense to take the transpose of a vector—but remember that $\boldsymbol{v}$ is to be interpreted as a one-column matrix, and the transpose of a one-column matrix is a one-row matrix.

The reason for interpreting a vector as a column vector instead of as a row vector is that matrix-vector multiplication is more common than vector-matrix multiplication.

Example 4.11.17: According to the column-vector convention, the matrix-vector product $\left[\begin{array}{ccc} 1 & 2 & 3 \\ 10 & 20 & 30 \end{array}\right] * [7, 0, 4]$ from Example 4.5.2 (Page 156) is written as

$$\left[\begin{array}{ccc} 1 & 2 & 3 \\ 10 & 20 & 30 \end{array}\right]\left[\begin{array}{c} 7 \\ 0 \\ 4 \end{array}\right]$$

Example 4.11.18: According to the column-vector convention, the vector-matrix product $[3, 4] * \left[\begin{array}{ccc} 1 & 2 & 3 \\ 10 & 20 & 30 \end{array}\right]$ is written as

$$\left[\begin{array}{c} 3 \\ 4 \end{array}\right]^T \left[\begin{array}{ccc} 1 & 2 & 3 \\ 10 & 20 & 30 \end{array}\right]$$

By interpreting vectors as matrices in expressions, we can make use of associativity of matrix-matrix multiplication. We will see this used, for example, in the next section but also in algorithms for solving matrix-vector equations, in Chapters 7 and 9.

From now on, we will employ the vector-as-column-vector convention, so will eschew the use of * in matrix-vector or vector-matrix multiplication—except in Python code, where we use * for matrix-vector, vector-matrix, and matrix-matrix multiplication.

4.11.7 Linear combinations of linear combinations revisited

In Example 3.2.11 (Page 121), I expressed each of three "old" vectors [3, 0, 0], [0, 2, 0], and [0, 0, 1] as a linear combination of three "new" vectors [1, 0, 0], [1, 1, 0], and [1, 1, 1]. I then illustrated that a linear combination of the old vectors could be transformed into a linear combination of the new vectors. In this section, we show that this transformation is a consequence of associativity of matrix-matrix multiplication.

As we saw in Example 4.11.6 (Page 182), we can use a matrix equation to express the statement that each of the old vectors is a linear combination of the new vectors:

$$\left[\begin{array}{c|c|c} 3 & 0 & 0 \\ 0 & 2 & 0 \\ 0 & 0 & 1 \end{array}\right] = \left[\begin{array}{c|c|c} 1 & 1 & 1 \\ 0 & 1 & 1 \\ 0 & 0 & 1 \end{array}\right] \left[\begin{array}{c|c|c} 3 & -2 & 0 \\ 0 & 2 & -1 \\ 0 & 0 & 1 \end{array}\right] \tag{4.9}$$

We express a linear combination of the old vectors as a matrix-vector product:

$$\begin{bmatrix} x \\ y \\ z \end{bmatrix} = \left[\begin{array}{c|c|c} 3 & 0 & 0 \\ 0 & 2 & 0 \\ 0 & 0 & 1 \end{array}\right] \begin{bmatrix} x/3 \\ y/2 \\ z \end{bmatrix}$$

Now we use Equation 4.9 to substitute for the matrix:

$$\begin{bmatrix} x \\ y \\ z \end{bmatrix} = \left(\left[\begin{array}{c|c|c} 1 & 1 & 1 \\ 0 & 1 & 1 \\ 0 & 0 & 1 \end{array}\right] \begin{bmatrix} 3 & -2 & 0 \\ 0 & 2 & -1 \\ 0 & 0 & 1 \end{bmatrix} \right) \begin{bmatrix} x/3 \\ y/2 \\ z \end{bmatrix}$$

By associativity, we can rewrite this as

$$\begin{bmatrix} x \\ y \\ z \end{bmatrix} = \left[\begin{array}{c|c|c} 1 & 1 & 1 \\ 0 & 1 & 1 \\ 0 & 0 & 1 \end{array}\right] \left(\begin{bmatrix} 3 & -2 & 0 \\ 0 & 2 & -1 \\ 0 & 0 & 1 \end{bmatrix} \begin{bmatrix} x/3 \\ y/2 \\ z \end{bmatrix} \right)$$

We simplify the expression in parentheses, $\begin{bmatrix} 3 & -2 & 0 \\ 0 & 2 & -1 \\ 0 & 0 & 1 \end{bmatrix} \begin{bmatrix} x/3 \\ y/2 \\ z \end{bmatrix}$, obtaining $\begin{bmatrix} x-y \\ y-z \\ z \end{bmatrix}$, and substitute:

$$\begin{bmatrix} x \\ y \\ z \end{bmatrix} = \left[\begin{array}{c|c|c} 1 & 1 & 1 \\ 0 & 1 & 1 \\ 0 & 0 & 1 \end{array}\right] \begin{bmatrix} x-y \\ y-z \\ z \end{bmatrix}$$

which shows that $[x, y, z]$ can be written as a linear combination of the new vectors.

As this example illustrates, the fact that linear combinations of linear combinations are linear combinations is a consequence of the associativity of matrix-matrix multiplication.

4.12 Inner product and outer product

Now that we can interpret vectors as matrices, we will see what happens when we multiply together two such vectors masquerading as matrices. There are two ways to do this.

4.12.1 Inner product

Let $\boldsymbol{u}$ and $\boldsymbol{v}$ be two D-vectors. Consider the "matrix-matrix" product $\boldsymbol{u}^T\boldsymbol{v}$. The first "matrix" has one row and the second matrix has one column. By the dot-product definition of matrix-matrix multiplication, the product consists of a single entry whose value is $\boldsymbol{u} \cdot \boldsymbol{v}$.

Example 4.12.1:

$$\begin{bmatrix} 1 & 2 & 3 \end{bmatrix} \begin{bmatrix} 3 \\ 2 \\ 1 \end{bmatrix} = \begin{bmatrix} 10 \end{bmatrix}$$

For this reason, the dot-product of $\boldsymbol{u}$ and $\boldsymbol{v}$ is often written $\boldsymbol{u}^T\boldsymbol{v}$. This product is often called an *inner product*. However, the term "inner product" has taken on another, related meaning, which we will discuss in Chapter 8.

4.12.2 Outer product

Now let $\boldsymbol{u}$ and $\boldsymbol{v}$ be any vectors (not necessarily sharing the same domain), and consider $\boldsymbol{u}\boldsymbol{v}^T$. For each element s of the domain of $\boldsymbol{u}$ and each element t of the domain of $\boldsymbol{v}$, the s,t element of $\boldsymbol{u}\boldsymbol{v}^T$ is $\boldsymbol{u}[s]\ \boldsymbol{v}[t]$.

Example 4.12.2:

$$\begin{bmatrix} u_1 \\ u_2 \\ u_3 \end{bmatrix} \begin{bmatrix} v_1 & v_2 & v_3 & v_4 \end{bmatrix} = \begin{bmatrix} u_1v_1 & u_1v_2 & u_1v_3 & u_1v_4 \\ u_2v_1 & u_2v_2 & u_2v_3 & u_2v_4 \\ u_3v_1 & u_3v_2 & u_3v_3 & u_3v_4 \end{bmatrix}$$

This kind of product is called the *outer product* of vectors $\boldsymbol{u}$ and $\boldsymbol{v}$.

4.13 From function inverse to matrix inverse

Let us return to the idea of a matrix defining a function. Recall that a matrix M gives rise to a function $f(\boldsymbol{x}) = M\boldsymbol{x}$. We study the case in which f has a functional inverse. Recall that g is the functional inverse of f if $f \circ g$ and $g \circ f$ are the identity functions on their domains.

4.13.1 The inverse of a linear function is linear

Lemma 4.13.1: If f is a linear function and g is its inverse then g is also a linear function.

Proof

We need to prove that

1. for every pair of vectors $\boldsymbol{y}_1, \boldsymbol{y}_2$ in the domain of g, $g(\boldsymbol{y}_1 + \boldsymbol{y}_2) = g(\boldsymbol{y}_1) + g(\boldsymbol{y}_2)$, and
2. for every scalar α and vector $\boldsymbol{y}$ in the domain of g, $g(\alpha\, \boldsymbol{y}) = \alpha\, g(\boldsymbol{y})$.

We prove Part 1. Let $\boldsymbol{y}_1$ and $\boldsymbol{y}_2$ be vectors in the domain of g. Let $\boldsymbol{x}_1 = g(\boldsymbol{y}_1)$ and $\boldsymbol{x}_2 = g(\boldsymbol{y}_2)$. By definition of inverse, $f(\boldsymbol{x}_1) = \boldsymbol{y}_1$ and $f(\boldsymbol{x}_2) = \boldsymbol{y}_2$.

$$\begin{aligned} g(\boldsymbol{y}_1 + \boldsymbol{y}_2) &= g(f(\boldsymbol{x}_1) + f(\boldsymbol{x}_2)) \\ &= g(f(\boldsymbol{x}_1 + \boldsymbol{x}_2)) && \text{by the linearity of } f \\ &= \boldsymbol{x}_1 + \boldsymbol{x}_2 && \text{because } g \text{ is the inverse of } f \\ &= g(\boldsymbol{y}_1) + g(\boldsymbol{y}_1) && \text{because } g \text{ is the inverse of } f \end{aligned}$$

The proof of Part 2 is similar and is left for the reader. □

Problem 4.13.2: Complete the proof of Lemma 4.13.1 by proving Part 2.

4.13.2 The matrix inverse

Definition 4.13.3: Let A be an $R \times C$ matrix over $\mathbb{F}$, and let B be a $C \times R$ matrix over $\mathbb{F}$. Define the function $f : \mathbb{F}^C \longrightarrow \mathbb{F}^R$ by $f_A(\boldsymbol{x}) = A\boldsymbol{x}$ and define the function $g : \mathbb{F}^R \longrightarrow \mathbb{F}^C$ by $g(\boldsymbol{y}) = B\boldsymbol{y}$. If f and g are functional inverses of each other, we say the matrices A and B are inverses of each other. If A has an inverse, we say A is an *invertible matrix*. It can be shown using the uniqueness of a functional inverse (Lemma 0.3.19) that a matrix has at most one inverse; we denote the inverse of an invertible matrix A by A^{-1}.

A matrix that is not invertible is often called a *singular* matrix. We don't use that term in this book.

Example 4.13.4: The 3×3 identity matrix $\mathbb{1} = \begin{bmatrix} 1 & 0 & 0 \\ 0 & 1 & 0 \\ 0 & 0 & 1 \end{bmatrix}$ corresponds to the identity function on $\mathbb{R}^3$. The inverse of the identity function is itself, so $\mathbb{1}$ is its own inverse.

Example 4.13.5: What is the inverse of the 3×3 diagonal matrix $\begin{bmatrix} 2 & 0 & 0 \\ 0 & 3 & 0 \\ 0 & 0 & 4 \end{bmatrix}$? This matrix corresponds to the function $f : \mathbb{R}^3 \longrightarrow \mathbb{R}^3$ defined by $f([x, y, z]) = [2x, 3y, 4z]$. The inverse of this function is the function $g([x, y, z]) = [\frac{1}{2}x, \frac{1}{3}y, \frac{1}{4}z]$, which corresponds to the matrix $\begin{bmatrix} \frac{1}{2} & 0 & 0 \\ 0 & \frac{1}{3} & 0 \\ 0 & 0 & \frac{1}{4} \end{bmatrix}$

Example 4.13.6: The 3×3 diagonal matrix $\begin{bmatrix} 2 & 0 & 0 \\ 0 & 0 & 0 \\ 0 & 0 & 4 \end{bmatrix}$ corresponds to the function $f([x, y, z]) = [2x, 0, 4z]$, which is not an invertible function, so this matrix does not have an inverse.

Example 4.13.7: Consider the following elementary row-addition matrix from Example 4.11.2 (Page 181):

$$A = \begin{bmatrix} 1 & 0 & 0 \\ 2 & 1 & 0 \\ 0 & 0 & 1 \end{bmatrix}$$

This matrix corresponds to the function $f([x_1, x_2, x_3]) = [x_1, x_2 + 2x_1, x_3])$. That is, the function adds twice the first entry to the second entry. The inverse is the function that subtracts the twice the first entry from the second entry: $f^{-1}([x_1, x_2, x_3]) = [x_1, x_2 - 2x_1, x_3]$. Thus the inverse of A is

$$A^{-1} = \begin{bmatrix} 1 & 0 & 0 \\ -2 & 1 & 0 \\ 0 & 0 & 1 \end{bmatrix}$$

This matrix is also a row-addition matrix.

Example 4.13.8: Here is another elementary row-addition matrix:

$$B = \begin{bmatrix} 1 & 0 & 5 \\ 0 & 1 & 0 \\ 0 & 0 & 1 \end{bmatrix}$$

This matrix corresponds to the function $f([x_1, x_2, x_3])$ that adds five times the third entry to the first entry. That is, $f([x_1, x_2, x_3]) = [x_1 + 5x_3, x_2, x_3]$. The The inverse of f is the function that subtracts five times the third entry from the first entry, $f^{-1}([x_1, x_2, x_3]) = [x_1 - 5x_3, x_2, x_3]$. The matrix corresponding to f^{-1}, the inverse of B, is

$$B^{-1} = \begin{bmatrix} 1 & 0 & -5 \\ 0 & 1 & 0 \\ 0 & 0 & 1 \end{bmatrix}$$

which is another elementary row-addition matrix.

It should be apparent that every elementary row-addition matrix is invertible, and that its inverse is also an elementary row-addition matrix. What about a matrix that adds different multiples of a single row to all the other rows?

Example 4.13.9: Consider the matrix from Example 4.11.10 (Page 187):

$$A = \begin{bmatrix} 1 & 0 & 0 & 0 \\ 2 & 1 & 0 & 0 \\ 3 & 0 & 1 & 0 \\ 4 & 0 & 0 & 1 \end{bmatrix}$$

This matrix adds two times the first row to the second row, three times the first row to the third row, and four times the first row to the fourth row: The inverse of this matrix is the matrix that *subtracts* two times the first row from the second row, three times the first row from the third row, and four times the first row from the fourth row:

$$A^{-1} = \begin{bmatrix} 1 & 0 & 0 & 0 \\ -2 & 1 & 0 & 0 \\ -3 & 0 & 1 & 0 \\ -4 & 0 & 0 & 1 \end{bmatrix}$$

Example 4.13.10: What about the matrix $\begin{bmatrix} 1 & 1 \\ 1 & 1 \end{bmatrix}$, which corresponds to the function $f([x_1, x_2]) = [x_1 + x_2, x_1 + x_2]$? The function maps both $[1, -1]$ and $[0, 0]$ to $[0, 0]$, so is not invertible. Therefore the matrix is not invertible.

4.13.3 Uses of matrix inverse

Lemma 4.13.11: If the $R \times C$ matrix A has an inverse A^{-1} then AA^{-1} is the $R \times R$ identity matrix.

Proof

Let $B = A^{-1}$. Define $f_A(\boldsymbol{x}) = A\boldsymbol{x}$ and $f_B(\boldsymbol{y}) = B\boldsymbol{y}$. By the Matrix-Multiplication Lemma (Lemma 4.11.8), the function $f_A \circ f_B$ satisfies $(f_A \circ f_B)(\boldsymbol{x}) = AB\boldsymbol{x}$ for every R-vector $\boldsymbol{x}$. On the other hand, $f_A \circ f_B$ is the identity function, so AB is the $R \times R$ identity matrix. □

Consider a matrix-vector equation

$$A\boldsymbol{x} = \boldsymbol{b}$$

If A has an inverse A^{-1} then by multiplying both sides of the equation on the left by A^{-1}, we obtain the equation

$$A^{-1}A\boldsymbol{x} = A^{-1}\boldsymbol{b}$$

Since $A^{-1}A$ is the identity matrix, we get

$$\mathbb{1}\boldsymbol{x} = A^{-1}\boldsymbol{b}$$

Since multiplication by the identity matrix is the identity function, we obtain

$$\boldsymbol{x} = A^{-1}\boldsymbol{b} \tag{4.10}$$

This tells us that, if the equation $A\boldsymbol{x} = \boldsymbol{b}$ has a solution, the solution must be $A^{-1}\boldsymbol{b}$. Conversely, if we let $\hat{\boldsymbol{x}} = A^{-1}\boldsymbol{b}$ then $A\hat{\boldsymbol{x}} = AA^{-1}\boldsymbol{b} = \mathbb{1}\boldsymbol{b} = \boldsymbol{b}$, which shows that $A^{-1}\boldsymbol{b}$ is indeed a solution to $A\boldsymbol{x} = \boldsymbol{b}$.

Summarizing,

Proposition 4.13.12: if A is invertible then, for any vector $\boldsymbol{b}$ with domain equal to the row-label set of A, the matrix-vector equation $A\boldsymbol{x} = \boldsymbol{b}$ has exactly one solution, namely $A^{-1}\boldsymbol{b}$.

This result turns out to be mathematically very useful. For example, the following proposition is used in our study of eigenvalues, in Chapter 12. Consider an upper-triangular matrix A.

Lemma 4.13.13: Suppose A is an upper-triangular matrix. Then A is invertible if and only if none of its diagonal elements is zero.

Proof

Suppose none of A's diagonal elements is zero. By Proposition 2.11.5, for any right-hand side vector $\boldsymbol{b}$, there is exactly one solution to $A\boldsymbol{x} = \boldsymbol{b}$. Thus the function $\boldsymbol{x} \mapsto A\boldsymbol{x}$ is one-to-one and onto.

On the other hand, suppose at least one of A's diagonal elements is zero. By Proposition 2.11.6, there is a vector $\boldsymbol{b}$ such that the equation $A\boldsymbol{x} = \boldsymbol{b}$ has no solution. If A had an inverse then by Proposition 4.13.12, $A^{-1}\boldsymbol{b}$ would be a solution. □

We will use Proposition 4.13.12 repeatedly in what is to come, culminating in the design of an algorithm to solve linear programs (Chapter 13).

In order to apply this result, however, we must develop a useful criterion for when a matrix is invertible. In the next section, we begin that process. We don't finish it until Chapter 6.

Equation 4.10 might lead one to think that the way to solve the matrix-vector equation $A\boldsymbol{x} = \boldsymbol{b}$ is to find the inverse of A and multiply it by $\boldsymbol{b}$. Indeed, this is traditionally suggested to math students. However, this turns out not to be a good idea when working with real numbers represented in a computer by floating-point numbers, since it can lead to much less accurate answers than can be computed by other methods.

4.13.4 The product of invertible matrices is an invertible matrix

Proposition 4.13.14: If A and B are invertible matrices and the matrix product AB is defined then AB is an invertible matrix, and $(AB)^{-1} = B^{-1}A^{-1}$.

Proof

Define the functions f and g by $f(\boldsymbol{x}) = A\boldsymbol{x}$ and $g(\boldsymbol{x}) = B\boldsymbol{x}$.

Suppose A and B are invertible matrices. Then the corresponding functions f and g are invertible. Therefore, by Lemma 0.3.20, $f \circ g$ is invertible and its inverse is $g^{-1} \circ f^{-1}$, so the matrix corresponding to $f \circ g$ (which is AB) is an invertible matrix, and its inverse is $B^{-1}A^{-1}$. □

Example 4.13.15: $A = \begin{bmatrix} 1 & 1 \\ 0 & 1 \end{bmatrix}$ and $B = \begin{bmatrix} 1 & 0 \\ 1 & 1 \end{bmatrix}$ correspond to functions $f : \mathbb{R}^2 \longrightarrow \mathbb{R}^2$ and $g : \mathbb{R}^2 \longrightarrow \mathbb{R}^2$

$$\begin{aligned} f\left(\begin{bmatrix} x_1 \\ x_2 \end{bmatrix}\right) &= \begin{bmatrix} 1 & 1 \\ 0 & 1 \end{bmatrix}\begin{bmatrix} x_1 \\ x_2 \end{bmatrix} \\ &= \begin{bmatrix} x_1 + x_2 \\ x_2 \end{bmatrix} \end{aligned}$$

f is an invertible function.

$$g\left(\begin{bmatrix} x_1 \\ x_2 \end{bmatrix}\right) = \begin{bmatrix} 1 & 0 \\ 1 & 1 \end{bmatrix}\begin{bmatrix} x_1 \\ x_2 \end{bmatrix} = \begin{bmatrix} x_1 \\ x_1 + x_2 \end{bmatrix}$$

g is an invertible function.

The functions f and g are invertible so the function $f \circ g$ is invertible.

By the Matrix-Multiplication Lemma, the function $f \circ g$ corresponds to the matrix product

$$AB = \begin{bmatrix} 1 & 1 \\ 0 & 1 \end{bmatrix}\begin{bmatrix} 1 & 0 \\ 1 & 1 \end{bmatrix} = \begin{bmatrix} 2 & 1 \\ 1 & 1 \end{bmatrix}$$

so that matrix is invertible.

Example 4.13.16: $A = \begin{bmatrix} 1 & 0 & 0 \\ 4 & 1 & 0 \\ 0 & 0 & 1 \end{bmatrix}$ and $B = \begin{bmatrix} 1 & 0 & 0 \\ 0 & 1 & 0 \\ 5 & 0 & 1 \end{bmatrix}$

Multiplication by the matrix A adds four times the first element to the second element:

$$f([x_1, x_2, x_3]) = [x_1, x_2 + 4x_1, x_3])$$

This function is invertible.

Multiplication by the matrix B adds five times the first element to the third element:

$$g([x_1, x_2, x_3]) = [x_1, x_2, x_3 + 5x_1]$$

This function is invertible.

By the Matrix Multiplication Lemma, multiplication by matrix AB corresponds to composition of functions $f \circ g$:

$$(f \circ g)([x_1, x_2, x_3]) = [x_1, x_2 + 4x_1, x_3 + 5x_1]$$

The function $f \circ g$ is also an invertible function, so AB is an invertible matrix.

Example 4.13.17: $A = \begin{bmatrix} 1 & 2 & 3 \\ 4 & 5 & 6 \\ 7 & 8 & 9 \end{bmatrix}$ and $B = \begin{bmatrix} 1 & 0 & 1 \\ 0 & 1 & 0 \\ 1 & 1 & 0 \end{bmatrix}$

The product is $AB = \begin{bmatrix} 4 & 5 & 1 \\ 10 & 11 & 4 \\ 16 & 17 & 7 \end{bmatrix}$ which is *not* invertible, so at least one of A and B is not invertible, and in fact $\begin{bmatrix} 1 & 2 & 3 \\ 4 & 5 & 6 \\ 7 & 8 & 9 \end{bmatrix}$ is not invertible.

4.13.5 More about matrix inverse

We saw that AA^{-1} is an identity matrix. It is tempting to conjecture that, conversely, for any matrix A, if B is a matrix such that AB is an identity matrix $\mathbb{1}$ then B is the inverse of A. This is not true.

Example 4.13.18: A simple counterexample is

$$A = \begin{bmatrix} 1 & 0 & 0 \\ 0 & 1 & 0 \end{bmatrix}, B = \begin{bmatrix} 1 & 0 \\ 0 & 1 \\ 0 & 0 \end{bmatrix}$$

since

$$AB = \begin{bmatrix} 1 & 0 & 0 \\ 0 & 1 & 0 \end{bmatrix} \begin{bmatrix} 1 & 0 \\ 0 & 1 \\ 0 & 0 \end{bmatrix} = \begin{bmatrix} 1 & 0 \\ 0 & 1 \end{bmatrix}$$

but the function $f_A : \mathbb{F}^3 \longrightarrow \mathbb{F}^2$ defined by $f_A(\boldsymbol{x}) = A\boldsymbol{x}$ and the function $f_B : \mathbb{F}^2 \longrightarrow \mathbb{F}^3$ defined by $f_B(\boldsymbol{y}) = B\boldsymbol{y}$ are not inverses of each other. Indeed, $A \begin{bmatrix} 0 \\ 0 \\ 1 \end{bmatrix}$ and $A \begin{bmatrix} 0 \\ 0 \\ 0 \end{bmatrix}$ are both $\begin{bmatrix} 0 \\ 0 \end{bmatrix}$, which demonstrates that the function f_A is not one-to-one, so is not an invertible function.

So $AB = I$ is *not* sufficient to ensure that A and B are inverses. However,

Corollary 4.13.19: Matrices A and B are inverses of each other if and only if both AB and BA are identity matrices.

Proof

- Suppose A and B are inverses of each other. Because B is the inverse of A, Lemma 4.13.11 implies that AB is an identity matrix. Because A is the inverse of B, the same lemma implies that BA is an identity matrix.
- Conversely, suppose AB and BA are both identity matrices. By the Matrix-Multiplication Lemma (Lemma 4.11.8), the functions corresponding to A and B are therefore functional inverses of each other, and therefore A and B are inverses of each other.

□

We finish with a Question:

Question 4.13.20: How can we tell if a matrix M is invertible?

We can relate this Question to others. By definition, M is an invertible matrix if the function $f(\boldsymbol{x}) = M\boldsymbol{x}$ is an invertible function, i.e. if the function is one-to-one and onto.

- *One-to-one:* Since the function is linear, we know by the One-to-One Lemma that the function is one-to-one if its kernel is trivial, i.e. if the null space of M is trivial.
- *Onto:* Question 4.10.16 asks: *how we can tell if a linear function is onto?*

If we knew how to tell if a linear function is onto, therefore, we would know how to tell if a matrix is invertible.

In the next two chapters, we will discover the tools to answer these Questions.

4.14 *Lab: Error-correcting codes*

In this lab, we work with vectors and matrices over $GF(2)$. So when you see 1's and 0's in this description, remember that each 1 is really the value `one` from the module `GF2`.

4.14.1 *The check matrix*

In Section 4.7.3, I introduced error-correcting codes. As we have seen, in a linear binary code, the set $\mathcal{C}$ of codewords is a vector space over $GF(2)$. In such a code, there is a matrix H, called the *check matrix*, such that $\mathcal{C}$ is the null space of H. When the Receiver receives the vector $\tilde{\boldsymbol{c}}$, she can check whether the received vector is a codeword by multiplying it by H and checking whether the resulting vector (called the *error syndrome*) is the zero vector.

4.14.2 *The generator matrix*

We have characterized the vector space $\mathcal{C}$ as the null space of the check matrix H. There is another way to specify a vector space: in terms of generators. The *generator matrix* for a linear code is a matrix G whose columns are generators for the set $\mathcal{C}$ of codewords.[a]

By the linear-combinations definition of matrix-vector multiplication, every matrix-vector product $G*\boldsymbol{p}$ is a linear combination of the columns of G, and is therefore a codeword.

4.14.3 *Hamming's code*

Hamming discovered a code in which a four-bit message is represented by a seven-bit *codeword.* The generator matrix is

$$G = \begin{bmatrix} 1 & 0 & 1 & 1 \\ 1 & 1 & 0 & 1 \\ 0 & 0 & 0 & 1 \\ 1 & 1 & 1 & 0 \\ 0 & 0 & 1 & 0 \\ 0 & 1 & 0 & 0 \\ 1 & 0 & 0 & 0 \end{bmatrix}$$

A four-bit message is represented by a 4-vector $\boldsymbol{p}$ over $GF(2)$. The encoding of $\boldsymbol{p}$ is the 7-vector resulting from the matrix-vector product $G * \boldsymbol{p}$.

Let f_G be the encoding function, the function defined by $f_G(\boldsymbol{x}) = G * \boldsymbol{p}$. The image of f_G, the set of all codewords, is the row space of G.

Task 4.14.1: Create an instance of `Mat` representing the generator matrix G. You can use the procedure `listlist2mat` in the `matutil` module. Since we are working over $GF(2)$, you should use the value `one` from the `GF2` module to represent 1.

[a]It is traditional to define the generator matrix so that its *rows* are generators for $\mathcal{C}$. We diverge from this tradition for the sake of simplicity of presentation.

What is the encoding of the message $[1, 0, 0, 1]$?

Task 4.14.2:

4.14.4 *Decoding*

Note that four of the rows of G are the standard basis vectors $\boldsymbol{e}_1, \boldsymbol{e}_2, \boldsymbol{e}_3, \boldsymbol{e}_4$ of $GF(2)^4$. What does that imply about the relation between words and codewords? Can you easily decode the codeword $[0, 1, 1, 1, 1, 0, 0]$ without using a computer?

Task 4.14.3: Think about the manual decoding process you just did. Construct a 4×7 matrix R such that, for any codeword $\boldsymbol{c}$, the matrix-vector product $R*\boldsymbol{c}$ equals the 4-vector whose encoding is $\boldsymbol{c}$. What should the matrix-matrix product RG be? Compute the matrix and check it against your prediction.

4.14.5 *Error syndrome*

Suppose Alice sends the codeword $\boldsymbol{c}$ across the noisy channel. Let $\tilde{\boldsymbol{c}}$ be the vector received by Bob. To reflect the fact that $\tilde{\boldsymbol{c}}$ might differ from $\boldsymbol{c}$, we write

$$\tilde{\boldsymbol{c}} = \boldsymbol{c} + \boldsymbol{e}$$

where $\boldsymbol{e}$ is the error vector, the vector with ones in the corrupted positions.

If Bob can figure out the error vector $\boldsymbol{e}$, he can recover the codeword $\boldsymbol{c}$ and therefore the original message. To figure out the error vector $\boldsymbol{e}$, Bob uses the check matrix, which for the Hamming code is

$$H = \begin{bmatrix} 0 & 0 & 0 & 1 & 1 & 1 & 1 \\ 0 & 1 & 1 & 0 & 0 & 1 & 1 \\ 1 & 0 & 1 & 0 & 1 & 0 & 1 \end{bmatrix}$$

As a first step towards figuring out the error vector, Bob computes the *error syndrome*, the vector $H * \tilde{\boldsymbol{c}}$, which equals $H * \boldsymbol{e}$.

Examine the matrix H carefully. What is special about the order of its columns?

Define the function f_H by $f_H(\boldsymbol{y}) = H * \boldsymbol{y}$. The image under f_H of any codeword is the zero vector. Now consider the function $f_H \circ f_G$ that is the composition of f_H with f_G. For any vector $\boldsymbol{p}$, $f_G(\boldsymbol{p})$ is a codeword $\boldsymbol{c}$, and for any codeword $\boldsymbol{c}$, $f_H(\boldsymbol{c}) = \boldsymbol{0}$. This implies that, for any vector $\boldsymbol{p}$, $(f_H \circ f_G)(\boldsymbol{p}) = \boldsymbol{0}$.

The matrix HG corresponds to the function $f_H \circ f_G$. Based on this fact, predict the entries of the matrix HG.

Task 4.14.4: Create an instance of `Mat` representing the check matrix H. Calculate the matrix-matrix product HG. Is the result consistent with your prediction?

4.14.6 *Finding the error*

Bob assumes that at most one bit of the codeword is corrupted, so at most one bit of $\boldsymbol{e}$ is nonzero, say the bit in position $i \in \{1, 2, \ldots, 7\}$. In this case, what is the value of $H * \boldsymbol{e}$? (Hint: this uses the special property of the order of H's rows.)

Task 4.14.5: Write a procedure `find_error` that takes an error syndrome and returns the corresponding error vector $\boldsymbol{e}$.

Task 4.14.6: Imagine that you are Bob, and you have received the *non*-codeword $\tilde{\boldsymbol{c}} = [1, 0, 1, 1, 0, 1, 1]$. Your goal is to derive the original 4-bit message that Alice intended to send. To do this, use `find_error` to figure out the corresponding error vector $\boldsymbol{e}$, and then add $\boldsymbol{e}$ to $\tilde{\boldsymbol{c}}$ to obtain the correct codeword. Finally, use the matrix R from Task 4.14.3 to derive the original 4-vector.

Task 4.14.7: Write a one-line procedure `find_error_matrix` with the following spec:

- *input:* a matrix S whose columns are error syndromes
- *output:* a matrix whose c^{th} column is the error corresponding to the cth column of S.

This procedure consists of a comprehension that uses the procedure `find_error` together with some procedures from the `matutil` module.

Test your procedure on a matrix whose columns are $[1, 1, 1]$ and $[0, 0, 1]$.

4.14.7 *Putting it all together*

We will now encode an entire string and will try to protect it against errors. We first have to learn a little about representing a text as a matrix of bits. Characters are represented using a variable-length encoding scheme called *UTF-8*. Each character is represented by some number of bytes. You can find the value of a character c using `ord(c)`. What are the numeric values of of the characters 'a', 'A' and space?

You can obtain the character from a numerical value using `chr(i)`. To see the string of characters numbered 0 through 255, you can use the following:

```
>>> s = ''.join([chr(i) for i in range(256)])
>>> print(s)
```

We have provided a module `bitutil` that defines some procedures for converting between lists of $GF(2)$ values, matrices over $GF(2)$, and strings. Two such procedures are `str2bits(str)` and `bits2str(L)`:

The procedure `str2bits(str)` has the following spec:

- *input:* a string
- *output:* a list of $GF(2)$ values (0 and one) representing the string

The procedure `bits2str(L)` is the inverse procedure:

- *input:* a list of $GF(2)$ values
- *output:* the corresponding string

Task 4.14.8: Try out `str2bits(str)` on the string `s` defined above, and verify that `bits2str(L)` gets you back the original string.

The Hamming code operates on four bits at a time. A four-bit sequence is called a *nibble* (sometimes *nybble*). To encode a list of bits (such as that produced by `str2bits`), we break the list into nibbles and encode each nibble separately.

To transform each nibble, we interpret the nibble as a 4-vector and we multiply it by the generating matrix G. One strategy is to convert the list of bits into a list of 4-vectors, and then use, say, a comprehension to multiply each vector in that list by G. In keeping with our current interest in matrices, we will instead convert the list of bits into a matrix B each column of which is a 4-vector representing a nibble. Thus a sequence of $4n$ bits is represented by a $4 \times n$ matrix P. The module `bitutil` defines a procedure `bits2mat(bits)` that transforms a list of bits into such a matrix, and a procedure `mat2bits(A)` that transforms such a matrix A back into a list of bits.

Task 4.14.9: Try converting a string to a list of bits to a matrix P and back to a string, and verify that you get the string you started with.

Task 4.14.10: Putting these procedures together, compute the matrix P which represents the string "I'm trying to free your mind, Neo. But I can only show you the door. You're the one that has to walk through it."

Imagine that you are transmitting the above message over a noisy communication channel. This channel transmits bits, but occasionally sends the wrong bit, so one becomes 0 and vice versa.

The module `bitutil` provides a procedure `noise(A, s)` that, given a matrix A and a probability parameter s, returns a matrix with the same row- and column-labels as A but with entries chosen from $GF(2)$ according the probability distribution `{one:s, 0:1-s}`.

For example, each entry of `noise(A, 0.02)` will be one with probability 0.02 and zero with probability 0.98

Task 4.14.11: To simulate the effects of the noisy channel when transmitting your matrix P, use `noise(P, 0.02)` to create a random matrix E. The matrix $E + P$ will introduce some errors. To see the effect of the noise, convert the perturbed matrix back to text.

Looks pretty bad, huh? Let's try to use the Hamming code to fix that. Recall that to encode a word represented by the row vector $\boldsymbol{p}$, we compute $G * \boldsymbol{p}$.

Task 4.14.12: Encode the words represented by the columns of the matrix P, obtaining a matrix C. You should not use any loops or comprehensions to compute C from P. How many bits represented the text before the encoding? How many after?

Task 4.14.13: Imagine that you send the encoded data over the noisy channel. Use `noise` to construct a noise matrix of the appropriate dimensions with error probability 0.02, and add it to C to obtain a perturbed matrix `CTILDE`. Without correcting the errors, decode `CTILDE` and convert it to text to see how garbled the received information is.

Task 4.14.14: In this task, you are to write a one-line procedure `correct(A)` with the following spec:

- *input:* a matrix A each column of which differs from a codeword in at most one bit
- *output:* a matrix whose columns are the corresponding valid codewords.

The procedure should contain no loops or comprehensions. Just use matrix-matrix multiplications and matrix-matrix additions together with a procedure you have written in this lab.

Task 4.14.15: Apply your procedure `correct(A)` to `CTILDE` to get a matrix of codewords. Decode this matrix of codewords using the matrix R from Task 4.14.3, obtaining a matrix whose columns are 4-vectors. Then derive the string corresponding to these 4-vectors.

Did the Hamming code succeed in fixing all of the corrupted characters? If not, can you explain why?

Task 4.14.16: Repeat this process with different error probabilities to see how well the Hamming code does under different circumstances.

4.15 *Lab: Transformations in 2D geometry*

4.15.1 *Our representation for points in the plane*

You are familiar with representing a point (x, y) in the plane by a `{'x','y'}`-vector $\begin{bmatrix} x \\ y \end{bmatrix}$. In this lab, for a reason that will become apparent, we will use an `{'x','y','u'}`-vector $\begin{bmatrix} x \\ y \\ u \end{bmatrix}$. This representation is called *homogeneous coordinates*. We will not be making use of homogeneous coordinates in their full generality; here, the u coordinate will always be 1.

4.15.2 *Transformations*

A geometric transformation will be represented by a matrix M. To apply the transformation to the location of a single point, use matrix-vector multiplication to multiply the matrix by the position vector representing the point.

For example, let's say you want to scale the point by two in the vertical direction. If we were representing points in the plane by 2-vectors $\begin{bmatrix} x \\ y \end{bmatrix}$, we would represent the transformation by the matrix

$$\begin{bmatrix} 1 & 0 \\ 0 & 2 \end{bmatrix}$$

You could apply the transformation to the vector by using matrix-vector multiplication:

$$\begin{bmatrix} 1 & 0 \\ 0 & 2 \end{bmatrix} \begin{bmatrix} x \\ y \end{bmatrix}$$

For example, if the point were located at $(12, 15)$, you would calculate

$$\begin{bmatrix} 1 & 0 \\ 0 & 2 \end{bmatrix} \begin{bmatrix} 12 \\ 15 \end{bmatrix} = \begin{bmatrix} 12 \\ 30 \end{bmatrix}$$

However, since here we instead represent points in the plane by 3-vectors $\begin{bmatrix} x \\ y \\ u \end{bmatrix}$ (with $u = 1$), you would instead represent the transformation by the matrix

$$\begin{bmatrix} 1 & 0 & 0 \\ 0 & 2 & 0 \\ 0 & 0 & 1 \end{bmatrix}$$

You would apply it to a point (x, y) by matrix-vector multiplication:

$$\begin{bmatrix} 1 & 0 & 0 \\ 0 & 2 & 0 \\ 0 & 0 & 1 \end{bmatrix} \begin{bmatrix} x \\ y \\ 1 \end{bmatrix}$$

For example, to apply it to the point $(12, 15)$, you would calculate

$$\begin{bmatrix} 1 & 0 & 0 \\ 0 & 2 & 0 \\ 0 & 0 & 1 \end{bmatrix} \begin{bmatrix} 12 \\ 15 \\ 1 \end{bmatrix} = \begin{bmatrix} 12 \\ 30 \\ 1 \end{bmatrix}$$

Note that the resulting vector also has a 1 in the u entry.

Suppose we want to apply such a transformation to many points at the same time. As illustrated in Examples 4.11.4 and 4.11.5, according to the matrix-vector definition of matrix-matrix multiplication, to apply the transformation to many points, we put the points together to form a position matrix and left-multiply that position matrix by the matrix representing the transformation:

$$\begin{bmatrix} 3 & 0 & 0 \\ 0 & 3 & 0 \\ 0 & 0 & 1 \end{bmatrix} \left[\begin{array}{c|c|c|c} 2 & 2 & -2 & -2 \\ 2 & -2 & 2 & 2 \\ 1 & 1 & 1 & 1 \end{array}\right] = \left[\begin{array}{c|c|c|c} 6 & 6 & -6 & -6 \\ 6 & -6 & 6 & 6 \\ 1 & 1 & 1 & 1 \end{array}\right]$$

4.15.3 *Image representation*

In this lab, we will be manipulating images using matrices in Python. In order to do this, we need to represent images as matrices. We represent an image by a set of colored points in the plane.

Colored points

To represent a colored point, we need to specify its location and its color. We will therefore represent a point using two vectors; the *location* vector with labels $\{'x','y','u'\}$ and the *color* vector with labels $\{'r','g','b'\}$. The location vector represents the location of the point in the usual way—as an (x, y) pair. The u entry is always 1 for now; later you will see how this is used. For example, the point $(12, 15)$ would be represented by the vector `Vec({'x','y','u'}, {'x':12, 'y':15, 'u':1})`.

The color vector represents the color of the point: the $'r'$, $'g'$, and $'b'$ entries give the intensities for the color channels *red*, *green*, and *blue*. For example, the color red is represented by the function $\{'r' : 1\}$.

Our scheme for representing images

Ordinarily, an image is a regular rectangular grid of rectangular pixels, where each pixel is assigned a color. Because we will be transforming images, we will use a slightly more general representation.

A *generalized image* consists of a grid of generalized pixels, where each generalized pixel is a quadrilateral (not necessarily a rectangle).

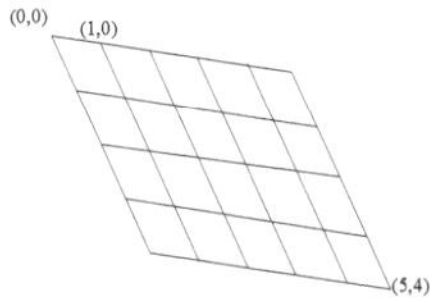

The points at the corners of the generalized pixels are identified by pairs (x, y) of integers, which are called *pixel coordinates*. The top-left corner has pixel coordinates (0,0), the corner directly to its right has pixel coordinates (1,0), and so on. For example, the pixel coordinates of the four corners of the top-left generalized pixel are (0,0), (0,1), (1,0), and (1,1).

Each corner is assigned a location in the plane, and each generalized pixel is assigned a color. The mapping of corners to points in the plane is given by a matrix, the *location matrix*. Each corner corresponds to a column of the location matrix, and the label of that column is the pair (x, y) of pixel coordinates of the corner. The column is a `{'x','y','u'}`-vector giving the location of the corner. Thus the row labels of the location matrix are `'x'`, `'y'`, and `'u'`.

The mapping of generalized pixels to colors is given by another matrix, the *color matrix*. Each generalized pixel corresponds to a column of the color matrix, and the label of that column is the pair of pixel coordinates of the top-left corner of that generalized pixel. The column is a `{'r','g','b'}`-vector giving the color of that generalized pixel.

For example, the image  consists of four generalized pixels, comprising a total of nine corners. This image is represented by the location matrix

	(0, 0)	(0, 1)	(0, 2)	(1, 2)	(1, 1)	(1, 0)	(2, 2)	(2, 0)	(2, 1)
x	0	0	0	1	1	1	2	2	2
y	0	1	2	2	1	0	2	0	1
u	1	1	1	1	1	1	1	1	1

and the color matrix

	(0, 0)	(0, 1)	(1, 1)	(1, 0)
b	225	125	75	175
g	225	125	75	175
r	225	125	75	175

By applying a suitable transformation to the location matrix, we can obtain

```
      (0, 0) (0, 1) (0, 2) (1, 2) (1, 1) (1, 0) (2, 2) (2, 0) (2, 1)
     ------------------------------------------------------------------
x  |      0      2      4     14     12     10     24     20     22
y  |      0     10     20     22     12      2     24      4     14
u  |      1      1      1      1      1      1      1      1      1
```

which, combined with the unchanged color matrix, looks like this: 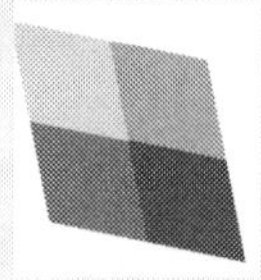

4.15.4 *Loading and displaying images*

We provide a module, `image_mat_util`, with some helpful procedures:

- `file2mat`
  - *input:* string giving the pathname of a `.png` image file
  - *output:* a 2-tuple (position matrix, color matrix) representing the image
- `mat2display`
  - *input:* a position matrix and a color matrix (two arguments)
  - *output:* Displays the image in a web browser

Task 4.15.1: Download a `.png` image file, then use `file2mat` to load it and `mat2display` to display it on the screen.

4.15.5 *Linear transformations*

You will be writing a module named `transform` that provides a number of simple linear transformations. Instead of writing procedures that operate on images, your methods will return the transformation matrix and you can apply it to a specific image using matrix multiplication. For each task, we want you to not just write the procedure but also test it on some of the images provided in `matrix_resources/images`.

Task 4.15.2: Write a procedure `identity()` which takes no arguments and returns an identity matrix for position vectors. Verify that this matrix works by applying it first to some points and then to an image, making sure that nothing changes. (Hint: *Think about the correct row and column labels.)*

4.15.6 *Translation*

Recall that a translation is a transformation that moves a point (x, y) to $(x + \alpha, y + \beta)$, where α and β are parameters of the transformation. Can you come up with a 2×2 matrix that represents a translation? That is, is there a matrix M such that $x' = Mx$, where x and x' are the coordinates of a point before and after the translation, respectively? (*Hint:* How does a translation act on the origin, or the zero vector?)

Now consider a representation of points in two dimensions by 3-dimensional vectors whose third coordinate is fixed to be 1. This is a special case of a representation known as *homogeneous coordinates.* Can you come up with a 3×3 matrix that describes a translation?

Task 4.15.3: Write a procedure `translation(alpha, beta)` that takes two translation parameters and returns the corresponding 3×3 translation matrix. Test it on some images.

4.15.7 *Scaling*

A scaling transformation transforms a point (x, y) into $(\alpha x, \beta y)$, where α and β are the $x-$ and y-scaling parameters, respectively. Can scaling be represented by a 2×2 matrix multiplying a vector $\begin{bmatrix} x \\ y \end{bmatrix}$? Can it be represented by a 3×3 matrix multiplying a vector $\begin{bmatrix} x \\ y \\ 1 \end{bmatrix}$?

Task 4.15.4: Write a procedure `scale(alpha, beta)` that takes $x-$ and $y-$scaling parameters and returns the corresponding 3×3 scaling matrix that multiplies a vector $\begin{bmatrix} x \\ y \\ 1 \end{bmatrix}$.

4.15.8 *Rotation*

- What point does the vector $(1, 0, 1)$ represent in homogeneous coordinates? What are the homogeneous coordinates of this point after rotating it about the origin by 30 degrees counterclockwise?
- Answer the same question for the vectors $(0, 1, 1)$ and $(0, 0, 1)$.
- What is the 3×3 matrix M that describes a counterclockwise rotation of 30 degrees about the origin? That is, a matrix M such that $x' = Mx$, where x and x' are the coordinates of a point before and after the rotation, respectively.
- What is the general matrix form for a counterclockwise rotation of θ radians about the origin? Compare this to the 2×2 rotation matrix derived in the book.

Task 4.15.5: Write a procedure `rotation(theta)` that takes an angle in radians and returns the corresponding rotation matrix. *Hint:* Both `sin(·)` and `cos(·)` are available in the `math` module.

4.15.9 *Rotation about a center other than the origin*

Task 4.15.6: Write a procedure `rotation_about(theta, x, y)` that takes three parameters—an angle `theta` in radians, an x coordinate, and a y coordinate—and returns the matrix that rotates counterclockwise about (x, y) by `theta`. *Hint:* Use procedures you've already written.

4.15.10 *Reflection*

A reflection about the y-axis transforms a point (x, y) into a point $(-x, y)$.

Task 4.15.7: Write a procedure `reflect_y()` that takes no parameters and returns the matrix which corresponds to a reflection about the y axis.

Task 4.15.8: Write a procedure `reflect_x()` which that takes no parameters and returns the matrix which corresponds to a reflection about the x axis.

4.15.11 *Color transformations*

Our image representation supports transformations on colors as well as positions. Such a transformation would be applied by multiplying the corresponding matrix times the color matrix.

Task 4.15.9: Write a procedure `scale_color` that takes r, g, and b scaling parameters and returns the corresponding scaling matrix.

Task 4.15.10: Write a procedure `grayscale()` that returns a matrix that converts a color image to a grayscale image. Note that both images are still represented in RGB. If a pixel in the original image had the values r, g, b in each of the color channels, then in the grayscale image it has the value $\frac{77r}{256} + \frac{151g}{256} + \frac{28b}{256}$ in all three color channels.

4.15.12 *Reflection more generally*

Task 4.15.11: Write a procedure `reflect_about(x1,y1, x2,y2)` that takes two points and returns the matrix that reflects about the line defined by the two points. (*Hint:* Use rotations, translations, and a simple reflection).

4.16 Review questions

- What is the transpose of a matrix?
- What is the sparsity of a matrix and why is it important in computation?
- What is the linear-combination definition of matrix-vector multiplication?
- What is the linear-combinations definition of vector-matrix multiplication?
- What is the dot-product definition of matrix-vector multiplication?
- What is the dot-product definition of vector-matrix multiplication?
- What is an identity matrix?
- What is an upper-triangular matrix?
- What is a diagonal matrix?
- What is a linear function?
- What are two ways that a linear function $f : \mathbb{F}^n \longrightarrow \mathbb{F}^m$ can be represented by a matrix?
- What are the kernel and image of a linear function?
- What are the null space, column space, and row space of a matrix?
- What is the matrix-vector definition of matrix-matrix multiplication?
- What is the vector-matrix definition of matrix-matrix multiplication?
- What is the dot-product definition of matrix-matrix multiplication?
- What is associativity of matrix-matrix multiplication?

- How can matrix-vector and vector-matrix multiplication be represented using matrix-matrix multiplication?
- What is an outer product?
- How can dot-product be represented using matrix-matrix multiplication?
- What is the inverse of a matrix?
- What is one criterion for whether two matrices are inverses of each other?

4.17 Problems

Matrix-Vector Multiplication

Problem 4.17.1: Compute the following matrix-vector products (I recommend you not use the computer to compute these):

1. $\begin{bmatrix} 1 & 1 \\ 1 & -1 \end{bmatrix} * [0.5, 0.5]$
2. $\begin{bmatrix} 0 & 0 \\ 0 & 1 \end{bmatrix} * [1.2, 4.44]$
3. $\begin{bmatrix} 1 & 2 & 3 \\ 2 & 3 & 4 \\ 3 & 4 & 5 \end{bmatrix} * [1, 2, 3]$

Problem 4.17.2: What 2×2 matrix M satisfies $M * [x, y] = [y, x]$ for all vectors $[x, y]$?

Problem 4.17.3: What 3×3 matrix M satisfies $M * [x, y, z] = [z + x, y, x]$ for all vectors $[x, y, z]$?

Problem 4.17.4: What 3-by-3 matrix M satisfies $M * [x, y, z] = [2x, 4y, 3z]$ for all vectors $[x, y, z]$?

Matrix-matrix multiplication: dimensions of matrices

Problem 4.17.5: For each of the following problems, answer whether the given matrix-matrix product is valid or not. If it is valid, give the number of rows and the number of columns of the resulting matrix (you need not provide the matrix itself).

1. $\begin{bmatrix} 1 & 1 & 0 \\ 1 & 0 & 1 \end{bmatrix} \begin{bmatrix} 2 & 1 & 1 \\ 3 & 1 & 2 \end{bmatrix}$
2. $\begin{bmatrix} 3 & 3 & 0 \end{bmatrix} \begin{bmatrix} 1 & 4 & 1 \\ 1 & 7 & 2 \end{bmatrix}$
3. $\begin{bmatrix} 3 & 3 & 0 \end{bmatrix} \begin{bmatrix} 1 & 4 & 1 \\ 1 & 7 & 2 \end{bmatrix}^T$
4. $\begin{bmatrix} 1 & 4 & 1 \\ 1 & 7 & 2 \end{bmatrix} \begin{bmatrix} 3 & 3 & 0 \end{bmatrix}^T$

5. $\begin{bmatrix} 1 & 4 & 1 \\ 1 & 7 & 2 \end{bmatrix} \begin{bmatrix} 3 & 3 & 0 \end{bmatrix}$

6. $\begin{bmatrix} 2 & 1 & 5 \end{bmatrix} \begin{bmatrix} 1 & 6 & 2 \end{bmatrix}^T$

7. $\begin{bmatrix} 2 & 1 & 5 \end{bmatrix}^T \begin{bmatrix} 1 & 6 & 2 \end{bmatrix}$

Matrix-matrix multiplication practice

Problem 4.17.6: Compute:

1. $\begin{bmatrix} 2 & 3 \\ 4 & 2 \end{bmatrix} \begin{bmatrix} 1 & 2 \\ 2 & 3 \end{bmatrix}$

2. $\begin{bmatrix} 2 & 4 & 1 \\ 3 & 0 & -1 \end{bmatrix} \begin{bmatrix} 1 & 2 & 0 \\ 5 & 1 & 1 \\ 2 & 3 & 0 \end{bmatrix}$

3. $\begin{bmatrix} 2 & 2 & 1 \end{bmatrix} \begin{bmatrix} 3 & 1 \\ -2 & 6 \\ 1 & -1 \end{bmatrix}$

4. $\begin{bmatrix} 1 & 2 & 3 \end{bmatrix} \begin{bmatrix} 1 \\ 2 \\ 3 \end{bmatrix}$

5. $\begin{bmatrix} 1 \\ 2 \\ 3 \end{bmatrix} \begin{bmatrix} 1 & 2 & 3 \end{bmatrix}$

6. $\begin{bmatrix} 4 & 1 & -3 \\ 2 & 2 & -2 \end{bmatrix}^T \begin{bmatrix} -1 & 1 \\ 1 & 0 \end{bmatrix}$ (Remember the superscript T means "transpose".)

Problem 4.17.7: Let

$$A = \begin{bmatrix} 2 & 0 & 1 & 5 \\ 1 & -4 & 6 & 2 \\ 3 & 0 & -4 & 2 \\ 3 & 4 & 0 & -2 \end{bmatrix}$$

For each of the following values of the matrix B, compute AB and BA. (I recommend you not use the computer to compute these.)

1. $B = \begin{bmatrix} 0 & 1 & 0 & 0 \\ 0 & 0 & 1 & 0 \\ 0 & 0 & 0 & 1 \\ 1 & 0 & 0 & 0 \end{bmatrix}$ 2. $B = \begin{bmatrix} 0 & 0 & 0 & 1 \\ 0 & 0 & 1 & 0 \\ 0 & 1 & 0 & 0 \\ 1 & 0 & 0 & 0 \end{bmatrix}$ 3. $B = \begin{bmatrix} 0 & 0 & 0 & 1 \\ 0 & 1 & 0 & 0 \\ 1 & 0 & 0 & 0 \\ 0 & 0 & 1 & 0 \end{bmatrix}$

Problem 4.17.8: Let a, b be numbers and let $A = \begin{bmatrix} 1 & a \\ 0 & 1 \end{bmatrix}$ and $B = \begin{bmatrix} 1 & b \\ 0 & 1 \end{bmatrix}$.

1. What is AB? Write it in terms of a and b.

2. Recall that, for a matrix M and a nonnegative integer k, we denote by M^k the k-fold product of M with itself, i.e.

$$\underbrace{MMM \ldots M}_{k \text{ times}}$$

Plug in 1 for a in A. What is A^2, A^3? What is A^n where n is a positive integer?

Problem 4.17.9: Let

$$A = \begin{bmatrix} 4 & 2 & 1 & -1 \\ 1 & 5 & -2 & 3 \\ 4 & 4 & 4 & 0 \\ -1 & 6 & 2 & -5 \end{bmatrix}$$

For each of the following values of the matrix B, compute AB and BA without using a computer. (To think about: Which definition of matrix-matrix multiplication is most useful here? What does a nonzero entry at position (i, j) in B contribute to the jth column of AB? What does it contribute to the ith row of BA?)

(a) $\begin{bmatrix} 0 & 0 & 0 & 0 \\ 0 & 0 & 1 & 0 \\ 0 & 0 & 0 & 0 \\ 0 & 0 & 0 & 0 \end{bmatrix}$ (b) $\begin{bmatrix} 0 & 0 & 0 & 0 \\ 0 & 1 & 0 & 0 \\ 0 & 0 & 0 & 0 \\ 0 & 0 & 1 & 0 \end{bmatrix}$ (c) $\begin{bmatrix} 1 & 0 & 0 & 0 \\ 1 & 0 & 0 & 0 \\ 0 & 0 & 0 & 0 \\ 0 & 0 & 0 & 0 \end{bmatrix}$

(d) $\begin{bmatrix} 0 & 1 & 0 & 1 \\ 0 & 0 & 0 & 0 \\ 0 & 0 & 0 & 0 \\ 0 & 1 & 0 & 0 \end{bmatrix}$ (e) $\begin{bmatrix} 0 & 0 & 0 & 2 \\ 0 & 0 & 0 & 0 \\ 0 & 0 & 0 & 0 \\ 0 & -3 & 0 & 0 \end{bmatrix}$ (f) $\begin{bmatrix} -1 & 0 & 0 & 0 \\ 0 & 2 & 0 & 0 \\ 0 & 0 & 2 & 0 \\ 0 & 0 & 0 & 3 \end{bmatrix}$

Matrix-matrix multiplication: dimensions of matrices (revisited)

Problem 4.17.10: For each of the following problems, answer whether the given matrix-matrix product is valid or not. If it is valid, give the number of rows and the number of columns of the resulting matrix (you need not provide the matrix itself).

1. $\begin{bmatrix} 1 & 1 & 0 \\ 1 & 0 & 1 \end{bmatrix} \begin{bmatrix} 2 & 1 & 1 \\ 3 & 1 & 2 \end{bmatrix}$
2. $\begin{bmatrix} 3 & 3 & 0 \end{bmatrix} \begin{bmatrix} 1 & 4 & 1 \\ 1 & 7 & 2 \end{bmatrix}$
3. $\begin{bmatrix} 3 & 3 & 0 \end{bmatrix} \begin{bmatrix} 1 & 4 & 1 \\ 1 & 7 & 2 \end{bmatrix}^T$
4. $\begin{bmatrix} 1 & 4 & 1 \\ 1 & 7 & 2 \end{bmatrix} \begin{bmatrix} 3 & 3 & 0 \end{bmatrix}^T$
5. $\begin{bmatrix} 1 & 4 & 1 \\ 1 & 7 & 2 \end{bmatrix} \begin{bmatrix} 3 & 3 & 0 \end{bmatrix}$
6. $\begin{bmatrix} 2 & 1 & 5 \end{bmatrix} \begin{bmatrix} 1 & 6 & 2 \end{bmatrix}^T$
7. $\begin{bmatrix} 2 & 1 & 5 \end{bmatrix}^T \begin{bmatrix} 1 & 6 & 2 \end{bmatrix}$

Column-vector and row-vector matrix multiplication

Problem 4.17.11: Compute the result of the following matrix multiplications.

(a) $\begin{bmatrix} 2 & 3 & 1 \\ 1 & 3 & 4 \end{bmatrix} \begin{bmatrix} 2 \\ 2 \\ 3 \end{bmatrix}$

(b) $\begin{bmatrix} 2 & 4 & 1 \end{bmatrix} \begin{bmatrix} 1 & 2 & 0 \\ 5 & 1 & 1 \\ 2 & 3 & 0 \end{bmatrix}$

(c) $\begin{bmatrix} 2 & 1 \end{bmatrix} \begin{bmatrix} 3 & 1 & 5 & 2 \\ -2 & 6 & 1 & -1 \end{bmatrix}$

(d) $\begin{bmatrix} 1 & 2 & 3 & 4 \\ 1 & 1 & 3 & 1 \end{bmatrix} \begin{bmatrix} 1 \\ 2 \\ 3 \\ 4 \end{bmatrix}$

(e) $\begin{bmatrix} 4 \\ 1 \\ -3 \end{bmatrix}^T \begin{bmatrix} -1 & 1 & 1 \\ 1 & 0 & 2 \\ 0 & 1 & -1 \end{bmatrix}$ (Remember the superscript T means "transpose".)

Matrix Class

Problem 4.17.12: You will write a module `mat` implementing a matrix class `Mat`. The data structure used for instances of `Mat` resembles that used for instances of `Vec`. The only difference is that the domain `D` will now store a pair (i.e., a 2-tuple) of sets instead of a single set. The keys of the dictionary `f` are pairs of elements of the Cartesian product of the two sets in `D`.

The operations defined for `Mat` include entry setters and getters, an equality test, addition and subtraction and negative, multiplication by a scalar, transpose, vector-matrix and matrix-vector multiplication, and matrix-matrix multiplication. Like `Vec`, the class `Mat` is defined to enable use of operators such as + and *. The syntax for using instances of `Mat` is as follows, where `A` and `B` are matrices, `v` is a vector, `alpha` is a scalar, `r` is a row label, and `c` is a column label:

operation	syntax
Matrix addition and subtraction	`A+B` and `A-B`
Matrix negative	`-A`
Scalar-matrix multiplication	`alpha*A`
Matrix equality test	`A == B`
Matrix transpose	`A.transpose()`
Getting and setting a matrix entry	`A[r,c]` and `A[r,c] = alpha`
Matrix-vector and vector-matrix multiplication	`v*A` and `A*v`
Matrix-matrix multiplication	`A*B`

You are required to write the procedures `getitem`, `setitem`, `mat_add`, `mat_scalar_mul`, `equal`, `transpose`, `vector_matrix_mul`, `matrix_vector_mul`, and `matrix_matrix_mul`.

Put the file `mat.py` in your working directory, and, for each procedure, replace the pass statement with a working version. Test your implementation using `doctest` as you did with `vec.py` in Problem 2.14.10. Make sure your implementation works with matrices whose row-label sets differ from their column-label sets.

Note: Use the sparse matrix-vector multiplication algorithm described in Section 4.8 (the one based on the "ordinary" definition") for matrix-vector multiplication. Use the analogous algorithm for vector-matrix multiplication. Do not use `transpose` in your multiplication algorithms. Do not use any external procedures or modules other than `vec`. In particular, do not use procedures from `matutil`. If you do, your Mat implementation is likely not to be efficient enough for use with large sparse matrices.

Matrix-vector and vector-matrix multiplication definitions in Python

You will write several procedures, each implementing matrix-vector multiplication using a *specified definition* of matrix-vector multiplication or vector-matrix multiplication.

- These procedures can be written and run after you write `getitem(M, k)` but before you make any other additions to `Mat`.
- These procedures must *not* be designed to exploit sparsity.
- Your code must *not* use the matrix-vector and vector-matrix multiplication operations that are a part of `Mat`.

- Your code should use procedures
  `mat2rowdict`, `mat2coldict`, `rowdict2mat(rowdict)`, and/or `coldict2mat(coldict)`
  from the `matutil` module.

Try reproducing the results below with the procedures you have written:

- $\begin{bmatrix} -1 & 1 & 2 \\ 1 & 2 & 3 \\ 2 & 2 & 1 \end{bmatrix} \begin{bmatrix} 1 \\ 2 \\ 0 \end{bmatrix} = \begin{bmatrix} 1 \\ 5 \\ 6 \end{bmatrix}$
- $\begin{bmatrix} 4 & 3 & 2 & 1 \end{bmatrix} \begin{bmatrix} -5 & 10 \\ -4 & 8 \\ -3 & 6 \\ -2 & 4 \end{bmatrix} = \begin{bmatrix} -40 & 80 \end{bmatrix}$

Problem 4.17.13: Write the procedure `lin_comb_mat_vec_mult(M,v)`, which multiplies `M` times `v` using the linear-combination definition. For this problem, the only operation on `v` you are allowed is getting the value of an entry using brackets: `v[k]`. The vector returned must be computed as a linear combination.

Problem 4.17.14: Write `lin_comb_vec_mat_mult(v,M)`, which multiply `v` times `M` using the linear-combination definition. For this problem, the only operation on `v` you are allowed is getting the value of an entry using brackets: `v[k]`. The vector returned must be computed as a linear combination.

Problem 4.17.15: Write `dot_product_mat_vec_mult(M,v)`, which multiplies `M` times `v` using the dot-product definition. For this problem, the only operation on `v` you are allowed is taking the dot-product of `v` and another vector and `v`: `u*v` or `v*u`. The entries of the vector returned must be computed using dot-product.

Problem 4.17.16: Write `dot_product_vec_mat_mult(v,M)`, which multiplies `v` times `M` using the dot-product definition. For this problem, the only operation on `v` you are allowed is taking the dot-product of `v` and another vector and `v`: `u*v` or `v*u`. The entries of the vector returned must be computed using dot-product.

Matrix-matrix multiplication definitions in Python

You will write several procedures, each implementing matrix-matrix multiplication using a *specified definition* of matrix-matrix multiplication.

- These procedures can be written and run only after you have written and tested the procedures in `mat.py` that perform matrix-vector and vector-matrix multiplication.
- These procedures must *not* be designed to exploit sparsity.
- Your code must *not* use the matrix-matrix multiplication that is a part of `Mat`. For this reason, you can write these procedures before completing that part of `Mat`.
- Your code should use procedures
  `mat2rowdict`, `mat2coldict`, `rowdict2mat(rowdict)`, and/or `coldict2mat(coldict)`
  from the `matutil` module.

Problem 4.17.17: `Mv_mat_mat_mult(A,B)`, using the matrix-vector multiplication definition of matrix-matrix multiplication. For this procedure, the only operation you are allowed to do on

A is matrix-vector multiplication, using the * operator: `A*v`. Do *not* use the named procedure `matrix_vector_mul` or any of the procedures defined in the previous problem.

Problem 4.17.18: `vM_mat_mat_mult(A,B)`, using the vector-matrix definition. For this procedure, the only operation you are allowed to do on A is vector-matrix-vector multiplication, using the * operator: `v*A`. Do *not* use the named procedure `vector_matrix_mul` or any of the procedures defined in the previous problem.

Dot products via matrix-matrix multiplication

Problem 4.17.19: Let A be a matrix whose column labels countries and whose row labels are votes taken in the United Nations (UN), where $A[i,j]$ is $+1$ or -1 or 0 depending on whether country j votes in favor of or against or neither in vote i.

As in the politics lab, we can compare countries by comparing their voting records. Let $M = A^T A$. Then M's row and column labels are countries, and $M[i,j]$ is the dot-product of country i's voting record with country j's voting record.

The provided file `UN_voting_data.txt` has one line per country. The line consists of the country name, followed by $+1$'s, -1's, and zeroes, separated by spaces. Read in the data and form the matrix A. Then form the matrix $M = A^T A$. (Note: this will take quite a while—from fifteen minutes to an hour, depending on your computer.)

Use M to answer the following questions.

1. Which pair of countries are most opposed? (They have the most negative dot-product.)
2. What are the ten most opposed pairs of countries?
3. Which pair of distinct countries are in the greatest agreement (have the most positive dot-product)?

Hint: the items in $M.f$ are key-value pairs where the value is the dot-product. You can use a comprehension to obtain a list of value-key pairs, and then sort by the value, using the expression `sorted([(value,key) for key,value in M.f.items()])`.

Comprehension practice

Problem 4.17.20: Write the one-line procedure `dictlist_helper(dlist, k)` with the following spec:

- *input:* a list `dlist` of dictionaries which all have the same keys, and a key `k`
- *output:* the list whose i^{th} element is the value corresponding to the key `k` in the ith dictionary of `dlist`
- *example:* With inputs `dlist=[{'a':'apple', 'b':'bear'}, {'a':1, 'b':2}]` and `k='a'`, the output is `['apple', 1]`

The procedure should use a comprehension.

Save your solution to this problem since you will use it later.

The inverse of a 2×2 matrix

Problem 4.17.21:

1. Use a formula given in the text to solve the linear system $\begin{bmatrix} 3 & 4 \\ 2 & 1 \end{bmatrix} \begin{bmatrix} x_1 \\ x_2 \end{bmatrix} = \begin{bmatrix} 1 \\ 0 \end{bmatrix}$.

2. Use the formula to solve the linear system $\begin{bmatrix} 3 & 4 \\ 2 & 1 \end{bmatrix} \begin{bmatrix} y_1 \\ y_2 \end{bmatrix} = \begin{bmatrix} 0 \\ 1 \end{bmatrix}$.

3. Use your solutions to find a 2×2 matrix M such that $\begin{bmatrix} 3 & 4 \\ 2 & 1 \end{bmatrix}$ times M is an identity matrix.

4. Calculate M times $\begin{bmatrix} 3 & 4 \\ 2 & 1 \end{bmatrix}$ and calculate $\begin{bmatrix} 3 & 4 \\ 2 & 1 \end{bmatrix}$ times M, and use Corollary 4.13.19 to decide whether M is the inverse of $\begin{bmatrix} 3 & 4 \\ 2 & 1 \end{bmatrix}$. Explain your answer.

Matrix inverse criterion

Problem 4.17.22: For each of the parts below, use Corollary 4.13.19 to demonstrate that the pair of matrices given are or are not inverse of each other.

1. matrices $\begin{bmatrix} 5 & 1 \\ 9 & 2 \end{bmatrix}, \begin{bmatrix} 2 & -1 \\ -9 & 5 \end{bmatrix}$ over $\mathbb{R}$

2. matrices $\begin{bmatrix} 2 & 0 \\ 0 & 1 \end{bmatrix}, \begin{bmatrix} \frac{1}{2} & 0 \\ 0 & 1 \end{bmatrix}$ over $\mathbb{R}$

3. matrices $\begin{bmatrix} 3 & 1 \\ 0 & 2 \end{bmatrix}, \begin{bmatrix} 1 & \frac{1}{6} \\ -2 & \frac{1}{2} \end{bmatrix}$ over $\mathbb{R}$

4. matrices $\begin{bmatrix} 1 & 0 & 1 \\ 0 & 1 & 0 \end{bmatrix}, \begin{bmatrix} 0 & 1 \\ 0 & 1 \\ 1 & 1 \end{bmatrix}$ over $GF(2)$

Problem 4.17.23: Specify a function f (by domain, co-domain, and rule) that is invertible but such that there is no matrix A such that $f(\boldsymbol{x}) = A\boldsymbol{x}$.

Chapter 5

The Basis

All your bases are belong to us.
Zero Wing, Sega Mega Drive version, 1991, misquoted

5.1 Coordinate systems

5.1.1 René Descartes' idea

In 1618, the French mathematician René Descartes had an idea that forever transformed the way mathematicians viewed geometry.

In deference to his father's wishes, he studied law in college. But something snapped during this time:

> I entirely abandoned the study of letters. Resolving to seek no knowledge other than that of which could be found in myself or else in the great book of the world, I spent the rest of my youth traveling, visiting courts and armies, mixing with people of diverse temperaments and ranks, gathering various experiences, testing myself in the situations which fortune offered me, and at all times reflecting upon whatever came my way so as to derive some profit from it.

After tiring of the Paris social scene, he joined the army of Prince Maurice of Nassau one year, then joined the opposing army of the duke of Bavaria the next year, although he never saw combat.

He had a practice of lying in bed in the morning, thinking about mathematics. He found the prevailing approach to geometry—the approach taken since the ancient Greeks—needlessly cumbersome.

His great idea about geometry came to him, according to one story, while lying in bed and watching a fly on the ceiling of his room, near a corner of the room. Descartes realized that the location of the fly could be described in terms of two numbers: its distance from the two walls it was near. Significantly, Descartes realized that this was true even if the two walls were not perpendicular. He further realized that geometrical analysis could thereby be reduced to algebra.

5.1.2 Coordinate representation

The two numbers characterizing the fly's location are what we now call *coordinates*. In vector analysis, a *coordinate system* for a vector space $\mathcal{V}$ is specified by generators $\boldsymbol{a}_1, \ldots, \boldsymbol{a}_n$ of $\mathcal{V}$. Every vector $\boldsymbol{v}$ in $\mathcal{V}$ can be written as a linear combination

$$\boldsymbol{v} = \alpha_1 \boldsymbol{a}_1 + \cdots + \alpha_n \boldsymbol{a}_n$$

We can therefore represent $\boldsymbol{v}$ by the vector $[\alpha_1, \ldots, \alpha_n]$ of coefficients. In this context, the coefficients are called *coordinates*, and the vector $[\alpha_1, \ldots, \alpha_n]$ is called the *coordinate representation* of $\boldsymbol{v}$ in terms of $\boldsymbol{a}_1, \ldots, \boldsymbol{a}_n$.

But assigning coordinates to points is not enough. In order to avoid confusion, we must ensure that each point is assigned coordinates in exactly one way. To ensure this, we must use care in selecting the generators $\boldsymbol{a}_1, \ldots, \boldsymbol{a}_n$. We address *existence and uniqueness of representation* in Section 5.7.1.

Example 5.1.1: The vector $\boldsymbol{v} = [1, 3, 5, 3]$ is equal to $1\,[1, 1, 0, 0] + 2\,[0, 1, 1, 0] + 3\,[0, 0, 1, 1]$, so the coordinate representation of $\boldsymbol{v}$ in terms of the vectors $[1, 1, 0, 0], [0, 1, 1, 0], [0, 0, 1, 1]$ is $[1, 2, 3]$.

Example 5.1.2: What is the coordinate representation of the vector $[6, 3, 2, 5]$ in terms of the vectors $[2, 2, 2, 3], [1, 0, -1, 0], [0, 1, 0, 1]$? Since

$$[6, 3, 2, 5] = 2\,[2, 2, 2, 3] + 2\,[1, 0, -1, 0] - 1\,[0, 1, 0, 1],$$

the coordinate representation is $[2, 2, -1]$.

Example 5.1.3: Now we do an example with vectors over $GF(2)$. What is the coordinate representation of the vector [0,0,0,1] in terms of the vectors [1,1,0,1], [0,1,0,1], and [1,1,0,0]? Since

$$[0, 0, 0, 1] = 1\,[1, 1, 0, 1] + 0\,[0, 1, 0, 1] + 1\,[1, 1, 0, 0]$$

the coordinate representation of $[0, 0, 0, 1]$ is $[1, 0, 1]$.

5.1.3 Coordinate representation and matrix-vector multiplication

Why put the coordinates in a vector? This actually makes a lot of sense in view of the linear-combinations definitions of matrix-vector and vector-matrix multiplication. Suppose the coordinate axes are $\boldsymbol{a}_1, \ldots, \boldsymbol{a}_n$. We form a matrix $A = \begin{bmatrix} \boldsymbol{a}_1 & \cdots & \boldsymbol{a}_n \end{bmatrix}$ whose columns are the generators.

- We can write the statement "$\boldsymbol{u}$ is the coordinate representation of $\boldsymbol{v}$ in $\boldsymbol{a}_1, \ldots, \boldsymbol{a}_n$" as the matrix-vector equation
$$A\boldsymbol{u} = \boldsymbol{v}$$
- Therefore, to go from a coordinate representation $\boldsymbol{u}$ to the vector being represented, we multiply A times $\boldsymbol{u}$.
- Moreover, to go from a vector $\boldsymbol{v}$ to its coordinate representation, we can solve the matrix-vector equation $A\boldsymbol{x} = \boldsymbol{v}$. Because the columns of A are generators for $\mathcal{V}$ and $\boldsymbol{v}$ belongs to $\mathcal{V}$, the equation must have at least one solution.

We will often use matrix-vector multiplication in the context of coordinate representations.

5.2 First look at lossy compression

In this section, I describe one application of coordinate representation. Suppose we need to store many 2000×1000 grayscale images. Each such image can be represented by a D-vector where $D = \{0, 1, \ldots, 19999\} \times \{0, 1, \ldots, 999\}$. However, we want to store the images more compactly. We consider three strategies.

5.2.1 Strategy 1: Replace vector with closest sparse vector

If an image vector has few nonzeroes, it can be stored compactly—but this will happen only rarely. We therefore consider a strategy that replaces an image with a different image, one that is sparse but that we hope will be perceptually similar. Such a compression method is said to be *lossy* since information in the original image is lost.

Consider replacing the vector with the closest k-sparse vector. This strategy raises a Question:

Question 5.2.1: Given a vector $\boldsymbol{v}$ and a positive integer k, what is the k-sparse vector closest to $\boldsymbol{v}$?

We are not yet in a position to say what "closest" means because we have not defined a distance between vectors. The distance between vectors over $\mathbb{R}$ is the subject of Chapter 8, where we will will discover that the closest k-sparse vector is obtained from $\boldsymbol{v}$ by simply replacing all but the k largest-magnitude entries by zeroes. The resulting vector will be k-sparse—and therefore, for, say, $k = 200,000$, can be represented more compactly. But is this a good way to compress an image?

Example 5.2.2: The image  consists of a single row of four pixels, with intensities 200, 200, 75, 75. The image is thus represented by four numbers. The closest 2-sparse image, which has intensities 200, 200, 0, 0, is .

Here is a realistic image:

and here is the result of suppressing all but 10% of the entries:

The result is far from the original image since so many of the pixel intensities have been set to zero. This approach to compression won't work well.

5.2.2 Strategy 2: Represent image vector by its coordinate representation

Here is another strategy, one that will incur no loss of fidelity to the original image.

- Before trying to compress any images, select a collection of vectors $\boldsymbol{a}_1, \ldots, \boldsymbol{a}_n$.
- Next, for each image vector, find and store its coordinate representation $\boldsymbol{u}$ in terms of $\boldsymbol{a}_1, \ldots, \boldsymbol{a}_n$.[1]
- To recover the original image from the coordinate representation, compute the corresponding linear combination.[2]

Example 5.2.3: We let $\boldsymbol{a}_1 =$  (an image with one row of pixels with intensities 255, 0, 255, 0) and $\boldsymbol{a}_2 =$  (an image with one row of pixels with intensities 0, 255, 0, 255).

Now suppose we want to represent the image  (with intensities 200, 75, 200, 75) in terms of $\boldsymbol{a}_1$ and $\boldsymbol{a}_2$.

 $= \frac{200}{255}\boldsymbol{a}_1 + \frac{75}{255}\boldsymbol{a}_2$

Thus this image is represented in compressed form by the coordinate representation $[\frac{200}{255}, \frac{75}{255}]$.

On the other hand, the image  (intensities 255, 200, 150, 90) cannot be written as a linear combination of $\boldsymbol{a}_1$ and $\boldsymbol{a}_2$, and so has no coordinate representation in terms of these vectors.

As the previous example suggests, for this strategy to work reliably, we need to ensure that every possible $2,000 \times 1,000$ image vector can be represented as a linear combination of $\boldsymbol{a}_1, \ldots, \boldsymbol{a}_n$. This comes down to asking whether $\mathbb{R}^D = \text{Span}\ \{\boldsymbol{a}_1, \ldots, \boldsymbol{a}_n\}$.

Formulated in greater generality, this is a Fundamental Question:

Question 5.2.4: For a given vector space $\mathcal{V}$, how can we tell if $\mathcal{V} = \text{Span}\ \{\boldsymbol{a}_1, \ldots, \boldsymbol{a}_n\}$?

Furthermore, the strategy will only be useful in compression if the number n of vectors used in linear combinations is much smaller than the number of pixels. Is it possible to select such vectors? What is the minimum number of vectors whose span equals $\mathbb{R}^D$?

Formulated in greater generality, this is another Fundamental Question:

Question 5.2.5: For a given vector space $\mathcal{V}$, what is the minimum number of vectors whose span equals $\mathcal{V}$?

It will turn out that our second strategy for image compression will fail: the minimum number of vectors required to span the set of all possible $2,000 \times 1,000$ images is not small enough to achieve any compression at all.

Strategy 3: A hybrid approach

The successful strategy will combine both of the previous two strategies: coordinate representation and closest k-sparse vector:

Step 1: Select vectors $\boldsymbol{a}_1, \ldots, \boldsymbol{a}_n$.

Step 2: For each image you want to compress, take the corresponding vector $\boldsymbol{v}$ and find its coordinate representation $\boldsymbol{u}$ in terms of $\boldsymbol{a}_1, \ldots, \boldsymbol{a}_n$.[3]

Step 3: Next, replace $\boldsymbol{u}$ with the closest k-sparse vector $\tilde{\boldsymbol{u}}$, and store $\tilde{\boldsymbol{u}}$.

[1] You could do this by solving a matrix-vector equation, as mentioned in Section 5.1.3.
[2] You could do this by matrix-vector multiplication, as mentioned in Section 5.1.3.
[3] You could do this by solving a matrix-vector equation, as mentioned in Section 5.1.3.

Step 4: To recover an image from $\tilde{\boldsymbol{u}}$, calculate the corresponding linear combination of $\boldsymbol{a}_1, \ldots \boldsymbol{a}_n$.[4]

How well does this method work? It all depends on which vectors we select in Step 1. We need this collection of vectors to have two properties:

- *Step 2 should always succeed.* It should be possible to express any vector $\boldsymbol{v}$ in terms of the vectors in the collection.
- *Step 3 should not distort the image much.* The image whose coordinate representation is $\tilde{\boldsymbol{u}}$ should not differ much from the original image, the image whose coordinate representation is $\boldsymbol{u}$.

How well does this strategy work? Following a well-known approach for selecting the vectors in Step 1 (described in detail in Chapter 10), we get the following nice result using only 10% of the numbers:

5.3 Two greedy algorithms for finding a set of generators

In this section, we consider two algorithms to address Question 5.2.5:

> *For a given vector space $\mathcal{V}$, what is the minimum number of vectors whose span equals $\mathcal{V}$?*

It will turn out that the ideas we discover will eventually help us answer many other questions, including Question 5.2.4.

5.3.1 Grow algorithm

How can we obtain a minimum number of vectors? Two natural approaches come to mind, the *Grow* algorithm and the *Shrink* algorithm. Here we present the *Grow* algorithm.

```
def GROW(𝒱)
  B = ∅
  repeat while possible:
      find a vector v in 𝒱 that is not in Span B, and put it in B.
```

The algorithm stops when there is no vector to add, at which time B spans all of $\mathcal{V}$. Thus, if the algorithm stops, it will have found a generating set. The question is: is it bigger than necessary?

Note that this algorithm is not very restrictive: we ordinarily have lots of choices of which vector to add.

[4]You could do this by matrix-vector multiplication, as mentioned in Section 5.1.3.

Example 5.3.1: We use the Grow algorithm to select a set of generators for $\mathbb{R}^3$. In Section 3.2.3, we defined the *standard* generators for $\mathbb{R}^n$. In the first iteration of the Grow algorithm we add to our set B the vector $[1, 0, 0]$ It should be apparent that $[0, 1, 0]$ is not in Span $\{[1, 0, 0]\}$. In the second iteration, we therefore add this vector to B. Likewise, in the third iteration we add $[0, 0, 1]$ to B. We can see that any vector $\boldsymbol{v} = (\alpha_1, \alpha_2, \alpha_3) \in \mathbb{R}^3$ is in Span $(\boldsymbol{e}_1, \boldsymbol{e}_2, \boldsymbol{e}_3)$ since we can form the linear combination

$$\boldsymbol{v} = \alpha_1 \boldsymbol{e}_1 + \alpha_2 \boldsymbol{e}_2 + \alpha_3 \boldsymbol{e}_3$$

Therefore there is no vector $\boldsymbol{v} \in \mathbb{R}^3$ to add to B, and the algorithm stops.

5.3.2 Shrink algorithm

In our continuing effort to find a minimum set of vectors that span a given vector space $\mathcal{V}$, we now present the *Shrink* algorithm.

```
def SHRINK(V)
  B = some finite set of vectors that spans V
  repeat while possible:
      find a vector v in B such that Span (B − {v}) = V, and remove v from B.
```

The algorithm stops when there is no vector whose removal would leave a spanning set. At every point during the algorithm, B spans $\mathcal{V}$, so it spans $\mathcal{V}$ at the end. Thus the algorithm certainly finds a generating set. The question is, again: is it bigger than necessary?

Example 5.3.2: Consider a simple example where B initially consists of the following vectors:

$$\begin{aligned} \boldsymbol{v}_1 &= [1, 0, 0] \\ \boldsymbol{v}_2 &= [0, 1, 0] \\ \boldsymbol{v}_3 &= [1, 2, 0] \\ \boldsymbol{v}_4 &= [3, 1, 0] \end{aligned}$$

In the first iteration, since $\boldsymbol{v}_4 = 3\boldsymbol{v}_1 + \boldsymbol{v}_2$, we can remove $\boldsymbol{v}_4$ from B in the first iteration without changing Span B. After this iteration, $B = \{\boldsymbol{v}_1, \boldsymbol{v}_2, \boldsymbol{v}_3\}$. In the second iteration, since $\boldsymbol{v}_3 = \boldsymbol{v}_1 + 2\boldsymbol{v}_2$, we remove $\boldsymbol{v}_3$ from B, resulting in $B = \{\boldsymbol{v}_1, \boldsymbol{v}_2\}$. Finally, note that Span $B = \mathbb{R}^3$ and that neither $\boldsymbol{v}_1$ nor $\boldsymbol{v}_2$ alone could generate $\mathbb{R}^3$. Therefore the Shrink algorithm stops.

Note: These are not algorithms that you can go and implement. They are abstract algorithms, algorithmic thought experiments:

- We don't specify how the input—a vector space—is specified.
- We don't specify how each step is carried out.
- We don't specify which vector to choose in each iteration.

In fact we later exploit the last property—the freedom to choose which vector to add or remove—in our proofs.

5.3.3 When greed fails

Before analyzing the Grow and Shrink algorithms for finding minimum generating set, I want to look at how similar algorithms perform on a different problem, a problem on graphs.

Dominating set A *dominating set* is a set of nodes such that every node in the graph is either in the set or is a neighbor (via a single edge) of some node in the set. The goal of the *minimum-dominating-set problem* is to find a dominating set of minimum size.

I like to think of a dominating set as a set of guards posted at intersections. Each intersection must be guarded by a guard at that intersection or a neighboring intersection.

Consider this graph:

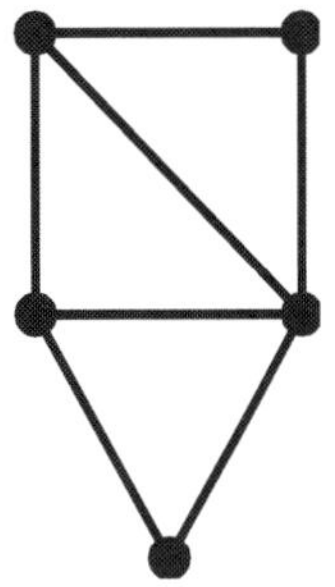

A dominating set is indicated here:

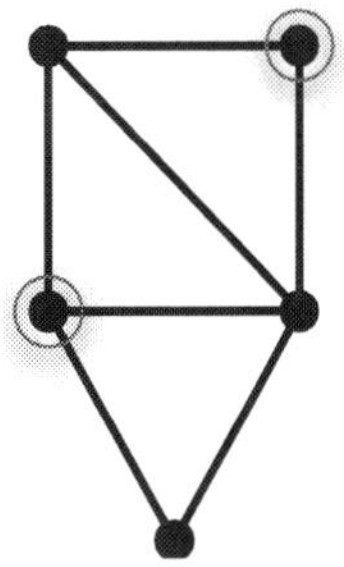

You might consider finding a dominating set using a Grow algorithm:

> *Grow Algorithm for Dominating Set:*
> Initialize B to be empty; while B is not a dominating set, add a node v to B

or a Shrink algorithm:

> *Shrink Algorithm for Dominating Set:*
> Initialize B to contain all nodes; while there is a node v such that $B - \{v\}$ is a dominating set, remove v from B

but either of these algorithms could, by unfortunate choices, end up selecting the dominating set shown above, whereas there is a smaller dominating set:

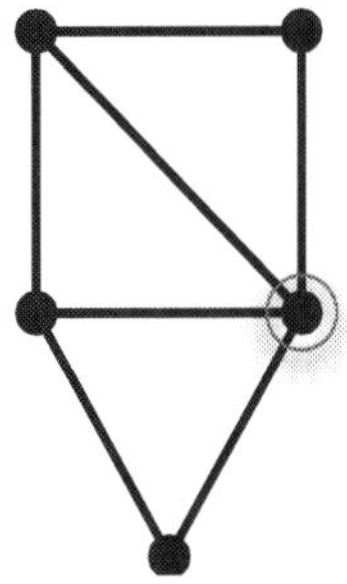

Grow and Shrink algorithms are called *greedy* algorithms because in each step the algorithm makes a choice without giving thought to the future. This example illustrates that greedy algorithms are not reliably good at finding the best solutions.

The Grow and Shrink algorithms for finding a smallest generating set for a vector space are remarkable: as we will see, they do in fact find the smallest solution.

5.4 Minimum Spanning Forest and $GF(2)$

I will illustrate the Grow and Shrink algorithms using a graph problem: *Minimum Spanning Forest*. Imagine you must replace the hot-water delivery network for the Brown University campus. You are given a graph with weights on edges:

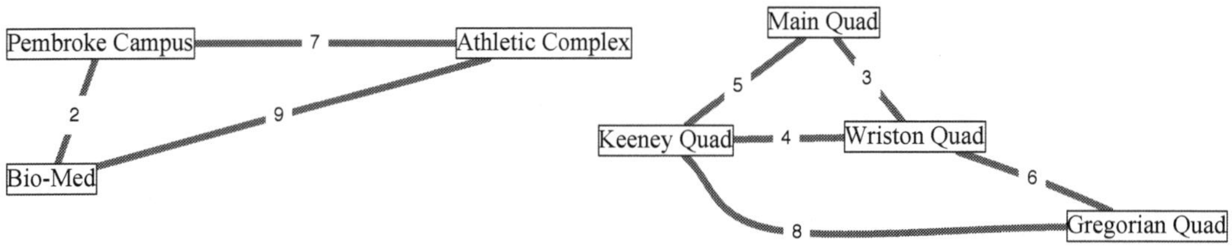

where there is a node for each campus area. An edge represents a possible hot-water pipe between different areas, and the edge's weight represents the cost of installing that pipe. Your goal is to select a set of pipes to install so every pair of areas that are connected in the graph are connected by the installed pipes, and to do so at minimum cost.

5.4.1 Definitions

Definition 5.4.1: For a graph G, a sequence of edges

$$[\{x_1, x_2\}, \{x_2, x_3\}, \{x_3, x_4\}, \ldots, \{x_{k-1}, x_k\}]$$

is called an x_1-to-x_k *path* (or a *path from x_1 to x_k*).

In this graph

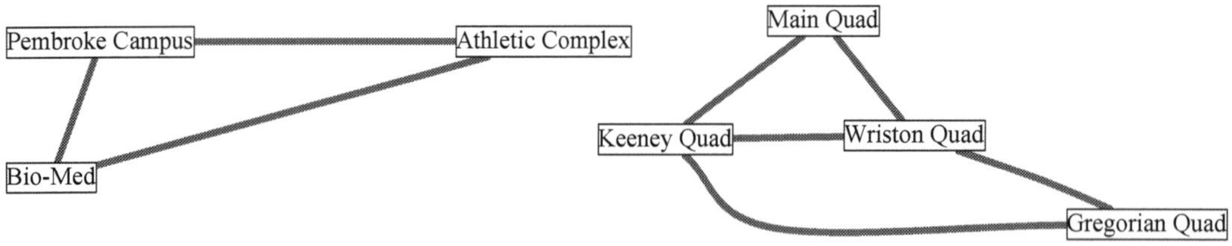

there is a path from "Main Quad" to "Gregorian Quad" but no path from "Main Quad" to "Athletic Complex".

Definition 5.4.2: A set S of edges is *spanning* for a graph G if, for every edge $\{x, y\}$ of G, there is an x-to-y path consisting of edges of S.

For example, the dark edges in the following diagram are spanning for the graph depicted:

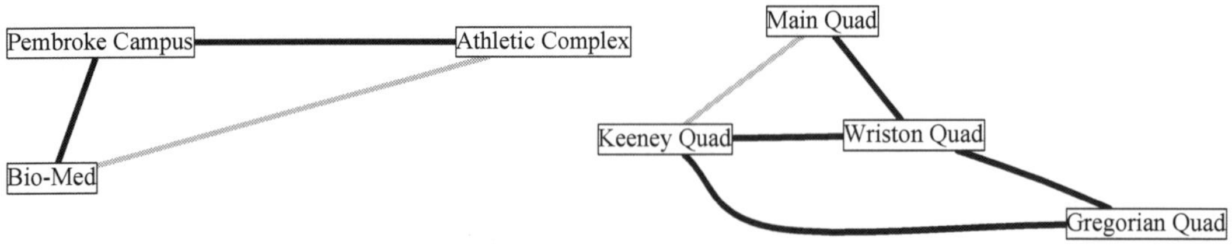

We will soon see a connection between this sense of "spanning" and the sense in which we use the term in linear algebra.

Definition 5.4.3: A *forest* is a set of edges containing no cycles (loops possibly consisting of several edges).

For example, the dark edges in the earlier diagram do *not* form a forest because there are three dark edges that form a cycle. On the other hand, the dark edges in the following diagram *do* form a forest:

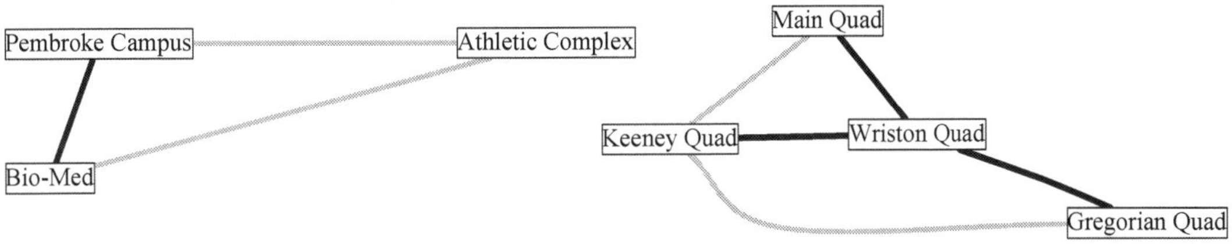

A graph-theoretical forest resembles a biological forest, i.e. collection of trees, in that a tree's branches do not diverge and then rejoin to form a cycle.

We will give two algorithms for a computational problem, *Minimum Spanning Forest*,[5] abbreviated MSF.

- *input:* a graph G, and an assignment of real-number *weights* to the edges of G.
- *output:* a minimum-weight set B of edges that is spanning and a forest.

The reason for the term "forest" is that the solution need not contain any cycles (as we will see), so the solution resembles a collection of trees. (A tree's branches do not diverge and then rejoin to form a cycle.)

5.4.2 The Grow algorithm and the Shrink algorithm for *Minimum Spanning Forest*

There are many algorithms for *Minimum Spanning Forest* but I will focus on two: a Grow algorithm and a Shrink algorithm. First, the Grow algorithm:

```
def Grow(G)
  B := ∅
  consider the edges in order, from lowest-weight to highest-weight
  for each edge e:
       if e's endpoints are not yet connected via edges in B:
         add e to B.
```

This algorithm exploits the freedom we have in the Grow algorithm to select which vector to add.

The weights in increasing order are: 2, 3, 4, 5, 6, 7, 8, 9. The solution obtained, which consists of the edges with weights 2, 3, 4, 6, 7, is this:

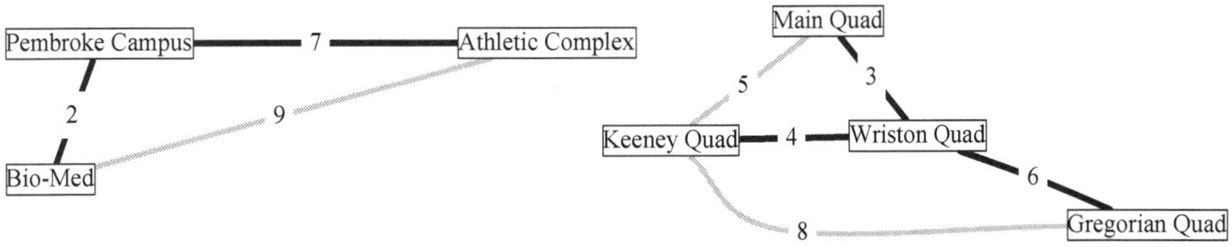

Here is the Shrink algorithm:

[5]The problem is also called *minimum-weight spanning forest*. The problem *maximum-weight spanning forest* can be solved by the same algorithms by just negating the weights.

```
def SHRINK(G)
  B = {all edges}
  consider the edges in order, from highest-weight to lowest-weight
  for each edge e:
      if every pair of nodes are connected via B − {e}:
        remove e from B.
```

This algorithm exploits the freedom in the Shrink algorithm to select which vector to remove. The weights in decreasing order are: 9, 8, 7, 6, 5, 4, 3, 2. The solution consists of the edges with weights 7, 6, 4, 3, and 2.

The Grow algorithm and the Shrink algorithm came up with the same solution, the correct solution.

5.4.3 Formulating *Minimum Spanning Forest* in linear algebra

It is no coincidence that the Grow and Shrink algorithms for minimum spanning forest resemble those for finding a set of generators for a vector space. In this section, we describe how to model a graph using vectors over $GF(2)$.

Let $D = \{\text{Pembroke}, \text{Athletic}, \text{Bio-Med}, \text{MainKeeney}, \text{Wriston}, \text{Gregorian}\}$ be the set of nodes. A subset of D is represented by the vector with ones in the corresponding entries and zeroes elsewhere. For example, the subset {`Pembroke, Main, Gregorian`} is represented by the vector whose dictionary is {`Pembroke:one, Main:one, Gregorian:one`}, which we can write as

Pembroke	Athletic	Bio-Med	Main	Keeney	Wriston	Gregorian
1			1			1

Each edge is a two-element subset of D, so it is represented by a vector, namely the vector that has a one at each of the endpoints of e and zeroes elsewhere. For example, the edge connecting Pembroke and Athletic is represented by the vector `{'Pembroke':one, 'Athletic':one}`.

Here are the vectors corresponding to all the edges in our graph:

edge	vector						
	Pem.	Athletic	Bio-Med	Main	Keeney	Wriston	Greg.
{Pem., Athletic}	1	1					
{Pem., Bio-Med}	1		1				
{Athletic, Bio-Med}		1	1				
{Main, Keeney}				1	1		
{Main, Wriston}				1		1	
{Keeney, Wriston}					1	1	
{Keeney, Greg.}					1		1
{Wriston, Greg.}						1	1

The vector representing {`Keeney, Gregorian`},

Pembroke	Athletic	Bio-Med	Main	Keeney	Wriston	Gregorian
				1		1

is the sum, for example, of the vectors representing {`Keeney, Main`}, {`Main, Wriston`}, and {`Wriston, Gregorian`} :

Pembroke	Athletic	Bio-Med	Main	Keeney	Wriston	Gregorian
			1	1		
			1		1	
					1	1

because the 1's in entries `Main` and `Wriston` cancel out, leaving 1's just in entries `Keeney` and `Gregorian`.

In general, a vector with 1's in entries x and y is the sum of vectors corresponding to edges that form an x-to-y path in the graph. Thus, for these vectors, it is easy to tell whether one vector is in the span of some others.

Example 5.4.4: The span of the vectors representing

`{Pembroke, Bio-Med}`, `{Main, Wriston}`, `{Keeney, Wriston}`, `{Wriston, Gregorian }`

contains the vector corresponding to `{Main, Keeney}` but not the vector corresponding to `{Athletic, Bio-Med }` or the vector corresponding to `{Bio-Med, Main}`.

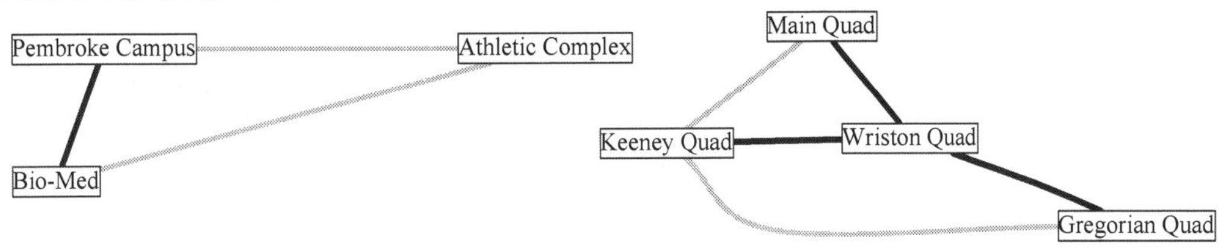

Example 5.4.5: The span of the vectors representing

`{Athletic, Bio-Med}`, `{Main, Keeney }`, `{Keeney, Wriston}`, `{Main, Wriston}`

does not contain `{Pembroke, Keeney}` or `{Main, Gregorian}` or `{Pembroke, Gregorian}`:

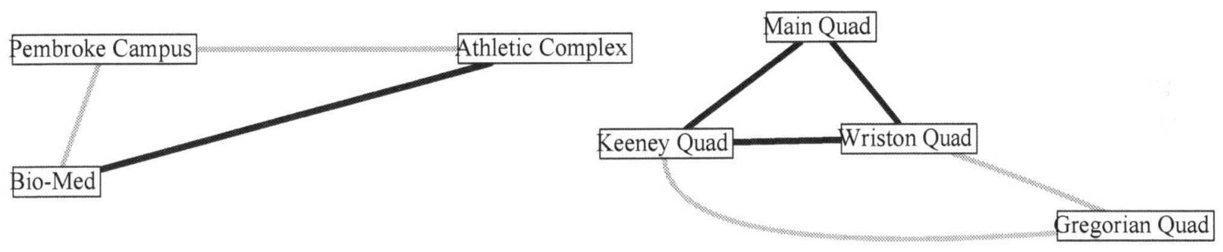

We see that the conditions used in the MSF algorithms to decide whether to add an edge (in the Grow algorithm) or remove an edge (in the Shrink algorithm) are just testing a span condition, exactly as in the vector Grow and Shrink algorithms.

5.5 Linear dependence

5.5.1 The Superfluous-Vector Lemma

To better understand the Grow and Shrink algorithms, we need to understand what makes it possible to omit a vector from a set of generators without changing the span.

Lemma 5.5.1 (Superfluous-Vector Lemma): For any set S and any vector $\boldsymbol{v} \in S$, if $\boldsymbol{v}$ can be written as a linear combination of the other vectors in S then Span $(S - \{\boldsymbol{v}\}) =$ Span S

Proof

Let $S = \{\boldsymbol{v}_1, \ldots, \boldsymbol{v}_n\}$, and suppose

$$\boldsymbol{v}_n = \alpha_1 \boldsymbol{v}_1 + \alpha_2 \boldsymbol{v}_2 + \cdots + \alpha_{n-1} \boldsymbol{v}_{n-1} \tag{5.1}$$

Our goal is to show that every vector in Span S is also in Span $(S - \{\boldsymbol{v}\})$. Every vector $\boldsymbol{v}$ in Span S can be written as

$$\boldsymbol{v} = \beta_1 \boldsymbol{v}_1 + \cdots \beta_n \boldsymbol{v}_n$$

Using Equation 5.1 to substitute for $\boldsymbol{v}_n$, we obtain

$$\begin{aligned} \boldsymbol{v} &= \beta_1 \boldsymbol{v}_1 + \beta_2 \boldsymbol{v}_2 + \cdots + \beta_n (\alpha_1 \boldsymbol{v}_1 + \alpha_2 \boldsymbol{v}_2 + \cdots + \alpha_{n-1} \boldsymbol{v}_{n-1}) \\ &= (\beta_1 + \beta_n \alpha_1)\boldsymbol{v}_1 + (\beta_2 + \beta_n \alpha_2)\boldsymbol{v}_2 + \cdots + (\beta_{n-1} + \beta_n \alpha_{n-1})\boldsymbol{v}_{n-1} \end{aligned}$$

which shows that an arbitrary vector in Span S can be written as a linear combination of vectors in $S - \{\boldsymbol{v}_n\}$ and is therefore in Span $(S - \{\boldsymbol{v}_n\})$. □

5.5.2 Defining linear dependence

The concept that connects the Grow algorithm and the Shrink algorithm, shows that each algorithm produces an optimal solution, resolves many other questions, and generally saves the world is.... linear dependence.

Definition 5.5.2: Vectors $\boldsymbol{v}_1, \ldots, \boldsymbol{v}_n$ are *linearly dependent* if the zero vector can be written as a **nontrivial** linear combination of the vectors:

$$\mathbf{0} = \alpha_1 \boldsymbol{v}_1 + \cdots + \alpha_n \boldsymbol{v}_n$$

In this case, we refer to the linear combination as a *linear dependency* in $\boldsymbol{v}_1, \ldots, \boldsymbol{v}_n$.

On the other hand, if the *only* linear combination that equals the zero vector is the trivial linear combination, we say $\boldsymbol{v}_1, \ldots, \boldsymbol{v}_n$ are linearly *in*dependent.

Remember that a nontrivial linear combination is one in which at least one coefficient is nonzero.

Example 5.5.3: The vectors $[1, 0, 0]$, $[0, 2, 0]$, and $[2, 4, 0]$ are linearly dependent, as shown by the following equation:

$$2\,[1, 0, 0] + 2\,[0, 2, 0] - 1\,[2, 4, 0] = [0, 0, 0]$$

Thus $2\,[1, 0, 0] + 2\,[0, 2, 0] - 1\,[2, 4, 0]$ is a linear dependency in $[1, 0, 0]$, $[0, 2, 0]$, and $[2, 4, 0]$.

Example 5.5.4: The vectors $[1, 0, 0]$, $[0, 2, 0]$, and $[0, 0, 4]$ are linearly independent. This is easy to see because of the particularly simple form of these vectors: each has a nonzero entry in a position in which the others have zeroes. Consider any nontrivial linear combination

$$\alpha_1\,[1, 0, 0] + \alpha_2\,[0, 2, 0] + \alpha_3\,[0, 0, 4]$$

i.e., one in which at least one of the coefficients is nonzero. Suppose α_1 is nonzero. Then the first entry of $\alpha_1\,[1, 0, 0]$ is nonzero. Since the first entry of $\alpha_2\,[0, 2, 0]$ is zero and the first entry of $\alpha_3\,[0, 0, 4]$ is zero, adding these other vectors to $\alpha_1\,[1, 0, 0]$ cannot affect the first entry, so it remains nonzero in the sum. This shows that we cannot obtain the zero vector from any linear combination in which the first coefficient is nonzero. A similar argument applies when the second coefficient is nonzero and when the third coefficient is nonzero. Thus there is *no* way to obtain the zero vector from a nontrivial linear combination.

Example 5.5.3 (Page 226) uses vectors for which it is easy to find an equation showing linear dependence. Example 5.5.4 (Page 226) uses a very simple argument to show linear independence. Most of the time, it is not so easy to tell!

Computational Problem 5.5.5: *Testing linear dependence*

- *input:* a list $[\boldsymbol{v}_1, \ldots, \boldsymbol{v}_n]$ of vectors
- *output:* DEPENDENDENT if the vectors are linearly dependent, and INDEPENDENT otherwise.

This Computational Problem is a restatement of two old Questions:

- Let $A = \left[\begin{array}{c|c|c} \boldsymbol{v}_1 & \cdots & \boldsymbol{v}_n \end{array}\right]$. The vectors $\boldsymbol{v}_1, \ldots, \boldsymbol{v}_n$ are linearly dependent if and only if there is a nonzero vector $\boldsymbol{u}$ such that $A\boldsymbol{u} = \mathbf{0}$, i.e. if and only if the null space of A contains a nonzero vector. This is Question 4.7.7: *How can we tell if the null space of a matrix consist solely of the zero vector?*

- As noted in Section 4.7.2, that Question is equivalent to Question 3.6.5: *How can we tell if a homogeneous linear system has only the trivial solution?*

Problem 5.5.6: Show that no independent set contains the zero vector.

5.5.3 Linear dependence in *Minimum Spanning Forest*

What about linear dependence in *Minimum Spanning Forest*? We can get the zero vector by adding together vectors corresponding to edges that form a cycle: in such a sum, for each entry x, there are exactly two vectors having 1's in position x. For example, the vectors corresponding to

$$\{\texttt{Main, Keeney}\}, \{\texttt{Keeney, Wriston}\}, \{\texttt{Main, Wriston}\}$$

are as follows:

Pembroke	Athletic	Bio-Med	Main	Keeney	Wriston	Gregorian
			1	1		
				1	1	
			1		1	

The sum of these vectors is the zero vector.

Therefore, if S is a collection of vectors corresponding to edges, and some subset of these edges form a cycle, we can get the zero vector as a nontrivial linear combination by assigning coefficient 1 to the vectors corresponding to the edges in the subset.

Example 5.5.7: The vectors corresponding to

$$\{\texttt{Main, Keeney}\}, \{\texttt{Main, Wriston}\}, \{\texttt{Keeney, Wriston }\}, \{\texttt{Wriston, Gregorian}\}$$

are linearly dependent because these edges include a cycle.

The zero vector is equal to the nontrivial linear combination

		Pembroke	Athletic	Bio-Med	Main	Keeney	Wriston	Gregorian
1	*				1	1		
+ 1	*				1		1	
+ 1	*					1	1	
+ 0	*						1	1

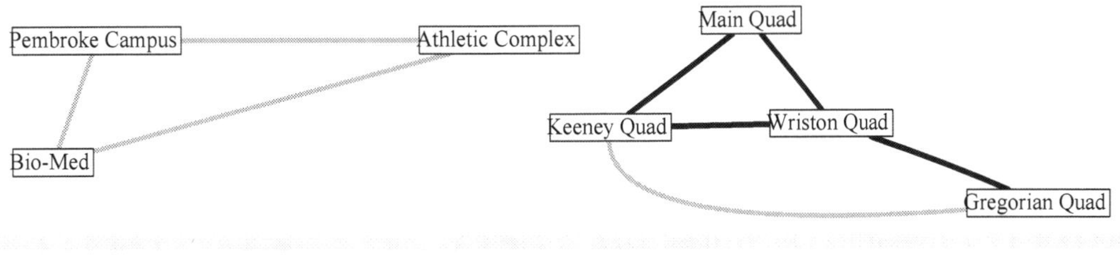

Conversely, if a set of edges contains no cycle (i.e. is a forest) then the corresponding set of vectors is linearly independent, as in the following diagram:

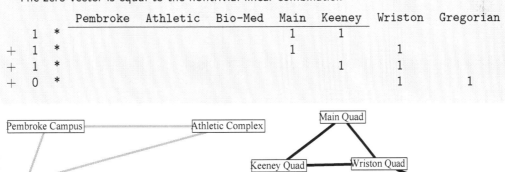

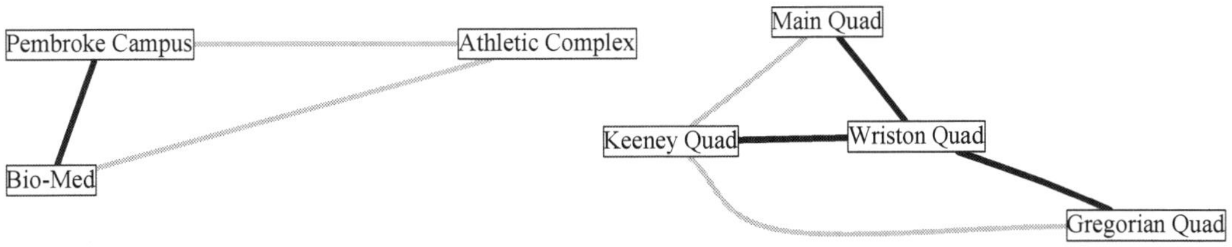

5.5.4 Properties of linear (in)dependence

Lemma 5.5.8: A subset of a linearly independent set is linearly independent.

For example, in the context of MSF, the set of vectors corresponding to a spanning forest is linearly independent, so any subset is also linearly independent.

Proof

Let S and T be subsets of vectors, and suppose S is a subset of T. Our goal is to prove that if T is linearly independent then S is linearly independent. This is equivalent to the contrapositive: if S is linearly dependent then T is linearly dependent.

It is easy to see that this is true: if the zero vector can be written as a nontrivial linear combination of some vectors, it can be so written even if we allow some additional vectors to be in the linear combination.

Here is a more formal proof. Write $T = \{\boldsymbol{s}_1, \ldots, \boldsymbol{s}_n, \boldsymbol{t}_1, \ldots, \boldsymbol{t}_k\}$ where $S = \{\boldsymbol{s}_1, \ldots, \boldsymbol{s}_n\}$. Suppose that S is linearly dependent. Then there are coefficients $\alpha_1, \ldots, \alpha_n$, not all zero, such that

$$\mathbf{0} = \alpha_1 \boldsymbol{s}_1 + \cdots + \alpha_n \boldsymbol{s}_n$$

Therefore

$$\mathbf{0} = \alpha_1 \boldsymbol{s}_1 + \cdots + \alpha_n \boldsymbol{s}_n + 0\, \boldsymbol{t}_1 + \cdots 0\, \boldsymbol{t}_k$$

which shows that the zero vector can be written as a nontrivial linear combination of thc vectors of T, i.e. that T is linearly dependent. □

Lemma 5.5.9 (Span Lemma): Let $\boldsymbol{v}_1, \ldots, \boldsymbol{v}_n$ be vectors. A vector $\boldsymbol{v}_i$ is in the span of the other vectors if and only if the zero vector can be written as a linear combination of $\boldsymbol{v}_1, \ldots, \boldsymbol{v}_n$ in which the coefficient of $\boldsymbol{v}_i$ is nonzero.

In the context of graphs, the Span Lemma states that an edge e is in the span of other edges if there is a cycle consisting of e and a subset of the other edges.

Proof

Like many "if and only if" proofs, this one has two directions.

First suppose $\boldsymbol{v}_i$ is in the span of the other vectors. That is, there exist coefficients $\alpha_1, \ldots, \alpha_{n-1}$ such that

$$\boldsymbol{v}_i = \alpha_1 \boldsymbol{v}_1 + \cdots + \alpha_{i-1} \boldsymbol{v}_{i-1} + \alpha_{i+1} \boldsymbol{v}_{i+1} + \cdots \alpha_n \boldsymbol{v}_n$$

Moving $\boldsymbol{v}_i$ to the other side, we can write

$$\mathbf{0} = \alpha_1 \boldsymbol{v}_1 + \cdots + (-1)\, \boldsymbol{v}_i + \cdots + \alpha_n \boldsymbol{v}_n$$

which shows that the all-zero vector can be written as a linear combination of $\boldsymbol{v}_1, \ldots, \boldsymbol{v}_n$ in which the coefficient of $\boldsymbol{v}_i$ is nonzero.

Now for the other direction. Suppose there are coefficients $\alpha_1, \ldots, \alpha_n$ such that

$$\mathbf{0} = \alpha_1 \boldsymbol{v}_1 + \cdots + \alpha_i \boldsymbol{v}_i + \cdots + \alpha_n \boldsymbol{v}_n$$

and such that $\alpha_i \neq 0$.

Subtracting $\alpha_i \boldsymbol{v}_i$ from both sides and dividing by $-\alpha_i$ yields

$$1\, \boldsymbol{v}_i = (\alpha_1 / - \alpha_i)\, \boldsymbol{v}_1 + \cdots + (\alpha_{i-1} / - \alpha_i)\, \boldsymbol{v}_{i-1} + (\alpha_{i+1} / - \alpha_i)\, \boldsymbol{v}_{i+1} + \cdots + (\alpha_n / - \alpha_i)\, \boldsymbol{v}_n$$

which shows that $\boldsymbol{v}_i$ is in the span of the other vectors. □

5.5.5 Analyzing the Grow algorithm

Corollary 5.5.10 (Grow-Algorithm Corollary): The vectors obtained by the Grow algorithm are linearly independent.

Proof

For $n = 1, 2, \ldots$, let $\boldsymbol{v}_n$ be the vector added to B in the nth iteration of the Grow algorithm. We show by induction that $\boldsymbol{v}_1, \boldsymbol{v}_2, \ldots, \boldsymbol{v}_n$ are linearly independent.

For $n = 0$, there are no vectors, so the claim is trivially true. Assume the claim is true for $n = k - 1$. We prove it for $n = k$.

The vector $\boldsymbol{v}_k$ added to B in the kth iteration is not in the span of $\boldsymbol{v}_1, \ldots, \boldsymbol{v}_{k-1}$. Therefore, by the Span Lemma, for any coefficients $\alpha_1, \ldots, \alpha_k$ such that

$$\mathbf{0} = \alpha_1 \boldsymbol{v}_1 + \cdots + \alpha_{k-1} \boldsymbol{v}_{k-1} + \alpha_k \boldsymbol{v}_k$$

it must be that α_k equals zero. We may therefore write

$$\mathbf{0} = \alpha_1 \boldsymbol{v}_1 + \cdots + \alpha_{k-1} \boldsymbol{v}_{k-1}$$

By the claim for $n = k - 1$, however, $\boldsymbol{v}_1, \ldots, \boldsymbol{v}_{k-1}$ are linearly independent, so $\alpha_1, \ldots, \alpha_{k-1}$ are all zero. We have proved that the only linear combination of $\boldsymbol{v}_1, \ldots, \boldsymbol{v}_k$ that equals the zero vector is the *trivial* linear combination, i.e. that $\boldsymbol{v}_1, \ldots, \boldsymbol{v}_k$ are linearly independent. This proves the claim for $n = k$. □

In the Grow algorithm for *Minimum Spanning Forest*, when considering whether to add an edge $\{x, y\}$, we only add it if there is no x-to-y path using previously selected edges, i.e. if the vector corresponding to $\{x, y\}$ is not spanned by the vectors corresponding to previously selected edges. If we do add the edge, therefore, it does not form a cycle with previously added edges, i.e. the set of corresponding vectors remains linearly independent.

5.5.6 Analyzing the Shrink algorithm

Corollary 5.5.11 (Shrink-Algorithm Corollary): The vectors obtained by the Shrink algorithm are linearly independent.

Proof

Let $B = \{\boldsymbol{v}_1, \ldots, \boldsymbol{v}_n\}$ be the set of vectors obtained by the Shrink algorithm. Assume for a contradiction that the vectors are linearly dependent. Then $\mathbf{0}$ can be written as a nontrivial linear combination

$$\mathbf{0} = \alpha_1 \boldsymbol{v}_1 + \cdots + \alpha_n \boldsymbol{v}_n$$

where at least one of the coefficients is nonzero. Let α_i be a nonzero coefficient. By the Linear-Dependence Lemma, $\boldsymbol{v}_i$ can be written as a linear combination of the other vectors. Hence by the Superfluous-Vector Lemma (Lemma 5.5.1), Span $(B - \{\boldsymbol{v}_i\})$ = Span B, so the Shrink algorithm should have removed $\boldsymbol{v}_i$. □

In the MSF Shrink algorithm, when considering whether to remove an edge $\{x, y\}$, we remove it if there would still be an x-to-y path using the remaining edges. If the edge $\{x, y\}$ is part of a cycle then the algorithm can safely remove it since the other edges of the cycle form an x-to-y path.

5.6 Basis

The Grow algorithm and the Shrink algorithm each find a set of vectors spanning the vector space $\mathcal{V}$. In each case, the set of vectors found is linearly independent.

5.6.1 Defining basis

We have arrived at the most important concept in linear algebra.

Definition 5.6.1: Let $\mathcal{V}$ be a vector space. A *basis* for $\mathcal{V}$ is a linearly independent set of generators for $\mathcal{V}$.

The plural of *basis* is *bases*, pronounced "basees".

Thus a set B of vectors of $\mathcal{V}$ is a *basis* for $\mathcal{V}$ if B satisfies two properties:

Property B1 (*Spanning*) Span $B = \mathcal{V}$, and

Property B2 (*Independent*) B is linearly independent.

Example 5.6.2: Define $\mathcal{V}$ to be the vector space spanned by $[1,0,2,0]$, $[0,-1,0,-2]$, and $[2,2,4,4]$. Then the set $\{[1,0,2,0],[0,-1,0,-2],[2,2,4,4]\}$ is *not* a basis for $\mathcal{V}$ because it is not linearly independent. For example,

$$1\,[1,0,2,0] - 1\,[0,-1,0,-2] - \frac{1}{2}\,[2,2,4,4] = \mathbf{0}$$

However, the set $\{[1,0,2,0],[0,-1,0,-2]\}$ *is* a basis:

- You can tell that these two vectors are linearly independent because each has a nonzero entry in a position in which the other has a zero. The argument for linear independence is that given in Example 5.5.4 (Page 226).
- You can tell that these two vectors span $\mathcal{V}$ by using the Superfluous-Vector Lemma (Lemma 5.5.1): the third vector in the definition of $\mathcal{V}$, namely $[2,2,4,4]$, is superfluous because it can be written as a linear combination of the first two:

$$[2,2,4,4] = 2\,[1,0,2,0] - 2\,[0,-1,0,-2] \tag{5.2}$$

Since $\{[1,0,2,0],[0,-1,0,-2]\}$ spans $\mathcal{V}$ and is linearly independent, it is a basis.

Example 5.6.3: Also, $\{[1,0,2,0],[2,2,4,4]\}$ is a basis for the same vector space $\mathcal{V}$:

- To show linear independence, consider any nontrivial linear combination

$$\alpha_1\,[1,0,2,0] + \alpha_2\,[2,2,4,4]$$

  If α_2 is nonzero then the sum has a nonzero entry in, for example, the second position. If α_2 is zero but α_1 is nonzero then the sum has a nonzero entry in the first position. $\{[1,0,2,0],[2,2,4,4]\}$

- To show that these vectors span $\mathcal{V}$, we again use the Superfluous-Vector Lemma: the vector $[0,-1,0,-2]$ is superfluous because it can be written as a linear combination of the others. A little manipulation of Equation 5.2 yields

$$[0,-1,0,-2] = -1\,[1,0,2,0] + \frac{1}{2}\,[2,2,4,4]$$

Example 5.6.4: What about the vector space $\mathbb{R}^3$? One basis for $\mathbb{R}^3$ is $[1,0,0],[0,1,0],[0,0,1]$. How do we know this is a basis?

- Every vector $[x,y,z] \in \mathbb{R}^3$ can be expressed as $x[1,0,0]+y[0,1,0]+z[0,0,1]$. This shows the vectors span $\mathbb{R}^3$.
- How do we know the vectors are linearly independent? We just need to show that none of these three can be expressed as a linear combination of the other two. Consider $[1,0,0]$. Since none of other two vectors have a nonzero first entry, $[1,0,0]$ cannot be expressed as a linear combination of them. The argument can be used for $[0,1,0]$ and $[0,0,1]$ as well. This shows that the dimension of $\mathbb{R}^3$ is 3.

Example 5.6.5: Another basis for $\mathbb{R}^3$ is $[1,1,1],[1,1,0],[0,1,1]$. How do we know these vectors span $\mathbb{R}^3$? We already know that the vectors $[1,0,0],[0,1,0],[0,0,1]$ are spanning, so if these vectors are in the span of $[1,1,1],[1,1,0],[0,1,1]$ then we know the latter vectors span all of $\mathbb{R}^3$.

$$\begin{aligned}
[1,0,0] &= [1,1,1]-[0,1,1] \\
[0,1,0] &= [1,1,0]+[0,1,1]-[1,1,1] \\
[0,0,1] &= [1,1,1]-[1,1,0]
\end{aligned}$$

How do we know the vectors $[1,1,1],[1,1,0],[0,1,1]$ are linearly independent? Suppose they were linearly dependent. By the Linear-Dependence Lemma, one of them could be written in terms of the other two vectors. There are three cases.

- Can $[1,1,1]$ be written as a linear combination $\alpha\,[1,1,0]+\beta\,[0,1,1]$? In order for the first entry of the linear combination to be 1, α would have to be 1. In order for the third entry to be 1, β would have to be 1. That would mean the second entry would be 2, so $[1,1,1]$ cannot be written as a linear combination.
- Can $[1,1,0]$ be written as a linear combination $\alpha\,[1,1,1]+\beta\,[0,1,1]$? In order for the first entry of the linear combination to be 1, α would have to be 1. Therefore, in order for the second entry to be 1, β would have to be 0 but in order for the third entry to be 0, β would ahve to be -1.
- Can $[0,1,1]$ be written as a linear combination $\alpha\,[1,1,1]+\beta\,[1,1,0]$? Adapting the argument used in the previous case shows that the answer is no.

Example 5.6.6: Does a trivial vector space, consisting only of a zero vector, have a basis? Of course: the empty set. We know from Quiz 3.2.4 that the span of an empty set is the set consisting of the zero vector. An empty set is linearly independent since there is no nontrivial linear combination of an empty set of vectors. (For every linear combination of an empty set, there is no nonzero coefficient.)

Example 5.6.7: In the context of a graph G, a basis of the set of edges forming the graph corresponds to a set B of edges that is spanning for G (in the sense of Definition 5.4.2) and is a forest (Definition 5.4.3). Thus a basis is precisely a spanning forest. Here are two examples:

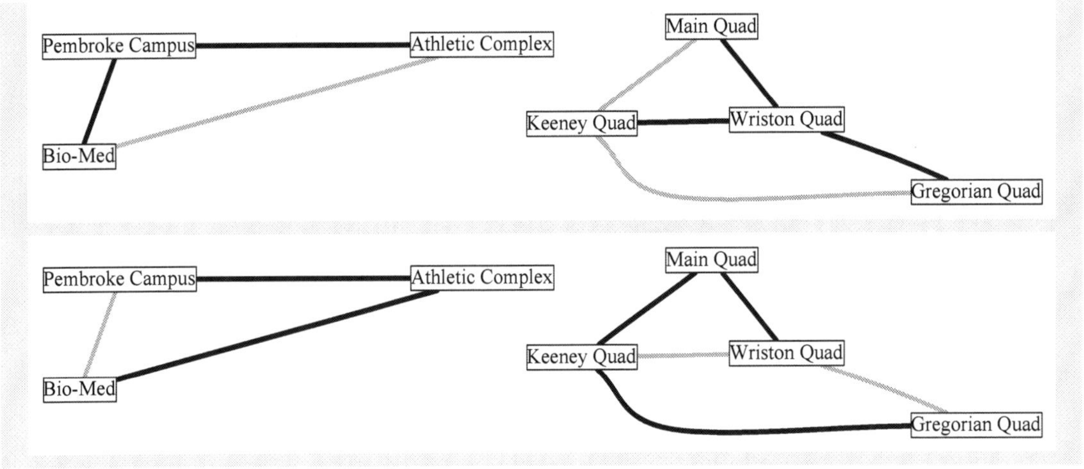

Example 5.6.8: Let T be a subset of edges of a graph, e.g. the solid edges in this graph:

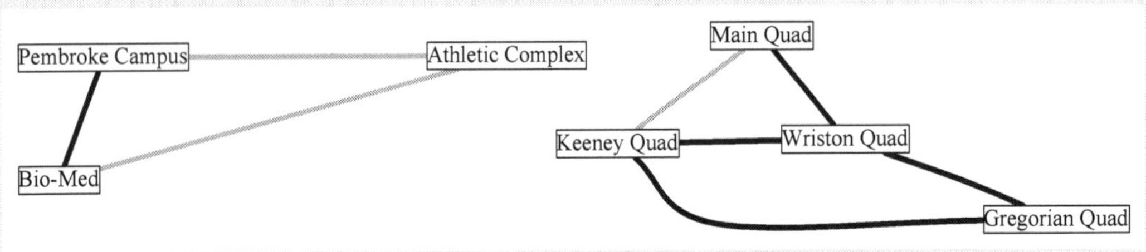

In the following diagram, the set B of dark edges form a basis for Span T:

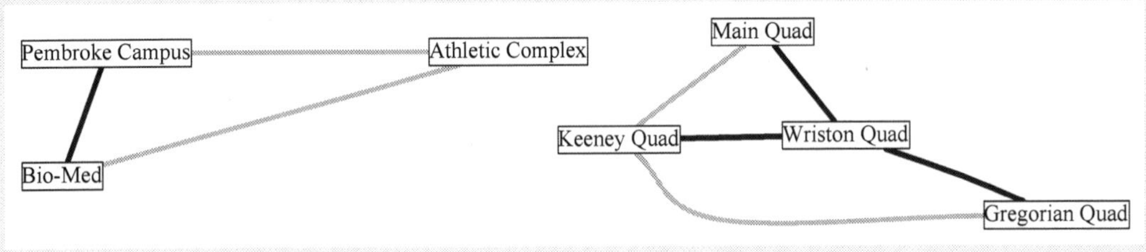

Let's verify that B is a basis for Span T: the edges of B do not form any cycles—so B is linearly independent—and for every edge of T, the endpoints are connected via edges of B, so Span B = Span T.

In the following diagram, the dark edges form another example of a basis for Span T.

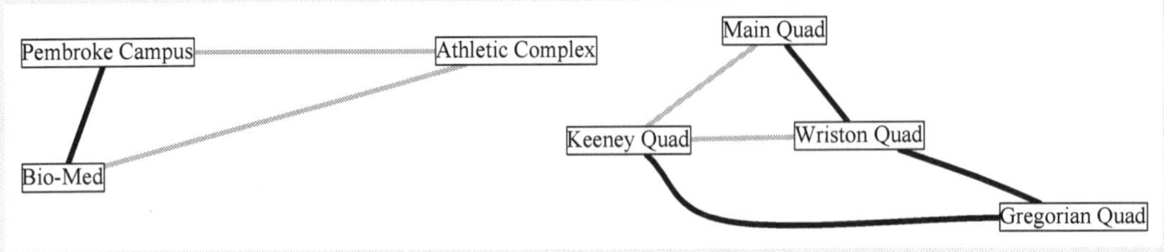

5.6.2 The standard basis for $\mathbb{F}^D$

In Section 3.2.5, we defined a set of generators for $\mathbb{F}^D$, the *standard* generators. The following lemma shows that a better name for these generators (and in fact the traditional name) for these vectors is *standard* **basis** *vectors* for $\mathbb{F}^D$:

Lemma 5.6.9: The standard generators for $\mathbb{F}^D$ form a basis.

Problem 5.6.10: Prove Lemma 5.6.9. The argument is a generalization of that in Example 5.6.4 (Page 231) showing $[1, 0, 0], [0, 1, 0], [0, 0, 1]$ is a basis.

5.6.3 Towards showing that every vector space has a basis

We would like to prove that every vector space $\mathcal{V}$ has a basis. The Grow algorithm and the Shrink algorithm each provides a way to prove this, but we are not there yet:

- The Grow-Algorithm Corollary implies that, if the Grow algorithm terminates, the set of vectors it has selected is a basis for the vector space $\mathcal{V}$. However, *we have not yet shown that it always terminates!*
- The Shrink-Algorithm Corollary implies that, if we can run the Shrink algorithm starting with a finite set of vectors that spans $\mathcal{V}$, upon termination it will have selected a basis for $\mathcal{V}$. However, *we have not yet shown that every vector space $\mathcal{V}$ is spanned by some finite set of vectors.*

These are not mathematically trivial matters. The issues above will be resolved in the next chapter—for us. That is because in this book we have confined our attention to D-vectors where D is a finite set. In the wider world of mathematics, however, D can be infinite, raising various difficulties that we avoid.

5.6.4 Any finite set of vectors contains a basis for its span

If $\mathcal{V}$ is specified as the span of a finite set of vectors then we can show that $\mathcal{V}$ has a basis—one consisting of a subset of that set.

Lemma 5.6.11 (Subset-Basis Lemma): Any finite set T of vectors contains a subset B that is a basis for Span T.

This result corresponds in graphs to the fact that every graph contains a spanning forest.

Proof

Let $\mathcal{V} =$ Span T. We use a version of the Grow algorithm.

def subset_basis(T):
  Initialize B to the empty set.
  Repeat while possible: select a vector $\boldsymbol{v}$ in T that is not in Span B, and put it in B.

This algorithm differs from the generic Grow algorithm in that the vector selected $\boldsymbol{v}$ is required to be in the set T. Is this algorithm nevertheless an instantiation of the Grow algorithm? This algorithm stops when Span B contains every vector in T, whereas the original Grow algorithm stops only once Span B contains every vector in $\mathcal{V}$. However, that's okay: when Span B contains all the vectors in T, Span B also contains all linear combinations of vectors in T, so at this point Span $B = \mathcal{V}$. This shows that our version of the original Grow algorithm is a true instantiation of it, so our version produces a basis. □

The reader will notice that the proof is really a procedure, a version of the Grow algorithm that can actually be implemented.

Example 5.6.12: Let $T = \{[1, 0, 2, 0], [0, -1, 0, -2], [2, 2, 4, 4]\}$. The procedure must find a subset B that is a basis for Span T:

- Initialize $B = \emptyset$.
- Choose a vector in T that is not in Span $\emptyset$, and add it to S. Since Span $\emptyset$ consists of

just the zero vector, the first vector chosen need only be nonzero. Suppose $[1, 0, 2, 0]$ is chosen.

- Choose a vector in T that is not in Span $\{[1, 0, 2, 0]\}$. Suppose $[0, -1, 0, -2]$ is chosen.
- Choose a vector in T that is not in Span $\{[1, 0, 2, 0], [0, -1, 0, -2]\}$. Oops, there is no such vector. Every vector in T is in Span $\{[1, 0, 2, 0], [0, -1, 0, -2]\}$. Therefore the procedure is finished.

Problem 5.6.13: Give an alternative proof of Lemma 5.6.11 that is based instead on the Shrink algorithm.

5.6.5 Can any linearly independent subset of vectors belonging to $\mathcal{V}$ be extended to form a basis for $\mathcal{V}$?

Analogous to the Subset-Basis Lemma, we would like to prove a *Superset*-Basis Lemma, a lemma that states:

> For any vector space $\mathcal{V}$ and any linearly independent set T of vectors, $\mathcal{V}$ has a basis that contains all of T.

Perhaps we can adapt the Grow algorithm to find such a basis, as follows:

def superset_basis(T, $\mathcal{V}$):
  Initialize B to be equal to T.
  Repeat while possible: select a vector in $\mathcal{V}$ that is not in Span B, and put it in B.
  Return B

Initially, B contains all of T (in fact, is equal to T). By the Grow-Algorithm Corollary, the set B is linearly independent throughout the algorithm. When the algorithm terminates, Span $B = \mathcal{V}$. Hence upon termination B is a basis for $\mathcal{V}$. Furthermore, B still contains all of T since the algorithm did not remove any vectors from B.

There is just one catch with this reasoning. As in our attempt in Section 5.6.3 to show that every vector space has a basis, *we have not yet shown that the algorithm terminates!* This issue will be resolved in the next chapter.

5.7 Unique representation

As discussed in Section 5.1, in a coordinate system for $\mathcal{V}$, specified by generators $\boldsymbol{a}_1, \ldots, \boldsymbol{a}_n$, each vector $\boldsymbol{v}$ in $\mathcal{V}$ has a coordinate representation $[\alpha_1, \ldots, \alpha_n]$, which consists of the coefficients with which $\boldsymbol{v}$ can be represented as a linear combination:

$$\boldsymbol{v} = \alpha_1 \boldsymbol{a}_1 + \cdots + \alpha_n \boldsymbol{a}_n$$

But we need the axes to have the property that each vector $\boldsymbol{v}$ has a *unique* coordinate representation. How can we ensure that?

5.7.1 Uniqueness of representation in terms of a basis

We ensure that by choosing the axis vectors so that they form a basis for $\mathcal{V}$.

Lemma 5.7.1 (Unique-Representation Lemma): Let $\boldsymbol{a}_1, \ldots, \boldsymbol{a}_n$ be a basis for a vector space $\mathcal{V}$. For any vector $\boldsymbol{v} \in \mathcal{V}$, there is exactly one representation of $\boldsymbol{v}$ in terms of the basis vectors.

In a graph G, this corresponds to the fact that, for any spanning forest F of G, for any pair x, y of vertices, if G contains an x-to-y path then F contains exactly one such path.

Proof

Because Span $\{\boldsymbol{a}_1, \ldots, \boldsymbol{a}_n\} = \mathcal{V}$, every vector $\boldsymbol{v} \in \mathcal{V}$ has at least one representation in terms of $\boldsymbol{a}_1, \ldots, \boldsymbol{a}_n$. Suppose that there are two representations;

$$\boldsymbol{v} = \alpha_1 \, \boldsymbol{a}_1 + \cdots + \alpha_n \, \boldsymbol{a}_n = \beta_1 \, \boldsymbol{a}_1 + \cdots + \beta_n \, \boldsymbol{a}_n$$

Then we can get the zero vector by subtracting one linear combination from the other:

$$\begin{aligned} \boldsymbol{0} &= \alpha_1 \, \boldsymbol{a}_1 + \cdots + \alpha_n \, \boldsymbol{a}_n - (\beta_1 \, \boldsymbol{a}_1 + \cdots + \beta_n \, \boldsymbol{a}_n) \\ &= (\alpha_1 - \beta_1)\boldsymbol{a}_1 + \cdots + (\alpha_n - \beta_n)\boldsymbol{a}_n \end{aligned}$$

Since the vectors $\boldsymbol{a}_1, \ldots, \boldsymbol{a}_n$ are linearly independent, the coefficients $\alpha_1 - \beta_1, \ldots, \alpha_n - \beta_n$ must all be zero, so the two representations are really the same. □

5.8 Change of basis, first look

Change of basis consists in changing from a vector's coordinate representation in terms of one basis to the same vector's coordinate representation in terms of another basis.

5.8.1 The function from representation to vector

Let $\boldsymbol{a}_1, \ldots, \boldsymbol{a}_n$ form a basis for a vector space $\mathcal{V}$ over a field $\mathbb{F}$. Define the function $f : \mathbb{F}^n \mapsto \mathcal{V}$ by

$$f([x_1, \ldots, x_n]) = x_1 \, \boldsymbol{a}_1 + \cdots + x_n \, \boldsymbol{a}_n$$

That is, f maps the representation in $\boldsymbol{a}_1, \ldots, \boldsymbol{a}_n$ of a vector to the vector itself. The Unique-Representation Lemma tells us that every vector in $\mathcal{V}$ has exactly one representation in terms of $\boldsymbol{a}_1, \ldots, \boldsymbol{a}_n$, so the function f is both onto and one-to-one, so it is invertible.

Example 5.8.1: I assert that one basis for the vector space $\mathbb{R}^3$ consists of $\boldsymbol{a}_1 = [2, 1, 0], \boldsymbol{a}_2 = [4, 0, 2], \boldsymbol{a}_3 = [0, 1, 1]$. The matrix with these vectors as columns is

$$A = \left[\begin{array}{c|c|c} 2 & 4 & 0 \\ 1 & 0 & 1 \\ 0 & 2 & 1 \end{array}\right]$$

Then the function $f : \mathbb{R}^3 \longrightarrow \mathbb{R}^3$ defined by $f(\boldsymbol{x}) = A\boldsymbol{x}$ maps the representation of a vector in terms of $\boldsymbol{a}_1, \ldots, \boldsymbol{a}_3$ to the vector itself. Since I have asserted that $\boldsymbol{a}_1, \boldsymbol{a}_2, \boldsymbol{a}_3$ form a basis, every vector has a unique representation in terms of these vectors, so f is an invertible function. Indeed, the inverse function is the function $g : \mathbb{R}^3 \longrightarrow \mathbb{R}^3$ defined by $g(\boldsymbol{y}) = M\boldsymbol{y}$ where

$$M = \begin{bmatrix} \frac{1}{4} & \frac{1}{2} & -\frac{1}{5} \\ \frac{1}{8} & -\frac{1}{4} & \frac{1}{4} \\ -\frac{1}{4} & \frac{1}{2} & \frac{1}{2} \end{bmatrix}$$

Thus M is the inverse of A.

5.8.2 From one representation to another

Now suppose $\boldsymbol{a}_1, \ldots, \boldsymbol{a}_n$ form one basis for $\mathcal{V}$ and $\boldsymbol{b}_1, \ldots, \boldsymbol{b}_m$ form another basis. Define $f : \mathbb{F}^n \longrightarrow \mathcal{V}$ and $g : \mathbb{F}^m \longrightarrow \mathcal{V}$ by

$$f([x_1, \ldots, x_n]) = x_1 \, \boldsymbol{a}_1 + \cdots + x_n \, \boldsymbol{a}_n \text{ and } g([y_1, \ldots, y_m]) = y_1 \, \boldsymbol{b}_1 + \cdots + y_m \, \boldsymbol{b}_m$$

By the linear-combinations definition of matrix-vector definition, each of these functions can be represented by matrix-vector multiplication:

$$f(\boldsymbol{x}) = \left[\begin{array}{c|c|c} & & \\ \boldsymbol{a}_1 & \cdots & \boldsymbol{a}_n \\ & & \end{array}\right]\left[\begin{array}{c} \boldsymbol{x} \end{array}\right] \text{ and } g(\boldsymbol{y}) = \left[\begin{array}{c|c|c} & & \\ \boldsymbol{b}_1 & \cdots & \boldsymbol{b}_m \\ & & \end{array}\right]\left[\begin{array}{c} \boldsymbol{y} \end{array}\right]$$

Furthermore, by the reasoning in Section 5.8.1, the functions f and g are both invertible. By Lemma 4.13.1, their inverses are linear functions

Now consider the function $g^{-1} \circ f$. It is the composition of linear functions so it is a linear function. Its domain is the domain of f, which is $\mathbb{F}^n$, and its co-domain is the domain of g, which is $\mathbb{F}^m$. Therefore, by Lemma 4.10.19, there is a matrix C such that $C\boldsymbol{x} = (g^{-1} \circ f)(\boldsymbol{x})$.

The matrix C is a *change-of-basis* matrix:

- Multiplying by C converts from a vector's coordinate representation in terms of $\boldsymbol{a}_1, \ldots, \boldsymbol{a}_n$ to the same vector's coordinate representation in terms of $\boldsymbol{b}_1, \ldots, \boldsymbol{b}_n$.

Since $g^{-1} \circ f$ is the composition of invertible functions, it too is an invertible function. By the same reasoning, there is a matrix D such that $D\boldsymbol{y} = (f^{-1} \circ g)(\boldsymbol{y})$.

- Multiplying by D converts from a vector's coordinate representation in terms of $\boldsymbol{b}_1, \ldots, \boldsymbol{b}_k$ to the same vector's coordinate representation in terms of $\boldsymbol{a}_1, \ldots, \boldsymbol{a}_n$.

Finally, since $f^{-1} \circ g$ and $g^{-1} \circ f$ are inverses of each other, the matrices C and D are inverses of each other.

Why would you want functions that map between different representations of a vector? There are many reasons. In the next section, we'll explore one: dealing with perspective in images. Lab 5.12 will similarly deal with perspective using change of basis. Change of basis is crucially important in Chapters 10, 11, and 12.

5.9 Perspective rendering

As an application of coordinate representation, we show how to synthesize a camera view from a set of points in three dimensions, taking into account perspective. The mathematics underlying this task will be useful in a lab, where we will go in the opposite direction, removing perspective from a real image.

5.9.1 Points in the world

We start with the points making up a wire-frame cube with coordinates as shown:

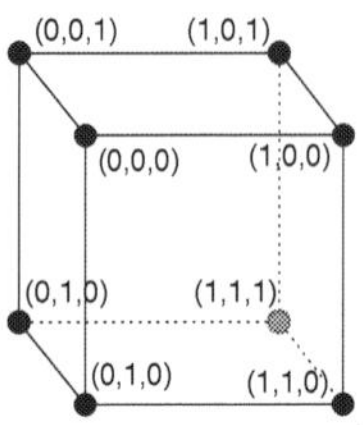

The coordinates might strike you as a bit odd: the point (0,1,0) is vertically *below* the point (0,0,0). We are used to y-coordinates that increase as you move up. We use this coordinate system in order to be consistent with the way pixel coordinates work.

The list of points making up the wire-frame cube can be produced as follows:

```
>>> L = [[0,0,0],[1,0,0],[0,1,0],[1,1,0],[0,0,1],[1,0,1],[0,1,1],[1,1,1]]
>>> corners = [list2vec(v) for v in L]
```

```
>>> def line_segment(pt1, pt2, samples=100):
    return [(i/samples)*pt1 + (1-i/samples)*pt2 for i in range(samples+1)]

>>> line_segments = [line_segment(corners[i], corners[j]) for i,j in
 [(0,1),(2,3), (0,2),(1,3),(4,5),(6,7),(4,6),(5,7),(0,4),(1,5),(2,6), (3,7)]]

>>> pts = sum(line_segments, [])
```

Imagine that a camera takes a picture of this cube: what does the picture look like? Obviously that depends on where the camera is located and which direction it points. We will choose to place the camera at the location (-1,-1, -8) facing straight at the plane containing the front face of the cube.

5.9.2 The camera and the image plane

We present a simplified model of a camera, a *pinhole* camera. Assume the location and orientation of the camera are fixed. The pinhole is a point called the *camera center.*

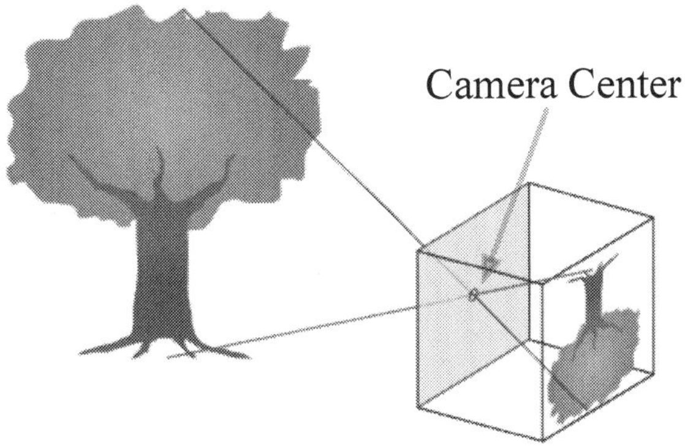

There is an image sensor array in the back of the camera. Photons bounce off objects in the scene and travel through the camera center to the image sensor array. A photon from the scene only reaches the image sensor array if it travels in a straight line through the camera center. The image ends up being reversed.

An even simpler model is usually adopted to make the math easier. In this model, the image sensor array is between the camera center and the scene.

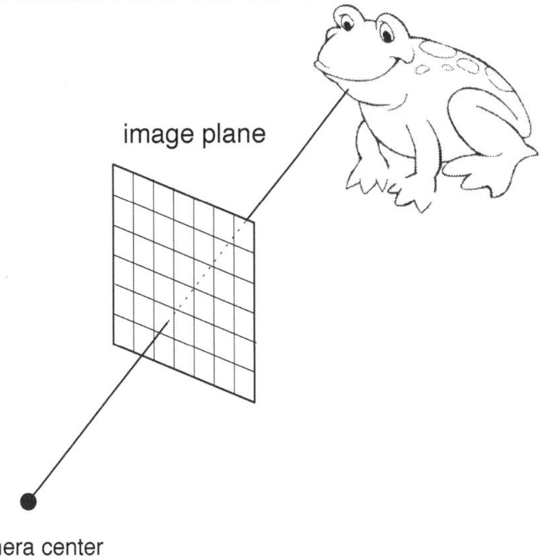

We retain the rule that a photon is only sensed by the sensor array if the photon is traveling in a line through the camera center.

The image sensor array is located in a plane, called the *image plane.* A photon bounces off an object in the scene, in this case the chin of a frog, and heads in a straight line towards the camera center. On its way, it bumps into the image plane where it encounters the sensor array.

The sensor array is a grid of rectangular sensor elements. Each element of the image sensor array measures the amount of red, green, and blue light that hits it, producing three numbers. Which sensor element is struck by the photon? The one located at the intersection between the image plane and the line along which the photon is traveling.

The result of all this sensing is an image, which is a grid of rectangular picture elements (pixels), each assigned a color. The pixels correspond to the sensor elements.

The pixels are assigned coordinates, as we have seen before:

(0,0)					(5,0)
(0,3)					(5,3)

5.9.3 The camera coordinate system

For each point $\boldsymbol{q}$ in the sensor array, the light that hits $\boldsymbol{q}$ is light that travels in a straight line towards the camera center. Thus the color detected by the sensor at $\boldsymbol{q}$ is located at some point $\boldsymbol{p}$ in the world such that the line through $\boldsymbol{p}$ and the origin intersects the image plane at $\boldsymbol{q}$.

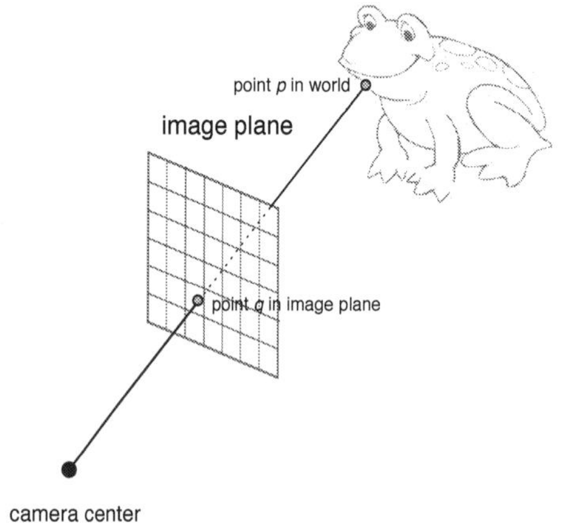

To synthesize an image of the wire-frame cube, we need to define a function that maps points $\boldsymbol{p}$ in the world to the pixel coordinates of the corresponding point $\boldsymbol{q}$ in the image plane.

There is a particularly convenient basis that enables us to simply express this function. We call it the *camera coordinate system.*

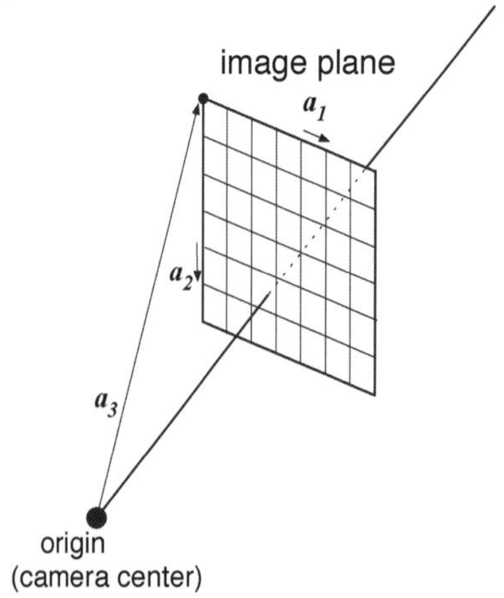

The origin is defined to be the camera center. The first basis vector $\boldsymbol{a}_1$ goes horizontally from the top-left corner of a sensor element to the top-right corner. The second vector $\boldsymbol{a}_2$ goes vertically from the top-left corner of a sensor element down to the bottom-left corner. The third vector $\boldsymbol{a}_3$ goes from the origin (the camera center) to the top-left corner of sensor element (0,0).

There's something nice about this basis. Let $\boldsymbol{q}$ be a point in the image plane

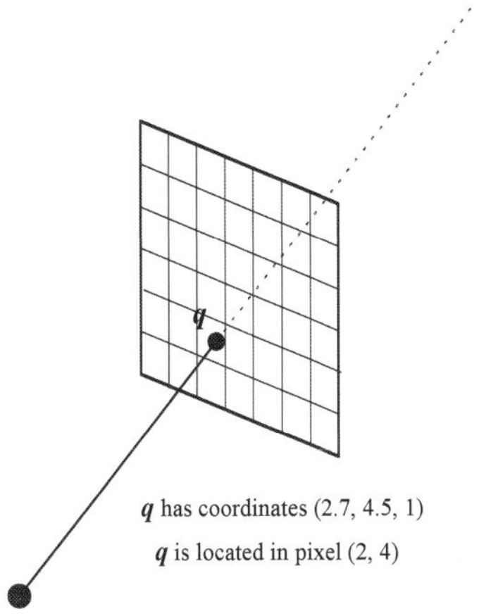

and let the coordinates of $\boldsymbol{q}$ in this coordinate system be $\boldsymbol{x} = (x_1, x_2, x_3)$, so $\boldsymbol{q} = x_1\,\boldsymbol{a}_1 + x_2\,\boldsymbol{a}_2 + x_3\,\boldsymbol{a}_3$.

Then the third coordinate x_3 is 1, and the first and second x_1, x_2 tell us which pixel contains the point $\boldsymbol{q}$.

```
def pixel(x): return (x[0], x[1])
```

We can round x_1 and x_2 down to integers i, j to get the coordinates of that pixel (if there exists an i, j pixel).

5.9.4 From the camera coordinates of a point in the scene to the camera coordinates of the corresponding point in the image plane

Let's take a side view from a point in the image plane, so we only see the edge of the sensor array:

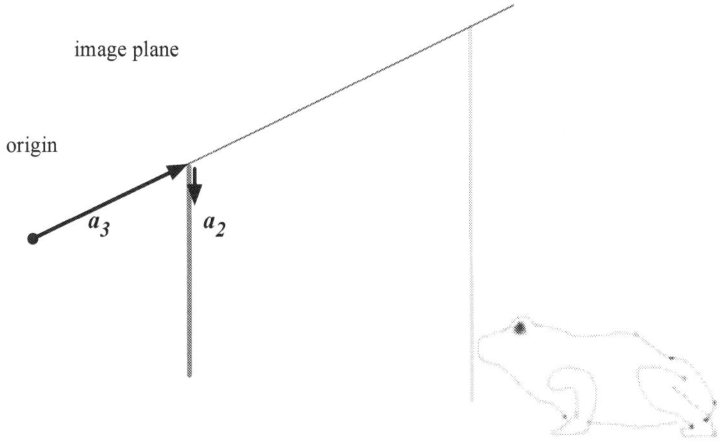

In this view, we can see the basis vectors $\boldsymbol{a}_2$ and $\boldsymbol{a}_3$ but not $\boldsymbol{a}_1$ since it is pointed directly at us.

Now suppose $\boldsymbol{p}$ is a point in the scene, way beyond the image plane.

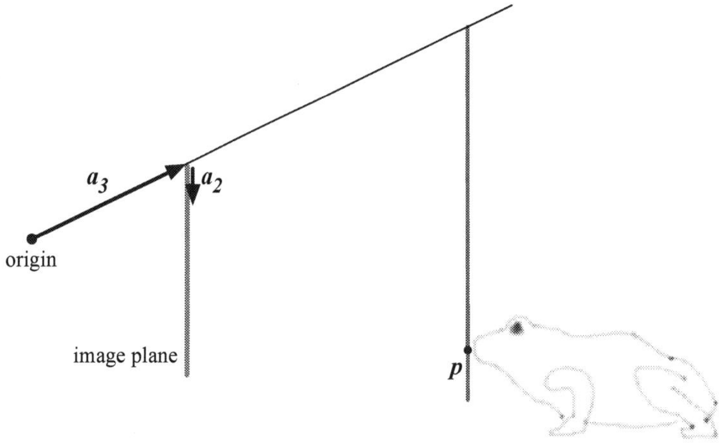

You can see the edge of the plane through $\boldsymbol{p}$ that is parallel to the image plane. We write $\boldsymbol{p}$ as a linear combination of the vectors of the camera basis:

$$\boldsymbol{p} = x_1\, \boldsymbol{a}_1 + x_2\, \boldsymbol{a}_2 + x_3\, \boldsymbol{a}_3$$

Think of the vector $x_3\, \boldsymbol{a}_3$ extending through the bottom-left corner of the sensor array all the way to the plane through $\boldsymbol{p}$ that is parallel to the image plane.

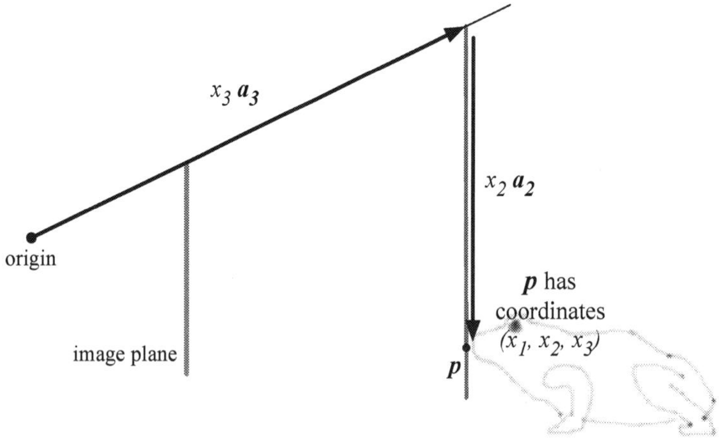

The vector $x_2\, \boldsymbol{a}_2$ extends vertically downward, and the vector $x_1\, \boldsymbol{a}_1$, which is not visible, extends horizontally towards us.

Let $\boldsymbol{q}$ be the point where the line through $\boldsymbol{p}$ and the origin intersects the image plane.

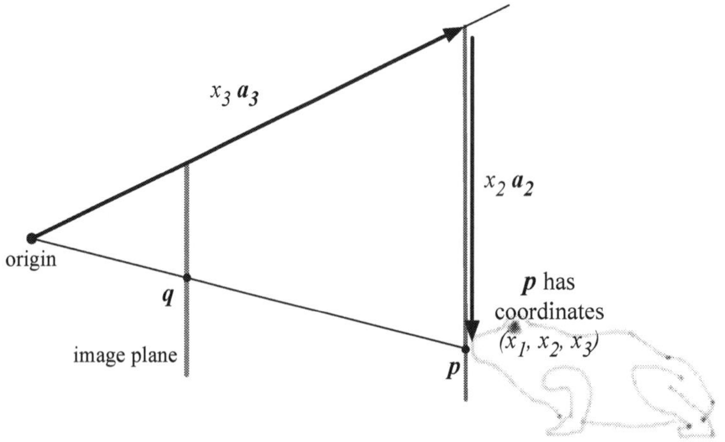

What are the coordinates of $\boldsymbol{q}$?

We see that the triangle formed by the origin, the head of $\boldsymbol{a}_3$ (when the tail is located at the origin), and $\boldsymbol{q}$ is a scaled-down version of the triangle formed by the origin, the head of $x_3\, \boldsymbol{a}_3$, and the point $\boldsymbol{p}$. Since the side formed by $\boldsymbol{a}_3$ has length $1/x_3$ times that of the side formed by $x_3\, \boldsymbol{a}_3$, a little geometric intution tells us that the coordinates of $\boldsymbol{q}$ are $1/x_3$ times each of the coordinates of $\boldsymbol{p}$, i.e. that $\boldsymbol{q}$'s coordinates are $(x_1/x_3, x_2/x_3, x_3/x_3)$.

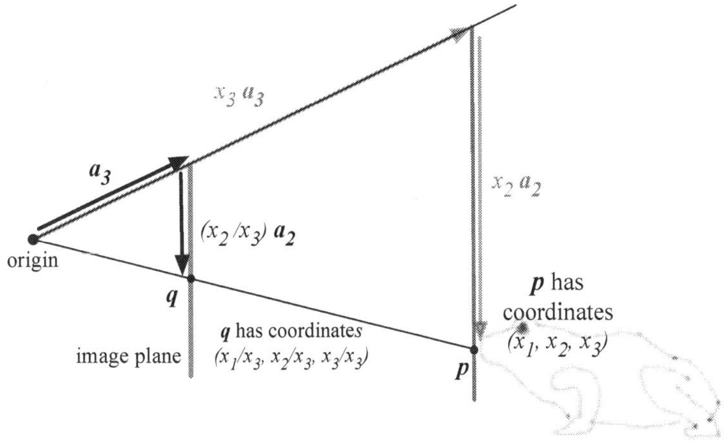

Thus, when the camera basis is used, it is easy to go from the representation of $\boldsymbol{p}$ to the representation of $\boldsymbol{q}$: just divide each of the entries by the third entry:

```
def scale_down(x): return list2vec([x[0]/x[2], x[1]/x[2], 1])
```

5.9.5 From world coordinates to camera coordinates

We now know how to map

- from the representation in camera coordinates of a point in the world
- to the coordinates of the pixel that "sees" that point.

However, to map the points of our wire-frame cube, we need to map from the coordinates of a point of the cube to the representation of that same point in camera coordinates.

First we write down the camera basis vectors.

We do this in two steps. The first step accounts for the fact that we are locating the camera center at (-1,-1,-8) in world coordinates. In order to use the camera coordinate system, we need to locate the camera at (0,0,0), so we translate the points of the wire-frame cube by adding (1,1,8) to each of them:

```
>>> shifted_pts = [v+list2vec([1,1,8]) for v in pts]
```

In the second step, we must do a change of basis. For each point in `shifted_pts`, we obtain its coordinate representation in terms of the camera basis.

To do this, we first write down the camera basis vectors. Imagine we have 100 horizontal pixels and 100 vertical pixels making up an image sensor array of dimensions 1×1. Then $a_1 = [1/100, 0, 0]$ and $\boldsymbol{a}_2 = [0, 1/100, 0]$. For the third basis vector $\boldsymbol{a}_3$, we decide that the sensor array is positioned so that the camera center lines up with the center of the sensor array. Remember that $\boldsymbol{a}_2$ points from the camera center to the top-left corner of the sensor array, so $\boldsymbol{a}_2 = [0, 0, 1]$.

```
>>> cb = [list2vec([1/xpixels,0,0]),
          list2vec([0,1/ypixels,0]),
          list2vec([0,0,1])]
```

We find the coordinates in the camera basis of the points in `shifted_pts`:

```
>>> reps = [vec2rep(cb, v) for v in shifted_pts]
```

5.9.6 ... to pixel coordinates

Next we obtain the projections of these points onto the image plane:

```
>>> in_camera_plane = [scale_down(u) for u in reps]
```

Now that these points lie in the camera plane, their third coordinates are all 1, and their first and second coordinates can be interpreted as pixel coordinates:

```
>>> pixels = [pixel(u) for u in in_camera_plane]
```

To see the result, we can use the `plot` procedure from the `plotting` module.

```
>>> plot(pixels, 30, 1)
```

However, keep in mind that increasing second pixel coordinate corresponds to moving downwards, whereas our `plot` procedure interprets the second coordinate in the usual mathematical way, so the plot will be a vertical inversion of what you would see in an image:

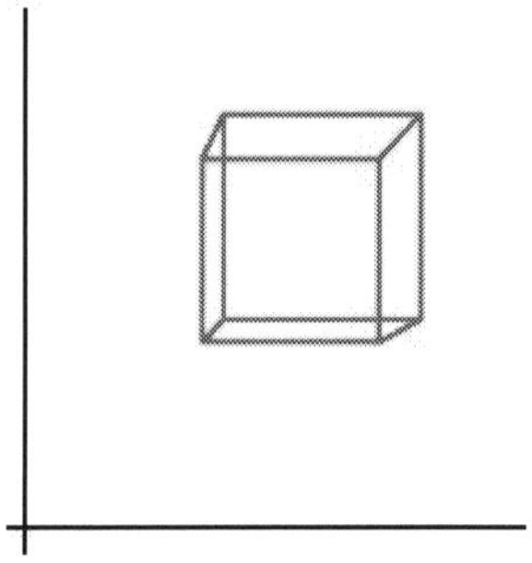

5.10 Computational problems involving finding a basis

Bases are quite useful. It is important for us to have implementable algorithms to find a basis for a given vector space. But a vector space can be huge—even infinite—how can it be the input to a procedure? There are two natural ways to specify a vector space $\mathcal{V}$:

1. Specifying generators for $\mathcal{V}$. This is equivalent to specifying a matrix A such that $\mathcal{V} =$ Col A.
2. Specifying a homogeneous linear system whose solution set is $\mathcal{V}$. This is equivalent to specifying a matrix A such that $\mathcal{V} =$ Null A.

For each of these ways to specify $\mathcal{V}$, we consider the Computational Problem of finding a basis.

Computational Problem 5.10.1: *Finding a basis of the vector space spanned by given vectors*

- *input:* a list $[\boldsymbol{v}_1, \ldots, \boldsymbol{v}_n]$ of vectors
- *output:* a list of vectors that form a basis for Span $\{\boldsymbol{v}_1, \ldots, \boldsymbol{v}_n\}$.

You might think we could use the approach of the Subset-Basis Lemma (Lemma 5.6.11) and the procedure `subset_basis(T)` of Problem 5.14.17, but this approach depends on having a way to tell if a vector is in the span of other vectors, which is itself a nontrivial problem.

Computational Problem 5.10.2: *Finding a basis of the solution set of a homogeneous linear system*

- *input:* a list $[\boldsymbol{a}_1, \ldots, \boldsymbol{a}_m]$ of vectors

- *output:* a list of vectors that form a basis for the set of solutions to the system $\boldsymbol{a}_1 \cdot \boldsymbol{x} = 0, \ldots, \boldsymbol{a}_m \cdot \boldsymbol{x} = 0$

This problem can be restated as

Given a matrix

$$A = \begin{bmatrix} \quad \boldsymbol{a}_1 \quad \\ \hline \vdots \\ \hline \boldsymbol{a}_m \end{bmatrix},$$

find a basis for the null space of A.

An algorithm for this problem would help us with several Questions. For example, having a basis would tell us whether the solution set was trivial: if the basis is nonempty, the solution set is nontrivial.

In Chapters 7 and 9, we will discuss efficient algorithms for solving these problems.

5.11 The Exchange Lemma

Do the Minimum Spanning Tree algorithms find a truly minimum-weight spanning tree? We have seen one example of a computational problem—Minimum Dominating Set–for which greedy algorithms sometimes fail to find the best answer. What makes Minimum Spanning Tree different?

5.11.1 The lemma

We present a lemma, the Exchange Lemma, that applies to vectors. In Section 5.4.3 we saw the close connection between MSF and vectors. In Section 5.11.2, we use the Exchange Lemma to prove the correctness of the Grow algorithm for MSF.

Lemma 5.11.1 (Exchange Lemma): Suppose S is a set of vectors and A is a subset of S. Suppose $\boldsymbol{z}$ is a vector in Span S such that $A \cup \{\boldsymbol{z}\}$ is linearly independent. Then there is a vector $\boldsymbol{w} \in S - A$ such that Span S = Span $(\{\boldsymbol{z}\} \cup S - \{\boldsymbol{w}\})$.

It's called the *Exchange* Lemma because it says you can inject a vector $\boldsymbol{z}$ and eject another vector without changing the span. The set A is used to keep certain vectors from being ejected.

Proof

Write $S = \{\boldsymbol{v}_1, \ldots, \boldsymbol{v}_k, \boldsymbol{w}_1, \ldots, \boldsymbol{w}_\ell\}$ and $A = \{\boldsymbol{v}_1, \ldots, \boldsymbol{v}_k\}$. Since $\boldsymbol{z}$ is in Span S, it can be expressed as a linear combination of vectors in S:

$$\boldsymbol{z} = \alpha_1 \boldsymbol{v}_1 + \cdots + \alpha_k \boldsymbol{v}_k + \beta_1 \boldsymbol{w}_1 + \cdots + \beta_\ell \boldsymbol{w}_\ell \tag{5.3}$$

If the coefficients $\beta_1, \ldots, \beta_\ell$ were all zero then we would have $\boldsymbol{z} = \alpha_1 \boldsymbol{v}_1 + \cdots + \alpha_k \boldsymbol{v}_k$, contradicting the linear independence of $A \cup \{\boldsymbol{z}\}$. Thus the coefficients $\beta_1, \ldots, \beta_\ell$ cannot all be zero. Let β_j be a nonzero coefficient. Then Equation (5.3) can be rewritten as

$$\begin{aligned} \boldsymbol{w}_j = (1/\beta_j)\, \boldsymbol{z} + (-\alpha_1/\beta_j)\, v_1 + \cdots + (-\alpha_k/\beta_j)\, \boldsymbol{v}_k + (-\beta_1/\beta_j)\, \boldsymbol{w}_1 + \ldots + (-\beta_{j-1}/\beta_j)\, \boldsymbol{w}_{j-1} \\ +(-\beta_{j+1}/\beta_j)\, \boldsymbol{w}_{j+1} + \cdots + (-\beta_\ell/\beta_j)\, \boldsymbol{w}_\ell \end{aligned} \tag{5.4}$$

By the Superfluous-Vector Lemma (Lemma 5.5.1),

$$\text{Span } (\{z\} \cup S - \{\boldsymbol{w}_j\}) = \text{Span } (\{z\} \cup S) = \text{Span } S$$

□

We use the Exchange Lemma in the next section to prove the correctness of the Grow algorithm for MSF. In the next chapter, we use the Exchange Lemma in a more significant and relevant way: to show that all bases for a vector space $\mathcal{V}$ have the same size. This is the central result in linear algebra.

5.11.2 Proof of correctness of the Grow algorithm for MSF

We show that the algorithm $\text{Grow}(G)$ returns a minimum-weight spanning tree for G. We assume for simplicity that all edge-weights are distinct. Let T^* be the true minimum-weight spanning tree for G, and let T be the set of edges chosen by the algorithm. Let $e_1, e_2, \ldots, e_m$ be the edges of G in increasing order. Let e_k be the minimum-weight edge that is not in both T and T^*. This means that $e_1, \ldots, e_{k-1}$ are all in both T and T^*. There are two cases.

- *Case 1:* e_k is in T^* but not in T. Since also $e_1, \ldots, e_{k-1}$ are in T^* and T^* is a spanning tree, e_k does not form a cycle with edges from $e_1, \ldots, e_{k-1}$, so the endpoints of e_k are not connected via edges in $\{e_1, \ldots, e_{k-1}\}$. Therefore the algorithm should have added e_k to its solution, a contradiction.

- *Case 2:* e_k is in T but not T^*. Let A consist of the vectors corresponding to $e_1, \ldots, e_{k-1}$ and let S consist of the vectors corresponding to all edges in T^*. Let $\boldsymbol{z}$ be the vector corresponding to e_k. Since the algorithm included e_k, the set $A \cup \{\boldsymbol{z}\}$ is linearly independent. Therefore, by the Exchange Lemma, there is a vector $\boldsymbol{w}$ in $S - A$ such that

$$\text{Span } S = \text{Span } (S \cup \{\boldsymbol{z}\} - \{\boldsymbol{w}\}]) \tag{5.5}$$

Since the vector $\boldsymbol{w}$ is in S, it corresponds to an edge e_n in T^*. Therefore $S \cup \{\boldsymbol{z}\} - \{\boldsymbol{w}\}$ corresponds to the set $T^* \cup \{e_k\} - \{e_n\}$ of edges. Equation 5.5 implies that every pair of nodes in G is connected via edges in $T^* \cup \{e_k\} - \{e_n\}$. However, this set of edges is obtained from T^* by replacing a heavy edge with a lighter one, showing that T^* is not a minimum-weight spanning tree, a contradiction.

This completes the proof of the correctness of $\text{Grow}(G)$.

5.12 *Lab: Perspective rectification*

The goal for this lab is to remove perspective from an image of a flat surface. Consider the following image (stored in the file `board.png`) of the whiteboard in my office:

Looks like there's some interesting linear algebra written on the board!

We will synthesize a new image. This new image has never been captured by a camera; in fact, it would be impossible using a traditional camera since it completely lacks perspective:

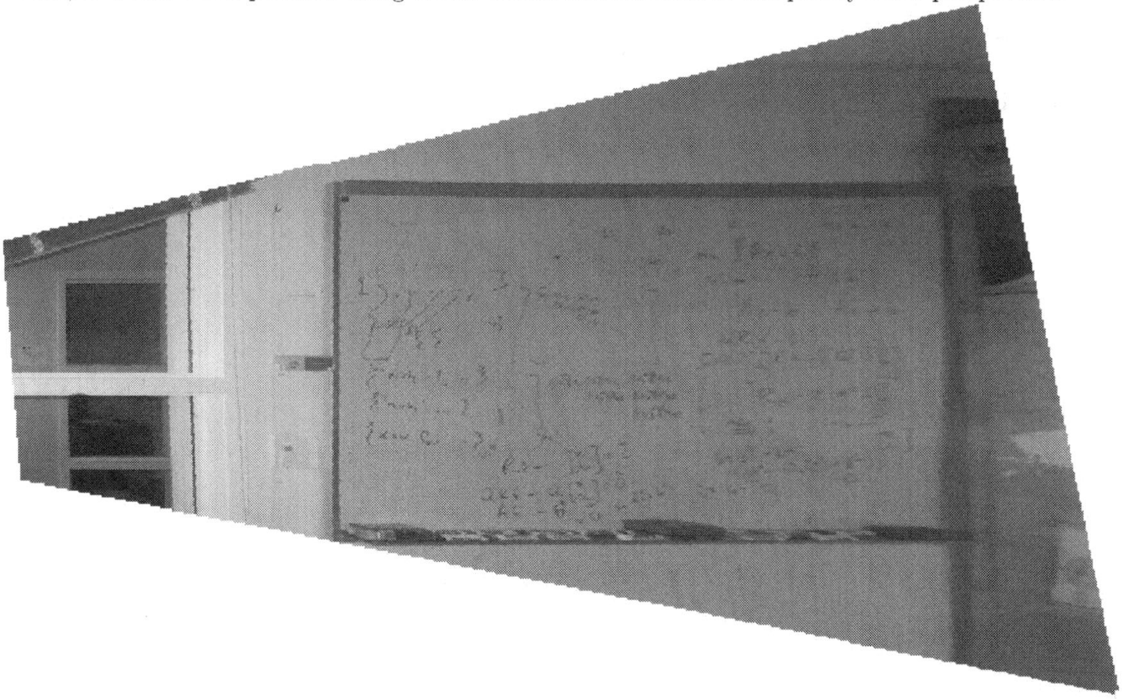

The technique for carrying out this transformation makes use of the idea of a coordinate system. Actually, it requires us to consider *two* coordinate systems and to transform between a representation within one system and a representation within the other.

Think of the original image as a grid of rectangles, each assigned a color. (The rectangles correspond to the pixels.) Each such rectangle in the image corresponds to a parallelogram in the plane of the whiteboard. The perspective-free image is created by painting each such parallelogram the color of the corresponding rectangle in the original image.

Forming the perspective-free image is easy once we have a function that maps pixel coordinates to the coordinates of the corresponding point in the plane of the whiteboard. How can we derive such a function?

The same problem arises in using a wiimote light pen. The light from the light pen strikes a particular sensor element of the wiimote, and the wiimote reports the coordinates of this sensor element to the computer. The computer needs to compute the corresponding location on the screen in order to move the mouse to that location. We therefore need a way to derive the function that maps coordinates of a sensor element to the coordinates in the computer screen.

The basic approach to derive this mapping is by example. We find several input-output pairs—points in the image plane and corresponding points in the whiteboard plane—and we derive the function that agrees with this behavior.

5.12.1 *The camera basis*

We use the camera basis $\boldsymbol{a}_1, \boldsymbol{a}_2, \boldsymbol{a}_3$ where:

- The origin is the camera center.
- The first vector $\boldsymbol{a}_1$ goes horizontally from the top-left corner of the whiteboard element to the top-right corner.
- The second vector $\boldsymbol{a}_2$ goes vertically from the top-left corner of whiteboard to the bottom-left corner.
- The third vector $\boldsymbol{a}_3$ goes from the origin (the camera center) to the top-left corner of sensor element (0,0).

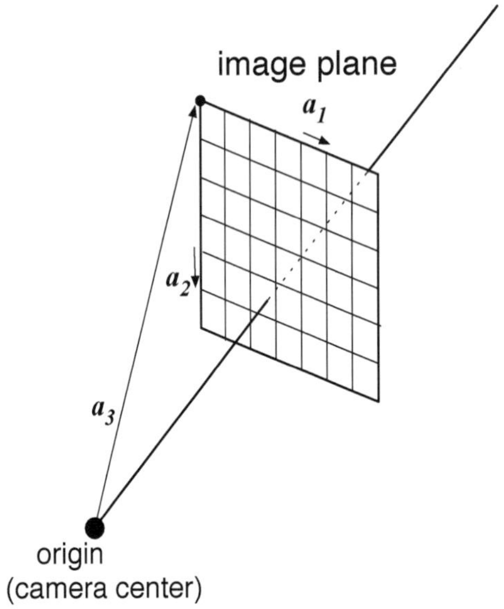

This basis has the advantage that the top-left corner of sensor element (x_1, x_2) has coordinate representation $(x_1, x_2, 1)$.

5.12.2 *The whiteboard basis*

In addition, we define a *whiteboard basis* $\boldsymbol{c}_1, \boldsymbol{c}_2, \boldsymbol{c}_3$ where:

- The origin is the camera center.
- The first vector $\boldsymbol{c}_1$ goes horizontally from the top-left corner of whiteboard to top-right corner.
- The second vector $\boldsymbol{c}_2$ goes vertically from the top-left corner of whiteboard to the bottom-left corner.

- The third vector c_3 goes from the origin (the camera center) to the top-right corner of whiteboard.

As we will see, this basis has the advantage that, given the coordinate representation (y_1, y_2, y_3) of a point $\boldsymbol{q}$, the intersection of the line through the origin and $\boldsymbol{q}$ with the whiteboard plane has coordinates $(y_1/y_3, y_2/y_3, y_3/y_3)$.

5.12.3 *Mapping from pixels to points on the whiteboard*

Our goal is to derive the function that maps the representation in camera coordinates of a point in the image plane to the representation in whiteboard coordinates of the corresponding point in the whiteboard plane.

At the heart of the function is a change of basis. We have two coordinate systems to think about, the camera coordinate system, defined by the basis $\boldsymbol{a}_1, \boldsymbol{a}_2, \boldsymbol{a}_3$, and the whiteboard coordinate system, defined by the basis $\boldsymbol{c}_1, \boldsymbol{c}_2, \boldsymbol{c}_3$. This gives us two representations for a point. Each of these representations is useful:

1. *It is easy to go from pixel coordinates to camera coordinates:* the point with pixel coordinates (x_1, x_2) has camera coordinates $(x_1, x_2, 1)$.

2. *It is easy to go from the whiteboard coordinates of a point $\boldsymbol{q}$ in space to the whiteboard coordinates of the corresponding point $\boldsymbol{p}$ on the whiteboard:* if $\boldsymbol{q}$ has whiteboard coordinates (y_1, y_2, y_3) then $\boldsymbol{p}$ has whiteboard coordinates $(y_1/y_3, y_2/y_3, y_3/y_3)$.

In order to construct the function that maps from pixel coordinates to whiteboard coordinates, we need to add a step in the middle: mapping from camera coordinates of a point $\boldsymbol{q}$ to whiteboard coordinates of the same point.

To help us keep track of whether a vector is the coordinate representation in terms of camera coordinates or is the coordinate representation in terms of whiteboard coordinates, we will use different domains for these two kinds of vectors. A coordinate representation in terms of camera coordinates will have domain C=`{'x1','x2','x3'}`. A coordinate representation in terms of whiteboard coordinates will have domain R=`{'y1','y2','y3'}`.

Our aim is to derive the function $f : \mathbb{R}^C \longrightarrow \mathbb{R}^R$ with the following spec:

- *input:* the coordinate representation $\boldsymbol{x}$ in terms of camera coordinates of a point $\boldsymbol{q}$
- *output:* the coordinate representation $\boldsymbol{y}$ in terms of whiteboard coordinates of the point $\boldsymbol{p}$ such that the line through the origin and $\boldsymbol{q}$ intersects the whiteboard plane at $\boldsymbol{p}$.

There is a little problem here; if $\boldsymbol{q}$ lies in the plane through the origin that is parallel to the whiteboard plane then the line through the origin and $\boldsymbol{q}$ does not intersect the whiteboard plane. We'll disregard this issue for now.

We will write f as the composition of two functions $f = g \circ h$, where

- $h : \mathbb{R}^C \longrightarrow \mathbb{R}^R$ is defined thus:
  - *input:* a point's coordinate representation with respect to the camera basis
  - *output:* the same point's coordinate representation with respect to the whiteboard basis
- $g : \mathbb{R}^R \longrightarrow \mathbb{R}^R$ is defined thus:
  - *input:* the coordinate representation in terms of whiteboard coordinates of a point $\boldsymbol{q}$
  - *output:* the coordinate representation in terms of whiteboard coordinates of the point $\boldsymbol{p}$ such that the line through the origin and $\boldsymbol{q}$ intersects the whiteboard plane at $\boldsymbol{p}$.

5.12.4 *Mapping a point not on the whiteboard to the corresponding point on the whiteboard*

In this section, we develop a procedure for the function g.

We designed the whiteboard coordinate system in such a way that a point on the whiteboard has coordinate y_3 equal to 1. For a point that is closer to the camera, the y_3 coordinate is less than 1.

Suppose $\boldsymbol{q}$ is a point that is not on the whiteboard, e.g. a point closer to the camera. Consider the line through the origin and $\boldsymbol{q}$. It intersects the whiteboard plane at some point $\boldsymbol{p}$. How do we compute the point $\boldsymbol{p}$ from the point $\boldsymbol{q}$?

This figure shows a top-view of the situation.

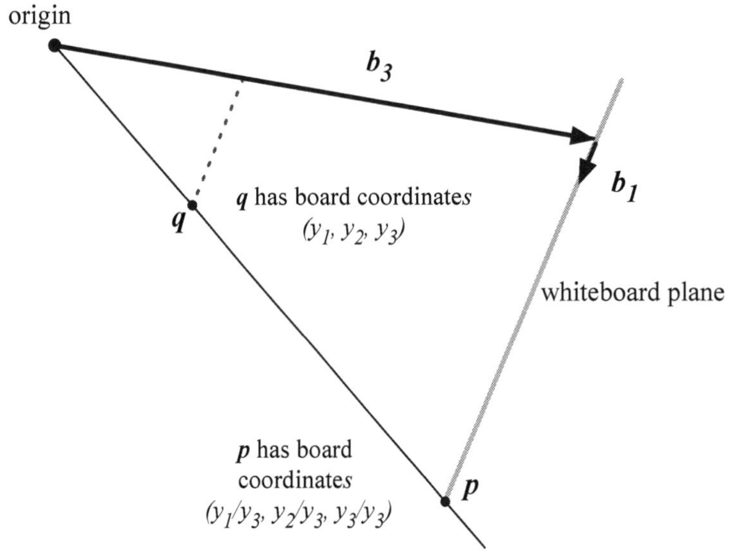

In this view, we see the top edge of the whiteboard, a point $\boldsymbol{q}$ not on the whiteboard, and the point $\boldsymbol{p}$ on the whiteboard that corresponds to $\boldsymbol{q}$ (in the sense that the line between the origin and $\boldsymbol{q}$ intersects the whiteboard plane at $\boldsymbol{p}$).

Let the whiteboard-coordinate representation of $\boldsymbol{q}$ be (y_1, y_2, y_3). In this figure, y_3 is less than 1. Elementary geometric reasoning (similar triangles) shows that the whiteboard-coordinate representation of the point $\boldsymbol{p}$ is $(y_1/y_3, y_2/y_3, y_3/y_3)$. Note that the third coordinate is 1, as required of a point in the whiteboard plane.

Task 5.12.1: Write a procedure `move2board(y)` with the following spec:

- *input:* a `{'y1','y2','y3'}`-vector $\boldsymbol{y}$, the coordinate representation in whiteboard coordinates of a point $\boldsymbol{q}$
  (Assume $\boldsymbol{q}$ is not in the plane through the origin that is parallel to the whiteboard plane, i.e. that the y3 entry is nonzero.)
- *output:* a `{'y1','y2','y3'}`-vector $\boldsymbol{z}$, the coordinate representation in whiteboard coordinates of the point $\boldsymbol{p}$ such that the line through the origin and $\boldsymbol{q}$ intersects the whiteboard

plane at $\boldsymbol{p}$.

5.12.5 *The change-of-basis matrix*

You have developed a procedure for g. Now we begin to address the procedure for h.

Writing a point $\boldsymbol{q}$ in terms of both the camera coordinate system $\boldsymbol{a}_1, \boldsymbol{a}_2, \boldsymbol{a}_3$ and the whiteboard coordinate system $\boldsymbol{c}_1, \boldsymbol{c}_2, \boldsymbol{c}_3$, and using the linear-combinations definition of matrix-vector multiplication, we have

$$\left[\begin{array}{c} \boldsymbol{q} \end{array}\right] = \left[\begin{array}{c|c|c} \boldsymbol{a}_1 & \boldsymbol{a}_2 & \boldsymbol{a}_3 \end{array}\right] \begin{bmatrix} x_1 \\ x_2 \\ x_3 \end{bmatrix} = \left[\begin{array}{c|c|c} \boldsymbol{c}_1 & \boldsymbol{c}_2 & \boldsymbol{c}_3 \end{array}\right] \begin{bmatrix} y_1 \\ y_2 \\ y_3 \end{bmatrix}$$

Let $A = \left[\begin{array}{c|c|c} \boldsymbol{a}_1 & \boldsymbol{a}_2 & \boldsymbol{a}_3 \end{array}\right]$ and let $C = \left[\begin{array}{c|c|c} \boldsymbol{c}_1 & \boldsymbol{c}_2 & \boldsymbol{c}_3 \end{array}\right]$. Since the function from $\mathbb{R}^3$ to $\mathbb{R}^3$ defined by $\boldsymbol{y} \mapsto C\boldsymbol{y}$ is an invertible function, the matrix C has an inverse C^{-1}. Let $H = C^{-1}A$. Then a little algebra shows

$$\left[\begin{array}{c} H \end{array}\right] \begin{bmatrix} x_1 \\ x_2 \\ x_3 \end{bmatrix} = \begin{bmatrix} y_1 \\ y_2 \\ y_3 \end{bmatrix}$$

This is just a recapitulation of the argument that change of basis is matrix-multiplication.

5.12.6 *Computing the change-of-basis matrix*

Now that we know a change-of-basis matrix H exists, we don't use the camera basis or the whiteboard basis to compute it because we don't know those bases! Instead, we will compute H by observing how it behaves on known points, setting up a linear system based on these observations, and solving the linear system to find the entries of H.

Write $H = \begin{bmatrix} h_{y_1,x_1} & h_{y_1,x_2} & h_{y_1,x_3} \\ h_{y_2,x_1} & h_{y_2,x_2} & h_{y_2,x_3} \\ h_{y_3,x_1} & h_{y_3,x_2} & h_{y_3,x_3} \end{bmatrix}$.

Let $\boldsymbol{q}$ be a point on the image plane. If $\boldsymbol{q}$ is the top-left corner of pixel x_1, x_2 then its camera coordinates are $(x_1, x_2, 1)$, and

$$\begin{bmatrix} y_1 \\ y_2 \\ y_3 \end{bmatrix} = \begin{bmatrix} h_{y_1,x_1} & h_{y_1,x_2} & h_{y_1,x_3} \\ h_{y_2,x_1} & h_{y_2,x_2} & h_{y_2,x_3} \\ h_{y_3,x_1} & h_{y_3,x_2} & h_{y_3,x_3} \end{bmatrix} \begin{bmatrix} x_1 \\ x_2 \\ 1 \end{bmatrix}$$

where (y_1, y_2, y_3) are the whiteboard coordinates of $\boldsymbol{q}$.

Multiplying out, we obtain

$$y_1 = h_{y_1,x_1}x_1 + h_{y_1,x_2}x_2 + h_{y_1,x_3} \tag{5.6}$$
$$y_2 = h_{y_2,x_1}x_1 + h_{y_2,x_2}x_2 + h_{y_2,x_3} \tag{5.7}$$
$$y_3 = h_{y_3,x_1}x_1 + h_{y_3,x_2}x_2 + h_{y_3,x_3} \tag{5.8}$$

If we had a point with known camera coordinates and known whiteboard coordinates, we could plug in these coordinates to get three linear equations in the unknowns, the entries of H. By using three such points, we would get nine linear equations and could solve for the entries of H.

For example, by inspecting the image of the board, you can find the pixel coordinates of the bottom-left corner of the whiteboard. You can do this using an image viewer such as The GIMP by opening the image and pointing with your cursor at the corner and reading off the pixel coordinates.[6] I get the coordinates $x_1 = 329$, $x_2 = 597$. Therefore the sensor element that detected the light from this corner is located at $(x_1, x_2, x_3) = (329, 597, 1)$ in camera coordinates.

[6]There are simpler programs that can be used for the same purpose. Under Mac OS, you don't need to install anything: the tool for making a screenshot of a rectangular region can be used.

Plugging in these values for x_1, x_2, we get

$$y_1 = h_{y_1,x_1}329 + h_{y_1,x_2}597 + h_{y_1,x_3} \tag{5.9}$$
$$y_2 = h_{y_2,x_1}329 + h_{y_2,x_2}597 + h_{y_2,x_3} \tag{5.10}$$
$$y_3 = h_{y_3,x_1}329 + h_{y_3,x_2}597 + h_{y_3,x_3} \tag{5.11}$$

Pretend we knew that the whiteboard coordinates for the same point were $(0.2, 0.1, 0.3)$. Then we could plug these values into the equations and obtain three equations in the unknown values of the entries of H. By considering other points in the image, we could get still more equations, and eventually get enough equations to allow us to solve for the entries of H. This approach consists in learning the function h from input-output pairs $(\boldsymbol{x}, \boldsymbol{y})$ such that $h(\boldsymbol{x}) = \boldsymbol{y}$.

The bad news is that we don't know the whiteboard coordinates for these points. The good news is that we can use a similar strategy, learning the function f from input-output pairs such that $f(\boldsymbol{x}) = \boldsymbol{y}$. For example, if $\boldsymbol{x}$ is $(329, 597, 1)$ then $\boldsymbol{y} = f(\boldsymbol{x})$ is $(0, 1, 1)$.

How can we use the knowledge of input-output pairs for f to calculate the entries of H? We need to do a bit of algebra. Let (y_1, y_2, y_3) be the whiteboard coordinates of the point $\boldsymbol{q}$ whose camera coordinates are $(329, 597, 1)$. We don't know the values of y_1, y_2, y_3 but we do know (from the discussion in Section 5.12.4) that

$$0 = y_1/y_3$$
$$1 = y_2/y_3$$

so

$$0y_3 = y_1$$
$$1y_3 = y_2$$

The first equation tells us that $y_1 = 0$. Combining this with Equation 5.9 gives us a linear equation:

$$h_{y_1,x_1}329 + h_{y_1,x_2}597 + h_{y_1,x_3} = 0$$

The second equation tells us that $y_3 = y_2$. Therefore, combining Equations 5.10 and 5.11 gives us

$$h_{y_3,x_1}329 + h_{y_3,x_2}597 + h_{y_3,x_3} = h_{y_2,x_1}329 + h_{y_2,x_2}597 + h_{y_2,x_3}$$

Thus we have obtained two linear equations in the unknown entries of H.

By considering the other three corners of the whiteboard, we can get six more equations. In general, suppose we know numbers x_1, x_2, w_1, w_2 such that

$$f([x_1, x_2, 1]) = [w_1, w_2, 1]$$

Let $[y_1, y_2, y_3]$ be the whiteboard coordinates of the point whose camera coordinates are $[x_1, x_2, 1]$. According to Section 5.12.4, the whiteboard-coordinate representation of the original point $\boldsymbol{p}$ is $(y_1/y_3, y_2/y_3, 1)$. This shows

$$w_1 = y_1/y_3$$
$$w_2 = y_2/y_3$$

Multiplying through by y_3, we obtain

$$w_1 y_3 = y_1$$
$$w_2 y_3 = y_2$$

Combining these equations with Equations 5.6, 5.7, and 5.8, we obtain

$$w_1(h_{y_3,x_1}x_1 + h_{y_3,x_2}x_2 + h_{y_3,x_3}) = h_{y_1,x_1}x_1 + h_{y_1,x_2}x_2 + h_{y_1,x_3}$$
$$w_2(h_{y_3,x_1}x_1 + h_{y_3,x_2}x_2 + h_{y_3,x_3}) = h_{y_2,x_1}x_1 + h_{y_2,x_2}x_2 + h_{y_2,x_3}$$

Multiplying through and moving everything to the same side, we obtain

$$(w_1x_1)h_{y_3,x_1} + (w_1x_2)h_{y_3,x_2} + w_1h_{y_3,x_3} - x_1h_{y_1,x_1} - x_2h_{y_1,x_2} - 1h_{y_1,x_3} = 0 \tag{5.12}$$
$$(w_2x_1)h_{y_3,x_1} + (w_2x_2)h_{y_3,x_2} + w_2h_{y_3,x_3} - x_1h_{y_2,x_1} - x_2h_{y_2,x_2} - 1h_{y_2,x_3} = 0 \tag{5.13}$$

Because we started with numbers for x_1, x_2, w_1, w_2, we obtain two linear equations with known coefficients. Recall that a linear equation can be expressed as an equation stating the value of the dot-product of a coefficient vector—a vector whose entries are the coefficients—and a vector of unknowns.

Task 5.12.2: Define the domain $D = R \times C$.

Write a procedure `make_equations(x1, x2, w1, w2)` that outputs a list $[\boldsymbol{u}, \boldsymbol{v}]$ consisting of two D-vectors $\boldsymbol{u}$ and $\boldsymbol{v}$ such that Equations 5.12 and 5.13 are expressed as

$$\boldsymbol{u} \cdot \boldsymbol{h} = 0$$
$$\boldsymbol{v} \cdot \boldsymbol{h} = 0$$

where $\boldsymbol{h}$ is the D-vector of unknown entries of H.

By using the four corners of the whiteboard, we obtain eight equations. However, no matter how many points we use, we cannot hope to exactly pin down H using only input-output pairs of f.

Here is the reason. Suppose $\hat{H}$ were a matrix that satisfied all such equations: for any input vector $\boldsymbol{x} = [x_1, x_2, 1]$, $g(\hat{H}\boldsymbol{x}) = f(\boldsymbol{x})$. For any scalar α, an algebraic property of matrix-vector multiplication is

$$(\alpha\hat{H})\boldsymbol{x} = \alpha(\hat{H}\boldsymbol{x})$$

Let $[y_1, y_2, y_3] = H\boldsymbol{x}$. Then $\alpha(\hat{H}\boldsymbol{x}) = [\alpha y_1, \alpha y_2, \alpha y_3]$. But since g divides the first and second entries by the third, multiplying all three entries by α does not change the output of g:

$$g(\alpha\hat{H}\boldsymbol{x}) = g([\alpha y_1, \alpha y_2, \alpha y_3]) = g([y_1, y_2, y_3])$$

This shows that if $\hat{H}$ is a suitable matrix for H then so is $\alpha\hat{H}$.

This mathematical result corresponds to the fact that we cannot recover the scale of the whiteboard from the image. It could be a huge whiteboard that is very far away, or a tiny whiteboard that is very close. Fortunately, the math also shows it doesn't matter to the function f.

In order to pin down *some* matrix H, we impose a scaling equation. We simply require that some entry, say the `('y1', 'x1')` entry, be equal to 1. We will write this as $\boldsymbol{w} \cdot \boldsymbol{h} = 1$

Task 5.12.3: Write the D-vector $\boldsymbol{w}$ with a 1 in the `('y1', 'x1')` entry.

Now we have a linear system consisting of nine equations. To solve it, we construct a $\{0, 1, \ldots, 8\} \times D$ matrix L whose rows are the coefficient vectors and we construct a $\{0, 1, \ldots, 8\}$-vector $\boldsymbol{b}$ whose entries are all zero except for a 1 in the position corresponding to the scaling equation.

Task 5.12.4: Here are the pixel coordinates for the corners of the whiteboard in the image.

top left	$x_1 = 358, x_2 = 36$
bottom left	$x_1 = 329, x_2 = 597$
top right	$x_1 = 592, x_2 = 157$
bottom right	$x_1 = 580, x_2 = 483$

Assign to L the $\{0, 1, \ldots, 8\} \times D$ matrix whose rows are, in order,

- the vector $\boldsymbol{u}$ and the vector $\boldsymbol{v}$ from `make_equations(x1, x2, w1, w2)` applied to the top-left corner,
- the vector $\boldsymbol{u}$ and the vector $\boldsymbol{v}$ from `make_equations(x1, x2, w1, w2)` applied to the bottom-left corner,
- the vector $\boldsymbol{u}$ and the vector $\boldsymbol{v}$ from `make_equations(x1, x2, w1, w2)` applied to the

top-right corner,

- the vector $\boldsymbol{u}$ and the vector $\boldsymbol{v}$ from `make_equations(x1, x2, w1, w2)` applied to the bottom-right corner,
- the vector $\boldsymbol{w}$ from the above ungraded task.

Assign to $\boldsymbol{b}$ the $\{0, 1, \ldots, 8\}$-vector $\boldsymbol{b}$ whose entries are all zero except for a 1 in position 8.

Assign to $\boldsymbol{h}$ the solution obtained by solving the equation $L\boldsymbol{h} = \boldsymbol{b}$. Verify for yourself that it is indeed a solution to this equation.

Finally, assign to H the matrix whose entries are given by the vector $\boldsymbol{h}$.

5.12.7 *Image representation*

Recall the image representation used in the 2D geometry lab. A *generalized image* consists of a grid of generalized pixels, where each generalized pixel is a quadrilateral (not necessarily a rectangle).

The points at the corners of the generalized pixels are identified by pairs (x, y) of integers, the pixel coordinates.

Each corner is assigned a location in the plane, and each generalized pixel is assigned a color. The mapping of corners to points in the plane is given by a matrix, the *location matrix.* Each corner corresponds to a column of the location matrix, and the label of that column is the pair (x, y) of pixel coordinates of the corner. The column is a `{'x','y','u'}`-vector giving the location of the corner. Thus the row labels of the location matrix are `'x'`, `'y'`, and `'u'`.

The mapping of generalized pixels to colors is given by another matrix, the *color matrix.* Each generalized pixel corresponds to a column of the color matrix, and the label of that column is the pair of pixel coordinates of the top-left corner of that generalized pixel. The column is a `{'r','g','b'}`-vector giving the color of that generalized pixel.

The module `image_mat_util` defines the procedures

- `file2mat(filename, rowlabels)`, which, given a path to a `.png` image file and optionally a tuple of row labels, returns the pair `(points, colors)` of matrices representing the image, and
- and `mat2display(pts, colors, row_labels)`, which displays an image given by a matrix `pts` and a matrix `colors` and optionally a tuple of row labels. There are a few additional optional parameters that we will use in this lab.

As in the 2D geometry lab, you will apply a transformation to the locations to obtain new locations, and view the resulting image.

5.12.8 *Synthesizing the perspective-free image*

Now we present the tasks involved in using H to create the synthetic image.

Task 5.12.5: Construct the generalized image from the image file `board.png`:

```
(X_pts, colors) = image_mat_util.file2mat('board.png', ('x1','x2','x3'))
```

Task 5.12.6: The columns of the matrix `X_pts` are the camera-coordinates representations of points in the image. We want to obtain the board-coordinates representations of these points. To apply the transformation `H` to each column of `X_pts`, we use matrix-matrix multiplication:

```
Y_pts = H * X_pts
```

Task 5.12.7: Each column of `Y_pts` gives the whiteboard-coordinate representation (y_1, y_2, y_3) of a point $\boldsymbol{q}$ in the image. We need to construct another matrix `Y_board` in which each column gives the whiteboard-coordinate representation $(y_1/y_3, y_2/y_3, 1)$ of the corresponding point $\boldsymbol{p}$ in the plane containing the whiteboard.

Write a procedure `mat_move2board(Y)` with the following spec:

- *input:* a Mat each column of which is a 'y1','y2','y3'-vector giving the whiteboard coordinates of a point q
- *output:* a Mat each column of which is the corresponding point in the whiteboard plane (the point of intersection with the whiteboard plane of the line through the origin and q

Here's a small example:

```
>>> Y_in = Mat(({'y1', 'y2', 'y3'}, {0,1,2,3}),
    {('y1',0):2, ('y2',0):4, ('y3',0):8,
     ('y1',1):10, ('y2',1):5, ('y3',1):5,
     ('y1',2):4, ('y2',2):25, ('y3',2):2,
     ('y1',3):5, ('y2',3):10, ('y3',3):4})
>>> print(Y_in)

        0  1  2  3
      ------------
 y1  |  2 10  4  5
 y2  |  4  5 25 10
 y3  |  8  5  2  4

>>> print(mat_move2board(Y_in))

          0 1    2    3
      ------------------
 y1  |  0.25 2    2 1.25
 y2  |   0.5 1 12.5  2.5
 y3  |     1 1    1    1
```

Once your `mat_move2board` procedure is working, use it to derive the matrix `Y_board` from `Y_pts`:

```
>>> Y_board = mat_move2board(Y_pts)
```

One simple way to implement `mat_move2board(Y)` is to convert the Mat to a column dictionary (coldict), call your `move2board(y)` procedure for each column, and convert the resulting column dictionary back to a matrix.

Task 5.12.8: Finally, display the result of

```
>>> image_mat_util.mat2display(Y_board, colors, ('y1', 'y2', 'y3'),
scale=100, xmin=None, ymin=None)
```

Task 5.12.9: If you have time, repeat with `cit.png`. This is a picture of Brown University's Computer Science building. Select a rectangle on a wall of the building (e.g. one of the windows), and define a coordinate system that assigns coordinates $(0,0,1), (1,0,1), (0,1,1), (1,1,1)$ to the corners of the rectangle. Then find out the pixels corresponding to these points, and so on.

5.13 Review questions

- What is coordinate representation?
- How can you express conversion between a vector and its coordinate representation using

matrices?

- What is linear dependence?
- How would you prove a set of vectors are linearly independent?
- What is the Grow algorithm?
- What is the Shrink algorithm?
- How do the concepts of linear dependence and spanning apply to subsets of edges of graphs?
- Why is the output of the growing algorithm a set of linearly independent vectors?
- Why is the output of the shrinking algorithm a set of linearly independent vectors?
- What is a basis?
- What is unique representation?
- What is change of basis?
- What is the Exchange Lemma?

5.14 Problems

Span of vectors over $\mathbb{R}$

Problem 5.14.1: Let $\mathcal{V} = \text{Span }\{[2,0,4,0],[0,1,0,1],[0,0,-1,-1]\}$. For each of the following vectors, show it belongs to $\mathcal{V}$ by writing it as a linear combination of the generators of $\mathcal{V}$.

(a) $[2,1,4,1]$

(b) $[1,1,1,0]$

(c) $[0,1,1,2]$

Problem 5.14.2: Let $\mathcal{V} = \text{Span }\{[0,0,1],[2,0,1],[4,1,2]\}$. For each of the following vectors, show it belongs to $\mathcal{V}$ by writing it as a linear combination of the generators of $\mathcal{V}$.

(a) $[2,1,4]$

(b) $[1,1,1]$

(c) $[5,4,3]$

(d) $[0,1,1]$

Span of vectors over $GF(2)$

Problem 5.14.3: Let $\mathcal{V} = \text{Span }\{[0,1,0,1],[0,0,1,0],[1,0,0,1],[1,1,1,1]\}$ where the vectors are over $GF(2)$. For each of the following vectors over $GF(2)$, show it belongs to $\mathcal{V}$ by writing it as a linear combination of the generators of $\mathcal{V}$.

(a) $[1,1,0,0]$

(b) $[1,0,1,0]$

(c) $[1,0,0,0]$

Problem 5.14.4: The vectors over $GF(2)$ representing the graph

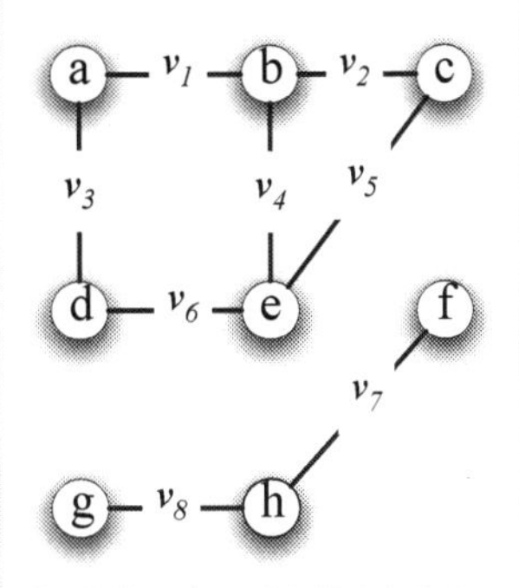

are

	a	b	c	d	e	f	g	h
v_1	1	1						
v_2		1	1					
v_3	1			1				
v_4		1			1			
v_5			1		1			
v_6				1	1			
v_7						1		1
v_8							1	1

For each of the following vectors over $GF(2)$, show it belongs to the span of the above vectors by writing it as a linear combination of the above vectors.

(a) $[0,0,1,1,0,0,0,0]$

(b) $[0,0,0,0,0,1,1,0]$

(c) $[1,0,0,0,1,0,0,0]$

(d) $[0,1,0,1,0,0,0,0]$

Linear dependence over $\mathbb{R}$

Problem 5.14.5: For each of the parts below, show the given vectors over $\mathbb{R}$ are linearly dependent by writing the zero vector as a nontrivial linear combination of the vectors.

(a) $[1,2,0],[2,4,1],[0,0,-1]$

(b) $[2,4,0],[8,16,4],[0,0,7]$

(c) $[0,0,5],[1,34,2],[123,456,789],[-3,-6,0],[1,2,0.5]$

Problem 5.14.6: For each of the parts below, show the given vectors over $\mathbb{R}$ are linearly dependent by writing the zero vector as a nontrivial linear combination of the vectors.

(a) $[1,2,3],[4,5,6],[1,1,1]$

(b) $[0,-1,0,-1],[\pi,\pi,\pi,\pi],[-\sqrt{2},\sqrt{2},-\sqrt{2},\sqrt{2}]$

(c) $[1,-1,0,0,0],[0,1,-1,0,0],[0,0,1,-1,0],[0,0,0,1,-1],[-1,0,0,0,1]$

Problem 5.14.7: Show that one of the vectors is superfluous by expressing it as a linear combination of the other two.

$$\begin{aligned} \boldsymbol{u} &= [3,9,6,5,5] \\ \boldsymbol{v} &= [4,10,6,6,8] \\ \boldsymbol{w} &= [1,1,0,1,3] \end{aligned}$$

Problem 5.14.8: Give four vectors that are linearly dependent but such that any three are linearly independent.

Linear dependence over $GF(2)$

Problem 5.14.9: For each of the parts below, show the given vectors over $GF(2)$ are linearly dependent by writing the zero vector as a nontrivial linear combination of the vectors.

(a) $[1,1,1,1], [1,0,1,0], [0,1,1,0], [0,1,0,1]$

(b) $[0,0,0,1], [0,0,1,0], [1,1,0,1], [1,1,1,1]$

(c) $[1,1,0,1,1], [0,0,1,0,0], [0,0,1,1,1], [1,0,1,1,1], [1,1,1,1,1]$

Problem 5.14.10: Each of the parts below specifies some of the vectors over $GF(2)$ specified in Problem 5.14.4. Show that these vectors are linearly dependent by giving a subset of those vectors whose sum is the zero vector. (Hint: Looking at the graph will help.)

(a) $\{\boldsymbol{v}_1, \boldsymbol{v}_2, \boldsymbol{v}_3, \boldsymbol{v}_4, \boldsymbol{v}_5\}$

(b) $\{\boldsymbol{v}_1, \boldsymbol{v}_2, \boldsymbol{v}_3, \boldsymbol{v}_4, \boldsymbol{v}_5, \boldsymbol{v}_7, \boldsymbol{v}_8\}$

(c) $\{\boldsymbol{v}_1, \boldsymbol{v}_2, \boldsymbol{v}_3, \boldsymbol{v}_4, \boldsymbol{v}_6\}$

(d) $\{\boldsymbol{v}_1, \boldsymbol{v}_2, \boldsymbol{v}_3, \boldsymbol{v}_5, \boldsymbol{v}_6, \boldsymbol{v}_7, \boldsymbol{v}_8\}$

Exchange Lemma for vectors over $\mathbb{R}$

Problem 5.14.11: Let $S = \{[1,0,0,0,0], [0,1,0,0,0], [0,0,1,0,0], [0,0,0,1,0], [0,0,0,0,1]\}$, and let
$A = \{[1,0,0,0,0], [0,1,0,0,0]\}$. For each of the following vectors $\boldsymbol{z}$, find a vector $\boldsymbol{w}$ in $S - A$ such that Span $S =$ Span $(S \cup \{\boldsymbol{z}\} - \{\boldsymbol{w}\})$.

(a) $\boldsymbol{z} = [1,1,1,1,1]$

(b) $\boldsymbol{z} = [0,1,0,1,0]$

(c) $\boldsymbol{z} = [1,0,1,0,1]$

Exchange Lemma for vectors over $GF(2)$

Problem 5.14.12: We refer in this problem to the vectors over $GF(2)$ specified in Problem 5.14.4.

Let $S = \{\boldsymbol{v}_1, \boldsymbol{v}_2, \boldsymbol{v}_3, \boldsymbol{v}_4\}$. Each of the following parts specifies a subset A of S and a vector $\boldsymbol{z}$ such that $A \cup \{\boldsymbol{z}\}$ is linearly independent. For each part, specify a vector $\boldsymbol{w}$ in $S - A$ such that Span $S =$ Span $(S \cup \{\boldsymbol{z}\} - \{\boldsymbol{w}\})$. (Hint: Drawing subgraphs of the graph will help.)

(a) $A = \{\boldsymbol{v}_1, \boldsymbol{v}_4\}$ and $\boldsymbol{z}$ is

a	b	c	d	e	f	g	h
			1	1			

(b) $A = \{v_2, v_3\}$ and z is

a	b	c	d	e	f	g	h
		1	1				

(c) $A = \{v_2, v_3\}$ and z is

a	b	c	d	e	f	g	h
1				1			

Problem 5.14.13: Write and test a procedure `rep2vec(u, veclist)` with the following spec:

- *input:* a vector u and a list *veclist* of Vecs $[a_0, \ldots, a_{n-1}]$. The domain of u should be $\{0, 1, 2, n-1\}$ where n is the length of `veclist`.
- *output:* the vector v such that u is the coordinate representation of v with respect to $a_0, \ldots, a_{n-1}$, where entry i of u is the coefficient of a_i for $i = 0, 1, 2 \ldots, n-1$.

Your procedure should not use any loops or comprehensions but of course can use the operations on instances of `Mat` and `Vec` and can also use procedures from the `matutil` module. Note that the procedures `coldict2mat` and `rowdict2mat` (defined in `matutil`) can accept lists, not just dictionaries.

Here is an illustration of how the procedure is used.

```
>>> a0 = Vec({'a','b','c','d'}, {'a':1})
>>> a1 = Vec({'a','b','c','d'}, {'b':1})
>>> a2 = Vec({'a','b','c','d'}, {'c':1})
>>> rep2vec(Vec({0,1,2}, {0:2, 1:4, 2:6}), [a0,a1,a2])
Vec({'a', 'c', 'b', 'd'},{'a': 2, 'c': 6, 'b': 4, 'd': 0})
```

Test your procedure with the following examples.

- $u = [5, 3, -2]$, $\texttt{veclist} = [[1, 0, 2, 0], [1, 2, 5, 1], [1, 5, -1, 3]]$ over $\mathbb{R}$
- $u = [1, 1, 0]$, $\texttt{veclist} = [[1, 0, 1], [1, 1, 0], [0, 0, 1]]$ over $GF(2)$

Note that the vectors in the examples above are given in math notation, but the procedure requires vectors to be represented as instances of `Vec`. You can use the procedure `list2vec` (defined in the module `vecutil`) to convert the example vectors from list representation to `Vec` for use as inputs to the procedure.

Also note that, in trying out the GF2 example, you should use the value one defined in the module `GF2` in place of the number 1.

In the following problem and others to come, you should make use of the `solve` procedure of the `solver` module, described in Section 4.5.4.

Problem 5.14.14: Write and test a procedure `vec2rep(veclist, v)` with the following spec:

- *input:* a list `veclist` of vectors $[a_0, \ldots, a_{n-1}]$, and a vector v with the domain $\{0, 1, 2, \ldots, n-1\}$ where n is the length of `veclist`. You can assume v is in Span $\{a_0, \ldots, a_{n-1}\}$.
- *output:* the vector u whose coordinate representation in tems of $a_0, \ldots, a_{n-1}$ is v.

As in Problem 5.14.13, your procedure should use no loops or comprehensions directly but can use procedures defined in `matutil`.

Here is an illustration of how the procedure is used.

```
>>> a0 = Vec({'a','b','c','d'}, {'a':1})
>>> a1 = Vec({'a','b','c','d'}, {'b':1})
>>> a2 = Vec({'a','b','c','d'}, {'c':1})
>>> vec2rep([a0,a1,a2], Vec({'a','b','c','d'}, {'a':3, 'c':-2}))
```

```
Vec({0, 1, 2},{0: 3.0, 1: 0.0, 2: -2.0})
```

Test your procedure with the following examples:

- $v = [6, -4, 27, -3]$, veclist $= [[1, 0, 2, 0], [1, 2, 5, 1], [1, 5, -1, 3]]$ in $\mathbb{R}$
- $v = [0, 1, 1]$, veclist $= [[1, 0, 1], [1, 1, 0], [0, 0, 1]]$ in $GF(2)$

As in Problem 5.14.13, the examples above are given in math notation but the procedure expects vectors to be represented as instances of Vec, and the value one from the module GF2 should be used in the GF(2) examples.

Problem 5.14.15: Write and test a procedure is_superfluous(L, i) with the following spec:

- *input:* a list L of vectors, and an integer i in $\{0, 1, \ldots, n-1\}$ where $n = \text{len}(L)$
- *output:* True if the span of the vectors in L equals the span of

$$L[0], L[1], \ldots, L[i-1], L[i+1], ..., L[n-1]$$

Your procedure should not use loops or comprehensions but can use procedures defined in the module matutil and can use the procedure solve(A,b) defined in solver module. Your procedure will most likely need a special case for the case where len(L) is 1.

Note that the solve(A,b) always returns a vector $\boldsymbol{u}$. It is up to you to check that $\boldsymbol{u}$ is in fact a solution to the equation $A\boldsymbol{x} = \boldsymbol{b}$. Moreover, over $\mathbb{R}$, even if a solution exists, the solution returned by solve is approximate due to roundoff error. To check whether the vector $\boldsymbol{u}$ returned is a solution, you should compute the residual $\boldsymbol{b} - A * \boldsymbol{u}$, and test if it is close to the zero vector:

```
>>> residual = b - A*u
>>> residual * residual
1.819555009546577e-25
```

If the sum of squares of the entries of the residual (the dot-product of the residual with itself) is less than, say 10^{-14}, it is pretty safe to conclude that $\boldsymbol{u}$ is indeed a solution.
Here is an illustration of how is_superfluous(L, v) is used.

```
>>> a0 = Vec({'a','b','c','d'}, {'a':1})
>>> a1 = Vec({'a','b','c','d'}, {'b':1})
>>> a2 = Vec({'a','b','c','d'}, {'c':1})
>>> a3 = Vec({'a','b','c','d'}, {'a':1,'c':3})
>>> is_superfluous(L, 3)
True
>>> is_superfluous([a0,a1,a2,a3], 3)
True
>>> is_superfluous([a0,a1,a2,a3], 0)
True
>>> is_superfluous([a0,a1,a2,a3], 1)
False
```

Test your procedure with the following examples:

- $L = [[1, 2, 3]], \boldsymbol{v} = [1, 2, 3]$ over $\mathbb{R}$
- $L = [[2, 5, 5, 6], [2, 0, 1, 3], [0, 5, 4, 3]], \boldsymbol{v} = [0, 5, 4, 3]$ over R

- $L = [[1,1,0,0],[1,1,1,1],[0,0,0,1]]$, $v = [0,0,0,1]$ over $GF(2)$

As in Problems 5.14.13 and 5.14.14, the examples are written in mathese and you have to translate into our Python representation:

Problem 5.14.16: Write and test a procedure `is_independent(L)` with the following spec:

- *input:* a list L of vectors
- *output:* `True` if the vectors form a linearly independent list.

Your algorithm for this procedure should be based on the Span Lemma (Lemma 5.5.9). You can use as a subroutine any one of the following:

- the procedure `is_superfluous(L, b)` from Problem 5.14.15, or
- the `solve(A,b)` procedure from the `solver` module (but see the provisos in Problem 5.14.15).

You will need a loop or comprehension for this procedure.

Here is an illustration of how the procedure is used:

```
>>> a0 = Vec({'a','b','c','d'}, {'a':1})
>>> a1 = Vec({'a','b','c','d'}, {'b':1})
>>> a2 = Vec({'a','b','c','d'}, {'c':1})
>>> a3 = Vec({'a','b','c','d'}, {'a':1,'c':3})
>>> is_independent([a0, a1, a2])
True
>>> is_independent([a0, a2, a3])
False
>>> is_independent([a0, a1, a3])
True
>>> is_independent([a0, a1, a2, a3])
False
```

Note: There is a slight technical difference between a set being linearly independent and a list being linearly independent. A list can contain some vector v twice, in which case the list is considered linearly dependent. Your code should *not* deal specially with this case; it will be handled naturally, so don't think about this case until after you have written your procedure.

Test your procedure with the following examples:

- $[[2,4,0],[8,16,4],[0,0,7]]$ over $\mathbb{R}$
- $[[1,3,0,0],[2,1,1,0],[0,0,1,0],[1,1,4,-1]]$ over $\mathbb{R}$
- $[[1,0,1,0],[0,1,0,0],[1,1,1,1],[1,0,0,1]]$ over $GF(2)$

As usual, the examples are given in math notation but the procedure expects vectors represented as `Vec` instances, and the 1 in $GF(2)$ vectors should be replaced by the value `one` defined in the module `GF2`.

Problem 5.14.17: Write and test a procedure `subset_basis(T)` with the following spec:

- *input:* a list T of vectors
- *output:* a list S consisting of vectors of T such that S is a basis for the span of T.

Your procedure should be based on either a version of the Grow algorithm or a version of the Shrink algorithm. Think about each one to see which is easier for you. You will need a loop or comprehension for this procedure. You can use as a subroutine any one of the following:

- the procedure is_superfluous(L, b) from Problem 5.14.15, , or
- the procedure is_independent(L) from Problem 5.14.16 or from the module independence we provide, or
- the procedure solve(A,b) from the solver module (but see the provisos in Problem 5.14.15).

Here is an illustration of how the procedure is used.

```
>>> a0 = Vec({'a','b','c','d'}, {'a':1})
>>> a1 = Vec({'a','b','c','d'}, {'b':1})
>>> a2 = Vec({'a','b','c','d'}, {'c':1})
>>> a3 = Vec({'a','b','c','d'}, {'a':1,'c':3})
>>> subset_basis([a0,a1,a2,a3])
[Vec({'a', 'c', 'b', 'd'},{'a': 1}), Vec({'a', 'c', 'b', 'd'},{'b':
1}),
 Vec({'a', 'c', 'b', 'd'},{'c': 1})]
>>> subset_basis([a0,a3,a1,a2])
[Vec({'a', 'c', 'b', 'd'},{'a': 1}),
 Vec({'a', 'c', 'b', 'd'},{'a': 1, 'c': 3}),
 Vec({'a', 'c', 'b', 'd'},{'b': 1})]
```

Note that the order in which vectors appear in T is likely to affect the returned list. Note also that there are different valid outputs.

Test your procedure with the following examples:

- $[1,1,2,1], [2,1,1,1], [1,2,2,1], [2,2,1,2], [2,2,2,2]$ over $\mathbb{R}$
- $[1,1,0,0], [1,1,1,1], [0,0,1,1], [0,0,0,1], [0,0,1,0]$ over $GF(2)$

As usual, the examples are given in math notation but the procedure expects vectors represented as Vec instances, and the 1 in $GF(2)$ vectors should be replaced by the value one defined in the module GF2.

Problem 5.14.18: Write and test a procedure superset_basis(T, L) with the following spec:

- *input:* a linearly independent list T of vectors, and a list L of vectors such that every vector in T is in the span of L.
- *output:* a linearly independent list S containing all vectors in T such that the span of S equals the span of L (i.e. S is a basis for the span of L).

Your procedure should be based on either a version of the Grow algorithm or a version of the Shrink algorithm. Think about each one to see which is easier for you. You will need a loop or comprehension for this procedure. You can use as a subroutine any one of the following:

- the procedure is_superfluous(L, b) from Problem 5.14.15, or
- the procedure is_independent(L) from Problem 5.14.16, or
- the procedure solve(A,b) from the solver module (but see the provisos in Problem 5.14.15).

Here is an illustration of how the procedure is used.

```
>>> a0 = Vec({'a','b','c','d'}, {'a':1})
>>> a1 = Vec({'a','b','c','d'}, {'b':1})
>>> a2 = Vec({'a','b','c','d'}, {'c':1})
>>> a3 = Vec({'a','b','c','d'}, {'a':1,'c':3})
```

```
>>> superset_basis([a0, a3], [a0, a1, a2])
[Vec({'b', 'c', 'd', 'a'},{'a': 1}), Vec({'b', 'c', 'd', 'a'},{'c': 3, 'a': 1}),
 Vec({'b', 'c', 'd', 'a'},{'b': 1})]
```

Test your procedure with the following examples:

- $T = [[0, 5, 3], [0, 2, 2], [1, 5, 7]], L = [[1, 1, 1], [0, 1, 1], [0, 0, 1]]$ over $\mathbb{R}$
- $T = [[0, 5, 3], [0, 2, 2]], L = [[1, 1, 1], [0, 1, 1], [0, 0, 1]]$ over $\mathbb{R}$
- $T = [[0, 1, 1, 0], [1, 0, 0, 1]], L = [[1, 1, 1, 1], [1, 0, 0, 0], [0, 0, 0, 1]]$ over $GF(2)$

As in Problems 5.14.13 and 5.14.14, the examples are written in mathese and you have to translate into our Python representation.

Problem 5.14.19: Write and test a procedure `exchange(S, A, z)` with the following spec:

- *input:* A list S of vectors, a list A of vectors that are all in S (such that len(A) < len(S)), and a vector z such that $A + [z]$ is linearly independent
- *output:* a vector $\boldsymbol{w}$ in S but not in A such that

$$\text{Span } S = \text{Span } (\{\boldsymbol{z}\} \cup S - \{\boldsymbol{w}\})$$

Your procedure should follow the proof of the Exchange Lemma (Lemma 5.11.1). You should use the `solver` module or the procedure `vec2rep(veclist, u)` from Problem 5.14.14. You can test whether a vector is in a list using the expression `v in L`.

Here is an illustration of how the procedure is used:

```
>>> S=[list2vec(v) for v in [[0,0,5,3] , [2,0,1,3],[0,0,1,0],[1,2,3,4]]]
>>> A=[list2vec(v) for v in [[0,0,5,3],[2,0,1,3]]]
>>> z=list2vec([0,2,1,1])
>>> print(exchange(S, A, z))

 0 1 2 3
--------
 0 0 1 0
```

Test your procedure with the following examples:

- $S = [[0, 0, 5, 3], [2, 0, 1, 3], [0, 0, 1, 0], [1, 2, 3, 4]], A = [[0, 0, 5, 3], [2, 0, 1, 3]], \boldsymbol{z} = [0, 2, 1, 1]$ in $\mathbb{R}$
- $S = [[0, 1, 1, 1], [1, 0, 1, 1], [1, 1, 0, 1], [1, 1, 1, 0]], A = [[0, 1, 1, 1], [1, 1, 0, 1]], \boldsymbol{z} = [1, 1, 1, 1]$ in $GF(2)$

Chapter 6

Dimension

> I hope that posterity will judge me kindly, not only as to the things which I have explained, but also to those which I have intentionally omitted so as to leave to others the pleasure of discovery.
>
> *René Descartes*, **Geometry**

The crucial fact about bases is that all bases for a vector space have the same size. I prove this fact in Section 6.1. In Section 6.2, I use this fact to define the *dimension* of a vector space. This concept opens the door to many important insights about vector spaces, about homogeneous linear systems, about linear functions, and about matrices.

6.1 The size of a basis

At the heart of linear algebra is the next result. (The name I give for it is, uh, nonstandard.)

6.1.1 The Morphing Lemma and its implications

Lemma 6.1.1 (Morphing Lemma): Let $\mathcal{V}$ be a vector space. Suppose S is a set of generators for $\mathcal{V}$, and B is a linearly independent set of vectors belonging to $\mathcal{V}$. Then $|S| \geq |B|$.

We will prove it presently. For now, let us see how it can be used to prove what is perhaps the most important result in linear algebra.

Theorem 6.1.2 (Basis Theorem): Let $\mathcal{V}$ be a vector space. All bases for $\mathcal{V}$ have the same size.

Proof

Let B_1 and B_2 be two bases for $\mathcal{V}$. Applying the Morphing Lemma with $S = B_1$ and $B = B_2$, we infer that $|B_1| \geq |B_2|$. Appying it with $S = B_2$ and $B = B_1$, we infer that $|B_2| \geq |B_1|$. Putting these inequalities together, we infer $|B_1| = |B_2|$. □

Theorem 6.1.3: Let $\mathcal{V}$ be a vector space. Then a set of generators for $\mathcal{V}$ is a *smallest* set of generators for $\mathcal{V}$ if and only if the set is a basis for $\mathcal{V}$.

Proof

Let T be a set of generators for $\mathcal{V}$. We must prove (1) if T is a basis for $\mathcal{V}$ then T is a smallest set of generators for $\mathcal{V}$, and (2) if T is not a basis for $\mathcal{V}$ then there is a set of generators smaller than T.

1. Suppose T is a basis. Let S be a smallest set of generators for $\mathcal{V}$. By the Morphing Lemma, $|T| \leq |S|$, so T is also a smallest set of generators.

2. Suppose T is not a basis. A basis is a linearly independent set of generators. Since T *is* a set of generators, it must be linearly dependent. By the Linear-Dependence Lemma (Lemma 5.5.9), there is some vector in T that is in the span of the other vectors. Therefore, by the Superfluous-Vector Lemma (Lemma 5.5.1), there is some vector whose removal from T leaves a set of generators for $\mathcal{V}$. Thus T is not a smallest set of generators.

□

6.1.2 Proof of the Morphing Lemma

The proof of the Morphing Lemma is algorithmic. We give an algorithm to convert S to a set S' that still spans $\mathcal{V}$, and that has the same cardinality as S, but that includes all the elements of B. This shows that $|S|$ is at least $|B|$.

Why is it called the "Morphing Lemma"? The algorithm modifies S step by step to include more and more vectors of B, while preserving its cardinality and the fact that it spans $\mathcal{V}$, until it includes all of B. In each step, the algorithm "injects" a vector of B into S and, in order to preserve the cardinality of S, it "ejects" a vector from S. The choice of vector to eject is the responsibility of a subroutine, the Exchange Lemma of Section 5.11.1.

For the reader's convenience, I restate the Morphing Lemma (Lemma 6.1.1):

> Let $\mathcal{V}$ be a vector space. Suppose S is a set of generators for $\mathcal{V}$, and B is a linearly independent set of vectors belonging to $\mathcal{V}$. Then $|B| \leq |S|$.

Proof

Let $B = \{\boldsymbol{b}_1, \ldots, \boldsymbol{b}_n\}$. We will transform S step by step to include more and more vectors in B without ever increasing the size of the set. For $k = 0, 1, \ldots, |B|$, we use S_k to denote the set obtained after k steps. We will prove by induction on k that there is a set S_k that spans $\mathcal{V}$ and contains $\boldsymbol{b}_1, \ldots, \boldsymbol{b}_k$ but has the same cardinality as S.

Consider the case $k = 0$. The set S_0 is just S, which spans $\mathcal{V}$ by assumption, and which has the same cardinality as itself. It is not required to contain any of the vectors of B, so the base case holds.

Now we prove the induction step. For $k = 1, \ldots, n$, we obtain S_k from S_{k-1} as follows. Let $A_k = \{\boldsymbol{b}_1, \ldots, \boldsymbol{b}_{k-1}\}$. Since $\boldsymbol{b}_k \cup A_k$ is linearly independent, we can apply the Exchange Lemma to A_k and S_{k-1}. There is a vector $\boldsymbol{w} \in S_{k-1} - A_k$ such that $\{\boldsymbol{b}_k\} \cup (S_{k-1} - \{\boldsymbol{w}\})$ spans $\mathcal{V}$. We define $S_k = \{\boldsymbol{b}_k\} \cup S_{k-1} - \{\boldsymbol{w}\}$. Then $|S_k| = |S_{k-1}|$ and S_k spans $\mathcal{V}$ and S_k contains $\boldsymbol{b}_1, \ldots, \boldsymbol{b}_k$, so we have proved the induction step. □

Example 6.1.4: Let S be the set of vectors corresponding to dark edges in the following graph:

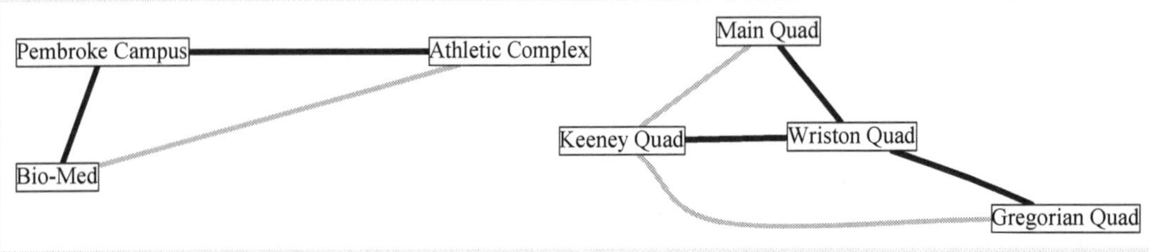

How can we morph S into the set B of vectors corresponding to the dark edges in the next diagram?

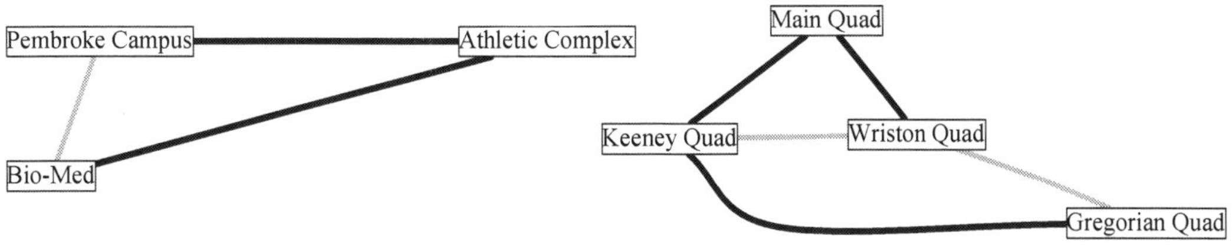

We insert edges of B into S edge by edge, removing edges from S in each step in accordance with the Exchange Lemma. First we inject, say, the edge with endpoints Bio-Med and Athletic:

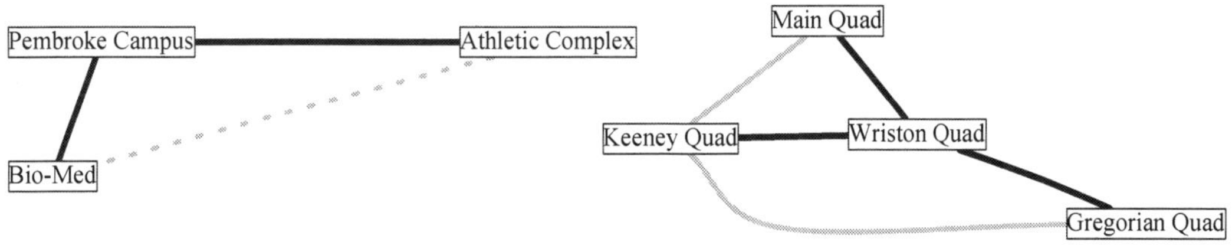

which entails ejecting an edge, say the edge with endpoints Bio-Med and Pembroke:

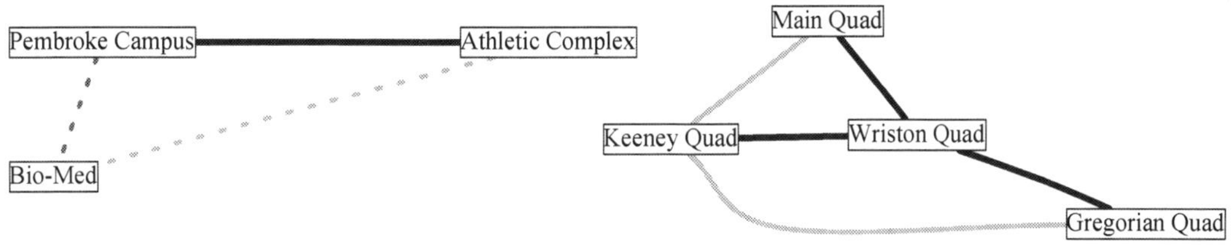

resulting in the forest

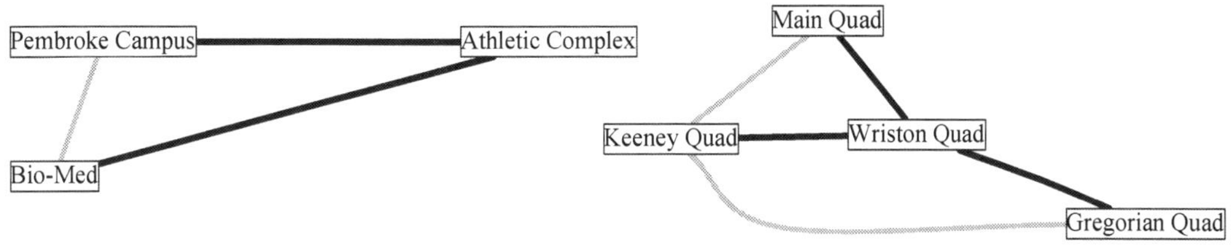

Next we inject, say, the edge with endpoints Keeney and Main, and eject the edge with endpoints Keeney and Wriston:

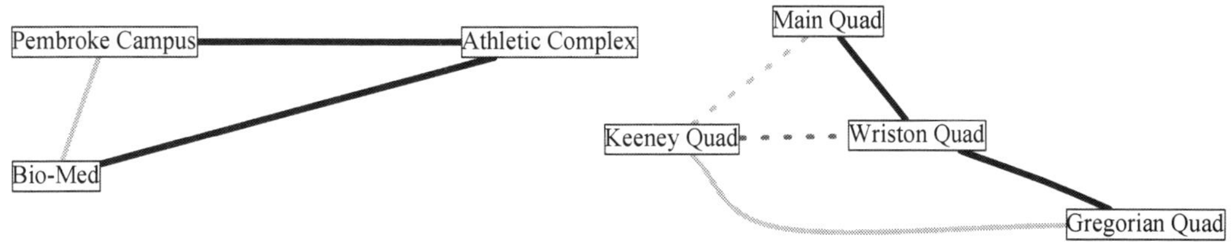

resulting in

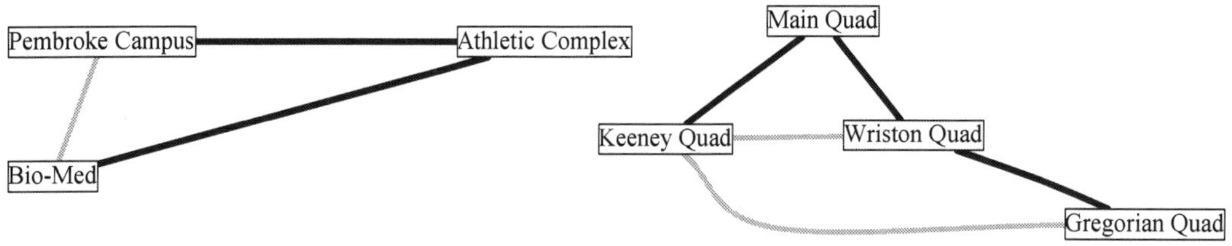

Finally, we inject the edge with endpoints Keeney and Gregorian, ejecting the edge with endpoints Wriston and Gregorian:

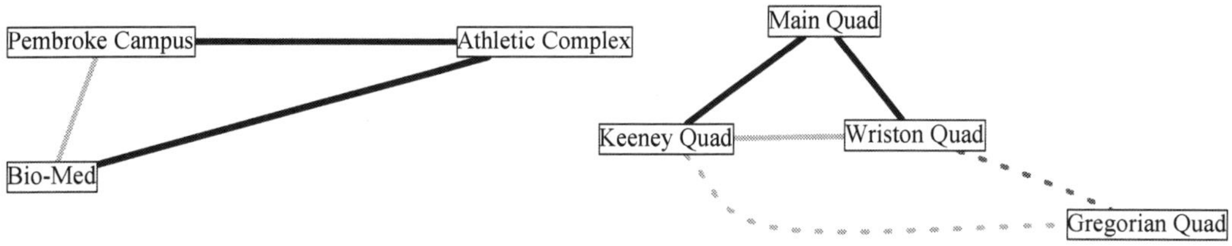

which results in the target spanning forest B:

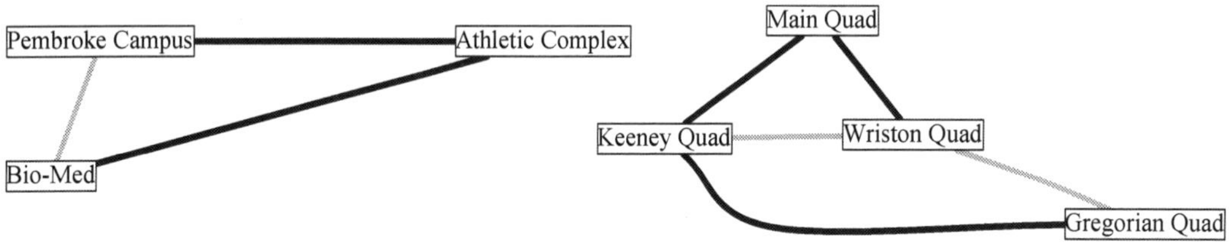

6.2 Dimension and rank

We have shown that all bases of $\mathcal{V}$ have the same size.

6.2.1 Definitions and examples

Definition 6.2.1: We define the *dimension* of a vector space to be the size of a basis for that vector space. The dimension of a vector space $\mathcal{V}$ is written $\dim \mathcal{V}$.

Example 6.2.2: One basis for $\mathbb{R}^3$ is the standard basis: $\{[1, 0, 0], [0, 1, 0], [0, 0, 1]\}$. Therefore the dimension of $\mathbb{R}^3$ is 3.

Example 6.2.3: More generally, for any field $\mathbb{F}$ and any finite set D, one basis for $\mathbb{F}^D$ is the standard basis, which consists of $|D|$ vectors. Therefore $\mathbb{F}^D$ has dimension $|D|$.

Example 6.2.4: Let $S = \{[-0.6, -2.1, -3.5, -2.2], [-1.3, 1.5, -0.9, -0.5], [4.9, -3.7, 0.5, -0.3], [2.6, -3.5, -1.2, -2.0], [-1.5, -2.5, -3.5, 0.94]\}$. What can we say about the dimension of Span S?

If we knew that S were linearly independent, we would know that S is a basis for Span S, so $\dim$ Span S would equal the cardinality of S, namely five.

But let's say we don't know whether S is linearly independent. By the Subset-Basis Lemma (Lemma 5.6.11), we know that S contains a basis. That basis must have size at most $|S|$, so we can be sure that $\dim$ Span S is less than or equal to five.

Since S contains nonzero vectors, so does Span S. We can therefore be sure that $\dim$ Span S

is greater than zero.

Definition 6.2.5: We define the *rank* of a set S of vectors as the dimension of Span S. We write rank S for the rank of S.

Example 6.2.6: In Example 5.5.3 (Page 226), we showed that the vectors $[1,0,0], [0,2,0], [2,4,0]$ are linearly dependent. Therefore their rank is less than three. Any two of these vectors form a basis for the span of all three, so the rank is two.

Example 6.2.7: The set S given in Example 6.2.4 (Page 266) has rank between one and five.

As illustrated in Example 6.2.4 (Page 266), the Subset-Basis Lemma (Lemma 5.6.11) shows the following.

Proposition 6.2.8: For any set S of vectors, rank $S \leq |S|$.

Definition 6.2.9: For a matrix M, the *row rank* of M is the rank of its rows, and the *column rank* of M is the rank of its columns.

Equivalently, the row rank of M is the dimension of Row M, and the column rank of M is the dimension of Col M.

Example 6.2.10: Consider the matrix

$$M = \begin{bmatrix} 1 & 0 & 0 \\ 0 & 2 & 0 \\ 2 & 4 & 0 \end{bmatrix}$$

whose rows are the vectors of Example 6.2.6 (Page 267). We saw that the set consisting of these vectors has rank two, so the row rank of M is two.

The columns of M are $[1,0,2]$, $[0,2,4]$, and $[0,0,0]$. Since the third vector is the zero vector, it is not needed for spanning the column space. Since each of the first two vectors has a nonzero where the other has a zero, these two are linearly independent, so the column rank is two.

Example 6.2.11: Consider the matrix

$$M = \begin{bmatrix} 1 & 0 & 0 & 5 \\ 0 & 2 & 0 & 7 \\ 0 & 0 & 3 & 9 \end{bmatrix}$$

Each of the rows has a nonzero where the others have zeroes, so the three rows are linearly independent. Thus the row rank of M is three.

The columns of M are $[1,0,0]$, $[0,2,0]$, $[0,0,3]$, and $[5,7,9]$. The first three columns are linearly independent, and the fourth can be written as a linear combination of the first three, so the column rank is three.

In both of these examples, the row rank equals the column rank. This is not a coincidence; we will show this is true in any matrix.

Example 6.2.12: Consider the set of vectors $S = \{[1,0,3], [0,4,0], [0,0,3], [2,1,3]\}$. By computation, one can show that the first three vectors of S are linearly independent. Thus, the rank of S is three. On the other hand, consider the matrix whose rows are the vectors of S.

$$M = \begin{bmatrix} 1 & 0 & 3 \\ 0 & 4 & 0 \\ 0 & 0 & 3 \\ 2 & 1 & 3 \end{bmatrix}$$

Because S has rank three, we also infer that the row rank of M is three. Moreover, since each of the columns has a nonzero where the others have zeroes, three columns are linearly independent. Thus the column rank of M is three.

6.2.2 Geometry

We asked in Section 3.3.1 about predicting the dimensionality of the geometric object formed by the span of given vectors.

We can now understand geometry in terms of coordinate systems. The dimensionality of such a geometric object is the minimum number of coordinates that must be assigned to its points. The number of coordinates is the size of the basis, and the size of the basis is the rank of the set of given vectors.

For example,

- Span $\{[1, 2, -2]\}$ is a line, a one-dimensional object, whereas Span $\{[0, 0, 0]\}$ is a point, a one-dimensional structure—-the first vector space has dimension one and the second has dimension zero.
- Span $\{[1, 2], [3, 4]\}$ consists of all of $\mathbb{R}^2$, a two-dimensional object, whereas Span $\{[1, 3], [2, 6]\}$ is a line, a one-dimensional object—the first has dimension two and the second has dimension one.
- Span $\{[1, 0, 0], [0, 1, 0], [0, 0, 1]\}$ is all of $\mathbb{R}^3$, a three-dimensional object, whereas Span $\{[1, 0, 0], [0, 1, 0], [1, 1, 0]\}$ is a plane, a two-dimensional object—the first has dimension three and the second has dimension two.

6.2.3 Dimension and rank in graphs

In Chapter 5, we described the concepts of spanning, linear independence, and basis as they apply to a graph such as:

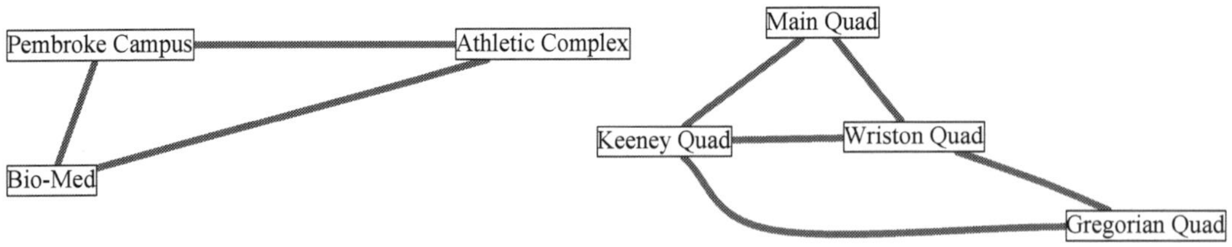

In Example 5.6.8 (Page 232), we saw examples of bases of the span of subsets of edges. Let T be the set of dark edges in the following diagram:

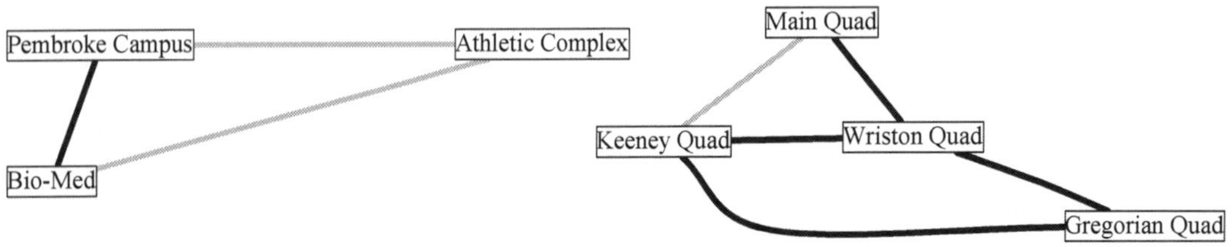

In each of the following two diagrams, the dark edges form a basis for T:

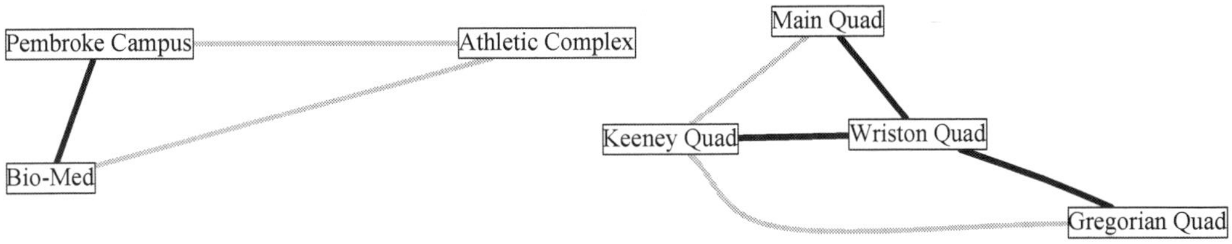

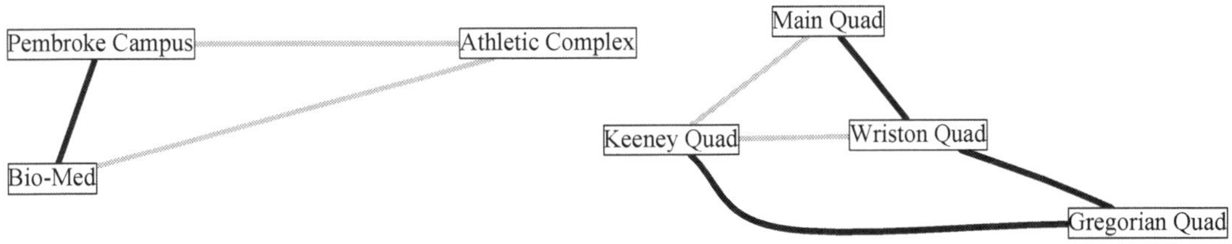

Each basis has size four, so dim Span $T = 4$ so rank $T = 4$.

A set T of edges of a graph forms a *connected subgraph* if every pair of edges of T belong to some path consisting of edges of T. An easy induction shows that the rank of a connected subgraph T equals the number of nodes that are endpoints of edges in T, minus one. For example, the set consisting of a single edge has rank one, and a set consisting of three edges forming a cycle has rank two.

In each of the diagrams above, the dark edges comprise two connected subgraphs that share no nodes. The connected subgraph on the left, which consists of a single edge, has rank one. The connected subgraph on the right, which has four nodes, has rank three. The rank of the set of all dark edges is one plus three, i.e. four.

6.2.4 The cardinality of a vector space over $GF(2)$

In Section 3.6.2, we saw that the number of solutions to a linear system

$$\begin{aligned} \boldsymbol{a}_1 \cdot \boldsymbol{x} &= \beta_1 \\ &\vdots \\ \boldsymbol{a}_m \cdot \boldsymbol{x} &= \beta_m \end{aligned}$$

over $GF(2)$ equals the number of vectors in the vector space V consisting of solutions to the corresponding system of homogeneous linear equations.

In Section 3.6.4, in order to calculate the probability of undetected corruption of a file, we similarly needed to know the cardinality of a vector space $\mathcal{V}$ over $GF(2)$.

Let d be the dimension of $\mathcal{V}$, and suppose $\boldsymbol{b}_1, \ldots, \boldsymbol{b}_d$ is a basis for V. By the Unique Representation Lemma (Lemma 5.7.1), each vector in V has a unique representation as a linear combination of the basis vectors. Thus the number of vectors in V equals the number of linear combinations of the basis vectors. Since there are d basis vectors, there are d coefficients in each linear combination. Each coefficient can be zero or one. Therefore there are 2^d different linear combinations.

6.2.5 Any linearly independent set of vectors belonging to $\mathcal{V}$ can be extended to form a basis for $\mathcal{V}$

Now that we have the notion of dimension, we can prove the Superset-Basis Lemma stated in Section 5.6.5.

Lemma 6.2.13 (Superset-Basis Lemma): For any vector space $\mathcal{V}$ and any linearly independent set A of vectors, $\mathcal{V}$ has a basis that contains all of A.

Proof

We started the proof in Section 5.6.5 but were unable to complete it because we lacked the notion of dimension. Let's try again.

We use a version of the Grow algorithm:

def superset_basis($T, \mathcal{V}$):
  Initialize B to be equal to T.
  Repeat while possible: select a vector in $\mathcal{V}$ that is not in Span B, and put it in B.
  Return B

Initially, B contains all of T (in fact, is equal to T) and is linearly independent. By the Grow-Algorithm Corollary, the set B remains linearly independent throughout the algorithm. If the algorithm terminates, Span $B = \mathcal{V}$. Hence upon termination B is a basis for $\mathcal{V}$. Furthermore, B still contains all of T since the algorithm did not remove any vectors from B.

How do we show that the algorithm terminates? For some field $\mathbb{F}$ and some set D, the vector space $\mathcal{V}$ consists of vectors in $\mathbb{F}^D$. In this book, we assume D is finite. Therefore there is a standard basis for $\mathbb{F}^D$, which consists of $|D|$ vectors.

By the Morphing Lemma, since B is a linearly independent set of vectors belonging to $\mathbb{F}^D$, the cardinality of B is at most the cardinality of the standard basis for $\mathbb{F}^D$. However, each iteration of the Grow algorithm increases the cardinality of B by one, so the algorithm cannot continue forever (in fact, it cannot continue for more than $|D|$ iterations). □

This proof crucially uses the fact that, for the purposes of this book, every vector space is a subspace of $\mathbb{F}^D$ where D is a finite set. Things get trickier when D is allowed to be infinite!

6.2.6 The Dimension Principle

The Superset-Basis Lemma is used in the proof of the following principle.

Lemma 6.2.14 (Dimension Principle): If $\mathcal{V}$ is a subspace of $\mathcal{W}$ then

Property D1: $\dim \mathcal{V} \leq \dim \mathcal{W}$, and

Property D2: if $\dim \mathcal{V} = \dim \mathcal{W}$ then $\mathcal{V} = \mathcal{W}$.

Proof

Let $\boldsymbol{v}_1, \ldots, \boldsymbol{v}_k$ be a basis for $\mathcal{V}$. By the Superset-Basis Lemma (Lemma 6.2.13), there is a basis B for $\mathcal{W}$ that contains $\boldsymbol{v}_1, \ldots, \boldsymbol{v}_k$. Thus the cardinality of B is at least k. This proves 1. Moreover, if the cardinality of B is exactly k then B contains no vectors other than $\boldsymbol{v}_1 \ldots, \boldsymbol{v}_k$, which shows that the basis of $\mathcal{V}$ is also a basis for $\mathcal{W}$, which proves 2. □

Example 6.2.15: Suppose $\mathcal{V} = \text{Span } \{[1, 2], [2, 1]\}$. Clearly $\mathcal{V}$ is a subspace of $\mathbb{R}^2$. However, the set $\{[1, 2], [2, 1]\}$ is linearly independent, so $\dim \mathcal{V} = 2$. Since $\dim \mathbb{R}^2 = 2$, 1 shows that $\mathcal{V} = \mathbb{R}^2$.

Example 6.2.16: In Example 6.2.4 (Page 266), we considered the set
$S = \{[-0.6, -2.1, -3.5, -2.2],$
$[-1.3, 1.5, -0.9, -0.5], [4.9, -3.7, 0.5, -0.3], [2.6, -3.5, -1.2, -2.0], [-1.5, -2.5, -3.5, 0.94]\}$. We observed that, because $|S| = 5$, we know $\dim \text{Span } S \leq 5$. We can say more. Since every vector in S is a 4-vector, Span S is a subspace of $\mathbb{R}^4$, so $\dim \text{Span } S \leq 4$.

Using the argument in Example 6.2.16 (Page 270), we can obtain the following:

Proposition 6.2.17: Any set of D-vectors has rank at most $|D|$.

6.2.7 The Grow algorithm terminates

In Section 5.6.3, I pointed out that the Grow algorithm *almost* shows that every vector space has a basis—but we had not yet shown that the algorithm terminates. We can now do that, using the Dimension Principle.

Recall the Grow algorithm from Section 5.3.1:

```
def Grow(V)
  S = ∅
  repeat while possible:
      find a vector v in V that is not in Span S, and put it in S.
```

Lemma 6.2.18 (Grow Algorithm Termination Lemma): If $\dim \mathcal{V}$ is finite then $\text{GROW}(\mathcal{V})$ terminates.

Proof

In each iteration, a new vector is added to S. Therefore after k iterations, $|S|$ equals k. The Grow-Algorithm Corollary (Corollary 5.5.10) ensures that, at each point in the execution of the algorithm, the set S is linearly independent, so after k iterations, rank $S = k$. Every vector added to S belongs to $\mathcal{V}$ so Span S is a subspace of $\mathcal{V}$. Suppose $\dim \mathcal{V}$ is a finite number. Then after $\dim \mathcal{V}$ iterations, $\dim \text{Span } S = \dim \mathcal{V}$ so, by Property D2 of the Dimension Principle, Span $S = \mathcal{V}$. Therefore, the algorithm must terminate. At this point, S is a basis for $\mathcal{V}$. □

How can we be sure that $\dim \mathcal{V}$ is a finite number? Suppose $\mathcal{V}$ is a vector space consisting of D-vectors over a field $\mathbb{F}$, where D is a finite set. Then $\mathcal{V}$ is a subspace of $\mathbb{F}^D$, so, by Property D1 of the Dimension Principle, $\dim \mathcal{V} \leq |D|$. We therefore obtain:

Corollary 6.2.19: For finite D, any subspace of $\mathbb{F}^D$ has a basis.

In this book, we only consider D-vectors where D is a finite set. As mentioned in Section 5.6.3, in the wider world of mathematics, one must on occasion consider D-vectors where D is an infinite set. For such vectors, the notion of dimension is more complicated and we do not consider it in this book, except to mention without proof that, even for such vectors, every vector space has a basis (albeit possibly one of infinite size).

6.2.8 The Rank Theorem

We observed earlier that, in a couple of examples, the row rank happen to be equal to the column rank. Here we show that this was not a coincidence.

Theorem 6.2.20 (Rank Theorem): For any matrix, the row rank equals the column rank.

Proof

We show that, for any matrix A, the row rank of A is less than or equal to the column rank of A. Applying the same argument to A^T proves that the row rank of A^T is less than or equal to the column rank of A^T, i.e. that the column rank of A is less than or equal to to the row rank of A. Putting these two inequalities together shows that the row rank of A equals the column rank of A.

Let A be a matrix. Write A in terms of its columns: $A = \left[\begin{array}{c|c|c} \boldsymbol{a}_1 & \cdots & \boldsymbol{a}_n \end{array}\right]$. Let r be the column rank of A, and let $\boldsymbol{b}_1, \ldots, \boldsymbol{b}_r$ be a basis for the column space of A.

For each column $\boldsymbol{a}_j$ of A, let $\boldsymbol{u}_j$ be the coordinate representation of $\boldsymbol{a}_j$ in terms of $\boldsymbol{b}_1, \ldots, \boldsymbol{b}_r$. Then by the linear-combinations definition of matrix-vector multiplication,

$$\left[\begin{array}{c} \boldsymbol{a}_j \end{array}\right] = \left[\begin{array}{c|c|c} \boldsymbol{b}_1 & \cdots & \boldsymbol{b}_r \end{array}\right] \left[\begin{array}{c} \boldsymbol{u}_j \end{array}\right]$$

By the matrix-vector definition of matrix-matrix multiplication, therefore,

$$\left[\begin{array}{c|c|c} \boldsymbol{a}_1 & \cdots & \boldsymbol{a}_n \end{array}\right] = \left[\begin{array}{c|c|c} \boldsymbol{b}_1 & \cdots & \boldsymbol{b}_r \end{array}\right] \left[\begin{array}{c|c|c} \boldsymbol{u}_1 & \cdots & \boldsymbol{u}_n \end{array}\right]$$

which we write as

$$A = BU$$

Note that B has r columns and U has r rows.

Now we switch interpretations, and interpret A and B as consisting of rows instead of columns:

$$\left[\begin{array}{c} \bar{\boldsymbol{a}}_1 \\ \hline \vdots \\ \hline \bar{\boldsymbol{a}}_m \end{array}\right] = \left[\begin{array}{c} \bar{\boldsymbol{b}}_1 \\ \hline \vdots \\ \hline \bar{\boldsymbol{b}}_m \end{array}\right] U$$

By the vector-matrix definition of matrix-matrix multiplication, row i of A, $\bar{\boldsymbol{a}}_i$, is the product of row i of B, $\bar{\boldsymbol{b}}_i$, times the matrix U:

$$\left[\begin{array}{c} \bar{\boldsymbol{a}}_i \end{array}\right] = \left[\begin{array}{c} \bar{\boldsymbol{b}}_i \end{array}\right] \left[\begin{array}{c} U \end{array}\right]$$

By the linear-combinations definition of vector-matrix multiplication, therefore, every row of A is some linear combination of the rows of U. Therefore the row space of A is a subspace of the row space of U. By Proposition 6.2.8, the dimension of the row space of U is at most r, the number of rows of U. Therefore, by 1 of the Dimension Principle, the row rank of A is at most r.

We have shown that, for any matrix A, the row rank of A is at most the column rank of A. For any matrix M, applying this result to M shows that

$$\text{row rank of } M \leq \text{column rank of } M$$

Applying the result to M^T shows that

$$\text{row rank of } M^T \leq \text{column rank of } M^T$$

which means

$$\text{column rank of} M \leq \text{row rank of } M$$

This shows that the row rank of M equals the column rank of M. □

Definition 6.2.21: We define the *rank* of a matrix to be its column rank, which is also equal to its row rank.

6.2.9 Simple authentication revisited

Recall the simple authentication scheme of Section 2.9.6. The password is an n-vector $\hat{\boldsymbol{x}}$ over $GF(2)$. The computer issues challenges to the human and the human responds:

- **Challenge:** Computer sends random n-vector $\boldsymbol{a}$.
- **Response:** Human sends back $\boldsymbol{a} \cdot \hat{\boldsymbol{x}}$.

until the computer is convinced that the human knows the password $\hat{\boldsymbol{x}}$.

Suppose Eve eavesdrops on the communication, and learns m pairs $\boldsymbol{a}_1, b_1, \ldots, \boldsymbol{a}_m, b_m$ such that b_i is the right response to challenge $\boldsymbol{a}_i$. We saw in Section 2.9.9 that Eve can calculate the right response to any challenge in Span $\{\boldsymbol{a}_1, \ldots, \boldsymbol{a}_m\}$:

Indeed, let $\boldsymbol{a} = \alpha_1\, \boldsymbol{a}_1 + \cdots + \alpha_m\, \boldsymbol{a}_m$. Then the right right response is $\alpha_1 b_1 + \cdots + \alpha_m b_m$

Using probability theory, it is possible to show:

Fact 6.2.22: Probably rank $[\boldsymbol{a}_1, \ldots, \boldsymbol{a}_m]$ is not much less than $\min\{m, n\}$.

You can try this for yourself using Python. Set $n = 100$, say, and generate some number m of random n-vectors over $GF(2)$.

```
>>> from vec import Vec
>>> from random import randint
>>> from GF2 import one
>>> def rand_GF2(): return one if randint(0,1)==1 else 0
>>> def rand_GF2_vec(D): return Vec(D, {d:rand_GF2() for d in D})
>>> D = set(range(100))
```

We provide a procedure `rank(L)` in the module `independence`.

```
>>> L = [rand_GF2_vec(D) for i in range(50)]
>>> from independence import rank
>>> rank(L)
50
```

Once $m > n$, probably Span $\{\boldsymbol{a}_1, \ldots, \boldsymbol{a}_m\}$ is all of $GF(2)^n$

so Eve can respond to *any* challenge.

Also, the password $\hat{\boldsymbol{x}}$ is a solution to the linear system
$$\underbrace{\begin{bmatrix} \boldsymbol{a}_1 \\ \hline \vdots \\ \hline \boldsymbol{a}_m \end{bmatrix}}_{A} \begin{bmatrix} \boldsymbol{x} \end{bmatrix} = \underbrace{\begin{bmatrix} b_1 \\ \vdots \\ b_m \end{bmatrix}}_{\boldsymbol{b}}$$

The solution set of $A\boldsymbol{x} = \boldsymbol{b}$ is $\hat{\boldsymbol{x}} + \text{Null } A$.

Once rank A reaches n, the columns of A are linearly independent so Null A is trivial. This implies that the only solution is the password $\hat{\boldsymbol{x}}$, which means Eve can compute the password by using `solver`.

6.3 Direct sum

We are familiar with the idea of adding vectors—now we learn about adding vector spaces. The ideas presented here will be useful in proving a crucial theorem in the next section—the Kernel-Image Theorem—and also prepare the way for a concept to be presented a couple of chapters hence: orthogonal complement.

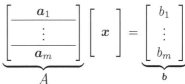

6.3.1 Definition

Let $\mathcal{U}$ and $\mathcal{V}$ be two vector spaces consisting of D-vectors over a field $\mathbb{F}$.

Definition 6.3.1: If $\mathcal{U}$ and $\mathcal{V}$ share only the zero vector then we define the *direct sum* of $\mathcal{U}$ and $\mathcal{V}$ to be the set

$$\{u + v \ : u \in \mathcal{U}, v \in \mathcal{V}\}$$

written $\mathcal{U} \oplus \mathcal{V}$

That is, $\mathcal{U} \oplus \mathcal{V}$ is the set of all sums of a vector in $\mathcal{U}$ and a vector in $\mathcal{V}$.

In Python, we can compute the list of vectors comprising the direct sum of `U` and `V` as follows:

```
>>> {u+v for u in U for v in V}
```

Recall that the Cartesian product of `U` and `V` is written `[(u,v) for u in U for v in V]` (except that the Python computes a list, not a set). This shows that the direct sum is similar to a Cartesian product—you add the two vectors instead of turning them into a tuple.

What if $\mathcal{U}$ and $\mathcal{V}$ share a nonzero vector? In that case, it is considered an error to form their direct sum!

Here is an example using vectors over $GF(2)$.

Example 6.3.2: Let $\mathcal{U}$ = Span $\{1000, 0100\}$ and let $\mathcal{V}$ = Span $\{0010\}$.

- Every nonzero vector in $\mathcal{U}$ has a one in the first or second position (or both) and nowhere else.
- Every nonzero vector in $\mathcal{V}$ has a one in the third position and nowhere else.

Therefore the only vector in both $\mathcal{U}$ and $\mathcal{V}$ is the zero vector. Therefore $\mathcal{U} \oplus \mathcal{V}$ is defined

$\mathcal{U} \oplus \mathcal{V} = \{0000+0000, 1000+0000, 0100+0000, 1100+0000, 0000+0010, 1000+0010, 0100+0010, 1100+0010\}$

which is equal to $\{0000, 1000, 0100, 1100, 0010, 1010, 0110, 1110\}$.

A a couple of examples using vectors over $\mathbb{R}$:

Example 6.3.3: Let $\mathcal{U}$ = Span $\{[1, 2, 1, 2], [3, 0, 0, 4]\}$ and let $\mathcal{V}$ be the null space of $\begin{bmatrix} 0 & 1 & -1 & 0 \\ 1 & 0 & 0 & -1 \end{bmatrix}$.

- The vector $[2, -2, -1, 2]$ is in $\mathcal{U}$ because it is $[3, 0, 0, 4] - [1, 2, 1, 2]$
- It is also in $\mathcal{V}$ because

$$\begin{bmatrix} 0 & 1 & -1 & 0 \\ 1 & 0 & 0 & -1 \end{bmatrix} \begin{bmatrix} 2 \\ -2 \\ -1 \\ 2 \end{bmatrix} = \begin{bmatrix} 0 \\ 0 \end{bmatrix}$$

Therefore we cannot form $V \oplus W$.

Example 6.3.4: Let $\mathcal{U}$ = Span $\{[4, -1, 1]\}$, and let $\mathcal{V}$ = Span $\{[0, 1, 1]\}$. Each of $\mathcal{U}$ and $\mathcal{V}$ is the span of a single vector, and so forms a line:

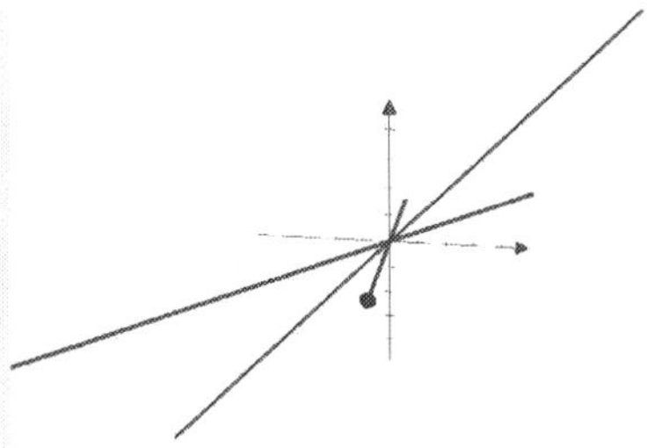

The only intersection is at the origin, so $\mathcal{U} \oplus \mathcal{V}$ is defined. It is is just Span $\{[4, -1, 1], [0, 1, 1]\}$, which we know is the plane containing the two lines.

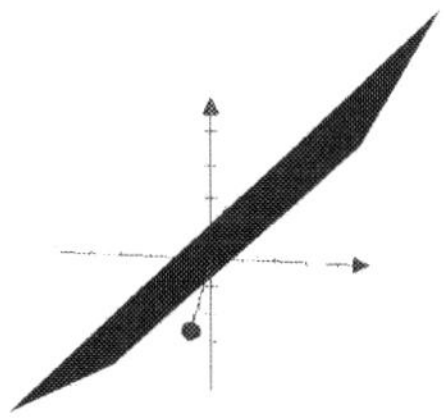

Proposition 6.3.5: The direct sum $\mathcal{U} \oplus \mathcal{V}$ is a vector space.

The proof is left as an exercise.

6.3.2 Generators for the direct sum

In Example 6.3.4 (Page 274), one set of generators for the direct sum $\mathcal{U} \oplus \mathcal{V}$ is obtained by taking the union of a set of generators for $\mathcal{U}$ and a set of generators for $\mathcal{V}$. This is true generally:

Lemma 6.3.6: The union of

- a set of generators of $\mathcal{V}$, and
- a set of generators of $\mathcal{W}$

is a set of generators for $\mathcal{V} \oplus \mathcal{W}$.

Proof

Suppose $\mathcal{V} = \text{Span}\ \{\boldsymbol{v}_1, \ldots, \boldsymbol{v}_m\}$ and $\mathcal{W} = \text{Span}\ \{\boldsymbol{w}_1, \ldots, \boldsymbol{w}_n\}$.
Then

- every vector in $\mathcal{V}$ can be written as $\alpha_1\, \boldsymbol{v}_1 + \cdots + \alpha_m\, \boldsymbol{v}_m$, and
- every vector in $\mathcal{W}$ can be written as $\beta_1\, \boldsymbol{w}_1 + \cdots + \beta_n\, \boldsymbol{w}_n$

so every vector in $\mathcal{V} \oplus \mathcal{W}$ can be written as

$$\alpha_1\, \boldsymbol{v}_1 + \cdots + \alpha_m\, \boldsymbol{v}_m \; + \; \beta_1\, \boldsymbol{w}_1 + \cdots + \beta_n\, \boldsymbol{w}_n$$

□

Example 6.3.7: Let $\mathcal{U}$ be the set of points in $\mathbb{R}^3$ comprising a plane that contains the origin, and let $\mathcal{V}$ be the set of points comprising a line that contains the origin:

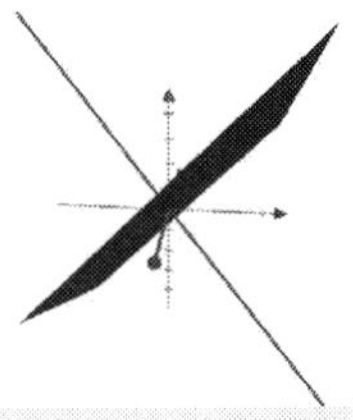

As long as the line is not in the plane, their intersection consists just of the origin, so their direct sum is defined, and consists of all of $\mathbb{R}^3$.

6.3.3 Basis for the direct sum

Lemma 6.3.8 (Direct Sum Basis Lemma): The union of a basis of $\mathcal{U}$ and a basis of $\mathcal{V}$ is a basis of $\mathcal{U} \oplus \mathcal{V}$.

Proof

Let $\{\boldsymbol{u}_1, \ldots, \boldsymbol{u}_m\}$ be a basis for $\mathcal{U}$. Let $\{\boldsymbol{v}_1, \ldots, \boldsymbol{v}_n\}$ be a basis for $\mathcal{V}$. Since a basis is a set of generators, we already know from the previous lemma that $\{\boldsymbol{u}_1, \ldots, \boldsymbol{u}_m, \boldsymbol{v}_1, \ldots, \boldsymbol{v}_n\}$ is a set of generators for $\mathcal{U} \oplus \mathcal{V}$. To show it is a basis, we need only show is is linearly independent.

Suppose

$$\mathbf{0} = \alpha_1\, \boldsymbol{u}_1 + \cdots + \alpha_m \boldsymbol{u}_m + \beta_1\, \boldsymbol{v}_1 + \cdots + \beta_n\, \boldsymbol{v}_n \tag{6.1}$$

Then

$$\underbrace{\alpha_1\, \boldsymbol{u}_1 + \cdots + \alpha_m\, \boldsymbol{u}_m}_{\text{in } \mathcal{U}} = \underbrace{(-\beta_1)\, \boldsymbol{v}_1 + \cdots + (-\beta_n)\, \boldsymbol{v}_n}_{\text{in } \mathcal{V}}$$

The left-hand side is a vector in $\mathcal{U}$, and the right-hand side is a vector in $\mathcal{V}$.

By definition of $\mathcal{U} \oplus \mathcal{V}$, the only vector in both $\mathcal{U}$ and $\mathcal{V}$ is the zero vector. This shows:

$$\mathbf{0} = \alpha_1\, \boldsymbol{u}_1 + \cdots + \alpha_m\, \boldsymbol{u}_m$$

and

$$\mathbf{0} = (-\beta_1)\, \boldsymbol{v}_1 + \cdots + (-\beta_n)\, \boldsymbol{v}_n$$

By linear independence, the linear combinations must be trivial, so the original linear combination in Equation 6.1 must also be trivial. This completes the proof that the union of bases is linearly independent. □

The definition of basis gives us an immediate corollary, which will be used in the proof of the Kernel-Image Theorem.

Corollary 6.3.9 (Direct-Sum Dimension Corollary): $\dim \mathcal{U} + \dim \mathcal{V} = \dim \mathcal{U} \oplus \mathcal{V}$

6.3.4 Unique decomposition of a vector

By definition, $\mathcal{U} \oplus \mathcal{V} = \{\boldsymbol{u} + \boldsymbol{v} \ : \ \boldsymbol{u} \in \mathcal{U}, \boldsymbol{v} \in \mathcal{V}\}$. Can the same vector occur in two different ways as the sum of a vector in $\mathcal{U}$ and a vector in $\mathcal{V}$?

We next see that the answer is no. If I obtain $\boldsymbol{w}$ by adding a vector $\boldsymbol{u}$ in $\mathcal{U}$ and a vector $\boldsymbol{v}$ in $\mathcal{V}$ and give you $\boldsymbol{w}$, you can figure out from $\boldsymbol{w}$ exactly which $\boldsymbol{u}$ and $\boldsymbol{v}$ I started with.

Corollary 6.3.10 (Direct-Sum Unique Representation Corollary): Any vector in $\mathcal{U} \oplus \mathcal{V}$ has a unique representation as $\boldsymbol{u} + \boldsymbol{v}$ where $\boldsymbol{u} \in \mathcal{U}, \boldsymbol{v} \in \mathcal{V}$.

Proof

Let $\{\boldsymbol{u}_1, \ldots, \boldsymbol{u}_m\}$ be a basis for $\mathcal{U}$. Let $\{\boldsymbol{v}_1, \ldots, \boldsymbol{v}_n\}$ be a basis for $\mathcal{V}$. Then $\{\boldsymbol{u}_1, \ldots, \boldsymbol{u}_m, \boldsymbol{v}_1, \ldots, \boldsymbol{v}_n\}$ is a basis for $\mathcal{U} \oplus V$.

Let $\boldsymbol{w}$ be any vector in $\mathcal{U} \oplus \mathcal{V}$. Write $\boldsymbol{w}$ as

$$\boldsymbol{w} = \underbrace{\alpha_1\, \boldsymbol{u}_1 + \cdots + \alpha_m\, \boldsymbol{u}_m}_{\text{in } \mathcal{U}} + \underbrace{\beta_1\, \boldsymbol{v}_1 + \cdots + \beta_n\, \boldsymbol{v}_n}_{\text{in } \mathcal{V}} \tag{6.2}$$

Consider any way of writing $\boldsymbol{w}$ as $\boldsymbol{w} = \boldsymbol{u} + \boldsymbol{v}$ where $\boldsymbol{u}$ is in $\mathcal{U}$ and $\boldsymbol{v}$ is in $\mathcal{V}$. Writing $\boldsymbol{u}$ in terms of the basis of $\mathcal{U}$ and writing $\boldsymbol{v}$ in terms of the basis of $\mathcal{V}$, we have

$$\boldsymbol{w} = \gamma_1\, \boldsymbol{u}_1 + \cdots + \gamma_m\, \boldsymbol{u}_m + \delta_1\, \boldsymbol{v}_1 + \cdots + \delta_n\, \boldsymbol{v}_n$$

By the Unique-Representation Lemma, $\gamma_1 = \alpha_1, \ldots, \gamma_m = \alpha_m, \delta_1 = \beta_1, \ldots, \delta_n = \beta_n$, which shows that Equation 6.2 specifies the unique way in which $\boldsymbol{w}$ can be written as a sum of a vector in $\mathcal{U}$ and a vector in $\mathcal{V}$. □

6.3.5 Complementary subspaces

Definition 6.3.11: If $\mathcal{U} \oplus \mathcal{V} = \mathcal{W}$, we say that $\mathcal{U}$ and $\mathcal{V}$ are *complementary subspaces* of $\mathcal{W}$.

Example 6.3.12: Suppose $\mathcal{U}$ is a plane in $\mathbb{R}^3$:

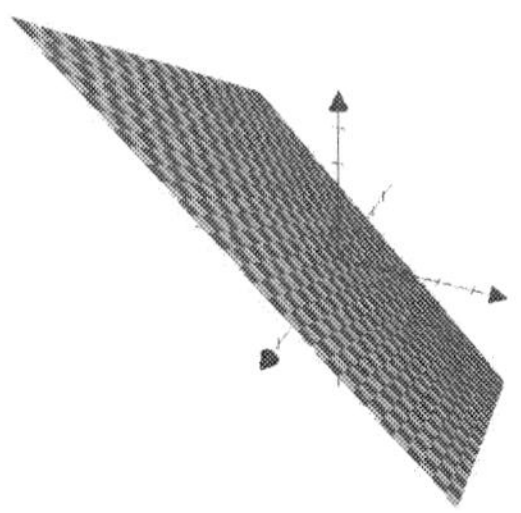

Then $\mathcal{U}$ and any line through the origin that does not lie in $\mathcal{U}$ are complementary subspaces of $\mathbb{R}^3$, e.g.:

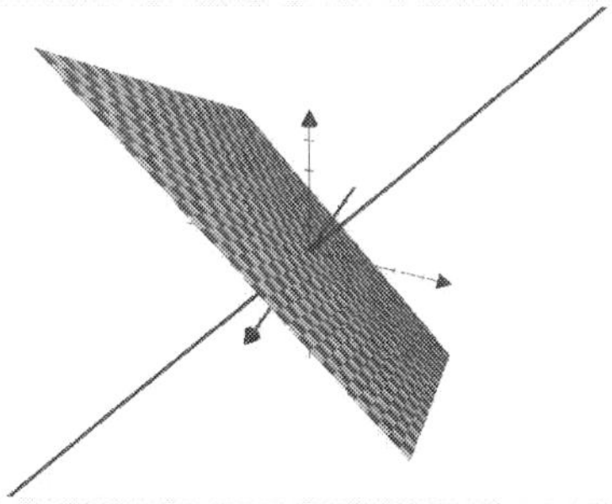

or, alternatively,

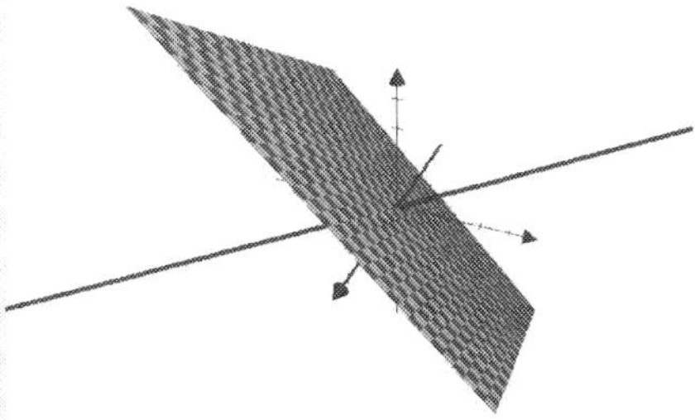

The example illustrates that, for a given subspace $\mathcal{U}$ of $\mathcal{W}$, there can be many different subspaces $\mathcal{V}$ such that $\mathcal{U}$ and $\mathcal{V}$ are complementary.

Problem 6.3.13: In each part, give a set T of vectors over $\mathbb{R}$ such that (Span S)$\oplus$(Span T) = $\mathbb{R}^3$, and show how to express a generic vector $[x, y, z]$ as a linear combination of the vectors in $S \cup T$.

1. $S = \{[2, 1, 2], [1, 1, 1]\}$
2. $S = \{[0, 1, -1], [0, 0, 0]\}$

Hint: To express $[x, y, z]$, you might first try to express the standard basis vectors in terms of $S \cup T$.

Problem 6.3.14: In each part, give a set T of vectors over $GF(2)$ such that (Span S) $\oplus$ (Span T) = $GF(2)^3$, and show how to express a generic vector $[x, y, z]$ as a linear combination of the vectors in $S \cup T$.

1. $S = \{[1, 1, 0], [0, 1, 1]\}$
2. $S = \{[1, 1, 1]\}$

Hint: To express $[x, y, z]$, you might first try to express the standard basis vectors in terms of $S \cup T$.

Proposition 6.3.15: For any vector space $\mathcal{W}$ and any subspace $\mathcal{U}$ of $\mathcal{W}$, there is a subspace $\mathcal{V}$ of $\mathcal{W}$ such that $\mathcal{W} = \mathcal{U} \oplus \mathcal{V}$.

Proof

Let $u_1, \ldots, \boldsymbol{u}_k$ be a basis for $\mathcal{U}$. By the Superset-Basis Lemma (Lemma 6.2.13), there is a basis for $\mathcal{W}$ that includes $\boldsymbol{u}_1, \ldots, \boldsymbol{u}_k$. Write this basis as $\{\boldsymbol{u}_1, \ldots, \boldsymbol{u}_k, \boldsymbol{v}_1, \ldots, \boldsymbol{v}_r\}$. Let $\mathcal{V} = \text{Span } \{\boldsymbol{v}_1, \ldots, \boldsymbol{v}_r\}$. Any vector $\boldsymbol{w}$ in $\mathcal{W}$ can be written in terms of its basis,

$$\boldsymbol{w} = \underbrace{\alpha_1\, \boldsymbol{u}_1 + \cdots + \alpha_k\, \boldsymbol{u}_k}_{\text{in } \mathcal{U}} + \underbrace{\beta_1\, \boldsymbol{v}_1 + \cdots + \beta_r\, \boldsymbol{v}_r}_{\text{in } \mathcal{V}}$$

so we will have shown that $\mathcal{W} = \mathcal{U} \oplus \mathcal{V}$ if we can only show that the direct sum is legal, i.e. that the only vector in both $\mathcal{U}$ and $\mathcal{V}$ is the zero vector.

Suppose some vector $\boldsymbol{v}$ belongs to both $\mathcal{U}$ and $\mathcal{V}$. That is,

$$\alpha_1\, \boldsymbol{u}_1 + \cdots + \alpha_k\, \boldsymbol{u}_k = \beta_1\, \boldsymbol{v}_1 + \cdots + \beta_r\, \boldsymbol{v}_r$$

Then

$$\mathbf{0} = \alpha_1\, \boldsymbol{u}_1 + \cdots + \alpha_k\, \boldsymbol{u}_k - \beta_1\, \boldsymbol{v}_1 - \cdots - \beta_r\, \boldsymbol{v}_r$$

which implies that $\alpha_1 = \cdots = \alpha_k = \beta_1 = \cdots = \beta_r = 0$, which shows that $\boldsymbol{v}$ is the zero vector. □

6.4 Dimension and linear functions

We will develop a criterion for whether a linear function is invertible. That in turn will provide a criterion for whether a matrix is invertible. These criteria will build on an important theorem, the Kernel-Image Theorem, that will also help us answer other questions.

6.4.1 Linear function invertibility

How can we tell if a linear function $f : \mathcal{V} \longrightarrow \mathcal{W}$ is invertible? We need to know (i) whether f is one-to-one and (ii) whether it is onto.

By the One-to-One Lemma (Lemma 4.10.15), we know that f is one-to-one iff its kernel is trivial. We asked in Question 4.10.16 if there is a similarly nice criterion for whether a linear function is onto.

Recall that the image of f is Im $f = \{f(\boldsymbol{v}) \ : \boldsymbol{v} \in \mathcal{V}\}$. Thus f is onto iff Im $f = \mathcal{W}$.

We can show that Im f is a supspace of $\mathcal{W}$. By the Dimension Principle (Lemma 6.2.14), therefore, f is onto iff $\dim \text{Im } f = \dim \mathcal{W}$.

We can conclude: a linear function $f : U \longrightarrow W$ is invertible if $\dim \text{Ker } f = 0$ and $\dim \text{Im } f = \dim W$.

How does this relate to the dimension of the domain? You might think that, for f to be invertible, we would need $\dim U = \dim W$. You would be right. We will obtain this as a result of a more powerful and useful theorem.

6.4.2 The largest invertible subfunction

Let $f : \mathcal{V} \longrightarrow \mathcal{W}$ be a linear function that is not necessarily invertible.

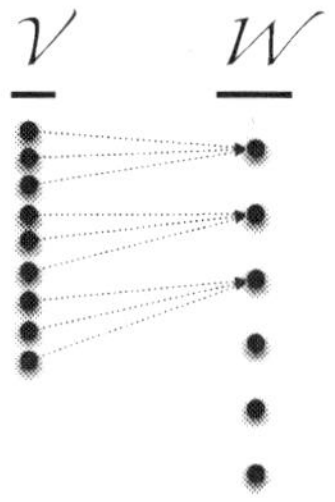

Let's try to define a subfunction $f^* : \mathcal{V}^* \longrightarrow \mathcal{W}^*$ that *is* invertible.

By "subfunction", I mean that $\mathcal{V}^*$ is a subset of $\mathcal{V}$, $\mathcal{W}^*$ is a subset of $\mathcal{W}$, and f^* agrees with f on all elements of $\mathcal{V}^*$. Along the way, we will also select a basis for $\mathcal{V}^*$ and a basis for $\mathcal{W}^*$.

First, we choose $\mathcal{W}^*$ so as to ensure that f^* is onto. This step is easy: we define $\mathcal{W}^*$ to be the image of f, i.e. the elements of $\mathcal{W}$ that are images of domain elements. Let $\boldsymbol{w}_1, \ldots, \boldsymbol{w}_r$ be a basis for $\mathcal{W}^*$.

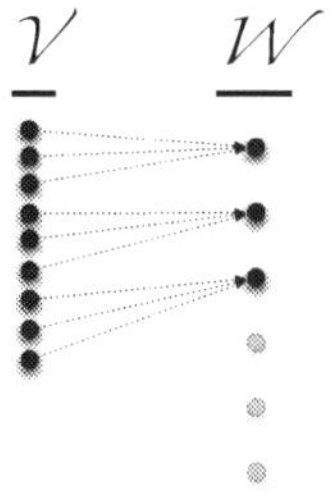

Next, let $\boldsymbol{v}_1, \ldots, \boldsymbol{v}_r$ be pre-images of $\boldsymbol{w}_1, \ldots, \boldsymbol{w}_r$. That is, select any vectors $\boldsymbol{v}_1, \ldots, \boldsymbol{v}_r$ in $\mathcal{V}$ such that $f(\boldsymbol{v}_1) = \boldsymbol{w}_1, \ldots, f(\boldsymbol{v}_r) = \boldsymbol{w}_r$. Now define $\mathcal{V}^*$ to be Span $\{\boldsymbol{v}_1, \ldots, \boldsymbol{v}_r\}$.

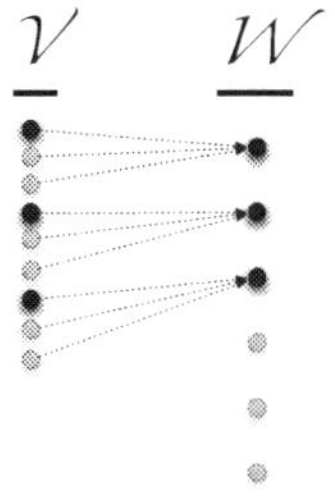

We now define $f^* : \mathcal{V}^* \longrightarrow \mathcal{W}^*$ by the rule $f^*(\boldsymbol{x}) = f(x)$.

Lemma 6.4.1: f^* is onto.

Proof

Let $\boldsymbol{w}$ be any vector in co-domain $\mathcal{W}^*$. There are scalars $\alpha_1, \ldots, \alpha_r$ such that

$$\boldsymbol{w} = \alpha_1\, \boldsymbol{w}_1 + \cdots + \alpha_r\, \boldsymbol{w}_r$$

Because f is linear,

$$\begin{aligned} & f(\alpha_1\, \boldsymbol{v}_1 + \cdots + \alpha_r\, \boldsymbol{v}_r) \\ = \; & \alpha_1\, f(\boldsymbol{v}_1) + \cdots + \alpha_r\, f(\boldsymbol{v}_r) \\ = \; & \alpha_1\, \boldsymbol{w}_1 + \cdots + \alpha_r\, \boldsymbol{w}_r \end{aligned}$$

so $\boldsymbol{w}$ is the image of $\alpha_1\, \boldsymbol{v}_1 + \cdots + \alpha_r\, \boldsymbol{v}_r \in \mathcal{V}^*$ □

Lemma 6.4.2: f^* is one-to-one

Proof

By the One-to-One Lemma, we need only show that the kernel of f^* is trivial. Suppose $\boldsymbol{v}^*$ is in $\mathcal{V}^*$ and $f(\boldsymbol{v}^*) = \boldsymbol{0}$. Because $\mathcal{V}^* = \text{Span}\,\{\boldsymbol{v}_1, \ldots, \boldsymbol{v}_r\}$, there are scalars $\alpha_1, \ldots, \alpha_r$ such that

$$\boldsymbol{v}^* = \alpha_1\, \boldsymbol{v}_1 + \cdots + \alpha_r\, \boldsymbol{v}_r$$

Applying f to both sides,

$$\begin{aligned} \boldsymbol{0} &= f(\alpha_1\, \boldsymbol{v}_1 + \cdots + \alpha_r\, \boldsymbol{v}_r) \\ &= \alpha_1\, \boldsymbol{w}_1 + \cdots + \alpha_r\, \boldsymbol{w}_r \end{aligned}$$

Because $\boldsymbol{w}_1, \ldots, \boldsymbol{w}_r$ are linearly independent, $\alpha_1 = \cdots = \alpha_r = 0$, so $\boldsymbol{v}^* = \boldsymbol{0}$. □

Lemma 6.4.3: $\boldsymbol{v}_1, \ldots, \boldsymbol{v}_r$ form a basis for $\mathcal{V}^*$

Proof

Since $\mathcal{V}^*$ is defined to be the span of $\boldsymbol{v}_1, \ldots, \boldsymbol{v}_r$, we need only show that these vectors are linearly independent.

Suppose

$$\boldsymbol{0} = \alpha_1\, \boldsymbol{v}_1 + \cdots + \alpha_r\, \boldsymbol{v}_r$$

Applying f to both sides,

$$\begin{aligned} \boldsymbol{0} &= f(\alpha_1\, \boldsymbol{v}_1 + \cdots + \alpha_r\, \boldsymbol{v}_r) \\ &= \alpha_1\, \boldsymbol{w}_1 + \cdots + \alpha_r\, \boldsymbol{w}_r \end{aligned}$$

Because $\boldsymbol{w}_1, \ldots, \boldsymbol{w}_r$ are linearly independent, we infer $\alpha_1 = \cdots = \alpha_r = 0$. □

Example 6.4.4: Let $A = \left[\begin{array}{c|c|c} 1 & 2 & 1 \\ 2 & 1 & 1 \\ 1 & 2 & 1 \end{array}\right]$, and define $\boldsymbol{f} : \mathbb{R}^3 \longrightarrow \mathbb{R}^3$ by $f(\boldsymbol{x}) = A\boldsymbol{x}$.

Define $\mathcal{W}^* = \mathsf{Im}\ f = \mathsf{Col}\ A = \mathsf{Span}\ \{[1,2,1],[2,1,2],[1,1,1]\}$. One basis for $\mathcal{W}^*$ is $\boldsymbol{w}_1 = [0,1,0]$, $\boldsymbol{w}_2 = [1,0,1]$

Now we select pre-images for $\boldsymbol{w}_1$ and $\boldsymbol{w}_2$. We select

$$\boldsymbol{v}_1 = \left[\frac{1}{2}, -\frac{1}{2}, \frac{1}{2}\right]$$
$$\boldsymbol{v}_2 = \left[-\frac{1}{2}, \frac{1}{2}, \frac{1}{2}\right]$$

for then $A\boldsymbol{v}_1 = \boldsymbol{w}_1$ and $A\boldsymbol{v}_2 = \boldsymbol{w}_2$.

Let $\mathcal{V}^* = \mathsf{Span}\ \{\boldsymbol{v}_1, \boldsymbol{v}_2\}$. Then the function $f^* : \mathcal{V}^* \longrightarrow \mathsf{Im}\ f$ defined by $f^*(\boldsymbol{x}) = f(\boldsymbol{x})$ is onto and one-to-one.

6.4.3 The Kernel-Image Theorem

The construction of an invertible subfunction $f^* : \mathcal{V}^* \longrightarrow \mathcal{W}^*$ from a linear function f allows us to relate the domain of the subfunction to the kernel of the original linear function f:

Lemma 6.4.5: $\mathcal{V} = \mathsf{Ker}\ f \oplus \mathcal{V}^*$

Proof

We must prove two things:

1. Ker f and $\mathcal{V}^*$ share only zero vector, and
2. every vector in $\mathcal{V}$ is the sum of a vector in Ker f and a vector in $\mathcal{V}^*$

We already showed kernel of f^* is trivial. This shows that the only vector of Ker f in $\mathcal{V}^*$ is zero, which proves 1.

Let $\boldsymbol{v}$ be any vector in $\mathcal{V}$, and let $\boldsymbol{w} = f(\boldsymbol{v})$. Since f^* is onto, its domain $\mathcal{V}^*$ contains a vector $\boldsymbol{v}^*$ such that $f(\boldsymbol{v}^*) = \boldsymbol{w}$. Therefore $f(\boldsymbol{v}) = f(\boldsymbol{v}^*)$, so $f(\boldsymbol{v}) - f(\boldsymbol{v}^*) = \mathbf{0}$, so $f(\boldsymbol{v} - \boldsymbol{v}^*) = \mathbf{0}$. Thus $\boldsymbol{u} = \boldsymbol{v} - \boldsymbol{v}^*$ is in Ker f, and $\boldsymbol{v} = \boldsymbol{u} + \boldsymbol{v}^*$. This proves 2. □

Example 6.4.6: Following up on Example 6.4.4 (Page 280), let $A = \left[\begin{array}{c|c|c} 1 & 2 & 1 \\ 2 & 1 & 1 \\ 1 & 2 & 1 \end{array}\right]$, and define $f : \mathbb{R}^3 \longrightarrow \mathbb{R}^3$ by $f(\boldsymbol{x}) = A\boldsymbol{x}$. Recall that the basis for $\mathcal{V}^*$ consists of $\boldsymbol{v}_1 = [\frac{1}{2}, -\frac{1}{2}, \frac{1}{2}]$ and $\boldsymbol{v}_2 = [-\frac{1}{2}, \frac{1}{2}, \frac{1}{2}]$.

The kernel of f is $\mathsf{Span}\ \{[1,1,-3]\}$. Therefore $\mathcal{V} = (\mathsf{Span}\ \{[1,1,-3]\}) \oplus (\mathsf{Span}\ \{\boldsymbol{v}_1, \boldsymbol{v}_2\})$

Now at last we state and prove the Kernel-Image Theorem:

Theorem 6.4.7 (Kernel-Image Theorem): For any linear function $f : \mathcal{V} \to W$,

$$\dim \mathsf{Ker}\ f + \dim \mathsf{Im}\ f = \dim \mathcal{V}$$

Proof

Lemma 6.4.5 shows that $\mathcal{V} = \text{Ker}\ f \oplus \mathcal{V}^*$. By the Direct-Sum Dimension Corollary,

$$\dim \mathcal{V} = \dim \text{Ker}\ f + \dim \mathcal{V}^*$$

Since $\boldsymbol{v}_1, \ldots, \boldsymbol{v}_r$ form a basis for $\mathcal{V}^*$, and the number r of these vectors equals the cardinality of a basis for Im f,

$$\dim \mathcal{V}^* = r = \dim \text{Im } f$$

This proves the theorem. □

6.4.4 Linear function invertibility, revisited

Now we can give a more appealing criterion for linear function invertibility.

Theorem 6.4.8 (Linear-Function Invertibility Theorem): Let $f : \mathcal{V} \longrightarrow \mathcal{W}$ be a linear function. Then f is invertible if and only if $\dim \text{Ker } f = 0$ and $\dim \mathcal{V} = \dim \mathcal{W}$.

Proof

We saw in Section 6.4.1 that f is invertible if and only if $\dim \text{Ker } f = 0$ and $\dim \text{Im } f = \dim \mathcal{W}$. By the Kernel-Image Theorem, $\dim \text{Ker } f = 0$ and $\dim \text{Im } f = \dim \mathcal{W}$ if and only if $\dim \text{Ker } f = 0$ and $\dim \mathcal{V} = \dim \mathcal{W}$. □

6.4.5 The Rank-Nullity Theorem

For an $R \times C$ matrix A, define $f : \mathbb{F}^C \longrightarrow \mathbb{F}^R$ by $f(\boldsymbol{x}) = A\boldsymbol{x}$. The Kernel-Image Theorem states that $\dim \mathbb{F}^C = \dim \text{Ker } f + \dim \text{Im } f$. The kernel of f is just the null space of A, and, by the linear-combinations definition of matrix-vector multiplication, the image of f is the column space of A, so we obtain

$$\dim \mathbb{F}^C = \dim \text{Null } A + \dim \text{Col } A$$

Continuing, the dimension of $\mathbb{F}^C$ is just $|C|$, the number of columns of A, and the dimension of the column space of A is called the rank of A. Finally, the dimension of the null space of a matrix A is called the *nullity* of A. We therefore obtain:

Theorem 6.4.9 (Rank-Nullity Theorem): For any n-column matrix A,

$$\text{rank } A + \text{nullity } A = n$$

6.4.6 Checksum problem revisited

Recall our toy checksum function maps n-vectors over $GF(2)$ to 64-vectors over $GF(2)$:

$$\boldsymbol{x} \mapsto [\boldsymbol{a}_1 \cdot \boldsymbol{x}, \ldots, \boldsymbol{a}_{64} \cdot \boldsymbol{x}]$$

We represent the original "file" by an n-vector $\boldsymbol{p}$, and we represent the transmission error by an n-vector $\boldsymbol{e}$, so the corrupted file is $\boldsymbol{p} + \boldsymbol{e}$.

If the error is chosen according to the uniform distribution,

$$\text{Probability}\,(\boldsymbol{p} + \boldsymbol{e} \text{ has same checksum as } \boldsymbol{p}) = \frac{2^{\dim \mathcal{V}}}{2^n}$$

where $\mathcal{V}$ is the null space of the matrix

$$A = \begin{bmatrix} \quad \boldsymbol{a}_1 \quad \\ \hline \vdots \\ \hline \boldsymbol{a}_{64} \end{bmatrix}$$

Suppose each of the vectors $\boldsymbol{a}_1, \ldots, \boldsymbol{a}_{64}$ defining the checksum function is chosen according to the uniform distribution. Fact 6.2.22 tells us that (assuming $n > 64$) probably rank $A = 64$.

By the Rank-Nullity Theorem,

$$\begin{array}{rcrcl} \text{rank } A & + & \text{nullity } A & = & n \\ 64 & + & \dim \mathcal{V} & = & n \\ & & \dim \mathcal{V} & = & n - 64 \end{array}$$

Therefore

$$\text{Probability} = \frac{2^{n-64}}{2^n} = \frac{1}{2^{64}}$$

Thus there is only a *very* tiny chance that the change is undetected.

6.4.7 Matrix invertibility

Question 4.13.20 asks for necessary and sufficient conditions for A to be invertible.

Corollary 6.4.10: Let A be an $R \times C$ matrix. Then A is invertible if and only if $|R| = |C|$ and the columns of A are linearly independent.

Proof

Let $\mathbb{F}$ be the field. Define $f : \mathbb{F}^C \longrightarrow \mathbb{F}^R$ by $f(\boldsymbol{x}) = A\boldsymbol{x}$. Then A is an invertible matrix if and only if f is an invertible function.

By Theorem 6.4.8, f is invertible iff $\dim \operatorname{Ker} f = 0$ and $\dim \mathbb{F}^C = \dim \mathbb{F}^R$, iff $\dim \operatorname{Null} A = 0$ and $|C| = |R|$. Moreover, $\dim \operatorname{Null} A = 0$ iff the only linear combination of the columns that equals the zero vector is the trivial linear combination, i.e. if the columns are linearly independent. □

Corollary 6.4.11: The transpose of an invertible matrix is invertible.

Proof

Suppose A is an invertible matrix. Then A is square and its columns are linearly independent. Let n be the number of columns. Write

$$A = \left[\begin{array}{c|c|c} \boldsymbol{v}_1 & \cdots & \boldsymbol{v}_n \end{array}\right] = \left[\begin{array}{c} \boldsymbol{a}_1 \\ \hline \vdots \\ \hline \boldsymbol{a}_n \end{array}\right]$$

and

$$A^T = \left[\begin{array}{c|c|c} \boldsymbol{a}_1 & \cdots & \boldsymbol{a}_n \end{array}\right]$$

Because the columns of A are linearly independent, the rank of A is n. Because A is square, it has n rows. By the Rank Theorem, the row rank of A is n, so its rows are linearly independent.

The columns of the transpose A^T are the rows of A, so the columns of A^T are linearly independent. Since A^T is square and its columns are linearly independent, we infer from Corollary 6.4.10 that A^T is invertible. □

Lemma 4.13.11 showed that *if A has an inverse A^{-1} then AA^{-1} is an identity matrix.* We saw in Example 4.13.18 (Page 197) that the converse was not always true: there are matrices A and B such that AB is an identity matrix but A and B are not inverses of each other. Now we know the missing ingredient: the matrices need to be square.

Corollary 6.4.12: Suppose A and B are square matrices such that BA is the identity matrix. Then A and B are inverses of each other.

Proof

Suppose A is an $R \times C$ matrix. We then require that B is a $C \times R$ matrix. It follows that BA is the the $C \times C$ identity matrix, I_C.

We first show that the columns of A are linearly independent. Let $\boldsymbol{u}$ be any vector such that $A\boldsymbol{u} = \mathbf{0}$. Then $B(A\boldsymbol{u}) = B\mathbf{0} = \mathbf{0}$. On the other hand, $(BA)\boldsymbol{u} = I_C\boldsymbol{u} = \boldsymbol{u}$, so $\boldsymbol{u} = \mathbf{0}$.

By Corollary 6.4.10, A is invertible. We denote its inverse by A^{-1}. By Lemma 4.13.11, AA^{-1} is the $R \times R$ identity matrix, $\mathbb{1}_R$.

$$\begin{aligned} BA &= I_C \\ BAA^{-1} &= \mathbb{1}_R A^{-1} \text{ by multiplying on the right by } B^{-1} \\ BAA^{-1} &= A^{-1} \\ BI_R &= A^{-1} \text{ by Lemma 4.13.11} \\ B &= A^{-1} \end{aligned}$$

□

Example 6.4.13: $\begin{bmatrix} 1 & 2 & 3 \\ 4 & 5 & 6 \end{bmatrix}$ is not square so cannot be invertible.

Example 6.4.14: $\begin{bmatrix} 1 & 2 \\ 3 & 4 \end{bmatrix}$ is square and its columns are linearly independent so it is invertible.

Example 6.4.15: $\left[\begin{array}{c|c|c} 1 & 1 & 2 \\ 2 & 1 & 3 \\ 3 & 1 & 4 \end{array}\right]$ is square but its columns are not linearly independent so it is not invertible

6.4.8 Matrix invertibility and change of basis

We saw in Section 5.8 that, for basis $\boldsymbol{a}_1, \ldots, \boldsymbol{a}_n$ and a basis $\boldsymbol{b}_1, \ldots, \boldsymbol{b}_m$ of the same space, there is an $m \times n$ matrix C such that multiplication by C converts from a vector's coordinate representation in terms of $\boldsymbol{a}_1, \ldots, \boldsymbol{a}_n$ to the same vector's coordinate representation in terms of $\boldsymbol{b}_1, \ldots, \boldsymbol{b}_m$. We saw that the matrix C is invertible.

Now we know that the two bases must have the same size, so in fact C is a square matrix, as is required by Corollary 6.4.10.

6.5 The annihilator

We saw in Section 3.3.3 that two representations of a vector space

- as the span of a finite set of vectors, and
- as the solution set of a homogeneous linear system

were each useful. We saw in Section 3.5.5 that the analogous representations of an affine space

- as the affine hull of a finite set of vectors, and
- as the solution set of a linear system

were similarly each quite useful. For these different representations to fulfill their computational potential, it is essential that we have computational methods for converting between these representations.

6.5.1 Conversions between representations

It would seem that we would need computational methods for four different Conversion Problems:

Conversion Problem 1: Given a homogeneous linear system $A\boldsymbol{x} = \boldsymbol{0}$, find vectors $\boldsymbol{w}_1, \ldots, \boldsymbol{w}_k$ whose span is the solution set of the system.

Conversion Problem 2: Given vectors $\boldsymbol{w}_1, \ldots, \boldsymbol{w}_k$, find a homogeneous linear system $A\boldsymbol{x} = \boldsymbol{0}$ whose solution set equals Span $\{\boldsymbol{w}_1, \ldots, \boldsymbol{w}_k\}$.

Conversion Problem 3: Given a linear system $A\boldsymbol{x} = \boldsymbol{b}$, find vectors $\boldsymbol{u}_1, \ldots, \boldsymbol{u}_k$ whose affine hull is the solution set of the system (if the solution set is nonempty).

Conversion Problem 4: Given vectors $\boldsymbol{w}_1, \ldots, \boldsymbol{w}_k$, find a linear system $A\boldsymbol{x} = \boldsymbol{0}$ whose solution set equals the affine hull of $\{\boldsymbol{w}_1, \ldots, \boldsymbol{w}_k\}$.

Conversion Problem 1 can be restated thus:

Given a matrix A, find generators for the null space of A.

This is precisely Computational Problem 5.10.2 as described in Section 5.10. We will give algorithms for this problem in Chapters 7 and 9. It turns out, in fact, that using any algorithm for Conversion Problem 1 as a subroutine, we can also solve Conversion Problems 2 through 4.

First, consider Conversion Problem 2. It turns out that this problem can be solved by an algorithm for Conversion Problem 1. The solution is quite suprising and elegant. In the next section, we develop the math underlying it.

Next, consider Conversion Problem 3. Given a linear system $A\boldsymbol{x} = \boldsymbol{b}$, we can find a solution $\boldsymbol{u}_1$ (if one exists) using a method for solving a matrix-vector equation (such as that provided by the `solver` module. Next, use the algorithm for Conversion Problem 1 to obtain generators $\boldsymbol{w}_1, \ldots, \boldsymbol{w}_k$ for the solution set of the corresponding homogeneous linear system $A\boldsymbol{x} = \boldsymbol{0}$. As we saw in Section 3.5.3, the solution set of $A\boldsymbol{x} = \boldsymbol{b}$ is then

$$\boldsymbol{u}_1 + \text{Span } \{\boldsymbol{w}_1, \ldots, \boldsymbol{w}_k\}$$

which is in turn the affine hull of $\boldsymbol{u}_1, \boldsymbol{w}_1 - \boldsymbol{u}_1, \ldots, \boldsymbol{w}_k - \boldsymbol{u}_k$.

Finally, consider Conversion Problem 4. The goal is to represent the affine hull of $\boldsymbol{w}_1, \ldots, \boldsymbol{w}_k$ as the solution set of a linear system. As we saw in Section 3.5.3, this affine hull equals

$$\boldsymbol{w}_1 + \text{Span } \{\boldsymbol{w}_2 - \boldsymbol{w}_1, \ldots, \boldsymbol{w}_k - \boldsymbol{w}_1\}$$

Use the algorithm for Conversion Problem 2 to find a homogeneous linear system $A\boldsymbol{x} = \boldsymbol{0}$ whose solution set is Span $\{\boldsymbol{w}_2 - \boldsymbol{w}_1, \ldots, \boldsymbol{w}_k - \boldsymbol{w}_1\}$. Let $\boldsymbol{b} = A\boldsymbol{w}_1$. This ensures that $\boldsymbol{w}_1$ is one solution to the matrix-vector equation $A\boldsymbol{x} = \boldsymbol{b}$, and Lemma 3.6.1 ensures that the solution set is

$$\boldsymbol{w}_1 + \text{Span } \{\boldsymbol{w}_2 - \boldsymbol{w}_1, \ldots, \boldsymbol{w}_k - \boldsymbol{w}_1\}$$

which is the affine hull of $\boldsymbol{w}_1, \ldots, \boldsymbol{w}_k$.

Example 6.5.1: We are given the plane $\{[x, y, z] \in \mathbb{R}^3 : [4, -1, 1] \cdot [x, y, z] = 0\}$.

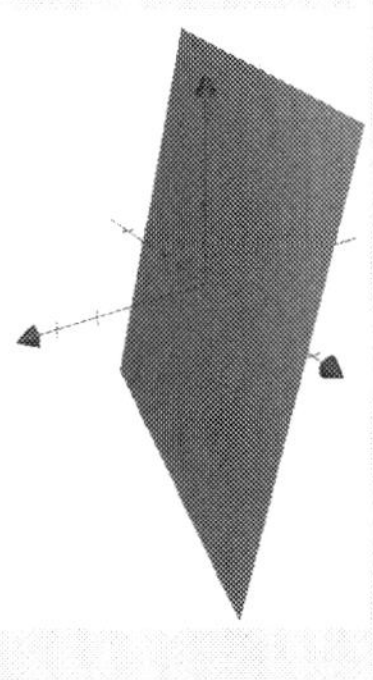

An algorithm for Conversion Problem 1 would tell us that this plane can also be written as Span $\{[1, 2, -2], [0, 1, 1]\}$.

Example 6.5.2: We are given the line $\{[x, y, z] \in \mathbb{R}^3 : [1, 2, -2] \cdot [x, y, z] = 0, [0, 1, 1] \cdot [x, y, z] = 0\}$.

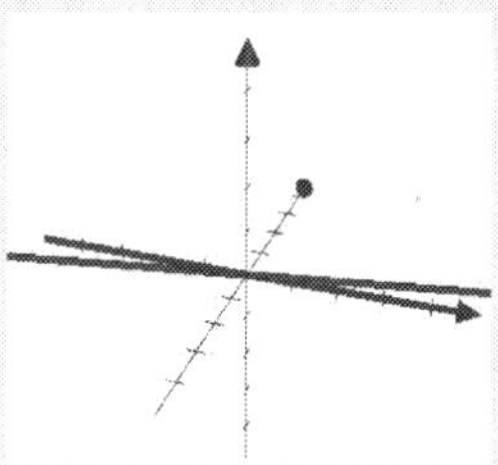

An algorithm for Conversion Problem 1 would tell us that this line can also be written as Span $\{[4, -1, 1]\}$.

Example 6.5.3: We are given the plane Span $\{[1, 2, -2], [0, 1, 1]\}$. An algorithm for Conversion Problem 2 would tell us that this plane can also be written as $\{[x, y, z] \in \mathbb{R}^3 : [4, -1, 1] \cdot [x, y, z] = 0\}$.

Example 6.5.4: We are given the line Span $\{[4, -1, 1]\}$. An algorithm for Conversion Problem 2 would tell us that this line can also be written as $\{[x, y, z] \in \mathbb{R}^3 : [1, 2, -2] \cdot [x, y, z] = 0, [0, 1, 1] \cdot [x, y, z] = 0\}$.

Conversion Problems 1 and 2 are clearly inverses of each other. This is illustrated by the fact that Example 6.5.4 (Page 286) is in a sense the inverse of Example 6.5.2 (Page 286) and that Example 6.5.3 (Page 286) is the inverse of Example 6.5.4 (Page 286).

However, the observant reader might notice something else, something tantalizing. In Examples 6.5.1 and 6.5.3, the plane is specified either as the solution set of a homogeneous linear system consisting of the equation $[4, -1, 1] \cdot [x, y, z] = 0$ or as the span of $[1, 2, -2]$ and $[0, 1, 1]$. In Examples 6.5.2 and 6.5.4, the *same* vectors play opposite roles: the line is specified either as the solution set of a homogeneous linear system $[1, 2, -2] \cdot [x, y, z] = 0, [0, 1, 1] \cdot [x, y, z] = 0$ or as Span $\{[4, -1, 1]\}$. In Section 6.5.2, we will start to learn what is going on here.

Example 6.5.5: A line is given as $\{[x, y, z] \in \mathbb{R}^3 : [5, 2, 4] \cdot [x, y, z] = 13, [0, 2, -1] \cdot [x, y, z] = 3\}$.

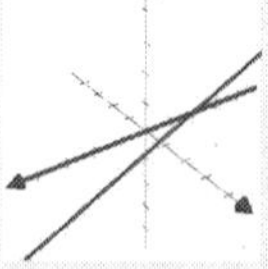

An algorithm for Conversion Problem 3 would tell us that the line is the affine hull of $[3, 1, -1]$ and $[1, 2, 1]$

Example 6.5.6: In Section 3.5.5, we illustrated the use of multiple representations by showing how to find the intersection of a ray with a triangle. One key step is: given the vertices of a triangle, find the equation for the plane containing this triangle. The vertices are $[1, 1, 1]$, $[2, 2, 3]$, and $[-1, 3, 0]$.

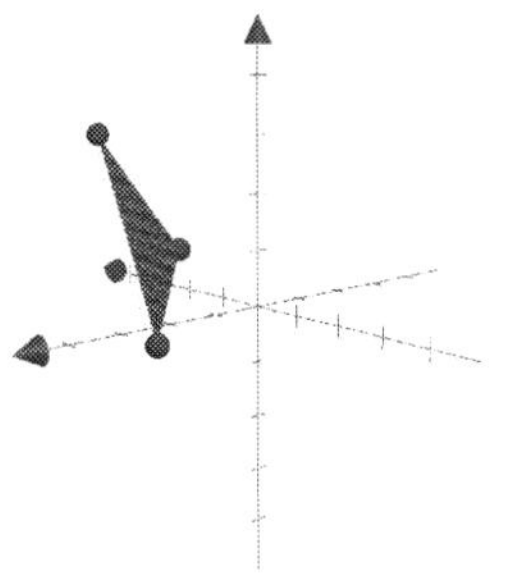

The plane containing the triangle is the affine hull of the vertices. An algorithm for Conversion Problem 4 would tell us that the plane is the solution set of $[5, 3, -4] \cdot [x, y, z] = 4$.

6.5.2 The annihilator of a vector space

Now we develop the mathematical ideas that permit an algorithm for Conversion Problem 1 to be used for Conversion Problem 2.

Definition 6.5.7: For a subspace $\mathcal{V}$ of $\mathbb{F}^n$, the *annihilator* of $\mathcal{V}$, written $\mathcal{V}^o$, is

$$\mathcal{V}^o = \{\boldsymbol{u} \in \mathbb{F}^n \ : \ \boldsymbol{u} \cdot \boldsymbol{v} = 0 \text{ for every vector } \boldsymbol{v} \in \mathcal{V}\}$$

Conversion Problem 1 concerns finding generators for the null space of a matrix. What does the annihilator of a vector space have to do with null space?

Lemma 6.5.8: Let $\boldsymbol{a}_1, \ldots, \boldsymbol{a}_m$ be generators for $\mathcal{V}$, and let

$$A = \begin{bmatrix} \boldsymbol{a}_1 \\ \hline \vdots \\ \hline \boldsymbol{a}_m \end{bmatrix}$$

Then $\mathcal{V}^o = \text{Null } A$.

Proof

Let $\boldsymbol{v}$ be a vector in $\mathbb{F}^n$. Then

$$\begin{aligned} \boldsymbol{v} \text{ is in Null } A \quad & \text{if and only if} \quad \boldsymbol{a}_1 \cdot \boldsymbol{v} = 0, \ldots, \boldsymbol{a}_m \cdot \boldsymbol{v} = 0 \\ & \text{if and only if} \quad \boldsymbol{a} \cdot \boldsymbol{v} = 0 \text{ for every vector } \boldsymbol{a} \in \text{Span } \{\boldsymbol{a}_1, \ldots, \boldsymbol{a}_m\} \\ & \text{if and only if} \quad \boldsymbol{v} \text{ is in } \mathcal{V}^o \end{aligned}$$

□

Example 6.5.9 (Example over $\mathbb{R}$): Let $\mathcal{V} = \text{Span } \{[1, 0, 1], [0, 1, 0]\}$. I show that $\mathcal{V}^o = \text{Span } \{[1, 0, -1]\}$:

- Note that $[1, 0, -1] \cdot [1, 0, 1] = 0$ and $[1, 0, -1] \cdot [0, 1, 0] = 0$.
  Therefore $[1, 0, -1] \cdot \boldsymbol{v} = 0$ for every vector $\boldsymbol{v}$ in Span $\{[1, 0, 1], [0, 1, 0]\}$.
- For any scalar β,
  $$\beta\,[1, 0, -1] \cdot \boldsymbol{v} = \beta\,([1, 0, -1] \cdot \boldsymbol{v}) = 0$$
  for every vector $\boldsymbol{v}$ in Span $\{[1, 0, 1], [0, 1, 0]\}$.
- Which vectors $\boldsymbol{u}$ satisfy $\boldsymbol{u} \cdot \boldsymbol{v} = 0$ for every vector $\boldsymbol{v}$ in Span $\{[1, 0, 1], [0, 1, 0]\}$? Only scalar multiples of $[1, 0, -1]$.

Note that in this case $\dim \mathcal{V} = 2$ and $\dim \mathcal{V}^o = 1$, so

$$\dim \mathcal{V} + \dim \mathcal{V}^o = 3$$

Example 6.5.10 (Example over $GF(2)$): Let $\mathcal{V} = \text{Span } \{[1,0,1],[0,1,0]\}$. I show that $\mathcal{V}^o = \text{Span } \{[1,0,1]\}$:

- Note that $[1,0,1] \cdot [1,0,1] = 0$ (remember, $GF(2)$ addition) and $[1,0,1] \cdot [0,1,0] = 0$.
- Therefore $[1,0,1] \cdot \boldsymbol{v} = 0$ for every vector $\boldsymbol{v}$ in Span $\{[1,0,1],[0,1,0]\}$.
- Of course $[0,0,0] \cdot \boldsymbol{v} = 0$ for every vector $\boldsymbol{v}$ in Span $\{[1,0,1],[0,1,0]\}$.
- $[1,0,1]$ and $[0,0,0]$ are the only such vectors.

Note that in this case again $\dim \mathcal{V} = 2$ and $\dim \mathcal{V}^o = 1$ so

$$\dim \mathcal{V} + \dim \mathcal{V}^o = 3$$

Example 6.5.11 (Example over $\mathbb{R}$): Let $\mathcal{V} = \text{Span } \{[1,0,1,0],[0,1,0,1]\}$. One can show that $\mathcal{V}^o = \text{Span } \{[1,0,-1,0],[0,1,0,-1]\}$. Note that in this case, $\dim \mathcal{V} = 2$ and $\dim \mathcal{V}^o = 2$, so

$$\dim \mathcal{V} + \dim \mathcal{V}^o = 4$$

Remark 6.5.12: In the traditional, abstract approach to linear algebra, alluded to in Section 3.4.4, the annihilator is defined differently, but in a way that is consistent with our definition.

6.5.3 The Annihilator Dimension Theorem

It is not a coincidence that, in each of the above examples, the sum of the dimension of $\mathcal{V}$ and that of its annihilator equals the dimension of the underlying space.

Theorem 6.5.13 (Annihilator Dimension Theorem): If $\mathcal{V}$ and $\mathcal{V}^o$ are subspaces of $\mathbb{F}^n$ then

$$\dim \mathcal{V} + \dim \mathcal{V}^o = n$$

Proof

Let A be a matrix whose row space is $\mathcal{V}$. By Lemma 6.5.8, $\mathcal{V}^o = \text{Null } A$. The Rank-Nullity Theorem states that rank $A +$ nullity $A = n$, which implies $\dim \mathcal{V} + \dim \mathcal{V}^o = n$. □

Example 6.5.14: Let's find a basis for the null space of $A = \begin{bmatrix} 1 & 0 & 2 & 4 \\ 0 & 5 & 1 & 2 \\ 0 & 2 & 5 & 6 \end{bmatrix}$ Let $\mathcal{V} =$ Row A. By Lemma 6.5.8, the null space of A is the annihilator $\mathcal{V}^o$. Since the three rows of A are linearly independent, we know dim Row A is 3, so, by the Annihilator Dimension Theorem, $\dim \mathcal{V}^o$ is $4-3$, which is 1. The vector $[1, \frac{1}{10}, \frac{13}{20}, \frac{-23}{40}]$ has a dot-product of zero with every row of A, so this vector forms a basis for the annihilator, and thus a basis for the null space of A.

6.5.4 From generators for $\mathcal{V}$ to generators for $\mathcal{V}^o$, and vice versa

Lemma 6.5.8 shows that an algorithm for finding generators for the null space of a matrix with rows $\boldsymbol{a}_1, \ldots, \boldsymbol{a}_m$ is an algorithm for finding generators for the annihilator of Span $\{\boldsymbol{a}_1, \ldots, \boldsymbol{a}_m\}$.

Let us suppose we have such an algorithm; call it *Algorithm X*. If we give it generators for a vector space $\mathcal{V}$, it outputs generators for the annihilator $\mathcal{V}^o$:

generators for a vector space $\mathcal{V}$
↓
Algorithm X
↓
generators for annihilator $\mathcal{V}^o$

What if we give it generators for the annihilator $\mathcal{V}^o$? It should output generators for the annihilator of the annihilator:

generators for annihilator $\mathcal{V}^o$
↓
Algorithm X
↓
generators for annihilator of annihilator $(\mathcal{V}^o)^o$

In the next section, we learn that the annihilator of the annihilator is the original space. That means that if we give Algorithm X generators for the annihilator $\mathcal{V}^o$ of $\mathcal{V}$, it will output generators for the original space $\mathcal{V}$:

generators for annihilator $\mathcal{V}^o$
↓
Algorithm Y
↓
generators for original space $\mathcal{V}$

Drawing on the connection between null space and annihilator, this means that the matrix whose rows are the output vectors has as its null space the span of the input vectors. Thus Algorithm X, by virtue of solving Conversion Problem 1, also solves Conversion Problem 2. The two problems, which are apparently different, are in fact the same.

6.5.5 The Annihilator Theorem

Theorem 6.5.15 (Annihilator Theorem): $(\mathcal{V}^o)^o = \mathcal{V}$ (The annihilator of the annihilator is the original space.)

Proof

Let $\boldsymbol{a}_1, \ldots, \boldsymbol{a}_m$ be a basis for $\mathcal{V}$. Let $\boldsymbol{b}_1, \ldots, \boldsymbol{b}_k$ be a basis for $\mathcal{V}^o$. Since $\boldsymbol{b}_1 \cdot \boldsymbol{v} = 0$ for every vector $\boldsymbol{v}$ in $\mathcal{V}$,

$$\boldsymbol{b}_1 \cdot \boldsymbol{a}_1 = 0, \boldsymbol{b}_1 \cdot \boldsymbol{a}_2 = 0, \ldots, \boldsymbol{b}_1 \cdot \boldsymbol{a}_m = 0$$

Similarly $\boldsymbol{b}_i \cdot \boldsymbol{a}_1 = 0, \boldsymbol{b}_i \cdot \boldsymbol{a}_2 = 0, \ldots, \boldsymbol{b}_i \cdot \boldsymbol{a}_m = 0$ for $i = 1, 2, \ldots, k$.

Reorganizing,

$$\boldsymbol{a}_1 \cdot \boldsymbol{b}_1 = 0, \boldsymbol{a}_1 \cdot \boldsymbol{b}_2 = 0, \ldots, \boldsymbol{a}_1 \cdot \boldsymbol{b}_k = 0$$

which implies that $\boldsymbol{a}_1 \cdot \boldsymbol{u} = 0$ for every vector $\boldsymbol{u}$ in Span $\{\boldsymbol{b}_1, \ldots, \boldsymbol{b}_k\}$, which is $\mathcal{V}^o$. This shows $\boldsymbol{a}_1$ is in $(\mathcal{V}^o)^o$.

Similarly $\boldsymbol{a}_2$ is in $(\mathcal{V}^o)^o$, $\boldsymbol{a}_3$ is in $(\mathcal{V}^o)^o$, ..., $\boldsymbol{a}_m$ is in $(\mathcal{V}^o)^o$. Therefore every vector in Span $\{\boldsymbol{a}_1, \boldsymbol{a}_2, \ldots, \boldsymbol{a}_m\}$ is in $(V^o)^o$.

Thus Span $\{\boldsymbol{a}_1, \boldsymbol{a}_2, \ldots, \boldsymbol{a}_m\}$, which is $\mathcal{V}$, is a subspace of $(\mathcal{V}^o)^o$. It remains to show that $\dim \mathcal{V} = \dim(\mathcal{V}^o)^o$, for then the Dimension Principle proves that $\mathcal{V}$ and $\mathcal{V}^o$ are equal.

By the Annihilator Dimension Theorem, $\dim \mathcal{V} + \dim \mathcal{V}^o = n$. By the Annihilator Dimension Theorem applied to $\mathcal{V}^o$, $\dim \mathcal{V}^o + \dim(\mathcal{V}^o)^o = n$.

Together these equations show $\dim \mathcal{V} = \dim(\mathcal{V}^o)^o$. □

6.6 Review questions

- Can a vector space have bases of different sizes?
- What is the rank of a set of vectors?
- What is the rank of a matrix?
- What is the difference between dimension and rank?
- How do dimension and rank apply to graphs?
- What is the Rank Theorem?
- What is the Dimension Principle?
- When can two vector spaces form a direct sum?
- How does the dimension of a direct sum of two vector spaces relate to their dimension?
- How can dimension be used in a criterion for a linear function to be invertible?
- What is the Kernel-Image Theorem?
- What is the Rank-Nullity Theorem?
- How can dimension be used to give a criterion for matrix invertibility?
- What is the annihilator of a vector space?
- What is the Annihilator Theorem?

6.7 Problems

Morphing using the Exchange Lemma

Problem 6.7.1: You will practice using the Exchange Lemma to transform one spanning forest into another.

Consider the campus map in Figure 6.1(a). Use the Exchange Lemma for spanning trees to transform a spanning forest F_0 = {(W, K), (W,M), (P,W), (K,A)} in Figure 6.1(a), into the spanning forest F_4 = {(P,K), (P,M), (P,A), (W,A)} in Figure 6.1(b). You should draw forests F_0, F_1, F_2, F_3 and F_4 to show each step of your transformation.

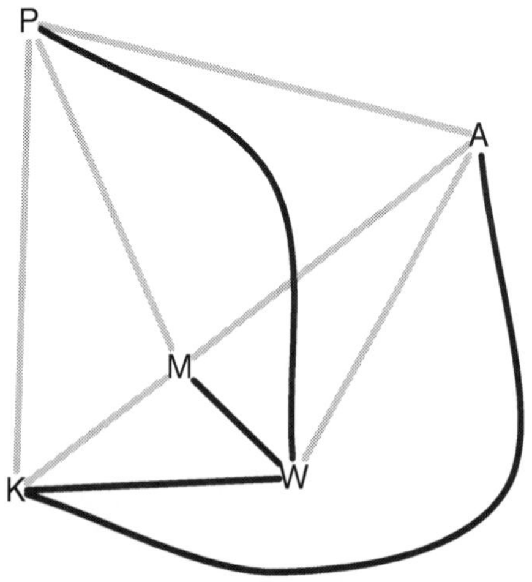

(a) Campus map with spanning forest F_0.

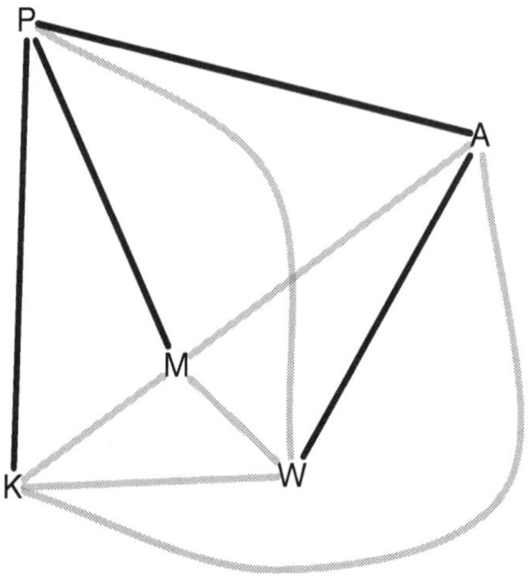

(b) Campus map with spanning forest F_4.

For the next two problems, use the Exchange Lemma iteratively to transform a set $S = \{\boldsymbol{w}_0, \boldsymbol{w}_1, \boldsymbol{w}_2\}$ into a set $B = \{\boldsymbol{v}_0, \boldsymbol{v}_1, \boldsymbol{v}_2\}$. In each step, one vector of B is injected, and one vector of S is ejected. Be careful to ensure that the ejection does not change the set of vectors spanned.

You might find the following table useful in keeping track of the iterations.

	S_i	A	$\boldsymbol{v}$ to inject	$\boldsymbol{w}$ to eject
i = 0	$\{\boldsymbol{w}_0, \boldsymbol{w}_1, \boldsymbol{w}_2\}$	$\emptyset$		
i = 1				
i = 2				
i = 3	$\{\boldsymbol{v}_0, \boldsymbol{v}_1, \boldsymbol{v}_2\}$	$\{\boldsymbol{v}_0, \boldsymbol{v}_1, \boldsymbol{v}_2\}$	-	

You are to specify the list of vectors comprising S_1 (after one iteration) and S_2 (after two iterations) in the process of transforming from $\{\boldsymbol{w}_0, \boldsymbol{w}_1, \boldsymbol{w}_2\}$ to $\{\boldsymbol{v}_0, \boldsymbol{v}_1, \boldsymbol{v}_2\}$.

Problem 6.7.2: Vectors over $\mathbb{R}$:

$$\boldsymbol{w}_0 = [1,0,0] \qquad \boldsymbol{v}_0 = [1,2,3]$$
$$\boldsymbol{w}_1 = [0,1,0] \qquad \boldsymbol{v}_1 = [1,3,3]$$
$$\boldsymbol{w}_2 = [0,0,1] \qquad \boldsymbol{v}_2 = [0,3,3]$$

Problem 6.7.3: Vectors over $GF(2)$:

$$\boldsymbol{w}_0 = [0, one, 0] \qquad \boldsymbol{v}_0 = [one, 0, one]$$
$$\boldsymbol{w}_1 = [0, 0, one] \qquad \boldsymbol{v}_1 = [one, 0, 0]$$
$$\boldsymbol{w}_2 = [one, one, one] \qquad \boldsymbol{v}_2 = [one, one, 0]$$

Problem 6.7.4: In this problem, you will write a procedure to achieve the following goal:

- *input:* a list S of vectors, and a list B of linearly independent vectors such that Span $S =$ Span B
- *output:* a list T of vectors that includes B and possibly some vectors of S such that
  - $|T| = |S|$, and
  - Span $T =$ Span S

This is not useful in its own sake, and indeed there is a trivial implementation in which T is defined to consist of the vectors in B together with enough vectors of S to make $|T| = |S|$. The point of writing this procedure is to illustrate your understanding of the proof of the Morphing Lemma. The procedure should therefore mimic that proof: T should be obtained step by step from S by, in each iteration, injecting a vector of B and ejecting a vector of $S - B$ using the Exchange Lemma. The procedure must return the list of pairs (injected vector, ejected vector) used in morphing S into T.

The procedure is to be called `morph(S, B)`. The spec is as follows:

- *input:* a list S of distinct vectors, and a list B of linearly independent vectors such that Span $S =$ Span B
- *output:* a k-element list $[(\boldsymbol{z}_1, \boldsymbol{w}_1), (\boldsymbol{z}_2, \boldsymbol{v}_2), \ldots, (\boldsymbol{z}_k, \boldsymbol{w}_k)]$ of pairs of vectors such that, for $i = 1, 2, \ldots, k$,

$$\text{Span } S = \text{Span } S \cup \{\boldsymbol{z}_1, \boldsymbol{z}_2, \ldots, \boldsymbol{z}_i\} - \{\boldsymbol{w}_1, \boldsymbol{w}_2, \ldots, \boldsymbol{w}_k\}$$

where $k = |B|$.

This procedure uses a loop. You can use the procedure exchange(S, A, z) from Problem 5.14.19 or the procedure vec2rep(veclist, u) from Problem 5.14.14 or the solver module.

Here is an illustration of how the procedure is used.

```
>>> S = [list2vec(v) for v in [[2,4,0],[1,0,3],[0,4,4],[1,1,1]]]
>>> B = [list2vec(v) for v in [[1,0,0],[0,1,0],[0,0,1]]]
>>> for (z,w) in morph(S, B):
...   print("injecting ", z)
...   print("ejecting ", w)
...   print()
...
injecting
 0 1 2
------
 1 0 0
ejecting
 0 1 2
------
 2 4 0

injecting
 0 1 2
------
 0 1 0
ejecting
 0 1 2
------
 1 0 3

injecting
 0 1 2
------
 0 0 1
ejecting
 0 1 2
------
 0 4 4
```

Test your procedure with the above example. Your results need not exactly match the results above.

Dimension and rank

Problem 6.7.5: For each of the following matrices, (a) give a basis for the row space (b) give a basis for the column space, and (c) verify that the row rank equals the column rank. Justify your answers.

1. $\begin{bmatrix} 1 & 2 & 0 \\ 0 & 2 & 1 \end{bmatrix}$

2. $\begin{bmatrix} 1 & 4 & 0 & 0 \\ 0 & 2 & 2 & 0 \\ 0 & 0 & 1 & 1 \end{bmatrix}$

3. $\begin{bmatrix} 1 \\ 2 \\ 3 \end{bmatrix}$

4. $\begin{bmatrix} 1 & 0 \\ 2 & 1 \\ 3 & 4 \end{bmatrix}$

Problem 6.7.6: In this problem you will again write an independence-testing procedure. Write and test a procedure `my_is_independent(L)` with the following spec:

- *input:* a list L of vectors
- *output:* `True` if the vectors form a linearly independent list.

Vectors are represented as instances of `Vec`. We have provided a module `independence` that provides a procedure `rank(L)`. You should use this procedure to write `my_is_independent(L)`. No loop or comprehension is needed. This is a very simple procedure.

Here is an illustration of how the procedure is used.

```
>>> my_is_independent([list2vec(v) for v in [[2,4,0],[8,16,4],[0,0,7]]])
False
>>> my_is_independent([list2vec(v) for v in [[2,4,0],[8,16,4]]])
True
```

Test your procedure with the following examples (written in Mathese, not in Python):

- $[[2, 4, 0], [8, 16, 4], [0, 0, 7]]$ over $\mathbb{R}$
- $[[1, 3, 0, 0], [2, 1, 1, 0], [0, 0, 1, 0], [1, 1, 4, -1]]$ over $\mathbb{R}$
- $[[one, 0, one, 0], [0, one, 0, 0], [one, one, one, one], [one, 0, 0, one]]$ over $GF(2)$

Problem 6.7.7: Write and test a procedure `my_rank(L)` with the following spec:

- *input:* a list L of Vecs
- *output:* The rank of L

You can use the procedure `subset_basis(T)` from Problem 5.14.17, in which case no loop is needed. Alternatively, you can use the procedure `is_independent(L)` from Problem 5.14.16 or from the module `independence` we provide; in this case, the procedure requires a loop.

Here is an illustration of how the procedure is used:

```
>>> my_rank([list2vec(v) for v in [[1,2,3],[4,5,6],[1.1,1.1,1.1]]])
2
```

Test your procedure with the following examples:

- $[[1, 2, 3], [4, 5, 6], [1.1, 1.1, 1.1]]$ over $\mathbb{R}$ (rank is 2)
- $[[1, 3, 0, 0], [2, 0, 5, 1], [0, 0, 1, 0], [0, 0, 7, -1]]$ over $\mathbb{R}$
- $[[one, 0, one, 0], [0, one, 0, 0], [one, one, one, one], [0, 0, 0, one]]$ over $GF(2)$

Problem 6.7.8: Prove that if a vector space has dimension n then any $n + 1$ of its vectors are linearly dependent.

Direct sum

Problem 6.7.9: Each of the following subproblems specifies two subspaces $\mathcal{U}$ and $\mathcal{V}$ of a vector space. For each subproblem, check whether $\mathcal{U} \cap \mathcal{V} = \{\mathbf{0}\}$.

1. Subspaces of $GF(2)^4$: let $\mathcal{U}$ = Span $\{1010, 0010\}$ and let $\mathcal{V}$ = Span $\{0101, 0001\}$.
2. Subspaces of $\mathbb{R}^3$: let $\mathcal{U}$ = Span $\{[1,2,3],[1,2,0]\}$ and let $\mathcal{V}$ = Span $\{[2,1,3],[2,1,3]\}$.
3. Subspaces of $\mathbb{R}^4$: let $\mathcal{U}$ = Span $\{[2,0,8,0],[1,1,4,0]\}$ and let $\mathcal{V}$ = Span $\{[2,1,1,1],[0,1,1,1]\}$

Problem 6.7.10: Proposition 6.3.5 states that the direct sum $\mathcal{U} \oplus \mathcal{V}$ is a vector space. Prove this using the Properties V1, V2, and V3 that define vector spaces.

Direct sum unique representation

Problem 6.7.11: Write and test a procedure `direct_sum_decompose(U_basis, V_basis, w)` with the following spec:

- *input:* A list *U_basis* containing a basis for a vector space $\mathcal{U}$, a list *V_basis* containing a basis for a vector space $\mathcal{V}$, and a vector $\boldsymbol{w}$ that belongs to the direct sum $\mathcal{U} \oplus \mathcal{V}$
- *output:* a pair $(\boldsymbol{u}, \boldsymbol{v})$ such that $\boldsymbol{w} = \boldsymbol{u} + \boldsymbol{v}$ and $\boldsymbol{u}$ belongs to $\mathcal{U}$ and $\boldsymbol{v}$ belongs to $\mathcal{V}$.

All vectors are represented as instances of `Vec`. Your procedure should use the fact that a basis of $\mathcal{U}$ joined with a basis of $\mathcal{V}$ is a basis for $\mathcal{U} \oplus \mathcal{V}$. It should use the *solver* module or the procedure `vec2rep(veclist, u)` from Problem 5.14.14.

Over R: Given `U_basis` $= \{[2,1,0,0,6,0],[11,5,0,0,1,0],[3,1.5,0,0,7.5,0]\}$, `V_basis` $= \{[0,0,7,0,0,1],[0,0,15,0,0,2]\}$, test your procedure with each of the following vectors $\boldsymbol{w}$:

1. $w = [2,5,0,0,1,0]$
2. $w = [0,0,3,0,0,-4]$
3. $w = [1,2,0,0,2,1]$
4. $w = [-6,2,4,0,4,5]$

Over $GF(2)$: Given `U_basis` $= \{[one, one, 0, one, 0, one],[one, one, 0, 0, 0, one],[one, 0, 0, 0, 0, 0]\}$, `V_basis` $= \{[one, one, one, 0, one, one]\}$, test your procedure with each of the the following vectors $\boldsymbol{w}$:

1. $w = [0,0,0,0,0,0]$
2. $w = [one, 0, 0, one, 0, 0]$
3. $w = [one, one, one, one, one, one]$

Testing invertibility

Problem 6.7.12: Write and test a procedure `is_invertible(M)` with the following spec:

- *input:* an instance `M` of `Mat`
- *output:* *True* if `M` is an invertible matrix, *False* otherwise.

Your procedure should not use any loops or comprehensions. It can use procedures from the `matutil` module and from the `independence` module.

Test your procedure with the following examples:

Over $\mathbb{R}$:

$\begin{bmatrix} 1 & 2 & 3 \\ 3 & 1 & 1 \end{bmatrix}$ False

$\begin{bmatrix} 1 & 0 & 1 & 0 \\ 0 & 2 & 1 & 0 \\ 0 & 0 & 3 & 1 \\ 0 & 0 & 0 & 4 \end{bmatrix}$ True

$\begin{bmatrix} 1 & 0 \\ 0 & 1 \\ 2 & 1 \end{bmatrix}$ False

$\begin{bmatrix} 1 & 0 \\ 0 & 1 \end{bmatrix}$ True

$\begin{bmatrix} 1 & 0 & 1 \\ 0 & 1 & 1 \\ 1 & 1 & 0 \end{bmatrix}$ True

Over $GF(2)$:

$\begin{bmatrix} \text{one} & 0 & \text{one} \\ 0 & \text{one} & \text{one} \\ \text{one} & \text{one} & 0 \end{bmatrix}$ False

$\begin{bmatrix} \text{one} & \text{one} \\ 0 & \text{one} \end{bmatrix}$ True

Note the resemblance between two matrices, one over $\mathbb{R}$ and one over $GF(2)$, and note that one is invertible and the other is not.

Finding the matrix inverse

Problem 6.7.13: Write a procedure `find_matrix_inverse(A)` with the following spec:

- *input:* an invertible matrix A over $GF(2)$ (represented as a `Mat`)
- *output:* the inverse of A (also represented as a `Mat`)

Note that the input and output matrices are over $GF(2)$.

Test your inverse procedure by printing AA^{-1} and $A^{-1}A$.

Try out your procedure on the following matrices over $GF(2)$. (Note: once your procedure outputs a matrix, you should test that it is indeed the inverse of the input matrix by multiplying the input and output matrices together.

- $\begin{bmatrix} 0 & one & 0 \\ one & 0 & 0 \\ 0 & 0 & one \end{bmatrix}$
- $\begin{bmatrix} one & one & one & one \\ one & one & one & 0 \\ 0 & one & 0 & one \\ 0 & 0 & one & 0 \end{bmatrix}$
- $\begin{bmatrix} one & one & 0 & 0 & 0 \\ 0 & one & one & 0 & 0 \\ 0 & 0 & one & one & 0 \\ 0 & 0 & 0 & one & one \\ 0 & 0 & 0 & 0 & one \end{bmatrix}$

Your procedure should use as a subroutine the `solve` procedure of the `solver` module. Since we are using $GF(2)$, you need not worry about rounding errors. Your procedure should be based on the following result:

> Suppose A and B are square matrices such that AB is the identity matrix. Then A and B are inverses of each other.

In particular, your procedure should try to find a square matrix B such that AB is an identity matrix:

$$\begin{bmatrix} & & \\ & A & \\ & & \end{bmatrix} \begin{bmatrix} & & \\ & B & \\ & & \end{bmatrix} = \begin{bmatrix} 1 & & \\ & \ddots & \\ & & 1 \end{bmatrix}$$

To do this, consider B and the identity matrix as consisting of columns.

$$\left[\begin{array}{c} \\ A \\ \\ \end{array}\right]\left[\begin{array}{c|c|c} & & \\ \boldsymbol{b}_1 & \cdots & \boldsymbol{b}_n \\ & & \end{array}\right] = \left[\begin{array}{c|c|c} 1 & & \\ & \cdots & \\ & & 1 \end{array}\right]$$

Using the matrix-vector definition of matrix-matrix multiplication, you can interpret this matrix-matrix equation as a collection of n matrix-vector equations: one for $\boldsymbol{b}_1$, ..., one for $\boldsymbol{b}_n$. By solving these equations, you can thus obtain the columns of B.

Remember: If A is an $R \times C$ matrix then AB must be an $R \times R$ matrix so the inverse B must be a $C \times R$ matrix.

Problem 6.7.14: You will write a procedure for finding the inverse of an upper-triangular matrix.

`find_triangular_matrix_inverse(A)`

- *input:* an instance M of `Mat` representing an upper triangular matrix with nonzero diagonal elements.

  You can assume that the row-label set and column-label set are of the form $\{0, 1, 2 \ldots, n-1\}$.

- *output:* a Mat representing the inverse of M

This procedure should use `triangular_solve` which is defined in the module `triangular`. It can also use the procedures in `matutil`, but that's all.

Try out your procedure on

```
>>> A = listlist2mat([[1, .5, .2, 4],[0, 1, .3, .9],[0,0,1,.1],[0,0,0,1]])
```

Chapter 7

Gaussian elimination

> Thou, nature, art my goddess; to thy laws my services are bound.
>
> Carl Friedrich Gauss

In this chapter, we finally present a sophisticated algorithm for solving computational problems in linear algebra. The method is commonly called *Gaussian elimination* but was illustrated in Chapter Eight of a Chinese text, *The Nine Chapters on the Mathematical Art*, that was written roughly two thousand years before Gauss.

Many years later, the method was rediscovered in Europe; it was described in full by Isaac Newton and by Michel Rolle, and elaborated by many others. Gauss got into the act in adapting the method to address another computational problem, one we address in Chapter 8.

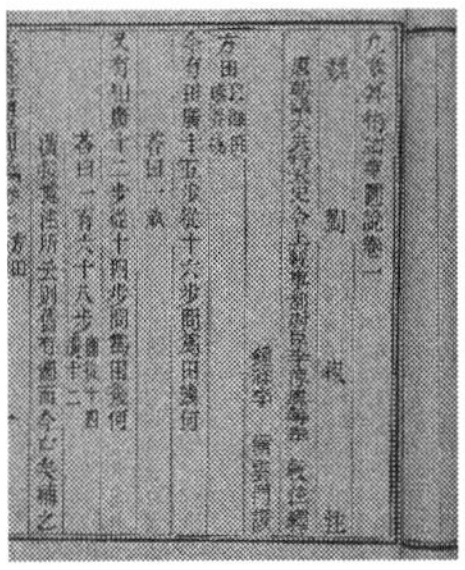

Gauss referred to the method as *eliminiationem vulgarem* ("common elimination"). He did, however, introduce a convenient notation for expressing the computation, and this is likely the reason that an author of a survey published in 1944, Henry Jensen, called it the "Gauss'ian algorithm", starting a tradition that has persisted to this day.

The algorithm clearly predated the concept of matrices, but was reformulated, by Jensen and others, in terms of matrices. The algorithm is related in spirit to backward substitution, the algorithm described in Section 2.11.

Gaussian elimination is most often applied to solving a system of linear equations—and we will see how—but it has other uses as well. Formulating the algorithm in terms of matrices helps to make clear its broader applicability.

Traditionally, Gaussian elimination is applied to matrices over the field $\mathbb{R}$ of real numbers. When it is actually carried it out on a computer using floating-point arithmetic, there are some subtleties involved in ensuring that the outputs are accurate. In this chapter, we outline the process applied to matrices over $\mathbb{R}$ but we focus on Gaussian elimination applied to matrices over $GF(2)$. We apply this algorithm to:

- *Finding a basis for the span of given vectors.* This additionally gives us an algorithm for rank and therefore for testing linear dependence.
- *Finding a basis for the null space of a matrix.*
- *Solving a matrix equation* (Computational Problem 4.5.13), which is the same as *expressing a given vector as a linear combination of other given vectors* (Computational Prob-

lem 3.1.8), which is the same as *solving a system of linear equations* (Computational Problem 2.9.12) and 2.8.7

7.1 Echelon form

Echelon form is a generalization of triangular matrices. Here is an example of a matrix in echelon form:

$$\begin{bmatrix} 0 & 2 & 3 & 0 & 5 & 6 \\ 0 & 0 & 1 & 0 & 3 & 4 \\ 0 & 0 & 0 & 0 & 1 & 2 \\ 0 & 0 & 0 & 0 & 0 & 9 \end{bmatrix}$$

Note that

- the first nonzero entry in row 0 is in column 1,
- the first nonzero entry in row 1 is in column 2,
- the first nonzero entry in row 2 is in column 4, and
- the first nonzero entry in row 4 is in column 5.

Definition 7.1.1: An $m \times n$ matrix A is in *echelon form* if it satisfies the following condition: for any row, if that row's first nonzero entry is in position k then every previous row's first nonzero entry is in some position less than k.

This definition implies that, as you iterate through the rows of A, the first nonzero entries per row move strictly right, forming a sort of staircase that descends to the right:

2	1	0	4	1	3	9	7
0	6	0	1	3	0	4	1
0	0	0	0	2	1	3	2
0	0	0	0	0	0	0	1

A triangular matrix such as $\begin{bmatrix} 4 & 1 & 3 & 0 \\ 0 & 3 & 0 & 1 \\ 0 & 0 & 1 & 7 \\ 0 & 0 & 0 & 9 \end{bmatrix}$ is a special case: the first nonzero entry in row i is in column i.

If a row of a matrix in echelon form is all zero then every subsequent row must also be all zero, e.g.

$$\begin{bmatrix} 0 & 2 & 3 & 0 & 5 & 6 \\ 0 & 0 & 1 & 0 & 3 & 4 \\ 0 & 0 & 0 & 0 & 0 & 0 \\ 0 & 0 & 0 & 0 & 0 & 0 \end{bmatrix}$$

7.1.1 From echelon form to a basis for row space

What good is it having a matrix in echeleon form?

Lemma 7.1.2: If a matrix is in echelon form, the nonzero rows form a basis for the row space.

For example, a basis for the row space of

$$\begin{bmatrix} 0 & 2 & 3 & 0 & 5 & 6 \\ 0 & 0 & 1 & 0 & 3 & 4 \\ 0 & 0 & 0 & 0 & 0 & 0 \\ 0 & 0 & 0 & 0 & 0 & 0 \end{bmatrix}$$

is $\{\begin{bmatrix} 0 & 2 & 3 & 0 & 5 & 6 \end{bmatrix}, \begin{bmatrix} 0 & 0 & 1 & 0 & 3 & 4 \end{bmatrix}\}$.

In particular, if every row is nonzero, as in each of the matrices

$$\begin{bmatrix} 0 & 2 & 3 & 0 & 5 & 6 \\ 0 & 0 & 1 & 0 & 3 & 4 \\ 0 & 0 & 0 & 0 & 1 & 2 \\ 0 & 0 & 0 & 0 & 0 & 9 \end{bmatrix}, \begin{bmatrix} 2 & 1 & 0 & 4 & 1 & 3 & 9 & 7 \\ 0 & 6 & 0 & 1 & 3 & 0 & 4 & 1 \\ 0 & 0 & 0 & 0 & 2 & 1 & 3 & 2 \\ 0 & 0 & 0 & 0 & 0 & 0 & 0 & 1 \end{bmatrix}, \begin{bmatrix} 4 & 1 & 3 & 0 \\ 0 & 3 & 0 & 1 \\ 0 & 0 & 1 & 7 \\ 0 & 0 & 0 & 9 \end{bmatrix}$$

then the rows form a basis of the row space.

To prove Lemma 7.1.2, note that it is obvious that the nonzero rows span the row space; we need only show that these vectors are linearly independent.

Before giving the formal argument, I illustrate it using the matrix

$$\begin{bmatrix} 4 & 1 & 3 & 0 \\ 0 & 3 & 0 & 1 \\ 0 & 0 & 1 & 7 \\ 0 & 0 & 0 & 9 \end{bmatrix}$$

Recall the Grow algorithm:

def GROW($\mathcal{V}$)
  $S = \emptyset$
  repeat while possible:
    find a vector $\boldsymbol{v}$ in $\mathcal{V}$ that is not in Span S, and put it in S

We imagine the Grow algorithm adding to S each of the rows of the matrix, in reverse order:

- Initially $S = \emptyset$
- Since Span $\emptyset$ does not include $\begin{bmatrix} 0 & 0 & 0 & 9 \end{bmatrix}$, the algorithm adds this vector to S.
- Now $S = \{\begin{bmatrix} 0 & 0 & 0 & 9 \end{bmatrix}\}$. Since every vector in Span S has zeroes in the first three positions, Span S does not contain $[0, 0, 1, 7]$, so the algorithm adds this vector to S.
- Now $S = \{\begin{bmatrix} 0 & 0 & 0 & 9 \end{bmatrix}, \begin{bmatrix} 0 & 0 & 1 & 7 \end{bmatrix}\}$. Since every vector in Span S has zeroes in the first two positions, Span S does not contain $[0, 3, 0, 1]$, so the algorithm adds this vector to S.
- Now $S = \{\begin{bmatrix} 0 & 0 & 0 & 9 \end{bmatrix}, \begin{bmatrix} 0 & 0 & 1 & 7 \end{bmatrix}, \begin{bmatrix} 0 & 3 & 0 & 1 \end{bmatrix}\}$. Since every vector in Span S has a zero in the first position, Span S does not contain $\begin{bmatrix} 4 & 1 & 3 & 0 \end{bmatrix}$, so the algorithm adds this vector to S, and we are done.

By the Grow-Algorithm Corollary (Corollary 5.5.10), the set S is linearly independent.

Now we present the same argument more formally:

Proof

Let $\boldsymbol{a}_1, \ldots, \boldsymbol{a}_m$ be the vectors of a matrix in echelon form, in order from top to bottom. To show that these vectors are linearly independent, imagine we run the Grow algorithm on Span $\{\boldsymbol{a}_1, \ldots, \boldsymbol{a}_m\}$. We direct the Grow algorithm to add the $\boldsymbol{a}_m, \boldsymbol{a}_{m-1}, \ldots, \boldsymbol{a}_2, \boldsymbol{a}_1$ to S, in that order.

After the Grow algorithm has added $\boldsymbol{a}_m, \boldsymbol{a}_{m-1}, \ldots, \boldsymbol{a}_i$ to S, we want it to add $\boldsymbol{a}_{i-1}$. How can we be sure $\boldsymbol{a}_{i-1}$ is not already in Span S? Suppose the first nonzero entry of $\boldsymbol{a}_{i-1}$ is in the $k+1^{st}$ position. Then, by the definition of echelon form, the first k entries of $\boldsymbol{a}_{i-1}$ are zero, so the first $k+1$ entries of $\boldsymbol{a}_i, \boldsymbol{a}_{i+1}, \ldots, \boldsymbol{a}_m$ are zero. Therefore the first $k+1$

entries of every vector in Span S are zero. Since the $k+1^{th}$ entry of $\boldsymbol{a}_{i-1}$ is nonzero, this vector is not in Span S, so the algorithm is allowed to add it. □

7.1.2 Rowlist in echelon form

Since echelon form has a lot to do with rows, it is convenient in working with echelon form to represent a matrix not as an instance of `Mat` but as a row-list—a list of vectors.

Since we want to handle vectors over arbitrary finite domains D rather than just sets of the form $\{0, 1, 2, \ldots, n-1\}$, we have to decide on an ordering of the labels (the column-labels of the matrix). For this purpose, we use

```
col_label_list = sorted(rowlist[0].D, key=hash)
```

which sorts the labels.

7.1.3 Sorting rows by position of the leftmost nonzero

Of course, not every rowlist is in echelon form. Our goal is to develop an algorithm that, given a matrix represented by a row-list, transforms the matrix into one in echelon form. We will later see exactly what kind of transformations are permitted.

To start, let's simply find a reordering of the vectors in `rowlist` that has a chance of being in echelon form. The definition of echelon form implies that the shows should be ordered according to the positions of their leftmost nonzero entries. We will use a naive sorting algorithm: try to find a row with a nonzero in the first column, then a row with a nonzero in the second column, and so on. The algorithm will accumulate the rows found in a list `new_rowlist`, initially empty:

```
new_rowlist = []
```

The algorithm maintains the set of indices of rows remaining to be sorted, `rows_left`, initially consisting of all the row indices:

```
rows_left = set(range(len(rowlist)))
```

It iterates through the column labels in order, finding a list of indices of the remaining rows that have nonzero entries in the current column. It takes one of these rows, adds it to `new_rowlist`, and removes its index from `rows_left`:

```
for c in col_label_list:
    rows_with_nonzero = [r for r in rows_left if rowlist[r][c] != 0]
    pivot = rows_with_nonzero[0]
    new_rowlist.append(rowlist[pivot])
    rows_left.remove(pivot)
```

The row added to `new_rowlist` is called the *pivot row*, and the element of the pivot row in column `c` is called the *pivot element*.

Okay, let's try out our algorithm with the matrix

$$\begin{bmatrix} 0 & 2 & 3 & 4 & 5 \\ 0 & 0 & 0 & 3 & 2 \\ 1 & 2 & 3 & 4 & 5 \\ 0 & 0 & 0 & 6 & 7 \\ 0 & 0 & 0 & 9 & 9 \end{bmatrix}$$

Things start out okay. After the iterations $c = 0$ and $c = 1$, `new_rowlist` is

$$\left[\begin{bmatrix} 1 & 2 & 3 & 4 & 5 \end{bmatrix}, \begin{bmatrix} 0 & 2 & 3 & 4 & 5 \end{bmatrix}\right]$$

and `rows_left` is $\{1, 2, 4\}$. The algorithm runs into trouble in iteration $c = 2$ since none of the remaining rows have a nonzero in column 2. The code above raises a `list index out of range` exception. How can we correct this flaw?

When none of the remaining rows have a nonzero in the current column, the algorithm should just move on to the next column without changing `new_rowlist` or `rows_left`. We amend the code accordingly:

```
for c in col_label_list:
    rows_with_nonzero = [r for r in rows_left if rowlist[r][c] != 0]
    if rows_with_nonzero != []:
        pivot = rows_with_nonzero[0]
        new_rowlist.append(rowlist[pivot])
        rows_left.remove(pivot)
```

With this change, the code does not raise an exception but at termination `new_rowlist` is

$$\begin{bmatrix} 1 & 2 & 3 & 4 & 5 \\ 0 & 2 & 3 & 4 & 5 \\ 0 & 0 & 0 & 3 & 2 \\ 0 & 0 & 0 & 6 & 7 \end{bmatrix}$$

which violates the definition of echelon form: the fourth row's first nonzero entry occurs in the fourth position, which means that every previous row's first nonzero entry must be strictly to the left of the third position—but the third row's first nonzero entry is in the fourth position.

7.1.4 Elementary row-addition operations

There is a way to repair the algorithm, however. If, in the iteration corresponding to some column-label c, there is a row other than the pivot row that has a nonzero element in the corresponding column then the algorithm must perform an *elementary row-addition operation* to make that element into a zero.

For example, given the matrix

$$\begin{bmatrix} 0 & 2 & 3 & 4 & 5 \\ 0 & 0 & 0 & 3 & 2 \\ 1 & 2 & 3 & 4 & 5 \\ 0 & 0 & 0 & 6 & 7 \\ 0 & 0 & 0 & 9 & 8 \end{bmatrix}$$

in the iteration corresponding to the fourth column, the algorithm subtracts twice the second row

$$2 \begin{bmatrix} 0 & 0 & 0 & 3 & 2 \end{bmatrix}$$

from the fourth

$$\begin{bmatrix} 0 & 0 & 0 & 6 & 7 \end{bmatrix}$$

gettting new fourth row

$$\begin{bmatrix} 0 & 0 & 0 & 6 & 7 \end{bmatrix} - 2 \begin{bmatrix} 0 & 0 & 0 & 3 & 2 \end{bmatrix} = \begin{bmatrix} 0 & 0 & 0 & 6-6 & 7-4 \end{bmatrix} = \begin{bmatrix} 0 & 0 & 0 & 0 & 3 \end{bmatrix}$$

During the same iteration, the algorithm also subtracts thrice the second row

$$3 \begin{bmatrix} 0 & 0 & 0 & 3 & 2 \end{bmatrix}$$

from the fifth

$$\begin{bmatrix} 0 & 0 & 0 & 9 & 9 \end{bmatrix}$$

getting new fifth row

$$\begin{bmatrix} 0 & 0 & 0 & 9 & 9 \end{bmatrix} - 3 \begin{bmatrix} 0 & 0 & 0 & 3 & 2 \end{bmatrix} = \begin{bmatrix} 0 & 0 & 0 & 0 & 3 \end{bmatrix}$$

The resulting matrix is

$$\begin{bmatrix} 0 & 2 & 3 & 4 & 5 \\ 0 & 0 & 0 & 3 & 2 \\ 1 & 2 & 3 & 4 & 5 \\ 0 & 0 & 0 & 0 & 3 \\ 0 & 0 & 0 & 0 & 2 \end{bmatrix}$$

In the iteration corresponding to the fifth column, the algorithm selects $\begin{bmatrix} 0 & 0 & 0 & 3 \end{bmatrix}$ as the pivot row, and adds it to new_rowlist. Next, the algorithm subtracts two-thirds times the fourth row from the fifth row, getting new fifth row

$$\begin{bmatrix} 0 & 0 & 0 & 0 & 2 \end{bmatrix} - \frac{2}{3} \begin{bmatrix} 0 & 0 & 0 & 0 & 3 \end{bmatrix} = \begin{bmatrix} 0 & 0 & 0 & 0 & 0 \end{bmatrix}$$

There are no more columns, and the algorithms stops. At this point, new_rowlist is

$$\begin{bmatrix} 1 & 2 & 3 & 4 & 5 \\ 0 & 2 & 3 & 4 & 5 \\ 0 & 0 & 0 & 3 & 2 \\ 0 & 0 & 0 & 0 & 3 \end{bmatrix}$$

The code for the procedure is

```
   for c in col_label_list:
       rows_with_nonzero = [r for r in rows_left if rowlist[r][c] != 0]
       if rows_with_nonzero != []:
           pivot = rows_with_nonzero[0]
           rows_left.remove(pivot)
           new_rowlist.append(rowlist[pivot])
=>         for r in rows_with_nonzero[1:]:
=>             multiplier = rowlist[r][c]/rowlist[pivot][c]
=>             rowlist[r] -= multiplier*rowlist[pivot]
```

The only change is the addition of a loop in which the appropriate multiple of the pivot row is subtracted from the other remaining rows.

We will prove that, when the algorithm completes, new_rowlist is a basis for the row space of the original matrix.

7.1.5 Multiplying by an elementary row-addition matrix

Subtracting a multiple of one row from another can be performed by multiplying the matrix by a *elementary row-addition matrix*

$$\begin{bmatrix} 1 & 0 & 0 & 0 \\ 0 & 1 & 0 & 0 \\ 0 & 0 & 1 & 0 \\ 0 & 0 & -2 & 1 \end{bmatrix} \begin{bmatrix} 1 & 2 & 3 & 4 & 5 \\ 0 & 2 & 3 & 4 & 5 \\ 0 & 0 & 0 & 3 & 2 \\ 0 & 0 & 0 & 6 & 7 \end{bmatrix} = \begin{bmatrix} 1 & 2 & 3 & 4 & 5 \\ 0 & 2 & 3 & 4 & 5 \\ 0 & 0 & 0 & 3 & 2 \\ 0 & 0 & 0 & 0 & 3 \end{bmatrix}$$

As we noticed in Chapter 4, such a matrix is invertible:

$$\begin{bmatrix} 1 & 0 & 0 & 0 \\ 0 & 1 & 0 & 0 \\ 0 & 0 & 1 & 0 \\ 0 & 0 & -2 & 1 \end{bmatrix} \text{ and } \begin{bmatrix} 1 & 0 & 0 & 0 \\ 0 & 1 & 0 & 0 \\ 0 & 0 & 1 & 0 \\ 0 & 0 & 2 & 1 \end{bmatrix}$$

are inverses.

7.1.6 Row-addition operations preserve row space

Our nominal goal in transforming a matrix into echelon form is to obtain a basis for the row space of the matrix. We will prove that row-addition operations do not change the row space. Therefore a basis for the row space of the transformed matrix is a basis for the original matrix.

Lemma 7.1.3: For matrices A and N, Row $NA \subseteq$ Row A.

Proof

Let $\boldsymbol{v}$ be any vector in Row NA. That is, $\boldsymbol{v}$ is a linear combination of the rows of NA. By the linear-combinations definition of vector-matrix multiplication, there is a vector $\boldsymbol{u}$ such that

$$\boldsymbol{v} = \left[\quad \boldsymbol{u}^T \quad\right]\left(\left[\quad N \quad\right]\left[\quad A \quad\right]\right)$$

$$= \left(\left[\quad \boldsymbol{u}^T \quad\right]\left[\quad N \quad\right]\right)\left[\quad A \quad\right] \qquad \text{by associativity}$$

which shows that $\boldsymbol{v}$ can be written as a linear combination of the rows of A. □

Corollary 7.1.4: For matrices A and M, if M is invertible then Row MA = Row A.

Proof

By applying Lemma 7.1.3 with $N = M$, we obtain Row $MA \subseteq$ Row A. Let $B = MA$. Since M is invertible, it has an inverse M^{-1}. Applying the lemma with $N = M^{-1}$, we obtain Row $M^{-1}B \subseteq$ Row B. Since $M^{-1}B = M^{-1}(MA) = (M^{-1}M)A = IA = A$, this proves Row $A \subseteq$ Row MA. □

Example 7.1.5: We return to the example in Section 7.1.4. Let $A = \begin{bmatrix} 0 & 2 & 3 & 4 & 5 \\ 0 & 0 & 0 & 3 & 2 \\ 1 & 2 & 3 & 4 & 5 \\ 0 & 0 & 0 & 6 & 7 \\ 0 & 0 & 0 & 9 & 8 \end{bmatrix}$ and let $M = \begin{bmatrix} 1 & 0 & 0 & 0 & 0 \\ 0 & 1 & 0 & 0 & 0 \\ 0 & 0 & 1 & 0 & 0 \\ 0 & 0 & -2 & 1 & 0 \\ 0 & 0 & 0 & 0 & 1 \end{bmatrix}$. Multiplying M by A yields $MA = \begin{bmatrix} 0 & 2 & 3 & 4 & 5 \\ 0 & 0 & 0 & 3 & 2 \\ 1 & 2 & 3 & 4 & 5 \\ 0 & 0 & 0 & 0 & 3 \\ 0 & 0 & 0 & 9 & 8 \end{bmatrix}$.

We will use the argument of Lemma 7.1.3 to show that Row $MA \subseteq$ Row A and Row $A \subseteq$ Row MA.

Every vector $\boldsymbol{v}$ in Row MA can be written as

$$\begin{aligned}
\boldsymbol{v} &= \begin{bmatrix} u_1 & u_2 & u_3 & u_4 \end{bmatrix} MA \\
&= \begin{bmatrix} u_1 & u_2 & u_3 & u_4 \end{bmatrix} \begin{bmatrix} 0 & 2 & 3 & 4 & 5 \\ 0 & 0 & 0 & 3 & 2 \\ 1 & 2 & 3 & 4 & 5 \\ 0 & 0 & 0 & 0 & 3 \\ 0 & 0 & 0 & 9 & 8 \end{bmatrix} \\
&= \begin{bmatrix} u_1 & u_2 & u_3 & u_4 \end{bmatrix} \left(\begin{bmatrix} 1 & 0 & 0 & 0 & 0 \\ 0 & 1 & 0 & 0 & 0 \\ 0 & 0 & 1 & 0 & 0 \\ 0 & 0 & -2 & 1 & 0 \\ 0 & 0 & 0 & 0 & 1 \end{bmatrix} \begin{bmatrix} 0 & 2 & 3 & 4 & 5 \\ 0 & 0 & 0 & 3 & 2 \\ 1 & 2 & 3 & 4 & 5 \\ 0 & 0 & 0 & 6 & 7 \\ 0 & 0 & 0 & 9 & 8 \end{bmatrix} \right) \\
&= \left(\begin{bmatrix} u_1 & u_2 & u_3 & u_4 \end{bmatrix} \begin{bmatrix} 1 & 0 & 0 & 0 & 0 \\ 0 & 1 & 0 & 0 & 0 \\ 0 & 0 & 1 & 0 & 0 \\ 0 & 0 & -2 & 1 & 0 \\ 0 & 0 & 0 & 0 & 1 \end{bmatrix} \right) \begin{bmatrix} 0 & 2 & 3 & 4 & 5 \\ 0 & 0 & 0 & 3 & 2 \\ 1 & 2 & 3 & 4 & 5 \\ 0 & 0 & 0 & 6 & 7 \\ 0 & 0 & 0 & 9 & 8 \end{bmatrix}
\end{aligned}$$

showing that $\boldsymbol{v}$ can be written as a vector times the matrix A. This shows that $\boldsymbol{v}$ is Row A. Since every vector in Row MA is also in Row A, we have shown Row $MA \subseteq$ Row A.

We also need to show that Row $A \subseteq$ Row MA. Since $A = M^{-1}MA$, it suffices to show that Row $M^{-1}MA \subseteq$ Row MA.

Every vector $\boldsymbol{v}$ in Row $M^{-1}MA$ can be written as

$$\begin{aligned}
\boldsymbol{v} &= \begin{bmatrix} u_1 & u_2 & u_3 & u_4 \end{bmatrix} M^{-1}MA \\
&= \begin{bmatrix} u_1 & u_2 & u_3 & u_4 \end{bmatrix} \left(\begin{bmatrix} 1 & 0 & 0 & 0 & 0 \\ 0 & 1 & 0 & 0 & 0 \\ 0 & 0 & 1 & 0 & 0 \\ 0 & 0 & 2 & 1 & 0 \\ 0 & 0 & 0 & 0 & 1 \end{bmatrix} \begin{bmatrix} 1 & 0 & 0 & 0 & 0 \\ 0 & 1 & 0 & 0 & 0 \\ 0 & 0 & 1 & 0 & 0 \\ 0 & 0 & -2 & 1 & 0 \\ 0 & 0 & 0 & 0 & 1 \end{bmatrix} \begin{bmatrix} 0 & 2 & 3 & 4 & 5 \\ 0 & 0 & 0 & 3 & 2 \\ 1 & 2 & 3 & 4 & 5 \\ 0 & 0 & 0 & 6 & 7 \\ 0 & 0 & 0 & 9 & 8 \end{bmatrix} \right) \\
&= \left(\begin{bmatrix} u_1 & u_2 & u_3 & u_4 \end{bmatrix} \begin{bmatrix} 1 & 0 & 0 & 0 & 0 \\ 0 & 1 & 0 & 0 & 0 \\ 0 & 0 & 1 & 0 & 0 \\ 0 & 0 & 2 & 1 & 0 \\ 0 & 0 & 0 & 0 & 1 \end{bmatrix} \right) \begin{bmatrix} 1 & 0 & 0 & 0 & 0 \\ 0 & 1 & 0 & 0 & 0 \\ 0 & 0 & 1 & 0 & 0 \\ 0 & 0 & -2 & 1 & 0 \\ 0 & 0 & 0 & 0 & 1 \end{bmatrix} \begin{bmatrix} 0 & 2 & 3 & 4 & 5 \\ 0 & 0 & 0 & 3 & 2 \\ 1 & 2 & 3 & 4 & 5 \\ 0 & 0 & 0 & 6 & 7 \\ 0 & 0 & 0 & 9 & 8 \end{bmatrix}
\end{aligned}$$

showing that $\boldsymbol{v}$ can be written as a vector times the matrix MA. This shows that $\boldsymbol{v}$ is in Row MA.

7.1.7 Basis, rank, and linear independence through Gaussian elimination

The program we have written has been incorporated into a procedure `row_reduce(rowlist)` that, given a list `rowlist` of vectors, mutates the list, performing the row-addition operations, and returns a list of vectors in echelon form with the same span as `rowlist`. The list of vectors returned includes no zero vectors, so is a basis for the span of `rowlist`.

Now that we have a procedure for finding a basis for the span of given vectors, we can easily write procedures for rank and linear independence. But are they correct?

7.1.8 When Gaussian elimination fails

We have shown that the algorithm for obtaining a basis is mathematically correct. However, Python carries out its computations using floating-point numbers, and arithmetic operations are only approximately correct. As a consequence, it can be tricky to use the result to decide on the rank of a set of vectors.

Consider the example

$$A = \begin{bmatrix} 10^{-20} & 0 & 1 \\ 1 & 10^{20} & 1 \\ 0 & 1 & -1 \end{bmatrix}$$

The rows of A are linearly independent. However, when we call `row_reduce` on these rows, the result is just two rows, which might lead us to conclude that the row rank is three.

First, for column $c = 0$, the algorithm selects the first row, $\begin{bmatrix} 10^{-20} & 0 & 1 \end{bmatrix}$, as the pivot row. It subtracts 10^{20} times this row from the second row, $\begin{bmatrix} 1 & 10^{20} & 1 \end{bmatrix}$, which should result in

$$\begin{bmatrix} 1 & 10^{20} & 1 \end{bmatrix} - 10^{20} \begin{bmatrix} 10^{-20} & 0 & 1 \end{bmatrix} = \begin{bmatrix} 0 & 10^{20} & 1 - 10^{20} \end{bmatrix}$$

However, let's see how Python computes the last entry:

```
>>> 1 - 1e+20
-1e+20
```

The `1` is swamped by the `- 1e+20`, and is lost. Thus, according to Python, the matrix after the row-addition operation is

$$\begin{bmatrix} 10^{-20} & 0 & 1 \\ 0 & 10^{20} & -10^{20} \\ 0 & 1 & -1 \end{bmatrix}$$

Next, for column $c = 1$, the algorithm selects the second row $\begin{bmatrix} 0 & 10^{20} & -10^{20} \end{bmatrix}$ as the pivot row, and subtracts 10^{20} times this row from the third row, resulting in the matrix

$$\begin{bmatrix} 10^{-20} & 0 & 1 \\ 0 & 10^{20} & -10^{20} \\ 0 & 0 & 0 \end{bmatrix}$$

The only remaining row, the third row, has a zero in column $c = 2$, so no pivot row is selected, and the algorithm completes.

7.1.9 Pivoting, and numerical analysis

While errors in calculation cannot be avoided when using inexact floating-point arithmetic, disastrous scenarios can be avoided by modifying Gaussian elimination. *Pivoting* refers to careful selection of the pivot element. Two strategies are employed:

- *Partial pivoting:* Among rows with nonzero entries in column c, choose row with entry having *largest* absolute value.

- *Complete pivoting:* Instead of selecting order of columns beforehand, choose each column on the fly to maximize pivot element.

Usually partial pivoting is used in practice because it is easy to implement and it runs quickly, but theoretically it can still get things disastrously bad for big matrices. Complete pivoting keeps those errors under control.

The field of *numerical analysis* provides tools for the mathematical analysis of errors resulting from using algorithms such as Gaussian elimination with inexact arithmetic. I don't cover numerical analysis in this text; I merely want the reader to be aware of the pitfalls of numerical algorithms and of the fact that mathematical analysis can help guide the development of algorithms that sidestep these pitfalls.

Using inexact arithmetic to compute the rank of a matrix is notoriously tricky. The accepted approach uses the *singular value decomposition* of the matrix, a concept covered in a later chapter.

7.2 Gaussian elimination over $GF(2)$

Gaussian elimination can be carried out on vectors over $GF(2)$, and in this case all the arithmetic is exact, so no numerical issues arise.

Here is an example. We start with the matrix

	A	B	C	D
1	0	0	one	one
2	one	0	one	one
3	one	0	0	one
4	one	one	one	one

The algorithm iterates through the columns in the order A, B, C, D. For column A, the algorithm selects row 2 as the pivot row. Since rows 3 and 4 also have nonzeroes in column A, the algorithm perform row-addition operations to add row 2 to rows 3 and 4, obtaining the matrix

	A	B	C	D
1	0	0	one	one
2	one	0	one	one
3	0	0	one	0
4	0	one	0	0

Now the algorithm handles column B. The algorithm selects row 4 as the pivot row. Since the other remaining rows (1 and 3) have zeroes in column B, no row operations need be performed for this iteration, so the matrix does not change.

Now the algorithm handles column c. It selects row 1 as the pivot row. The only other remaining row is row 3, and the algorithm performs a row-addition operation to add row 1 to row 3, obtaining the matrix

	A	B	C	D
1	0	0	one	one
2	one	0	one	one
3	0	0	0	one
4	0	one	0	0

Finally, the algorithm handles column d. The only remaining row is row 3, the algorithm selects it as the pivot row. There are no other rows, so no row-addition operations need to be performed. We have completed all the iterations for all columns. The matrix represented by `new_rowlist` is

	A	B	C	D
0	one	0	one	one
1	0	one	0	0
2	0	0	one	one
3	0	0	0	one

You can find a couple of example matrices in the file `gaussian_examples.py`

7.3 Using Gaussian elimination for other problems

We have learned that the nonzero rows of a matrix in echelon form are a basis for the row space of the matrix. We have learned how to use Gaussian elimination to transform a matrix into echelon form without changing the row space. This gives us an algorithm for finding a basis of the row space of a matrix.

However, Gaussian elimination can be used to solve other problems as well:

- Solving linear systems, and
- Finding a basis for the null space.

Over $GF(2)$, the algorithm for solving a linear system can be used, for example, to solve an instance of *Lights Out.* It can be used by Eve to find the secret password used in the simple authentication scheme. More seriously, it can even be used to predict the next random numbers coming from Python's random-number generator `random`. (See `resources.codingthematrix.com`.)

Over $GF(2)$, finding a basis for the null space can be used to find a way to corrupt a file that will not be detected by our naive checksum function. More seriously, it can be used to help factor integers, a notoriously difficult computational problem whose difficulty is at the heart of the cryptographic scheme, RSA, commonly used in protecting credit-card numbers transmitted via web browsers.

7.3.1 There is an invertible matrix M such that MA is in echelon form

The key idea in using Gaussian elimination to solve these other problems is to keep track of the elementary row-addition operations used to bring the input matrix into echelon form.

Remember that you can apply an elementary row-addition operation to a matrix by multiplying an elementary row-addition matrix M times the matrix. Starting with the matrix A,

- the algorithm performs one row-addition operation, resulting in the matrix $M_1 A$,
- then performs another row-addition operation on that matrix, resulting in the matrix $M_2 M_1 A$

  ⋮

and so on, resulting in the end in the matrix

$$M_k M_{k-1} \cdots M_2 M_1 A$$

if k is the total number of row-addition operations. Let $\bar{M}$ be the product of M_k through M_1. Then the final matrix resulting from applying Gaussian elimination to A is $\bar{M}A$.

In our code, the final matrix resulting is not in echelon form because its rows are not in correct order. By reordering the rows of $\bar{M}$, we can obtain a matrix M such that MA is a matrix in echelon form.

Moreover, since each of the matrices M_k through M_1 is invertible, so is their product $\bar{M}$. Thus $\bar{M}$ is square and its rows are linearly independent. Since M is obtained from $\bar{M}$ by reordering rows, M is also square and its rows are linearly independent. Informally, we have proved the following:

Proposition 7.3.1: For any matrix A, there is an invertible matrix M such that MA is in echelon form.

7.3.2 Computing M without matrix multiplications

Actually computing M does not require all these matrix-matrix multiplications however. There is a much slicker approach. The procedure maintains two matrices, each represented by a row-list:

- the matrix undergoing the transformation, represented in our code by `rowlist`, and
- the transforming matrix, which we will represent in code by `M_rowlist`.

The algorithm maintains the invariant that the transforming matrix times the input matrix equals the matrix represented by `rowlist`:

$$\texttt{M\_rowlist}\,(\text{initial matrix}) = \texttt{rowlist} \tag{7.1}$$

Performing the ith row-addition operation consists in subtracting some multiple of one row of `rowlist` from another. This is equivalent to multiplying the matrix `rowlist` by a row-addition matrix M_i. To maintain the invariant (Equation 7.1), we multiply both sides of the equation by M_i:

$$M_i\,(\texttt{M\_rowlist})\,(\text{initial matrix}) = M_i\,(\texttt{rowlist})$$

On the right-hand side, the procedure carries out the operation by subtracting a multiple of the pivot row from another row.

What about on the left-hand side? To update `M_rowlist` to be the product of M_i with `M_rowlist`, the procedure similarly carries out the row-addition operation on `M_rowlist`, subtracting the same multiple of the corresponding row from the corresponding row.

Example 7.3.2: Let's run through an example using the matrix

$$A = \begin{bmatrix} 0 & 2 & 3 & 4 & 5 \\ 0 & 0 & 0 & 3 & 2 \\ 1 & 2 & 3 & 4 & 5 \\ 0 & 0 & 0 & 6 & 7 \\ 0 & 0 & 0 & 9 & 8 \end{bmatrix}$$

Initially, `rowlist` consists of the rows of A. To make the invariant (Equation 7.1) true, the algorithm initializes `M_rowlist` to be the identity matrix. Now we have

$$\begin{bmatrix} 1 & & & & \\ & 1 & & & \\ & & 1 & & \\ & & & 1 & \\ & & & & 1 \end{bmatrix} \begin{bmatrix} 0 & 2 & 3 & 4 & 5 \\ 0 & 0 & 0 & 3 & 2 \\ 1 & 2 & 3 & 4 & 5 \\ 0 & 0 & 0 & 6 & 7 \\ 0 & 0 & 0 & 9 & 8 \end{bmatrix} = \begin{bmatrix} 0 & 2 & 3 & 4 & 5 \\ 0 & 0 & 0 & 3 & 2 \\ 1 & 2 & 3 & 4 & 5 \\ 0 & 0 & 0 & 6 & 7 \\ 0 & 0 & 0 & 9 & 8 \end{bmatrix}$$

The first row-addition operation is to subtract twice the second row from the fourth row. The algorithm applies this operation to the transforming matrix (the first matrix on the left-hand side) and the matrix being transformed (the matrix on the right-hand side), resulting in:

$$\begin{bmatrix} 1 & & & & \\ & 1 & & & \\ & & 1 & & \\ & -2 & & 1 & \\ & & & & 1 \end{bmatrix} \begin{bmatrix} 0 & 2 & 3 & 4 & 5 \\ 0 & 0 & 0 & 3 & 2 \\ 1 & 2 & 3 & 4 & 5 \\ 0 & 0 & 0 & 6 & 7 \\ 0 & 0 & 0 & 9 & 8 \end{bmatrix} = \begin{bmatrix} 0 & 2 & 3 & 4 & 5 \\ 0 & 0 & 0 & 3 & 2 \\ 1 & 2 & 3 & 4 & 5 \\ 0 & 0 & 0 & 0 & 3 \\ 0 & 0 & 0 & 9 & 8 \end{bmatrix}$$

Since the same matrix M_1 has multiplied the left-hand side and the right-hand side, the invariant is still true. The next row-addition operation is to subtract three times the second row from the fifth row. The algorithm applies this operation to the transforming matrix and the matrix being transformed, resulting in:

$$\begin{bmatrix} 1 & & & & \\ & 1 & & & \\ & & 1 & & \\ & -2 & & 1 & \\ & -3 & & & 1 \end{bmatrix} \begin{bmatrix} 0 & 2 & 3 & 4 & 5 \\ 0 & 0 & 0 & 3 & 2 \\ 1 & 2 & 3 & 4 & 5 \\ 0 & 0 & 0 & 6 & 7 \\ 0 & 0 & 0 & 9 & 8 \end{bmatrix} = \begin{bmatrix} 0 & 2 & 3 & 4 & 5 \\ 0 & 0 & 0 & 3 & 2 \\ 1 & 2 & 3 & 4 & 5 \\ 0 & 0 & 0 & 0 & 3 \\ 0 & 0 & 0 & 0 & 2 \end{bmatrix}$$

The third and final row-addition operation is to subtract two-thirds times the fourth row from the fifth row. The algorithm must apply this operation to the transforming matrix and the matrix being transformed. The fourth row of the transforming matrix is $\begin{bmatrix} 0 & -2 & 0 & 1 & 0 \end{bmatrix}$, and two-thirds of this row is $\begin{bmatrix} 0 & -1\frac{1}{3} & 0 & \frac{2}{3} & 0 \end{bmatrix}$. The fifth row of the transforming matrix is $\begin{bmatrix} 0 & -3 & 0 & 0 & 1 \end{bmatrix}$, and subtracting two-thirds of the fourth row yields $\begin{bmatrix} 0 & -1\frac{1}{3} & 0 & -\frac{2}{3} & 0 \end{bmatrix}$. Thus the equation becomes

$$\begin{bmatrix} 1 & & & & \\ & 1 & & & \\ & & 1 & & \\ & -2 & & 1 & \\ & -1\frac{1}{3} & & -\frac{2}{3} & \end{bmatrix} \begin{bmatrix} 0 & 2 & 3 & 4 & 5 \\ 0 & 0 & 0 & 3 & 2 \\ 1 & 2 & 3 & 4 & 5 \\ 0 & 0 & 0 & 6 & 7 \\ 0 & 0 & 0 & 9 & 8 \end{bmatrix} = \begin{bmatrix} 0 & 2 & 3 & 4 & 5 \\ 0 & 0 & 0 & 3 & 2 \\ 1 & 2 & 3 & 4 & 5 \\ 0 & 0 & 0 & 0 & 3 \\ 0 & 0 & 0 & 0 & 0 \end{bmatrix} \tag{7.2}$$

To incorporate this strategy into our code, we need to make two changes:

- initialize the variable `M_rowlist` to represent the identity matrix (using 1 or `GF2.one` as appropriate):

```
M_rowlist = [Vec(row_labels, {row_label_list[i]:one}) for i in range(m)]
```

- and, whenever a row-addition operation is performed on `rowlist`, perform the same row-addition operation on `M_rowlist`.

Here's the main loop, which shows the second change:

```
    for c in sorted(col_labels, key=hash):
        rows_with_nonzero = [r for r in rows_left if rowlist[r][c] != 0]
        if rows_with_nonzero != []:
            pivot = rows_with_nonzero[0]
            rows_left.remove(pivot)
            for r in rows_with_nonzero[1:]:
                multiplier = rowlist[r][c]/rowlist[pivot][c]
                rowlist[r] -= multiplier*rowlist[pivot]
=>              M_rowlist[r] -= multiplier*M_rowlist[pivot]
```

To make this useful for solving the other problems mentioned in Section 7.3, we need to finally produce the matrix M such that multiplying M by the input matrix gives a matrix in echelon form. Note that in Equation 7.2, the matrix in the right-hand side is not in echelon form because the rows are in the wrong order. Thus the matrix

$$\begin{bmatrix} 1 & & & \\ & 1 & & \\ & & 1 & \\ & -2 & & 1 \\ & -1\frac{1}{3} & & -\frac{2}{3} \end{bmatrix}$$

represented by `M_rowlist` is not quite M. It has the correct rows but those rows need to be reordered.

Here is a simple way to get the rows in the correct order. Recall our initial efforts in sorting rows by position of the leftmost nonzero (Section 7.1.3). There we accumulated the pivot rows in a list called `new_rowlist`. We use the same idea but this time, instead of accumulating the pivot rows, we accumulate the corresponding rows of `M_rowlist` in a list called `new_M_rowlist`:

```
=>  new_M_rowlist = []
    for c in sorted(col_labels, key=hash):
        rows_with_nonzero = [r for r in rows_left if rowlist[r][c] != 0]
        if rows_with_nonzero != []:
            pivot = rows_with_nonzero[0]
            rows_left.remove(pivot)
=>          new_M_rowlist.append(M_rowlist[pivot])
            for r in rows_with_nonzero[1:]:
                multiplier = rowlist[r][c]/rowlist[pivot][c]
                rowlist[r] -= multiplier*rowlist[pivot]
                M_rowlist[r] -= multiplier*M_rowlist[pivot]
```

One problem with this approach: it fails to append the rows of `M_rowlist` corresponding to zero rows of `rowlist` since no zero row becomes a pivot row. I therefore put another loop at the end to append these rows to `new_M_rowlist`:

```
    for c in sorted(col_labels, key=hash):
        rows_with_nonzero = [r for r in rows_left if rowlist[r][c] != 0]
        if rows_with_nonzero != []:
            pivot = rows_with_nonzero[0]
            rows_left.remove(pivot)
            new_M_rowlist.append(M_rowlist[pivot])
            for r in rows_with_nonzero[1:]:
                multiplier = rowlist[r][c]/rowlist[pivot][c]
                rowlist[r] -= multiplier*rowlist[pivot]
                M_rowlist[r] -= multiplier*M_rowlist[pivot]
 => for r in rows_left: new_M_rowlist.append(M_rowlist[r])
```

The module `echelon` contains a procedure `transformation(A)` that returns an invertible matrix M such that MA is in echelon form. It uses the above code.

Example 7.3.3: Here is another example of maintaining the transforming matrix.

$$\begin{bmatrix} 1 & 0 & 0 & 0 \\ 0 & 1 & 0 & 0 \\ 0 & 0 & 1 & 0 \\ 0 & 0 & 0 & 1 \end{bmatrix} \begin{bmatrix} 0 & 2 & 4 & 2 & 8 \\ 2 & 1 & 0 & 5 & 4 \\ 4 & 1 & 2 & 4 & 2 \\ 5 & 0 & 0 & 2 & 8 \end{bmatrix} = \begin{bmatrix} 0 & 2 & 4 & 2 & 8 \\ 2 & 1 & 0 & 5 & 4 \\ 4 & 1 & 2 & 4 & 2 \\ 5 & 0 & 0 & 2 & 8 \end{bmatrix}$$

$$\begin{bmatrix} 1 & 0 & 0 & 0 \\ 0 & 1 & 0 & 1 \\ 0 & -2 & 1 & 0 \\ 0 & 0 & 0 & 1 \end{bmatrix} \begin{bmatrix} 0 & 2 & 4 & 2 & 8 \\ 2 & 1 & 0 & 5 & 4 \\ 4 & 1 & 2 & 4 & 2 \\ 5 & 0 & 0 & 2 & 8 \end{bmatrix} = \begin{bmatrix} 0 & 2 & 4 & 2 & 8 \\ 2 & 1 & 0 & 5 & 4 \\ 0 & -1 & 2 & -6 & -6 \\ 5 & 0 & 0 & 2 & 8 \end{bmatrix}$$

$$\begin{bmatrix} 1 & 0 & 0 & 0 \\ 0 & 1 & 0 & 0 \\ 0 & -2 & 1 & 0 \\ 0 & -2.5 & 0 & 1 \end{bmatrix} \begin{bmatrix} 0 & 2 & 4 & 2 & 8 \\ 2 & 1 & 0 & 5 & 4 \\ 4 & 1 & 2 & 4 & 2 \\ 5 & 0 & 0 & 2 & 8 \end{bmatrix} = \begin{bmatrix} 0 & 2 & 4 & 2 & 8 \\ 2 & 1 & 0 & 5 & 4 \\ 0 & -1 & 2 & -6 & -6 \\ 0 & -2.5 & 0 & -10.5 & -2 \end{bmatrix}$$

$$\begin{bmatrix} 1 & 0 & 0 & 0 \\ 0 & 1 & 0 & 0 \\ .5 & -2 & 1 & 0 \\ 0 & -2.5 & 0 & 1 \end{bmatrix} \begin{bmatrix} 0 & 2 & 4 & 2 & 8 \\ 2 & 1 & 0 & 5 & 4 \\ 4 & 1 & 2 & 4 & 2 \\ 5 & 0 & 0 & 2 & 8 \end{bmatrix} = \begin{bmatrix} 0 & 2 & 4 & 2 & 8 \\ 2 & 1 & 0 & 5 & 4 \\ 0 & 0 & 4 & -5 & -2 \\ 0 & -2.5 & 0 & -10.5 & -2 \end{bmatrix}$$

7.4 Solving a matrix-vector equation using Gaussian elimination

Suppose you want to solve a matrix-vector equation

$$A\boldsymbol{x} = \boldsymbol{b} \tag{7.3}$$

Compute an invertible matrix M such that MA is a matrix U in echelon form, and multiply both sides of Equation 7.3 by M, obtaining the equation

$$MA\boldsymbol{x} = M\boldsymbol{b} \tag{7.4}$$

This shows that if the original equation (Equation 7.3) has a solution $\boldsymbol{u}$, the same solution satisfies the new equation (Equation 7.4). Conversely, suppose $\boldsymbol{u}$ is a solution to the new equation. We then have $MA\boldsymbol{u} = M\boldsymbol{b}$. Multiplying both sides by the inverse M^{-1}, we obtain $M^{-1}MA\boldsymbol{u} = M^{-1}M\boldsymbol{b}$, which implies $A\boldsymbol{u} = \boldsymbol{b}$, showing that $\boldsymbol{u}$ is therefore a solution to the original equation.

The new equation $MA\boldsymbol{x} = M\boldsymbol{b}$ is easier to solve than the original equation because the matrix MA on the left-hand side is in echelon form.

7.4.1 Solving a matrix-vector equation when the matrix is in echelon form—the invertible case

Can we give an algorithm to solve a matrix-vector equation $U\boldsymbol{x} = \boldsymbol{b}$ where U is in echelon form?

Consider first the case in which U is an invertible matrix. In this case, U is square and its diagonal elements are nonzero. It is upper triangular, and we can solve the equation $U\boldsymbol{x} = \bar{\boldsymbol{b}}$ using backward substitution, the algorithm described in Section 2.11.2 and embodied in the procedure `triangular_solve(A, b)` defined in the module `triangular`.

7.4.2 Coping with zero rows

Now consider the general case. There are two ways in which U can fail to be triangular:

- There can be rows that are all zero, and
- there can be columns of U for which no row has its leftmost nonzero entry in this column.

The first issue is easy to cope with: just ignore zero rows.

Consider an equation $\boldsymbol{a}_i \cdot \boldsymbol{x} = b_i$ where $\boldsymbol{a}_i = \boldsymbol{0}$.

- If $b_i = 0$ then the equation is true regardless of the choice of $\boldsymbol{x}$.
- If $b_i \neq 0$ then the equation is false regardless of the choice of $\boldsymbol{x}$.

Thus the only disadvantage of ignoring the rows that are zero is that the algorithm will not notice if the equations cannot be solved.

7.4.3 Coping with irrelevant columns

Assume therefore that U has no zero rows. Consider the following example:

$$\begin{array}{c|ccccc} & A & B & C & D & E \\ \hline 0 & 1 & & 1 & & \\ 1 & & 2 & & 3 & \\ 2 & & & & 1 & 9 \end{array} * \begin{bmatrix} x_a & x_b & x_c & x_d & x_e \end{bmatrix} = \begin{bmatrix} 1 \\ 1 \\ 1 \end{bmatrix}$$

There are no nonzero rows. Every row therefore has a leftmost nonzero entry. Discard every column c such that no row has a leftmost nonzero in that column. (In the example, we discard columns C and E.) The resulting system looks like this:

$$\begin{array}{c|ccc} & A & B & D \\ \hline 0 & 1 & & \\ 1 & & 2 & 3 \\ 2 & & & 1 \end{array} * \begin{bmatrix} x_a & x_b & x_d \end{bmatrix} = \begin{bmatrix} 1 \\ 1 \\ 1 \end{bmatrix}$$

This system is triangular, so can be solved using backward substitution. The solution assigns numbers to the variables x_a, x_b, x_d. What about the variables x_c and x_e corresponding to discarded columns? We just set these to zero. Using the linear-combinations definition of matrix-vector multiplication, the effect is that the discarded columns contribute nothing to the linear combination. This shows this assignment to the variables remains a solution when the discarded columns are reinserted.

In Problem 7.9.6, you will write a procedure to try to find a solution to a matrix-vector equation where the matrix is in echelon form. A straightforward but somewhat cumbersome approach is to form a new matrix by deleting the zero rows and irrelevant columns and to then use `triangular_solve`.

There is also a simpler, shorter, and more elegant solution; the code is a slightly modified version of that for `triangular_solve`.

7.4.4 Attacking the simple authentication scheme, and improving it

Recall the simple authentication scheme of Section 2.9.7:

- The password is an n-vector $\hat{\boldsymbol{x}}$ over $GF(2)$.
- As a *challenge*, Computer sends random n-vector $\boldsymbol{a}$.
- As the *response*, Human sends back $\boldsymbol{a} \cdot \hat{\boldsymbol{x}}$.
- The challenge-response interaction is repeated until Computer is convinced that Human knowns password $\hat{\boldsymbol{x}}$.

Eve eavesdrops on communication, and learns m pairs $\boldsymbol{a}_1, b_1, \ldots, \boldsymbol{a}_m, b_m$ such that b_i is the correct response to challenge $\boldsymbol{a}_i$. Then the password $\hat{\boldsymbol{x}}$ is a solution to

$$\underbrace{\begin{bmatrix} \quad \boldsymbol{a}_1 \quad \\ \hline \vdots \\ \hline \boldsymbol{a}_m \end{bmatrix}}_{A} \begin{bmatrix} \boldsymbol{x} \end{bmatrix} = \underbrace{\begin{bmatrix} b_1 \\ \vdots \\ b_m \end{bmatrix}}_{\boldsymbol{b}}$$

Once rank A reaches n, the solution is unique, and Eve can use Gaussian elimination to find it, obtaining the password.

Making the scheme more secure by introducing mistakes The way to make the scheme more secure is to introduce *mistakes.*

- In about 1/6 of the rounds, randomly, Human sends the *wrong* dot-product.
- Computer is convinced if Human gets the right answers 75% of the time.

Even if Eve knows that Human is making mistakes, she doesn't know *which* rounds involve mistakes. Gaussian elimination does *not* find the solution when some of the right-hand side values b_i are wrong. In fact, we don't know *any* efficient algorithm Eve can use to find the solution, even if Eve observes many, many rounds. Finding an "approximate" solution to a large matrix-vector equation over $GF(2)$ is considered a difficult computational problem.

In contrast, in the next couple of chapters we will learn how to find approximate solutions to matrix-vector equations over $\mathbb{R}$.

7.5 Finding a basis for the null space

Given a matrix A, we describe an algorithm to find a basis for the vector space $\{\boldsymbol{v} : \boldsymbol{v} * A = \boldsymbol{0}\}$. This is the null space of A^T.

The first step is to find an invertible matrix M such that $MA = U$ is in echelon form. In order to use the vector-matrix definition of matrix-matrix multiplication, interpret M and U as consisting of rows:

$$\begin{bmatrix} \quad \boldsymbol{b}_1 \quad \\ \hline \vdots \\ \hline \boldsymbol{b}_m \end{bmatrix} \begin{bmatrix} & & \\ & A & \\ & & \end{bmatrix} = \begin{bmatrix} \quad \boldsymbol{u}_1 \quad \\ \hline \vdots \\ \hline \boldsymbol{u}_m \end{bmatrix}$$

For each row $\boldsymbol{u}_i$ of U that is a zero vector, the corresponding row $\boldsymbol{b}_i$ of M has the property that $\boldsymbol{b}_i * A = \boldsymbol{0}$.

Example 7.5.1: Suppose A is the following matrix over $GF(2)$:

$A =$

	A	B	C	D	E
a	0	0	0	one	0
b	0	0	0	one	one
c	one	0	0	one	0
d	one	0	0	0	one
e	one	0	0	0	0

Using `transformation(A)`, we obtain the transforming matrix M such that $MA = U$:

	a	b	c	d	e
0	0	0	one	0	0
1	one	0	0	0	0
2	one	one	0	0	0
3	0	one	one	one	0
4	one	0	one	0	one

$*$

	A	B	C	D	E
a	0	0	0	one	0
b	0	0	0	one	one
c	one	0	0	one	0
d	one	0	0	0	one
e	one	0	0	0	0

$=$

	A	B	C	D	E
0	one	0	0	one	0
1	0	0	0	one	0
2	0	0	0	0	one
3	0	0	0	0	0
4	0	0	0	0	0

Since rows 3 and 4 of the right-hand side matrix are zero vectors, rows 3 and 4 of M, the first matrix on the left-hand side, belong to the vector space $\{\boldsymbol{v} : \boldsymbol{v} * A = \boldsymbol{0}\}$. We will show that in fact these two vectors form a basis for that vector space.

Thus the second step of our algorithm is to find out which rows of U are zero, and select the corresponding rows of M.

To show that the selected rows of M are a basis for the vector space $\{\boldsymbol{v} : \boldsymbol{v} * A = \boldsymbol{0}\}$, we must prove two things:

- they are linearly independent, and
- they span the vector space.

Since M is an invertible matrix, it is square and its columns are linearly independent. Therefore its rank equals the number of columns, which is the same as the number of rows. Therefore its rows are linearly independent (Corollary 6.4.11). Therefore any subset of its rows are linearly independent (Lemma 5.5.8).

To show that the selected rows span the vector space $\{\boldsymbol{v} : \boldsymbol{v} * A = \boldsymbol{0}\}$, we take an oblique approach. Let s be the number of selected rows. These rows belong to the vector space and so their span is a subspace. If we can show that the rank of the selected rows equals the dimension of the vector space, the Dimension Principle (Lemma 6.2.14), 2, will show that the span of the selected rows in fact equals the vector space. Because the selected rows are linearly independent, their rank equals s.

Let m be the number of rows of A. Note that U has the same number of rows. U has two kinds of rows: nonzero rows and zero rows. We saw in Section 7.1.1 that the nonzero rows form a basis for Row A, so the number of nonzero rows is rank A.

$$\begin{aligned} m &= \text{(number of nonzero rows of } U) &+& \text{(number of zero rows of } U) \\ &= \text{rank } A &+& s \end{aligned}$$

By the Rank-Nullity Theorem (the matrix version of the Kernel-Image Theorem),

$$m = \text{rank } A + \text{nullity } A^T$$

Therefore $s = \text{nullity } A^T$.

7.6 Factoring integers

We begin with a quotation from Gauss, writing more than two hundred years ago.

> The problem of distinguishing prime numbers from composite numbers and of resolving the latter into their prime factors is known to be one of the most important and useful in arithmetic. It has engaged the industry and wisdom of ancient and modern geometers to such an extent that it would be superfluous to discuss the problem at length. Further, the dignity of the science itself seems to require solution of a problem so elegant and so celebrated. (Carl Friedrich Gauss, *Disquisitiones Arithmeticae*, 1801)

Recall that a prime number is an integer greater than 1 whose only divisors are 1 and itself. A composite number is an integer greater than 1 that is not prime, i.e. a positive integer that has a divisor greater than one. A fundamental theorem of number theory is this:

Theorem 7.6.1 (Prime Factorization Theorem): For every positive integer N, there is a unique bag of primes whose product is N.

For example, 75 is the product of the elements in the bag $\{3, 5, 5\}$, and 126 is the product of the elements in the bag $\{2, 3, 3, 7\}$, and 23 is the product of the elements in the bag $\{23\}$. All the elements in a bag must be prime. If N is itself prime, the bag for N is just $\{N\}$.

Factoring a number N means finding the corresponding bag of primes. Gauss really spoke of two problems: (1) distinguishing prime numbers from composite numbers, and (2) factoring integers. The first problem has been solved. The second, factoring, has not, although there has been tremendous progress in algorithms for factoring since Gauss's time.

In Gauss's day, the problems were of mathematical interest. In our day, primality and factorization lie at the heart of the RSA cryptosystem, which we use every day to securely transfer credit-card numbers and other secrets. In your web browser, when you navigate to a secure website,

the browser communicates with the server using the protocol HTTPS (Secure HTTP), which is based on RSA. To quote from a book by a current-day expert, Bill Gates:

> Because both the system's privacy and the security of digital money depend on encryption, a breakthrough in mathematics or computer science that defeats the cryptographic system could be a disaster. The obvious mathematical breakthrough would be the development of an easy way to factor large prime numbers (Bill Gates, *The Road Ahead*, 1995).

Okay, Bill got it slightly wrong. Factoring a large prime number is easy. Don't worry—this was corrected in the next release of his book.

Factoring a composite number N into primes isn't the hard part. Suppose you had an algorithm `factor(N)` that, given a composite number N, found *any* integers a and b bigger than 1 such that $N = ab$. You could then obtain the prime factorization by recursively factoring a and b:

```
def prime_factorize(N):
    if is_prime(N):
      return [N]
    a,b = factor(N)
    return prime_factorize(a)+prime_factorize(b)
```

The challenge is implementing `factor(N)` to run quickly.

7.6.1 First attempt at factoring

Let's consider algorithms that involve *trial divisions*. A *trial division* is testing, for a particular integer b, whether N is divisible by b. Trial division is not a trivial operation—it's much slower than operations on floats because it has to be done in exact arithmetic—but it's not too bad.

Consider some obvious methods: The most obvious method of finding a factor of N is to try all numbers between 2 and $N - 1$. This requires $N - 2$ trial divisions. If your budget is one billion trial divisions, you can factor numbers up to a billion, i.e. 9 digits.

```
def find_divisor(N):
  for i in range(2, N):
    if N % i == 0:
       return i
```

We can get a slight improvement using the following claim:

Claim: If N is composite, it has a nontrivial divisor that is at most $\sqrt{N}$.

Proof

Suppose N is composite, and let b be a nontrivial divisor. If b is no more than $\sqrt{N}$, the claim holds. If $b > \sqrt{N}$ then $N/b < \sqrt{N}$ and N/b is an integer such that $b \cdot (N/b) = N$. □

By the claim, it suffices to look for a divisor of N that is less than or equal to $\sqrt{N}$. Thus we need only carry out $\sqrt{N}$ trial divisions. With the same billion trial divisions, you can now handle numbers up to a billion squared, i.e. 18 digits.

The next refinement you might consider it to do trial division only by primes less than or equal to $\sqrt{N}$. The Prime Number Theorem states essentially that the number of primes less than or equal to a number K is roughly $K/\ln(K)$, where $\ln(K)$ is the natural log of K. It turns out, therefore, that this refinement saves you about a factor of fifty, so now you can handle numbers with about 19 digits.

Okay, but it's easy for RSA to add another ten digits to its numbers, increasing the time for this method by a factor of ten thousand or so. What else you got?

In an upcoming lab, you will explore a more sophisticated method for factoring, the quadratic sieve. At its heart, it is based on (you guessed it) linear algebra. There is a still more sophisticated methods for factoring, but it uses linear algebra in a similar way.

7.7 *Lab: Threshold Secret-Sharing*

Recall we had a method for splitting a secret into two pieces so that both were required to recover the secret. The method used $GF(2)$. We could generalize this to split the secret among, say, four teaching assistants (TAs), so that jointly they could recover the secret but any three cannot. However, it is risky to rely on all four TAs showing up for a meeting.

We would instead like a *threshold* secret-sharing scheme, a scheme by which, say, we could share a secret among four TAs so that any three TAs could jointly recover the secret, but any two TAs could not. There are such schemes that use fields other than $GF(2)$, but let's see if we can do it using $GF(2)$.

7.7.1 *First attempt*

Here's a (doomed) attempt. I work with five 3-vectors over $GF(2)$: $\boldsymbol{a}_0, \boldsymbol{a}_1, \boldsymbol{a}_2, \boldsymbol{a}_3, \boldsymbol{a}_4$. These vectors are supposed to satisfy the following requirement:

> **Requirement:** every set of three are linearly independent.

These vectors are part of the scheme; they are known to everybody. Now suppose I want to share a one-bit secret s among the TAs. I randomly select a 3-vector $\boldsymbol{u}$ such that $\boldsymbol{a}_0 \cdot \boldsymbol{u} = s$. I keep $\boldsymbol{u}$ secret, but I compute the other dot-products:

$$\begin{aligned} \beta_1 &= \boldsymbol{a}_1 \cdot \boldsymbol{u} \\ \beta_2 &= \boldsymbol{a}_2 \cdot \boldsymbol{u} \\ \beta_3 &= \boldsymbol{a}_3 \cdot \boldsymbol{u} \\ \beta_4 &= \boldsymbol{a}_4 \cdot \boldsymbol{u} \end{aligned}$$

Now I give the bit β_1 to TA 1, I give β_2 to TA 2, I give β_3 to TA 3, and I give β_4 to TA 4. The bit given to a TA is called the TA's *share*.

First I argue that this scheme allows any three TAs to combine their shares to recover the secret.

Suppose TAs 1, 2, and 3 want to recover the secret. They solve the matrix-vector equation

$$\begin{bmatrix} \quad \boldsymbol{a}_1 \quad \\ \hline \boldsymbol{a}_2 \\ \hline \boldsymbol{a}_3 \end{bmatrix} \begin{bmatrix} x_1 \\ x_2 \\ x_3 \end{bmatrix} = \begin{bmatrix} \beta_1 \\ \beta_2 \\ \beta_3 \end{bmatrix}$$

The three TAs know the right-hand side bits, so can construct this matrix-vector equation. Since the vectors $\boldsymbol{a}_1, \boldsymbol{a}_2, \boldsymbol{a}_3$ are linearly independent, the rank of the matrix is three, so the columns are also linearly independent. The matrix is square and its columns are linearly independent, so it is invertible, so there is a unique solution. The solution must therefore be the secret vector $\boldsymbol{u}$. The TAs use `solve` to recover $\boldsymbol{u}$, and take the dot-product with $\boldsymbol{a}_0$ to get the secret $\boldsymbol{s}$.

Similarly, any three TAs can combine their shares to recover the secret vector $\boldsymbol{u}$ and thereby get the secret.

Now suppose two rogue TAs, TA 1 and TA 2, decide they want to obtain the secret without involving either of the other TAs. They know β_1 and β_2. Can they use these to get the secret s? The answer is no: their information is consistent with both $s = 0$ and $s =$ one: Since the matrix

$$\begin{bmatrix} \boldsymbol{a}_0 \\ \hline \boldsymbol{a}_1 \\ \hline \boldsymbol{a}_2 \end{bmatrix}$$

is invertible, each of the two matrix equations

$$\begin{bmatrix} \boldsymbol{a}_0 \\ \hline \boldsymbol{a}_1 \\ \hline \boldsymbol{a}_2 \end{bmatrix} \begin{bmatrix} x_0 \\ xvec_1 \\ xvec_2 \end{bmatrix} = \begin{bmatrix} 0 \\ \beta_1 \\ \beta_2 \end{bmatrix}$$

$$\begin{bmatrix} \boldsymbol{a}_0 \\ \hline \boldsymbol{a}_1 \\ \hline \boldsymbol{a}_2 \end{bmatrix} \begin{bmatrix} x_0 \\ xvec_1 \\ xvec_2 \end{bmatrix} = \begin{bmatrix} \text{one} \\ \beta_1 \\ \beta_2 \end{bmatrix}$$

has a unique solution. The solution to the first equation is a vector $\boldsymbol{v}$ such that $\boldsymbol{a}_0 \cdot \boldsymbol{v} = 0$, and the solution to the second equation is a vector $\boldsymbol{v}$ such that $\boldsymbol{a}_0 \cdot \boldsymbol{v} =$ one.

7.7.2 *Scheme that works*

So the scheme seems to work. What's the trouble?

The trouble is that there are no five 3-vectors satisfying the requirement. There are just not enough 3-vectors over $GF(2)$ to make it work.

Instead, we go to bigger vectors. We will seek ten 6-vectors $\boldsymbol{a}_0, \boldsymbol{b}_0, \boldsymbol{a}_1, \boldsymbol{b}_1, \boldsymbol{a}_2, \boldsymbol{b}_2, \boldsymbol{a}_3, \boldsymbol{b}_3, \boldsymbol{a}_4, \boldsymbol{b}_4$ over $GF(2)$. We think of them as forming five pairs:

- Pair 0 consists of $\boldsymbol{a}_0$ and $\boldsymbol{b}_0$,
- Pair 1 consists of $\boldsymbol{a}_1$ and $\boldsymbol{b}_1$,
- Pair 2 consists of $\boldsymbol{a}_2$ and $\boldsymbol{b}_2$,
- Pair 3 consists of $\boldsymbol{a}_3$ and $\boldsymbol{b}_3$, and
- Pair 4 consists of $\boldsymbol{a}_4$ and $\boldsymbol{b}_4$.

The requirement is as follows:

> **Requirement:** For any three pairs, the corresponding six vectors are linearly independent.

To use this scheme to share two bits s and t, I choose a secret 6-vector $\boldsymbol{u}$ such that $\boldsymbol{a}_0 \cdot \boldsymbol{u} = s$ and $\boldsymbol{b}_0 \cdot \boldsymbol{u} = t$. I then give TA 1 the two bits $\beta_1 = \boldsymbol{a}_1 \cdot \boldsymbol{u}$ and $\gamma_1 = \boldsymbol{b}_1 \cdot \boldsymbol{u}$, I give TA 2 the two bits $\beta_2 = \boldsymbol{a}_2 \cdot \boldsymbol{u}$ and $\gamma_2 = \boldsymbol{b}_2 \cdot \boldsymbol{u}$, and so on. Each TA's share thus consists of a pair of bits.

Recoverability: Any three TAs jointly can solve a matrix-vector equation with a 6×6 matrix to obtain $\boldsymbol{u}$, whence they can obtain the secret bits s and t. Suppose, for example, TAs 1, 2, and 3 came together. Then they would solve the equation

$$\begin{bmatrix} \boldsymbol{a}_1 \\ \hline \boldsymbol{b}_1 \\ \hline \boldsymbol{a}_2 \\ \hline \boldsymbol{b}_2 \\ \hline \boldsymbol{a}_3 \\ \hline \boldsymbol{b}_3 \end{bmatrix} \begin{bmatrix} \\ \\ \boldsymbol{x} \\ \\ \\ \end{bmatrix} = \begin{bmatrix} \beta_1 \\ \gamma_1 \\ \beta_2 \\ \gamma_2 \\ \beta_3 \\ \gamma_3 \end{bmatrix}$$

to obtain $\boldsymbol{u}$ and thereby obtain the secret bits. Since the vectors $\boldsymbol{a}_1, \boldsymbol{b}_1, \boldsymbol{a}_2, \boldsymbol{b}_2, \boldsymbol{a}_3, \boldsymbol{b}_3$ are linearly independent, the matrix is invertible, so there is a unique solution to this equation.

Secrecy: However, for any two TAs, the information they possess is consistent with any assignment to the two secret bits s and t. Suppose TAs 1 and 2 go rogue and try to recover s and t. They possess the bits $\beta_1, \gamma_1, \beta_2, \gamma_2$. Are these bits consistent with $s = 0$ and $t = \text{one}$? They are if there is a vector $\boldsymbol{u}$ that solves the equation

$$\begin{bmatrix} \boldsymbol{a}_0 \\ \hline \boldsymbol{b}_0 \\ \hline \boldsymbol{a}_1 \\ \hline \boldsymbol{b}_1 \\ \hline \boldsymbol{a}_2 \\ \hline \boldsymbol{b}_2 \end{bmatrix} \begin{bmatrix} \\ \\ \boldsymbol{x} \\ \\ \\ \end{bmatrix} = \begin{bmatrix} 0 \\ \text{one} \\ \beta_1 \\ \gamma_1 \\ \beta_2 \\ \gamma_2 \end{bmatrix}$$

where the first two entries of the right-hand side are the guessed values of s and t.

Since the vectors $\boldsymbol{a}_0, \boldsymbol{b}_0, \boldsymbol{a}_1, \boldsymbol{b}_1, \boldsymbol{a}_2, \boldsymbol{b}_2$ are linearly independent, the matrix is invertible, so there is a unique solution. Similarly, no matter what you put in the first two entries of the right-hand side, there is exactly one solution. This shows that the shares of TAs 1 and 2 tell them nothing about the true values of s and t.

7.7.3 *Implementing the scheme*

To make thing simple, we will define $\boldsymbol{a}_0 = [\text{one}, \text{one}, 0, \text{one}, 0, \text{one}]$ and $\boldsymbol{b}_0 = [\text{one}, \text{one}, 0, 0, 0, \text{one}]$:

```
>>> a0 = list2vec([one, one, 0, one, 0, one])
>>> b0 = list2vec([one, one, 0, 0, 0, one])
```

Remember, `list2vec` is defined in the module `vecutil` and `one` is defined in `GF2`.

7.7.4 *Generating mathbfu*

Task 7.7.1: Write a procedure `choose_secret_vector(s,t)` with the following spec:

- *input:* $GF(2)$ field elements s and t (i.e. bits)
- *output:* a random 6-vector $\boldsymbol{u}$ such that $\boldsymbol{a}_0 \cdot \boldsymbol{u} = s$ and $\boldsymbol{b}_0 \cdot \boldsymbol{u} = t$

Why must the output be random? Suppose that the procedure was not random: the output vector $\boldsymbol{u}$ was determined by the two secret bits. The TA could use the information to make a good guess as to the value of $\boldsymbol{u}$ and therefore the values of s and t.

For this task, you can use Python's `random` module to generate pseudorandom elements of $GF(2)$.

```
>>> import random
>>> def randGF2(): return random.randint(0,1)*one
```

However, be warned: Don't use this method if you really intend to keep a secret. Python's `random` module does not generate cryptographically secure pseudorandom bits. In particular, a rogue TA could use his shares to actually figure out the state of the pseudorandom-number generator, predict future pseudorandom numbers, and break the security of the scheme. (Figuring out the state of the pseudorandom-number generator uses—you guessed it—linear algebra over $GF(2)$.)

7.7.5 *Finding vectors that satisfy the requirement*

Task 7.7.2: We have decided that $\boldsymbol{a}_0 = [\text{one}, \text{one}, 0, \text{one}, 0, \text{one}]$ and $\boldsymbol{b}_0 = [\text{one}, \text{one}, 0, 0, 0, \text{one}]$. Your goal is to select vectors $\boldsymbol{a}_1, \boldsymbol{b}_1, \boldsymbol{a}_2, \boldsymbol{b}_2, \boldsymbol{a}_3, \boldsymbol{b}_3, \boldsymbol{a}_4, \boldsymbol{b}_4$ over $GF(2)$ so that the requirement is satisfied:

For any three pairs, the corresponding six vectors are linearly independent.

Include in your answer any code you used, and the vectors you came up with. Your solution to this problem should be submitted electronically and on paper.

Hint: try selecting eight random vectors and testing whether they satisfy the requirement. Repeat until you succeed. Use the `independence` module.

7.7.6 *Sharing a string*

Now that we can share two bits, we can share an arbitrarily long string.

The module `bitutil` defines procedures

- `str2bits(str)`, which converts a string to a list of $GF(2)$ values
- `bits2str(bitlist)`, the inverse of `str2bits`
- `bits2mat(bitlist, nrows)`, which uses the bits in `bitlist` to populate a matrix with `nrows` rows.
- `mat2bits(M)`, which is the inverse of `bits2mat`

You can use `str2bits` to transform a string, say `"Rosebud"`, into a list of bits, and use `bits2mat` to transform the list of bits to a $2 \times n$ matrix.

For each column of this matrix, you can use the procedure `choose_secret_vector(s,t)` of Task 7.7.1 to obtain a corresponding secret vector $\boldsymbol{u}$, constructing a matrix U whose columns are the secret vectors.

To compute the shares of the TAs, multiply the matrix

$$\begin{bmatrix} \boldsymbol{a}_0 \\ \hline \boldsymbol{b}_0 \\ \hline \boldsymbol{a}_1 \\ \hline \boldsymbol{b}_1 \\ \hline \boldsymbol{a}_2 \\ \hline \boldsymbol{b}_2 \\ \hline \boldsymbol{b}_3 \\ \hline \boldsymbol{a}_4 \\ \hline \boldsymbol{b}_4 \end{bmatrix}$$

times U. The second and third rows of the product form the share for TA 1, and so on.

7.8 *Lab: Factoring integers*

7.8.1 *First attempt to use square roots*

In one step towards a modern factorization algorithm, suppose you could find integers a and b such that

$$a^2 - b^2 = N$$

for then

$$(a - b)(a + b) = N$$

so $a - b$ and $a + b$ are divisors of N. We hope that they happen to be nontrivial divisors (ie. that $a - b$ is neither 1 nor N).

Task 7.8.1: To find integers a and b such that $a^2 - b^2 = N$, write a procedure `root_method(N)` to implement the following algorithm:

- Initialize integer a to be an integer greater than $\sqrt{N}$
- Check if $\sqrt{a^2 - N}$ is an integer.
- If so, let $b = \sqrt{a^2 - N}$. Success! Return $a - b$.
- If not, repeat with the next greater value of a.

The module `factoring_support` provides a procedure `intsqrt(x)` with the following spec:

- *input:* an integer x
- *output:* an integer y such that $y * y$ is close to x and, if x happens to be a perfect square, $y * y$ is exactly x.

You should use `intsqrt(x)` in your implementation of the above algorithm. Try it out with 55, 77, 146771, and 118. Hint: the procedure might find just a trivial divisor or it might run forever.

7.8.2 *Euclid's algorithm for greatest common divisor*

In order to do better, we turn for help to a lovely algorithm that dates back some 2300 years: Euclid's algorithm for greatest common divisor. Here is code for it:

```
def gcd(x,y): return x if y == 0 else gcd(y, x % y)
```

Task 7.8.2: Enter the code for `gcd` or import it from the module `factoring_support` that we provide. Try it out. Specifically, use Python's pseudo-random-number generator (use the procedure `randint(a,b)` in the module `random`) or use pseudo-random whacking at your keyboard to generate some very big integers r, s, t. Then set $a = r * s$ and $b = s * t$, and find the greatest common divisor d of a and b. Verify that d has the following properties:

- a is divisible by d (verify by checking that $a\%d$ equals zero)
- b is divisible by d, and
- $d \geq s$

7.8.3 *Using square roots revisited*

It's too hard to find integers a and b such that $a^2 - b^2$ equals N. We will lower our standards a bit, and seek integers a and b such that $a^2 - b^2$ is divisible by N. Suppose we find such integers. Then there is another integer k such that

$$a^2 - b^2 = kN$$

That means

$$(a - b)(a + b) = kN$$

Every prime in the bag of primes whose product is kN

- belongs either to the the bag of primes whose product is k or the bag of primes whose product is N, and
- belongs either to the the bag of primes whose product is $a - b$ or the bag of primes whose product is $a + b$.

Suppose N is the product of two primes, p and q. If we are even a little lucky, one of these primes will belong to the bag for $a-b$ and the other will belong to the bag for $a+b$. If this happens, the greatest common divisor of $a-b$ with N will be nontrivial! And, thanks to Euclid's algorithm, we can actually compute it.

Task 7.8.3: Let $N = 367160330145890434494322103$, let $a = 67469780066325164$, and let $b = 9429601150488992$, and verify that $a*a-b*b$ is divisible by N. That means that the greatest common divisor of $a-b$ and N has a chance of being a nontrivial divisor of N. Test this using the gcd procedure, and report the nontrivial divisor you found.

But how can we find such a pair of integers? Instead of hoping to get lucky, we'll take matters into our own hands. We'll try to create a and b. This method starts by creating a set `primeset` consisting of the first thousand or so primes. We say an integer x factors over *primeset* if you can multiply together some of the primes in S (possibly using a prime more than once) to form x.

For example:

- 75 factors over $\{2,3,5,7\}$ because $75 = 3 \cdot 5 \cdot 5$.
- 30 factors over $\{2,3,5,7\}$ because $30 = 2 \cdot 3 \cdot 5$.
- 1176 factors over $\{2,3,5,7\}$ because $1176 = 2 \cdot 2 \cdot 2 \cdot 7 \cdot 7$.

We can represent a factorization of an integer over a set of primes by a list of pairs (prime, exponent). For example:

- We can represent the factorization of 75 over $\{2,3,5,7\}$ by the list of pairs $[(3,1),(5,2)]$, indicating that 75 is obtained by multiplying a single 3 and two 5's.
- We can represent the factorization of 30 by the list $[(2,1),(3,1),(5,1)]$, indicating that 30 is obtained by multiplying 2, 3, and 5.
- We can represent the factorization of 1176 by the list $[(2,3),(5,2)]$, indicating that 1176 is obtained by multipying together three 2's and two 5's.

The first number in each pair is a prime in the set *primeset* and the second number is its exponent:

$$\begin{aligned} 75 &= 3^1 5^2 \\ 30 &= 2^1 3^1 5^1 \\ 1176 &= 2^3 5^2 \end{aligned}$$

The module `factoring_support` defines a procedure `dumb_factor(x, primeset)` with the following spec:

- *input:* an integer x and a set *primeset* of primes
- *output:* if there are primes $p_1, \ldots, \boldsymbol{p}_s$ in *primeset* and positive integers $e_1, e_2, \ldots, \boldsymbol{e}_s$ (the exponents) such that $x = p_1^{e_1} p_2^{e_2} \cdots p_s^{e_s}$ then the procedure returns the list $[(p_1,e_1),(p_2,e_2),\ldots,(p_s,e_s)]$ of pairs (prime, exponent). If not, the procedure returns the empty list.

Here are some examples:

```
>>> dumb_factor(75, {2,3,5,7})
[(3, 1), (5, 2)]
>>> dumb_factor(30, {2,3,5,7})
[(2, 1), (3, 1), (5, 1)]
>>> dumb_factor(1176, {2,3,5,7})
[(2, 3), (3, 1), (7, 2)]
```

```
>>> dumb_factor(2*17, {2,3,5,7})
[]
>>> dumb_factor(2*3*5*19, {2,3,5,7})
[]
```

Task 7.8.4: Define *primeset*=$\{2,3,5,7,11,13\}$. Try out `dumb_factor(x, primeset)` on integers $x = 12, x = 154, x = 2*3*3*3*11*11*13, x = 2*17, x = 2*3*5*7*19$. Report the results.

Task 7.8.5: From the `GF2` module, import the value `one`. Write a procedure `int2GF2(i)` that, given an integer i, returns `one` if i is odd and 0 if i is even.

```
>>> int2GF2(3)
one
>>> int2GF2(4)
0
```

The module `factoring_support` defines a procedure `primes(P)` that returns a set consisting of the prime numbers less than P.

Task 7.8.6: From the module `vec`, import `Vec`. Write a procedure `make_Vec(primeset, factors)` with the following spec:

- *input:* a set of primes *primeset* and a list *factors*=$[(p_1,a_1),(p_2,a_2),\ldots,(p_s,a_s)]$ such as produced by `dumb_factor`, where every p_i belongs to *primeset*
- *output:* a *primeset*-vector v over $GF(2)$ with domain *primeset* such that $v[p_i] =$ `int2GF2`(a_i) for $i = 1,\ldots,s$

For example,

```
>>> make_Vec({2,3,5,7,11}, [(3,1)])
Vec({3, 2, 11, 5, 7},{3: one})
>>> make_Vec({2,3,5,7,11}, [(2,17), (3, 0), (5,1), (11,3)])
Vec({3, 2, 11, 5, 7},{11: one, 2: one, 3: 0, 5: one})
```

Now comes the interesting part.

Task 7.8.7: Suppose you want to factor the integer $N = 2419$ (easy but big enough to demonstrate the idea).

Write a procedure `find_candidates(N, primeset)` that, given an integer N to factor and a set *primeset* of primes, finds `len(primeset)+1` integers a for which $a \cdot a - N$ can be factored completely over `primeset` The procedure returns two lists:

- the list `roots` consisting of $a_0, a_1, a_2, ...$ such that $a_i \cdot a_i - N$ can be factored completely over `primeset`, and
- the list `rowlist` such that element i is the `primeset`-vector over $GF(2)$ corresponding to a_i (that is, the vector produced by `make_vec`).

The algorithm should initialize

```
roots = []
rowlist = []
```

and then iterate for $x =$ `intsqrt(N)+2, intsqrt(N)+3`, ..., and for each value of x,

- if $x \cdot x - N$ can be factored completely over *primeset*,
  - append x to `roots`,
  - append to `rowlist` the vector corresponding to the factors of $x \cdot x - N$

continuing until at least *len(primeset)*+1 roots and vectors have been accumulated.

Try out your procedure on $N = 2419$ by calling `find_candidates(N, primes(32))`.

Here's a summary of the result of this computation:

x	x^2-N	factored	result of `dumb_factor`	vector.f
51	182	$2 \cdot 7 \cdot 13$	$[(2,1),(7,1),(13,1)]$	{2 : one, 13 : one, 7 : one}
52	285	$3 \cdot 5 \cdot 19$	$[(3,1),(5,1),(19,1)]$	{19 : one, 3 : one, 5 : one}
53	390	$2 \cdot 3 \cdot 5 \cdot 13$	$[(2,1),(3,1),(5,1),(13,1)]$	{2 : one, 3 : one, 5 : one, 13 : one}
58	945	$3^3 \cdot 5 \cdot 7$	$[(3,3),(5,1),(7,1)]$	{3 : one, 5 : one, 7 : one}
61	1302	$2 \cdot 3 \cdot \cdot 7 \cdot 13$	$[(2,1),(3,1),(7,1),(31,1)]$	{31 : one, 2 : one, 3 : one, 7 : one}
62	1425	$3 \cdot 5^2 \cdot 19$	$[(3,1),(5,2),(19,1)]$	{19 : one, 3 : one, 5 : 0}
63	1550	$2 \cdot 5^2 \cdot 31$	$[(2,1),(5,2),(31,1)]$	{2 : one, 5 : 0, 31 : one}
67	2070	$2 \cdot 3^2 \cdot 5 \cdot 23$	$[[(2,1),(3,2),(5,1),(23,1)]$	{2 : one, 3 : 0, 5 : one, 23 : one}
68	2205	$3^2 \cdot 5 \cdot 7^2$	$[(3,2),(5,1),(7,2)]$	{3 : 0, 5 : one, 7 : 0}
71	2622	$2 \cdot 3 \cdot 19 \cdot 23$	$[(2,1),(3,1),(19,1),(23,1)]$	{19 : one, 2 : one, 3 : one, 23 : one}
77	3510	$2 \cdot 3^3 \cdot 5 \cdot 13$	$[(2,1),(3,3),(5,1),(13,1)]$	{2 : one, 3 : one, 5 : one, 13 : one}
79	3822	$2 \cdot 3 \cdot 7^2$	$[(2,1),(3,1),(7,1)]$	{2 : one, 3 : one, 13 : one, 7 : 0}

Thus, after the loop completes, the value of `roots` should be the list

$$[51, 52, 53, 58, 61, 62, 63, 67, 68, 71, 77, 79]$$

and the value of `rowlist` should be the list

```
[Vec({2,3,5, ..., 31},{2: one, 13: one, 7: one}),
                          ⋮
 Vec({2,3,5, ...  , 31},{2: one, 3: one, 5: one, 13: one}),
  Vec({2,3,5, ..., 31}, {2: one, 3: one, 13: one, 7: 0})]
```

Now we use the results to find a nontrivial divisor of N.

Examine the table rows corresponding to 53 and 77. The factorization of $53 * 53 - N$ is $2 \cdot 3 \cdot 5 \cdot 13$. The factorization of $77 * 77 - N$ is $2 \cdot 3^3 \cdot 5 \cdot 13$. Therefore the factorization of the product $(53 * 53 - N)(77 * 77 - N)$ is

$$(2 \cdot 3 \cdot 5 \cdot 13)(2 \cdot 3^3 \cdot 5 \cdot 13) = 2^2 \cdot 3^4 \cdot 5^2 \cdot 13^2$$

Since the exponents are all even, the product is a perfect square: it is the square of

$$2 \cdot 3^2 \cdot 5 \cdot 13$$

Thus we have derived

$$\begin{aligned}(53^2 - N)(77^2 - N) &= (2 \cdot 3^2 \cdot 5 \cdot 13)^2 \\ 53^2 \cdot 77^2 - kN &= (2 \cdot 3^2 \cdot 5 \cdot 13)^2 \\ (53 \cdot 77)^2 - kN &= (2 \cdot 3^2 \cdot 5 \cdot 13)^2\end{aligned}$$

Task 7.8.8: To try to find a factor, let $a = 53 \cdot 77$ and let $b = 2 \cdot 3^2 \cdot 5 \cdot 13$, and compute gcd$(a - b, N)$. Did you find a proper divisor of N?

Similarly, examine the table rows corresponding to 52, 67, and 71. The factorizations of $x * x - N$ for these values of x are

$$\begin{gathered}3 \cdot 5 \cdot 19 \\ 2 \cdot 3^2 \cdot 5 \cdot 23 \\ 2 \cdot 3 \cdot 19 \cdot 23\end{gathered}$$

Therefore the factorization of the product $(52 * 52 - N)(67 * 67 - N)(71 * 71 - N)$ is

$$(3 \cdot 5 \cdot 19)(2 \cdot 3^2 \cdot 5 \cdot 23)(2 \cdot 3 \cdot 19 \cdot 23) = 2^2 \cdot 3^4 \cdot 5^2 \cdot 19^2 \cdot 23^2$$

which is again a perfect square; it is the square of

$$2 \cdot 3^2 \cdot 5 \cdot 19 \cdot 23$$

Task 7.8.9: To again try to find a factor of N (just for practice), let $a = 52 \cdot 67 \cdot 71$ and let $b = 2 \cdot 3^2 \cdot 5 \cdot 19 \cdot 23$, and compute gcd$(a - b, N)$. Did you find a proper divisor of N?

How did I notice that the rows corresponding to 52, 67, and 71 combine to provide a perfect square? That's where the linear algebra comes in. The sum of the vectors in these rows is the zero vector. Let A be the matrix consisting of these rows. Finding a nonempty set of rows of A whose $GF(2)$ sum is the zero vector is equivalent, by the linear-combinations definition of vector-matrix multiplication, to finding a nonzero vector $\boldsymbol{v}$ such that $\boldsymbol{v} * A$ is the zero vector. That is, $\boldsymbol{v}$ is a nonzero vector in the null space of A^T.

How do I know such a vector exists? Each vector in `rowlist` is a `primeset`-vector and so lies in a K-dimensional space where K = *len(primelist)*. Therefore the rank of these vectors is at most K. But `rowlist` consists of at least $K + 1$ vectors. Therefore the rows are linearly dependent.

How do I find such a vector? When I use Gaussian elimination to transform the matrix into echelon form, the last one is guaranteed to be zero.

More specifically, I find a matrix M representing a transformation that reduced the vectors in `rowlist` to echelon form. The last row of M, multiplied by the original matrix represented by `rowlist`, yields the last row of the matrix in echelon form, which is a zero vector.

To compute M, you can use the procedure `transformation_rows(rowlist_input)` defined in the module `echelon` we provide.

Given a matrix A (represented by as a list `rowlist_input` of rows), this procedure returns a matrix M (also represented as a list of rows) such that MA is in echelon form.

Since the last row of MA must be a zero vector, by the vector-matrix definition of matrix-vector multiplication, the last row of M times A is the zero vector. By the linear-combinations definition of vector-matrix multiplication, the zero vector is a linear combination of the rows of A where the coefficients are given by the entries of the last row of M. The last row of M is

```
Vec({0, 1, 2, 3, 4, 5, 6, 7, 8, 9, 10, 11},{0: 0, 1: one, 2: one, 4: 0,
    5: one, 11: one})
```

Note that entries 1, 2, 5, and 11 are nonzero, which tells us that the sum of the corresponding rows of `rowlist` is the zero vector. That tells us that these rows correspond to the factorizations of numbers whose product is a perfect square. The numbers are: 285, 390, 1425, and 3822. Their product is 605361802500, which is indeed a perfect square: it is the square of 778050. We therefore set $b = 778050$. We set a to be the product of the corresponding values of x (52, 53, 62, and 79), which is 1395 498888. The greatest common divisor of $a - b$ and N is, uh, 1. Oops, we were unlucky–it didn't work.

Was all that work for nothing? It turns out we were not so unlucky. The rank of the matrix A could have been *len(rowlist)* but turned out to be somewhat less. Consequently, the second-to-last row of MA is also a zero vector. The second-to-last vector of M is

```
Vec({0, 1, 2, 3, 4, 5, 6, 7, 8, 9, 10, 11},{0: 0, 1: 0, 10: one, 2: one})
```

Note that entries 10 and 2 are nonzero, which tells us that combining row 2 of `rowlist` (the row corrresponding to 53) with row 10 of `rowlist` (the row corresponding to 77) will result in a perfect square.

Task 7.8.10: Define a procedure `find_a_and_b(v, roots, N)` that, given a vector $\boldsymbol{v}$ (one of the rows of M), the list `roots`, and the integer N to factor, computes a pair (a, b) of integers such that $a^2 - b^2$ is a multiple of N.

Your procedure should work as follows:

- Let `alist` be the list of elements of `roots` corresponding to nonzero entries of the vector $\boldsymbol{v}$. (Use a comprehension.)
- Let a be the product of these. (Use the procedure `prod(alist)` defined in the module `factoring`.)
- Similarly, let c be the product of $\{x \cdot x - N : x \in \text{alist}\}$.
- Let b be `intsqrt(c)`.
- Verify using an assertion that `b*b == c`
- Return the pair (a, b).

Try out your procedure with $\boldsymbol{v}$ being the last row of M. See if $a - b$ and N have a nontrivial common divisor. If it doesn't work, try it with $\boldsymbol{v}$ being the second-to-last row of M, etc.

Finally, you will try the above strategy on larger integers.

Task 7.8.11: Let $N = 2461799993978700679$, and try to factor N

- Let *primelist* be the set of primes up to 10000.
- Use `find_candidates(N, primelist)` to compute the lists `roots` and `rowlist`.
- Use `echelon.transformation_rows(rowlist)` to get a matrix M.
- Let v be the last row of M, and find a and b using `find_a_and_b(v, roots, N)`.
- See if $a - b$ has a nontrivial common divisor with N. If not, repeat with v being the second-to-last row of M or the third-to-last row....

Give a nontrivial divisor of N.

Task 7.8.12: Let $N = 20672783502493917028427$, and try to factor N. This time, since N is a lot bigger, finding $K + 1$ rows will take a lot longer, perhaps six to ten minutes depending on your computer. Finding M could take a few minutes.

Task 7.8.13: Here is a way to speed up finding M: The procedure `echelon.transformation_rows` takes an optional third argument, a list of column-labels. The list instructs the procedure in which order to handle column-labels. The procedure works much faster if the list consists of the primes of `primeset` in descending order:

```
>>> M_rows = echelon.transformation_rows(rowlist,
                                         sorted(primeset, reverse=True))
```

Why should the order make a difference? Why does this order work well? *Hint:* a large prime is less likely than a small prime to belong to the factorization of an integer.

7.9 Review questions

- What is echelon form?
- What can we learn about the rank of a matrix in echelon form?
- How can a matrix be converted into echelon form by multiplication by an invertible matrix?
- How can Gaussian elimination be used to find a basis for the null space of a matrix?
- How can Gaussian elimination be used to solve a matrix-vector equation when the matrix is invertible?

Problems

Practice on Gaussian elimination

Problem 7.9.1: Carry out Gaussian elimination by hand for the following matrix over $GF(2)$. Handle the columns in the order A,B,C,D. For each of the column-labels, tell us

- which row you select as the pivot row,
- which rows the pivot row is added to (if any), and
- what the resulting matrix is.

Finally, reorder the rows of the resulting matrix to obtain a matrix in echelon form.

Note: Remember that each row is only used once as a pivot-row, and that the pivot-row for column c must have a nonzero value for that column. Remember that the matrix is over $GF(2)$.

	A	B	C	D
0	*one*	*one*	0	0
1	*one*	0	*one*	0
2	0	*one*	*one*	*one*
3	*one*	0	0	0

Recognizing echelon form

Problem 7.9.2: Each of the matrices given below is almost in echelon form; replace the *MINIMUM* number of elements with zeroes to obtain a matrix in echelon form. You are not allowed to reorder the rows or columns of the matrix. (Note: you don't need to actually do any steps of Gaussian elimination for this problem.)

Example: Given the matrix

$$\begin{bmatrix} 1 & 2 & 3 & 4 \\ 9 & 2 & 3 & 4 \\ 0 & 0 & 3 & 4 \\ 0 & 8 & 0 & 4 \end{bmatrix}$$

you would replace the 9 and 8 with zeroes, so you answer would be the matrix

$$\begin{bmatrix} 1 & 2 & 3 & 4 \\ 0 & 2 & 3 & 4 \\ 0 & 0 & 3 & 4 \\ 0 & 0 & 0 & 4 \end{bmatrix}$$

Okay, here are the problems:

1. $\begin{bmatrix} 1 & 2 & 0 & 2 & 0 \\ 0 & 1 & 0 & 3 & 4 \\ 0 & 0 & 2 & 3 & 4 \\ 1 & 0 & 0 & 2 & 0 \\ 0 & 3 & 0 & 0 & 4 \end{bmatrix}$

2. $\begin{bmatrix} 0 & 4 & 3 & 4 & 4 \\ 6 & 5 & 4 & 2 & 0 \\ 0 & 0 & 0 & 0 & 1 \\ 0 & 0 & 0 & 0 & 2 \end{bmatrix}$

3. $\begin{bmatrix} 1 & 0 & 0 & 1 \\ 1 & 0 & 0 & 1 \\ 0 & 0 & 0 & 1 \end{bmatrix}$

4. $\begin{bmatrix} 1 & 0 & 0 & 0 \\ 0 & 1 & 0 & 0 \\ 1 & 1 & 0 & 0 \\ 0 & 0 & 0 & 1 \end{bmatrix}$

Problem 7.9.3: Write a procedure `is_echelon(A)` that takes a matrix in list of row lists and returns `True` if it is echelon form, and `False` otherwise.

Try out your procedure on the following matrices:

$$\begin{bmatrix} 2 & 1 & 0 \\ 0 & -4 & 0 \\ 0 & 0 & 1 \end{bmatrix} \text{(True)}, \begin{bmatrix} 2 & 1 & 0 \\ -4 & 0 & 0 \\ 0 & 0 & 1 \end{bmatrix} \text{(False)}, \begin{bmatrix} 2 & 1 & 0 \\ 0 & 3 & 0 \\ 1 & 0 & 1 \end{bmatrix} \text{(False)}, \begin{bmatrix} 1 & 1 & 1 & 1 & 1 \\ 0 & 2 & 0 & 1 & 3 \\ 0 & 0 & 0 & 5 & 3 \end{bmatrix} \text{(True)}$$

Solving a matrix-vector equation when the matrix is in echelon form

Consider solving the matrix-vector equation

	a	b	c	d	e
0	1		1		
1		2		3	
2				1	9

$$* [x_a, x_b, x_c, x_d, x_e] = [1, 1, 1]$$

where the matrix is in echelon form. The algorithm for solving this is very similar to the algorithm for solving a triangular system. The difference is that the algorithm disregards some of the columns. In particular, any column not containing the leftmost nonzero entry of some row is disregarded.

For the above example, the algorithm should disregard columns c and e. Column a contains the leftmost nonzero in row 0, column b contains the leftmost nonzero in row 1, and column d contains the leftmost nonzero in row 2.

Problem 7.9.4: For each of the following matrix-vector equations, find the solution:

(a) $\begin{bmatrix} 10 & 2 & -3 & 53 \\ 0 & 0 & 1 & 2013 \end{bmatrix} * [x_1, x_2, x_3, x_4] = [1, 3]$

(b) $\begin{bmatrix} 2 & 0 & 1 & 3 \\ 0 & 0 & 5 & 3 \\ 0 & 0 & 0 & 1 \end{bmatrix} * [x_1, x_2, x_3, x_4] = [1, -1, 3]$

(c) $\begin{bmatrix} 2 & 2 & 4 & 3 & 2 \\ 0 & 0 & -1 & 11 & 1 \\ 0 & 0 & 0 & 0 & 5 \end{bmatrix} * [x_1, x_2, x_3, x_4, x_5] = [2, 0, 10]$

The examples above have no rows that are zero. What do we do when the matrix has some rows that are zero? Ignore them!

Consider an equation $\boldsymbol{a}_i \cdot \boldsymbol{x} = b_i$ where $\boldsymbol{a}_i = \boldsymbol{0}$.

- If $b_i = 0$ then the equation is true regardless of the choice of $\boldsymbol{x}$.
- If $b_i \neq 0$ then the equation is false regardless of the choice of $\boldsymbol{x}$.

Thus the only disadvantage of ignoring the rows that are zero is that the algorithm will not notice that the equations cannot be solved.

Problem 7.9.5: For each of the following matrix-vector equations, say whether the equation has a solution. If so, compute the solution.

(a) $\begin{bmatrix} 1 & 3 & -2 & 1 & 0 \\ 0 & 0 & 2 & -3 & 0 \\ 0 & 0 & 0 & 0 & 0 \end{bmatrix} * [x_1, x_2, x_3, x_4, x_5] = [5, 3, 2]$

(b) $\begin{bmatrix} 1 & 2 & -8 & -4 & 0 \\ 0 & 0 & 2 & 12 & 0 \\ 0 & 0 & 0 & 0 & 0 \\ 0 & 0 & 0 & 0 & 0 \end{bmatrix} * [x_1, x_2, x_3, x_4, x_5] = [5, 4, 0, 0]$

Problem 7.9.6: Give a procedure echelon_solve(rowlist, label_list, b) with the following spec:

- *input:* for some integer n, a matrix in echelon form represented by a list rowlist of n vectors, a list of column-labels giving the order of the columns of the matrix (i.e. the domain of the vectors), and a length-n list $\boldsymbol{b}$ of field elements
- *output:* a vector $\mathbf{x}$ such that, for $i = 0, 1, \ldots, n-1$, the dot-product of rowlist[i] with $\mathbf{x}$ equals $\boldsymbol{b}[i]$ if rowlist[i] is not a zero vector

Obviously your code should not use the solver module.

If you wanted to use this procedure with floating-point numbers, the procedure would have to interpret very small numbers as zero. To avoid this issue, you should assume that the field is $GF(2)$. (The interested student can modify the solution to work for $\mathbb{R}$.)

Hints for your implementation:

- The slickest way to write this procedure is to adapt the code of the procedure triangular_solve(rowlist, label_list, b) in module triangular. As in that procedure, initialize a vector x to zero, then iterate through the rows of rowlist from last row to first row; in each iteration, assign to an entry of x. In this procedure, however, you must assign to the variable corresponding to the column containing the first nonzero entry in that row. (If there are no nonzero entries in that row, the iteration should do nothing.)

  This approach leads to a very simple implementation consisting of about seven lines. The code closely resembles that for triangular_solve.

- For those for whom the above approach does not make sense, here is an alternative approach that leads to about twice as many lines of code. Form a new matrix-vector equation by removing

  - zero rows and
  - irrelevant columns,

  and then use the procedure triangular_solve in module triangular to solve the new matrix-vector equation.

  Removing zero rows: When you remove zero rows from the matrix, you must remove the corresponding entries from the right-hand side vector b. Because the matrix is in echelon form, all the zero rows come at the end. Find out which rows are zero, and form a new rowlist from the old that omits those rows; form a new right-hand side vector b by removing the corresponding entries.

  Removing irrelevant columns: For each of the remaining rows, find out the position of the leftmost nonzero entry. Then remove the columns that do not contain any leftmost nonzero entries.

  Let $A\boldsymbol{x} = \boldsymbol{b}$ be the original matrix-vector equation, and let $\hat{A}\hat{\boldsymbol{x}} = \hat{\boldsymbol{b}}$ be the one resulting from these operations. Finally, solve $\hat{A}\hat{\boldsymbol{x}} = \hat{\boldsymbol{b}}$ using triangular_solve, and let $\hat{\boldsymbol{u}}$ be the solution. The domain of $\hat{\boldsymbol{u}}$ equals the column-label set of $\hat{A}$ rather than that of A. From $\hat{\boldsymbol{u}}$, construct a solution $\boldsymbol{u}$ to $A\boldsymbol{x} = \boldsymbol{b}$. The domain of $\boldsymbol{u}$ is the column-label set of A, and the extra entries of $\boldsymbol{u}$ (the ones not given by $\hat{\boldsymbol{u}}$) are set to zero (easy because of our sparsity convention).

Here are some examples to test your procedure.

'A'	'B'	'C'	'D'	'E'
one	0	one	one	0
0	one	0	0	one
0	0	one	0	one
0	0	0	0	one

  and b = [one, 0, one, one].

  The solution is

'A'	'B'	'C'	'D'	'E'
one	one	0	0	one

'A'	'B'	'C'	'D'	'E'
one	one	0	one	0
0	one	0	one	one
0	0	one	0	one
0	0	0	0	0

  and b = [one, 0, one, 0].

  The solution is

'A'	'B'	'C'	'D'	'E'
one	0	one	0	0

Problem 7.9.7: Now that you have developed a procedure for solving a matrix-vector equation where the matrix is in echelon form, you can use this in a procedure for the general case. We have already described the method. The code for the method is

```
def solve(A, b):
    M = echelon.transformation(A)
    U = M*A
    col_label_list = sorted(A.D[1])
    U_rows_dict = mat2rowdict(U)
    rowlist = [U_rows_dict[i] for i in U_rows_dict]
    return echelon_solve(rowlist,col_label_list, M*b)
```

(Sorting is problematic when the column-labels include values of different types, e.g. ints and strings.)

Suppose you have the matrix

$A =$

	A	B	C	D
a	one	one	0	one
b	one	0	0	one
c	one	one	one	one
d	0	0	one	one

and the right-hand side vector g=

a	b	c	d
one	0	one	0

You are interested in finding the solution to the matrix-vector equation $Ax = g$. The first step in using Gaussian elimination to find the solution is to find a matrix M such that MA is in echelon form.

In this case, $M =$

	a	b	c	d
0	one	0	0	0
1	one	one	0	0
2	one	0	one	0
3	one	0	one	one

and $MA =$

	A	B	C	D
0	one	one	0	one
1	0	one	0	0
2	0	0	one	0
3	0	0	0	one

Use the above data and above procedure to figure out not the solution but what actual arguments should be provided to `echelon_solve` in order to obtain the solution to the original matrix-vector equation.

Finding a basis for $\{u \ : \ u * A = 0\} = A^T$

Problem 7.9.8: We consider matrices over $GF(2)$. Let

$A =$

	A	B	C	D	E
a	0	0	0	one	0
b	0	0	0	one	one
c	one	0	0	one	0
d	one	0	0	0	one
e	one	0	0	0	0

.

Then the matrix

$M =$

	a	b	c	d	e
0	0	0	one	0	0
1	one	0	0	0	0
2	one	one	0	0	0
3	0	one	one	one	0
4	one	0	one	0	one

has the property that MA is a matrix in echelon form, namely

$MA =$

	A	B	C	D	E
0	one	0	0	one	0
1	0	0	0	one	0
2	0	0	0	0	one
3	0	0	0	0	0
4	0	0	0	0	0

List the rows $\boldsymbol{u}$ of M such that $\boldsymbol{u} * A = \mathbf{0}$. (Note that these are vectors in the null space of the transpose A^T.)

Problem 7.9.9: We consider matrices over $GF(2)$. Let

$A =$

	A	B	C	D	E
a	0	0	0	one	0
b	0	0	0	one	one
c	one	0	0	one	0
d	one	one	one	0	one
e	one	0	0	one	0

Then the matrix

$M =$

	a	b	c	d	e
0	0	0	one	0	0
1	0	0	one	one	0
2	one	0	0	0	0
3	one	one	0	0	0
4	0	0	one	0	one

has the property that MA is a matrix in echelon form, namely

$MA =$

A	B	C	D	E	
0	one	0	0	one	0
1	0	one	one	one	one
2	0	0	0	one	0
3	0	0	0	0	one
4	0	0	0	0	0

List the rows $\boldsymbol{u}$ of M such that $\boldsymbol{u} * A = \mathbf{0}$. (Note that these are vectors in the null space of the transpose A^T.)

Chapter 8

The Inner Product

> The purpose of computing is insight, not numbers.
>
> Richard Hamming

In this chapter, we learn how the notions of *length* and *perpendicular* are interpreted in Mathese. We study the problem of finding the point on a given line closest to a given point. In the next chapter, we study a generalization of this problem.

8.1 The *fire engine* problem

There is a burning house located at coordinates $[2, 4]$!A street runs near the house, along the line through the origin and through $[6, 2]$—but it is near enough? The fire engine has a hose only three and a half units long. If we can navigate the fire engine to the point on the line nearest the house, will the distance be small enough to save the house?

We're faced with two questions: what point along the line is closest to the house, and how far is it?

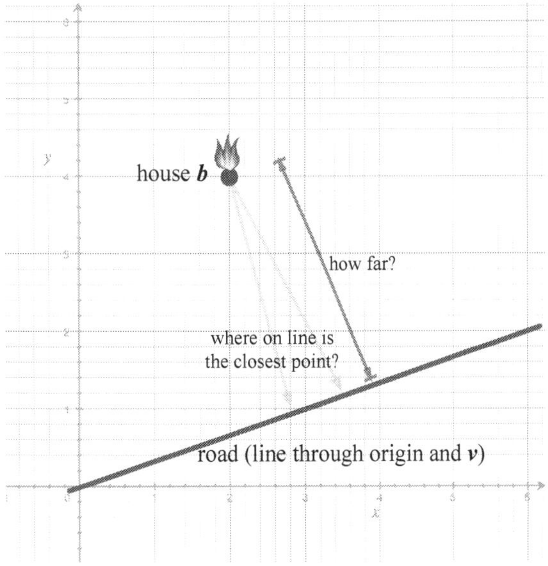

Let's formulate this as a computational problem. Recall from Section 2.5.3 that a line through the origin can be represented as the set of scalar multiples of a vector; in our example, the street runs along the line $\{\alpha[3, 1] \ : \ \alpha \in \mathbb{R}\}$. The *fire engine* problem can therefore be formulated as follows.

Computational Problem 8.1.1: *The vector in the span of one given vector closest to another given vector,* a.k.a. *fire-engine problem:*

- *input:* vectors $\boldsymbol{v}$ and $\boldsymbol{b}$
- *output:* the point on the line $\{\alpha \boldsymbol{v} \ : \ \alpha \in \mathbb{R}\}$ that is closest to $\boldsymbol{b}$

This problem is not yet completely formulated because we have not said what we mean by *closest*.

8.1.1 Distance, length, norm, inner product

We will define the distance between two vectors $\boldsymbol{p}$ and $\boldsymbol{b}$ to be the length of the difference $\boldsymbol{p} - \boldsymbol{b}$. This means that we must define the length of a vector. Instead of using the term "length" for vectors, we typically use the term *norm*. The norm of a vector $\boldsymbol{v}$ is written $\|\boldsymbol{v}\|$. Since it plays the role of length, it should satisfy the following *norm properties*:

Property N1: For any vector $\boldsymbol{v}$, $\|\boldsymbol{v}\|$ is a nonnegative real number.

Property N2: For any vector $\boldsymbol{v}$, $\|\boldsymbol{v}\|$ is zero if and only if $\boldsymbol{v}$ is a zero vector.

Property N3: For any vector $\boldsymbol{v}$ and any scalar α, $\|\alpha\, \boldsymbol{v}\| = |\alpha| \|\boldsymbol{v}\|$.

Property N4: For any vectors $\boldsymbol{u}$ and $\boldsymbol{v}$, $\|\boldsymbol{u} + \boldsymbol{v}\| \leq \|\boldsymbol{u}\| + \|\boldsymbol{v}\|$

One way to define vector norm is to define an operation on vectors called *inner product*. The notation for the inner product of vectors $\boldsymbol{u}$ and $\boldsymbol{v}$ is

$$\langle \boldsymbol{u}, \boldsymbol{v} \rangle$$

The inner product must satisfy certain axioms, which we outline later.

It turns out, however, that there is no way to define inner product for $GF(2)$ so that it satisfies the axioms. We will therefore regretfully leave aside $GF(2)$ for the remainder of the book.

For the real numbers and complex numbers, one has some flexibility in defining the inner product—but there is one most natural and convenient way of defining it, one that leads to the norm of a vector over the reals being the length (in the geometrical sense) of the arrow representing the vector. (Some advanced applications, not covered in this book, require more complicated inner products.)

Once we have defined an inner product, the norm of a vector $\boldsymbol{u}$ is defined by

$$\|\boldsymbol{v}\| = \sqrt{\langle \boldsymbol{v}, \boldsymbol{v} \rangle} \tag{8.1}$$

8.2 The inner product for vectors over the reals

Our inner product for vectors over $\mathbb{R}$ is defined as the dot-product:

$$\langle \boldsymbol{u}, \boldsymbol{v} \rangle = \boldsymbol{u} \cdot \boldsymbol{v}$$

Some algebraic properties of the inner product for vectors over the reals follow easily from properties of the dot-product (bilinearity, homogeneity, symmetry):

- *linearity in the first argument:* $\langle \boldsymbol{u} + \boldsymbol{v}, \boldsymbol{w} \rangle = \langle \boldsymbol{u}, \boldsymbol{w} \rangle + \langle \boldsymbol{v}, \boldsymbol{w} \rangle$
- *symmetry:* $\langle \boldsymbol{u}, \boldsymbol{v} \rangle = \langle \boldsymbol{v}, \boldsymbol{u} \rangle$
- *homogeneity:* $\langle \alpha\, \boldsymbol{u}, \boldsymbol{v} \rangle = \alpha\ \langle \boldsymbol{u}, \boldsymbol{v} \rangle$

8.2.1 Norms of vectors over the reals

Let's see what the resulting norm function looks like:

$$\|\boldsymbol{v}\| = \sqrt{\langle \boldsymbol{v}, \boldsymbol{v} \rangle}$$

Suppose $\boldsymbol{v}$ is an n-vector, and write $\boldsymbol{v} = [v_1, \ldots, v_n]$. Then

$$\begin{aligned}\|\boldsymbol{v}\|^2 &= \langle \boldsymbol{v}, \boldsymbol{v} \rangle = \boldsymbol{v} \cdot \boldsymbol{v} \\ &= v_1^2 + \cdots + v_n^2\end{aligned}$$

More generally, if $\boldsymbol{v}$ is a D-vector,

$$\|\boldsymbol{v}\|^2 = \sum_{i \in D} v_i^2$$

so $\|\boldsymbol{v}\| = \sqrt{\sum_{i \in D} v_i^2}$.

Does this norm satisfy the *norm properties* of Section 8.1.1?

1. The first property states that $\|\boldsymbol{v}\|$ is a real number. Is this true for every vector $\boldsymbol{v}$ over the reals? Every entry v_i is real, so its square v_i^2 is a nonnegative real number. The sum of squares is a nonnegative real number, so $\|\boldsymbol{v}\|$ is the square root of a nonnegative real number, so $\|\boldsymbol{v}\|$ is a nonnegative real number.
2. The second property states that $\|\boldsymbol{v}\|$ is zero if and only if $\boldsymbol{v}$ is a zero vector. If $\boldsymbol{v}$ is a zero vector then every entry is zero, so the sum of squares of entries is also zero. On the other hand, if $\boldsymbol{v}$ is not a zero vector then there is at least one nonzero entry v_i. Since $\|\boldsymbol{v}\|^2$ is the sum of squares, there is no cancellation—since at least one of the terms is positive, the sum is positive. Therefore $\|\boldsymbol{v}\|$ is positive in this case.
3. The third property states that, for any scalar α, $\|\alpha\, \boldsymbol{v}\| = |\alpha| \|\boldsymbol{v}\|$. Let's check this property:

$$\begin{aligned}\|\alpha\, \boldsymbol{v}\|^2 &= \langle \alpha\, \boldsymbol{v}, \alpha\, \boldsymbol{v} \rangle && \text{by definition of norm} \\ &= \alpha\, \langle \boldsymbol{v}, \alpha\, \boldsymbol{v} \rangle && \text{by homogeneity of inner product} \\ &= \alpha\, (\alpha\, \langle \boldsymbol{v}, \boldsymbol{v} \rangle) && \text{by symmetry and homogeneity (again) of inner product} \\ &= \alpha^2 \|\boldsymbol{v}\|^2 && \text{by definition of norm}\end{aligned}$$

Thus $\|\alpha\, \boldsymbol{v}\| = \alpha\, \|\boldsymbol{v}\|$.

Example 8.2.1: Consider the example of 2-vectors. What is the length of the vector $\boldsymbol{u} = [u_1, u_2]$? Remember the Pythagorean Theorem: for a right triangle with side-lengths a, b, c, where c is the length of the hypotenuse,

$$a^2 + b^2 = c^2 \tag{8.2}$$

We can use this equation to calculate the length of $\boldsymbol{u}$:

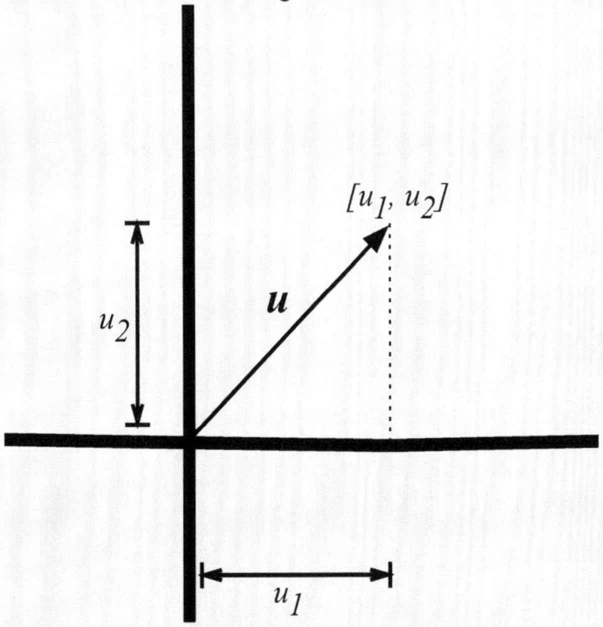

$$(\text{length of } \boldsymbol{u})^2 = u_1^2 + u_2^2$$

So this notion of length agrees with the one we learned in grade school, at least for vectors in $\mathbb{R}^2$.

8.3 Orthogonality

Orthogonal is Mathese for *perpendicular.*

Before giving the definition, I'll motivate it. We will use the Pythagorean Theorem in reverse: we will define the notion of orthogonality so that the Pythagorean Theorem holds. o Let $\boldsymbol{u}$ and $\boldsymbol{v}$ be vectors. Their lengths are $\|\boldsymbol{u}\|$ and $\|\boldsymbol{v}\|$. Think of these vectors as translations, and place the tail of $\boldsymbol{v}$ at the head of $\boldsymbol{u}$. Then the "hypotenuse" is the vector from the tail of $\boldsymbol{u}$ to the head of $\boldsymbol{v}$, which is $\boldsymbol{u}+\boldsymbol{v}$. (The triangle is not necessarily a right angle.)

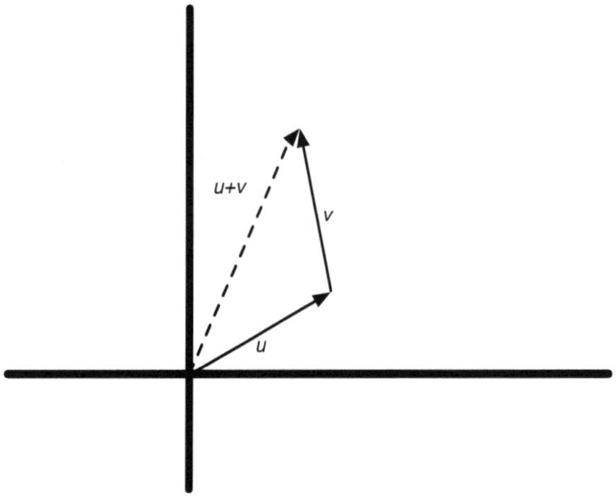

The squared length of the vector $\boldsymbol{u}+\boldsymbol{v}$ (the "hypotenuse") is

$$\begin{aligned}
\|\boldsymbol{u}+\boldsymbol{v}\|^2 &= \langle \boldsymbol{u}+\boldsymbol{v}, \boldsymbol{u}+\boldsymbol{v}\rangle \\
&= \langle \boldsymbol{u}, \boldsymbol{u}+\boldsymbol{v}\rangle + \langle \boldsymbol{v}, \boldsymbol{u}+\boldsymbol{v}\rangle && \text{by linearity of inner product in the first argument} \\
&= \langle \boldsymbol{u}, \boldsymbol{u}\rangle + \langle \boldsymbol{u}, \boldsymbol{v}\rangle + \langle \boldsymbol{v}, \boldsymbol{u}\rangle + \langle \boldsymbol{v}, \boldsymbol{v}\rangle && \text{by symmetry and linearity} \\
&= \|\boldsymbol{u}\|^2 + 2\langle \boldsymbol{u}, \boldsymbol{v}\rangle + \|\boldsymbol{v}\|^2 && \text{by symmetry}
\end{aligned}$$

The last expression is $\|\boldsymbol{u}\|^2 + \|\boldsymbol{v}\|^2$ if and only if $\langle \boldsymbol{u}, \boldsymbol{v}\rangle = 0$.

We therefore define $\boldsymbol{u}$ and $\boldsymbol{v}$ to be *orthogonal* if $\langle \boldsymbol{u}, \boldsymbol{v}\rangle = 0$. From the reasoning above, we obtain:

Theorem 8.3.1 (Pythagorean Theorem for vectors over the reals): If vectors $\boldsymbol{u}$ and $\boldsymbol{v}$ over the reals are orthogonal then

$$\|\boldsymbol{u}+\boldsymbol{v}\|^2 = \|\boldsymbol{u}\|^2 + \|\boldsymbol{v}\|^2$$

8.3.1 Properties of orthogonality

To solve the *fire engine* problem, we will use the Pythagorean Theorem in conjunction with the following simple observations:

Lemma 8.3.2 (Orthogonality Properties): For any vectors $\boldsymbol{u}$ and $\boldsymbol{v}$ and any scalar α,

Property O1: If $\boldsymbol{u}$ is orthogonal to $\boldsymbol{v}$ then $\alpha\,\boldsymbol{u}$ is orthogonal to $\alpha\,\boldsymbol{v}$ for every scalar α.

Property O2: If $\boldsymbol{u}$ and $\boldsymbol{v}$ are both orthogonal to $\boldsymbol{w}$ then $\boldsymbol{u}+\boldsymbol{v}$ is orthogonal to $\boldsymbol{w}$.

Proof

1. $\langle \boldsymbol{u}, \alpha\, \boldsymbol{v}\rangle = \alpha\, \langle \boldsymbol{u}, \boldsymbol{v}\rangle = \alpha\, 0 = 0$
2. $\langle \boldsymbol{u}+\boldsymbol{v}, \boldsymbol{w}\rangle = \langle \boldsymbol{u}, \boldsymbol{w}\rangle + \langle \boldsymbol{v}, \boldsymbol{w}\rangle = 0+0$

□

Lemma 8.3.3: If $\boldsymbol{u}$ is orthogonal to $\boldsymbol{v}$ then, for any scalars α, β,

$$\|\alpha\, \boldsymbol{u} + \beta\, \boldsymbol{v}\|^2 = \alpha_1^2 \|\boldsymbol{u}\|^2 + \beta^2 \|\boldsymbol{v}\|^2$$

Proof

$$\begin{aligned}(\alpha\, \boldsymbol{u} + \beta\, \boldsymbol{v}) \cdot (\alpha\, \boldsymbol{u} + \beta\, \boldsymbol{v}) &= \alpha\, \boldsymbol{u} \cdot \alpha\, \boldsymbol{u} + \beta\, \boldsymbol{v} \cdot \beta\, \boldsymbol{v} + \alpha\, \boldsymbol{u} \cdot \beta\, \boldsymbol{v} + \beta\, \boldsymbol{v} \cdot \alpha\, \boldsymbol{u} \\ &= \alpha\, \boldsymbol{u} \cdot \alpha\, \boldsymbol{u} + \beta\, \boldsymbol{v} \cdot \beta\, \boldsymbol{v} + \alpha\beta\, (\boldsymbol{u} \cdot \boldsymbol{v}) + \beta\alpha\, (\boldsymbol{v} \cdot \boldsymbol{u}) \\ &= \alpha\, \boldsymbol{u} \cdot \alpha\, \boldsymbol{u} + \beta\, \boldsymbol{v} \cdot \beta\, \boldsymbol{v} + 0 + 0 \\ &= \alpha^2\, \|\boldsymbol{u}\|^2 + \beta^2\, \|\boldsymbol{v}\|^2\end{aligned}$$

□

Problem 8.3.4: Demonstrate using a numerical example that Lemma 8.3.3 would not be true if we remove the requirement that $\boldsymbol{u}$ and $\boldsymbol{v}$ are orthogonal.

Problem 8.3.5: Using induction and Lemma 8.3.3, prove the following generalization: Suppose $\boldsymbol{v}_1, \ldots, \boldsymbol{v}_n$ are mutually orthogonal. For any coefficients $\alpha_1, \ldots, \alpha_n$,

$$\|\alpha_1 \boldsymbol{v}_1 + \cdots + \alpha_n \boldsymbol{v}_n\|^2 = \alpha_1^2 \|\boldsymbol{v}_1\|^2 + \cdots + \alpha_n^2 \|\boldsymbol{v}_n\|^2$$

8.3.2 Decomposition of b into parallel and perpendicular components

In order to state the solution to the *fire engine* problem, we first introduce a key concept.

Definition 8.3.6: For any vector $\boldsymbol{b}$ and any vector $\boldsymbol{v}$, define vectors $\boldsymbol{b}^{\|\boldsymbol{v}}$ and $\boldsymbol{b}^{\perp \boldsymbol{v}}$ to be, respectively, the *projection of* $\boldsymbol{b}$ *along* $\boldsymbol{v}$ and the *projection of* $\boldsymbol{b}$ *orthogonal to* $\boldsymbol{v}$ if

$$\boldsymbol{b} = \boldsymbol{b}^{\|\boldsymbol{v}} + \boldsymbol{b}^{\perp \boldsymbol{v}} \tag{8.3}$$

and, for some scalar $\sigma \in R$,

$$\boldsymbol{b}^{\|\boldsymbol{v}} = \sigma\, \boldsymbol{v} \tag{8.4}$$

and

$$\boldsymbol{b}^{\perp \boldsymbol{v}} \text{ is orthogonal to } \boldsymbol{v} \tag{8.5}$$

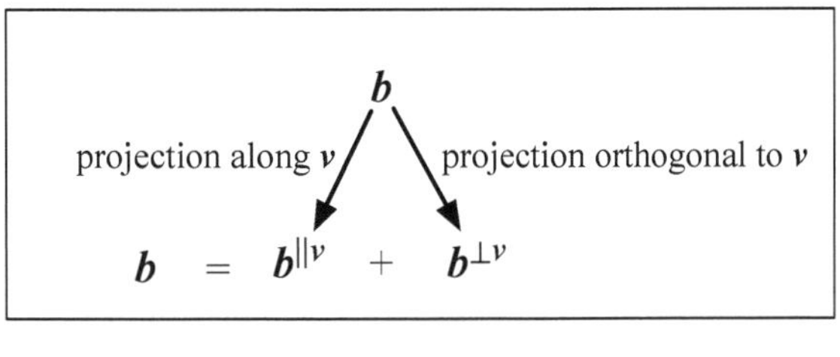

Example 8.3.7: Suppose we are working on the plane, and the line is the x-axis, i.e. the set $\{(x, y) : \ y = 0\}$. Say $\boldsymbol{b}$ is (b_1, b_2), and $\boldsymbol{v}$ is $(1, 0)$. The projection of $\boldsymbol{b}$ along $\boldsymbol{v}$ is $(b_1, 0)$, and the projection of $\boldsymbol{b}$ orthogonal to $\boldsymbol{v}$ is $(0, b_2)$.

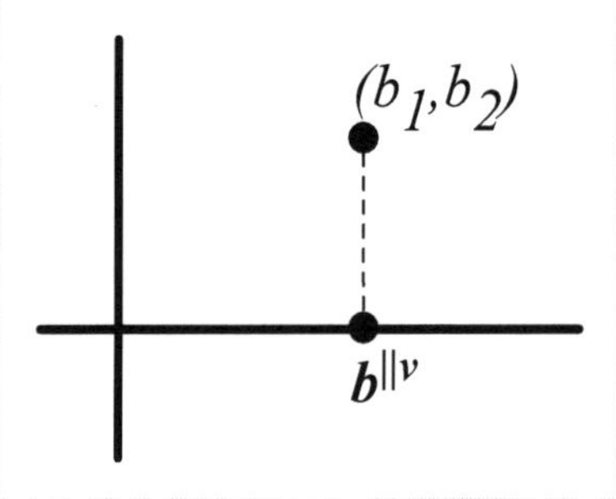

To verify this, consider the equations.

- Equation 8.3 requires that $(b_1, b_2) = (b_1, 0) + (0, b_2)$, which is certainly true.
- Equation 8.4 requires that, for some scalar σ, we have $(b_1, 0) = \sigma\,(1, 0)$, which is true when we choose σ to be b_1.
- Equation 8.5 requires that $(0, b_1)$ is orthogonal to $(1, 0)$, which is clearly true.

When $\boldsymbol{v}$ is the zero vector

What if $\boldsymbol{v}$ is the zero vector? In this case, the only vector $\boldsymbol{b}^{||\boldsymbol{v}}$ satisfying Equation 8.4 is the zero vector. According to Equation 8.3, this would mean that $\boldsymbol{b}^{\perp}$ must equal $\boldsymbol{b}$. Fortunately, this choice of $\boldsymbol{b}^{\perp}$ does satisfy Equation 8.5, i.e. $\boldsymbol{b}^{\perp}$ is orthogonal to $\boldsymbol{v}$. Indeed, *every* vector is orthogonal to $\boldsymbol{v}$ when $\boldsymbol{v}$ is the zero vector.

8.3.3 Orthogonality property of the solution to the *fire engine* problem

Orthogonality helps us solve the *fire engine* problem.

Lemma 8.3.8 (Fire Engine Lemma): Let $\boldsymbol{b}$ and $\boldsymbol{v}$ be vectors. The point in Span $\{\boldsymbol{v}\}$ closest to $\boldsymbol{b}$ is $\boldsymbol{b}^{||\boldsymbol{v}}$, and the distance is $||\boldsymbol{b}^{\perp\boldsymbol{v}}$.

Example 8.3.9: Continuing with Example 8.3.7 (Page 338) Then the lemma states that the point on the line Span $\{(1, 0)\}$ closest to (b_1, b_2) is $\boldsymbol{b}^{||\boldsymbol{v}} = (b_1, 0)$.

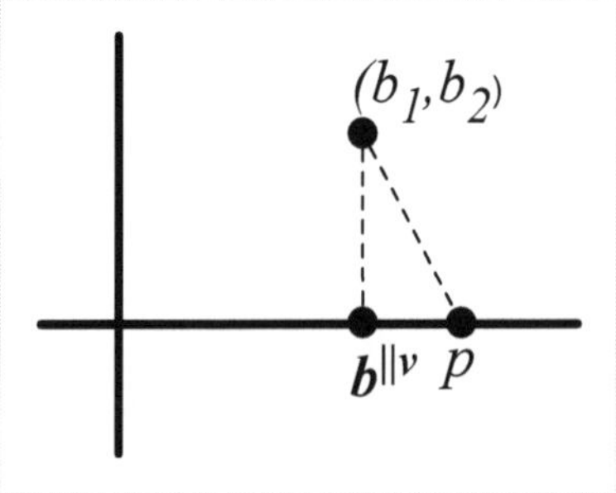

For any other point $\boldsymbol{p}$, the points (b_1, b_2), $\boldsymbol{b}^{||\boldsymbol{v}}$, and $\boldsymbol{p}$ form a right triangle. Since $\boldsymbol{p}$ is different from $\boldsymbol{b}^{||\boldsymbol{v}}$, the base is nonzero, and so, by the Pythagorean Theorem, the hypotenuse's lengh is greater than the height. This shows that $\boldsymbol{p}$ is farther from (b_1, b_2) than $\boldsymbol{b}^{||\boldsymbol{v}}$ is.

The proof uses the same argument as was used in the example. The proof works in any-dimensional space but here is a figure in $\mathbb{R}^2$:

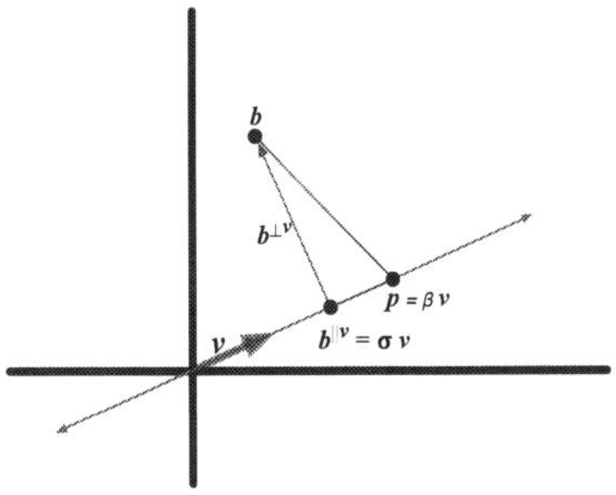

Proof

Let $\boldsymbol{p}$ be any point on $L = \text{Span }\{\boldsymbol{v}\}$. The three points $\boldsymbol{p}$, $\boldsymbol{b}^{||\boldsymbol{v}}$, and $\boldsymbol{b}$ form a triangle. The arrow from $\boldsymbol{p}$ to $\boldsymbol{b}^{||\boldsymbol{v}}$ is $\boldsymbol{b}^{||\boldsymbol{v}} - \boldsymbol{p}$. The arrow from $\boldsymbol{b}^{||\boldsymbol{v}}$ to $\boldsymbol{b}$ is $\boldsymbol{b} - \boldsymbol{b}^{||\boldsymbol{v}}$, which is $\boldsymbol{b}^{\perp \boldsymbol{v}}$. The arrow from $\boldsymbol{p}$ to $\boldsymbol{b}$ is $\boldsymbol{b} - \boldsymbol{p}$.

Since $\boldsymbol{b}^{||\boldsymbol{v}}$ and $\boldsymbol{p}$ are both on L, they are both multiples of $\boldsymbol{v}$, so their difference $\boldsymbol{b}^{||\boldsymbol{v}} - \boldsymbol{p}$ is also a multiple of $\boldsymbol{v}$. Since $\boldsymbol{b} - \boldsymbol{b}^{||\boldsymbol{v}}$ is orthogonal to $\boldsymbol{v}$, therefore, it is also orthogonal to $\boldsymbol{b}^{||\boldsymbol{v}} - \boldsymbol{p}$ by Orthogonality Property 1 (Lemma 8.3.2).

Hence by the Pythagorean Theorem,

$$\|\boldsymbol{b} - \boldsymbol{p}\|^2 = \|\boldsymbol{b}^{||\boldsymbol{v}} - \boldsymbol{p}\|^2 + \|\boldsymbol{b} - \boldsymbol{b}^{||\boldsymbol{v}}\|^2$$

If $\boldsymbol{p} \neq \boldsymbol{b}^{||\boldsymbol{v}}$ then $\|\boldsymbol{b}^{||\boldsymbol{v}} - \boldsymbol{p}\|^2 > 0$ so $\|\boldsymbol{b} - \boldsymbol{b}^{||\boldsymbol{v}}\| < \|\boldsymbol{b} - \boldsymbol{p}\|$.

We have shown that the distance from $\boldsymbol{b}^{||\boldsymbol{v}}\|$ to $\boldsymbol{b}$ is less than the distance to $\boldsymbol{b}$ from any other point on L. The distance is the length of the vector $\boldsymbol{b} - \boldsymbol{b}^{||\boldsymbol{v}}$, which is $\|\boldsymbol{b}^{\perp \boldsymbol{v}}\|$. □

Example 8.3.10: What about the case where $\boldsymbol{v}$ is the zero vector? In this case, L is not a line at all: it is the set consisting just of the zero vector. It is clear that the point in L closest to $\boldsymbol{b}$ is the *only* point in L, namely the zero vector, which is indeed $\boldsymbol{b}^{||\boldsymbol{v}}$ in this case (as discussed at the end of Section 8.3.2). The vector $\boldsymbol{b}^{\perp \boldsymbol{v}}$ is just $\boldsymbol{b}$ in this case, and the distance is $\|\boldsymbol{b}\|$.

8.3.4 Finding the projection and the closest point

Now that we have characterized the solution and given it a new name, how can we actually compute it? It suffices to compute the scalar σ in Equation 8.4. If $\boldsymbol{v}$ is the zero vector, we saw that σ must be zero. Otherwise, we can derive what σ must be from the other two equations. Equation 8.5 requires that $\langle \boldsymbol{b}^{\perp \boldsymbol{v}}, \boldsymbol{v} \rangle = 0$. Using Equation 8.3 to substitute for $\boldsymbol{b}^{\perp \boldsymbol{v}}$, we see that this requires that $\langle \boldsymbol{b} - \boldsymbol{b}^{||\boldsymbol{v}}, \boldsymbol{v} \rangle = 0$. Using Equation 8.4 to substitute for $\boldsymbol{b}^{||\boldsymbol{v}}$, we see that this requires that $\langle \boldsymbol{b} - \sigma\, \boldsymbol{v}, \boldsymbol{v} \rangle = 0$. Using linearity and homogeneity of inner product (Section 8.2), this requires that

$$\langle \boldsymbol{b}, \boldsymbol{v} \rangle - \sigma\, \langle \boldsymbol{v}, \boldsymbol{v} \rangle = 0 \tag{8.6}$$

Solving for σ, we obtain

$$\sigma = \frac{\langle \boldsymbol{b}, \boldsymbol{v} \rangle}{\langle \boldsymbol{v}, \boldsymbol{v} \rangle} \tag{8.7}$$

In the special case in which $\|\boldsymbol{v}\| = 1$, Equation 8.7 can be simplified: the denominator $\langle \boldsymbol{v}, \boldsymbol{v} \rangle = 1$ so

$$\sigma = \langle \boldsymbol{b}, \boldsymbol{v} \rangle \tag{8.8}$$

Beware! We have shown that if $\boldsymbol{b}$, $\boldsymbol{b}^{||\boldsymbol{v}}$, and $\boldsymbol{b}^{\perp \boldsymbol{v}}$ satisfy Definition 8.3.6 then σ must satisfy Equation 8.7. Formally, we must also prove the converse: Equation 8.7 implies that $\boldsymbol{b}^{||\boldsymbol{v}} = \sigma\, \boldsymbol{v}$ and $\boldsymbol{b}^{\perp \boldsymbol{v}} = \boldsymbol{b} - \boldsymbol{b}^{||\boldsymbol{v}}$ satisfy Definition 8.3.6.

The proof is just the reverse of the derivation: Equation 8.7 implies Equation 8.6, which implies that $\langle \boldsymbol{b} - \sigma\, \boldsymbol{v}, \boldsymbol{v} \rangle = 0$, which means that $\boldsymbol{b}^{\perp \boldsymbol{v}}$ is orthogonal to $\boldsymbol{v}$ as required by the definition.

We summarize our conclusions in a lemma.

Lemma 8.3.11: For any vector $\boldsymbol{b}$ and any vector $\boldsymbol{v}$ over the reals.

1. There is a scalar σ such that $\boldsymbol{b} - \sigma\,\boldsymbol{v}$ is orthogonal to $\boldsymbol{v}$.
2. The point $\boldsymbol{p}$ on Span $\{\boldsymbol{v}\}$ that minimizes $\|\boldsymbol{b} - \boldsymbol{p}\|$ is $\sigma\,\boldsymbol{v}$.
3. The value of σ is $\frac{\langle \boldsymbol{b},\boldsymbol{v}\rangle}{\langle \boldsymbol{v},\boldsymbol{v}\rangle}$.

Quiz 8.3.12: In Python, write a procedure `project_along(b, v)` that returns the projection of $\boldsymbol{b}$ onto the span of $\boldsymbol{v}$.

Answer

```
def project_along(b, v):
 sigma = ((b*v)/(v*v)) if v*v != 0 else 0
 return sigma * v
```

The one-line version is

```
def project_along(b, v): return (((b*v)/(v*v)) if v*v != 0 else 0) * v
```

Mathematically, this implementation of `project_along` is correct. However, because of floating-point roundoff error, it is crucial that we make a slight change.

Often the vector $\boldsymbol{v}$ will be not a truly zero vector but for practical purposes will be zero. If the entries of $\boldsymbol{v}$ are tiny, the procedure should treat $\boldsymbol{v}$ as a zero vector: `sigma` should be assigned zero. We will consider $\boldsymbol{v}$ to be a zero vector if its squared norm is no more than, say, 10^{-20}. So here is our revised implementation of `project_along`:

```
def project_along(b, v):
 sigma = (b*v)/(v*v) if v*v > 1e-20 else 0
 return sigma * v
```

Now we build on `project_along` to write a procedure to find the orthogonal projection of $\boldsymbol{b}$.

Quiz 8.3.13: In Python, write a procedure `project_orthogonal_1(b, v)` that returns the projection of $\boldsymbol{b}$ orthogonal to $\boldsymbol{v}$

Answer

```
def project_orthogonal_1(b, v): return b - project_along(b, v)
```

These procedures are defined in the module `orthogonalization`.

8.3.5 Solution to the *fire engine* problem

Example 8.3.14: We return to the *fire engine* problem posed at the beginning of the chapter. In that example, $\boldsymbol{v} = [6, 2]$ and $\boldsymbol{b} = [2, 4]$. The closest point on the line $\{\alpha\,\boldsymbol{v}\ :\ \alpha \in \mathbb{R}\}$ is the point $\sigma\,\boldsymbol{v}$ where

$$\begin{aligned}\sigma &= \frac{\boldsymbol{v}\cdot\boldsymbol{b}}{\boldsymbol{v}\cdot\boldsymbol{v}}\\ &= \frac{6\cdot 2+2\cdot 4}{6\cdot 6+2\cdot 2}\\ &= \frac{20}{40}\\ &= \frac{1}{2}\end{aligned}$$

Thus the point closest to $\boldsymbol{b}$ is $\frac{1}{2}\,[6,2]=[3,1]$. The distance to $\boldsymbol{b}$ is $\|[2,4]-[3,1]\| = \|[-1,3]\| = \sqrt{10}$, which is just under 3.5, the length of the firehose. The house is saved!

The *fire engine* problem can be restated as finding the vector on the line that "best approximates" the given vector $\boldsymbol{b}$. By "best approximation", we just mean closest. This notion of "best approximates" will come up several times in future chapters:

- in least-squares/regression, a fundamental data analysis technique,
- image compression,
- in principal component analysis, another data analysis technique,
- in latent semantic analysis, an information retrieval technique, and
- in compressed sensing.

8.3.6 *Outer product and projection

Recall that the *outer product* of vectors $\boldsymbol{u}$ and $\boldsymbol{v}$ is defined as the matrix-matrix product $\boldsymbol{u}\boldsymbol{v}^T$:

$$\begin{bmatrix} \\ \boldsymbol{u} \\ \\ \end{bmatrix}\begin{bmatrix} & \boldsymbol{v}^T & \end{bmatrix}$$

We use this idea to represent projection in terms of a matrix-vector product.

For a nonzero vector $\boldsymbol{v}$, define the *projection function* $\pi_{\boldsymbol{v}} : \mathbb{R}^n \longrightarrow \mathbb{R}^n$ by

$$\pi_{\boldsymbol{v}}(\boldsymbol{x}) = \text{projection of } \boldsymbol{x} \text{ along } \boldsymbol{v}$$

Can we represent this function as a matrix-vector product? Is it a linear function?

Let us assume for now that $\|\boldsymbol{v}\| = 1$, for, as we have seen, the formula for projection is simpler under this assumption. Then

$$\pi_{\boldsymbol{v}}(\boldsymbol{x}) = (\boldsymbol{v}\cdot\boldsymbol{x})\,\boldsymbol{v}$$

The first step is to represent the function in terms of matrix-matrix multiplications, using the idea of row and column vectors. The dot-product is replaced by a row vector times a column vector.

$$\pi_{\boldsymbol{v}}(\boldsymbol{x}) = \begin{bmatrix} \\ \boldsymbol{v} \\ \\ \end{bmatrix}\left(\begin{bmatrix} & \boldsymbol{v}^T & \end{bmatrix}\begin{bmatrix} \\ \boldsymbol{x} \\ \\ \end{bmatrix}\right) = \underbrace{\left(\begin{bmatrix} \\ \boldsymbol{v} \\ \\ \end{bmatrix}\begin{bmatrix} & \boldsymbol{v}^T & \end{bmatrix}\right)}_{\text{matrix}}\underbrace{\begin{bmatrix} \\ \boldsymbol{x} \\ \\ \end{bmatrix}}_{\text{vector}}$$

This shows that the projection function $\pi_{\boldsymbol{v}}(\boldsymbol{x})$ is a matrix times a vector. Of course, it follows (by Proposition 4.10.2) that therefore the function is linear.

We will later use outer product again, when we consider approximations to matrices.

Problem 8.3.15: Write a Python procedure `projection_matrix(v)` that, given a vector $\boldsymbol{v}$, returns the matrix M such that $\pi_{\boldsymbol{v}}(\boldsymbol{x}) = M\boldsymbol{x}$. Your procedure should be correct even if $\|\boldsymbol{v}\| \neq 1$.

Problem 8.3.16: Suppose $\boldsymbol{v}$ is a nonzero n-vector. What is the rank of the matrix M such that $\pi_{\boldsymbol{v}}(\boldsymbol{x}) = M\boldsymbol{x}$? Explain your answer using appropriate interpretations of matrix-vector or matrix-matrix multiplication.

Problem 8.3.17: Suppose $\boldsymbol{v}$ is a nonzero n-vector. Let M be the matrix such that $\pi_{\boldsymbol{v}}(\boldsymbol{x}) = M\boldsymbol{v}$.

1. How many scalar-scalar multiplications (i.e. ordinary multiplications) are required to multiply M times $\boldsymbol{v}$? Answer the question with a simple formula in terms of n, and justify your answer.
2. Suppose $\boldsymbol{x}$ is represented by a column vector, i.e. an $n \times 1$ matrix. There are two matrices M_1 and M_2 such that computing $\pi_{\boldsymbol{v}}(\boldsymbol{x})$ by computing $M_1(M_2\boldsymbol{x})$ requires only $2n$ scalar-scalar multiplications. Explain.

8.3.7 Towards solving the higher-dimensional version

A natural generalization of the *fire engine* problem is to find the vector in the span of several given vectors that is closest to a given vector $\boldsymbol{b}$. In the next lab, you will explore one approach to this computational problem, based on *gradient descent*. In the next chapter, we will develop an algorithm based on orthogonality and projection.

8.4 *Lab: machine learning*

In this lab you will use a rudimentary machine-learning algorithm to learn to diagnose breast cancer from features.

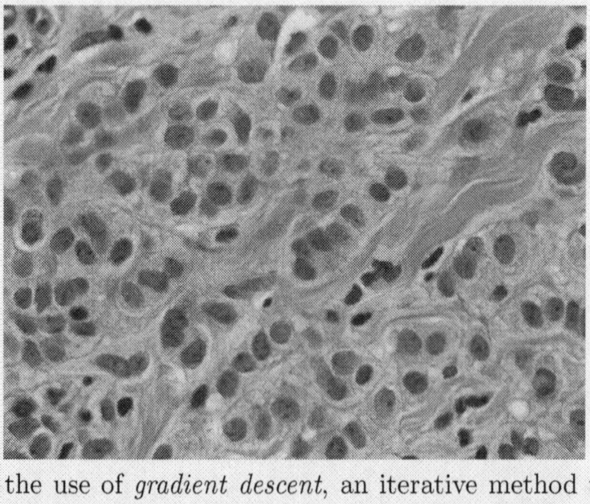

The core idea is the use of *gradient descent*, an iterative method to find a "best" hypothesis. Gradient descent is useful in finding a point that nearly minimizes a nonlinear function. In each iteration, it approximates the function by a linear function.

Disclaimer: For this particular function, there is a much faster and more direct way of finding the best point. We will learn it in Orthogonalization. However, gradient descent is useful more generally and is well worth knowing.

8.4.1 *The data*

You are given part of the Wisconsin Diagnostic Breast Cancer (WDBC) dataset. For each patient, you are given a vector $\boldsymbol{a}$ giving features computed from digitized images of a fine needle aspirate of a breast mass for that patient. The features describe characteristics of the cell nuclei present in the image. The goal is to decide whether the cells are malignant or benign.

Here is a brief description of the way the features were computed. Ten real-valued quantities are computed for each cell nucleus:

- radius (mean of distances from center to points on the perimeter)
- texture (standard deviation of gray-scale values)
- perimeter
- area
- smoothness (local variation in radius lengths)
- compactness (perimeter^2/ area)
- concavity (severity of concave portions of the contour)
- concave points (number of concave portions of the contour)
- symmetry
- fractal dimension ("coastline approximation")

The mean, standard error, and a measure of the largest (mean of the three largest values) of these features were computed for each image. Thus each specimen is represented by a vector $\boldsymbol{a}$ with thirty entries. The domain D consists of thirty strings identifying these features, e.g. `"radius (mean)"`, `"radius (stderr)"`, `"radius (worst)"`, `"area (mean)"`, and so on.

We provide two files containing data, `train.data` and `validate.data`.

The procedure `read_training_data` in the `cancer_data` module takes a single argument, a string giving the pathname of a file. It reads the data in the specified file and returns a pair $(A, \boldsymbol{b})$ where:

- A is a Mat whose row labels are patient identification numbers and whose column-label set is D
- $\boldsymbol{b}$ is a vector whose domain is the set of patient identification numbers, and $\boldsymbol{b}[r]$ is 1 if the specimen of patient r is malignant and is -1 if the specimen is benign.

Task 8.4.1: Use `read_training_data` to read the data in the file `train.data` into the variables $A, \boldsymbol{b}$.

8.4.2 *Supervised learning*

Your goal is to write a program to select a *classifier*, a function $C(\boldsymbol{y})$ that, given a feature vector $\boldsymbol{a}$, predicts whether the tissue is malignant or benign. To enable the program to select a classifier that is likely to be accurate, the program is provided with *training data* consisting of *labeled examples* $(\boldsymbol{a}_1, b_1), \ldots, (\boldsymbol{a}_m, b_m)$. Each labeled example consists of a feature vector $\boldsymbol{a}_i$ and the corresponding label b_i, which is +1 or -1 (+1 for *malignant*, -1 for *benign*). Once the program has selected a classifier, the classifier is tested for its accuracy on unlabeled feature vectors $\boldsymbol{a}$ for which the correct answers are known.

8.4.3 *Hypothesis class*

A classifier is selected from a set of possible classifiers (the *hypothesis class*). In this case (as is often the case in machine learning), the hypothesis class consists of linear functions $h(\cdot)$ from the space $\mathbb{R}^D$ of feature vectors to $\mathbb{R}$. The classifier is defined in terms of such a function as follows:

$$C(\boldsymbol{y}) = \begin{cases} +1 & \text{if } h(\boldsymbol{y}) \geq 0 \\ -1 & \text{if } h(\boldsymbol{y}) < 0 \end{cases}$$

For each linear function $h : \mathbb{R}^D \longrightarrow \mathbb{R}$, there is a D-vector $\boldsymbol{w}$ such that

$$h(\boldsymbol{y}) = \boldsymbol{w} \cdot \boldsymbol{y}$$

Thus selecting such a linear function amounts to selecting a D-vector $\boldsymbol{w}$. We refer to $\boldsymbol{w}$ as a *hypothesis vector* since choosing $\boldsymbol{w}$ is equivalent to choosing the hypothesis h.

You will write a procedure that calculates, for a given hypothesis vector $\boldsymbol{w}$, the number of labeled examples incorrectly predicted by the classifier that uses function $h(\boldsymbol{y}) = \boldsymbol{w} \cdot \boldsymbol{y}$. To make this easier, you will first write a simple utility procedure.

Task 8.4.2: Write the procedure `signum(u)` with the following spec:

- *input:* a Vec $\boldsymbol{u}$
- *output:* the Vec $\boldsymbol{v}$ with the same domain as $\boldsymbol{u}$ such that

$$\boldsymbol{v}[d] = \begin{cases} +1 & \text{if } \boldsymbol{u}[d] \geq 0 \\ -1 & \text{if } \boldsymbol{u}[d] < 0 \end{cases}$$

For example, `signum(Vec({'A','B'}, {'A':3, 'B':-2}))` is
`Vec({'A', 'B'},{'A': 1, 'B': -1})`

Task 8.4.3: Write the procedure `fraction_wrong(A, b, w)` with the following spec:

- *input:* An $R \times C$ matrix A whose rows are feature vectors, an R-vector $\boldsymbol{b}$ whose entries are $+1$ and -1, and a C-vector $\boldsymbol{w}$
- *output:* The fraction of of row labels r of A such that the sign of (row r of A) $\cdot\, \boldsymbol{w}$ differs from that of $\boldsymbol{b}[r]$.

(Hint: There is a clever way to write this without any explicit loops using matrix-vector multiplication and dot-product and the `signum` procedure you wrote.)

Pick a simple hypothesis vector such as $[1, 1, 1, ..., 1]$ or a random vector of $+1$'s and -1's, and see how well it classifies the data.

8.4.4 *Selecting the classifier that minimizes the error on the training data*

How should the function h be selected? We will define a way of measuring the error of a particular choice of h with respect to the training data, and the program will select the function with the minimum error among all classifiers in the hypothesis class.

The obvious way of measuring the error of a hypothesis is by using the fraction of labeled examples the hypothesis gets wrong, but it is too hard to find the solution that is best with respect to this criterion, so other ways of measuring the error are used. In this lab, we use a very rudimentary measure of error. For each labeled example $(\boldsymbol{a}_i, b_i)$, the error of h on that example is $(h(\boldsymbol{a}_i) - b_i)^2$. If $h(\boldsymbol{a}_i)$ is close to b_i then this error is small. The overall error on the training data is the sum of the errors on each of the labeled examples:

$$(h(\boldsymbol{a}_1) - b_1)^2 + (h(\boldsymbol{a}_2) - b_2)^2 + \cdots + (h(\boldsymbol{a}_m) - b_m)^2$$

Recall that choosing a function $h(\cdot)$ is equivalent to choosing a D-vector $\boldsymbol{w}$ and defining $h(\boldsymbol{y}) = \boldsymbol{y} \cdot \boldsymbol{w}$. The corresponding error is

$$(\boldsymbol{a}_1 \cdot \boldsymbol{w} - b_1)^2 + (\boldsymbol{a}_2 \cdot \boldsymbol{w} - b_2)^2 + \cdots + (\boldsymbol{a}_m \cdot \boldsymbol{w} - b_m)^2$$

Now we can state our goal for the learning algorithm. We define a function $L : \mathbb{R}^D \longrightarrow \mathbb{R}$ by the rule

$$L(\boldsymbol{x}) = (\boldsymbol{a}_1 \cdot \boldsymbol{x} - b_1)^2 + (\boldsymbol{a}_2 \cdot \boldsymbol{x} - b_2)^2 + \cdots + (\boldsymbol{a}_m \cdot \boldsymbol{x} - b_m)^2$$

This function is the *loss* function on the training data. It is used to measure the error of a particular choice of the hypothesis vector $\boldsymbol{w}$. The goal of the learning algorithm is to select the hypothesis vector $\boldsymbol{w}$ that makes $L(\boldsymbol{w})$ as small as possible (in other words, the *minimizer* of the function L).

One reason we chose this particular loss function is that it can be related to the linear algebra we are studying. Let A be the matrix whose rows are the training examples $\boldsymbol{a}_1, \ldots, \boldsymbol{a}_m$. Let $\boldsymbol{b}$ be the m-vector whose i^{th} entry is b_i. Let $\boldsymbol{w}$ be a D-vector. By the dot-product definition of matrix-vector multiplication, entry i of the vector $A\boldsymbol{w} - \boldsymbol{b}$ is $\boldsymbol{a}_i \cdot \boldsymbol{w} - b_i$. The squared norm of this vector is therefore $(\boldsymbol{a}_1 \cdot \boldsymbol{w} - b_1)^2 + \cdots + (\boldsymbol{a}_m \cdot \boldsymbol{w} - b_m)^2$. It follows that our goal is to select the vector $\boldsymbol{w}$ minimizing $\|A\boldsymbol{w} - \boldsymbol{b}\|^2$.

In Orthogonalization, we learn that this computational problem can be solved by an algorithm that uses orthogonality and projection.

Task 8.4.4: Write a procedure `loss(A, b, w)` that takes as input the training data $A, \boldsymbol{b}$ and a hypothesis vector $\boldsymbol{w}$, and returns the value $L(\boldsymbol{w})$ of the loss function for input $\boldsymbol{w}$. (Hint: You should be able to write this without any loops, using matrix multiplication and dot-product.)

Find the value of the loss function at a simple hypothesis vector such as the all-ones vector or a random vector of +1's and -1's.

8.4.5 *Nonlinear optimization by hill-climbing*

In this lab, however, we use a generic and commonly used heuristic for finding the minimizer of a function, *hill-climbing.* I call it *generic* because it can be used for a very broad class of functions; however, I refer to it as a *heuristic* because in general it is not guaranteed to find the true minimum (and often fails to do so). Generality of applicability comes at a price.

Hill-climbing maintains a solution $\boldsymbol{w}$ and iteratively makes small changes to it, in our case using vector addition. Thus it has the general form

initialize $\boldsymbol{w}$ to something
repeat as many times as you have patience for:
  $\boldsymbol{w} := \boldsymbol{w} + \textit{change}$
return $\boldsymbol{w}$

where *change* is a small vector that depends on the current value of $\boldsymbol{w}$. The goal is that each iteration improves the value of the function being optimized.

Imagine that the space of solutions forms a plane. Each possible solution $\boldsymbol{w}$ is assigned a value by the function being optimized. Interpret the value of each solution as the altitude. One can visualize the space as a three dimensional terrain.

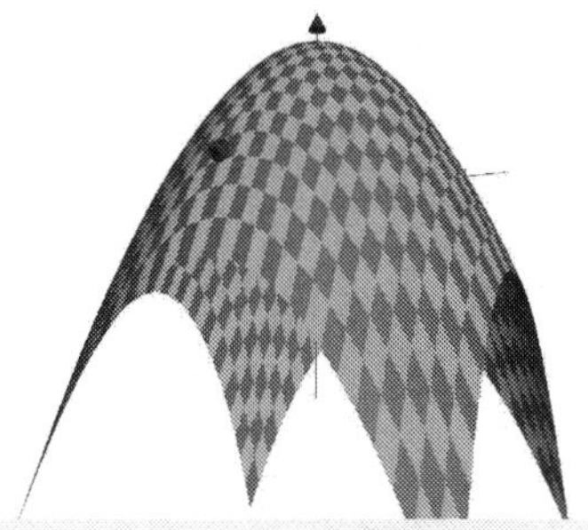

If we were trying to find a *maximizer* of the function, the algorithm gradually move the solution $\boldsymbol{w}$ towards the top of the terrain, thus the name *hill-climbing.*

In our case, the goal is to find the lowest point, so it's better to visualize the situation thus:

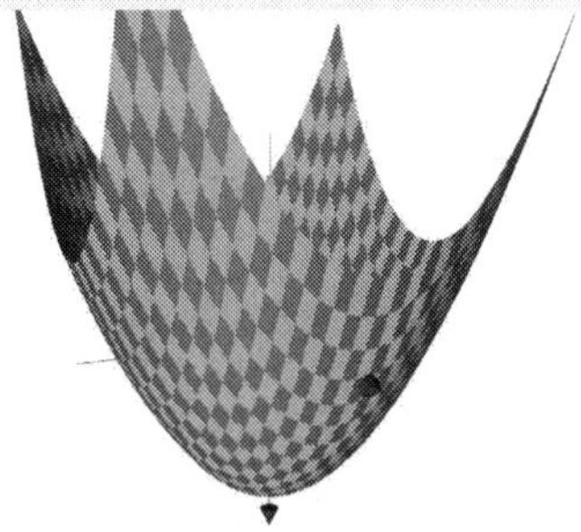

In this case, the algorithm tries to climb *down* the hill.

The strategy of hill-climbing works okay when the terrain is simple, but it is often applied to much more complicated terrains, e.g.

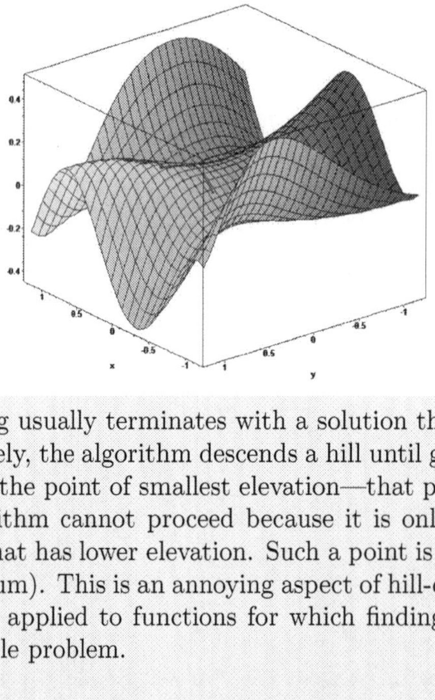

In such cases, hill-climbing usually terminates with a solution that is not truly a minimum for that function. Intuitively, the algorithm descends a hill until gets to the lowest point in a valley. It has not reached the point of smallest elevation—that point is somewhere far from the valley—but the algorithm cannot proceed because it is only allowed to descend, and there is nowhere nearby that has lower elevation. Such a point is called a *local* minimum (as opposed to a *global* minimum). This is an annoying aspect of hill-climbing but it is inevitable since hill-climbing can be applied to functions for which finding the global minimum is a computationally intractable problem.

8.4.6 *Gradient*

How should the *change* vector be selected in each iteration?

Example 8.4.5: Suppose the function to be minimized were a linear function, say $f(\boldsymbol{w}) = \boldsymbol{c} \cdot \boldsymbol{w}$. Suppose we change $\boldsymbol{w}$ by adding some vector $\boldsymbol{u}$ for which $\boldsymbol{c} \cdot \boldsymbol{u} < 0$. It follows that $f(\boldsymbol{w} + \boldsymbol{u}) < f(\boldsymbol{w})$ so we will make progress by assigning $\boldsymbol{w} + \boldsymbol{u}$ to $\boldsymbol{w}$. Moving in the direction $\boldsymbol{u}$ decreases the function's value.

In this lab, however, the function to be minimized is not a linear function. As a consequence, the right direction depends on where you are. for each particular point $\boldsymbol{w}$, there is in fact a direction of steepest descent *from that point.* We should move in that direction! Of course, once we've moved a little bit, the direction of steepest descent will have changed, so we recompute it and move a little more. We can move only a little bit in each iteration before we have to recompute the direction to move.

For a function $f : \mathbb{R}^n \longrightarrow \mathbb{R}$, the *gradient* of f, written ∇f, is a function from $\mathbb{R}^n$ to $\mathbb{R}^n$. Note that it outputs a vector, not a single number. For any particular input vector $\boldsymbol{w}$, the direction of steepest ascent of $f(\boldsymbol{x})$ for inputs near $\boldsymbol{w}$ is $\nabla f(\boldsymbol{w})$, the value of the function ∇f applied to $\boldsymbol{w}$. The direction of steepest *de*scent is the negative of $\nabla f(\boldsymbol{w})$.

The definition of the gradient is the one place in this course where we use calculus. If you don't know calculus, the derivation won't make sense but you can still do the lab.

Definition 8.4.6: The gradient of $f([x_1, \ldots, x_n])$ is defined to be

$$\left[\frac{\partial f}{\partial x_1}, \ldots, \frac{\partial f}{\partial x_n}\right]$$

Example 8.4.7: Let's return once more to the simple case where f is a linear function: $f(\boldsymbol{x}) = \boldsymbol{c} \cdot \boldsymbol{x}$. That means, of course, $f([x_1, \ldots, x_n]) = c_1 x_1 + \cdots + c_n x_n$. The partial derivative of f with respect to x_i is just c_i. Therefore $\nabla f([x_1, \ldots, x_n]) = [c_1, \ldots, c_n]$. This function disregards its argument; the gradient is the same everywhere.

Example 8.4.8: Let's take a function that is not linear. For a vector $\boldsymbol{a}$ and a scalar b, define $f(\boldsymbol{x}) = (\boldsymbol{a} \cdot \boldsymbol{x} - b)^2$. Write $\boldsymbol{x} = [x_1, \ldots, x_n]$. Then, for $j = 1, \ldots, n$,

$$\begin{aligned}\frac{\partial f}{\partial x_j} &= 2(\boldsymbol{a} \cdot \boldsymbol{x} - b)\frac{\partial}{\partial x_j}(\boldsymbol{a} \cdot \boldsymbol{x} - b) \\ &= 2(\boldsymbol{a} \cdot \boldsymbol{x} - b)a_j\end{aligned}$$

One reason for our choice of loss function

$$L(\boldsymbol{x}) = \sum_{i=1}^{m}(\boldsymbol{a}_i \cdot \boldsymbol{x} - b_i)^2$$

is that partial derivatives of this function exist and are easy to compute (if you remember a bit of calculus). The partial derivative of $L(\boldsymbol{x})$ with respect to x_j is

$$\begin{aligned}\frac{\partial L}{\partial x_j} &= \sum_{i=1}^{m}\frac{\partial}{\partial x_j}(\boldsymbol{a}_i \cdot \boldsymbol{x} - b_i)^2 \\ &= \sum_{i=1}^{m} 2(\boldsymbol{a}_i \cdot \boldsymbol{x} - b_i)a_{ij}\end{aligned}$$

where a_{ij} is entry j of $\boldsymbol{a}_i$.

Thus the value of the gradient function for a vector $\boldsymbol{w}$ is a vector whose entry j is

$$\sum_{i=1}^{m} 2(\boldsymbol{a}_i \cdot \boldsymbol{w} - b_i)a_{ij}$$

That is, the vector is

$$\nabla L(\boldsymbol{w}) = \left[\sum_{i=1}^{m} 2(\boldsymbol{a}_i \cdot \boldsymbol{w} - b_i)a_{i1}, \ldots, \sum_{i=1}^{m} \nabla L(\boldsymbol{w}) = 2(\boldsymbol{a}_i \cdot \boldsymbol{w} - b_i)a_{in}\right]$$

which can be rewritten using vector addition as

$$\sum_{i=1}^{m} 2(\boldsymbol{a}_i \cdot \boldsymbol{w} - b_i)\boldsymbol{a}_i \tag{8.9}$$

Task 8.4.9: Write a procedure `find_grad(A, b, w)` that takes as input the training data $A, \boldsymbol{b}$ and a hypothesis vector $\boldsymbol{w}$ and returns the value of the gradient of L at the point $\boldsymbol{w}$, using Equation 8.9. (Hint: You can write this without any loops, by using matrix multiplication and transpose and vector addition/subtraction.)

8.4.7 *Gradient descent*

The idea of gradient descent is to update the vector $\boldsymbol{w}$ iteratively; in each iteration, the algorithm adds to $\boldsymbol{w}$ a small scalar multiple of the negative of the value of the gradient at $\boldsymbol{w}$. The scalar is called the *step size*, and we denote it by σ.

Why should the step size be a small number? You might think that a big step allows the algorithm to make lots of progress in each iteration, but, since the gradient changes every time the hypothesis vector changes, it is safer to use a small number to as not to overshoot. (A more sophisticated method might adapt the step size as the computation proceeds.)

The basic algorithm for gradient descent is then

Set σ to be a small number
Initialize $\boldsymbol{w}$ to be some D-vector
repeat some number of times:
  $\boldsymbol{w} := \boldsymbol{w} + \sigma(\nabla L(\boldsymbol{w}))$ return $\boldsymbol{w}$

Task 8.4.10: Write a procedure `gradient_descent_step(A, b, w, sigma)` that, given the training data $A, \boldsymbol{b}$ and the current hypothesis vector $\boldsymbol{w}$, returns the next hypothesis vector.

The next hypothesis vector is obtained by computing the gradient, multiplying the gradient by the step size, and subtracting the result from the current hypothesis vector. (Why subtraction? Remember, the gradient is the direction of steepest ascent, the direction in which the function increases.)

Task 8.4.11: Write a procedure `gradient_descent(A, b, w, sigma, T)` that takes as input the training data A, $\boldsymbol{b}$, an initial value $\boldsymbol{w}$ for the hypothesis vector, a step size σ, and a number T of iterations. The procedure should implement gradient descent as described above for T iterations, and return the final value of $\boldsymbol{w}$. It should use `gradient_descent_step` as a subroutine.

Every thirty iterations or so, the procedure should print out the value of the loss function and the fraction wrong for the current hypothesis vector.

Task 8.4.12: Try out your gradient descent code on the training data! Notice that the fraction wrong might go up even while the value of the loss function goes down. Eventually, as the value of the loss function continues to decrease, the fraction wrong should also decrease (up to a point).

The algorithm is sensitive to the step size. While in principle the value of `loss` should go down in each iteration, that might not happen if the step size is too big. On the other hand, if the step size is too small, the number of iterations could be large. Try a step size of $\sigma = 2 \cdot 10^{-9}$, then try a step size of $\sigma = 10^{-9}$.

The algorithm is also sensitive to the initial value of $\boldsymbol{w}$. Try starting with the all-ones vector. Then try starting with the zero vector.

Task 8.4.13: After you have used your gradient descent code to find a hypothesis vector $\boldsymbol{w}$, see how well this hypothesis works for the data in the file `validate.data`. What is the percentage of samples that are incorrectly classified? Is it greater or smaller than the success rate on the training data? Can you explain the difference in performance?

8.5 Review questions

- What is an inner product for vectors over $\mathbb{R}$?
- How is norm defined in terms of dot-product?
- What does it mean for two vectors to be orthogonal?
- What is the Pythagorean Theorem for vectors?
- What is parallel-perpendicular decomposition of a vector?
- How does one find the projection of a vector $\boldsymbol{b}$ orthogonal to another vector $\boldsymbol{v}$?
- How can linear algebra help in optimizing a nonlinear function?

8.6 Problems

Norm

Problem 8.6.1: For each of the following problem, compute the norm of given vector $\boldsymbol{v}$:

(a) $\boldsymbol{v} = [2, 2, 1]$

(b) $\boldsymbol{v} = [\sqrt{2}, \sqrt{3}, \sqrt{5}, \sqrt{6}]$

(c) $\boldsymbol{v} = [1, 1, 1, 1, 1, 1, 1, 1, 1]$

Closest vector

Problem 8.6.2: For each of the following $\boldsymbol{a}, \boldsymbol{b}$, find the vector in Span $\{\boldsymbol{a}\}$ that is closest to $\boldsymbol{b}$:

1. $\boldsymbol{a} = [1, 2], \boldsymbol{b} = [2, 3]$
2. $\boldsymbol{a} = [0, 1, 0], \boldsymbol{b} = [1.414, 1, 1.732]$
3. $\boldsymbol{a} = [-3, -2, -1, 4], \boldsymbol{b} = [7, 2, 5, 0]$

Projection orthogonal to $\boldsymbol{a}$ and onto $\boldsymbol{a}$

Problem 8.6.3: For each of the following $\boldsymbol{a}, \boldsymbol{b}$, find $\boldsymbol{b}^{\perp \boldsymbol{a}}$ and $\boldsymbol{b}^{||\boldsymbol{a}}$.

1. $\boldsymbol{a} = [3, 0], \boldsymbol{b} = [2, 1]$
2. $\boldsymbol{a} = [1, 2, -1], \boldsymbol{b} = [1, 1, 4]$
3. $\boldsymbol{a} = [3, 3, 12], \boldsymbol{b} = [1, 1, 4]$

Chapter 9

Orthogonalization

> There are two kinds of geniuses, the "ordinary" and the "magicians." An ordinary genius is a fellow that you and I would be just as good as if we were only many times better.....It is different with the magicians. they are, to use mathematical jargon, in the orthogonal complement of where we are....
>
> Mark Kac, *Enigmas of Chance*

In this chapter, our first goal is to give an algorithm for the following problem

Computational Problem 9.0.4: *(Closest point in the span of several vectors)* Given a vector $\boldsymbol{b}$ and vectors $\boldsymbol{v}_1, \ldots, \boldsymbol{v}_n$ over the reals, find the vector in Span $\{\boldsymbol{v}_1, \ldots, \boldsymbol{v}_n\}$ closest to $\boldsymbol{b}$.

Example 9.0.5:

Let $\boldsymbol{v}_1 = [8, -2, 2]$ and $\boldsymbol{v}_2 = [4, 2, 4]$. These span a plane. Let $\boldsymbol{b} = [5, -5, 2]$.

Our goal is to find the point in Span $\{\boldsymbol{v}_1, \boldsymbol{v}_2\}$ closest to $\boldsymbol{b}$:

The closest point is $[6, -3, 0]$.

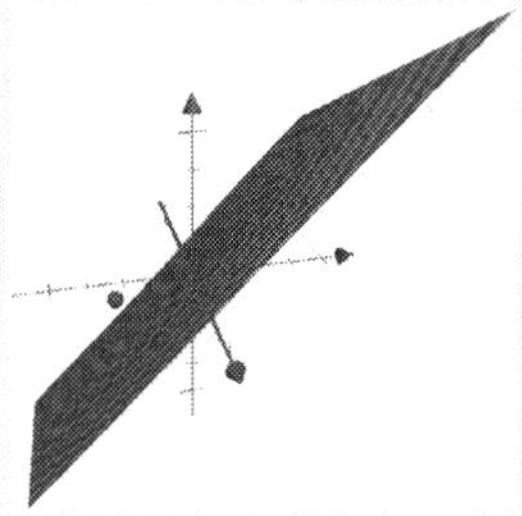

Solving Computational Problem 9.0.4 is an important goal in its own right, but through studying this goal we will also develop the techniques to solve several other computational problems.

In a significant modification, we will seek not just the closest point in Span $\{\boldsymbol{v}_1, \ldots, \boldsymbol{v}_n\}$ but also the coefficients linear of the combination with which to express that closest point.

Let $A = \begin{bmatrix} \boldsymbol{v}_1 & \cdots & \boldsymbol{v}_m \end{bmatrix}$. By the linear-combinations definition of matrix-vector multiplication, the set of vectors in Span $\{\boldsymbol{v}_1, \ldots, \boldsymbol{v}_m\}$ is exactly the set of vectors that can be written as $A\boldsymbol{x}$. As a consequence, finding the coefficients is equivalent to finding the vector $\boldsymbol{x}$ that minimizes $\|\boldsymbol{b} - A\boldsymbol{x}\|$. This is the *least-squares* problem.

If the matrix-vector equation $A\boldsymbol{x} = \boldsymbol{b}$ has a solution then of course the closest vector is $\boldsymbol{b}$ itself, and the solution to least squares is the solution to the matrix-vector equation. The advantage of an algorithm for least-squares is that it can be used even when the matrix-vector equation has no solution, as is often the case when the equation is formed based on real-world measurements.

Along the way to our solution to the least-squares problem, we will also discover algorithms

- for testing linear independence,
- for rank, for finding a basis of the span of given vectors, and
- for finding a basis of the null space, which as we know from Section 6.5 is equivalent to finding a basis of the annihilator.

We will also learn about *orthogonal complement*, which for vectors over $\mathbb{R}$ brings together the concepts of direct sum and annihilator.

9.1 Projection orthogonal to multiple vectors

Since the closest-point problem is a generalization of the Fire Engine Problem, you might think that it could be solved using the same concepts as we used to solve the latter, namely orthogonality and projection. You would be right.

9.1.1 Orthogonal to a set of vectors

Before stating the generalization of the Fire Engine Lemma, we need to extend the notion of orthogonality. So far, we have defined what it means for a vector to be orthogonal to another vector; now we define what it means for a vector to be orthogonal to a set of vectors.

Definition 9.1.1: A vector $\boldsymbol{v}$ is orthogonal to a set $\mathcal{S}$ of vectors if $\boldsymbol{v}$ is orthogonal to every vector in $\mathcal{S}$.

Example 9.1.2: The vector $[2, 0, -1]$ is orthogonal to the set $\{[0, 1, 0], [1, 0, 2]\}$ because it is orthogonal to $[0, 1, 0]$ and to $[1, 0, -2]$. Moreover, it is orthogonal to the infinite set $\mathcal{V} =$ Span $\{[0, 1, 0], [1, 0, 2]\}$ because every vector in $\mathcal{V}$ has the form $\alpha\,[0, 1, 0] + \beta\,[1, 0, 2]$, and

$$\begin{aligned}\langle [2, 0, -1], \alpha\,[0, 1, 0] + \beta\,[1, 0, 2]\rangle &= \alpha\,\langle [2, 0, -1], [0, 1, 0]\rangle + \beta\,\langle [2, 0, -1], [1, 0, 2]\rangle \\ &= \alpha\,0 + \beta\,0\end{aligned}$$

The argument used in Example 9.1.2 (Page 352) is quite general:

Lemma 9.1.3: A vector $\boldsymbol{v}$ is orthogonal to each of the vectors $\boldsymbol{a}_1, \ldots, \boldsymbol{a}_n$ if and only if it is orthogonal to every vector in Span $\{\boldsymbol{a}_1, \ldots, \boldsymbol{a}_n\}$.

Proof

Suppose $\boldsymbol{v}$ is orthogonal to $\boldsymbol{a}_1, \ldots, \boldsymbol{a}_n$. Let $\boldsymbol{w}$ be any vector in Span $\{\boldsymbol{a}_1, \ldots, \boldsymbol{a}_n\}$. We show that $\boldsymbol{v}$ is orthogonal to $\boldsymbol{w}$. By definition of span, there are coefficients $\alpha_1, \ldots, \alpha_n$ such that

$$\boldsymbol{w} = \alpha_1\,\boldsymbol{a}_1 + \cdots + \alpha_n\,\boldsymbol{a}_n$$

Therefore, using orthogonality properties (Lemma 8.3.2),

$$\begin{aligned}\langle \boldsymbol{v}, \boldsymbol{w}\rangle &= \langle \boldsymbol{v}, \alpha_1\,\boldsymbol{a}_1 + \cdots + \alpha_n\,\boldsymbol{a}_n\rangle \\ &= \alpha_1\,\langle \boldsymbol{v}, \boldsymbol{a}_1\rangle + \cdots + \alpha_n\,\langle \boldsymbol{v}, \boldsymbol{a}_n\rangle \\ &= \alpha_1\,0 + \cdots + \alpha_n\,0 \\ &= 0\end{aligned}$$

Thus $\boldsymbol{v}$ is orthogonal to $\boldsymbol{w}$.

Now suppose $\boldsymbol{v}$ is orthogonal to every vector in Span $\{\boldsymbol{a}_1, \ldots, \boldsymbol{a}_n\}$. Since the span includes $\boldsymbol{a}_1, \ldots, \boldsymbol{a}_n$, we infer that $\boldsymbol{v}$ is orthogonal to $\boldsymbol{a}_1, \ldots, \boldsymbol{a}_n$. □

Because of Lemma 9.1.3, we tend to blur the between a vector being orthogonal to a vector space and being orthogonal to a set of generators for that vector space.

9.1.2 Projecting onto and orthogonal to a vector space

It is similarly natural to generalize the notion of projection.

Definition 9.1.4: For a vector $\boldsymbol{b}$ and a vector space $\mathcal{V}$, we define the projection of $\boldsymbol{b}$ onto $\mathcal{V}$ (written $\boldsymbol{b}^{||\mathcal{V}}$) and the projection of $\boldsymbol{b}$ orthogonal to $\mathcal{V}$ (written $\boldsymbol{b}^{\perp\mathcal{V}}$) so that

$$\boldsymbol{b} = \boldsymbol{b}^{||\mathcal{V}} + \boldsymbol{b}^{\perp\mathcal{V}} \tag{9.1}$$

and $\boldsymbol{b}^{||\mathcal{V}}$ is in $\mathcal{V}$, and $\boldsymbol{b}^{\perp\mathcal{V}}$ is orthogonal to every vector in $\mathcal{V}$.

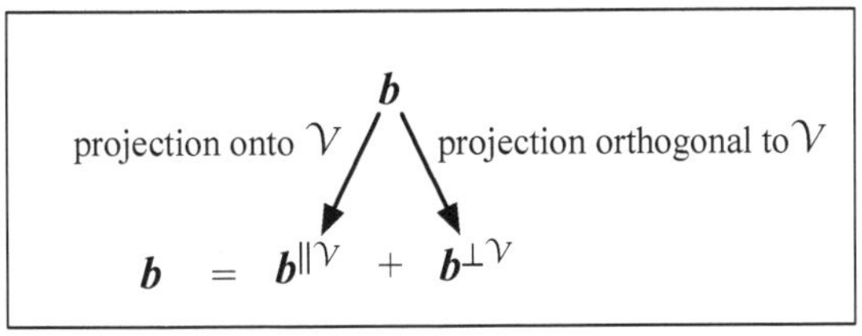

Example 9.1.5: Returning to Example 9.0.5 (Page 351), let $\mathcal{V} = \text{Span}\;\{[8,-2,2],[4,2,4]\}$ and let $\boldsymbol{b} = [5,-5,2]$. I claim that the projection of $\boldsymbol{b}$ onto $\mathcal{V}$ is $\boldsymbol{b}^{||\mathcal{V}} = [6,-3,0]$, and the projection of $\boldsymbol{b}$ orthogonal to $\mathcal{V}$ is $\boldsymbol{b}^{\perp\mathcal{V}} = [-1,-2,2]$. To prove this claim, I can show that these vectors satisfy the requirements:

- $\boldsymbol{b} = \boldsymbol{b}^{||\mathcal{V}} + \boldsymbol{b}^{\perp\mathcal{V}}$? Yes, $[5,-5,2] = [-1,-2,2] + [6,-3,0]$. ✓
- $\boldsymbol{b}^{||\mathcal{V}}$ is in $\mathcal{V}$? Yes, $\boldsymbol{b}^{||\mathcal{V}} = 1\,[8,-2,2] - \frac{1}{2}\,[4,2,4]$. ✓
- $\boldsymbol{b}\perp\mathcal{V}$ is orthogonal to $\mathcal{V}$? Yes, $[-1,-2,2]\cdot[8,-2,2] = 0$ and $[-1,-2,2]\cdot[4,2,4] = 0$. ✓

So this is the solution. But how can you calculate that solution? We need to do a bit more work before we can answer that question.

Now we can state the generalization of the Fire Engine Lemma:

Lemma 9.1.6 (Generalized Fire Engine Lemma): Let $\mathcal{V}$ be a vector space, and let $\boldsymbol{b}$ be a vector. The point in $\mathcal{V}$ closest to $\boldsymbol{b}$ is $\boldsymbol{b}^{||\mathcal{V}}$, and the distance is $\|\boldsymbol{b}^{\perp\mathcal{V}}\|$.

Proof

The proof is a straightforward generalization of that of the Fire Engine Lemma (Lemma 8.3.8). Clearly the distance between $\boldsymbol{b}$ and $\boldsymbol{b}^{||\mathcal{V}}$ is $\|\boldsymbol{b} - \boldsymbol{b}^{||\mathcal{V}}\|$ which is $\|\boldsymbol{b}^{\perp\mathcal{V}}\|$. Let $\boldsymbol{p}$ be any point in $\mathcal{V}$. We show that $\boldsymbol{p}$ is no closer to $\boldsymbol{b}$ than $\boldsymbol{b}^{||\mathcal{V}}$.

We write

$$\boldsymbol{b} - \boldsymbol{p} = \left(\boldsymbol{b} - \boldsymbol{b}^{||\mathcal{V}}\right) + \left(\boldsymbol{b}^{||\mathcal{V}} - \boldsymbol{p}\right)$$

The first summand on the right-hand side is $\boldsymbol{b}^{\perp\mathcal{V}}$. The second lies in $\mathcal{V}$ because it is the difference of two vectors that are both in $\mathcal{V}$. Since $\boldsymbol{b}^{\perp\mathcal{V}}$ is orthogonal to $\mathcal{V}$, the Pythagorean Theorem (Theorem 8.3.1) implies that

$$\|\boldsymbol{b} - \boldsymbol{p}\|^2 = \|\boldsymbol{b} - \boldsymbol{b}^{||\mathcal{V}}\|^2 + \|\boldsymbol{b}^{||\mathcal{V}} - \boldsymbol{p}\|^2$$

This shows that $\|\boldsymbol{b} - \boldsymbol{p}\| > \|\boldsymbol{b} - \boldsymbol{b}^{||\mathcal{V}}\|$ if $\boldsymbol{p} \neq \boldsymbol{b}^{||\mathcal{V}}$. □

Our goal is now to give a procedure to find these projections. It suffices to find $\boldsymbol{b}^{\perp\mathcal{V}}$, for we can then obtain $\boldsymbol{b}^{||\mathcal{V}}$ using Equation 9.1.

9.1.3 First attempt at projecting orthogonal to a list of vectors

Our first goal is a procedure `project_orthogonal` with the following spec:

- *input:* a vector $\boldsymbol{b}$, and a list *vlist* of vectors
- *output:* the projection of $\boldsymbol{b}$ orthogonal to Span *vlist*

Recall the procedure we used in Section 8.3.4 to find the projection of $\boldsymbol{b}$ orthogonal to a single vector $\boldsymbol{v}$:

```
def project_orthogonal_1(b, v): return b - project_along(b, v)
```

To project orthogonal to a list of vectors, let's try a procedure that mimics `project_orthogonal_1`:

```
def project_orthogonal(b, vlist):
  for v in vlist:
     b = b - project_along(b, v)
  return b
```

Short, elegant—and flawed. It doesn't satisfy the specification, as we now show. Consider a list *vlist* consisting of the vectors $[1, 0]$ and $[\frac{\sqrt{2}}{2}, \frac{\sqrt{2}}{2}]$. Let $\boldsymbol{b}$ be the vector $[1, 1]$.

Let $\boldsymbol{b}_i$ be the value of the variable $\boldsymbol{b}$ after i iterations. Then $\boldsymbol{b}_0$ denotes the initial value of $\boldsymbol{b}$, which is $[1, 1]$. The procedure carries out the following calculations

$$\begin{aligned} \boldsymbol{b}_1 &= \boldsymbol{b}_0 - (\text{projection of } [1,1] \text{ along } [1,0]) \\ &= \boldsymbol{b}_0 - [1,0] \\ &= [0,1] \end{aligned}$$

$$\begin{aligned} \boldsymbol{b}_2 &= \boldsymbol{b}_1 - (\text{projection of } [0,1] \text{ along } [\frac{\sqrt{2}}{2}, \frac{\sqrt{2}}{2}]) \\ &= \boldsymbol{b}_1 - [\frac{1}{2}, \frac{1}{2}] \\ &= [-\frac{1}{2}, \frac{1}{2}] \end{aligned}$$

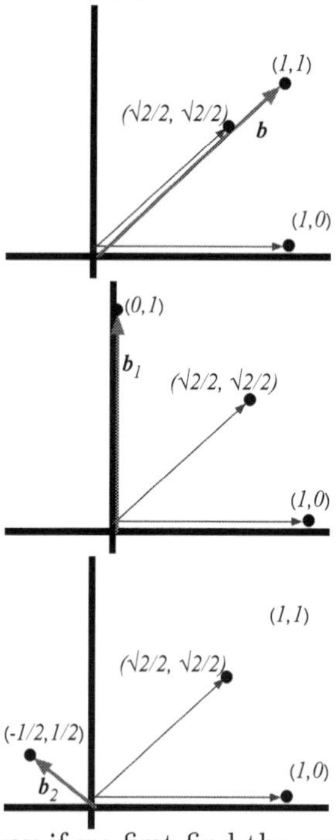

At the end, the procedure returns $\boldsymbol{b}_2$, which is $[-\frac{1}{2}, \frac{1}{2}]$. Unfortunately, this vector is not orthogonal to $[1, 0]$, the first vector in `vlist`, which shows that the procedure does not obey the specification.

How can we amend this flaw? Maybe the problem will go away if we first find the projection of b along each of the vectors in `vlist`, and only then subtract these projections all from b. Here is the procedure implementing that algorithm:

```
def classical_project_orthogonal(b, vlist):
  w = all-zeroes-vector
  for v in vlist:
    w = w + project_along(b, v)
  return b - w
```

Alas, this procedure also does not work: for the inputs specified earlier, the output vector is $[-1, 0]$, which is in fact orthogonal to neither of the two vectors in `vlist`.

9.2 Projecting b orthogonal to a list of *mutually orthogonal* vectors

Instead of abandoning this approach, let's consider a special case in which it works.

Example 9.2.1:

Let $\boldsymbol{v}_1 = [1, 2, 1]$ and $\boldsymbol{v}_2 = [-1, 0, -1]$, and let $\boldsymbol{b} = [1, 1, 2]$. Again, let $\boldsymbol{b}_i$ be the value of the variable $\boldsymbol{b}$ after i iterations. Then

$$\begin{aligned} \boldsymbol{b}_1 &= \boldsymbol{b} - \frac{\boldsymbol{b}\cdot\boldsymbol{v}_1}{\boldsymbol{v}_1\cdot\boldsymbol{v}_1}\boldsymbol{v}_1 \\ &= [1,1,2] - \frac{5}{6}[1,2,1] \\ &= \left[\frac{1}{6}, -\frac{4}{6}, \frac{7}{6}\right] \\ \boldsymbol{b}_2 &= \boldsymbol{b}_1 - \frac{\boldsymbol{b}_1\cdot\boldsymbol{v}_2}{\boldsymbol{v}_2\cdot\boldsymbol{v}_2}\boldsymbol{v}_2 \\ &= \left[\frac{1}{6}, -\frac{4}{6}, \frac{7}{6}\right] - \frac{1}{2}[-1,0,1] \\ &= \left[\frac{2}{3}, -\frac{2}{3}, \frac{2}{3}\right] \end{aligned}$$

and note that $\boldsymbol{b}_2$ is orthogonal to both $\boldsymbol{v}_1$ and $\boldsymbol{v}_2$.

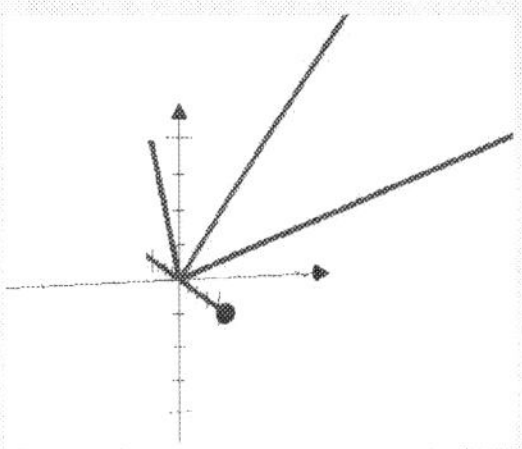

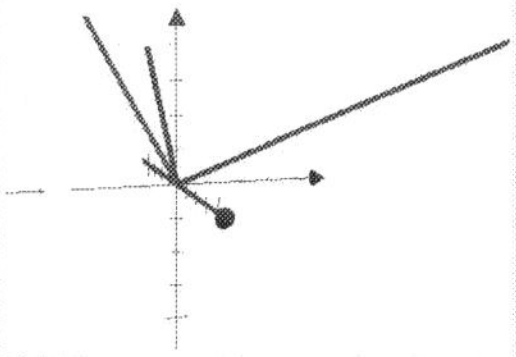

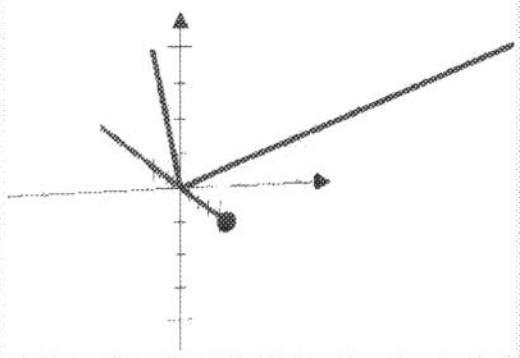

Suppose $\boldsymbol{v}_2$ is orthogonal to $\boldsymbol{v}_1$. Then the projection of $\boldsymbol{b}$ along $\boldsymbol{v}_2$ is also orthogonal to $\boldsymbol{v}_1$ (since the projection is just a scalar multiple of $\boldsymbol{v}_2$). If $\boldsymbol{b}$ is orthogonal to $\boldsymbol{v}_1$ (which is true after the first iteration), then subtracting off something else orthogonal to $\boldsymbol{v}_1$ preserves orthogonality. Let's see that formally:

Assume $\langle \boldsymbol{v}_1, \boldsymbol{b}\rangle = 0$ and $\langle \boldsymbol{v}_1, \boldsymbol{v}_2\rangle = 0$. Then

$$\begin{aligned} \langle \boldsymbol{v}_1, \boldsymbol{b} - \sigma\, \boldsymbol{v}_2\rangle &= \langle \boldsymbol{v}_1, \boldsymbol{b}\rangle - \langle \boldsymbol{v}_1, \sigma\, \boldsymbol{v}_2\rangle \\ &= \langle \boldsymbol{v}_1, \boldsymbol{b}\rangle - \sigma\, \langle \boldsymbol{v}_1, \boldsymbol{v}_2\rangle \\ &= 0 + 0 \end{aligned}$$

Returning to our flawed procedure `project_orthogonal(b,vlist)`, instead of trying to fix the flaw by changing the procedure, we will change the spec we expect the procedure to fulfill. We will only expect the correct answer when `vlist` consists of *mutually orthogonal* vectors, i.e. when the ith vector in the list is orthogonal to the jth vector in the list for every $i \neq j$. That is, the new spec will be:

- *input:* a vector $\boldsymbol{b}$, and a list *vlist* of *mutually orthogonal* vectors
- *output:* the projection $\boldsymbol{b}^{\perp}$ of $\boldsymbol{b}$ orthogonal to the vectors in *vlist*

With this restriction on the input, the procedure is correct!

Problem 9.2.2: Using hand-calculation, show the steps carried out when `project_orthogonal` is called with `b`$=[1,1,1]$ and `vlist`$= [\boldsymbol{v}_1, \boldsymbol{v}_2]$ where $\boldsymbol{v}_1 = [0, 2, 2]$ and $\boldsymbol{v}_2 = [0, 1, -1]$.

9.2.1 Proving the correctness of `project_orthogonal`

Theorem 9.2.3 (Correctness of `project_orthogonal`): For a vector $\boldsymbol{b}$ and a list `vlist` of mutually orthogonal vectors, the procedure `project_orthogonal(`$\boldsymbol{b}$`,vlist)` returns a vector $\boldsymbol{b}^{\perp}$ such that $\boldsymbol{b}^{\perp}$ is orthogonal to the vectors in `vlist` and $\boldsymbol{b} - \boldsymbol{b}^{\perp}$ is in the span of the vectors in `vlist`.

A procedure with a loop does not become correct all of a sudden at the end of a loop. To prove such a procedure is correct, we show that, after i iterations, some statement involving i is true, for $i = 0, 1, 2, \ldots$. A statement used in this way is called a *loop invariant.* The following lemma gives the loop invariant for the proof of correctness of `project_orthogonal`. Intuitively, the loop invariant states that, after i iterations, the current value of the variable `b` would be the correct return value if `vlist` had only i vectors.

Lemma 9.2.4 (Loop Invariant for `project_orthogonal`): Let k =`len(vlist)`. For $i = 0, \ldots, k$, let $\boldsymbol{b}_i$ be the value of the variable `b` after i iterations. Then

- $\boldsymbol{b}_i$ is orthogonal to the first i vectors of `vlist`, and
- $\boldsymbol{b} - \boldsymbol{b}_i$ is in the span of the first i vectors of `vlist`

The loop invariant implies correctness of the procedure

Note that the vector $\boldsymbol{b}^{\perp}$ returned by `project_orthogonal` is the value of `b` after all k iterations, which is denoted $\boldsymbol{b}_k$. By plugging in k for i in the loop invariant, we obtain

> $\boldsymbol{b}_k$ is orthogonal to the first k vectors of `vlist`, and $\boldsymbol{b} - \boldsymbol{b}_k$ is in the span of the first k vectors sof `vlist`.

Since `vlist` has exactly k vectors, this is equivalent to

> $\boldsymbol{b}_k$ is orthogonal to all vectors in `vlist`, and $\boldsymbol{b} - \boldsymbol{b}_k$ is in Span `vlist`

This is what is claimed by the theorem (taking into account that $\boldsymbol{b}^{\perp} = \boldsymbol{b}_k$).

The proof of the loop invariant

Proof

The proof is by induction on i. For $i = 0$, the loop invariant is trivially true: $\boldsymbol{b}_0$ is orthogonal to each of the first 0 vectors (every vector is), and $\boldsymbol{b} - \boldsymbol{b}_0$ is in the span of the first 0 vectors (because $\boldsymbol{b} - \boldsymbol{b}_0$ is the zero vector).

Assume the invariant holds for $i - 1$ iterations. We prove it holds for i iterations. We write *vlist* as $[\boldsymbol{v}_1, \ldots, \boldsymbol{v}_k]$.

In the ith iteration, the procedure computes

$$\boldsymbol{b}_i = \boldsymbol{b}_{i-1} - \texttt{project\_along}(b_{i-1}, \boldsymbol{v}_i)$$

Using our knowledge of the procedure `project_along`, we can rewrite this as

$$\boldsymbol{b}_i = \boldsymbol{b}_{i-1} - \alpha_i\, \boldsymbol{v}_i \tag{9.2}$$

where $\alpha_i = \frac{\langle \boldsymbol{b}_{i-1}, \boldsymbol{v}_i \rangle}{\langle \boldsymbol{v}_i, \boldsymbol{v}_i \rangle}$. The induction hypothesis states that $\boldsymbol{b}_{i-1}$ is the projection of $\boldsymbol{b}_0$ orthogonal to the first $i - 1$ vectors.

We need to prove that $\boldsymbol{b}_i$ is orthogonal to each vector in $\{\boldsymbol{v}_1, \ldots, \boldsymbol{v}_{i-1}, \boldsymbol{v}_i\}$. The choice of α_i ensures that $\boldsymbol{b}_i$ is orthogonal to $\boldsymbol{v}_I^*$. We still have to prove that $\boldsymbol{b}_i$ is orthogonal to $\boldsymbol{v}_j^*$ for each $j < i$.

$$\begin{aligned}\langle \boldsymbol{b}_i, \boldsymbol{v}_j\rangle &= \langle \boldsymbol{b}_{i-1} - \alpha_i \boldsymbol{v}_i, \boldsymbol{v}_j\rangle \\ &= \langle \boldsymbol{b}_{i-1}, \boldsymbol{v}_j\rangle - \alpha_i \langle \boldsymbol{v}_i, \boldsymbol{v}_j\rangle \\ &= 0 - \alpha_i \langle \boldsymbol{v}_i, \boldsymbol{v}_j\rangle \text{ by the inductive hypothesis} \\ &= 0 - \alpha_i 0 \text{ by mutual orthogonality}\end{aligned}$$

We also need to prove that $\boldsymbol{b}_0 - \boldsymbol{b}_i$ is in the span of the first i vectors of `vlist`. By the inductive hypothesis, $\boldsymbol{b}_0 - \boldsymbol{b}_{i-1}$ is in the span of the first $i-1$ vectors. Therefore

$$\begin{aligned}\boldsymbol{b}_0 - \boldsymbol{b}_i &= \boldsymbol{b}_0 - (\boldsymbol{b}_{i-1} - \alpha_i\, \boldsymbol{v}_i) \text{ by Equation 9.2} \\ &= (\boldsymbol{b}_0 - \boldsymbol{b}_{i-1}) + \alpha_i\, \boldsymbol{v}_i \\ &= (\text{a vector in the span of the first } i-1 \text{ vectors}) + \alpha_i\, \boldsymbol{v}_i \\ &= \text{a vector in the span of the first } i \text{ vectors}\end{aligned}$$

This completes the proof of the loop invariant. □

We have shown that `project_orthogonal` satisfies the new spec.

Having completed the proof of the loop invariant, we have completed the proof of Theorem 9.2.3, showing the correctness of the procedure.

Problem 9.2.5: In Section 9.1.3, we gave two procedures, `project_orthogonal` and `classical_project_orthogonal`. In this section, we proved that `project_orthogonal(b,vlist)` correctly computes the projection of `b` orthogonal to the vectors in `vlist` if those vectors are mutually orthogonal. In this problem, you will show an analogous result for `classical_project_orthogonal(b, vlist)`.

As in the proof of Lemma 9.2.4, let $\boldsymbol{b}_i$ be the value of the variable `b` in `project_orthogonal` after i iterations. Let $\boldsymbol{w}_i$ be the value of `w` in `classical_project_orthogonal` after i iterations. Use induction on i to prove the following claim, which shows that the procedures should return the same vector:

Claim: For $i = 0, 1, 2, \ldots,$

$$\boldsymbol{b}_i = \boldsymbol{b} - \boldsymbol{w}_i$$

Problem 9.2.5 shows that `project_orthogonal` and `classical_project_orthogonal` are mathematically equivalent. That is, they would produce identical answers *if arithmetic was perfectly accurate.*

Since computers use finite-precision (i.e., approximate) arithmetic, the answers you get in reality are not the same. The classical version is slightly worse in accuracy.

9.2.2 Augmenting `project_orthogonal`

We will now change to zero-based indexing for a while, in order to be more in tune with Python.

The fact that $\boldsymbol{b} - \boldsymbol{b}^{\perp}$ is in the span of the vectors $\boldsymbol{v}_0, \ldots, \boldsymbol{v}_{k-1}$ can be written as

$$\boldsymbol{b} = \sigma_0\, \boldsymbol{v}_0 + \cdots + \sigma_{k-1}\, \boldsymbol{v}_{k-1} + 1\, \boldsymbol{b}^{\perp} \tag{9.3}$$

The values of the coefficients $\sigma_0, \ldots, \sigma_{k-1}$ in this equation are the coefficients specified in Equation 9.2: σ_i is the coefficient such that $\sigma_i\, \boldsymbol{v}_i$ is the projection of $\boldsymbol{b}_{i-1}$ along $\boldsymbol{v}_i$.

Writing Equation 9.3 in matrix form, we have

$$\left[\begin{array}{c} \\ \boldsymbol{b} \\ \\ \end{array}\right] = \left[\begin{array}{c|c|c|c} & & & \\ \boldsymbol{v}_0 & \cdots & \boldsymbol{v}_{k-1} & \boldsymbol{b}^{\perp} \\ & & & \end{array}\right] \left[\begin{array}{c} \sigma_0 \\ \sigma_2 \\ \ddots \\ \sigma_{k-1} \\ 1 \end{array}\right] \tag{9.4}$$

We now write an augmented version of `project_orthogonal(b, vlist)`, called `aug_project_orthogonal(b, vlist)`, with the following spec:

- *input:* a vector $\boldsymbol{b}$ and a list $[\boldsymbol{v}_0, \ldots, \boldsymbol{v}_{k-1}]$ of mutually orthogonal vectors over the reals
- *output:* the pair $(\boldsymbol{b}^{\perp}$, *sigmadict*)) such that
  - the first element in the pair is the projection $\boldsymbol{b}^{\perp}$ of $\boldsymbol{b}$ orthogonal to Span $\{\boldsymbol{v}_0, \ldots, \boldsymbol{v}_{k-1}\}$, and
  - the second element in the pair is the dictionary *sigmadict* $= \{0 : \sigma_0, 1 : \sigma_1, \ldots, (k-1) : \sigma_{k-1}, k : 1\}$ such that

$$\boldsymbol{b} = \sigma_0\, \boldsymbol{v}_0 + \sigma_1\, \boldsymbol{v}_1 + \ldots + \sigma_{k-1}\, \boldsymbol{v}_{k-1} + 1\, \boldsymbol{b}^{\perp} \tag{9.5}$$

We are building on two prior procedures.

```
def project_along(b, v):
 sigma = ((b*v)/(v*v)) if v*v != 0 else 0
 return sigma * v
```

and

```
def project_orthogonal(b, vlist):
  for v in vlist:
     b = b - project_along(b, v)
  return b
```

The first procedure tells us how to compute the projection of a vector $\boldsymbol{b}$ onto another vector $\boldsymbol{b}$. Reviewing it reminds us of the formula:

$$\boldsymbol{b}^{||\boldsymbol{v}} = \sigma\, \boldsymbol{v}$$

where σ equals $\frac{\boldsymbol{b}\cdot\boldsymbol{v}}{\boldsymbol{v}\cdot\boldsymbol{v}}$ if $\boldsymbol{v} \neq \boldsymbol{0}$ and equals 0 if $\boldsymbol{v}$ is the zero vector.

Remember that, for practical purposes in working with floating-point numbers, the first procedure should really be implemented to assign zero to `sigma` when the vector `v` is very close to the zero vector:

```
def project_along(b, v):
 sigma = ((b*v)/(v*v)) if v*v > 1e-20 else 0
 return sigma * v
```

The second procedure tells us that the projection of $\boldsymbol{b}$ orthogonal to a list of mutually orthogonal vectors is obtained by subtracting off the projection onto each of the vectors in the list.

The procedure `aug_project_orthogonal(b, vlist)` is based on `project_orthogonal(b, vlist)` except that it must create and populate a dictionary, `sigmadict`, with the values of the coefficients in Equation 9.5. There is one coefficient for each vector in *vlist*, and one more, the coefficient of $\boldsymbol{b}^{\perp}$, which is 1. The procedure therefore initializes `sigmadict` to consist just of the entry for the coefficient of 1:

```
def aug_project_orthogonal(b, vlist):
    sigmadict = {len(vlist):1}
    ...
```

Because the procedure needs to create entries in `sigmadict` with indices matching the indices into *vlist*, we use `enumerate(vlist)` to iterate over the pairs i, v where i is an index into `vlist` and v is the corresponding element:

```
def aug_project_orthogonal(b, vlist):
 sigmadict = {len(vlist):1}
 for i,v in enumerate(vlist):
   sigma = (b*v)/(v*v) if v*v > 1e-20 else 0
   sigmadict[i] = sigma
   b = b - sigma*v
 return (b, sigmadict)
```

A more traditional but less Pythonic approach simply iterates over the indices:

```
def aug_project_orthogonal(b, vlist):
    sigmadict = {len(vlist):1}
    for i in range(len(vlist)):
        v = vlist[i]
        sigma = (b*v)/(v*v) if v*v > 1e-20 else 0
        sigmadict[i] = sigma
        b = b - sigma*v
    return (b, sigmadict)
```

Remember, the procedures are mathematically correct if we replace `1e-20` with zero, but the procedures given here are more likely to give correct solutions when working with floating-point numbers.

The procedure `aug_project_orthogonal` is defined in the module `orthogonalization`.

Having completed the definition of this procedure, we return to one-based addressing to be more consistent with mathematical convention. Sorry!

9.3 Building an orthogonal set of generators

Our goal was to project $\boldsymbol{b}$ orthogonal to the span of a set of arbitrary vectors $\boldsymbol{v}_1, \ldots, \boldsymbol{v}_n$, but so far we have succeeded only in projecting $\boldsymbol{b}$ orthogonal to the span of a set of mutually orthogonal vectors. In order to project $\boldsymbol{b}$ orthogonal to the span of vectors $\boldsymbol{v}_1, \ldots, \boldsymbol{v}_n$ that are not necessarily mutually orthogonal, we would have to first find mutually orthogonal generators for Span $\{\boldsymbol{v}_1, \ldots, \boldsymbol{v}_n\}$.

We therefore consider a new problem, *orthogonalization*:

- *input:* A list $[\boldsymbol{v}_1, \ldots, \boldsymbol{v}_n]$ of vectors over the reals
- *output:* A list of mutually orthogonal vectors $\boldsymbol{v}_1^*, \ldots, \boldsymbol{v}_n^*$ such that

$$\text{Span } \{\boldsymbol{v}_1^*, \ldots, \boldsymbol{v}_n^*\} = \text{Span } \{\boldsymbol{v}_1, \ldots, \boldsymbol{v}_n\}$$

9.3.1 The `orthogonalize` procedure

The idea for solving this problem is to use `project_orthogonal` iteratively to make a longer and longer list of mutually orthogonal vectors. First consider $\boldsymbol{v}_1$. We define $\boldsymbol{v}_1^* := \boldsymbol{v}_1$ since the set $\{\boldsymbol{v}_1^*\}$ is trivially a set of mutually orthogonal vectors. Next, we define $\boldsymbol{v}_2^*$ to be the projection of $\boldsymbol{v}_2$ orthogonal to $\boldsymbol{v}_1^*$. Now $\{\boldsymbol{v}_1^*, \boldsymbol{v}_2^*\}$ is a set of mutually orthogonal vectors. In the next step, we define $\boldsymbol{v}_3^*$ to be the projection of $\boldsymbol{v}_3$ orthogonal to $\boldsymbol{v}_1^*$ and $\boldsymbol{v}_2^*$, and so on. In each step, we use `project_orthogonal` to find the next orthogonal vector. In the ith iteration, we project $\boldsymbol{v}_i$ orthogonal to $\boldsymbol{v}_1^*, \ldots, \boldsymbol{v}_{i-1}^*$ to find $\boldsymbol{v}_i^*$.

```
def orthogonalize(vlist):
  vstarlist = []
  for v in vlist:
    vstarlist.append(project_orthogonal(v, vstarlist))
  return vstarlist
```

A simple induction using Theorem 9.2.3 proves the following lemma.

Lemma 9.3.1: Throughout the execution of `orthogonalize`, the vectors in `vstarlist` are mutually orthogonal.

In particular, the list `vstarlist` at the end of the execution, which is the list returned, consists of mutually orthogonal vectors.

Example 9.3.2: When orthogonalize is called on a vlist consisting of vectors

$$v_1 = [2,0,0], v_2 = [1,2,2], v_3 = [1,0,2]$$

it returns the list vstarlist consisting of

$$v_1^* = [2,0,0], v_2^* = [0,2,2], v_3^* = [0,-1,1]$$

(1) In the first iteration, when v is v_1, vstarlist is empty, so the first vector v_1^* added to vstarlist is v_1 itself.

(2) In the second iteration, when v is v_2, vstarlist consists only of v_1^*. The projection of v_2 orthogonal to v_1^* is

$$\begin{aligned} v_2 - \frac{\langle v_2, v_1^* \rangle}{\langle v_1^*, v_1^* \rangle} v_1^* &= [1,2,2] - \frac{2}{4}[2,0,0] \\ &= [0,2,2] \end{aligned}$$

(3) In the third iteration, when v is v_3, vstarlist consists of v_1^* and v_2^*. The projection of v_2 orthogonal to v_1^* is $[0,0,2]$, and the projection of $[0,0,2]$ orthogonal to v_2^* is

$$[0,0,2] - \frac{1}{2}[0,2,2] = [0,-1,1]$$

Example 9.3.3: Returning to the problem posed in Examples 9.0.5 and 9.1.5, we need to run orthogonalize on the list of vectors $[v_1, v_2]$ where $v_1 = [8,-2,2]$ and $v_2 = [4,2,4]$.

We set $v_1^* = v_1$. Next, we compute v_2^* as the projection of v_2 orthogonal to v_1^*:

$$\begin{aligned} v_2^* &= v_2 - \texttt{project\_along}(v_2, v_1^*) \\ &= v_2 - \frac{\langle v_2, v_1^* \rangle}{\langle v_1^*, v_1^* \rangle} v_1^* \\ &= v_2 - \frac{36}{72} v_1^* \\ &= v_2 - \frac{1}{2}[8,-2,2] \\ &= [0,3,3] \end{aligned}$$

We end up with $[v_1^*, v_2^*] = [[8,-2,2],[0,3,3]]$.

Problem 9.3.4: Using hand-calculation, show the steps carried out when orthogonalize is applied to $[v_1, v_2, v_3]$ where $v_1 = [1,0,2]$, $v_2 = [1,0,2]$, and $v_3 = [2,0,0]$.

9.3.2 Proving the correctness of `orthogonalize`

To show that `orthogonalize` satisfies its specification, we must also show that the span of the list of vectors returned equals the span of the list of vectors provided as input.

We use the following loop invariant:

Lemma 9.3.5: Consider orthogonalize applied to an n-element list $[v_1, \ldots, v_n]$. After i iterations of the algorithm, Span vstarlist = Span $\{v_1, \ldots, v_i\}$.

Proof

The proof is by induction on i. The case $i = 0$ is trivial. After $i - 1$ iterations, `vstarlist` consists of vectors $\boldsymbol{v}_1^*, \ldots, \boldsymbol{v}_{i-1}^*$. Assume the lemma holds at this point.

This means that

$$\text{Span } \{\boldsymbol{v}_1^*, \ldots, \boldsymbol{v}_{i-1}^*\} = \text{Span } \{\boldsymbol{v}_1, \ldots, \boldsymbol{v}_{i-1}\}$$

By adding the vector $\boldsymbol{v}_i$ to sets on both sides, we obtain

$$\text{Span } \{\boldsymbol{v}_1^*, \ldots, \boldsymbol{v}_{i-1}^*, \boldsymbol{v}_i\} = \text{Span } \{\boldsymbol{v}_1, \ldots, \boldsymbol{v}_{i-1}, \boldsymbol{v}_i\}$$

It therefore remains only to show that Span $\{\boldsymbol{v}_1^*, \ldots, \boldsymbol{v}_{i-1}^*, \boldsymbol{v}_i^*\} =$ Span $\{\boldsymbol{v}_1^*, \ldots, \boldsymbol{v}_{i-1}^*, \boldsymbol{v}_i\}$.

The ith iteration computes $\boldsymbol{v}_i^*$ using `project_orthogonal(`$\boldsymbol{v}_i$`, [`$\boldsymbol{v}_1^*, \ldots, \boldsymbol{v}_{i-1}^*$`])`. By Equation 9.3, there are scalars $\sigma_{i1}, \sigma_{i2}, \ldots, \sigma_{i,i-1}$ such that

$$\boldsymbol{v}_i = \sigma_{1i}\boldsymbol{v}_1^* + \cdots + \sigma_{i-1,i}\boldsymbol{v}_{i-1}^* + \boldsymbol{v}_i^* \tag{9.6}$$

This equation shows that any linear combination of

$$\boldsymbol{v}_1^*, \boldsymbol{v}_2^* \ldots, \boldsymbol{v}_{i-1}^*, \boldsymbol{v}_i$$

can be transformed into a linear combination of

$$\boldsymbol{v}_1^*, \boldsymbol{v}_2^* \ldots, \boldsymbol{v}_{i-1}^*, \boldsymbol{v}_i^*$$

and vice versa. □

The process of orthogonalization is often called *Gram-Schmidt orthogonalization* after the mathematicians Jørgen Pedersen Gram and Erhard Schmidt.

Remark 9.3.6: Order matters! Suppose you run the procedure `orthogonalize` twice, once with a list of vectors and once with the reverse of that list. The output lists will not be the reverses of each other. This contrasts with `project_orthogonal(b, vlist)`. The projection of a vector $\boldsymbol{b}$ orthogonal to a vector space is unique, so in principle[a] the order of vectors in `vlist` doesn't affect the output of `project_orthogonal(b, vlist)`.

[a] Note, however, that the exact vector returned might depend on the order due to the fact that numerical calculations are not exact.

Following up on the matrix equation (9.4), we can write (9.6) in matrix form:

$$\left[\begin{array}{c|c|c|c|c} & & & & \\ \boldsymbol{v}_1 & \boldsymbol{v}_2 & \boldsymbol{v}_3 & \cdots & \boldsymbol{v}_n \\ & & & & \end{array}\right] = \left[\begin{array}{c|c|c|c|c} & & & & \\ \boldsymbol{v}_1^* & \boldsymbol{v}_2^* & \boldsymbol{v}_3^* & \cdots & \boldsymbol{v}_n^* \\ & & & & \end{array}\right] \begin{bmatrix} 1 & \sigma_{12} & \sigma_{13} & & \sigma_{1n} \\ & 1 & \sigma_{23} & & \sigma_{2n} \\ & & 1 & & \sigma_{3n} \\ & & & \ddots & \\ & & & & \sigma_{n-1,n} \\ & & & & 1 \end{bmatrix} \tag{9.7}$$

Note that the two matrices on the right-hand side are special. The first one has mutually orthogonal columns. The second one is square, and has the property that the ij entry is zero if $i < j$. Such a matrix, you will recall, is called an *upper-triangular matrix*.

We shall have more to say about both kinds of matrices.

Example 9.3.7: For `vlist` consisting of vectors $\boldsymbol{v}_1 = [2, 0, 0]$, $\boldsymbol{v}_2 = [1, 2, 2]$ and $\boldsymbol{v}_3 = [1, 0, 2]$, the corresponding list of orthogonal vectors `vstarlist` consists of $\boldsymbol{v}_1^* = [2, 0, 0]$, $\boldsymbol{v}_2^* = [0, 2, 2]$

and $v_3^* = [0, -1, 1]$. The corresponding matrix equation is

$$\left[\begin{array}{c|c|c} & & \\ v_1 & v_2 & v_3 \\ & & \end{array}\right] = \begin{bmatrix} 2 & 0 & 0 \\ 0 & 2 & -1 \\ 0 & 2 & 1 \end{bmatrix} \begin{bmatrix} 1 & 0.5 & 0.5 \\ & 1 & 0.5 \\ & & 1 \end{bmatrix}$$

9.4 Solving the Computational Problem *closest point in the span of many vectors*

We can now give an algorithm for Computational Problem 9.0.4: finding the vector in Span $\{v_1, \ldots, v_n\}$ that is closest to b.

According to the Generalized Fire Engine Lemma (Lemma 9.1.6), the closest vector is $b^{||\mathcal{V}}$, the projection of b onto $\mathcal{V} = $ Span $\{v_1, \ldots, v_n\}$, which is $b - b^{\perp|V}$, where $b^{\perp\mathcal{V}}$ is the projection of b orthogonal to $\mathcal{V}$.

There are two equivalent ways to find $b^{\perp\mathcal{V}}$,

- *One method:* First, apply `orthogonalize` to $v_1, \ldots, v_n$, and we obtain $v_1^*, \ldots, v_n^*$. Second, call
  project_orthogonal(b, [$v_1^*, \ldots, v_n^*$])
  and obtain $b^{\perp\mathcal{V}}$ as the result.

- *Another method:* Exactly the same computations take place when `orthogonalize` is applied to $[v_1, \ldots, v_n, b]$ to obtain $[v_1^*, \ldots, v_n^*, b^*]$. In the last iteration of `orthogonalize`, the vector b^* is obtained by projecting b orthogonal to $v_1^*, \ldots, v_n^*$. Thus $b^* = b^{\perp\mathcal{V}}$.

Then $b^{||\mathcal{V}} = b - b^{\perp\mathcal{V}}$ is the closest vector to b in Span $\{v_1, \ldots, v_n\}$.

Example 9.4.1: We return to the problem posed in Examples 9.0.5 and 9.1.5. Let $v_1 = [8, -2, 2]$ and $v_2 = [4, 2, 4]$. In Example 9.3.3 (Page 360), we found that the vectors $v_1^* = [8, -2, 2]$ and $v_2^*] = [0, 3, 3]$ span the same space and are orthogonal to each other. We can therefore find the projection of $b = [5, -5, 2]$ onto this space using project_orthogonal(b, [v_1^*, v_2^*]).

For $i = 0, 1, 2$, let b_i denote the value of the variable b in `project_orthogonal` after i iterations.

$$\begin{aligned} b_1 &= b_0 - \frac{\langle b_0, v_1^* \rangle}{\langle v_1^*, v_1^* \rangle} v_1^* \\ &= b_0 - \frac{3}{4} [8, -2, 2] \\ &= [-1, -3.5, 0.5] b_2 \qquad = b_1 - \frac{\langle b_0, v_2^* \rangle}{\langle v_2^*, v_2^* \rangle} v_2^* \\ &= b_1 - \frac{-1}{2} [0, 3, 3] \\ &= [-1, -2, 2] \end{aligned}$$

The resulting vector b_2 is the projection of b orthogonal to Span $\{v_1^*, v_2^*\}$ and therefore the projection of b orthogonal to Span $\{v_1, v_2\}$ since these two spans are the same set.

9.5 Solving other problems using orthogonalize

We've shown how `orthogonalize` can be used to find the vector in Span $\{v_1, \ldots, v_n\}$ closest to b, namely $b^{||\mathcal{V}}$. Later we will give an algorithm to find the coordinate representation of $b^{||}$ in terms of $\{v_1, \ldots, v_n\}$. First we will see how we can use orthogonalization to solve other computational problems.

We need to prove something about mutually orthogonal vectors:

Proposition 9.5.1: Mutually orthogonal nonzero vectors are linearly independent.

Proof

Let $\boldsymbol{v}_1^*, \boldsymbol{v}_2^*, \ldots, \boldsymbol{v}_n^*$ be mutually orthogonal nonzero vectors.

Suppose $\alpha_1, \alpha_2, \ldots, \alpha_n$ are coefficients such that

$$\mathbf{0} = \alpha_1\, \boldsymbol{v}_1^* + \alpha_2\, \boldsymbol{v}_2^* + \cdots + \alpha_n\, \boldsymbol{v}_n^*$$

We must show that therefore the coefficients are all zero.

To show that α_1 is zero, take inner product with $\boldsymbol{v}_1^*$ on both sides:

$$\begin{aligned}
\langle \boldsymbol{v}_1^*, \mathbf{0}\rangle &= \langle \boldsymbol{v}_1^*, \alpha_1\, \boldsymbol{v}_1^* + \alpha_2\, \boldsymbol{v}_2^* + \cdots + \alpha_n\, \boldsymbol{v}_n^*\rangle \\
&= \alpha_1\ \langle \boldsymbol{v}_1^*, \boldsymbol{v}_1^*\rangle + \alpha_2\ \langle \boldsymbol{v}_1^*, \boldsymbol{v}_2^*\rangle + \cdots + \alpha_n\ \langle \boldsymbol{v}_1^*, \boldsymbol{v}_n^*\rangle \\
&= \alpha_1 \|\boldsymbol{v}_1^*\|^2 + \alpha_2\, 0 + \cdots + \alpha_n\, 0 \\
&= \alpha_1 \|\boldsymbol{v}_1^*\|^2
\end{aligned}$$

The inner product $\langle \boldsymbol{v}_1^*, 0\rangle$ is zero, so $\alpha_1\ \|\boldsymbol{v}_1^*\|^2 = 0$. Since $\boldsymbol{v}_1^*$ is nonzero, its norm is nonzero, so the only solution is $\alpha_1 = 0$.

One can similarly show that $\alpha_2 = 0, \cdots, \alpha_n = 0$. □

9.5.1 Computing a basis

The `orthogonalize` procedure does not require that the vectors of `vlist` be linearly independent. What happens if they are not?

Let $\boldsymbol{v}_1^*, \ldots, \boldsymbol{v}_n^*$ be the vectors returned by `orthogonalize`($[\boldsymbol{v}_1, \ldots, \boldsymbol{v}_n]$). They are mutually orthogonal and span the same space as $\boldsymbol{v}_1, \ldots, \boldsymbol{v}_n$. Some of them, however, might be zero vectors. Let S be the subset of $\{\boldsymbol{v}_1^*, \ldots, \boldsymbol{v}_n^*\}$ that are nonzero vectors. Clearly Span S = Span $\{\boldsymbol{v}_1^*, \ldots, \boldsymbol{v}_n^*\}$. Moreover, by Proposition 9.5.1, the vectors of S are linearly independent. Therefore they form a basis for Span $\{\boldsymbol{v}_1^*, \ldots, \boldsymbol{v}_n^*\}$ and thus also for Span $\{\boldsymbol{v}_1, \ldots, \boldsymbol{v}_n\}$.

We have thus obtained an algorithm for Computational Problem 5.10.1: finding a basis of the vector space spanned by given vectors.

Here is pseudocode for the algorithm:

```
def find_basis([v_1, ..., v_n]):
  "Return the list of nonzero starred vectors"
  [v*_1, ..., v*_n] = orthogonalize([v_1, ..., v_n])
  return [v* for v* in [v*_1, ..., v*_n] if v* is not the zero vector]
```

As a bonus, we get algorithms for:

- finding the rank of a list of vectors, and
- testing whether vectors $\boldsymbol{v}_1, \ldots, \boldsymbol{v}_n$ are linearly dependent, Computational Problem 5.5.5.

9.5.2 Computing a subset basis

With a bit more cleverness, we can find a basis of Span $\{\boldsymbol{v}_1, \ldots, \boldsymbol{v}_n\}$ consisting of a subset of the original vectors $\boldsymbol{v}_1, \ldots, \boldsymbol{v}_n$. Let k be the number of nonzero orthogonal vectors, and let $i_1, i_2, \ldots, i_k$ be the indices, in increasing order, of the nonzero orthogonal vectors. That is, the nonzero orthogonal vectors are

$$\boldsymbol{v}_{i_1}^*, \boldsymbol{v}_{i_2}^*, \ldots, \boldsymbol{v}_{i_k}^*$$

Then, I claim, the corresponding original vectors

$$\boldsymbol{v}_{i_1}, \boldsymbol{v}_{i_2}, \ldots, \boldsymbol{v}_{i_k}$$

span the same space as the basis $\boldsymbol{v}^*_{i_1}, \boldsymbol{v}^*_{i_2}, \ldots, \boldsymbol{v}^*_{i_k}$. Since they have the same cardinality as the basis, the original k vectors must also be a basis.

To see why the claim is true, consider a thought experiment in which one calls

$$\texttt{orthogonalize}([\boldsymbol{v}_{i_1}, \boldsymbol{v}_{i_2}, \ldots, \boldsymbol{v}_{i_k}])$$

A simple induction shows that, for $j = 1, \ldots, k$, the vector added to `vstarlist` is $\boldsymbol{v}^*_{i_j}$. The reason is that `project_orthogonal(v, vstarlist)` effectively ignores the zero vectors within `vstarlist` in computing the projection.

Here is pseudocode for the algorithm:

def find_subset_basis($[\boldsymbol{v}_0, \ldots, \boldsymbol{v}_n]$):
  "Return the list of original vectors that correspond to nonzero starred vectors."
  $[\boldsymbol{v}^*_0, \ldots, \boldsymbol{v}^*_n] =$ `orthogonalize`$([\boldsymbol{v}_0, \ldots, \boldsymbol{v}_n])$
  Return [$\boldsymbol{v}_i$ for i in $\{0, \ldots, n\}$ if $\boldsymbol{v}^*_i$ is not the zero vector]

9.5.3 augmented_orthogonalize

Building on `aug_project_orthgonal(b, vlist)`, we will write a procedure `aug_orthogonalize(vlist)` with the following spec:

- *input:* a list $[\boldsymbol{v}_1, \ldots, \boldsymbol{v}_n]$ of vectors
- *output:* the pair $([\boldsymbol{v}^*_1, \ldots, \boldsymbol{v}^*_n], [\boldsymbol{u}_1, \ldots, \boldsymbol{u}_n])$ of lists of vectors such that
  - $\boldsymbol{v}^*_1, \ldots, \boldsymbol{v}^*_n$ are mutually orthogonal vectors whose span equals Span $\{\boldsymbol{v}_1, \ldots, \boldsymbol{v}_n\}$, and
  - for $i = 1, \ldots, n$,

$$\left[\begin{array}{c|c|c} \\ \boldsymbol{v}_1 & \cdots & \boldsymbol{v}_n \\ \\ \end{array}\right] = \left[\begin{array}{c|c|c} \\ \boldsymbol{v}^*_1 & \cdots & \boldsymbol{v}^*_n \\ \\ \end{array}\right] \left[\begin{array}{c|c|c} \\ \boldsymbol{u}_1 & \cdots & \boldsymbol{u}_n \\ \\ \end{array}\right]$$

```
def aug_orthogonalize(vlist):
    vstarlist = []
    sigma_vecs = []
    D = set(range(len(vlist)))
    for v in vlist:
        (vstar, sigmadict)= aug_project_orthogonal(v, vstarlist)
        vstarlist.append(vstar)
        sigma_vecs.append(Vec(D, sigmadict))
    return vstarlist, sigma_vecs
```

9.5.4 Algorithms that work in the presence of rounding errors

We have given algorithms `find_basis` and `find_subset_basis` that are mathematically correct but will not work in practice; because of rounding errors, the vectors produced by `orthogonalize` that should be zero vectors will not really be zero vectors. We saw this before, in defining `project_along`. One solution is to consider a vector to be practically zero if its squared norm is very small, e.g. less than 10^{-20}.

9.6 Orthogonal complement

We have learned about projecting a vector $\boldsymbol{b}$ orthogonal to a vector space $\mathcal{V}$. Next we (in a sense) project a whole vector space orthogonal to another.

9.6.1 Definition of orthogonal complement

Definition 9.6.1: Let $\mathcal{W}$ be a vector space over the reals, and let $\mathcal{U}$ be a subspace of $\mathcal{W}$. The *orthogonal complement of $\mathcal{U}$ with respect to $\mathcal{W}$* is defined to be the set $\mathcal{V}$ such that

$$\mathcal{V} = \{\boldsymbol{w} \in \mathcal{W} \ : \ \boldsymbol{w} \text{ is orthogonal to every vector in } \mathcal{U}\}$$

The set $\mathcal{V}$ is by its definition a subset of $\mathcal{W}$, but we can say more:

Lemma 9.6.2: $\mathcal{V}$ is a subspace of $\mathcal{W}$.

Proof

For any two vectors $\boldsymbol{v}_1$ and $\boldsymbol{v}_2$ in $\mathcal{V}$, we want to show that $\boldsymbol{v}_1 + \boldsymbol{v}_2$ is also in $\mathcal{V}$. By definition of $\mathcal{V}$, the vectors $\boldsymbol{v}_1$ and and $\boldsymbol{v}_2$

1. are both in the vector space $\mathcal{W}$, and
2. are orthogonal to every vector in $\mathcal{U}$.

By 1, their sum is in $\mathcal{W}$. By combining 2 with Orthogonality Property 2 of Lemma 8.3.2, their sum is orthogonal to every vector in $\mathcal{U}$. Thus the sum is in $\mathcal{V}$.

Similarly, for any $\boldsymbol{v} \in \mathcal{V}$ and any scalar $\alpha \in \mathbb{R}$, we must show that $\alpha\,\boldsymbol{v}$ is in $\mathcal{V}$. Since $\boldsymbol{v}$ is in the vector space $\mathcal{W}$, it follows that $\alpha\,\boldsymbol{v}$ is also in $\mathcal{W}$. Since $\boldsymbol{v}$ is orthogonal to every vector in $\mathcal{U}$, it follows from Orthogonality Property 1 that $\alpha\,\boldsymbol{v}$ is also orthogonal to every vector in $\mathcal{U}$. Thus $\alpha\,\boldsymbol{v}$ is in $\mathcal{V}$. □

Example 9.6.3: Let $\mathcal{U} = \text{Span }\{[1,1,0,0],[0,0,1,1]\}$. Let $\mathcal{V}$ denote the orthogonal complement of $\mathcal{U}$ in $\mathbb{R}^4$. What vectors form a basis for $\mathcal{V}$?

Every vector in $\mathcal{U}$ has the form $[a,a,b,b]$. Therefore any vector of the form $[c,-c,d,-d]$ is orthogonal to every vector in $\mathcal{U}$.

Every vector in Span $\{[1,-1,0,0],[0,0,1,-1]\}$ is orthogonal to every vector in $\mathcal{U}$, so Span $\{[1,-1,0,0],[0,0,1,-1]\}$ is a subspace of $\mathcal{V}$, the orthogonal complement of $\mathcal{U}$ in $\mathbb{R}^4$. In fact, it is the whole thing, as we show using the Dimension Principle. We know $\mathcal{U} \oplus \mathcal{V} = \mathbb{R}^4$ so $\dim\mathcal{U} + \dim\mathcal{V} = 4$. We can tell that $\{[1,1,0,0],[0,0,1,1]\}$ is linearly independent so $\dim\mathcal{U} = 2$... so $\dim\mathcal{V} = 2$. We can tell that $\{[1,-1,0,0],[0,0,1,-1]\}$ is linearly independent so $\dim \text{Span }\{[1,-1,0,0],[0,0,1,-1]\}$ is also 2. By the Dimension Principle, therefore, Span $\{[1,-1,0,0],[0,0,1,-1]\}$ is equal to $\mathcal{V}$.

9.6.2 Orthogonal complement and direct sum

Now we see the connection between orthogonal complement and direct sum.

Lemma 9.6.4: Let $\mathcal{V}$ be the orthogonal complement of $\mathcal{U}$ with respect to $\mathcal{W}$. The only vector in $\mathcal{U} \cap \mathcal{V}$ is the zero vector.

Proof

A vector $\boldsymbol{u}$ that is in $\mathcal{V}$ is orthogonal to every vector in $\mathcal{U}$. If $\boldsymbol{u}$ is also in $\mathcal{U}$, then $\boldsymbol{u}$ is orthogonal to itself, i.e. $\langle \boldsymbol{u}, \boldsymbol{u} \rangle = 0$. By the second *norm property* (see Section 8.1.1), this implies that $\boldsymbol{u}$ is the zero vector. □

Suppose $\mathcal{V}$ is the orthogonal complement of $\mathcal{U}$ with respect to $\mathcal{W}$. By Lemma 9.6.4, we can form the direct sum $\mathcal{U} \oplus \mathcal{V}$ (see Section 6.3), which is defined to be the set

$$\{\boldsymbol{u} + \boldsymbol{v} \; : \; \boldsymbol{u} \in \mathcal{U}, \boldsymbol{v} \in \mathcal{V}\}$$

The following lemma shows that $\mathcal{W}$ is the direct sum of $\mathcal{U}$ and $\mathcal{V}$, so $\mathcal{U}$ and $\mathcal{V}$ are complementary subspaces of $\mathcal{W}$.

Lemma 9.6.5: If the orthogonal complement of $\mathcal{U}$ with respect to $\mathcal{W}$ is $\mathcal{V}$ then

$$\mathcal{U} \oplus \mathcal{V} = \mathcal{W}$$

Proof

The proof has two directions.

1. Every element of $\mathcal{U} \oplus \mathcal{V}$ has the form $\boldsymbol{u} + \boldsymbol{v}$ for $\boldsymbol{u} \in \mathcal{U}$ and $\boldsymbol{v} \in \mathcal{V}$. Since $\mathcal{U}$ and $\mathcal{V}$ are both subsets of the vector space $\mathcal{W}$, the sum $\boldsymbol{u} + \boldsymbol{v}$ is in $\mathcal{W}$. This shows $\mathcal{U} \oplus \mathcal{V} \subseteq \mathcal{W}$.
2. For any vector $\boldsymbol{b}$ in $\mathcal{W}$, write $\boldsymbol{b} = \boldsymbol{b}^{||\mathcal{U}} + \boldsymbol{b}^{\perp\mathcal{U}}$ where $\boldsymbol{b}^{||\mathcal{U}}$ is the projection of $\boldsymbol{b}$ onto $\mathcal{U}$ and $\boldsymbol{b}^{\perp\mathcal{U}}$ is the projection of $\boldsymbol{b}$ orthogonal to $\mathcal{U}$. Then $\boldsymbol{b}^{||\mathcal{U}}$ is in $\mathcal{U}$ and $\boldsymbol{b}^{\perp\mathcal{U}}$ is in $\mathcal{V}$, so $\boldsymbol{b}$ is the sum of a vector in $\mathcal{U}$ and a vector in $\mathcal{V}$. This shows $\mathcal{W} \subseteq \mathcal{U} \oplus \mathcal{V}$.

□

In Chapter 10, we use the link between orthogonal complement and direct sum in defining a wavelet basis, used in image compression.

9.6.3 Normal to a plane in $\mathbb{R}^3$ given as span or affine hull

You will often see the phrase "normal to a plane." Used in this way, "normal" means perpendicular. (Section 9.1.1).

Suppose the plane is specified as the span of two 3-vectors $\boldsymbol{u}_1$ and $\boldsymbol{u}_2$, in which case it is a vector space $\mathcal{U}$ of dimension 2. In this case, a vector is perpendicular to the plane if it is orthogonal to every vector in the plane.

Let $\boldsymbol{n}$ be a nonzero vector that is orthogonal to $\mathcal{U}$. Then Span $\{\boldsymbol{n}\}$ is a subspace of the orthogonal complement of $\mathcal{U}$ in $\mathbb{R}^3$. Moreover, by the Direct-Sum Dimension (Corollary 6.3.9), the dimension of the orthogonal complement is $\dim \mathbb{R}^3 - \dim \mathcal{U} = 1$, so, by the Dimension Principle (Lemma 6.2.14), the orthogonal complement is exactly Span $\{\boldsymbol{n}\}$. Thus any nonzero vector in Span $\{\boldsymbol{n}\}$ serves as a normal. Often the vector chosen to serve that role is the vector in Span $\{\boldsymbol{n}\}$ with norm one.

Example 9.6.6: As we learned in Example 9.4.1 (Page 362), one nonzero vector that is orthogonal to Span $\{[8, -2, 2], [0, 3, 3]\}$ is $[-1, -2, 2]$, so that is a normal. To get a normal with norm one, we divide by the norm of $[-1, -2, 2]$, getting $[\frac{-1}{9}, \frac{-2}{9}, \frac{2}{9}]$.

Similarly, one can give a vector that is normal to a line in the plane $\mathbb{R}^2$. Suppose a line in the plane is given as the span of a 2-vector $\boldsymbol{u}_1$. Any nonzero vector $\boldsymbol{n}$ that is orthogonal to $\boldsymbol{u}_1$ is a normal vector.

Now suppose a plane is specified as the affine hull of $\boldsymbol{u}_1$, $\boldsymbol{u}_2$, and $\boldsymbol{u}_3$. We know from Section 3.5.3 that we can rewrite it, for example, as

$$\boldsymbol{u}_1 + \text{Span } \{\boldsymbol{u}_2 - \boldsymbol{u}_1, \boldsymbol{u}_3 - \boldsymbol{u}_1\}$$

that is, as a translation of a plane that contains the origin. A bit of geometric intution tells us that a vector is perpendicular to this new plane if and only if it is perpendicular to the original plane. Therefore, we find a normal $\boldsymbol{n}$ to the vector space Span $\{\boldsymbol{u}_2 - \boldsymbol{u}_1, \boldsymbol{u}_3 - \boldsymbol{u}_1\}$ as described above, and it will also be a normal to the original plane.

9.6.4 Orthogonal complement and null space and annihilator

Let A be an $R \times C$ matrix over $\mathbb{R}$. Recall that the null space of A is the set of C-vectors $\boldsymbol{u}$ such that $A\boldsymbol{u}$ is a zero vector. By the dot-product definition of matrix-vector multiplication, this is the set of C-vectors $\boldsymbol{u}$ whose dot-product with each row of A is zero. Since our inner product for vectors over $\mathbb{R}$ is dot-product, this means that the orthogonal complement of Row A in $\mathbb{R}^C$ is Null A.

We already saw (in Section 6.5.2) the connection between null space and annihilator: the annihilator of Row A is Null A. This means that, for any subspace $\mathcal{U}$ of $\mathbb{R}^C$, the orthogonal complement of $\mathcal{U}$ in $\mathbb{R}^C$ is the annihilator $\mathcal{U}^o$.

The Annihilator Theorem tells us that the annihilator is the original space. The argument can be adapted to show that, for any vector space $\mathcal{W}$ over $\mathbb{R}$ (not just $\mathbb{R}^C$) and any subspace $\mathcal{U}$, the orthogonal complement of the orthogonal complement of $\mathcal{U}$ with respect to $\mathcal{W}$ is $\mathcal{U}$ itself.

9.6.5 Normal to a plane in $\mathbb{R}^3$ given by an equation

Returning to the problem of finding a normal to a plane, suppose the plane is given as the solution set for a linear equation:

$$\{[x, y, z] \in \mathbb{R}^3 \; : \; [a, b, c] \cdot [x, y, z] = d\}$$

As we saw in Section 3.6.1, the solution set is a translation of the solution set to the corresponding homogeneous linear equation:

$$\{[x, y, z] \in \mathbb{R}^3 \; : \; [a, b, c] \cdot [x, y, z] = 0\}$$

Let $\mathcal{U} = $ Span $\{[a, b, c]\}$. The solution set $\{[x, y, z] \in \mathbb{R}^3 \; : \; [a, b, c] \cdot [x, y, z] = 0\}$ is the annihilator $\mathcal{U}^o$. We want a normal to the plane consisting of the vectors in the annihilator $\mathcal{U}^o$. The set of vectors orthogonal to the annihilator $\mathcal{U}^o$ is the annihilator of the annihilator, i.e. $(\mathcal{U}^o)^o$, but the Annihilator Theorem (Theorem 6.5.15) tells us that the annihilator of the annihilatorl is the original space $\mathcal{U}$. Thus one candidate for the normal is the vector $[a, b, c]$ itself.

9.6.6 Computing the orthogonal complement

Suppose we have a basis $\boldsymbol{u}_1, \ldots, \boldsymbol{u}_k$ for $\mathcal{U}$ and a basis $\boldsymbol{w}_1, \ldots, \boldsymbol{w}_n$ for $\mathcal{W}$. How can we compute a basis for the orthogonal complement of $\mathcal{U}$ in $\mathcal{W}$?

We will give a method that uses `orthogonalize(vlist)` with

$$\text{vlist} = [\boldsymbol{u}_1, \ldots, \boldsymbol{u}_k, \boldsymbol{w}_1, \ldots, \boldsymbol{w}_n]$$

Write the list returned as $[\boldsymbol{u}_1^*, \ldots, \boldsymbol{u}_k^*, \boldsymbol{w}_1^*, \ldots, \boldsymbol{w}_n^*]$

These vectors span the same space as input vectors $\boldsymbol{u}_1, \ldots, \boldsymbol{u}_k, \boldsymbol{w}_1, \ldots, \boldsymbol{w}_n^*$, namely $\mathcal{W}$, which has dimension n. Therefore exactly n of the output vectors $\boldsymbol{u}_1^*, \ldots, \boldsymbol{u}_k^*, \boldsymbol{w}_1^*, \ldots, \boldsymbol{w}_n^*$ are nonzero.

The vectors $\boldsymbol{u}_1^*, \ldots, \boldsymbol{u}_k^*$ have same span as $\boldsymbol{u}_1, \ldots, \boldsymbol{u}_k$ and are all nonzero since $\boldsymbol{u}_1, \ldots, \boldsymbol{u}_k$ are linearly independent. Therefore exactly $n - k$ of the remaining vectors $\boldsymbol{w}_1^*, \ldots, \boldsymbol{w}_n^*$ are nonzero. Every one of them is orthogonal to $\boldsymbol{u}_1, \ldots, \boldsymbol{u}_n$, so they are orthogonal to every vector in $\mathcal{U}$, so they lie in the orthogonal complement of $\mathcal{U}$.

On the other hand, by the Direct-Sum Dimension Corollary (Corollary 6.3.9), the orthogonal complement has dimension $n - k$, so the remaining nonzero vectors are a basis for the orthogonal complement.

Here is pseudocode for the algorithm:

```
def find_orthogonal_complement(U_basis, W_basis):
    "Given a basis U_basis for 𝒰 and a basis W_basis for 𝒲,
    Returns a basis for the orthogonal complement of 𝒰 with respect to 𝒲"
    [u*_1, ..., u*_k, w*_1, ..., w*_n] = orthogonalize(U_basis, W_basis)
    Return [w_i for i in {1, ..., n} if w*_i is not the zero vector]
```

Example 9.6.7: Let's use this algorithm to find a basis for the orthogonal complement of Span $\{[8,-2,2],[0,3,3]\}$ in $\mathbb{R}^3$. We use the standard basis for $\mathbb{R}^3$, namely $[1,0,0]$, $[0,1,0]$, and $[0,0,1]$.

```
>>> L = [list2vec(v) for v in [[8,-2,2], [0,3,3], [1,0,0], [0,1,0], [0,0,1]]]
>>> Lstar = orthogonalize(L)
>>> print(Lstar[2])

     0     1     2
-------------------
 0.111 0.222 -0.222
>>> print(Lstar[3])

         0        1        2
----------------------------
 -8.33E-17 1.67E-16 5.55E-17
>>> print(Lstar[4])

        0        1        2
---------------------------
 8.33E-17 5.55E-17 1.67E-16
```

The third vector in Lstar, $\left[\frac{1}{9},\frac{2}{9},\frac{-2}{9}\right]$, is the projection of $[1,0,0]$ orthogonal to Span $\{[8,-2,2],[0,3,3]\}$. The fourth and fifth vectors in Lstar are zero vectors, so $\left[\frac{1}{9},\frac{2}{9},\frac{-2}{9}\right]$ is the sole vector in a basis for the orthogonal complement.

9.7 The QR factorization

We are now in a position to develop our first matrix factorization. Matrix factorizations play a mathematical role and a computational role:

- *Mathematical:* They provide insight into the nature of matrices—each factorization gives us a new way to think about a matrix.
- *Computational:* They give us ways to compute solutions to fundamental computational problems involving matrices.

We will use the QR algorithm for solving a square matrix equation and for the least-squares problem.

Equation 9.7 states that a matrix whose columns are $\boldsymbol{v}_1,\ldots,\boldsymbol{v}_n$ can be expressed as the product of two matrices:

- a matrix whose columns $\boldsymbol{v}_1^*,\ldots,\boldsymbol{v}_n^*$ are mutually orthogonal, and
- a triangular matrix.

9.7.1 Orthogonal and column-orthogonal matrices

Definition 9.7.1: Mutually orthogonal vectors are said to be *orthonormal* if they all have norm 1. A matrix is said to be *column-orthogonal* if the columns are orthonormal. A square column-orthogonal matrix is said to be an *orthogonal* matrix.

Yes, the terms are confusing. One would think that a matrix with orthonormal columns would be called an *orthonormal* matrix, but this is not the convention. Suppose Q is a column-orthogonal matrix, and write its columns as $\boldsymbol{q}_1^*,\ldots,\boldsymbol{q}_n^*$. Therefore the rows of Q^T are orthonormal. Let's see what happens when we take the matrix product $Q^T Q$, which we can write as

$$\begin{bmatrix} \quad q_1^* \quad \\ \hline \vdots \\ \hline q_n^* \end{bmatrix} \left[\begin{array}{c|c|c} q_1^* & \cdots & q_n^* \end{array}\right]$$

By the dot-product definition of matrix-matrix multiplication, the ij entry of the product is the dot-product of row i of the first matrix with column j of the second matrix. In this case, therefore the ij entry is $q_i^* \cdot q_j^*$. If $i = j$ then this is $q_i^* \cdot q_i^*$, which is the square of the norm of q_i^*, which is 1. If $i \neq j$ then this is the dot-product of two mutually orthogonal vectors, which is zero. Thus the ij entry of the product is 1 if $i = j$ and 0 otherwise. In other words, the product is an identity matrix.

Lemma 9.7.2: If Q is a column-orthogonal matrix then $Q^T Q$ is an identity matrix.

Now suppose in addition that Q is square, i.e. that Q is an orthogonal matrix. By Corollary 6.4.10, Q^T and Q are inverses of each other.

We have shown:

Corollary 9.7.3 (Inverse of Orthogonal Matrix): If Q is an orthogonal matrix then its inverse is Q^T.

9.7.2 Defining the QR factorization of a matrix

Definition 9.7.4: The QR factorization of an $m \times n$ matrix A (where $m \geq n$) is $A = QR$ where Q is an $m \times n$ column-orthogonal matrix Q and R is a triangular matrix:

$$\begin{bmatrix} \\ A \\ \\ \end{bmatrix} = \begin{bmatrix} \\ Q \\ \\ \end{bmatrix} \begin{bmatrix} R \end{bmatrix} \tag{9.8}$$

(What I have described is sometimes called the *reduced* QR factorization, as opposed to the *full* QR factorization.)

Later we see how having a QR factorization for a matrix A can help us solve computational problems. For now, let us consider the problem of computing a QR factorization of an input matrix A.

9.7.3 Requring A to have linearly independent columns

Let the columns of A be $v_1, \ldots, v_n$. The factorization of Equation 9.7 almost satisfies the definition of a QR factorization—it fails only in that the columns $v_1^*, \ldots, v_n^*$ do not generally have norm 1. To remedy this, we propose to *normalize* the columns, i.e. divide column j by $\|v_j^*\|$. To preserve the equality in Equation 9.7, we compensate by multiplying *row* j of the triangular matrix by $\|v_j^*\|$.

Here then is pseudocode for our proposed method for computing the QR factorization of a matrix A:

```
def qr.factor(A):
  apply aug_orthogonalize to the columns of A to get
    • mutually orthogonal vectors and
    • corresponding coefficients
  let Q = matrix with normalized versions of these vectors
```

let R = coefficient matrix with rows scaled return Q and R

What can go wrong when we do this? If some vector $\boldsymbol{v}_j^*$ is a zero vector, $\|\boldsymbol{v}_j^*\| = 0$ so we cannot divide by $\|\boldsymbol{v}_j^*\|$.

To avoid division by zero, we will impose a precondition on `QR`: the columns $\boldsymbol{v}_1, \ldots, \boldsymbol{v}_n$ of A are required to be linearly independent, i.e. they form a basis of Col$A$

No divisions by zero

The precondition implies that, by the Basis Theorem, there is no set of generators for Col$A$ that has fewer than n vectors.

Consider the vectors $\boldsymbol{v}_1^*, \ldots, \boldsymbol{v}_n^*$ returned by the orthogonalization procedure. They span Col$A$. If any of them were the zero vector, the remaining $n-1$ vectors would span Col A, a contradiction.

Diagonal elements of R are nonzero

Linear independence of the columns of A has another consequence. The coefficient matrix returned by the orthogonalization procedure is an upper-triangular matrix with diagonal entries equal to 1. The procedure `QR_special` obtains R from the coefficient matrix by multiplying its rows by the norms of the corresponding vectors $\boldsymbol{v}_i^*$. We just proved that these norms are nonzero (assuming the columns of A are linearly independent). It follows that (under the same assumption) the diagonal elements of R are nonzero. This will be important because we use backward substitution to solve the triangular system.

Col Q = Col A

The mutually orthogonal vectors returned by `aug_orthogonalize` span the same space as the columns of A. Normalizing them does not change them, so we obtain the following:

Lemma 9.7.5: In the QR factorization of A, if A's columns are linearly independent then Col Q = Col A.

9.8 Using the QR factorization to solve a matrix equation $A\mathrm{x} = \mathrm{b}$

9.8.1 The square case

Consider the matrix equation $A\boldsymbol{x} = \boldsymbol{b}$ over the reals. For the case where the matrix A is square and the columns of A are linearly independent, there is a method, based on QR factorization, for solving the equation.

Before giving the method and proving it obtains the correct answer, we give the intuition behind the method.

Suppose the columns of A are linearly independent, and $A = QR$ is the QR factorization of A. We are looking for a vector that satisfies the equation

$$A\boldsymbol{x} = \boldsymbol{b}$$

By substituting QR for A, we obtain

$$QR\boldsymbol{x} = \boldsymbol{b}$$

By left-multiplying both sides of the equation by Q^T, we obtain

$$Q^T QR\boldsymbol{x} = Q^T\boldsymbol{b}$$

Since the columns of Q are orthonormal, $Q^T Q$ is the identity matrix $\mathbb{1}$, so

$$\mathbb{1}R\boldsymbol{x} = Q^T\boldsymbol{b}$$

which is equivalent to

$$R\boldsymbol{x} = Q^T\boldsymbol{b}$$

We have shown that any vector $\hat{\boldsymbol{x}}$ that satisfies the equation $A\boldsymbol{x} = \boldsymbol{b}$ must also satisfy the equation $R\boldsymbol{x} = \mathbb{Q}^T\boldsymbol{b}$. This reasoning suggest the following method.

```
def QR_solve(A, b):
(assumes columns of A are linearly independent)
  find the QR factorization QR = A
  return the solution x̂ to Rx = Q^T b.
```

Let $\boldsymbol{b}' = Q^T\boldsymbol{b}$. Since R is an upper-triangular square matrix with nonzero diagonal elements, the solution to $R\boldsymbol{x} = \boldsymbol{b}'$ can be found using backward substitution (see Section 2.11.2 and the module `triangular`).

9.8.2 Correctness in the square case

Have we shown that `QR_solve` actually finds the solution to $A\boldsymbol{x} = \boldsymbol{b}$? No:

- We've shown that any solution to $A\boldsymbol{x} = \boldsymbol{b}$ is a solution to $R\boldsymbol{x} = Q^T\boldsymbol{b}$. (This argument applies even when A has more rows than columns. There might be *no* solutions to $A\boldsymbol{x} = \boldsymbol{b}$ in this case.)
- We must instead show that a solution to $R\boldsymbol{x} = Q^T\boldsymbol{b}$ is a solution to $A\boldsymbol{x} = \boldsymbol{b}$. (This is not necessarily true when A has more rows than columns. However, we will later see that a solution to $R\boldsymbol{x} = Q^T\boldsymbol{b}$ is, in a sense, a best *approximate* solution to $A\boldsymbol{x} = \boldsymbol{b}$.)

Theorem 9.8.1: Suppose A is a square matrix with linearly independent columns. The vector $\hat{\boldsymbol{x}}$ found by the above algorithm satisfies the equation $A\boldsymbol{x} = \boldsymbol{b}$.

Proof

We have

$$R\hat{\boldsymbol{x}} = Q^T\boldsymbol{b}$$

Multiply both sides on the left by Q. We get

$$QR\hat{\boldsymbol{x}} = QQ^T\boldsymbol{b}$$

which is equivalent to

$$A\hat{\boldsymbol{x}} = QQ^T\boldsymbol{b}$$

Because A is square, so is Q. Therefore Q is an orthogonal matrix (not just column-orthogonal) so by Corollary 9.7.3, its inverse is Q^T.

Therefore $QQ^T\boldsymbol{b} = \boldsymbol{b}$, so we obtain

$$A\hat{\boldsymbol{x}} = \boldsymbol{b}$$

□

We have given a solution to (a special case of) the computational problem *solving a matrix equation*, and therefore to *expressing a given vector as a linear combination of other given vectors.* We can solve $A\boldsymbol{x} = \boldsymbol{b}$ when

- the field is $\mathbb{R}$,
- the columns of A are linearly independent, and
- A is square.

We will see how to avoid the latter two assumptions. First we study the case where A is not square (but the other assumptions still hold).

9.8.3 The least-squares problem

We assume that the field is $\mathbb{R}$ and that the columns of A are linearly independent. First, let's consider what we can hope to achieve. Suppose A is an $R \times C$ matrix, and define the function $f_A : \mathbb{R}^C \longrightarrow \mathbb{R}^R$ by $f_A(\boldsymbol{x}) = A\boldsymbol{x}$. The domain is $\mathbb{R}^C$ so the dimension of the domain is $|C|$. The dimension of the co-domain is $|R|$. In the case where A has more rows than columns, therefore, the dimension of the co-domain is greater than that of the image. Therefore there are vectors in the co-domain that are not in the image, so f_A is *not* onto. Suppose the vector $\boldsymbol{b}$ is one of those vectors. Then there is *no* solution to $A\boldsymbol{x} = \boldsymbol{b}$.

What can we hope to find in this case? At the beginning of the chapter, we distinguished between two problems:

- finding the closest vector to $\boldsymbol{b}$ among the linear combinations of the columns of A, and
- finding the coefficients with which we can express that closest vector as such a linear combination.

Orthogonalization enables us to solve the first problem: the point closest to $\boldsymbol{b}$ is $\boldsymbol{b}^{||}$, the projection of $\boldsymbol{b}$ into the column space of A.

We addressed the second problem in Lab 8.4 but did not give a fully specified algorithm.

Computational Problem 9.8.2: *Least squares*

- *input:* an $R \times C$ matrix A and an R-vector $\boldsymbol{b}$ over the reals
- *output:* a vector $\hat{\boldsymbol{x}}$ that minimizes $\|A\boldsymbol{x} - \boldsymbol{b}\|$.

Recall from Section 4.5.4 that, for a given vector $\hat{\boldsymbol{x}}$, the vector $\boldsymbol{b} - A\hat{\boldsymbol{x}}$ is called the *residual vector.* The goal of the least-squares problem is to find a vector $\hat{\boldsymbol{x}}$ that minimizes the norm of the residual vector.

9.8.4 The coordinate representation in terms of the columns of a column-orthogonal matrix

Before we prove that `QR_solve` procedure solves the least-squares problem, we introduce a lemma that is used in that proof and in many others.

Lemma 9.8.3: Let Q be a column-orthogonal basis, and let $\mathcal{V} = \text{Col } Q$. Then, for any vector $\boldsymbol{b}$ whose domain equals Q's row-label set, $Q^T\boldsymbol{b}$ is the coordinate representation of $\boldsymbol{b}^{||\mathcal{V}}$ in terms of the columns of Q, and $QQ^T\boldsymbol{b}$ is $\boldsymbol{b}^{||\mathcal{V}}$ itself.

Proof

Write $\boldsymbol{b} = \boldsymbol{b}^{\perp\mathcal{V}} + \boldsymbol{b}^{||\mathcal{V}}$. Since $\boldsymbol{b}^{||\mathcal{V}}$ lies in $\mathcal{V}$, it can be expressed as a linear combination of the columns $\boldsymbol{q}_1, \ldots, \boldsymbol{q}_n$ of Q:

$$\boldsymbol{b}^{||\mathcal{V}} = \alpha_1 \boldsymbol{q}_1 + \cdots + \alpha_n \boldsymbol{q}_n \tag{9.9}$$

Then the coordinate representation of $\boldsymbol{b}^{||\mathcal{V}}$ is $[\alpha_1, \ldots, \alpha_n]$. We must show that this vector equals $Q^T\boldsymbol{b}$.

By the dot-product definition of matrix-vector multiplication, entry j of $Q^T\boldsymbol{b}$ is the dot-product of column j of Q with $\boldsymbol{b}$. Is this dot-product equal to α_j?

Column j of Q is $\boldsymbol{q}_i$, and the dot-product is our inner product. Let's use Equation 9.9

to calculate the inner product of $\boldsymbol{q}_j$ with $\boldsymbol{b}$:

$$\begin{aligned} \langle \boldsymbol{q}_j, \boldsymbol{b} \rangle &= \langle \boldsymbol{q}_j, \boldsymbol{b}^{\perp \mathcal{V}} + \boldsymbol{b}^{||\mathcal{V}} \rangle \\ &= \langle \boldsymbol{q}_j, \boldsymbol{b}^{\perp \mathcal{V}} \rangle + \langle \boldsymbol{q}_j, \boldsymbol{b}^{||\mathcal{V}} \rangle \\ &= 0 + \langle \boldsymbol{q}_j, \alpha_1 \boldsymbol{q}_1 + \cdots + \alpha_j \boldsymbol{q}_j + \cdots + \alpha_n \boldsymbol{q}_n \rangle \\ &= \alpha_1 \langle \boldsymbol{q}_j, \boldsymbol{q}_1 \rangle + \cdots + \alpha_j \langle \boldsymbol{q}_j, \boldsymbol{q}_j \rangle + \cdots + \alpha_n \langle \boldsymbol{q}_j, \boldsymbol{q}_n \rangle \\ &= \alpha_j \end{aligned}$$

We have shown that $\alpha_j = \langle \boldsymbol{q}_j, \boldsymbol{b} \rangle$ for $j = 1, \ldots, n$. This shows that $Q^T \boldsymbol{b}$ is the coordinate representation of $\boldsymbol{b}^{||\mathcal{V}}$ in terms of $\boldsymbol{q}_1, \ldots, \boldsymbol{q}_n$.

To go from the coordinate representation of a vector to the vector itself, we multiply by the matrix whose columns form the basis, which in this case is Q. Thus $QQ^T \boldsymbol{b}$ is $\boldsymbol{b}^{||\mathcal{V}}$ itself. □

9.8.5 Using `QR_solve` when A has more rows than columns

Here we see that the `QR_solve` procedure solves the least-squares problem. The goal is to find a vector $\hat{\boldsymbol{x}}$ that minimizes $A\boldsymbol{x} - \boldsymbol{b}$. We know from the Generalized Fire Engine Lemma (Lemma 9.1.6 that $A\hat{\boldsymbol{x}}$ should equal the projection $\boldsymbol{b}^{||\mathcal{V}}$ onto $\mathcal{V}$, where $\mathcal{V}$ is the column space of A. Is this true of the solution $\hat{\boldsymbol{x}}$ returned by `QR_solve`(A)?

`QR_solve`(A) returns a vector $\hat{\boldsymbol{x}}$ such that

$$R\hat{\boldsymbol{x}} = Q^T \boldsymbol{b}$$

Multiplying both sides of this equation by Q gives us

$$QR\hat{\boldsymbol{x}} = QQ^T \boldsymbol{b}$$

Substituting A for Q gives us

$$A\hat{\boldsymbol{x}} = QQ^T \boldsymbol{b}$$

By Lemma 9.7.5, $\mathcal{V}$ is also the column space of Q. By Lemma 9.8.3, therefore, $QQ^T \boldsymbol{b} = \boldsymbol{b}^{||\mathcal{V}}$, so we obtain

$$A\hat{\boldsymbol{x}} = \boldsymbol{b}^{||\mathcal{V}}$$

which proves that `QR_solve`(A) solves the least-squares problem.

9.9 Applications of least squares

9.9.1 Linear regression (Line-fitting)

An example application of least-squares is finding the line that best fits some two-dimensional data.

Suppose you collect some data on age versus brain mass. Here is some data from the Bureau of Made-up Numbers:

age	brain mass
45	4 lbs.
55	3.8
65	3.75
75	3.5
85	3.3

Let $f(x)$ be the function that best predicts brain mass for someone of age x. You hypothesize that, after age 45, brain mass decreases linearly with age, i.e. that $f(x) = a + cx$ for some numbers a, c. Our goal will be to find a, c to as to minimize the sum of squares of prediction errors. The observations are $(x_1, y_1) = (45, 4), (x_2, y_2) = (55, 3.8), \ldots, (x_5, y_5) = (85, 3.3)$. The prediction error on the the ith observation is $|f(x_i) - y_i|$. The sum of squares of prediction errors is $\sum_i (f(x_i) - y_i)^2$.

Here is a diagram (not a real plot of this data):

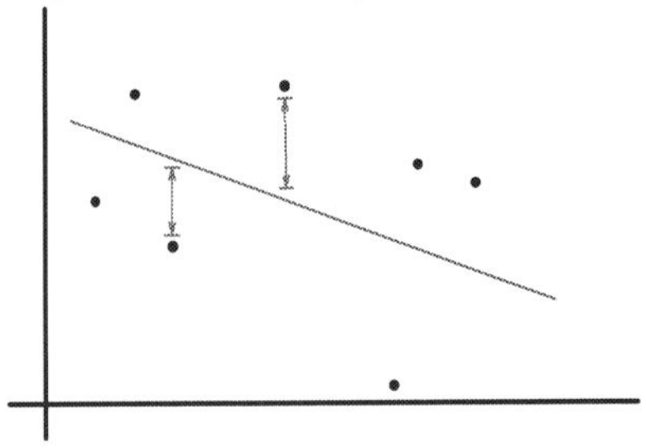

Note that, for each observation, we measure the difference between the predicted and observed y-value. In this application, this difference is measured in pounds.

We don't measure the distance from the point (x_i, y_i) to the line. Measuring the distance from the point to the line wouldn't make sense—in what units would you measure that distance? The vertical distance is measured in pounds and the horizontal distance is measured in years.

But why do we try to minimize the sum of *squares* of prediction errors? One reason is that we can handle it!A deeper reason is there is an interpretation based on probability theory for why this is a good measure. If the error is modeled by a certain probability distribution called a Gaussian (also called a *normal* distribution), minimizing this measure can be shown to be the best way to find a, b.

Be warned, however, that it's a bad model when there can be observations that are very far from the line. *Robust statistics* is a field that addresses the case when outliers are common.

Let A be the matrix whose rows are $(1, x_1), (1, x_2), \ldots, (1, x_5)$. The dot-product of row i with the vector (a, c) is $a + cx_i$, i.e. the value predicted by $f(y) = a + cx$ for the ith observation. Therefore, the vector of predictions is $A \cdot (a, c)$. The vector of differences between predictions and observed values is $A(a, c) - (y_1, y_2, \ldots, y_k)$, and the sum of squares of differences is the squared norm of this vector. Therefore the method of least squares can be used to find the pair (a, c) that minimizes the sum of squares, i.e. the line that best fits the data. The squared norm of the residual is a measure of how well the data fit the line model.

9.9.2 Fitting to a quadratic

Line-fitting is good when the data are expected to fit a line, but often you expect the data to follow a slightly more complicated pattern. We'll do an example of that.

Suppose you are trying to find occurences of some specific structure within an image, e.g. a tumor. You might have a linear filter that computes, for each pixel, some measure of how tumorlike the region centered on that pixel looks.

Considering the image to be a huge vector in $\mathbb{R}^n$, with one entry for each pixel, a linear filter is just a linear transformation from $\mathbb{R}^n$ to $\mathbb{R}^n$. The output of the filter assigns a signal strength to each pixel.

Our goal is to find the locations of tumor centers from the results of the filter. Keeping in mind that a pixel corresponds to an area on the sensor, not a single point, we don't want to know just which pixels are the centers of tumors, but exactly where in the pixel area. This is called sub-pixel accuracy.

First, consider a one-dimensional image.

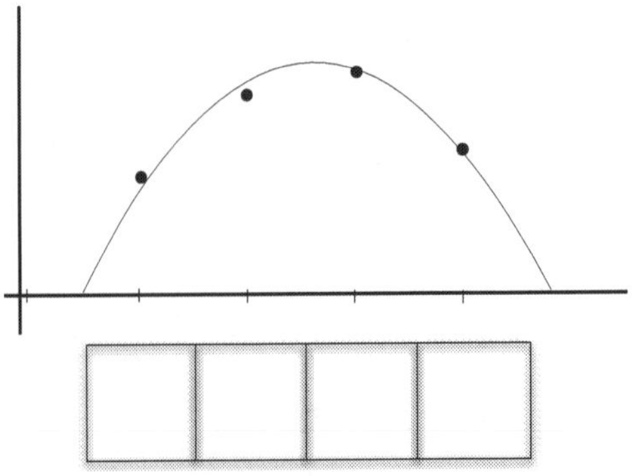

Suppose the pixel locations are $x_1, \ldots, x_5$ and the corresponding signal strengths are $y_1, \ldots, y_5$. We expect the strengths to form a peak whose maximum occurs at the exact center of the tumor.

The maximum might not occur at the center of a pixel. We therefore find the quadratic function $f(x) = u_0 + u_1 x + u_2 x^2$ that best fits the data. If the quadratic has the right form (i.e. it is concave, so it has a maximum), we conclude that the x-location of the maximum is the center of the tumor.

To find the best-fitting quadratic, we treat u_0, u_1, u_2 as unknowns in a least-squares problem in which we choose $\boldsymbol{u}$ to minimize the norm of the residual

$$\left\| \begin{bmatrix} 1 & x_1 & x_1^2 \\ 1 & x_2 & x_2^2 \\ 1 & x_3 & x_3^2 \\ 1 & x_4 & x_4^2 \\ 1 & x_5 & x_5^2 \end{bmatrix} \cdot \begin{bmatrix} u_0 \\ u_1 \\ u_2 \end{bmatrix} - \begin{bmatrix} y_1 \\ y_2 \\ y_3 \\ y_4 \\ y_5 \end{bmatrix} \right\|$$

9.9.3 Fitting to a quadratic in two variables

A similar technique works for a two-dimensional (or even three-dimensional) image. For a two-dimensional image, the data has the form $(x_1, y_1, z_1), \ldots, (x_m, y_m, z_m)$ where, for each i, z_i is the signal strength measured at pixel (x, y). We look for a quadratic in two variables:

$$f(x, y) = a + bx + cy + dxy + ex^2 + fy^2$$

To find the best-fitting such function, we minimize the norm of the residual

$$\left\| \begin{bmatrix} 1 & x_1 & y_1 & x_1 y_1 & x_1^2 & y_1^2 \\ 1 & x_2 & y_2 & x_2 y_2 & x_2^2 & y_2^2 \\ \vdots & & & & & \\ 1 & x_m & y_m & x_m y_m & x_m^2 & y_m^2 \end{bmatrix} \begin{bmatrix} a \\ b \\ c \\ d \\ e \\ f \end{bmatrix} - \begin{bmatrix} z_1 \\ z_2 \\ \vdots \\ z_m \end{bmatrix} \right\|$$

9.9.4 Coping with approximate data in the *industrial espionage* problem

Recall the *industrial espionage* problem. We were given a matrix M that specified how much of each resource was consumed per unit manufactured of each product:

	metal	concrete	plastic	water	electricity
garden gnome	0	1.3	.2	.8	.4
hula hoop	0	0	1.5	.4	.3
slinky	.25	0	0	.2	.7
silly putty	0	0	.3	.7	.5
salad shooter	.15	0	.5	.4	.8

The goal was to find the number of each product being produced from a vector $\boldsymbol{b}$ of measurements of the amount of each resource consumed:

$\boldsymbol{b} =$

metal	concrete	plastic	water	electricity
226.25	1300	677	1485	1409.5

To find the amount of each resource consumed, we can solve the vector-matrix equation $\boldsymbol{u}^T M = \boldsymbol{b}$, getting

gnome	hoop	slinky	putty	shooter
1000	175	860	590	75

In a more realistic scenario, we would get only approximate measurements of resources consumed:

$\tilde{\boldsymbol{b}} =$

metal	concrete	plastic	water	electricity
223.23	1331.62	679.32	1488.69	1492.64

Solving with these approximate quantities gives us

gnome	hoop	slinky	putty	shooter
1024.32	28.85	536.32	446.7	594.34

which are fairly inaccurate numbers. How can we improve accuracy of output without more accurate measurements? More measurements!

We have to measure something else, e.g. the amount of waste water produced. We would start with a slightly larger matrix M:

	metal	concrete	plastic	water	electricity	waste water
garden gnome	0	1.3	.2	.8	.4	.3
hula hoop	0	0	1.5	.4	.3	.35
slinky	.25	0	0	.2	.7	0
silly putty	0	0	.3	.7	.5	.2
salad shooter	.15	0	.5	.4	.8	.15

Now we have one additional measurement:

$\tilde{b} =$

metal	concrete	plastic	water	electricity	waste water
223.23	1331.62	679.32	1488.69	1492.64	489.19

Unfortunately, in adding a linear equation to our vector-matrix equation, we end up with an equation that has no solution.

However, we can still use least squares to find a best solution:

gnome	hoop	slinky	putty	shooter
1022.26	191.8	1005.58	549.63	41.1

which is considerably closer to the true amounts. We have achieved better output accuracy with same input accuracy.

9.9.5 Coping with approximate data in the *sensor node* problem

For another example, recall the *sensor node* problem: estimate current draw for each hardware component. Define `D = {'radio', 'sensor', 'memory', 'CPU'}`. Our goal was to compute a `D`-vector $\boldsymbol{u}$ that, for each hardware component, gives the current drawn by that component.

We used four test periods. In each test period, we measured the total mA-seconds in these test periods, obtaining $\boldsymbol{b} = [140, 170, 60, 170]$. For each test period, we have a vector specifying how long each hardware device was operating:

$\boldsymbol{duration}_1 =$ `{Vec(D, 'radio':0.1, 'CPU':0.3})`
$\boldsymbol{duration}_2 =$ `{Vec(D, 'sensor':0.2, 'CPU':0.4})`
$\boldsymbol{duration}_3 =$ `{Vec(D, 'memory':0.3, 'CPU':0.1})`
$\boldsymbol{duration}_4 =$ `{Vec(D, 'memory':0.5, 'CPU':0.4})`

To get $\boldsymbol{u}$, we solve $A\boldsymbol{x} = \boldsymbol{b}$ where

$$A = \begin{bmatrix} \boldsymbol{duration}_1 \\ \hline \boldsymbol{duration}_2 \\ \hline \boldsymbol{duration}_3 \\ \hline \boldsymbol{duration}_4 \end{bmatrix}$$

If measurement are exact, we get back the true current draw for each hardware component:

radio	sensor	CPU	memory
500	250	300	100

In a more realistic scenario, we would only get approximate measurements, such as

$$\tilde{\boldsymbol{b}} = [141.27, 160.59, 62.47, 181.25]$$

Solve $A\boldsymbol{x} = \tilde{\boldsymbol{b}}$ gives us the vector

radio	sensor	CPU	memory
421	142	331	98.1

How can we get more accurate results? Add more test periods and solve a least-squares problem. Suppose we use six test periods instead of four:

$duration_1 =$ `Vec(D, {'radio':0.1, 'CPU':0.3})`
$duration_2 =$ `Vec(D, {'sensor':0.2, 'CPU':0.4})`
$duration_3 =$ `Vec(D, {'memory':0.3, 'CPU':0.1})`
$duration_4 =$ `Vec(D, {'memory':0.5, 'CPU':0.4})`
$duration_5 =$ `Vec(D, {'radio':0.2, 'CPU':0.5})`
$duration_6 =$ `Vec(D, {'sensor':0.3, 'radio':0.8, 'CPU':0.9, 'memory':0.8})`
$duration_7 =$ `Vec(D, {'sensor':0.5, 'radio':0.3 'CPU':0.9, 'memory':0.5})`
$duration_8 =$ `Vec(D, {'radio':0.2 'CPU':0.6})`

Now, letting

$$A = \begin{bmatrix} \hline duration_1 \\ \hline duration_2 \\ \hline \vdots \\ \hline duration_8 \end{bmatrix}$$

and using as our measurement vector $\tilde{\boldsymbol{b}} = [141.27, 160.59, 62.47, 181.25, 247.74, 804.58, 609.10, 282.09]$, we get a matrix-vector equation $A\boldsymbol{x} = \tilde{\boldsymbol{b}}$ that has no solution.

However, the solution to least-squares problem is

radio	sensor	CPU	memory
451.40	252.07	314.37	111.66

which is much closer to the true values. Again, we achieved better output accuracy with same input accuracy by using more measurements.

9.9.6 Using the method of least squares in the machine-learning problem

In the breast-cancer machine-learning lab, the training data consisted of

- vectors $\boldsymbol{a}_1, \ldots, \boldsymbol{a}_m$ giving features of specimen, and
- values $b_1, \ldots, b_m$ specifying +1 (malignant) or -1 (benign)

Informally, the goal was to find a vector $\boldsymbol{w}$ such that sign of $\boldsymbol{a}_i \cdot \boldsymbol{w}$ predicts sign of b_i

We replaced that goal with a mathematically defined goal:

find the vector $\boldsymbol{w}$ that minimizes sum of squared errors $(\boldsymbol{b}[1] - \boldsymbol{a}_1 \cdot \boldsymbol{w})^2 + \cdots + (\boldsymbol{b}[1] - \boldsymbol{a}_m \cdot \boldsymbol{w})^2$

where $\boldsymbol{b} = [b_1, \ldots, \boldsymbol{b}_m]$.

In the machine-learning lab, we used gradient descent, which is very generally applicable but not always the best solution to a particular problem, and so it is with this problem.

It turns out that the optimization goal we posed could be addressed by a method for least squares. The mathematical goal stated above is equivalent to the goal

$$\text{find the vector } \boldsymbol{w} \text{ that minimizes } \left\| \begin{bmatrix} \boldsymbol{b} \end{bmatrix} - \begin{bmatrix} \boldsymbol{a}_1 \\ \hline \vdots \\ \hline \boldsymbol{a}_m \end{bmatrix} \begin{bmatrix} \boldsymbol{x} \end{bmatrix} \right\|^2$$

This is the least-squares problem. Using the algorithm based on QR factorization takes a fraction of the time of gradient descent, and is guaranteed to find the best solution, the one that truly minimizes the quantity. (When using gradient descent for this problem, getting the optimal solution depends on the step size.)

Machine learning provides even better solutions using more sophisticated techniques in linear algebra:

- using gradient descent but with loss functions that do a better job of modeling the learning problem,
- using an inner product that better reflects the variance of each of the features,
- using *linear programming*, a technique addressed in Chapter 13
- using *convex programming*, an even more general technique not addressed here.

9.10 Review questions

- What does it mean to normalize a vector?
- What does it mean for several vectors to be mutually orthogonal?
- What are orthonormal vectors? What is an orthonormal basis?
- How can one find the vector in Span $\{\boldsymbol{b}_1, \ldots, \boldsymbol{v}_n\}$ closest to $\boldsymbol{b}$?
- How does one find the projection of a vector $\boldsymbol{b}$ orthogonal to several mutually orthogonal vectors $\boldsymbol{v}_1, \ldots, \boldsymbol{v}_n$?
- How does one find vectors that (i) span the same space as $\boldsymbol{v}_1, \ldots, \boldsymbol{v}_n$ and that (ii) are mutually orthogonal?
- What is a column-orthogonal matrix? An orthogonal matrix?
- What is the inverse of an orthogonal matrix?
- How can you use matrix-vector multiplication to find the coordinate representation of a vector in terms of an orthonormal basis?
- What is the QR factorization of a matrix?
- How can the QR factorization be used to solve a matrix equation?
- How can the QR factorization be computed?
- How can the QR factorization be used to solve a least-squares problem?
- How can solving a least-squares problem help in fitting data to a line or a quadratic?
- How can solving a least-squares problem help to get more accurate output?
- What is the orthogonal complement?
- What is the connection between orthogonal complement and direct sum?

9.11 Problems

Orthogonal Complement

Problem 9.11.1: Find generators for the orthogonal complement of $\mathcal{U}$ with respect to $\mathcal{W}$ where

1. $\mathcal{U} = \text{Span}\ \{[0, 0, 3, 2]\}$ and $\mathcal{W} = \text{Span}\ \{[1, 2, -3, -1], [1, 2, 0, 1], [3, 1, 0, -1], [-1, -2, 3, 1]\}$.
2. $\mathcal{U} = \text{Span}\ \{[3, 0, 1]\}$ and $\mathcal{W} = \text{Span}\ \{[1, 0, 0], [1, 0, 1]\}$.
3. $\mathcal{U} = \text{Span}\ \{[[-4, 3, 1, -2], [-2, 2, 3, -1]\}$ and $\mathcal{W} = \mathbb{R}^4$

Problem 9.11.2: Explain why each statement cannot be true.

1. $\mathcal{U} = \text{Span}\ \{[0,0,1],[1,2,0]\}$ and $\mathcal{W} = \text{Span}\ \{[1,0,0],[1,0,1]\}$, and there is a vector space$\mathcal{V}$ that is the orthogonal complement of $\mathcal{U}$ in $\mathcal{W}$.
2. $\mathcal{U} = \text{Span}\ \{[3,2,1],[5,2,-3]\}$ and $\mathcal{W} = \text{Span}\ \{[1,0,0],[1,0,1],[0,1,1]\}$ and the orthogonal complement $\mathcal{V}$ of $\mathcal{U}$ in $\mathcal{W}$ contains the vector $[2,-3,1]$.

Problem 9.11.3: Let $A = \begin{bmatrix} -4 & -1 & -3 & -2 \\ 0 & 4 & 0 & -1 \end{bmatrix}$. Use orthogonal complement to find a basis for the null space of A.

Normal vector

Problem 9.11.4: Find a normal for each of the following lines in $\mathbb{R}^2$.

1. $\{\alpha\,[3,2] : \alpha \in \mathbb{R}\}$
2. $\{\alpha\,[3,5] : \alpha \in \mathbb{R}\}$

Problem 9.11.5: Find a normal for each of the following planes in $\mathbb{R}^3$.

1. Span $\{[0,1,0],[0,0,1]\}$
2. Span $\{[2,1,-3],[-2,1,1]\}$
3. affine hull of $[3,1,4]$, $[5,2,6]$, and $[2,3,5]$.

Problem 9.11.6: For each of the following vectors in $\mathbb{R}^2$, give a mathematical description of a line that has this vector as the normal.

1. $[0,7]$
2. $[1,2]$

Problem 9.11.7: For each of the following vectors, provide a set of vectors that span a plane in $\mathbb{R}^3$ for which the normal is the given vector.

1. $[0,1,1]$
2. $[0,1,0]$

Orthogonal complement and rank

Problem 9.11.8: In this problem, you will give an alternative proof of the Rank Theorem, a proof that works for matrices over the reals.

> **Theorem:** For a matrix A over the reals, the row rank equals the column rank.

Your proof should proceed as follows:

- The orthogonal complement of Row A is Null A.
- Using the connection between orthogonal complement and direct sum (Lemma 9.6.5) and

the Direct Sum Dimension Corollary (Corollary 6.3.9), show that

$$\dim \text{Row } A + \dim \text{Null } A = \text{number of columns of } A$$

- Using the Kernel-Image Theorem (Theorem 6.4.7), show that

$$\dim \text{Col } A + \dim \text{Null } A = \text{number of columns of } A$$

- Combine these equations to obtain the theorem.

QR factorization

Problem 9.11.9: Write a module `orthonormalization` that defines a procedure `orthonormalize(L)` with the following spec:

- *input:* a list L of linearly independent Vecs
- *output:* a list L^* of orthonormal Vecs such that, for $i = 1, \ldots, \text{len}(L)$, the first i Vecs of L^* and the first i Vecs of L span the same space.

Your procedure should follow this outline:

1. Call `orthogonalize(L)`,
2. Compute the list of norms of the resulting vectors, and
3. Return the list resulting from normalizing each of the vectors resulting from Step 1.

Be sure to test your procedure.

When the input consists of the list of Vecs corresponding to $[4, 3, 1, 2]$, $[8, 9, -5, -5]$, $[10, 1, -1, 5]$, your procedure should return the list of vecs corresponding approximately to $[0.73, 0.55, 0.18, 0.37]$, $[0.19, 0.40, -0.57, -0.69]$, $[0.53, -0.65, -0.51, 0.18]$.

Problem 9.11.10: Write a procedure `aug_orthonormalize(L)` in your `orthonormalization` module with the following spec:

- *input:* a list L of Vecs
- *output:* a pair `Qlist`, `Rlist` of lists of Vecs such that
  - `coldict2mat(L)` equals `coldict2mat(Qlist)` times `coldict2mat(Rlist)`, and
  - `Qlist` = `orthonormalize(L)`

Your procedure should start by calling the procedure `aug_orthogonalize(L)` defined in the module `orthogonalization`. I suggest that your procedure also use a subroutine `adjust(v, multipliers)` with the following spec:

- *input:* a Vec `v` with domain $\{0, 1, 2, \ldots, n-1\}$ and an n-element list `multipliers` of scalars
- *output:* a Vec `w` with the same domain as `v` such that `w[i] = multipliers[i]*v[i]`

Here is an example for testing `aug_orthonormalize(L)`:

```
>>> L = [list2vec(v) for v in [[4,3,1,2],[8,9,-5,-5],[10,1,-1,5]]]
>>> print(coldict2mat(L))

       0  1  2
     ---------
```

```
0   |  4  8 10
1   |  3  9  1
2   |  1 -5 -1
3   |  2 -5  5

>>> Qlist, Rlist = aug_orthonormalize(L)
>>> print(coldict2mat(Qlist))

             0      1      2
      ----------------------
0   |   0.73  0.187  0.528
1   |  0.548  0.403 -0.653
2   |  0.183 -0.566 -0.512
3   |  0.365 -0.695  0.181

>>> print(coldict2mat(Rlist))

            0    1      2
      ------------------
0   |  5.48 8.03   9.49
1   |     0 11.4 -0.636
2   |     0    0   6.04

>>> print(coldict2mat(Qlist)*coldict2mat(Rlist))

        0  1  2
      ---------
0   |  4  8 10
1   |  3  9  1
2   |  1 -5 -1
3   |  2 -5  5
```

Keep in mind, however, that numerical calculations are approximate:

```
>>> print(coldict2mat(Qlist)*coldict2mat(Rlist)-coldict2mat(L))

                 0 1        2
      -----------------------
0   |  -4.44E-16 0        0
1   |          0 0 4.44E-16
2   |  -1.11E-16 0        0
3   |  -2.22E-16 0        0
```

Problem 9.11.11: Compute the QR factorization for the following matrices. You can use a calculator or computer for the arithmetic.

1. $\begin{bmatrix} 6 & 6 \\ 2 & 0 \\ 3 & 3 \end{bmatrix}$

2. $\begin{bmatrix} 2 & 3 \\ 2 & 1 \\ 1 & 1 \end{bmatrix}$

Solving a matrix-vector equation with QR factorization

Problem 9.11.12: Write and test a procedure `QR_solve(A, b)`. Assuming the columns of A are linearly independent, this procedure should return the vector $\hat{x}$ that minimizes $\|\boldsymbol{b} - A\hat{\boldsymbol{x}}\|$.

The procedure should use

- `triangular_solve(rowlist, label_list, b)` defined in the module `triangular`, and
- the procedure `factor(A)` defined in the module `QR`, which in turn uses the procedure `aug_orthonormalize(L)` that you wrote in Problem 9.11.9

Note that `triangular_solve` requires its matrix to be represented as a list of rows. The row-labels of the matrix R returned by `QR.factor(R)` are 0,1,2,... so it suffices to use the dictionary returned by `mat2rowdict(R)`.

Note also that `triangular_solve` must be supplied with a list `label_list` of column-labels in order that it know how to interpret the vectors in `rowlist` as forming a triangular system. The column-labels of R are, of course, the column-labels of A. The ordering to provide here must match the ordering used in `QR.factor(A)`, which is `sorted(A.D[1], key=repr)`.

Demonstrate that your procedure works on some 3×2 and 3×3 matrices. Include the code and a transcript of your interaction with Python in testing it.

You can try your procedure on the examples given in Problem 9.11.13 and on the following example:

```
>>> A=Mat(({'a','b','c'},{'A','B'}), {('a','A'):-1, ('a','B'):2,
              ('b','A'):5, ('b','B'):3,('c','A'):1, ('c','B'):-2})
>>> print(A)

        A  B
     -------
 a  |  -1  2
 b  |   5  3
 c  |   1 -2

>>> Q, R = QR.factor(A)

>>> print(Q)

             0     1
     --------------
 a  |  -0.192  0.68
 b  |   0.962 0.272
 c  |   0.192 -0.68

>>> print(R)

          A    B
     ----------
 0  |   5.2 2.12
 1  |     0 3.54

>>> b = Vec({'a','b','c'}, {'a':1,'b':-1})
>>> x = QR_solve(A,b)
>>> x
Vec({'A', 'B'},{'A': -0.269..., 'B': 0.115...})
```

A good way to test your solution is to verify that the residual is (approximately) orthogonal to the columns of A:

```
>>> A.transpose()*(b-A*x)
Vec({'A', 'B'},{'A': -2.22e-16, 'B': 4.44e-16})
```

Least squares

Problem 9.11.13: In each of the following parts, you are given a matrix A and a vector $\boldsymbol{b}$. You are also given the approximate QR factorization of A. You are to

- find a vector $\hat{\boldsymbol{x}}$ that minimizes $||A\hat{\boldsymbol{x}} - \boldsymbol{b}||^2$,
- prove to yourself that the columns of A are (approximately) orthogonal to the residual $\boldsymbol{b} - A\hat{\boldsymbol{x}}$ by computing the inner products, and
- calculate the value of $||A\hat{\boldsymbol{x}} - \boldsymbol{b}||$.

1. $A = \begin{bmatrix} 8 & 1 \\ 6 & 2 \\ 0 & 6 \end{bmatrix}$ and $\boldsymbol{b} = [10, 8, 6]$

$$A = \underbrace{\begin{bmatrix} 0.8 & -0.099 \\ 0.6 & 0.132 \\ 0 & 0.986 \end{bmatrix}}_{Q} \underbrace{\begin{bmatrix} 10 & 2 \\ 0 & 6.08 \end{bmatrix}}_{R}$$

2. $A = \begin{bmatrix} 3 & 1 \\ 4 & 1 \\ 5 & 1 \end{bmatrix}$ and $\boldsymbol{b} = [10, 13, 15]$

$$A = \underbrace{\begin{bmatrix} 0.424 & .808 \\ 0.566 & 0.115 \\ 0.707 & -0.577 \end{bmatrix}}_{Q} \underbrace{\begin{bmatrix} 7.07 & 1.7 \\ 0 & 0.346 \end{bmatrix}}_{R}$$

Problem 9.11.14: For each of the following, find a vector $\hat{\boldsymbol{x}}$ that minimizes $||A\hat{\boldsymbol{x}} - \boldsymbol{b}||$. Use the algorithm based on the QR factorization.

1. $A = \begin{bmatrix} 8 & 1 \\ 6 & 2 \\ 0 & 6 \end{bmatrix}$ and $\boldsymbol{b} = (10, 8, 6)$
2. $A = \begin{bmatrix} 3 & 1 \\ 4 & 1 \end{bmatrix}$ and $\boldsymbol{b} = (10, 13)$

Linear regression

In this problem, you will find the "best" line through a given set of points, using QR factorization and solving a matrix equation where the matrix is upper triangular. You can use the `solver` module.

Module `read_data` defines a procedure `read_vectors(filename)` that takes a filename and reads a list of vectors from the named file.

The data we provide for this problem relates age to height for some young people in Kalam, an Egyptian village. Since children's heights vary a great deal, this dataset gives for each age from 18 to 29 the *average* height for the people of that age. The data is in the file `age-height.txt`

Problem 9.11.15: Use Python to find the values for parameters a and b defining the line $y = ax + b$ that best approximates the relationship between age (x) and height (y). Show your computational work (i.e., include a record of your interaction with Python).

Least squares in machine learning

Problem 9.11.16: Try using the least-squares approach on the problem addressed in the machine learning lab. Compare the quality of the solution with that you obtained using gradient descent.

Chapter 10

Special Bases

> There cannot be a language more universal and more simple, more free from errors and obscurities...more worthy to express the invariable relations of all natural things [than mathematics].
>
> Joseph Fourier

In this chapter, we discuss two special bases. Each of them consists of orthonormal vectors. For each, change of basis can be done much more quickly than by computing matrix-vector multiplication or solving a matrix-vector equation. Each of them is important in applications.

10.1 Closest k-sparse vector

Recall that we say a vector $\boldsymbol{b}$ is k-sparse if $\boldsymbol{b}$ has at most k nonzero entries. A k-sparse vector can be represented compactly. Suppose we want to compactly represent a vector $\boldsymbol{b}$ that is not k-sparse. We will have to give up accuracy. We represent not $\boldsymbol{b}$ but a vector that is similar to $\boldsymbol{b}$. This suggests the following computational problem:

- *input:* a vector $\boldsymbol{b}$, an integer k
- *output:* a k-sparse vector $\tilde{\boldsymbol{b}}$ that is closest to $\boldsymbol{b}$.

If our field is $\mathbb{R}$, the measure of closeness, as usual, is the norm of the difference $||\boldsymbol{b} - \tilde{\boldsymbol{b}}||$.

There is a simple procedure to solve this problem, *compression by suppression*: to get $\tilde{\boldsymbol{b}}$ from $\boldsymbol{b}$, suppress (i.e. zero out) all but the largest k entries. It can be proved that this procedure indeed finds the closest k-sparse vector. We will prove something more general that this, however, and more useful. Consider applying the *compression-through-suppression* procedure to an image.

photo by William Fehr, the top of the Creeper Trail in Virginia

The procedure zeroes out all but the k whitest pixels. Here is the result when k is one-quarter of the total number of pixels:

Catastrophic compression failure!

10.2 Closest vector whose representation with respect to a given basis is k-sparse

The problem is that, for a typical image vector, even the nearest k-sparse image vector looks nothing like the original image vector. It is part of the nature of images that suppressing lots of pixels destroys our perception of the image. Our perceptual system cannot fill in the missing data.

Here is the trick. Our usual way of storing or transmitting an image specifies each pixel; think of this format as the representation of the image vector in terms of the *standard* basis. Instead, consider storing or transmitting the representation of the image vector in terms of an alternative basis. If the representation in terms of the alternative basis is sparse, the representation requires few numbers so we achieve our goal of compression.

If the representation in terms of the standard basis is not sparse, we can make it so by suppressing entries of the representation that are close to zero.

Here is the new problem we face.

Computational Problem 10.2.1: *closest vector whose representation in a given basis is sparse*

- *input:* a D-vector $\boldsymbol{b}$, an integer k, a basis $\boldsymbol{u}_1, \ldots, \boldsymbol{u}_{|D|}$ for $\mathbb{R}^D$
- *output:* the vector $\tilde{\boldsymbol{b}}$ that is closest to $\boldsymbol{b}$ among all vectors whose representation with respect to $\boldsymbol{u}_1, \ldots, \boldsymbol{u}_{|D|}$ is k-sparse

10.2.1 Finding the coordinate representation in terms of an orthonormal basis

The first step in applying compression-by-suppression is converting from the original image vector to the representation in terms of the basis $\boldsymbol{u}_1, \ldots, \boldsymbol{u}_n$. Let $\boldsymbol{b}$ be the original image vector. Let $\boldsymbol{x}$ be the representation. Let Q be the matrix whose columns are $\boldsymbol{u}_1, \ldots, \boldsymbol{u}_n$. Then, by the linear-combinations interpretation of matrix-vector multiplication, $Q\boldsymbol{x} = \boldsymbol{b}$.

It appears that computing the representation $\boldsymbol{x}$ involves solving a matrix equation. We could in principle use `QR.solve` to solve such an equation. However, the number of scalar operations is roughly n^3. For a one-megapixel image, this is 10^{18} operations.

We will study the special case of Computational Problem 10.2.1 in which the basis $\boldsymbol{u}_1, \ldots, \boldsymbol{u}_{|D|}$ is orthonormal. Let $n = |D|$.

In this case, Q is an orthogonal matrix. By Corollary 9.7.3, the inverse of Q is Q^T. Therefore we can more easily solve the equation $Q\boldsymbol{x} = \boldsymbol{b}$. Multiplying this equation on the left by Q^T and invoking the lemma, we obtain $\boldsymbol{x} = Q^T\boldsymbol{b}$. Thus to compute the representation we need only carry out a matrix-vector multiplication. This requires only about n^2 scalar operations.

Furthermore, we plan to transmit the sparse representation of an image. Suppose a user downloads the sparse representation $\tilde{\boldsymbol{x}}$. In order to allow the user to view the image, the user's browser must convert the representation to an image vector. This involves computing $Q\tilde{\boldsymbol{x}}$, which also takes about n^2 scalar operations.

Even n^2 operations is somewhat impractical when dealing with megapixel images. Furthermore, suppose the real goal is to compress a movie, which is a long sequence of still images! The time would be even greater.

We will use a particular orthonormal basis for which there is a computational shortcut, a *wavelet* basis.

10.2.2 Multiplication by a column-orthogonal matrix preserves norm

Orthonormal bases have another nice property: they preserve norm.

Lemma 10.2.2: Let Q be a column-orthogonal matrix. Multiplication of vectors by Q preserves inner-products: For any vectors $\boldsymbol{u}$ and $\boldsymbol{v}$,

$$\langle Q\boldsymbol{u}, Q\boldsymbol{v}\rangle = \langle \boldsymbol{u}, \boldsymbol{v}\rangle$$

Proof

Our inner product is defined to be dot-product Recall that the dot-product of two vectors $\boldsymbol{a}$ and $\boldsymbol{b}$ can be written as

$$\begin{bmatrix} \quad \boldsymbol{a}^T \quad \end{bmatrix} \begin{bmatrix} \\ \boldsymbol{b} \\ \\ \end{bmatrix}$$

which we write as $\boldsymbol{a}^T\boldsymbol{b}$.

Thus $\langle Q\boldsymbol{u}, Q\boldsymbol{v}\rangle$ can be written as

$$\left(\begin{bmatrix} \\ \quad Q \quad \\ \\ \end{bmatrix}\begin{bmatrix} \\ \boldsymbol{u} \\ \\ \end{bmatrix}\right)^T \begin{bmatrix} \\ \quad Q \quad \\ \\ \end{bmatrix}\begin{bmatrix} \\ \boldsymbol{v} \\ \\ \end{bmatrix}$$

We can rewrite $\left(\begin{bmatrix} \\ \quad Q \quad \\ \\ \end{bmatrix}\begin{bmatrix} \\ \boldsymbol{u} \\ \\ \end{bmatrix}\right)^T$ as $\begin{bmatrix} \\ \boldsymbol{u} \\ \\ \end{bmatrix}^T \begin{bmatrix} \\ \quad Q^T \quad \\ \\ \end{bmatrix}$, getting

$$\begin{aligned}
&\left(\begin{bmatrix} \\ \quad Q \quad \\ \\ \end{bmatrix}\begin{bmatrix} \\ \boldsymbol{u} \\ \\ \end{bmatrix}\right)^T \begin{bmatrix} \\ \quad Q \quad \\ \\ \end{bmatrix}\begin{bmatrix} \\ \boldsymbol{v} \\ \\ \end{bmatrix} \\
&= \begin{bmatrix} \\ \boldsymbol{u} \\ \\ \end{bmatrix}^T \begin{bmatrix} \\ \quad Q^T \quad \\ \\ \end{bmatrix}\begin{bmatrix} \\ \quad Q \quad \\ \\ \end{bmatrix}\begin{bmatrix} \\ \boldsymbol{v} \\ \\ \end{bmatrix} \\
&= \begin{bmatrix} \quad \boldsymbol{u}^T \quad \end{bmatrix} \begin{bmatrix} \\ \quad Q^T \quad \\ \\ \end{bmatrix}\begin{bmatrix} \\ \quad Q \quad \\ \\ \end{bmatrix}\begin{bmatrix} \\ \boldsymbol{v} \\ \\ \end{bmatrix} \\
&= \begin{bmatrix} \quad \boldsymbol{u}^T \quad \end{bmatrix} \begin{bmatrix} \\ \boldsymbol{v} \\ \\ \end{bmatrix}
\end{aligned}$$

which is the dot-product of $\boldsymbol{u}$ and $\boldsymbol{v}$. □

Because vector norm is defined in terms of inner product, we obtain

Corollary 10.2.3: For any column-orthogonal matrix Q and vector $\boldsymbol{u}$, $||Q\boldsymbol{u}|| = ||\boldsymbol{u}||$.

Let $\boldsymbol{b}$ and $\tilde{\boldsymbol{b}}$ be two vectors, and let $\boldsymbol{x}$ and $\tilde{\boldsymbol{x}}$ be the representations of $\boldsymbol{b}$ and $\tilde{\boldsymbol{b}}$ with respect to an orthonormal basis $\boldsymbol{u}_1, \ldots, \boldsymbol{u}_n$. Let $Q = \left[\begin{array}{c|c|c} \\ \boldsymbol{u}_1 & \cdots & \boldsymbol{u}_n \\ \\ \end{array}\right]$. Since $Q\boldsymbol{x} = \boldsymbol{b}$ and $Q\tilde{\boldsymbol{x}} = \tilde{\boldsymbol{b}}$, the corollary imples that $||\boldsymbol{b} - \tilde{\boldsymbol{b}}|| = ||\boldsymbol{x} - \tilde{\boldsymbol{x}}||$. This means that finding a vector close to $\boldsymbol{b}$ is equivalent to finding a representation close to $\boldsymbol{x}$.

10.3 Wavelets

We'll discuss standard and alternative representations of signals such as images and sound. The representations are in terms of different bases for the same vector space.

For some purposes (including compression), it is convenient to use orthonormal bases.

Let's think about black-and-white images. A 512×512 image has an intensity–a number–for each pixel. (In a real image, the intensity is an integer, but we'll treat the intensities as real numbers.)

You might be familiar with the idea of downsampling an image. A 512×512 image can be downsampled to obtain a 256×256 image; the higher-dimensional image is divided into little 2×2 blocks of pixels, and each block is replaced by a pixel whose intensity is the average of the intensities of the pixels it replaces. The 256×256 image can be further downsampled, and so on, down to a 1×1 image. The intensity is the average of all the intensities of the original image. This idea of repeated subsampling gives rise to the notion of wavelets.

10.3.1 One-dimensional "images" of different resolutions

However, instead of directly studying wavelets for true images, we will study wavelets for one-dimensional "images". The traditional representation of an n-pixel one-dimensional image is as a sequence $x_0, x_1, \ldots, x_{n-1}$ of pixel intensities.

We will derive wavelets for an image by considering subsamples at different resolutions (different numbers of pixels). It is natural to consider a 512-pixel image as represented by a 512-vector, to consider a 256-pixel image as represented by a 256-vector, and so on. However, we will use a linear-algebra trick to enable us to view all such images as vectors in a single vector space, so that we can use the notion of orthogonal complement.

In order to carry out this trick, we select one basic resolution n, the highest resolution to be considered. To make things easy, we require that n be a power of two. In your lab, you will take n to be 512. In this lecture, we will work through an example with $n = 16$. We will use $\mathcal{V}_{16}$ to denote the space $\mathbb{R}^{16}$.

For our orthogonal basis for $\mathcal{V}_{16}$, we choose to use the standard basis. In this context, we refer to this basis as the *box* basis for $\mathcal{V}_{16}$, and we name the vectors $\boldsymbol{b}_0^{16}, \boldsymbol{b}_1^{16}, \ldots, \boldsymbol{b}_{15}^{16}$. Instead of the usual way of writing the vectors, I will write them using little squares to remind us that they represent images (albeit one-dimensional images). The basis vectors look like this:

$\boldsymbol{b}_0^{16} =$ | 1 | 0 | 0 | 0 | 0 | 0 | 0 | 0 | 0 | 0 | 0 | 0 | 0 | 0 | 0 | 0 |

$\boldsymbol{b}_1^{16} =$ | 0 | 1 | 0 | 0 | 0 | 0 | 0 | 0 | 0 | 0 | 0 | 0 | 0 | 0 | 0 | 0 |

$\vdots$

$\boldsymbol{b}_{15}^{16} =$ | 0 | 0 | 0 | 0 | 0 | 0 | 0 | 0 | 0 | 0 | 0 | 0 | 0 | 0 | 0 | 1 |

Any one-dimensional 16-pixel image can be represented as a linear combination of these basis vectors. For example, the image

can be represented as

$4 \times$ 

$+ \quad 5 \times$ 

$+$

$\vdots$

$+ \quad 0 \times$ 

What about 8-pixel images obtained from 16-pixel images by downsampling by a factor of 2? We want these 8-pixel images to "look" like 16-pixel images, only with less fine detail. For example, downsampling the image above yields

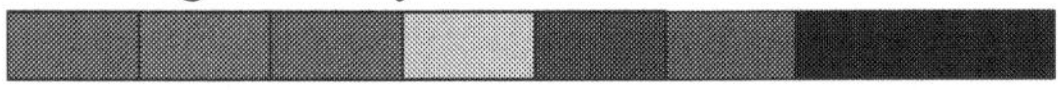

That is, we want to represent them as vectors in $\mathbb{R}^{16}$. We therefore define $\mathcal{V}_8$ to be the set of vectors in $\mathbb{R}^{16}$ such that intensity 0 equals intensity 1, intensity 2 equals intensity 3, intensity 4 equals intensity 5, and so on. The natural basis to use for $\mathcal{V}_8$ is

1	1	0	0	0	0	0	0	0	0	0	0	0	0	0	0

0	0	1	1	0	0	0	0	0	0	0	0	0	0	0	0

$\vdots$

0	0	0	0	0	0	0	0	0	0	0	0	0	0	1	1

We name these vectors $\boldsymbol{b}_0^8, \boldsymbol{b}_1^8, \ldots, \boldsymbol{b}_7^8$. We will call this the *box* basis for $\mathcal{V}_8$. Note that these are mutually orthogonal.

We similarly define $\mathcal{V}_4, \mathcal{V}_2, \mathcal{V}_1$. An image in $\mathcal{V}_4$ (the one obtained by downsampling an earlier image) is

An image in $\mathcal{V}_2$ looks like

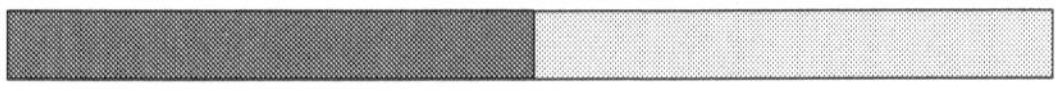

and, finally, an image in $\mathcal{V}_1$ looks like

You can probably figure out the box bases for $\mathcal{V}_4, \mathcal{V}_2, \mathcal{V}_1$. In general, the box basis for $\mathcal{V}_k$ consists of k vectors; vector i has ones in positions $ki, ki+1, ki+2, \ldots, ki+(k-1)$ and zeroes elsewhere.

10.3.2 Decomposing $\mathcal{V}_n$ as a direct sum

Wavelets arise from considering the orthogonal complements of subspaces of low-res images in the vector spaces of high-res images. For any positive integer $k < n$ that is a power of two, define the wavelet space $\mathcal{W}_k$ to be the orthogonal complement of $\mathcal{V}_k$ in $\mathcal{V}_{2k}$. By the Orthogonal-Complement Theorem,

$$\mathcal{V}_{2k} = \mathcal{V}_k \oplus \mathcal{W}_k \tag{10.1}$$

By plugging in $k = 8, 4, 2, 1$, we get

$$\begin{aligned} \mathcal{V}_{16} &= \mathcal{V}_8 \oplus \mathcal{W}_8 \\ \mathcal{V}_8 &= \mathcal{V}_4 \oplus \mathcal{W}_4 \\ \mathcal{V}_4 &= \mathcal{V}_2 \oplus \mathcal{W}_2 \\ \mathcal{V}_2 &= \mathcal{V}_1 \oplus \mathcal{W}_1 \end{aligned}$$

By using substitution repeatedly, we infer

$$\mathcal{V}_{16} = \mathcal{V}_1 \oplus \mathcal{W}_1 \oplus \mathcal{W}_2 \oplus \mathcal{W}_4 \oplus \mathcal{W}_8 \tag{10.2}$$

Therefore one basis for $\mathcal{V}_{16}$ is the union of:

- a basis for $\mathcal{V}_1$,
- a basis for $\mathcal{W}_2$,
- a basis for $\mathcal{W}_4$, and

- a basis for $\mathcal{W}_8$.

We will derive such a basis, the *Haar* basis. We will refer to the vectors forming this basis as *wavelet* vectors.

More generally,

$$\begin{aligned} \mathcal{V}_n &= \mathcal{V}_{n/2} \oplus \mathcal{W}_{n/2} \\ \mathcal{V}_{n/2} &= \mathcal{V}_{n/4} \oplus \mathcal{W}_{n/4} \\ &\vdots \\ \mathcal{V}_4 &= \mathcal{V}_2 \oplus \mathcal{W}_2 \\ \mathcal{V}_2 &= \mathcal{V}_1 \oplus \mathcal{W}_1 \end{aligned}$$

so

$$\mathcal{V}_n = \mathcal{V}_1 \oplus \mathcal{W}_1 \oplus \mathcal{W}_2 \oplus \mathcal{W}_4 \cdots \oplus \mathcal{W}_{n/2}$$

We get the *Haar* basis for $\mathcal{V}_n$ by choosing a particular basis for each of $\mathcal{V}_1, \mathcal{W}_1, \mathcal{W}_2, \mathcal{W}_4, \ldots, \mathcal{W}_{n/2}$, and taking the union.

10.3.3 The wavelet bases

We derive the bases for $\mathcal{W}_8, \mathcal{W}_4, \mathcal{W}_2, \mathcal{W}_1$. Recall that $\mathcal{W}_k$ is the orthogonal complement of $\mathcal{V}_k$ in $\mathcal{V}_{2k}$. Let's just use our method for computing generators for the orthogonal complement of a space for which we have an orthogonal basis. For example, to get generators for $\mathcal{W}_8$, we project the basis of $\mathcal{V}_{16}$ orthogonal to our basis for $\mathcal{V}_8$. We get a bunch of orthogonal vectors, of which half are zero; the other half form our basis for $\mathcal{W}_8$. We name them $\boldsymbol{w}_0^8, \boldsymbol{w}_1^8, \ldots, \boldsymbol{w}_7^8$. What form do they have?

The first wavelet vector is the projection of $\boldsymbol{b}_0^{16}$ orthogonal to the basis $\boldsymbol{b}_0^8, \ldots, \boldsymbol{b}_7^8$ of $\mathcal{V}_8$. That is, the projection of

$\boldsymbol{b}_0^{16} =$

1	0	0	0	0	0	0	0	0	0	0	0	0	0	0	0

orthogonal to

$\boldsymbol{b}_0^8 =$	1	1	0	0	0	0	0	0	0	0	0	0	0	0	0	0
$\boldsymbol{b}_1^8 =$	0	0	1	1	0	0	0	0	0	0	0	0	0	0	0	0
																⋮
$\boldsymbol{b}_7^8 =$	0	0	0	0	0	0	0	0	0	0	0	0	0	0	1	1

The projection of $\boldsymbol{b}_0^{16}$ along $\boldsymbol{b}_0^8$ is

$$((\boldsymbol{b}_0^{16} \cdot \boldsymbol{b}_0^8)/(\boldsymbol{b}_0^8 \cdot \boldsymbol{b}_0^8))\boldsymbol{b}_0^8$$

The numerator is 1 and the denominator is 2, so the projection of $\boldsymbol{b}_0^{16}$ along $\boldsymbol{b}_0^8$ is

.5	.5	0	0	0	0	0	0	0	0	0	0	0	0	0	0

Subtracting this vector from $\boldsymbol{b}_0^{16}$ yields

.5	-.5	0	0	0	0	0	0	0	0	0	0	0	0	0	0

Ordinarily, `project_orthogonal` would continue, projecting the result orthogonal to $\boldsymbol{b}_1^8, \ldots, \boldsymbol{b}_7^8$, but the result is already orthogonal to those vectors, so this is our first basis vector, $\boldsymbol{w}_0^8$.

Projecting $\boldsymbol{b}_1^{16}$ orthogonal to the box basis for $\mathcal{V}_8$ just gives the negative of $\boldsymbol{w}_0^8$, so we proceed to project $\boldsymbol{w}_2^{16}$ orthogonal to that basis, $\boldsymbol{b}_0^8, \ldots, \boldsymbol{b}_7^8$. Note that $\boldsymbol{w}_2^{16}$ is orthogonal to all these vectors except $\boldsymbol{b}_1^8$. The result is our second basis vector

$\boldsymbol{w}_1^8 =$

0	0	.5	-.5	0	0	0	0	0	0	0	0	0	0	0	0

We similarly obtain $\boldsymbol{w}_2^8, \boldsymbol{w}_3^8, \ldots, \boldsymbol{w}_7^8$. Each of these basis vectors has the same form: two adjacent entries having values .5 and -.5, and all other entries having value zero.

Note that the squared norm of each such vector is

$$(\frac{1}{2})^2 + (\frac{1}{2})^2 = \frac{1}{4} + \frac{1}{4} = \frac{1}{2}$$

The wavelet vectors $\boldsymbol{w}_0^8, \ldots, \boldsymbol{w}_7^8$ are an orthogonal basis for the orthogonal complement $\mathcal{W}_8$ of $\mathcal{V}_8$ in $\mathcal{V}_{16}$. Combining these vectors with the box vectors for $\mathcal{V}_8$ yields an orthogonal basis for $\mathcal{V}_{16}$.

We use the same approach to derive the vectors that comprise a basis for $\mathcal{W}_4$, the orthogonal complement of $\mathcal{V}_4$ in $\mathcal{V}_8$: project the basis vectors of $\mathcal{V}_8$ orthogonal to the basis vectors of $\mathcal{V}_4$. The resulting basis for $\mathcal{W}_4$ is

$\boldsymbol{w}_0^4$	=	.5	.5	-.5	-.5	0	0	0	0	0	0	0	0	0	0	0	0
$\boldsymbol{w}_1^4$	=	0	0	0	0	.5	.5	-.5	-.5	0	0	0	0	0	0	0	0
$\boldsymbol{w}_2^4$	=	0	0	0	0	0	0	0	0	.5	.5	-.5	-.5	0	0	0	0
$\boldsymbol{w}_3^4$	=	0	0	0	0	0	0	0	0	0	0	0	0	.5	.5	-.5	-.5

The squared norm of each such vector is

$$(.5)^2 + (.5)^2 + (.5)^2 + (.5^2 = 4(.25) = 1$$

The basis for $\mathcal{W}_2$ is

$\boldsymbol{w}_0^2$	=	.5	.5	.5	.5	-.5	-.5	-.5	-.5	0	0	0	0	0	0	0	0
$\boldsymbol{w}_1^2$	=	0	0	0	0	0	0	0	0	.5	.5	.5	.5	-.5	-.5	-.5	-.5

and the squared norms of these vectors is $8(.5)^2 = 2$.

The basis for $\mathcal{W}_1$ consists of the single vector

$\boldsymbol{w}_0^1$	=	.5	.5	.5	.5	.5	.5	.5	.5	-.5	-.5	-.5	-.5	-.5	-.5	-.5	-.5

which has squared norm $16(.5)^2 = 4$.

10.3.4 The basis for $\mathcal{V}_1$

The basis for $\mathcal{V}_1$, the space of one-pixel images, consists of the single vector

$\boldsymbol{b}_0^1$	=	1	1	1	1	1	1	1	1	1	1	1	1	1	1	1	1

and its squared norm is 16.

The vector $\boldsymbol{b}_0^1$, together with the wavelet vectors

$$\begin{array}{r} \boldsymbol{w}_0^8, \boldsymbol{w}_1^8, \boldsymbol{w}_2^8, \boldsymbol{w}_3^8, \boldsymbol{w}_4^8, \boldsymbol{w}_5^8, \boldsymbol{w}_6^8, \boldsymbol{w}_7^8, \\ \boldsymbol{w}_0^4, \boldsymbol{w}_1^4, \boldsymbol{w}_2^4, \boldsymbol{w}_3^4, \\ \boldsymbol{w}_0^2, \boldsymbol{w}_1^2, \\ \boldsymbol{w}_0^1 \end{array}$$

form the Haar wavelet basis for $\mathcal{V}^{16}$.

10.3.5 General n

We will want to consider the Haar wavelet basis for values of n other than 16.

For every power of two $s \leq n$, the basis vectors for $\mathcal{W}_k$ are written $\boldsymbol{w}_0^s, \ldots, \boldsymbol{w}_{k-1}^s$. The vector $\boldsymbol{w}_i^s$ has value one-half in entries $(n/k)i, (n/k)i + 1, \ldots, (n/k)i + n/2k - 1$, and value negative one-half in entries $(n/k)i + n/2k, (n/k)i + n/2k + 1, \ldots, (n/k)i + n/k - 1$, and value zero in all other entries.

The squared norm of each such vector is therefore

$$\|\boldsymbol{w}_i^s\|^2 = (n/k)\left(\frac{1}{2}\right)^2 = n/4s \tag{10.3}$$

The basis vector $\boldsymbol{b}_0^1$ for $\mathcal{V}_1$ has a one in each of the n entries. Its squared norm is therefore

$$\|\boldsymbol{b}_0^1\|^2 = n \tag{10.4}$$

10.3.6 The first stage of wavelet transformation

Our initial basis for $\mathcal{V}_{16}$ is the box basis. The decomposition $\mathcal{V}_{16} = \mathcal{V}_8 \oplus \mathcal{W}_8$ corresponds to another basis for $\mathcal{V}_{16}$, namely the union of the box basis for $\mathcal{V}_8$ and the wavelet basis for $\mathcal{W}_8$. Given a vector represented in terms of the initial basis, the first level of the transform produces a representation in terms of the other basis.

For example, let $\boldsymbol{v}$ be the one-dimensional image vector

$$\begin{bmatrix} 4 & 5 & 3 & 7 & 4 & 5 & 2 & 3 & 9 & 7 & 3 & 5 & 0 & 0 & 0 & 0 \end{bmatrix}$$

which looks like this: 

We represent this over the initial basis as

$$\boldsymbol{v} = 4\,\boldsymbol{b}_0^{16} + 5\,\boldsymbol{b}_1^{16} + 3\,\boldsymbol{b}_2^{16} + \cdots + 0\boldsymbol{b}_{15}^{16}$$

The input vector is represented as the list

```
>>> v = [4,5,3,7,4,5,2,3,9,7,3,5,0,0,0,0]
```

of coefficients in this linear combination.

Our goal is to represent this vector in terms of the basis $\boldsymbol{b}_0^8, \ldots, \boldsymbol{b}_7^8, \boldsymbol{w}_0^8, \ldots \boldsymbol{w}_7^8$. That is, we want $x_0, \ldots, x_7, y_0, \ldots, y_7$ such that

$$\boldsymbol{v} = x_0\,\boldsymbol{b}_0^8 + \cdots + x_7\,\boldsymbol{b}_7^8 + y_0\,\boldsymbol{w}_0^8 + \cdots + y_7\,\boldsymbol{w}_7^8$$

Since the vectors on the right-hand side are mutually orthogonal, each term on the right-hand side is the projection of $\boldsymbol{v}$ along the corresponding vector, so the coefficient can be found using the project-along formula:

$$\begin{aligned} x_i &= (\boldsymbol{v} \cdot \boldsymbol{b}_i^8)/(\boldsymbol{b}_i^8 \cdot \boldsymbol{b}_i^8) \\ y_i &= (\boldsymbol{v} \cdot \boldsymbol{w}_i^8)/(\boldsymbol{w}_i^8 \cdot \boldsymbol{w}_i^8) \end{aligned}$$

For example, consider $\boldsymbol{b}_0^8$. It has ones in entries 0 and 1, so its coefficient is

$$\begin{aligned} (\boldsymbol{v} \cdot \boldsymbol{b}_0^8)/(\boldsymbol{b}_0^8 \cdot \boldsymbol{b}_0^8) &= (4+5)/(1+1) \\ &= 4.5 \end{aligned}$$

That is, the coefficient of $\boldsymbol{b}_0^8$ is the average of the first two entries, which is 4.5. In general, for $i = 0, 1, 2, \ldots, 7$, the coefficient of $\boldsymbol{b}_i^8$ is the average of entries $2i$ and $2i+1$. Here's how the computation looks:

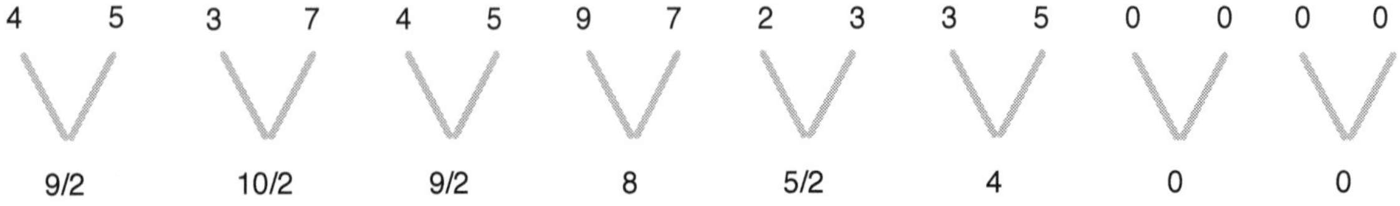

You could use a comprehension to obtain the list consisting of coefficients of $\boldsymbol{b}_0^8, \ldots, \boldsymbol{b}_7^8$:

```
vnew = [(v[2*i]+v[2*i+1])/2 for i in range(len(v)//2)]
```

Next, to find the coefficient of $\boldsymbol{w}_0^8$, we again use the project-along formula:

$$\begin{aligned}(\boldsymbol{v}\cdot\boldsymbol{w}_0^8)/(\boldsymbol{w}_0^8\cdot\boldsymbol{w}_0^8) &= \left(\frac{1}{2}4-\frac{1}{2}5\right)/\left(\frac{1}{2}\frac{1}{2}+\frac{1}{2}\frac{1}{2}\right)\\ &= 4-5\end{aligned}$$

because the $\frac{1}{2}$ in the denominator cancels out a $\frac{1}{2}$ in the numerator.

That is, the coefficient of $\boldsymbol{w}_0^8$ is entry 0 minus entry 1. In general, for $i = 0, 1, 2, \ldots, 7$, the coefficient of $\boldsymbol{w}_i^8$ is entry $2i$ minus entry $2i+1$.

Intuitively, the coefficients of the box vectors are the averages of pairs of intensities, and the coefficients of the wavelet vectors are the differences.

10.3.7 The subsequent levels of wavelet decomposition

We have so far described only one level of wavelet decomposition. We have shown

- how to get from a list of coefficients of box vectors $\boldsymbol{b}_0^{16}, \ldots, \boldsymbol{b}_{15}^{16}$ to a list of coefficients of box vectors $\boldsymbol{b}_0^8, \ldots, \boldsymbol{b}_7^8$, and

- how to calculate the corresponding eight wavelet coefficients, the coefficients of $\boldsymbol{w}_0^8, \ldots, \boldsymbol{w}_7^8$.

Just as the coefficients of box vectors $\boldsymbol{b}_0^{16}, \ldots, \boldsymbol{b}_{15}^{16}$ are the pixel intensities of a sixteen-pixel one-dimensional image, it is helpful to think of the coefficients of box vectors $\boldsymbol{b}_0^8, \ldots, \boldsymbol{b}_7^8$ as the pixel intensities of the eight-pixel one-dimensional image obtained by subsampling from the sixteen-pixel image:

The next level of wavelet decomposition consists of performing the same operation on the eight-pixel image. This results in a four-pixel image

and four more wavelet coefficients.

Another level of wavelet decomposition operates on the four-pixel image, yielding a two-pixel image

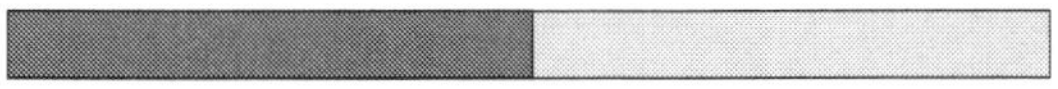

andtwo more wavelet coefficients.

A final level of wavelet decomposition operates on the two-pixel image, yielding a one-pixel image

and one more wavelet coefficient.

The computation of the box-vector coefficients is shown in the following diagram:

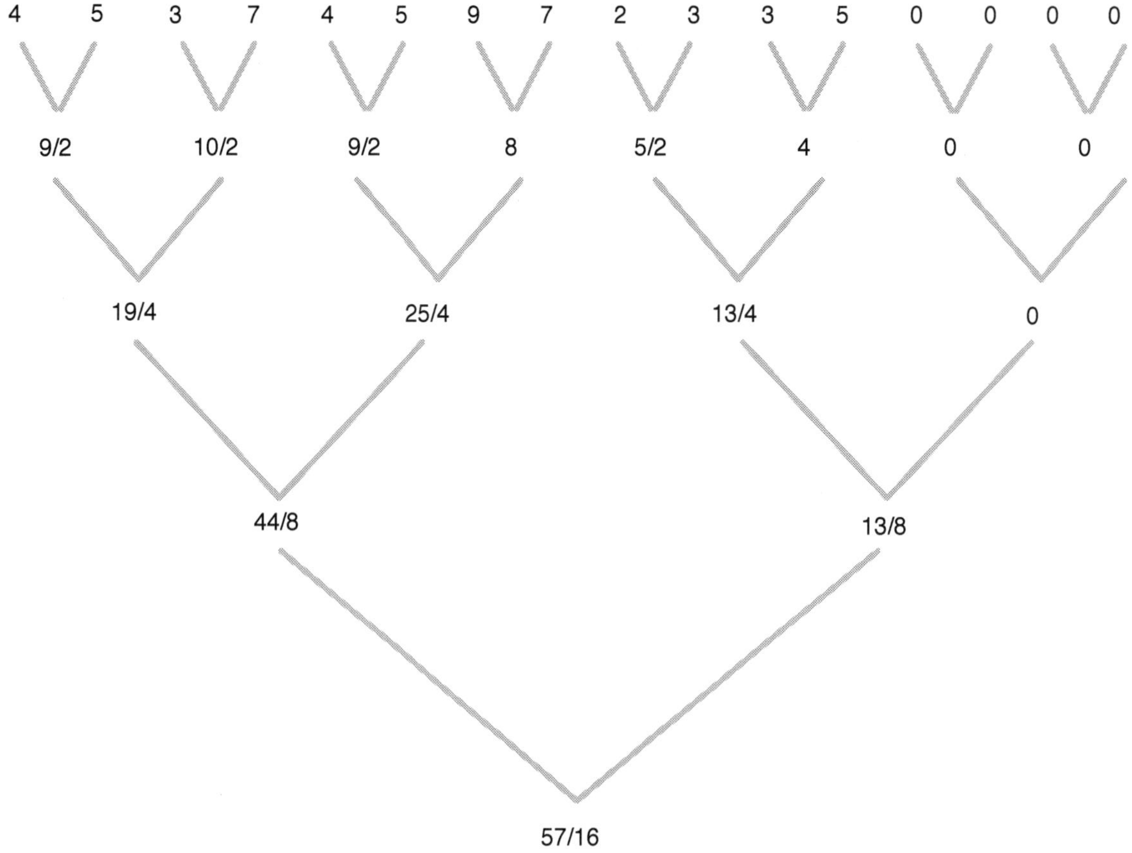

The next diagram shows the computation of the box-vector coefficients and also the wavelet coefficients (shown in ovals):

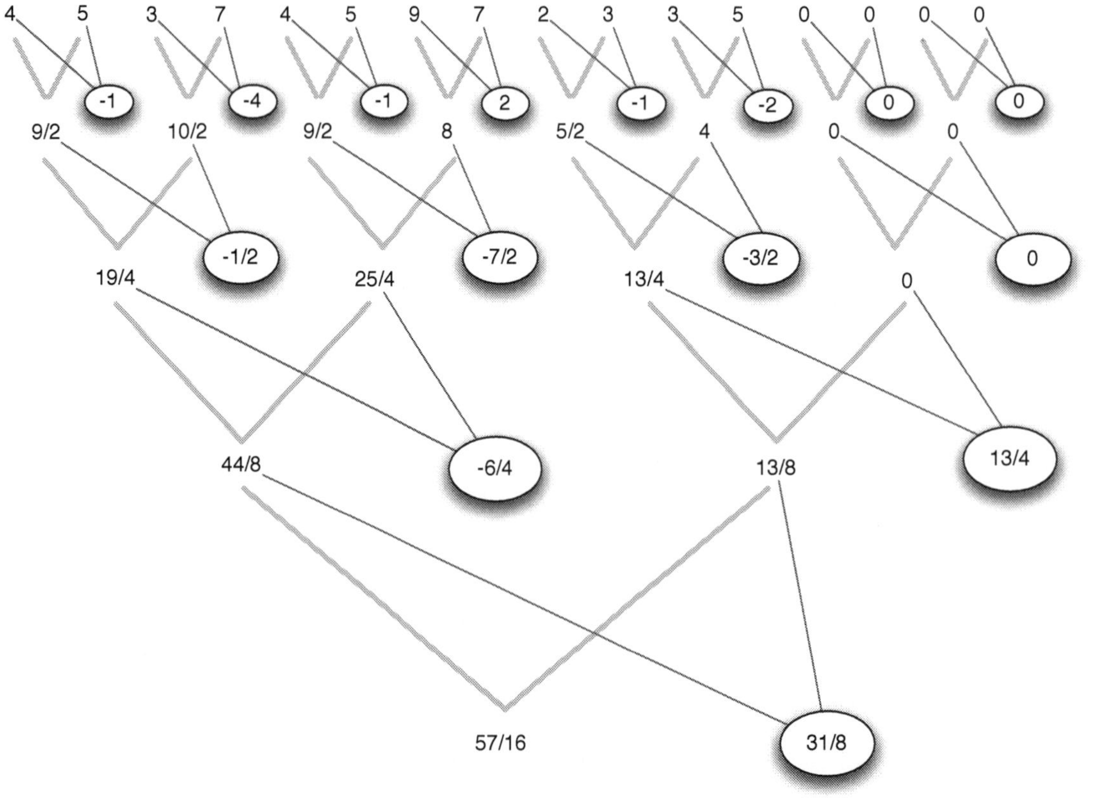

There are in total $8+4+2+1=15$ wavelet coefficients. Since the original vector space had dimension sixteen, we need one more number to represent the image uniquely: the intensity of the one pixel comprising the one-pixel image.

This, then, is the wavelet transform for $\mathcal{V}_{16}$. The input is the list of intensities of the original one-dimensional image. The output consists of

- the fifteen wavelet coefficients, the coefficients of the wavelet vectors

$$\begin{array}{r} \boldsymbol{w}_0^8, \boldsymbol{w}_1^8, \boldsymbol{w}_2^8, \boldsymbol{w}_3^8, \boldsymbol{w}_4^8, \boldsymbol{w}_5^8, \boldsymbol{w}_6^8, \boldsymbol{w}_7^8, \\ \boldsymbol{w}_0^4, \boldsymbol{w}_1^4, \boldsymbol{w}_2^4, \boldsymbol{w}_3^4, \\ \boldsymbol{w}_0^2, \boldsymbol{w}_1^2, \\ \boldsymbol{w}_0^1 \end{array}$$

- together with the intensity of the pixel of the one-pixel image (the overall average intensity).

For an n-pixel one-dimensional image (where n is a power of two), the input is the list of intensities of the original image, and the output is

- the wavelet coefficients, the number of which is

$$n/2+n/4+n/8+\cdots+2+1=n-1$$

- together with the overall average intensity.

10.3.8 Normalizing

The bases we have described are not orthonormal because the vectors do not have norm one. For the purpose of compression (e.g.), it is better to represent vectors in terms of an orthonormal basis. Normalizing a basis vector consists of dividing it by its norm. To compensate, the corresponding coefficient must be multiplied by the norm of the basis vector. That is, given a representation of a vector $\boldsymbol{v}$ in terms of vectors that are not necessarily unit-norm,

$$\boldsymbol{v}=\alpha_1\boldsymbol{v}_1+\cdots+\alpha_n\boldsymbol{v}_n$$

the representation in terms of the normalized versions is:

$$\boldsymbol{v}=(||\boldsymbol{v}_1||\alpha_1)\frac{\boldsymbol{v}_1}{||\boldsymbol{v}_1||}+\cdots+(||\boldsymbol{v}_n||\alpha_n)\frac{\boldsymbol{v}_n}{||\boldsymbol{v}_n||}$$

The procedure we have described for finding the wavelet coefficients produces coefficients of *unnormalized* basis vectors. There is a final step, therefore, in which the coefficients must be adjusted so that they are coefficients of normalized basis vectors.

- The squared norm of the unnormalized Haar wavelet basis vector $\boldsymbol{w}_i^s$ is $n/4s$, according to Equation 10.3, so the coefficient must be multiplied by $\sqrt{n/4s}$.
- The squared norm of the box vector $\boldsymbol{b}_0^1$ that forms the basis for $\mathcal{V}_1$ is n, according to Equation 10.4, so the coefficient must be multiplied by $\sqrt{n}$.

10.3.9 The backward transform

The wavelet transform's output consists of the coefficients of a set of vectors that form a basis for the original vector space. We have lost no information, so we can reverse the process. The backward transform uses the wavelet coefficients to find the coefficients of box basis of $\mathcal{V}_2, \mathcal{V}_4, \mathcal{V}_8, \ldots$ in succession.

10.3.10 Implementation

In the lab, you will implement the forward and backward transformations. You will also see how these can be used in tranforming real, 2-dimensional images, and experiment with lossy image compression.

10.4 Polynomial evaluation and interpolation

A *degree-d polynomial* is a single-variable function of the form

$$f(x) = a_0 1 + a_1 x^1 + a_2 x^2 + \cdots + a_d x^d$$

where $a_0, a_1, \ldots, a_d$ are scalar values. These are called the *coefficients* of the polynomial. You can specify a degree-d polynomial by the $d+1$-vector consisting the coefficients: $(a_0, a_1, a_2, \ldots, a_d)$.

The *value* of a polynomial at a number r is just the image of r under the function, i.e. plug r in for x, and find the value. *Evaluating* the polynomial means obtaining that value. For example, evaluating the polynomial $2 + 3x + x^2$ at 7 yields 72.

If for a given number r the value of the polynomial is zero, we call r a *root* of the polynomial.

Theorem 10.4.1: For any nonzero polynomial $f(x)$ of degree d, there are at most d values of x whose images under f is zero.

Unless all coefficients except a_1 are zero, a polynomial is not a linear function. However, there are linear functions lurking here: for a given number r, the function that takes a vector $(a_0, a_1, a_2, \ldots, a_d)$ of coefficients and outputs the value of the corresponding polynomial at r. For example, say r is 2 and d is 3. The corresponding linear function is

$$g((a_0, a_1, a_2, a_3)) = a_0 + a_1 2 + a_2 4 + a_3 8$$

The function can be written as a matrix-by-column-vector product:

$$\begin{bmatrix} 1 & 2 & 4 & 8 \end{bmatrix} \begin{bmatrix} a_0 \\ a_1 \\ a_2 \\ a_3 \end{bmatrix}$$

For an arbitrary value r, the matrix on the left would be

$$\begin{bmatrix} r^0 & r^1 & r^2 & r^3 \end{bmatrix}$$

More generally, suppose we have k numbers $r_0, r_1, \ldots, r_{k-1}$. The corresponding linear function takes a vector $(a_0, \ldots, a_d)$ of coefficients specifying a degree-d polynomial, and outputs the vector consisting of

- the value of the polynomial at r_0,
- the value of the polynomial at r_1,
- $\vdots$
- the value of the polynomial at r_{k-1}

For the case $d = 3$, here is this linear function as a matrix-by-column-vector product:

$$\begin{bmatrix} r_0^0 & r_0^1 & r_0^2 & r_0^3 \\ r_1^0 & r_1^1 & r_1^2 & r_1^3 \\ r_2^0 & r_2^1 & r_2^2 & r_2^3 \\ r_3^0 & r_3^1 & r_3^2 & r_3^3 \\ \vdots & & & \\ r_{k-1}^0 & r_{k-1}^1 & r_{k-1}^2 & r_{k-1}^3 \end{bmatrix} \begin{bmatrix} a_0 \\ a_1 \\ a_2 \\ a_3 \end{bmatrix}$$

For arbitrary d, the linear function can be written this way:

$$\begin{bmatrix} r_0^0 & r_0^1 & r_0^2 & \cdots & r_0^d \\ r_1^0 & r_1^1 & r_1^2 & \cdots & r_1^d \\ r_2^0 & r_2^1 & r_2^2 & \cdots & r_2^d \\ r_3^0 & r_3^1 & r_3^2 & \cdots & r_3^d \\ \vdots & & & & \\ r_{k-1}^0 & r_{k-1}^1 & r_{k-1}^2 & \cdots & r_{k-1}^d \end{bmatrix} \begin{bmatrix} a_0 \\ a_1 \\ a_2 \\ \vdots \\ a_d \end{bmatrix}$$

Theorem 10.4.2: In the case that $k = d + 1$ and the numbers $r_0, \ldots, r_{k-1}$ are all distinct, the function is invertible.

Proof

Suppose that there were two degree-d polynomials $f(x)$ and $g(x)$ for which $f(r_0) = g(r_0), f(r_1) = g(r_1), \ldots, f(r_d) = g(r_d)$. Define a third polynomial $h(x) = f(x) - g(x)$. Then h has degree at most d, and $h(r_0) = h(r_1) = \cdots = h(r_d) = 0$. By the earlier theorem, therefore, $h(x)$ is the zero polynomial, which shows $f(x) = g(x)$. □

There is a function that, given the values of a degree-d polynomial at $r_0, \ldots, r_d$, returns the coefficients of the polynomial. The process of obtaining the coefficients from the values is called *polynomial interpolation*. Thus polynomial evaluation and polynomial interpolation are inverse functions.

We defined degree-d polynomials in terms of coefficients, but an alternative representation is in terms of the values at $r_0, \ldots, r_d$. Each of these two representations has its strengths. For example:

- If you want to evaluate the polynomial at a completely new number, it's convenient to have the coefficients.
- If you want to multiply two polynomials, it's easier to do it using the values representation: just multiply the corresponding values.

In fact, the fastest algorithm known for multiplying two degree-d polynomials given in terms of their coefficients consists of (a) converting their representations to the value-based representation, (b) performing the multiplication using those representations, and (c) converting the result back to the coefficient representation.

For this to be a fast algorithm, however, one cannot evaluate the polynomials at just any numbers; the numbers must be carefully chosen. The key subroutine is the Fast Fourier Algorithm.

10.5 Fourier transform

A sound clip can be stored digitally as a sequence of amplitude samples. Connect a microphone to an analog-to-digital converter, and the converter will output a sequence of amplitudes represented digitally at a certain rate (say forty thousand samples per second). Say you have two seconds of sound. That's eighty-thousand numbers. You can represent that as an 80,000-vector. This is the representation in terms of the standard basis.

A pure tone is a sine wave. The Discrete Fourier basis consists of sine waves. If the sound sample is the result of sampling a pure tone and the frequency of the pure tone is carefully chosen, the representation of the sound sample in the Fourier basis will be very sparse—just one nonzero. More generally, if the sound sample consists of just a few pure tones added together, the Fourier representation will still be sparse.

Here is an example of a signal obtained by mixing two pure tones:

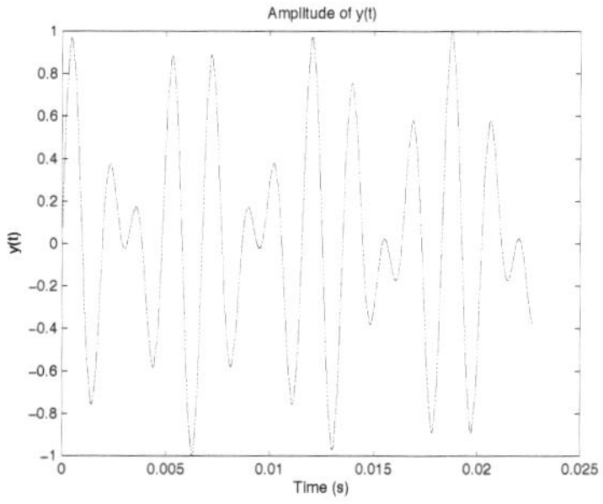

Here is a plot of the signal's coordinate representation in terms of the Fourier basis:

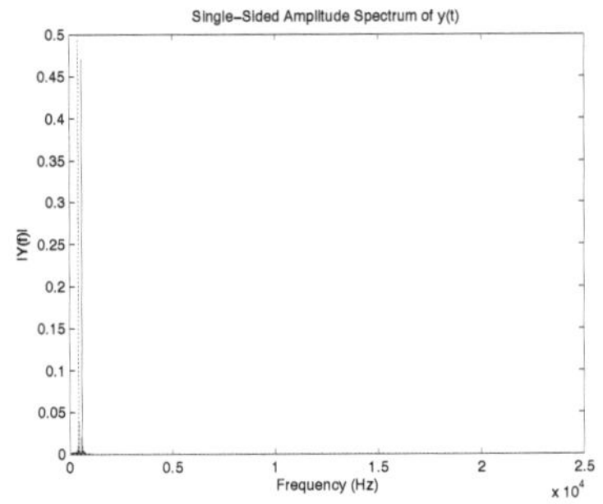

Here is a signal produced by a random-number generator. A randomly generated signal is called *noise*:

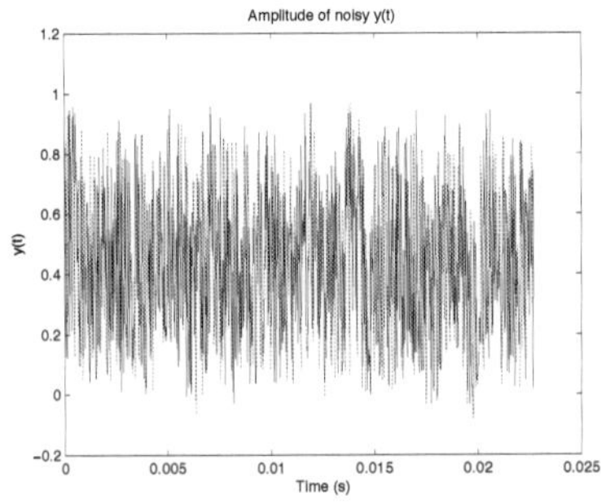

When the noise is added to the signal consisting of two pure tones, it still *looks* pretty random, but the ear can pick out the two tones about the noise. So can the Fourier transform; here's a plot of the noise-plus-signal's coordinate representation in terms of the Fourier basis:

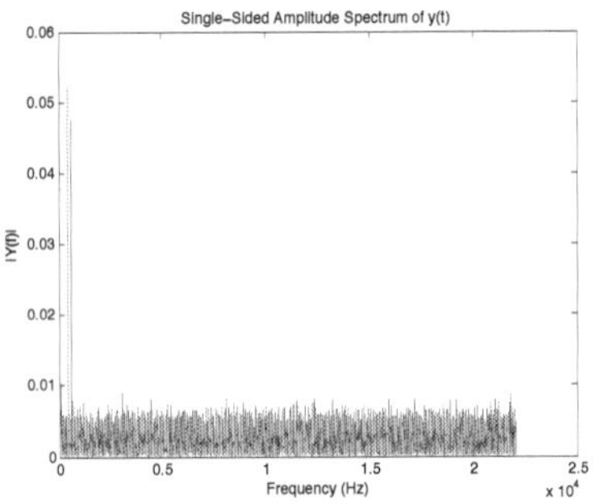

When we suppress the coefficients that are small, we obtain:

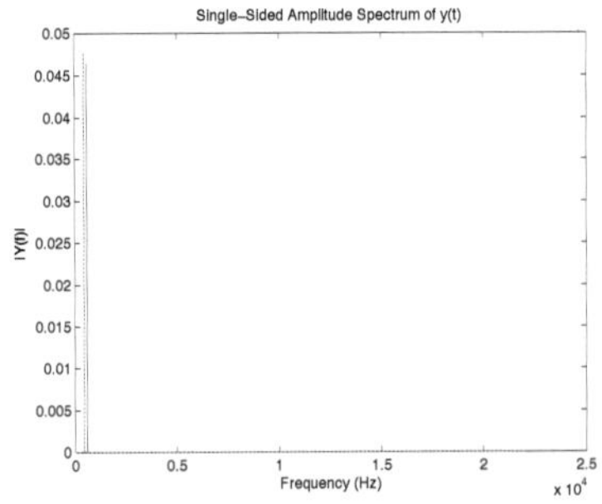

Transforming from this representation to the representation in terms of the standard basis, we obtain

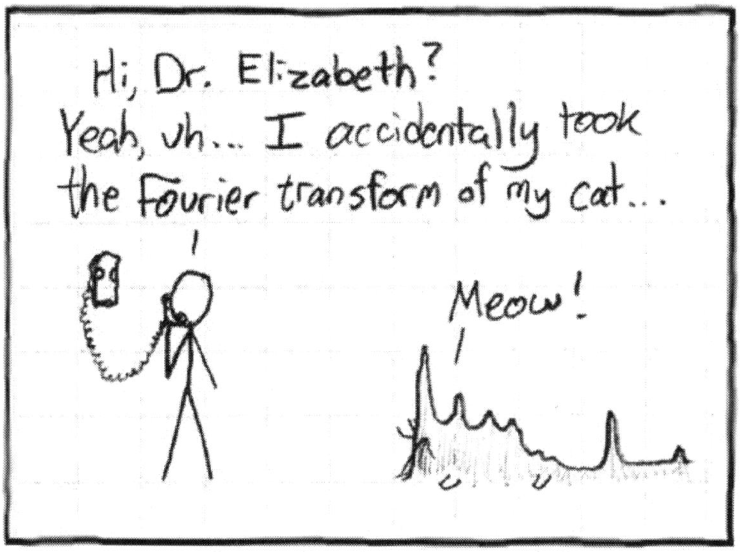

Figure 10.1: Fourier (`http://xkcd.com/26/`)

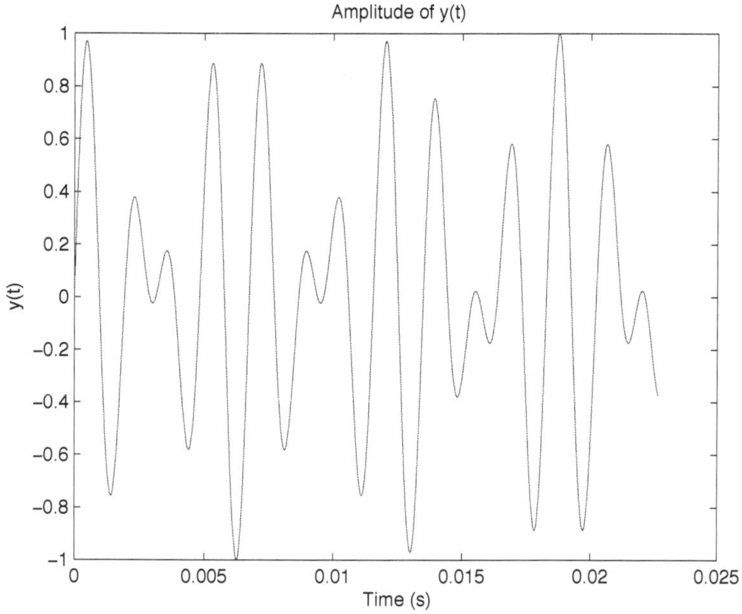

which looks (and sounds) just like the original signal.

10.6 Discrete Fourier transform

10.6.1 The Laws of Exponentiation

Exponentials are nice to work with because of the two familiar Laws of Exponentiation. We saw the First Law back in Section 1.4.9:

> $e^u e^v = e^{u+v}$ (multiplication of two exponentials (with the same base) is equivalent to *addition* of the exponents).

The Second Law of Exponentiation is

> $(e^u)^v = e^{uv}$ (composition of exponentiation is equivalent to *multiplication* of the exponents).

10.6.2 The n stopwatches

Let ω denote $e^{(2\pi/n)\mathbf{i}}$. Consider the function $F : \mathbb{R} \longrightarrow \mathbb{C}$ defined by $F(t) = \omega^t$. At $t = 0$, $F(t)$ is located at $1 + 0\mathbf{i}$. As t increases, $F(t)$ winds around the unit circle, returning to $1 + 0\mathbf{i}$ at $t = n, 2n, 3n, \ldots$. Thus $F(t)$ has a period of n (and therefore its frequency is $1/n$. Think of $F(t)$ as a stopwatch that runs counterclockwise.

Building on this idea, we define n stopwatches, each running at a different rate. However, these n stopwatches are much smaller than $F(t)$. Their radii of the circles are all $\frac{1}{\sqrt{n}}$ instead of 1. (We choose to have them be smaller so that, as you will see, a vector of values of these functions will have norm 1.)

We define $F_1 : \mathbb{R} \longrightarrow \mathbb{C}$ by

$$F_1(t) = \frac{1}{\sqrt{n}}\omega^t \tag{10.5}$$

so $F_1(t)$ is just like $F(t)$, only smaller. For example, $F(0) = \frac{1}{\sqrt{n}} + 0\mathbf{i}$. Note that the period of $F_1(t)$, like that of $F(t)$, is n.

For $k = 0, 1, 2, \ldots, n-1$, we define $F_k(t) = \frac{1}{\sqrt{n}}(\omega^k)^t$.

Each of the functions $F_0(t), F_1(t), F_2(t), \ldots, F_{n-1}(t)$ has value $\frac{1}{\sqrt{n}} + 0\mathbf{i}$ at $t = 0$, but as t increases they all go around the radius-$\frac{1}{\sqrt{n}}$ circle at different rates.

- $F_2(t)$ first returns to $\frac{1}{\sqrt{n}} + 0\mathbf{i}$ when t reaches $n/2$,
- $F_3(t)$ first returns when $t = n/3$, and so on.

Thus the period of $F_k(t)$ is n/k.

- The period of $F_0(t)$ is $n/0$, i.e. infinity, because $F_0(t)$ never moves—its hand perpetually points to $\frac{1}{\sqrt{n}} + 0\mathbf{i}$).

10.6.3 Discrete Fourier space: Sampling the basis functions

Fourier analysis allows us to represent functions as linear combinations of such clocks. Here we study *discrete* Fourier analysis, which represents a function $s(t)$ sampled at $t = 0, 1, 2, \ldots, n-1$

using a linear combination of the clocks clocks $F_0(t), F_1(t), F_2(t), \ldots, F_{n-1}(t)$ sampled at the same times.

The samples of the signal $s(t)$ are stored in a vector

$$\boldsymbol{s} = \begin{bmatrix} s(0) \\ s(1) \\ s(2) \\ \vdots \\ s(n-1) \end{bmatrix}$$

Similarly, we write the samples of clock $F_j(t)$ in a vector

$$\begin{bmatrix} F_j(0) \\ F_j(1) \\ F_j(2) \\ \vdots \\ F_j(n-1) \end{bmatrix}$$

Our goal is to write an equation showing how the vector $\boldsymbol{s}$ consisting of the samples of the signal can be represented as a linear combination of the vectors consisting of the samples of the clocks. The coefficients in this linear combination are called *Fourier coefficients.*

Fortunately, we know how to formulate this: we construct a matrix $\mathcal{F}$ with the clock-vectors as columns, we construct a vector ϕ of the Fourier coefficients $\phi_0, \phi_1, \phi_2, \ldots, \phi_{n-1}$ (one for each clock), and, drawing on the linear-combinations interpretation of matrix-vector multiplication, we write the matrix equation $\mathcal{F}_n \boldsymbol{f} = \boldsymbol{s}$. Written more explicitly, the equation is:

$$\left[\begin{array}{c|c|c|c|c} F_0(0) & F_1(0) & F_2(0) & & F_{n-1}(0) \\ F_0(1) & F_1(1) & F_2(1) & & F_{n-1}(1) \\ F_0(2) & F_1(2) & F_2(2) & & F_{n-1}(2) \\ F_0(3) & F_1(3) & F_2(3) & & F_{n-1}(3) \\ F_0(4) & F_1(4) & F_2(4) & & F_{n-1}(4) \\ F_0(5) & F_1(5) & F_2(5) & \cdots & F_{n-1}(5) \\ \vdots & \vdots & \vdots & & \vdots \\ F_0(n-1) & F_1(n-1) & F_2(n-1) & & F_{n-1}(n-1) \end{array}\right] \begin{bmatrix} \phi_0 \\ \phi_1 \\ \phi_2 \\ \vdots \\ \phi_{n-1} \end{bmatrix} = \begin{bmatrix} s(0) \\ s(1) \\ s(2) \\ s(3) \\ s(4) \\ s(5) \\ \vdots \\ s(n-1) \end{bmatrix}$$

This equation relates two representations of a signal: a representation in terms of samples of the signal, and a representation as a weighted sum of (sampled) clocks.

Each representation is useful.

- A `.wav` file represents a sound signal in terms of samples of the signal, and the representation in terms of clocks tells us which frequencies are most prominent in the sound. The signal is acquired in the first form (a microphone connected to an analog-to-digital converter produces samples) but often analyzing the signal involves converting to the second.
- In magnetic resonance imaging, the signal is acquired roughly as the vector ϕ of the Fourier coefficients, but creating a digital image requires the samples $\boldsymbol{s}$. image.

The equation $\mathcal{F}_n \phi = \boldsymbol{s}$ tells us how to convert between the two representations.

- Given $\boldsymbol{f}$, we can obtain $\boldsymbol{s}$ by left-multiplication by $\mathcal{F}_n$.
- Given $\boldsymbol{s}$, we can obtain ϕ by solving a matrix equation—or by left-multiplying $\boldsymbol{s}$ by the inverse matrix $\mathcal{F}_n^{-1}$.

10.6.4 The inverse of the Fourier matrix

Ordinarily, solving a linear system is preferable to computing a matrix inverse and multiplying by it. The Fourier matrix, however, is very special; its inverse is very similar to the Fourier matrix itself.

Our first step will be to describe the Fourier matrix as a scalar multiple of a special kind of matrix. Let ω be a complex number of the form $e^{\theta \mathbf{i}}$. We denote by $W(\omega, n)$ the matrix

- whose row-label and column-label set is $\{0, 1, 2, 3, \ldots, n-1\}$, and
- whose entry rc equals $\omega^{r \cdot c}$.

That is,

$$W(\omega, n) = \begin{bmatrix} \omega^{0\cdot 0} & \omega^{0\cdot 1} & \omega^{0\cdot 2} & & \omega^{0\cdot(n-1)} \\ \omega^{1\cdot 0} & \omega^{1\cdot 1} & \omega^{1\cdot 2} & & \omega^{1\cdot(n-1)} \\ \omega^{2\cdot 0} & \omega^{2\cdot 1} & \omega^{2\cdot 2} & & \omega^{2\cdot(n-1)} \\ \omega^{3\cdot 0} & \omega^{3\cdot 1} & \omega^{3\cdot 2} & \cdots & \omega^{3\cdot(n-1)} \\ \omega^{4\cdot 0} & \omega^{4\cdot 1} & \omega^{4\cdot 2} & & \omega^{4\cdot(n-1)} \\ \omega^{5\cdot 0} & \omega^{5\cdot 1} & \omega^{5\cdot 2} & & \omega^{5\cdot(n-1)} \\ \vdots & \vdots & \vdots & & \vdots \\ \omega^{(n-1)\cdot 0} & \omega^{(n-1)\cdot 1} & \omega^{(n-1)\cdot 2} & & \omega^{(n-1)\cdot(n-1)} \end{bmatrix}$$

In Python,

```
def W(w, n):
  R=set(range(n))
  return Mat((R,R), {(r,c):w**(r*c) for r in R for c in R})
```

although there is never any reason for actually representing this matrix explicitly in a computer. We can then write $\mathcal{F}_n = \frac{1}{\sqrt{n}} W(e^{2\pi \mathbf{i}/n}, n)$

Theorem 10.6.1 (Fourier Inverse Theorem): $\mathcal{F}_n^{-1} = \frac{1}{\sqrt{n}} W(e^{-2\pi \mathbf{i}/n}, n)$.

To prove the Fourier Inverse Theorem, it suffices to prove the following lemma.

Lemma 10.6.2: $W(e^{2\pi \mathbf{i}/n}, n) W(e^{-2\pi \mathbf{i}/n}, n) = n\,\mathbb{1}$.

Proof

Let ω denote $e^{2\pi \mathbf{i}/n}$. Note that $e^{-2\pi \mathbf{i}/n} = (e^{2\pi \mathbf{i}/n})^{-1} = \omega^{-1}$. By the dot-product definition of matrix-matrix multiplication, entry rc of this product is the dot-product of row r of $W(e^{2\pi \mathbf{i}/n}, n)$ with column c of $W(e^{-2\pi \mathbf{i}/n}, n)$, which is

$$\begin{aligned} &\omega^{r0}\omega^{-0c} + \omega^{r1}\omega^{-1c} + \omega^{r2}\omega^{-2c} + \cdots \omega^{r(n-1)}\omega^{-(n-1)c} & (10.6) \\ = \quad &\omega^{0(r-c)} + \omega^{1(r-c)} + \omega^{2(r-c)} + \cdots + \omega^{(n-1)(r-c)} \text{ by the addition-of-exponents Law} \end{aligned}$$

There are two possibilities. If $r = c$ then each of the exponents in the expression 10.6 equals 0, so the expression equals

$$\begin{aligned} & \omega^0 + \omega^0 + \omega^0 + \cdots + \omega^0 \\ = \quad & 1 + 1 + 1 + \cdots + 1 \\ = \quad & n \end{aligned}$$

Suppose $r \neq c$. By the multiplication-of-exponents Law, expression 10.6 equals

$$(\omega^{r-c})^0 + (\omega^{r-c})^1 + (\omega^{r-c})^2 + \cdots + (\omega^{r-c})^{n-1} \tag{10.7}$$

Let z be the value of this expression. We prove that z is zero. Since $r \neq c$ and both r and c are between 0 and $n-1$, we know ω^{r-c} is not equal to one. We show, however, that multiplying z by ω^{r-c} gives us z. That is,

$$\omega^{r-c} z = z$$

so

$$(\omega^{r-c} - 1) z = 0$$

so $z = 0$.

Now we complete the proof by showing that multiplying z by ω^{r-c} gives z:

$$\begin{aligned}
\omega^{r-c} z &= \omega^{r-c}((\omega^{r-c})^0 + (\omega^{r-c})^1 + (\omega^{r-c})^2 + \cdots + (\omega^{r-c})^{n-2} + (\omega^{r-c})^{n-1}) \\
&= (\omega^{r-c})^1 + (\omega^{r-c})^2 + (\omega^{r-c})^3 + \cdots + (\omega^{r-c})^{n-1} + (\omega^{r-c})^n \\
&= (\omega^{r-c})^1 + (\omega^{r-c})^2 + (\omega^{r-c})^3 + \cdots + (\omega^{r-c})^{n-1} + (\omega^n)^{r-c} \\
&= (\omega^{r-c})^1 + (\omega^{r-c})^2 + (\omega^{r-c})^3 + \cdots + (\omega^{r-c})^{n-1} + 1^{r-c} \\
&= (\omega^{r-c})^1 + (\omega^{r-c})^2 + (\omega^{r-c})^3 + \cdots + (\omega^{r-c})^{n-1} + (\omega^{r-c})^0 \\
&= z
\end{aligned}$$

□

10.6.5 The Fast Fourier Algorithm

In applications the number n of samples is usually very large, so we need to think about the time required by these computations. Left-multiplication by $\mathcal{F}$ or by $\mathcal{F}^{-1}$ would seem to take n^2 multiplications. However,

- we can multiply by $\mathcal{F}$ by multiplying by $W(e^{2\pi \mathbf{i}/n}, n)$ and then scaling by $\frac{1}{\sqrt{n}}$.
- Similarly, we can multiply by $\mathcal{F}^{-1}$ by first multiplying by $W(e^{-2\pi \mathbf{i}/n}, n)$ and then scaling.

There is a fast algorithm, called the *Fast Fourier transform* (FFT), for multiplying by $W(\omega, n)$ if the following preconditions hold:

FFT Preconditions

- n is a power of two, and
- $\omega^n = 1$.

The algorithm requires $O(n \log n)$ time, which for large n is much faster than naive matrix-vector multiplication. The FFT is a central part of modern digital signal processing.

10.6.6 Deriving the FFT

Here is the specification of FFT$(\omega, \boldsymbol{s})$:

- *input:*, a list $\boldsymbol{s} = [s_0, s_1, s_2, \ldots, s_{n-1}]$ of n complex numbers where n is a power of two, and a complex number ω such that $\omega^n = 1$
- *output:* the list $[z_0, z_1, z_2, \ldots, z_{n-1}]$ of complex numbers where

$$z_k = s_0(\omega^k)^0 + s_1(\omega^k)^1 + s_2(\omega^k)^2 + \cdots + s_{n-1}(\omega^k)^{n-1} \quad (10.8)$$

  for $k = 0, 1, 2, \ldots, n-1$

It is convenient to interpret the list $\boldsymbol{s}$ as representing a polynomial function:

$$s(x) = s_0 + s_1 x + s_2 x^2 + \cdots + s_{n-1} x^{n-1}$$

for then Equation 10.8 states that element k of the output list is the value of $s(x)$ on input ω^k.

The first step of the algorithm is to divide up the list $\boldsymbol{s} = [s_0, s_1, s_2, \ldots, s_{n-1}]$ into two lists, each consisting of half the elements:

- $\boldsymbol{s}_{even}$ consists of the even-numbered elements of $\boldsymbol{s}$:

$$\boldsymbol{s}_{even} = [s_0, s_2, s_4, \ldots, s_{n-2}]$$

  and

- $\boldsymbol{s}_{odd}$ consists of the odd-numbered elements of $\boldsymbol{s}$:

$$\boldsymbol{s}_{odd} = [s_1, s_2, s_3, \ldots, s_{n-1}]$$

We can interpret $\boldsymbol{s}_{even}$ and $\boldsymbol{s}_{odd}$ as representing polynomial functions:

$$\begin{aligned} s_{even}(x) &= s_0 + s_2 x + s_4 x^2 + s_6 x^3 + \cdots + s_{n-2} x^{\frac{n-2}{2}} \\ s_{odd}(x) &= s_1 + s_3 x + s_5 x^2 + s_7 x^3 + \cdots + s_{n-1} x^{\frac{n-2}{2}} \end{aligned}$$

The basis of the FFT is the following equation, in which we express the polynomial $s(x)$ in terms of the polynomials $s_{even}(x)$ and $s_{odd}(x)$:

$$s(x) = s_{even}(x^2) + x \cdot s_{odd}(x^2) \tag{10.9}$$

This equation allows us to achieve the goal of the FFT (evaluate $s(x)$ at $\omega^0, \omega^1, \omega^2, \ldots, \omega^{n-1}$) by evaluating $s_{even}(x)$ and s_{odd} at $(\omega^0)^2, (\omega^1)^2, (\omega^2)^2, \ldots, (\omega^{n-1})^2$ and combining corresponding values:

$$\begin{aligned} s(\omega^0) &= s_{even}((\omega^0)^2) + \omega^0 s_{odd}((\omega^0)^2) \\ s(\omega^1) &= s_{even}((\omega^1)^2) + \omega^1 s_{odd}((\omega^1)^2) \\ s(\omega^2) &= s_{even}((\omega^2)^2) + \omega^2 s_{odd}((\omega^2)^2) \\ s(\omega^3) &= s_{even}((\omega^3)^2) + \omega^3 s_{odd}((\omega^3)^2) \\ &\vdots \\ s(\omega^{n-1}) &= s_{even}((\omega^{n-1})^2) + \omega^{n-1} s_{odd}((\omega^{n-1})^2) \end{aligned} \tag{10.10}$$

It would seem as if the FFT would need to evaluate $s_{even}(x)$ and $s_{odd}(x)$ at n different values, but in fact

$$(\omega^0)^2, (\omega^1)^2, (\omega^2)^2, (\omega^3)^2, \ldots, (\omega^{n-1})^2$$

are not all distinct! Because of Precondition 10.6.5, $\omega^n = 1$ so

$$\begin{aligned} (\omega^0)^2 &= (\omega^0)^2(\omega^{\frac{n}{2}})^2 &= (\omega^{0+\frac{n}{2}})^2 \\ (\omega^1)^2 &= (\omega^1)^2(\omega^{\frac{n}{2}})^2 &= (\omega^{1+\frac{n}{2}})^2 \\ (\omega^2)^2 &= (\omega^2)^2(\omega^{\frac{n}{2}})^2 &= (\omega^{2+\frac{n}{2}})^2 \\ (\omega^3)^2 &= (\omega^3)^2(\omega^{\frac{n}{2}})^2 &= (\omega^{3+\frac{n}{2}})^2 \\ &\vdots \\ (\omega^{\frac{n}{2}-1})^2 &= (\omega^{\frac{n}{2}})^2(\omega^{\frac{n}{2}})^2 &= (\omega^{\frac{n}{2}-1+\frac{n}{2}})^2 \end{aligned} \tag{10.11}$$

which shows that there are only $n/2$ distinct numbers at which $s_{even}(x)$ and $s_{odd}(x)$ must be evaluated, namely

$$(\omega^0)^2, (\omega^1)^2, (\omega^2)^2, (\omega^3)^2, \ldots, (\omega^{\frac{n}{2}-1})^2$$

which are the same as

$$(\omega^2)^0, (\omega^2)^1, (\omega^2)^2, (\omega^2)^3, \ldots, (\omega^2)^{\frac{n}{2}-1}$$

Furthermore, the resulting values can be obtained using recursive calls to FFT:

- the values of $s_{even}(x)$ at each of these numbers can be obtained by calling

$$\texttt{f0} = \mathrm{FFT}(\omega^2, [s_0, s_2, s_4, \ldots, s_{n-2}])$$

  and

- the values of $s_{odd}(x)$ at each of these numbers can be obtained by calling

$$\texttt{f0} = \mathrm{FFT}(\omega^2, [s_1, s_3, s_5, \ldots, s_{n-1}])$$

After these statements have been executed,

$$\texttt{f0} = [s_{even}((\omega^2)^0),\ s_{even}((\omega^2)^1),\ s_{even}((\omega^2)^2),\ s_{even}((\omega^2)^3),\ \ldots,\ s_{even}((\omega^2)^{\frac{n}{2}-1})]$$

and

$$\texttt{f1} = [s_{odd}((\omega^2)^0),\ s_{odd}((\omega^2)^1),\ s_{odd}((\omega^2)^2),\ s_{odd}((\omega^2)^3),\ \ldots,\ s_{odd}((\omega^2)^{\frac{n}{2}-1})]$$

Once these values have been computed by the recursive call, the FFT combines them using Equations 10.10 to obtain

$$[s(\omega^0), s(\omega^1), s(\omega^2), s(\omega^3), \ldots, s(\omega^{\frac{n}{2}-1}), s(\omega^{\frac{n}{2}}), s(\omega^{\frac{n}{2}+1}), s(\omega^{\frac{n}{2}+2}), \ldots, s(\omega^{n-1})]$$

The first $n/2$ values are computed as you would expect:

$$\begin{aligned}[s(\omega^0),\ s(\omega^1),\ s(\omega^2),\ s(\omega^3),\ \ldots,\ s(\omega^{\frac{n}{2}-1})] = [& s_{even}(((\omega^2)^0) + \omega^0 \cdot s_{odd}((\omega^2)^0),\\ & s_{even}((\omega^2)^1) + \omega^1 \cdot s_{odd}((\omega^2)^1),\\ & s_{even}((\omega^2)^2 + \omega^2 \cdot s_{odd}((\omega^2)^2),\\ & s_{even}((\omega^2)^3) + \omega^3 \cdot s_{odd}((\omega^2)^3),\\ & \vdots\\ & s_{even}((\omega^2)^{\frac{n}{2}-1}) + \omega^{\frac{n}{2}-1} \cdot s_{odd}((\omega^2)^{\frac{n}{2}-1})]\end{aligned}$$

which can be computed using a comprehension: $[\texttt{f0}[j] + \omega^j * \texttt{f0}[j] \text{ for } j \text{ in range}(n//2)]$

The last $n/2$ values are computed similarly, except using the Equations 10.11:

$$\begin{aligned}[s(\omega^{\frac{n}{2}}),\ s(\omega^{\frac{n}{2}+1}),\ s(\omega^{\frac{n}{2}+2}),\ s(\omega^{\frac{n}{2}+3}),\ \ldots,\ s(\omega^{n-1})] = [& s_{even}((\omega^2)^0) + \omega^{\frac{n}{2}} \cdot s_{odd}((\omega^2)^0),\\ & s_{even}((\omega^2)^1) + \omega^{\frac{n}{2}+1} \cdot s_{odd}((\omega^2)^1),\\ & s_{even}((\omega^2)^2 + \omega^{\frac{n}{2}+2} \cdot s_{odd}((\omega^2)^2),\\ & s_{even}((\omega^2)^3) + \omega^{\frac{n}{2}+3} \cdot s_{odd}((\omega^2)^3),\\ & \vdots\\ & s_{even}((\omega^2)^{\frac{n}{2}-1}) + \omega^{n-1} \cdot s_{odd}((\omega^2)^{\frac{n}{2}-1})]\end{aligned}$$

which can also be computed using comprehension: $[\texttt{f0}[j] + \omega^{j+\frac{n}{2}} * \texttt{f0}[j] \text{ for } j \text{ in range}(n//2)]$.

10.6.7 Coding the FFT

Finally we give the Python code for FFT. There is a base case for the recursion, the case where the input list $\boldsymbol{s}$ is $[s_0]$. In this case, the polynomial $s(x)$ is just s_0, so the value of the polynomial at any number (and in particular at 1) is s_0. If the base case does not hold, FFT is called recursively on the even-numbered entries of the input list and the odd-numbered entries. The values returned are used in accordance with Equation 10.9 to compute the values of $s(x)$.

```
def FFT(w, s):
 n = len(s)
 if n==1: return [s[0]]
 f0 = FFT(w*w, [s[i] for i in range(n) if i % 2 == 0])
 f1 = FFT(w*w, [s[i] for i in range(n) if i % 2 == 1])
 return [f0[j]+w**j*f1[j] for j in range(n//2)] +
        [f0[j]-w**(j+n//2)*f1[j] for j in range(n//2)]
```

The analysis of the running time of FFT resembles that of, e.g., Mergesort (an algorithm commonly analyzed in introductory algorithms classes). The analysis shows that the number of operations is $O(n \log n)$.

Remark: For the case where the list length n is not a power of two, one option is to "pad" the input sequence $\boldsymbol{s}$ with extra zeroes until the length is a factor of two. FFT can also be tuned to specific values of n.

10.7 The inner product for the field of complex numbers

We have defined the inner product for the field of real numbers to be just dot-product. One requirement of inner product is that the inner product of a vector with itself is the norm, and one requirement for norm is that it be a nonnegative real number. However, the dot-product of a vector over the field $\mathbb{C}$ of complex numbers is not necessarily nonnegative! For example, the dot-product of the 1-vector $[\mathbf{i}]$ with itself is -1. Fortunately, there is a simple change to the definition of inner product that gives the same result as before if the vectors happen to have all real entries.

Recall that the conjugate of a complex number z, written $\bar{z}$, is defined as $z.\text{real} - z.\text{imag}$, and that the product of z and $\bar{z}$ is the absolute value of z, written $|z|$, which is a nonnegative real number.

Example 10.7.1: The value of $e^{\theta \mathbf{i}}$ is the point on the complex plane that is located on the unit circle and has argument θ, i.e. the complex number $\cos\theta + (\sin\theta)\mathbf{i}$. The conjugate is therefore $\cos\theta - (\sin\theta)\mathbf{i}$, which is the same as $\cos(-\theta) + (\sin(-\theta))\mathbf{i}$, which is $e^{-\theta\mathbf{i}}$. Thus conjugation is the same as negating the imaginary exponent.

For $\boldsymbol{z} = e^{\theta\mathbf{i}}$, what is the product of z and $\bar{z}$? Using the addition-of-exponents Law, $\boldsymbol{z\bar{z}} = e^{\theta\mathbf{i}}e^{-\theta\mathbf{i}} = e^{\theta\mathbf{i}-\theta\mathbf{i}} = e^0 = 1$

For a vector $\boldsymbol{v} = [z_1, \ldots, z_n]$ over $\mathbb{C}$, we denote by $\bar{\boldsymbol{v}}$ the vector derived from $\boldsymbol{v}$ by replacing each entry by its conjugate:

$$\bar{\boldsymbol{v}} = [\bar{z}_1, \ldots, \bar{z}_n]$$

Of course, if the entries of $\boldsymbol{v}$ happen to be real numbers then $\bar{\boldsymbol{v}}$ is the same as $\boldsymbol{v}$.

Definition 10.7.2: We define the inner product for vectors over the field of complex numbers as

$$\langle \boldsymbol{u}, \boldsymbol{v}\rangle = \bar{\boldsymbol{u}} \cdot \boldsymbol{v}$$

Of course, if the entries of $\boldsymbol{u}$ happen to be real numbers then the inner product is just the dot-product of $\boldsymbol{u}$ and $\boldsymbol{v}$.

This definition ensures that the inner product of a vector with itself is nonnegative. Suppose $\boldsymbol{v} = [z_1, \ldots, z_n]$ is an n-vector over $\mathbb{C}$. Then

$$\begin{aligned} \langle \boldsymbol{v}, \boldsymbol{v}\rangle &= [\bar{z}_1, \ldots, \bar{z}_n] \cdot [z_1, \ldots, z_n] && (10.12) \\ &= \bar{z}_1 z_1 + \cdots + \bar{z}_n z_n \\ &= |z_1|^2 + \cdots + |z_n|^2 && (10.13) \end{aligned}$$

which is a nonnegative real number

Example 10.7.3: Let ω be a complex number of the form $e^{\theta\mathbf{i}}$. Let $\boldsymbol{v}$ be column c of $W(\omega, n)$. Then

$$\boldsymbol{v} = [\omega^{0\cdot c}, \omega^{1\cdot c}, \ldots, \omega^{(n-1)\cdot c}]$$

so

$$\bar{\boldsymbol{v}} = [\omega^{-0\cdot c}, \omega^{-1\cdot c}, \ldots, \omega^{-(n-1)\cdot c}]$$

so

$$\begin{aligned} \langle \boldsymbol{v}, \boldsymbol{v}\rangle &= \omega^{0\cdot c}\omega^{-0\cdot c} + \omega^{1\cdot c}\omega^{-1\cdot c} + \cdots + \omega^{(n-1)\cdot c}\omega^{-(n-1)\cdot c} \\ &= \omega^0 + \omega^0 + \cdots + \omega^0 \\ &= 1 + 1 + \cdots + 1 \\ &= n \end{aligned}$$

This is the first case in the proof of Lemma 10.6.2.

Having defined an inner product, we define the norm of a vector just as we did in Equation 8.1:

$$\|\boldsymbol{v}\| = \sqrt{\langle \boldsymbol{v}, \boldsymbol{v} \rangle}$$

Equation 10.13 ensures that the norm of a vector is nonnegative (Norm Property N1), and is zero only if the vector is the zero vector (Norm Property N2).

Example 10.7.3 (Page 406) shows that the squared norm of a column of $W(\omega, n)$ is n, so the norm of a column of $\mathcal{F}_n$ is one.

Recal the convention for interpreting a vector as a column vector (a one-column matrix). Recall that the product $\boldsymbol{u}^T\boldsymbol{v}$ is therefore the matrix whose only entry is $\boldsymbol{u} \cdot \boldsymbol{v}$. Analogously, for vectors over $\mathbb{C}$, the product $\boldsymbol{u}^H\boldsymbol{v}$ is the matrix whose only entry is $\langle \boldsymbol{u}, \boldsymbol{v} \rangle$. In fact, the equation

$$\boldsymbol{u}^H\boldsymbol{v} = \left[\ \langle \boldsymbol{u}, \boldsymbol{v} \rangle\ \right] \tag{10.14}$$

holds whether the vectors are over $\mathbb{C}$ or over $\mathbb{R}$.

Whereas the inner product for vectors over $\mathbb{R}$ is symmetric, this is not true of the inner product for vectors over $\mathbb{C}$.

Example 10.7.4: Let $\boldsymbol{u} = [1 + 2\text{i}, 1]$ and $\boldsymbol{v} = [2, 1]$. Then

$$\begin{aligned}
\left[\ \langle \boldsymbol{u}, \boldsymbol{v} \rangle\ \right] &= \left[\ 1 - 2\text{i} \quad 1\ \right] \begin{bmatrix} 2 \\ 1 \end{bmatrix} \\
&= \left[\ 2 - 4\text{i} + 1\ \right] \\
\left[\ \langle \boldsymbol{v}, \boldsymbol{u} \rangle\ \right] &= \left[\ 2 \quad 1\ \right] \begin{bmatrix} 1 + 2\text{i} \\ 1 \end{bmatrix} \\
&= \left[\ 2 + 4\text{i} + 1\ \right]
\end{aligned}$$

We define orthogonality for vectors over $\mathbb{C}$ just as we did for vectors over $\mathbb{R}$: two vectors are orthogonal if their inner product is zero.

Example 10.7.5: Let $\boldsymbol{u} = [e^{0\cdot\pi\text{i}/2}, e^{1\cdot\pi\text{i}/2}, e^{2\cdot\pi\text{i}/2}, e^{3\cdot\pi\text{i}/2}]$ and let $\boldsymbol{v} = [e^{0\cdot\pi\text{i}}, e^{1\cdot\pi\text{i}}, e^{2\cdot\pi\text{i}}, e^{3\cdot\pi\text{i}}]$. Then

$$\begin{aligned}
\langle \boldsymbol{u}, \boldsymbol{v} \rangle &= \bar{\boldsymbol{u}} \cdot \boldsymbol{v} \\
&= [e^{-0\cdot\pi\text{i}/2}, e^{-1\cdot\pi\text{i}/2}, e^{-2\cdot\pi\text{i}/2}, e^{-3\cdot\pi\text{i}/2}] \cdot [e^{0\cdot\pi\text{i}}, e^{1\cdot\pi\text{i}}, e^{2\cdot\pi\text{i}}, e^{3\cdot\pi\text{i}}] \\
&= e^{-0\cdot\pi\text{i}/2}e^{0\cdot\pi\text{i}} + e^{-1\cdot\pi\text{i}/2}e^{1\cdot\pi\text{i}} + e^{-2\cdot\pi\text{i}/2}e^{2\cdot\pi\text{i}} + e^{-3\cdot\pi\text{i}/2}e^{3\cdot\pi\text{i}} \\
&= e^{0\cdot\pi\text{i}/2} + e^{1\cdot\pi\text{i}/2} + e^{2\cdot\pi\text{i}/2} + e^{3\cdot\pi\text{i}/2}
\end{aligned}$$

and the last sum is zero (try it out in Python or see the argument at the end of the proof of Lemma 10.6.2).

Example 10.7.6: More generally, the second case in the proof of Lemma 10.6.2 shows that two distinct columns of $W(e^{2\pi\text{i}/n}, n)$ are orthogonal.

Definition 10.7.7: The *Hermitian adjoint* of a matrix A over $\mathbb{C}$, written A^H, is the matrix obtained from A by taking the transpose and replacing each entry by its conjugate.

Definition 10.7.8: A matrix A over $\mathbb{C}$ is *unitary* if A is square and A^HA is an identity matrix.

The Fourier Inverse Theorem (Theorem 10.6.1) shows that the Fourier matrix is unitary.

Unitary matrices are the complex analogue of orthogonal matrices. Just as the tranpose of an orthogonal matrix is its inverse, we have the following lemma.

Lemma 10.7.9: The Hermitian adjoint of a unitary matrix is its inverse.

Moreover, just as multiplication by an orthogonal matrix preserve norms, so does multiplication by a unitary matrix.

The results we presented in Chapters 8 and 9 can be adapted to hold for vectors and matrices over $\mathbb{C}$, by replacing transpose with Hermitian adjoint.

10.8 Circulant matrices

For numbers (real or imaginary) a_0, a_1, a_2, a_3, consider the matrix

$$A = \begin{bmatrix} a_0 & a_1 & a_2 & a_3 \\ a_3 & a_0 & a_1 & a_2 \\ a_2 & a_3 & a_0 & a_1 \\ a_1 & a_2 & a_3 & a_0 \end{bmatrix}$$

Note that the second row can be obtained from the first row by a cyclic shift one position to the right. The third row is similarly obtained from the second row, and the fourth row is similarly obtained from the third row.

Definition 10.8.1: A $\{0, 1, \ldots, n-1\} \times \{0, 1, \ldots, n-1\}$ matrix A is called a *circulant* matrix if

$$A[i, j] = A[0, (i - j) \bmod n]$$

that is, if A has the form

$$\begin{bmatrix} a_0 & a_1 & a_2 & \cdots & a_{n-3} & a_{n-2} & a_{n-1} \\ a_{n-1} & a_0 & a_1 & \cdots & a_{n-4} & a_{n-3} & a_{n-2} \\ a_{n-2} & a_{n-1} & a_0 & \cdots & a_{n-3} & a_{n-2} & a_{n-3} \\ & & & \vdots & & & \\ a_2 & a_3 & a_4 & \cdots & a_{n-1} & a_0 & a_1 \\ a_1 & a_2 & a_3 & \cdots & a_{n-2} & a_{n-1} & a_0 \end{bmatrix}$$

Perhaps you think that a circulant matrix is a curiousity: nice to look at but irrelevant to applications. Consider Example 4.6.6 (Page 163), in which we are trying to compute many dot-products in order match an audio clip against a longer audio segment. As mentioned in that example, finding all these dot-products can be formulated as matrix-vector multiplication where the matrix is circulant. Each row is responsible for matching the short sequence against a particular subsequence of the long sequence.

We show in this section that the matrix-vector multiplication can be done much faster than the obvious algorithm. The obvious algorithm involves n^2 multiplications. We will see that you can do the computation with a couple of calls to the FFT at $O(n \log n)$ time per call, plus n multiplications. (There is an additional cost since the multiplications have to be done using complex numbers—but the sophisticated algorithm is still much faster when n is large.)

A similar phenomenon occurs when computing many dot-products of a pattern against an image, only in two dimensions. We won't go into it here, but it is not hard to extend the FFT-based algorithm to handle this.

10.8.1 Multiplying a circulant matrix by a column of the Fourier matrix

Let A be an $n \times n$ circulant matrix. Something interesting happens when we multiply A times a column of the matrix $W(\omega, n)$. Let's start with a small example, $n = 4$. Let $[a_0, a_1, a_2, a_3]$ be the first row of A. $W(\omega, 4)$ is

$$\begin{bmatrix} \omega^{0\cdot 0} & \omega^{0\cdot 1} & \omega^{0\cdot 2} & \omega^{0\cdot 3} \\ \omega^{1\cdot 0} & \omega^{1\cdot 1} & \omega^{1\cdot 2} & \omega^{1\cdot 3} \\ \omega^{2\cdot 0} & \omega^{2\cdot 1} & \omega^{2\cdot 2} & \omega^{2\cdot 3} \\ \omega^{3\cdot 0} & \omega^{3\cdot 1} & \omega^{3\cdot 2} & \omega^{3\cdot 3} \end{bmatrix}$$

For $j = 0, 1, 2, 3$,

column j of $W(\omega, 4)$ is the vector $[\omega^{0\cdot j}, \omega^{1\cdot j}, \omega^{2\cdot j}, \omega^{3\cdot j}]$ The first entry in the matrix-vector product is the dot-product of the first row of A with that column, which is is

$$a_0\omega^{j\cdot 0} + a_1\omega^{j\cdot 1} + a_2\omega^{j\cdot 2} + a_3\omega^{j\cdot 3} \tag{10.15}$$

The second row of A is $[a_3, a_0, a_1, a_2]$, so the second entry in the matrix-vector product is

$$a_3\omega^{j\cdot 0} + a_0\omega^{j\cdot 1} + a_1\omega^{j\cdot 2} + a_2\omega^{j\cdot 3} \tag{10.16}$$

Note that this second entry can be obtained from the first by multiplying by ω^j. Similarly, the third entry can be obtained from the first by multiplying by $\omega^{j\cdot 2}$ and the fourth entry can be obtained from the first by multiplying by $\omega^{j\cdot 3}$.

Letting λ_j denote the first entry of the matrix-vector product, that product has the form $[\lambda_j\omega^{j\cdot 0}, \lambda_j\omega^{j\cdot 1}, \lambda_j\omega^{j\cdot 2}, \lambda_j\omega^{j\cdot 3}]$. That is, the product of A with column j of $W(\omega, 4)$ is the scalar-vector product λ_j times column j of $W(\omega, 4)$.

This is a very special property, and one we will explore at greater length in Chapter 12. For now, let's write this result as an equation:

$$\lambda_j \begin{bmatrix} \omega^{0\cdot j} \\ \omega^{1\cdot j} \\ \omega^{2\cdot j} \\ \omega^{3\cdot j} \end{bmatrix} = \begin{bmatrix} a_0 & a_1 & a_2 & a_3 \\ a_3 & a_0 & a_1 & a_2 \\ a_2 & a_3 & a_0 & a_1 \\ a_1 & a_2 & a_3 & a_0 \end{bmatrix} \begin{bmatrix} \omega^{0\cdot j} \\ \omega^{1\cdot j} \\ \omega^{2\cdot j} \\ \omega^{3\cdot j} \end{bmatrix}$$

We have an equation for each column of $W(\omega, n)$. We will try to put these four equations together to form a single equation, using the matrix-vector definition of matrix-matrix multiplication.

The matrix whose columns are

$$\lambda_0 \begin{bmatrix} \omega^{0\cdot 0} \\ \omega^{1\cdot 0} \\ \omega^{2\cdot 0} \\ \omega^{3\cdot 0} \end{bmatrix} + \lambda_1 \begin{bmatrix} \omega^{0\cdot 1} \\ \omega^{1\cdot 1} \\ \omega^{2\cdot 1} \\ \omega^{3\cdot 1} \end{bmatrix} + \lambda_2 \begin{bmatrix} \omega^{0\cdot 2} \\ \omega^{1\cdot 2} \\ \omega^{2\cdot 2} \\ \omega^{3\cdot 2} \end{bmatrix} + \lambda_3 \begin{bmatrix} \omega^{0\cdot 3} \\ \omega^{1\cdot 3} \\ \omega^{2\cdot 3} \\ \omega^{3\cdot 3} \end{bmatrix}$$

can be written as

$$\begin{bmatrix} \omega^{0\cdot 0} & \omega^{0\cdot 1} & \omega^{0\cdot 2} & \omega^{0\cdot 3} \\ \omega^{1\cdot 0} & \omega^{1\cdot 1} & \omega^{1\cdot 2} & \omega^{1\cdot 3} \\ \omega^{2\cdot 0} & \omega^{2\cdot 1} & \omega^{2\cdot 2} & \omega^{2\cdot 3} \\ \omega^{3\cdot 0} & \omega^{3\cdot 1} & \omega^{3\cdot 2} & \omega^{3\cdot 3} \end{bmatrix} \begin{bmatrix} \lambda_0 & 0 & 0 & 0 \\ 0 & \lambda_1 & 0 & 0 \\ 0 & 0 & \lambda_2 & 0 \\ 0 & 0 & 0 & \lambda_3 \end{bmatrix}$$

so we write the equation as

$$\begin{bmatrix} \omega^{0\cdot 0} & \omega^{0\cdot 1} & \omega^{0\cdot 2} & \omega^{0\cdot 3} \\ \omega^{1\cdot 0} & \omega^{1\cdot 1} & \omega^{1\cdot 2} & \omega^{1\cdot 3} \\ \omega^{2\cdot 0} & \omega^{2\cdot 1} & \omega^{2\cdot 2} & \omega^{2\cdot 3} \\ \omega^{3\cdot 0} & \omega^{3\cdot 1} & \omega^{3\cdot 2} & \omega^{3\cdot 3} \end{bmatrix} \begin{bmatrix} \lambda_0 & 0 & 0 & 0 \\ 0 & \lambda_1 & 0 & 0 \\ 0 & 0 & \lambda_2 & 0 \\ 0 & 0 & 0 & \lambda_3 \end{bmatrix} = \begin{bmatrix} a_0 & a_1 & a_2 & a_3 \\ a_3 & a_0 & a_1 & a_2 \\ a_2 & a_3 & a_0 & a_1 \\ a_1 & a_2 & a_3 & a_0 \end{bmatrix} \begin{bmatrix} \omega^{0\cdot 0} & \omega^{0\cdot 1} & \omega^{0\cdot 2} & \omega^{0\cdot 3} \\ \omega^{1\cdot 0} & \omega^{1\cdot 1} & \omega^{1\cdot 2} & \omega^{1\cdot 3} \\ \omega^{2\cdot 0} & \omega^{2\cdot 1} & \omega^{2\cdot 2} & \omega^{2\cdot 3} \\ \omega^{3\cdot 0} & \omega^{3\cdot 1} & \omega^{3\cdot 2} & \omega^{3\cdot 3} \end{bmatrix}$$

or, more briefly, as

$$W(\omega, 4)\ \Lambda = A\ W(\omega, 4)$$

where Λ is the diagonal matrix $\begin{bmatrix} \lambda_0 & 0 & 0 & 0 \\ 0 & \lambda_1 & 0 & 0 \\ 0 & 0 & \lambda_2 & 0 \\ 0 & 0 & 0 & \lambda_3 \end{bmatrix}$ Since the Fourier matrix $\mathcal{F}_4$ is $\frac{1}{2}W(\omega, 4)$, we can write the equation as

$$\mathcal{F}_4\Lambda = A\ \mathcal{F}_4$$

where we have canceled out the $\frac{1}{2}$ on each side. Multiplying on the right on both sides of the equation by the inverse of $\mathcal{F}_4$, we obtain

$$\mathcal{F}_4 \, \Lambda \, \mathcal{F}_4^{-1} = A$$

Now suppose A is an $n \times n$ matrix. All the algebra we have done carries over to this case to show the following important equation:

$$\mathcal{F}_n \, \Lambda \, \mathcal{F}_n^{-1} = A \tag{10.17}$$

where Λ is some $n \times n$ diagonal matrix.

Suppose we want to compute the matrix-vector product $A\boldsymbol{v}$. We know from the equation that this is equal to $\mathcal{F}_n \, \Lambda \, \mathcal{F}_n^{-1} \, \boldsymbol{v}$, which can be written using associativity as

$$\mathcal{F}_n(\Lambda(\mathcal{F}_n^{-1}\boldsymbol{v}))$$

Thus we have replaced a single matrix-vector multiplication, $A\boldsymbol{v}$, with a series of three successive matrix-vector multiplications. What good is that? Well, each of the new matrix-vector multiplications involves a very special matrix. Multiplying a vector by $\mathcal{F}_n^{-1}$ or by $\mathcal{F}_n$ can be done using the FFT in $O(n \log n)$ time. Multiplying a vector by Λ is easy: each entry of the vector is multiplied by the corresponding diagonal entry of Λ. Thus the total time to carry out all these multiplications is $O(n \log n)$, which is much faster than the $O(n^2)$ required using the obvious matrix-vector multiplication algorithm.

10.8.2 Circulant matrices and change of basis

Equation 10.17 has an important interpretation. Transform the equation by multiplying it on the left by $\mathcal{F}_n^{-1}$ and multiplying it on the right by $\mathcal{F}$. The result is the equation

$$\Lambda = \mathcal{F}_n^{-1} A \mathcal{F}_n \tag{10.18}$$

The columns of $\mathcal{F}_n$ form a basis, the discrete Fourier basis. Multiplying a vector by $\mathcal{F}_n$ corresponds to `rep2vec`, to converting from the vector's coordinate representation in terms of this basis to the vector itself. Multiplying by $\mathcal{F}_n^{-1}$ corresponds to `vec2rep`, to converting from a vector to the vector's coordinate representation in terms of the discrete Fourier basis. Equation 10.18 states that the function $\boldsymbol{x} \mapsto A\boldsymbol{x}$ is very simple when viewed in terms of the discrete Fourier basis: it just multiplies each of the coordinates by some number.

Equation 10.18 is called a *diagonalization* of the matrix A. Diagonalization is the focus of Chapter 12.

10.9 *Lab: Using wavelets for compression*

In this lab, we will not use our vector class `Vec`. We will use lists and dictionaries to store vectors. For simplicity, we will require every vector's number of elements to be a power of two.

Note: Even though the goal of this lab is achieving sparse representations, for simplicity we will retain all values, even zeroes.

- The coordinate representation in terms of the standard basis will be stored as a list. In particular, each row of an image will be represented by a list.
- The coordinate representation in terms of the Haar wavelet basis will be represented by a dictionary.

For example, let $n = 16$. The representation of a vector $\boldsymbol{v}$ in terms of the standard basis looks like this

```
>>> v = [4,5,3,7,4,5,2,3,9,7,3,5,0,0,0,0]
```

in which v[i] is the coefficient of the standard basis vector b_i^{16}.

The (unnormalized) Haar wavelet basis consists of the vectors

$$\begin{aligned} \boldsymbol{w}_0^8, \boldsymbol{w}_1^8, \boldsymbol{w}_2^8, \boldsymbol{w}_3^8, \boldsymbol{w}_4^8, \boldsymbol{w}_5^8, \boldsymbol{w}_6^8, \boldsymbol{w}_7^8, \\ \boldsymbol{w}_0^4, \boldsymbol{w}_1^4, \boldsymbol{w}_2^4, \boldsymbol{w}_3^4, \\ \boldsymbol{w}_0^2, \boldsymbol{w}_1^2, \\ \boldsymbol{w}_0^1, \\ \boldsymbol{b}_0^1 \end{aligned}$$

For notational convenience, we use $\boldsymbol{w}_0^0$ to denote $\boldsymbol{b}_0^1$.

The representation of $\boldsymbol{v}$ in terms of the Haar wavelet basis is stored in a dictionary with the following keys:

$$\begin{aligned} (8,0),(8,1),(8,2),(8,3),(8,4),(8,5),(8,6),(8,7), \\ (4,0),(4,1),(4,2),(4,3), \\ (2,0),(2,1), \\ (1,0), \\ (0,0) \end{aligned}$$

where the value associated with key $(8,0)$ is the coefficient of $\boldsymbol{w}_0^8$, ..., the value associated with key $(1,0)$ is the coefficient of $\boldsymbol{w}_0^1$, and the value associated with key $(0,0)$ is the coefficient of $\boldsymbol{w}_0^0$ (i.e. the overall intensity average)

For example, for the vector $\boldsymbol{v}$ given earlier, the representation is

```
{(8, 3): -1, (0, 0): 3.5625, (8, 2): -1, (8, 1): -4, (4, 1): 2.0,
 (4, 3): 0.0, (8, 0): -1, (2, 1): 6.0, (2, 0): 1.25, (8, 7): 0,
(4, 2): 4.0, (8, 6): 0, (1, 0): 1.125, (4, 0): -0.5, (8, 5): -2,
 (8, 4): 2}
```

Remember, this is the representation with respect to *unnormalized basis vectors.* You will write a procedure to find this representation. You will then use this procedure in another procedure that returns the representation in terms of normalized basis vectors.

10.9.1 *Unnormalized forward transform*

Task 10.9.1: Write a procedure `forward_no_normalization(v)` with the following specification:

- *input:* a list representing a vector in $\mathbb{R}^n$ where n is some power of two.
- *output:* a dictionary giving the representation of the input vector in terms of the unnormalized Haar wavelet basis. The key for the coefficient of $\boldsymbol{w}_i^j$ should be the tuple (j, i).

Here is some pseudocode to get you started

```
def forward_no_normalization(v):
 D = {}
 while len(v) > 1:
    k = len(v)
    # v is a k-element list
    vnew = ... compute downsampled 1-d image of size k//2 from v ...
    # vnew is a k//2-element list
    w = ... compute unnormalized coefficients of basis for W(k/2) ...
    # w is a list of coefficients
    D.update( ...dictionary with keys (k//2, 0), (k//2, 1), ...,
                 (k//2, k//2-1) and values from w ...)
    v = vnew
 # v is a 1-element list
 D[(0,0)] = v[0]  #store the last coefficient
 return D
```

Here are some examples for you to test against.

```
>>> forward_no_normalization([1,2,3,4])
{(2, 0): -1, (1, 0): -2.0, (0, 0): 2.5, (2, 1): -1}
>>> v=[4,5,3,7,4,5,2,3,9,7,3,5,0,0,0,0]
>>> {(8,3): -1, (0,0): 3.5625, (8,2): -1, (8,1): -4, (4,1): 2.0, (4,3): 0.0,
 (8,0): -1, (2,1): 6.0, (2,0): 1.25, (8,7): 0, (4,2): 4.0, (8,6): 0,
 (1,0): 1.125, (4,0): -0.5, (8,5): -2, (8,4): 2}
```

Test it out on some small one-dimensional images in which there is a lot of similarity between nearby pixels, e.g. $[1, 1, 2, 2]$ and $[0, 1, 1, 1, -1, 1, 0, 1, 100, 101, 102, 100, 101, 100, 99, 100]$. Then test on some images in which nearby pixels vary wildly. Do you notice any difference in the magnitude of the coefficients that result?

10.9.2 *Normalization in the forward transform*

As we calculated in introducing the wavelet basis, the unnormalized squared norm of the wavelet basis vector $\boldsymbol{w}_i^s$ is $n/(4s)$. The special basis vector $\boldsymbol{w}_0^0$ corresponding to the overall average intensity is

$$[1, 1, \ldots, 1]$$

so its squared norm is $1^2 + 1^2 + \cdots + 1^2$, which is n.

As we have seen, to convert from the coefficient of an *unnormalized* basis vector to the coefficient of the corresponding *normalized* basis vector, we multiply by the norm of the unnormalized basis vector. Thus normalizing the coefficients means multiplying the coefficient of $\boldsymbol{w}_i^s$ by $\sqrt{n/(4s)}$, except that the coefficient of $\boldsymbol{w}_0^0$ must be multiplied by $\sqrt{n}$.

Task 10.9.2: Write a procedure `normalize_coefficients(n, D)` that, given the dimension n of the original space and a dictionary D of the form returned by `forward_no_normalization(v)`, returns the corresponding dictionary with the coefficients normalized.

Here are examples:

```
>>> normalize_coefficients(4, {(2,0):1, (2,1):1, (1,0):1, (0,0):1})
{(2, 0): 0.707..., (1, 0): 1.0, (0, 0): 2.0, (2, 1): 0.707...}
>>> normalize_coefficients(4, forward_no_normalization([1,2,3,4]))
(2, 0): -0.707, (1, 0): -2.0, (0, 0): 5.0, (2, 1): -0.707}
```

Task 10.9.3: Write a procedure `forward(v)` to find the representation with respect to the normalized Haar wavelet basis. This procedure should simply combine `forward_no_normalization` with `normalize_coefficients`.

As an example, try finding the forward transform of $[1, 2, 3, 4]$. The unnormalized coefficients of $\boldsymbol{w}_0^2$ and $\boldsymbol{w}_1^2$ are -1 and -1. The squared norms of these vectors are $1/2$, so these raw coefficients must be multiplied by $\sqrt{1/2}$.

The subsampled 2-pixel image is $[1.5, 3.5]$, which is the value assigned to v for the next iteration. The unnormalized coefficient of $\boldsymbol{w}_0^1$ is -2. The squared norm of $\boldsymbol{w}_0^1$ is 1, so the raw coefficient must be multiplied by $\sqrt{1}$.

The subsampled 1-pixel image is $[2.5]$, so the coefficient of $\boldsymbol{w}_0^0 = \boldsymbol{b}_0^1$ is 2.5. The squared norm of $\boldsymbol{w}_0^0$ is 4, so this coefficient must be multiplied by $\sqrt{4}$.

Therefore the output dictionary should be

```
{(2, 0): -sqrt(1/2), (2,1): -sqrt(1/2), (1, 0): -2, (0, 0): 5}
```

Again try your procedure out on one-dimensional images with small and large variations between nearby pixel values.

10.9.3 *Compression by suppression*

Our compression method is to zero out all coefficients whose absolute values are less than a given threshold.

Task 10.9.4: Write a procedure `suppress(D, threshold)` that, given a dictionary D giving the representation of a vector with respect to the normalized basis, returns a dictionary of the same form but where every value whose absolute value is less than `threshold` is replaced with zero. You should be able to use a simple comprehension for this.

Example:

```
>>> suppress(forward([1,2,3,4]), 1)
{(2, 0): 0, (1, 0): -2.0, (0, 0): 5.0, (2, 1): 0}
```

Task 10.9.5: Write a procedure `sparsity(D)` that, given such a dictionary, returns the percentage of its values that are nonzero. The smaller this value, the better the compression achieved.

```
>>> D = forward([1,2,3,4])
>>> sparsity(D)
1.0
>>> sparsity(suppress(D, 1))
0.5
```

10.9.4 *Unnormalizing*

We now have the procedures we need to compress a one-dimensional image, obtaining a sparse representation by suppressing values close to zero. However, we also want to recover the original image from the compressed representation. For that, we need a procedure `backward(D)` that calculates the list corresponding to a dictionary of wavelet coefficients.

The first step is undoing the normalization.

Task 10.9.6: Write a procedure `unnormalize_coefficients(n, D)` that corresponds to the functional inverse of `normalize_coefficients(n, D)`.

10.9.5 *Unnormalized backward transform*

Task 10.9.7: Write a procedure `backward_no_normalization(D)` that, given a dictionary of unnormalized wavelet coefficients, produces the corresponding list. It should be the inverse of `foward_no_normalization(v)`.

Here is pseudocode to get you started.

```
def backward_no_normalization(D):
  n = len(D)
  v =  (the one-element list whose entry is the coefficient of b_0^0)
  while len(v) < n:
    k = 2 * len(v)
    v =  (a k-element list)
  return v
```

Test your procedure to make sure it is the functional inverse of `forward_no_normalization(v)`.

10.9.6 *Backward transform*

Task 10.9.8: Write the procedure `backward(D)` that computes the inverse wavelet transform. This involves just combining `unnormalize_coefficients(n, D)` and `backward_no_normalization(D)`.

Test your procedure to make sure it is the inverse of `forward(v)`.

Handling two-dimensional images

There is a two-dimensional Haar wavelet basis. We will use an approach that differs slightly from the two-dimensional basis.

10.9.7 *Auxiliary procedure*

Back in Problem 4.17.20, you wrote the procedure `dictlist_helper(dlist, k)` with the following spec:

- *input:*
  - a list `dlist` of dictionaries which all have the same keys, and
  - a key `k`.
- *output:* a list whose i^{th} element is the value corresponding to the key `k` in the ith dictionary of `dlist`

In case you don't still have this procedure, it is a straightforward comprehension.

10.9.8 *Two-dimensional wavelet transform*

Our goal is to find a representation in terms of wavelets for a two-dimensional image.

We start with file of an image with m rows and n columns. Using the `image` module, we obtain a list-of-lists representation, an m-element list each element of which is an n-element list:

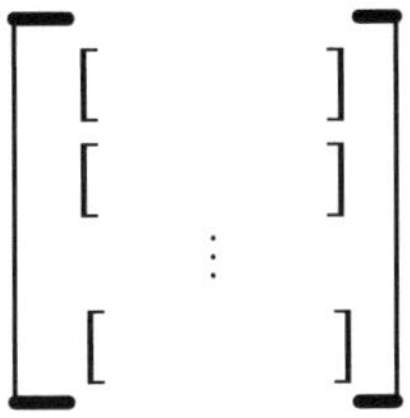

The `forward(v)` procedure will be used to apply the transformation to each row of the image. The output of `forward(v)` is a dictionary, so the result will be a dictlist, an m-element list of n-element dictionaries:

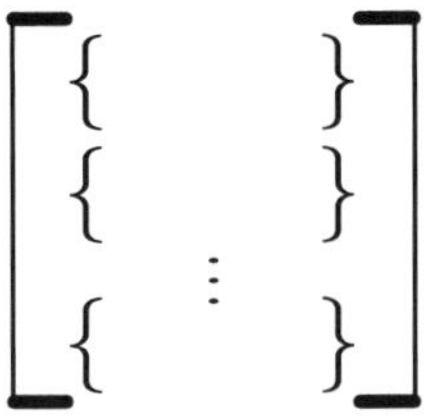

Next, our goal is to obtain columns to which we can again apply `forward(v)`. For each key k, each dictionary has a corresponding value; thus for each key k, we can extract an m-element list consisting of

- entry k of the first dictionary,
- entry k of the second dictionary,

  ⋮

- entry k of the mth dictionary.

and we create a dictionary mapping each key k to the corresponding m-element list. This is a listdict representation:

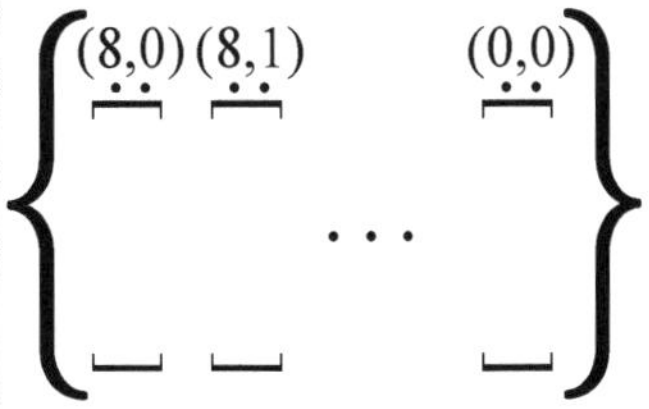

Finally, we will apply `forward(v)` to transform each list to a dictionary, resulting in a dictdict:

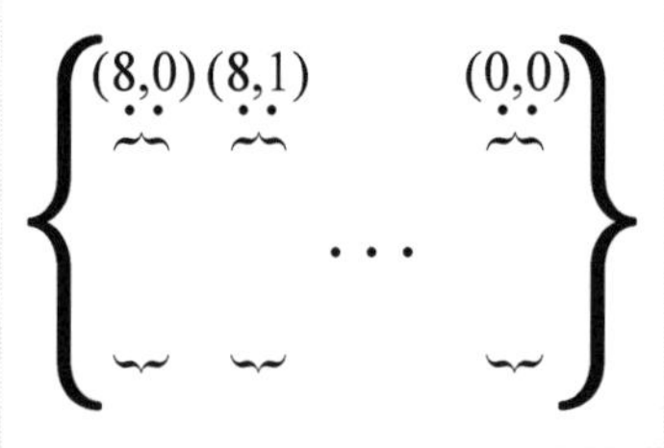

This is our representation of the wavelet transform of the image.

10.9.9 *Forward two-dimensional transform*

Now you will write the procedure `forward2d(listlist)` that transforms the listlist representation of an image into the dictdict representation of the wavelet coefficients.

The input *listlist* is an m-element list of n-element lists. Each inner list is a row of pixel intensities. We assume for simplicity that m and n are powers of two.

Step 1: Use `forward(v)` to transform each row of the image. For each row, the result is a dictionary. All the dictionaries have the same set of keys. Store these dictionaries in an m-element list `D_list`.

Step 2: This is a "transpose"-type step, to give you access to the columns. For each key k, construct an m-element list of consisting of

- entry k of the first dictionary,
- entry k of the second dictionary,

  ⋮

- entry k of the mth dictionary.

Store this list in a dictionary `L_dict` under key k. Thus `L_dict` is an n-element dictionary each element of which is an m-element list.

Step 3: Use `forward(v)` to transform each of those lists, obtaining a dictionary `D_dict` in which each list is replaced by a dictionary.

The result, then, is an n-element dictionary of m-element dictionaries.

Task 10.9.9: Write a procedure `forward2d(vlist)` to carry this out.

- Step 1 is a simple list comprehension.
- Step 2 is a dictionary comprehension that uses the procedure `dictlist_helper(dlist, k)`.
- Step 3 is a dictionary comprehension.

First, test your procedure on a 1×4 image such as $[[1, 2, 3, 4]]$. On such an image, it should agree with the one-dimensional transform. The first step should give you

```
[{(2,0): -0.707..., (1,0): -2.0, (0,0): 5.0, (2,1): -0.707...}]
```

The second step, the "transpose" step, should give you

```
{(2,0): [-0.707...], (1,0): [-2.0], (0,0): [5.0], (2,1): [-0.707...]}
```

The third and final step should give you

```
{(2,0): {(0,0): -0.707...}, (1,0): {(0,0): -2.0},
 (0,0): {(0,0): 5.0}, (2,1): {(0,0): -0.707...}}
```

```
{(2,0): {(0,0): -0.35...}, (1,0): {(0,0): -1.0},
 (0,0): {(0,0): 5.0}, (2,1): {(0,0): -0.35...}}
```

Next, test it on the 2×4 image $[[1, 2, 3, 4], [2, 3, 4, 3]]$. The first step should give you

```
{(2, 0): -0.707..., (1, 0): -2.0, (0, 0): 5.0, (2, 1): -0.707...},
 {(2, 0): -0.707..., (1, 0): -1.0, (0, 0): 6.0, (2, 1): 0.707...}]
```

The second step, the "transpose", should give you

```
{(2, 0): [-0.707..., -0.707...], (1, 0): [-2.0, -1.0],
 (0, 0): [5.0, 6.0], (2, 1): [-0.707..., 0.707...]}
```

The third and final step should give you

```
{(2, 0): {(1, 0): 0.0, (0, 0): -1}, (1, 0): {(1, 0): -0.707...,
 (0, 0): -2.121...}, (0, 0): {(1, 0): -0.707..., (0, 0): 7.778...},
 (2, 1): {(1, 0): -1, (0, 0): 0.0}}
```

Task 10.9.10: Write `suppress2d(D_dict, threshold)`, a two-dimensional version of `suppress(D,threshold)` that suppresses values with absolute value less than `threshold` in a dictionary of dictionaries such as is returned by `forward2d(vlist)`.

Task 10.9.11: Write `sparsity2d(D_dict)`, a two-dimensional version of `sparsity(D)`.

10.9.10 *More auxiliary procedures*

Task 10.9.12: Write a procedure `listdict2dict(L_dict, i)` with the following spec:

- *input:* a dictionary *L_dict* of lists, all the same length, and an index i into such a list
- *output:* a dictionary with the same keys as *L_dict*, in which key k maps to element i of L_dict[i]

Task 10.9.13: Write a procedure `listdict2dictlist(listdict)` that converts from a listdict representation (a dictionary of lists) to a dictlist representation (a list of dictionaries):

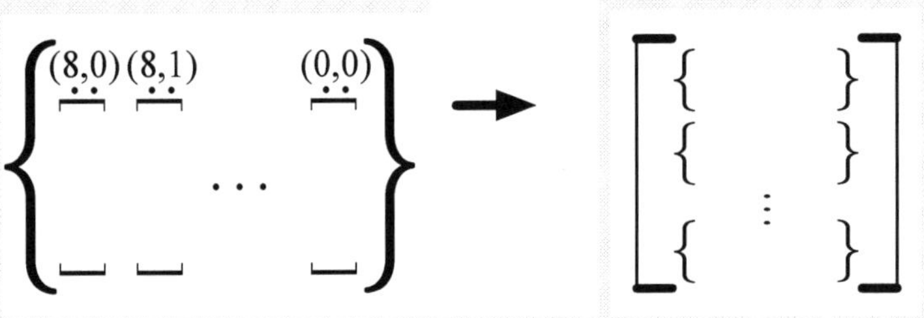

10.9.11 *Two-dimensional backward transform*

You will write the two-dimensional backward transform. To review the process:

Step 1: The input is a dictdict. This step applies `backward(D)` to transform each inner dictionary to a list, resulting in a listdict.

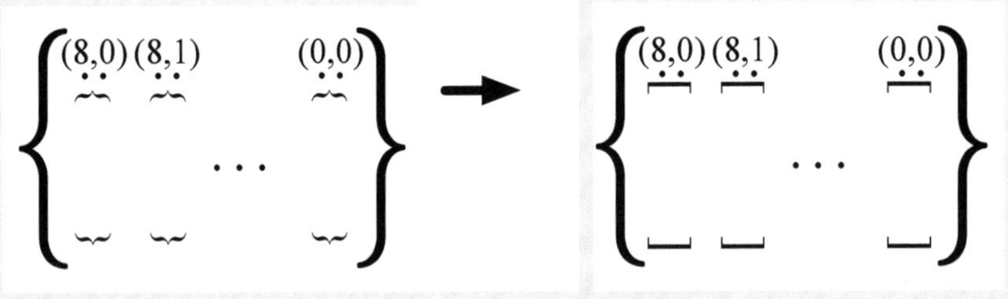

Step 2: This is a "transpose"-type step that transforms from a listdict to a dictlist:

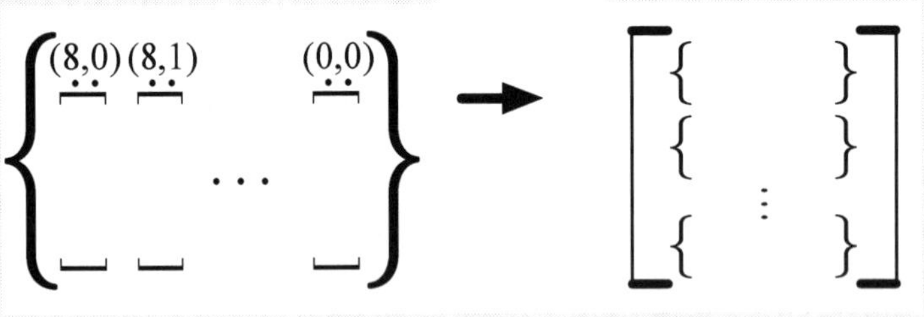

Step 3: This step applies `backward(D)` to each inner dictionary, resulting in a listlist:

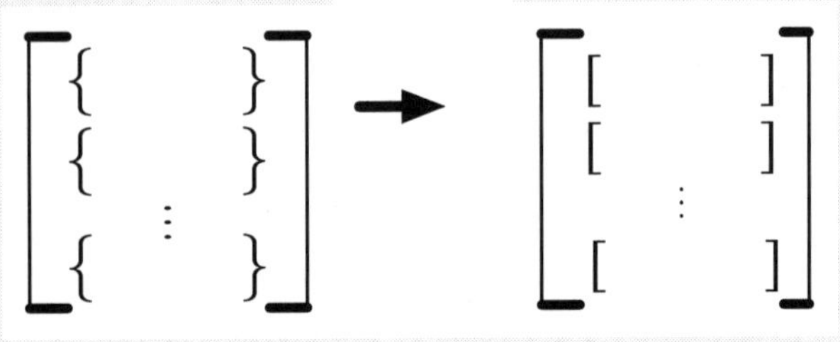

Task 10.9.14: Write `backward2d(dictdict)`, the functional inverse of `forward2d(vlist)`. Test it to make sure it is really the inverse.

10.9.12 *Experimenting with compression of images*

Task 10.9.15: Select a `.png` representing an image whose dimensions are powers of two.[a] We provide some example `.png` files at `resources.codingthematrix.com`. Use an application on your computer to display the image.

[a]You can, if you choose, adapt the code to "pad" an image, adding zeroes as necessary.

Task 10.9.16: Write a procedure `image_round(image)` with the following spec:

- *input:* a grayscale image, represented as a list of lists of floats
- *output:* the corresponding grayscale image, represented as a list of lists of integers, obtained by rounding the floats in the input image and taking their absolute values, and replacing numbers greater than 255 with 255.

Task 10.9.17: Import the procedures `file2image(filename)` and `color2gray(image)` from the `image` Use them in the expression `color2gray(file2image(filename))` to read the image into a list of lists.

Try using the procedure `image2display` from the `image` module to display the image. Compute the 2-D Haar wavelet transform using `forward2d`. Then apply `backward2d` to get a list-of-lists representation. Round this representation using `image_round`, and display the result either using `image2display` or by writing it to a file using the procedure `image2file(filename,vlist)` and then displaying the file you wrote. Make sure it looks like the original image.

Task 10.9.18: Finally, use `suppress2d(D_dict, threshold)` and `sparsity2d(D_dict)` to come up with a threshold that achieves the degree of compression you would like. Apply the backward transform and round the result using `image_round`, and then use `image2display` or `image2file` to view it to see how close it is to the original image.

Task 10.9.19: Try other images. See if the same threshold works well.

10.10 Review Questions

- Why does the function $\boldsymbol{x} \mapsto Q\boldsymbol{x}$ preserve dot-products when Q is a column-orthogonal matrix? Why does it preserve norms?
- How does one find the closest vector whose representation in a given basis is k-sparse? Why would you want to do so?
- Why might you want a basis such that there are fast algorithms for converting between a vector and the vector's coordinate representation?
- What is a wavelet basis? How does it illustrate the notions of direct sum and orthogonal complement?
- What is the process for computing a vector's representation in the wavelet basis?
- What is the inner product for vectors over $\mathbb{C}$?
- What is the Hermitian adjoint?
- What does the discrete Fourier basis have to do with circulant matrices?

10.11 Problems

Projections and representations in different bases

These problems seem to involve giving elaborate algorithms but your solutions must involve just simple operations: matrix-vector, vector-matrix, and matrix-matrix multiplication; dot-product, maybe transpose. Your code should not use any subroutines. (Of course, your code should make use of the operations defined on Mats and Vecs.)

You will find that in each problem the body of the procedure is very short. If the body of your procedure is at all complicated (e.g. involves a loop or even a comprehension), you're doing it wrong!

Try making very short solutions. In my solutions the average length of the body of the procedure (not counting `def ... return ....` is about five characters! :) I will admit, however, that one of my solutions involves cheating a little: I use an expression that would not be mathematically acceptable if the vectors were translated into row and column vectors.

Use your understanding of linear algebra to give solutions that are as simple and pure as possible.

Problem 10.11.1: Write a procedure `orthogonal_vec2rep(Q, b)` for the following:

- *input:* An orthogonal matrix Q, and a vector $\boldsymbol{b}$ whose label set equals the column-label set of Q
- *output:* the coordinate representation of $\boldsymbol{b}$ in terms of the rows of Q.

Your code should use the `mat` module and no other module and no other procedures.

Test case: For $Q = \begin{bmatrix} \frac{1}{\sqrt{2}} & \frac{1}{\sqrt{2}} & 0 \\ \frac{1}{\sqrt{3}} & -\frac{1}{\sqrt{3}} & \frac{1}{\sqrt{3}} \\ -\frac{1}{\sqrt{6}} & \frac{1}{\sqrt{6}} & \frac{2}{\sqrt{6}} \end{bmatrix}$, $\boldsymbol{b} = \begin{bmatrix} 10 & 20 & 30 \end{bmatrix}$, you should get $[21.213, 11.547, 28.577]$.

Problem 10.11.2: Write a procedure `orthogonal_change_of_basis(A, B, a)` for the following:

- *input:*
  - two orthogonal matrices A and B, such that the row-label set of A equals its column-label set which equals the row and column-label sets of B as well.
  - the coordinate representation a of a vector v in terms of the rows of A.
- *output:* the coordinate representation of v in terms of the columns of B.

Just for fun, try to limit your procedure's body to about five characters (not counting `return`).

Test case: For $A = B = \begin{bmatrix} \frac{1}{\sqrt{2}} & \frac{1}{\sqrt{2}} & 0 \\ \frac{1}{\sqrt{3}} & -\frac{1}{\sqrt{3}} & \frac{1}{\sqrt{3}} \\ -\frac{1}{\sqrt{6}} & \frac{1}{\sqrt{6}} & \frac{2}{\sqrt{6}} \end{bmatrix}$, $\boldsymbol{a} = [\sqrt{2}, \frac{1}{\sqrt{3}}, \frac{2}{\sqrt{6}}]$, you should get $[0.876, 0.538, 1.393]$.

Problem 10.11.3: Write a procedure `orthonormal_projection_orthogonal(W, b)` for the following spec.

- *input:* a matrix W whose rows are orthonormal, and a vector $\boldsymbol{b}$ whose label set is the column-label set of W
- *output:* the projection of $\boldsymbol{b}$ orthogonal to the row space of W.

Just for fun, try to limit your procedure's body to about seven characters (not counting `return`).

This is cheating, in that the expression would not be considered good mathematical syntax; without cheating, you can do it in nineteen characters.

(Hint: First find the projection of $\boldsymbol{b}$ onto the row space of W.)

Test case: For $W = \begin{bmatrix} \frac{1}{\sqrt{2}} & \frac{1}{\sqrt{2}} & 0 \\ \frac{1}{\sqrt{3}} & -\frac{1}{\sqrt{3}} & \frac{1}{\sqrt{3}} \end{bmatrix}$, $\boldsymbol{b} = [10, 20, 30]$, you should get $\begin{bmatrix} -11\frac{2}{3} & 11\frac{2}{3} & 23\frac{1}{3} \end{bmatrix}$.

Chapter 11

The Singular Value Decomposition

> One singular sensation...
> *A Chorus Line*, lyrics by Edward Kleban

In the last chapter, we studied special matrices—a Haar basis matrix, a Fourier matrix, a circulant matrix—such that multiplying by such a matrix was fast. Moreover, storage for such a matrix is very cheap. The Haar and Fourier matrices can be represented implicitly, by procedures, and a circulant matrix can be represented by storing just its first row, for the other rows can be derived from that one row.

11.1 Approximation of a matrix by a low-rank matrix

11.1.1 The benefits of low-rank matrices

A low-rank matrix has the same benefits. Consider a matrix whose rank is one. All the rows lie in a one-dimensional space. let $\{\boldsymbol{v}\}$ be a basis for that space. Every row of the matrix is some scalar multiple of $\boldsymbol{v}$. Let $\boldsymbol{u}$ be the vector whose entries are these scalar multiples. Then the matrix can be written as $\boldsymbol{u}\boldsymbol{v}^T$. Such a representation requires small storage—just $m+n$ numbers have to be stored for a rank-one $m \times n$ matrix. Moreover, to multiply the matrix $\boldsymbol{u}\boldsymbol{v}^T$ by a vector $\boldsymbol{w}$, we use the equation

$$\left(\begin{bmatrix} \\ \boldsymbol{u} \\ \\ \end{bmatrix}\begin{bmatrix} & \boldsymbol{v}^T & \end{bmatrix}\right)\begin{bmatrix} \\ \boldsymbol{w} \\ \\ \end{bmatrix} = \begin{bmatrix} \\ \boldsymbol{u} \\ \\ \end{bmatrix}\left(\begin{bmatrix} & \boldsymbol{v}^T & \end{bmatrix}\begin{bmatrix} \\ \boldsymbol{w} \\ \\ \end{bmatrix}\right)$$

which shows that the matrix-vector product can be computed by computing two dot-products.

Even if a matrix has rank more than one, if the rank of a matrix is small, we get some of the same benefits. A rank-two matrix, for example, can be written as

$$\left[\begin{array}{c|c} & \\ \boldsymbol{u}_1 & \boldsymbol{u}_2 \\ & \end{array}\right]\left[\begin{array}{c} \boldsymbol{v}_1^T \\ \hline \boldsymbol{v}_2 \end{array}\right]$$

so it can be stored compactly and can be multiplied by a vector quickly.

Unfortunately, most matrices derived from observed data do not have low rank. Fortunately, sometimes a low-rank approximation to a matrix will do nearly as well as the the matrix itself; sometimes even better! In this chapter, we will learn about how to find the best rank-k approximation to a given matrix, the rank-k matrix that is closest to the given matrix. There are a variety of applications, including two analytical methods, one called *principal components analysis* (PCA) and the other called *latent semantic indexing*.

11.1.2 Matrix norm

In order to define the problem of finding the rank-k matrix closest to a given matrix, we need to define a distance for matrices. For vectors, distance is given by the norm, which is in turn defined by the inner product. For vectors over $\mathbb{R}$, we defined inner product to be dot-product. We saw in Chapter 10 that the inner product for complex numbers was somewhat different. For this chapter, we will leave aside complex numbers and return to vectors and matrices over $\mathbb{R}$. Our inner product, therefore, is once again just dot-product, and so the norm of a vector is simply the square root of the sum of the squares of its entries. But how can we define the norm of a matrix?

Perhaps the most natural matrix norm arises from interpreting a matrix A as a vector. An $m \times n$ matrix is represented by an mn-vector, i.e. the vector has one entry for each entry of the matrix. The norm of a vector is the square root of the sum of the entries, and so that is how we measure the norm of a matrix A. This norm is called the *Frobenius* norm:

$$||A||_F = \sqrt{\sum_i \sum_j A[i,j]^2}$$

Lemma 11.1.1: The square of the Frobenius norm of A equals the sum of the squares of the rows of A.

Proof

Suppose A is an $m \times n$ matrix. Write A in terms of its rows: $A = \begin{bmatrix} \overline{\quad a_1 \quad} \\ \vdots \\ \overline{\quad a_m \quad} \end{bmatrix}$.

For each row label i,

$$||\boldsymbol{a}_i||^2 = \boldsymbol{a}_i[1]^2 + \boldsymbol{a}_i[2]^2 + \cdots + \boldsymbol{a}_i[n]^2 \qquad (11.1)$$

We use this equation to substitute in the definition of Frobenius norm:

$$\begin{aligned} ||A||_F^2 &= (A[1,1]^2 + A[1,2]^2 + \cdots + A[1,n]^2) + \cdots + (A[m,1]^2 + A[m,2]^2 + \cdots + A[m,n]^2) \\ &= ||\boldsymbol{a}_1||^2 + \cdots ||\boldsymbol{a}_m||^2 \end{aligned}$$

□

The analogous statement holds for columns as well.

11.2 The *trolley-line-location* problem

We start with a problem that is in a sense the opposite of the fire-engine problem. I call it the *trolley-line-location* problem. Given the locations of m houses, specified as vectors $\boldsymbol{a}_1, \ldots, \boldsymbol{a}_m$,

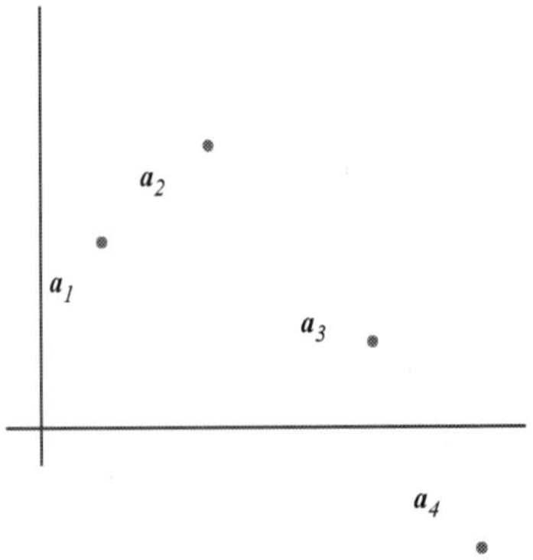

we must choose where to locate a trolley-line. The trolley-line is required to go through downtown (which we represent as the origin) and is required to be a straight line.

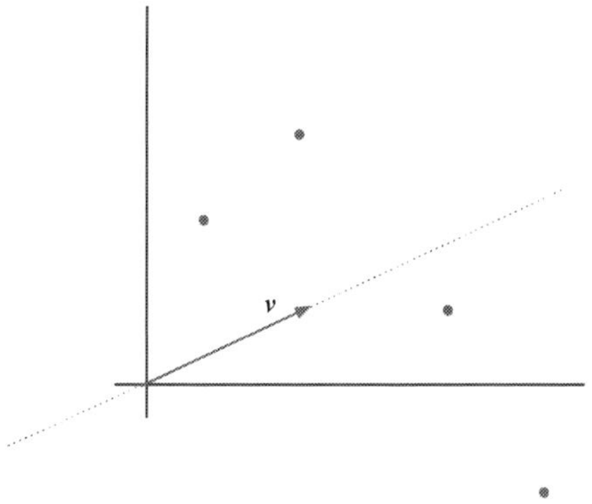

The goal is to locate the trolley-line so that it is as close as possible to the m houses.

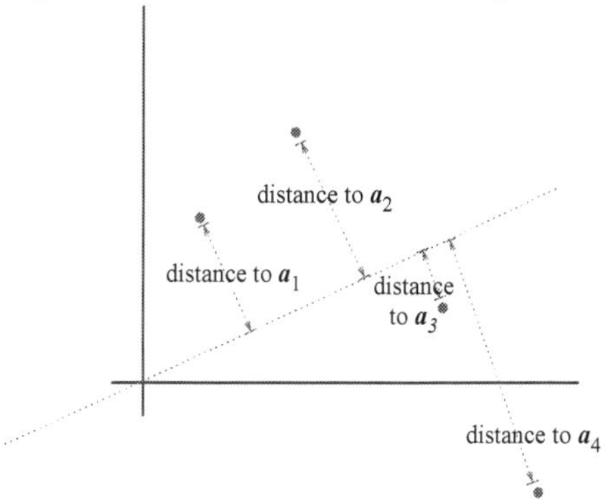

So far the problem is not fully specified. If there is only one house (i.e. one vector $\boldsymbol{a}_1$) then the solution is obvious: build a trolley-line along the line going through the origin and $\boldsymbol{a}_1$. In this case, the distance from the one house to the trolley-line is zero. If there are many vectors $\boldsymbol{a}_1, \ldots, \boldsymbol{a}_m$, how should we measure the distance from these vectors to the trolley-line? Each vector $\boldsymbol{a}_i$ has its own distance d_i from the trolley-line—-how should we combine the numbers $[d_1, \ldots, d_m]$ to get a single number to minimize? As in least squares, we minimize the norm of the vector $[d_1, \ldots, d_m]$. This is equivalent to minimizing the square of the norm of this vector, i.e. $d_1^2 + \cdots + d_m^2$.

And in what form should the output line be specified? By a unit-norm vector $\boldsymbol{v}$. The line of the trolley-line is then Span $\{\boldsymbol{v}\}$.

Computational Problem 11.2.1: *Trolley-Line-location problem:*

- *input:* vectors $\boldsymbol{a}_1, \ldots, \boldsymbol{a}_m$
- *output:* a unit vector $\boldsymbol{v}$ that minimizes

$$(\text{distance from } \boldsymbol{a}_1 \text{ to Span } \{\boldsymbol{v}\})^2 + \cdots + (\text{distance from } \boldsymbol{a}_m \text{ to Span } \{\boldsymbol{v}\})^2 \quad (11.2)$$

11.2.1 Solution to the trolley-line-location problem

For each vector $\boldsymbol{a}_i$, write $\boldsymbol{a}_i = \boldsymbol{a}_i^{\|\boldsymbol{v}} + \boldsymbol{a}_i^{\perp\boldsymbol{v}}$ where $\boldsymbol{a}_i^{\|\boldsymbol{v}}$ is the projection of $\boldsymbol{a}_i$ along $\boldsymbol{v}$ and $\boldsymbol{a}_i^{\perp\boldsymbol{v}}$ is the projection orthogonal to $\boldsymbol{v}$. Then

$$\begin{aligned} \boldsymbol{a}_1^{\perp\boldsymbol{v}} &= \boldsymbol{a}_1 - \boldsymbol{a}_1^{\|\boldsymbol{v}} \\ &\vdots \\ \boldsymbol{a}_m^{\perp\boldsymbol{v}} &= \boldsymbol{a}_m - \boldsymbol{a}_m^{\|\boldsymbol{v}} \end{aligned}$$

By the Pythagorean Theorem,

$$\begin{aligned} \|\boldsymbol{a}_1^{\perp\boldsymbol{v}}\|^2 &= \|\boldsymbol{a}_1\|^2 - \|\boldsymbol{a}_1^{\|\boldsymbol{v}}\|^2 \\ &\vdots \\ \|\boldsymbol{a}_m^{\perp\boldsymbol{v}}\|^2 &= \|\boldsymbol{a}_m\|^2 - \|\boldsymbol{a}_m^{\|\boldsymbol{v}}\|^2 \end{aligned}$$

Since the distance from $\boldsymbol{a}_i$ to Span $\{\boldsymbol{v}\}$ is $\|\boldsymbol{a}_i^{\perp \boldsymbol{v}}\|$, we have

$$\begin{aligned}
(\text{distance from } \boldsymbol{a}_1 \text{ to Span } \{\boldsymbol{v}\})^2 &= \|\boldsymbol{a}_1\|^2 - \|\boldsymbol{a}_1^{\|\boldsymbol{v}}\|^2 \\
&\vdots \\
(\text{distance from } \boldsymbol{a}_m \text{ to Span } \{\boldsymbol{v}\})^2 &= \|\boldsymbol{a}_m\|^2 - \|\boldsymbol{a}_m^{\|\boldsymbol{v}}\|^2
\end{aligned}$$

Adding vertically, we obtain

$$\begin{aligned}
\textstyle\sum_i(\text{distance from } \boldsymbol{a}_i \text{ to Span } \{\boldsymbol{v}\})^2 &= \|\boldsymbol{a}_1\|^2 + \cdots + \|\boldsymbol{a}_m\|^2 - \left(\|\boldsymbol{a}_1^{\|\boldsymbol{v}}\|^2 + \cdots + \|\boldsymbol{a}_m^{\|\boldsymbol{v}}\|^2\right) \\
&= \|A\|_F^2 - \left(\|\boldsymbol{a}_1^{\|\boldsymbol{v}}\|^2 + \cdots + \|\boldsymbol{a}_m^{\|\boldsymbol{v}}\|^2\right)
\end{aligned}$$

where A is the matrix whose rows are $\boldsymbol{a}_1, \ldots, \boldsymbol{a}_m$, by Lemma 11.1.1.

Using the fact that $\boldsymbol{a}_i^{\|\boldsymbol{v}} = \langle \boldsymbol{a}_i, \boldsymbol{v}\rangle\, \boldsymbol{v}$ because $\boldsymbol{v}$ is a norm-one vector, we have $\|\boldsymbol{a}_i^{\|\boldsymbol{v}}\|^2 = \langle \boldsymbol{a}_i, \boldsymbol{v}\rangle^2$, so

$$\textstyle\sum_i(\text{distance from } \boldsymbol{a}_i \text{ to Span } \{\boldsymbol{v}\})^2 = \|A\|_F^2 - \left(\langle \boldsymbol{a}_1, \boldsymbol{v}\rangle^2 + \langle \boldsymbol{a}_2, \boldsymbol{v}\rangle^2 + \cdots + \langle \boldsymbol{a}_m, \boldsymbol{v}\rangle^2\right) \tag{11.3}$$

Next, we show that $\left(\langle \boldsymbol{a}_1, \boldsymbol{v}\rangle^2 + \langle \boldsymbol{a}_2, \boldsymbol{v}\rangle^2 + \cdots + \langle \boldsymbol{a}_m, \boldsymbol{v}\rangle^2\right)$ can be replaced by $\|A\boldsymbol{v}\|^2$. By our dot-product interpretation of matrix-vector multiplication,

$$\begin{bmatrix} \quad \boldsymbol{a}_1 \quad \\ \hline \vdots \\ \hline \boldsymbol{a}_m \end{bmatrix} \begin{bmatrix} \\ \boldsymbol{v} \\ \\ \end{bmatrix} = \begin{bmatrix} \langle \boldsymbol{a}_1, \boldsymbol{v}\rangle \\ \vdots \\ \langle \boldsymbol{a}_m, \boldsymbol{v}\rangle \end{bmatrix} \tag{11.4}$$

so

$$\|A\boldsymbol{v}\|^2 = \left(\langle \boldsymbol{a}_1, \boldsymbol{v}\rangle^2 + \langle \boldsymbol{a}_2, \boldsymbol{v}\rangle^2 + \cdots + \langle \boldsymbol{a}_m, \boldsymbol{v}\rangle^2\right)$$

Substituting into Equation 11.3, we obtain

$$\textstyle\sum_i(\text{distance from } \boldsymbol{a}_i \text{ to Span } \{\boldsymbol{v}\})^2 = \|A\|_F^2 - \|A\boldsymbol{v}\|^2 \tag{11.5}$$

Therefore the best vector $\boldsymbol{v}$ is a unit vector that maximizes $\|A\boldsymbol{v}\|^2$ (equivalently, maximizes $\|A\boldsymbol{v}\|$). We now know a solution, at least in principle, for the trolley-line-location problem, Computational Problem 11.2.1:

```
def trolley_line_location(A):
  Given a matrix A, find the vector v1
  minimizing Σ_i(distance from row i of A to Span {v1})^2
  v1 = arg max{||Av|| : ||v|| = 1}
  σ1 = ||Av1||
  return v1
```

The arg max notation means the thing (in this case the norm-one vector $\boldsymbol{v}$) that results in the largest value of $\|A\boldsymbol{v}\|$.

So far, this is a solution only in *principle* since we have not specified how to actually compute $\boldsymbol{v}_1$. In Chapter 12, we will describe a method for approximating $\boldsymbol{v}_1$.

Definition 11.2.2: We refer to σ_1 as the *first singular value* of A, and we refer to $\boldsymbol{v}_1$ as the *first right singular vector*.

Example 11.2.3: Let $A = \begin{bmatrix} 1 & 4 \\ 5 & 2 \end{bmatrix}$, so $\boldsymbol{a}_1 = [1, 4]$ and $\boldsymbol{a}_2 = [5, 2]$. In this case, a unit vector maximizing $\|A\boldsymbol{v}\|$ is $\boldsymbol{v}_1 \approx \begin{bmatrix} 0.78 \\ 0.63 \end{bmatrix}$. We use σ_1 to denote $\|A\boldsymbol{v}_1\|$, which is about 6.1:

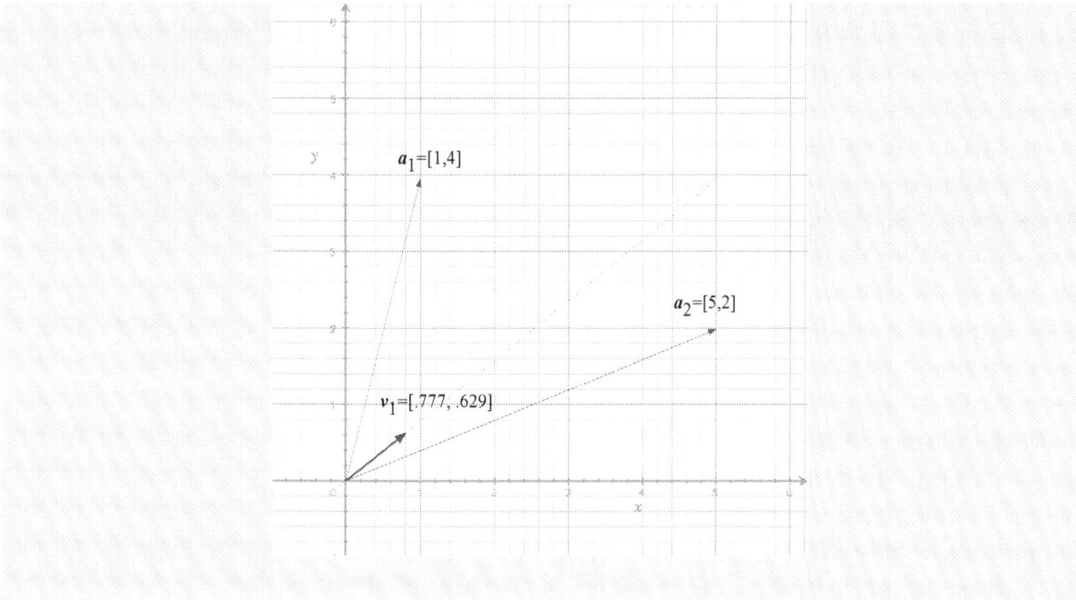

We have proved the following theorem, which states that `trolley_line_location`(A) finds the closest vector space.

Theorem 11.2.4: Let A be an $m \times n$ matrix over $\mathbb{R}$ with rows $\boldsymbol{a}_1, \ldots, \boldsymbol{a}_m$. Let $\boldsymbol{v}_1$ be the first right singular vector of A. Then Span $\{\boldsymbol{v}_1\}$ is the one-dimensional vector space $\mathcal{V}$ that minimizes

$$(\text{distance from } \boldsymbol{a}_1 \text{ to } \mathcal{V})^2 + \cdots + (\text{distance from } \boldsymbol{a}_m \text{ to } \mathcal{V})^2$$

How close is the closest vector space to the rows of A?

Lemma 11.2.5: The minimum sum of squared distances is $||A||_F^2 - \sigma_1^2$.

Proof

According to Equation 11.5, the squared distance is $\sum_i ||\boldsymbol{a}_i||^2 - \sum_i ||\boldsymbol{a}_i^{||\boldsymbol{v}}||^2$. By Lemma 11.1.1, the first sum is $||A||_F^2$. The second sum is the square of the quantity $||A\boldsymbol{v}_1||$, a quantity we have named σ_1. □

Example 11.2.6: Continuing with Example 11.2.3 (Page 426), we calculate the sum of squared distances.

First we find the projection of $\boldsymbol{a}_1$ orthogonal to $\boldsymbol{v}_1$:

$$\begin{aligned}\boldsymbol{a}_1 - \langle \boldsymbol{a}_1, \boldsymbol{v}_1 \rangle \, \boldsymbol{v}_1 &\approx [1,4] - (1 \cdot 0.78 + 4 \cdot 0.63)[0.78, 0.63] \\ &\approx [1,4] - 3.3\,[0.78, 0.63] \\ &\approx [-1.6, 1.9]\end{aligned}$$

The norm of this vector, about 2.5, is the distance from $\boldsymbol{a}_1$ to Span $\{\boldsymbol{v}_1\}$.

Next we find the projection of $\boldsymbol{a}_2$ orthogonal to $\boldsymbol{v}_1$:

$$\begin{aligned}\boldsymbol{a}_2 - \langle \boldsymbol{a}_1, \boldsymbol{v}_1 \rangle \, \boldsymbol{v}_1 &\approx [5,2] - (5 \cdot 0.78 + 2 \cdot 0.63)[0.78, 0.63] \\ &\approx [5,2] - 5.1\,[0.78, 0.63] \\ &\approx [1, -1.2]\end{aligned}$$

The norm of this vector, about 1.6, is the distance from $\boldsymbol{a}_2$ to Span $\{\boldsymbol{v}_1\}$.

Thus the sum of squared distances is about $2.5^2 + 1.6^2$, which is about 8.7.

According to Lemma 11.2.5, the sum of squared distances should be $||A||_F^2 - \sigma_1^2$. The squared Frobenius of A is $1^2 + 4^2 + 5^2 + 2^2 = 46$, and the first singular value is about 6.1, so $||A||_F^2 - \sigma_1^2$ is about 8.7. Lemma 11.2.5 is correct in this example!

Warning: We are measuring the error by distance to the subspace. The norm of a vector treats every entry equally. For this technique to be relevant, the units for the entries need to be appropriate.

Example 11.2.7: Let $\boldsymbol{a}_1, \ldots, \boldsymbol{a}_{100}$ be the voting records for US Senators (same data as you used in the politics lab). These are 46-vectors with ± 1 entries.

We find the unit-norm vector $\boldsymbol{v}$ that minimizes least-squares distance from $\boldsymbol{a}_1, \ldots, \boldsymbol{a}_{100}$ to Span $\{\boldsymbol{v}\}$, and we plot the projection along $\boldsymbol{v}$ of each of these vectors:

The results are not so meaningful. Moderates and conservatives have very similar projections:

Snowe	0.106605199	moderate Republican from Maine
Lincoln	0.106694552	moderate Republican from Rhode Island
Collins	0.107039376	moderate Republican from Maine
Crapo	0.107259689	conservative moderate Republican from Idaho
Vitter	0.108031374	conservative moderate Republican from Louisiana

There is that one outlier, way off to the left. That's Russ Feingold.

We'll later return to this data and try again....

11.2.2 Rank-one approximation to a matrix

Building on our solution to the trolley-line-location problem, we will obtain a solution to another computational problem: *finding the best rank-one approximation to a given matrix.* In finding the best k-sparse approximation to a vector (Chapter 10), "best" meant "closest to the original vector", where distance between vectors is measured in the usual way, by the norm. Here we would like to similarly measure the distance between the original matrix and its approximation. For that, we need a norm for matrices.

11.2.3 The best rank-one approximation

We are now in a position to define the problem *rank-one approximation.*

Computational Problem 11.2.8: *Rank-one approximation:*

- *input:* a nonzero matrix A
- *output:* the rank-one matrix $\tilde{A}$ that is closest to A according to Frobenius norm

Equivalently, the goal is to find the rank-one matrix $\tilde{A}$ that minimizes $||A - \tilde{A}||_F$:

$$\tilde{A} = \arg\min\{||A - B||_F \; : \; B \text{ has rank one}\}$$

Suppose we have some rank-one matrix $\tilde{A}$. How close is it to A? Let's look at the squared distance between A and $\tilde{A}$. By Lemma 11.1.1,

$$||A - \tilde{A}||_F^2 = \|\text{row 1 of } \; A - \tilde{A}\|^2 + \cdots + \|\text{row } m \text{ of } \; A - \tilde{A}\|^2 \tag{11.6}$$

This tells us that, in order to minimize the distance to A, we should choose each row of $\tilde{A}$ to be as close as possible to the corresponding row of A. On the other hand, we require that $\tilde{A}$

have rank one. That is, we require that, for some vector $\boldsymbol{v}$, each row of $\tilde{A}$ lies in Span $\{\boldsymbol{v}\}$. To minimize the distance to A, therefore, once $\boldsymbol{v}$ has been chosen, we should choose $\tilde{A}$ thus:

$$\tilde{A} = \begin{bmatrix} \text{vector in Span } \{\boldsymbol{v}\} \text{ closest to } \boldsymbol{a}_1 \\ \hline \vdots \\ \hline \text{vector in Span } \{\boldsymbol{v}\} \text{ closest to } \boldsymbol{a}_m \end{bmatrix} \tag{11.7}$$

Accordingly, for $i = 1, \ldots, m$,

$$\|\text{row } i \text{ of } A - \tilde{A}\|_F = \text{distance from } \boldsymbol{a}_i \text{ to Span } \{\boldsymbol{v}\}$$

Combining with Equation 11.6 tells us that, once we have chosen $\boldsymbol{v}$, the best approximation $\tilde{A}$ satisfies

$$\|A - \tilde{A}\|^2 = (\text{distance from } \boldsymbol{a}_1 \text{ to Span } \{\boldsymbol{v}\})^2 + \cdots + (\text{distance from } \boldsymbol{a}_m \text{ to Span } \{\boldsymbol{v}\})^2$$

Theorem 11.2.4 tells us that, to minimize the sum of squared distances to Span $\{\boldsymbol{v}\}$, we should choose $\boldsymbol{v}$ to be $\boldsymbol{v}_1$, the first right singular value. By Lemma 11.2.5, the sum of squared distances is then $\|A\|_F^2 - \sigma_1^2$. We therefore obtain

Theorem 11.2.9: The rank-one matrix $\tilde{A}$ that minimizes $\|A - \tilde{A}\|_F$ is

$$\tilde{A} = \begin{bmatrix} \text{vector in Span } \{\boldsymbol{v}_1\} \text{ closest to } \boldsymbol{a}_1 \\ \hline \vdots \\ \hline \text{vector in Span } \{\boldsymbol{v}_1\} \text{ closest to } \boldsymbol{a}_m \end{bmatrix} \tag{11.8}$$

and, for this choice, $\|A - \tilde{A}\|_F^2 = \|A\|_F^2 - \sigma_1^2$.

11.2.4 An expression for the best rank-one approximation

Equation 11.8 specifies $\tilde{A}$ but there is a slicker way of writing it. The vector in Span $\{\boldsymbol{v}_1\}$ closest to $\boldsymbol{a}_i$ is $\boldsymbol{a}_i^{\|\boldsymbol{v}_1}$, the projection of $\boldsymbol{a}_i$ onto Span $\{\boldsymbol{v}_1\}$. Using the formula $\boldsymbol{a}_i^{\|\boldsymbol{v}_1} = \langle \boldsymbol{a}_i, \boldsymbol{v}_1 \rangle \, \boldsymbol{v}_1$, we obtain

$$\tilde{A} = \begin{bmatrix} \langle \boldsymbol{a}_1, \boldsymbol{v}_1 \rangle \, \boldsymbol{v}_1^T \\ \hline \vdots \\ \hline \langle \boldsymbol{a}_m, \boldsymbol{v}_1 \rangle \, \boldsymbol{v}_1^T \end{bmatrix}$$

Using the linear-combinations interpretation of vector-matrix multiplication, we can write this as an outer product of two vectors:

$$\tilde{A} = \begin{bmatrix} \langle \boldsymbol{a}_1, \boldsymbol{v}_1 \rangle \\ \vdots \\ \langle \boldsymbol{a}_m, \boldsymbol{v}_1 \rangle \end{bmatrix} \begin{bmatrix} \quad \boldsymbol{v}_1^T \quad \end{bmatrix} \tag{11.9}$$

By Equation 11.4, the first vector in the outer product can be written as $A\boldsymbol{v}_1$. Substituting into Equation 11.9, we obtain

$$\tilde{A} = \begin{bmatrix} \\ A\boldsymbol{v}_1 \\ \\ \end{bmatrix} \begin{bmatrix} \quad \boldsymbol{v}_1^T \quad \end{bmatrix} \tag{11.10}$$

We have defined σ_1 to be the norm $\|A\boldsymbol{v}_1\|$. We define $\boldsymbol{u}_1$ to be the norm-one vector such that $\sigma_1 \, \boldsymbol{u}_1 = A\boldsymbol{v}_1$. Then we can rewrite Equation 11.10 as

$$\tilde{A} = \sigma_1 \begin{bmatrix} \\ \boldsymbol{u}_1 \\ \\ \end{bmatrix} \begin{bmatrix} \quad \boldsymbol{v}_1^T \quad \end{bmatrix} \tag{11.11}$$

Definition 11.2.10: The *first left singular vector* of A is defined to be the vector $\boldsymbol{u}_1$ such that $\sigma_1\,\boldsymbol{u}_1 = A\boldsymbol{v}_1$, where σ_1 and $\boldsymbol{v}_1$ are, respectively, the first singular value and the first right singular vector.

Theorem 11.2.11: The best rank-one approximation to A is $\sigma_1\,\boldsymbol{u}_1\boldsymbol{v}_1^T$ where σ_1 is the first singular value, $\boldsymbol{u}_1$ is the first left singular vector, and $\boldsymbol{v}_1$ is the first right singular vector of A.

Example 11.2.12: We saw in Example 11.2.3 (Page 426) that, for the matrix $A = \begin{bmatrix} 1 & 4 \\ 5 & 2 \end{bmatrix}$, the first right singular vector is $\boldsymbol{v}_1 \approx \begin{bmatrix} 0.78 \\ 0.63 \end{bmatrix}$ and the first singular value σ_1 is about 6.1. The first left singular vector is $\boldsymbol{u}_1 \approx \begin{bmatrix} 0.54 \\ 0.84 \end{bmatrix}$, meaning $\sigma_1\,\boldsymbol{u}_1 = A\boldsymbol{v}_1$.

We then have

$$\begin{aligned} \tilde{A} &= \sigma_1\,\boldsymbol{u}_1\boldsymbol{v}_1^T \\ &\approx 6.1\begin{bmatrix} 0.54 \\ 0.84 \end{bmatrix}\begin{bmatrix} 0.78 & 0.63 \end{bmatrix} \\ &\approx \begin{bmatrix} 2.6 & 2.1 \\ 4.0 & 3.2 \end{bmatrix} \end{aligned}$$

Then

$$\begin{aligned} A - \tilde{A} &\approx \begin{bmatrix} 1 & 4 \\ 5 & 2 \end{bmatrix} - \begin{bmatrix} 2.6 & 2.1 \\ 4.0 & 3.2 \end{bmatrix} \\ &\approx \begin{bmatrix} -1.56 & 1.93 \\ 1.00 & -1.23 \end{bmatrix} \end{aligned}$$

so the squared Frobenius norm of $A - \tilde{A}$ is

$$1.56^2 + 1.93^2 + 1^2 + 1.23^2 \approx 8.7$$

Does this agree with Theorem 11.2.9? That theorem states that $||A - \tilde{A}||_F^2 = ||A||_F^2 - \sigma_1^2$, which we calculated to be about 8.7 in Example 11.2.6 (Page 427).

11.2.5 The closest one-dimensional affine space

When we defined the trolley-line-location problem in Section 11.2, we stipulated that the trolley-line go through the origin. This was necessary in order that the trolley-line-location problem correspond to finding the closest one-dimensional *vector space*. A one-dimensional vector space is a line through the origin. Recall from Chapter 3 that an arbitrary line (one not necessarily passing through the origin) is an *affine* space.

We can adapt the trolley-line-location techniques to solve this problem as well. Given points $\boldsymbol{a}_1, \ldots, \boldsymbol{a}_m$, we choose a point $\bar{\boldsymbol{a}}$ and translate each of the input points by subtracing $\bar{\boldsymbol{a}}$:

$$\boldsymbol{a}_1 - \bar{\boldsymbol{a}}, \ldots, \boldsymbol{a}_m - \bar{\boldsymbol{a}}$$

We find the one-dimensional vector space closest to these translated points, and then translate that vector space by adding back $\bar{\boldsymbol{a}}$.

Whether the procedure we have just described correctly finds the *closest* affine space depends on how $\bar{\boldsymbol{a}}$ is chosen. The best choice of $\bar{\boldsymbol{a}}$, quite intuitively, is the *centroid* of the input points, the vector

$$\bar{\boldsymbol{a}} = \frac{1}{m}(\boldsymbol{a}_1 + \cdots + \boldsymbol{a}_m)$$

We omit the proof.

Finding the centroid of given points and then translating those points by subtracting off the centroid is called *centering* the points.

Example 11.2.13: We revisit Example 11.2.7 (Page 428), in which $a_1, \ldots, a_{100}$ were the voting records for US Senators. This time, we *center* the data, and only then find the closest one-dimensional vector space Span $\{v_1\}$.

Now projection along v gives a better spread:

Only three of the senators to the left of the origin are Republican:

```
>>> {r for r in senators if is_neg[r] and is_Repub[r]}
{'Collins', 'Snowe', 'Chafee'}
```

and these are perhaps the most moderate Republicans in the Senate at that time. Similarly, only three of the senators to the right of the origin are Democrat.

11.3 Closest dimension-k vector space

The generalization of the trolley-line-location problem to higher dimensions is this:

Computational Problem 11.3.1: *closest low-dimensional subspace:*

- *input:* Vectors $a_1, \ldots a_m$ and positive integer k
- *output:* basis for the k-dimensional vector space V_k that minimizes

$$\sum_i (\text{distance from } a_i \text{ to } V_k)^2$$

The trolley-line-location problem is merely the special case in which $k = 1$. In this special case, we seek the basis for a one-dimensional vector space. The solution, embodied in trolley_line_location(A), is a basis consisting of the unit-norm vector v that maximizes $||Av||$ where A is the matrix whose rows are $a_1, \ldots, a_m$.

11.3.1 A *Gedanken* algorithm to find the singular values and vectors

There is a natural generalization of this algorithm in which an *orthonormal* basis is sought. In the ith iteration, the vector v selected is the one that maximizes $||Av||$ subject to being orthogonal to all previously selected vectors:

- Let v_1 be the norm-one vector v maximizing $||Av||$,
- let v_2 be the norm-one vector v orthogonal to v_1 that maximizes $||Av||$,
- let v_3 be the norm-one vector v orthogonal to v_1 and v_2 that maximizes $||Av||$,

and so on.

Here is the same algorithm in pseudocode:

Given an $m \times n$ matrix A, find vectors $v_1, \ldots, v_{\text{rank } A}$ such that, for $k = 1, 2, \ldots, \text{rank}(A)$, the k-dimensional subspace $\mathcal{V}$ that minimizes $\sum_i (\text{distance from row } i \text{ of } A \text{ to } \mathcal{V}_k)^2$ is Span $\{v_1, \ldots, v_k\}$

```
def find_right_singular_vectors(A):
  for i = 1, 2, ...
```

$\quad\quad \boldsymbol{v}_i = \arg\max\{\|A\boldsymbol{v}\| : \|\boldsymbol{v}\| = 1, \boldsymbol{v} \text{ is orthogonal to } \boldsymbol{v}_1, \boldsymbol{v}_2, \ldots \boldsymbol{v}_{i-1}\}$
$\quad\quad \sigma_i = \|A\boldsymbol{v}_i\|$
until $A\boldsymbol{v} = \mathbf{0}$ for every vector $\boldsymbol{v}$ orthogonal to $\boldsymbol{v}_1, \ldots, \boldsymbol{v}_i$
let r be the final value of the loop variable i.
return $[\boldsymbol{v}_1, \boldsymbol{v}_2, \ldots, \boldsymbol{v}_r]$

Like the procedure `trolley_line_location(A)`, so far this procedure is not fully specified since we have not said how to compute each arg max. Indeed, this is not by any stretch the best algorithm for computing these vectors, but it is very helpful indeed to think about. It is a *Gedanken* algorithm.

Definition 11.3.2: The vectors $\boldsymbol{v}_1, \boldsymbol{v}_2, \ldots, \boldsymbol{v}_r$ are the *right singular vectors* of A, and the corresponding real numbers $\sigma_1, \sigma_2, \ldots, \sigma_r$ are the *singular values* of A.

11.3.2 Properties of the singular values and right singular vectors

The following property is rather obvious.

Proposition 11.3.3: The right singular vectors are orthonormal.

Proof

In iteration i, $\boldsymbol{v}_i$ is chosen from among vectors that have norm one and are orthogonal to $\boldsymbol{v}_1, \ldots, \boldsymbol{v}_{i-1}$. □

Example 11.3.4: We revisit the matrix $A = \begin{bmatrix} 1 & 4 \\ 5 & 2 \end{bmatrix}$ of Examples 11.2.3, 11.2.6 and 11.2.12. We saw that the first right singular vector is $\boldsymbol{v}_1 \approx \begin{bmatrix} 0.78 \\ 0.63 \end{bmatrix}$ and the first singular value σ_1 is about 6.1. The second right singular vector must therefore be chosen among the vectors orthogonal to $\begin{bmatrix} 0.78 \\ 0.63 \end{bmatrix}$. It turns out to be $\begin{bmatrix} 0.63 \\ -0.78 \end{bmatrix}$. The corresponding singular value is $\sigma_2 \approx 2.9$.

The vectors $\boldsymbol{v}_1$ and $\boldsymbol{v}_2$ vectors are obviously orthogonal. Notice that σ_2 is smaller than σ_1. It cannot be greater since the second maximization is over a smaller set of candidate solutions.

Since the vectors $\boldsymbol{v}_1$ and $\boldsymbol{v}_2$ are orthogonal and nonzero, we know they are linearly independent, and therefore that they span $\mathbb{R}^2$.

Here's another nearly obvious property.

Proposition 11.3.5: The singular values are nonnegative and in descending order.

Proof

Since each singular value is the norm of a vector, it is nonnegative. For each $i > 1$, the set of vectors from which $\boldsymbol{v}_i$ is chosen is a subset of the set of vectors from which $\boldsymbol{v}_{i-1}$ is chosen, so the maximum achieved in iteration i is no greater than the maximum achieved in iteration $i - 1$. This shows $\sigma_i \leq \sigma_{i-1}$. □

Now for something not at all obvious—a rather surprising fact that is at the heart of the notion of Singular Value Decomposition.

Lemma 11.3.6: Every row of A is in the span of the right singular vectors.

Proof

Let $\mathcal{V} = \text{Span }\{\boldsymbol{v}_1, \ldots, \boldsymbol{v}_r\}$. Let $\mathcal{V}^o$ be the annihilator of $\mathcal{V}$, and recall that $\mathcal{V}^o$ consists of all vectors orthogonal to $\mathcal{V}$. By the loop termination condition, for any vector $\boldsymbol{v}$ in $\mathcal{V}^o$, the product $A\boldsymbol{v}$ is the zero vector, so the rows of A are orthogonal to $\boldsymbol{v}$. The annihilator of the annihilator $(\mathcal{V}^o)^*$ consists of all vectors orthogonal to $\mathcal{V}^o$, so the rows of A are in $(\mathcal{V}^o)^*$. Theorem 6.5.15, the Annihilator Theorem, states that $(\mathcal{V}^o)^o$ equals $\mathcal{V}$. This shows that the rows of A are in $\mathcal{V}$. □

11.3.3 The singular value decomposition

Lemma 11.3.6 tells us that each row $\boldsymbol{a}_i$ of A is a linear combination of the right singular vectors:

$$\boldsymbol{a}_i = \sigma_{i1}\,\boldsymbol{v}_1 + \cdots + \sigma_{ir}\,\boldsymbol{v}_r$$

Since $\boldsymbol{v}_1, \ldots, \boldsymbol{v}_r$ are orthonormal, the jth summand $\sigma_{ij}\,\boldsymbol{v}_j$ is the projection of $\boldsymbol{a}_i$ along the jth right singular vector $\boldsymbol{v}_j$, and the coefficient σ_{ij} is just the inner product of $\boldsymbol{a}_i$ and $\boldsymbol{v}_j$:

$$\boldsymbol{a}_i = \langle \boldsymbol{a}_i, \boldsymbol{v}_1\rangle\ \boldsymbol{v}_1 + \cdots + \langle \boldsymbol{a}_i, \boldsymbol{v}_r\rangle\ \boldsymbol{v}_r$$

Using the dot-product definition of vector-matrix multiplication, we write this as

$$\boldsymbol{a}_i = \begin{bmatrix} \langle \boldsymbol{a}_i, \boldsymbol{v}_1\rangle & \cdots & \langle \boldsymbol{a}_i, \boldsymbol{v}_r\rangle \end{bmatrix} \begin{bmatrix} \boldsymbol{v}_1^T \\ \hline \vdots \\ \hline \boldsymbol{v}_r^T \end{bmatrix}$$

Combining all these equations and using the vector-matrix definition of matrix-matrix multiplication, we can express A as a matrix-matrix product:

$$\begin{bmatrix} \boldsymbol{a}_1^T \\ \hline \boldsymbol{a}_2^T \\ \hline \vdots \\ \hline \boldsymbol{a}_m^T \end{bmatrix} = \begin{bmatrix} \langle \boldsymbol{a}_1, \boldsymbol{v}_1\rangle & \cdots & \langle \boldsymbol{a}_1, \boldsymbol{v}_r\rangle \\ \langle \boldsymbol{a}_2, \boldsymbol{v}_1\rangle & \cdots & \langle \boldsymbol{a}_2, \boldsymbol{v}_r\rangle \\ & \vdots & \\ \langle \boldsymbol{a}_m, \boldsymbol{v}_1\rangle & \cdots & \langle \boldsymbol{a}_m, \boldsymbol{v}_r\rangle \end{bmatrix} \begin{bmatrix} \boldsymbol{v}_1^T \\ \hline \vdots \\ \hline \boldsymbol{v}_r^T \end{bmatrix}$$

We can further simplify this equation. The jth column of the first matrix on the right-hand side is

$$\begin{bmatrix} \langle \boldsymbol{a}_1, \boldsymbol{v}_j\rangle \\ \langle \boldsymbol{a}_2, \boldsymbol{v}_j\rangle \\ \vdots \\ \langle \boldsymbol{a}_m, \boldsymbol{v}_j\rangle \end{bmatrix}$$

which is, by the dot-product definition of linear combinations, simply $A\boldsymbol{v}_j$. It is convenient to have a name for these vectors.

Definition 11.3.7: The vectors $\boldsymbol{u}_1, \boldsymbol{u}_2, \ldots, \boldsymbol{u}_r$ such that $\sigma_j\,\boldsymbol{u}_j = A\boldsymbol{v}_j$ are the *left singular vectors* of A.

Proposition 11.3.8: The left singular vectors are orthonormal.

(The proof is given in Section 11.3.10.)

Using the definition of left singular vectors, we substitute $\sigma_j\,\boldsymbol{u}_j$ for $A\boldsymbol{v}_j$, resulting in the equation

$$\begin{bmatrix} \\ A \\ \\ \end{bmatrix} = \left[\begin{array}{c|c|c} \sigma_1\boldsymbol{u}_1 & \cdots & \sigma_r\boldsymbol{u}_r \end{array}\right] \begin{bmatrix} \boldsymbol{v}_1^T \\ \hline \vdots \\ \hline \boldsymbol{v}_r^T \end{bmatrix}$$

Finally, we separate out $\sigma_1, \ldots, \sigma_r$ into a diagonal matrix, obtaining the equation

$$\left[\begin{array}{c} \\ A \\ \\ \end{array}\right] = \left[\begin{array}{c|c|c} & & \\ \boldsymbol{u}_1 & \cdots & \boldsymbol{u}_r \\ & & \end{array}\right] \left[\begin{array}{ccc} \sigma_1 & & \\ & \ddots & \\ & & \sigma_r \end{array}\right] \left[\begin{array}{c} \boldsymbol{v}_1^T \\ \hline \vdots \\ \hline \boldsymbol{v}_r^T \end{array}\right] \tag{11.12}$$

Definition 11.3.9: The *singular value decomposition* of a matrix A is a factorization of A as $A = U\Sigma V^T$ in which the matrices U, Σ, and V have three properties:

Property S1: Σ is a diagonal matrix whose entries $\sigma_1, \ldots, \sigma_r$ are positive and in descending order.

Property S2: V is a column-orthogonal matrix.

Property S3: U is a column-orthogonal matrix.

(Sometimes this is called the *reduced* singular value decomposition.)

We have established the following theorem.

Theorem 11.3.10: Every matrix A over $\mathbb{R}$ has a singular value decomposition.

Proof

We have derived Equation 11.12, which shows the factorization of A into the product of matrices U, Σ, and V. Property S1 follows from Proposition 11.3.5. Property S2 follows from Proposition 11.3.3. Property S3 follows from Proposition 11.3.8. □

The procedure `def find_right_singular_vectors(`A`)` is not the most efficient way to find a singular value decomposition of A. The best algorithms are beyond the scope of this book, but we provide a module `svd` with a procedure `factor(A)` that, given a Mat A, returns a triple (U, Σ, V) such that $A = U\Sigma * V^T$.

It is worth noting that the singular value decomposition has a nice symmetry under transposition. By the properties of the transpose of a matrix product (Proposition 4.11.14),

$$\begin{aligned} A^T &= (U\Sigma V^T)^T \\ &= V\Sigma^T U^T \\ &= V\Sigma U^T \end{aligned}$$

because the transpose of Σ is Σ itself.

We see that the the SVD of A^T can be obtained from the SVD of A^T just by swapping U and V.

As we will see, the SVD is important as both a mathematical concept and a computational tool. One of the people who helped develop good algorithms for computing the SVD was Gene Golub, whose license plate reflected his interest in the topic:

Detail of a photograph due to Professor Pieter Kroonenberg

11.3.4 Using right singular vectors to find the closest k-dimensional space

Now we show how to use the right singular vectors to address Computational Problem 11.3.1. First we state how good a solution they provide.

Lemma 11.3.11: Let $\boldsymbol{v}_1, \ldots, \boldsymbol{v}_k$ be an orthonormal vector basis for a vector space $\mathcal{V}$. Then

$$(\text{distance from } \boldsymbol{a}_1 \text{ to } \mathcal{V})^2 + \cdots + (\text{distance from } \boldsymbol{a}_m \text{ to } \mathcal{V})^2$$

is $\|A\|_F^2 - \|A\boldsymbol{v}_1\|^2 - \|A\boldsymbol{v}_2\|^2 - \cdots - \|A\boldsymbol{v}_k\|^2$

Proof

The argument is the same as that given in Section 11.2.1. For each vector $\boldsymbol{a}_i$, write $\boldsymbol{a}_i = \boldsymbol{a}_i^{\|\mathcal{V}} + \boldsymbol{a}_i^{\perp\mathcal{V}}$. By the Pythagorean Theorem, $\|\boldsymbol{a}_1^{\perp\mathcal{V}}\|^2 = \|\boldsymbol{a}_1\|^2 - \|\boldsymbol{a}_1^{\|\mathcal{V}}\|^2$. Therefore the sum of squared distances is

$$\left(\|\boldsymbol{a}_1\|^2 - \|\boldsymbol{a}_1^{\|\mathcal{V}}\|^2\right) + \cdots + \left(\|\boldsymbol{a}_m\|^2 - \|\boldsymbol{a}_m^{\|\mathcal{V}}\|^2\right)$$

which equals

$$\left(\|\boldsymbol{a}_1\|^2 + \cdots + \|\boldsymbol{a}_m\|^2\right) + \left(\|\boldsymbol{a}_1^{\|\mathcal{V}}\|^2 + \cdots + \|\boldsymbol{a}_m^{\|\mathcal{V}}\|^2\right)$$

The first sum $\|\boldsymbol{a}_1\|^2 + \cdots + \|\boldsymbol{a}_m\|^2$ equals $\|A\|_F^2$. As for the second sum,

$$\begin{aligned}
&\|\boldsymbol{a}_1^{\|\mathcal{V}}\|^2 + \cdots + \|\boldsymbol{a}_m^{\|\mathcal{V}}\|^2 \\
&= \left(\|\boldsymbol{a}_1^{\|\boldsymbol{v}_1}\|^2 + \cdots + \|\boldsymbol{a}_1^{\|\boldsymbol{v}_k}\|^2\right) + \cdots + \left(\|\boldsymbol{a}_m^{\|\boldsymbol{v}_1}\|^2 + \cdots + \|\boldsymbol{a}_m^{\|\boldsymbol{v}_k}\|^2\right) \\
&= \left(\langle \boldsymbol{a}_1, \boldsymbol{v}_1\rangle^2 + \cdots + \langle \boldsymbol{a}_1, \boldsymbol{v}_k\rangle^2\right) + \cdots + \left(\langle \boldsymbol{a}_m, \boldsymbol{v}_1\rangle^2 + \cdots + \langle \boldsymbol{a}_m, \boldsymbol{v}_k\rangle^2\right)
\end{aligned}$$

Reorganizing all these squared inner products, we get

$$\begin{aligned}
&\left(\langle \boldsymbol{a}_1, \boldsymbol{v}_1\rangle^2 + \langle \boldsymbol{a}_2, \boldsymbol{v}_1\rangle^2 + \cdots + \langle \boldsymbol{a}_m, \boldsymbol{v}_1\rangle^2\right) + \cdots + \left(\langle \boldsymbol{a}_1, \boldsymbol{v}_k\rangle^2 + \langle \boldsymbol{a}_2, \boldsymbol{v}_k\rangle^2 + \cdots + \langle \boldsymbol{a}_m, \boldsymbol{v}_k\rangle^2\right) \\
&= \|A\boldsymbol{v}_1\|^2 + \cdots + \|A\boldsymbol{v}_k\|^2
\end{aligned}$$

□

The next theorem says that the span of the first k right singular vectors is the best solution.

Theorem 11.3.12: Let A be an $m \times n$ matrix, and let $\boldsymbol{a}_1, \ldots, \boldsymbol{a}_m$ be its rows. Let $\boldsymbol{v}_1, \ldots, \boldsymbol{v}_r$ be its right singular vectors, and let $\sigma_1, \ldots, \sigma_r$ be its singular values. For any positive integer $k \leq r$, Span $\{\boldsymbol{v}_1, \ldots, \boldsymbol{v}_k\}$ is the k-dimensional vector space $\mathcal{V}$ that minimizes

$$(\text{distance from } \boldsymbol{a}_1 \text{ to } \mathcal{V})^2 + \cdots + (\text{distance from } \boldsymbol{a}_m \text{ to } \mathcal{V})^2$$

and the minimum sum of squared distances is $\|A\|_F^2 - \sigma_1^2 - \sigma_2^2 - \cdots - \sigma_k^2$.

Proof

By Lemma 11.3.11, the sum of squared distances for the space $\mathcal{V} = \text{Span}\ \{\boldsymbol{v}_1, \ldots, \boldsymbol{v}_k\}$ is

$$\|A\|_F^2 - \sigma_1^2 - \sigma_2^2 - \cdots - \sigma_k^2 \tag{11.13}$$

To prove that this is the minimum, we need to show that any other k-dimensional vector space $\mathcal{W}$ leads to a sum of squares that is no smaller.

Any k-dimensional vector space $\mathcal{W}$ has an orthonormal basis. Let $\boldsymbol{w}_1, \ldots, \boldsymbol{w}_k$ be such a basis. Plugging these vectors into Lemma 11.3.11, we get that the sum of squared distances from $\boldsymbol{a}_1, \ldots, \boldsymbol{a}_m$ to $\mathcal{W}$ is

$$\|A\|_F^2 - \|A\boldsymbol{w}_1\|^2 - \|A\boldsymbol{w}_2\|^2 - \cdots - \|A\boldsymbol{w}_k\|^2 \tag{11.14}$$

In order to show that $\mathcal{V}$ is the closest, we need to show that the quantity in 11.14 is no less than the quantity in 11.13. This requires that we show that $\|A\boldsymbol{w}_1\|^2 + \cdots + \|A\boldsymbol{w}_k\|^2 \leq \sigma_1^2 + \cdots + \sigma_k^2$.

For conciseness, we work with the matrix W with columns $\boldsymbol{w}_1, \ldots, \boldsymbol{w}_k$. Then

$$\|AW\|_F^2 = \|A\boldsymbol{w}_1\|^2 + \cdots + \|A\boldsymbol{w}_k\|^2$$

by the column analogue of Lemma 11.1.1.

By Theorem 11.3.10, A can be factored as $A = U\Sigma V^T$ where U and V are column-orthogonal and Σ is diagonal. By substitution, $\|AW\|_F^2 = \|U\Sigma V^T W\|_F^2$. Since U is column-orthogonal, multiplication by U preserves norms, so $\|U\Sigma V^T W\|_F^2 = \|\Sigma V^T W\|_F^2$.

Now we consider, for $i = 1, \ldots, k$, the projection of $\boldsymbol{v}_i$ onto $\mathcal{W}$:

$$\boldsymbol{v}_i = \boldsymbol{v}_i^{||\mathcal{W}} + \boldsymbol{v}_i^{\perp\mathcal{W}}$$

Since the two vectors on the right-hand side are orthogonal,

$$\|\boldsymbol{v}_i\|^2 = \|\boldsymbol{v}_i^{||\mathcal{W}}\|^2 + \|\boldsymbol{v}_i^{\perp\mathcal{W}}\|^2$$

In particular, $\|\boldsymbol{v}_i^{||\mathcal{W}}\|^2 \leq \|\boldsymbol{v}_i\|^2 = 1$. Let $\boldsymbol{x}_i$ be the coordinate representation of $\boldsymbol{v}_i^{||\mathcal{W}}$ in terms of $\boldsymbol{w}_1, \ldots, \boldsymbol{w}_k$. Then $\boldsymbol{v}_i^{||\mathcal{W}} = W\boldsymbol{x}_i$. Since W is column-orthogonal, $\|W\boldsymbol{x}_i\| = \|\boldsymbol{x}_i\|$, so $\|\boldsymbol{x}_i\|^2 \leq 1$.

Since $\boldsymbol{w}_1, \ldots, \boldsymbol{w}_k$ are orthonormal,

$$\boldsymbol{x}_i = [\langle \boldsymbol{v}_i, \boldsymbol{w}_1 \rangle, \langle \boldsymbol{v}_i, \boldsymbol{w}_2 \rangle, \ldots, \langle \boldsymbol{v}_i, \boldsymbol{w}_k \rangle]$$

This vector is row i of $V^T W$, by the vector-matrix definition of matrix-matrix multiplication and the dot-product definition of vector-matrix multiplication.

$$\begin{aligned} \|\Sigma V^T W\|^2 &= \sigma_1^2 \|\text{row } 1 \text{ of } V^T W\|^2 + \cdots \sigma_k^2 \|\text{row } m \text{ of } V^T W\|^2 \\ &= \sigma_1^2 \|\boldsymbol{x}_1\|^2 + \cdots + \sigma_k^2 \|\boldsymbol{x}_k\|^2 \\ &\leq \sigma_1^2 + \cdots + \sigma_k^2 \end{aligned}$$

This completes the proof. □

11.3.5 Best rank-k approximation to A

We saw in Section 11.2.4 that the best rank-one approximation to A is $\sigma_1 \boldsymbol{u}_1 \boldsymbol{v}_1^T$. Now we generalize that formula:

Theorem 11.3.13: For $k \leq$ rank A, the best rank-at-most-k approximation to A is

$$\tilde{A} = \sigma_1 \, \boldsymbol{u}_1 \boldsymbol{v}_1^T + \cdots + \sigma_k \, \boldsymbol{u}_k \boldsymbol{v}_k^T \tag{11.15}$$

for which $\|A - \tilde{A}\|_F^2 = \|A\|_F^2 - \sigma_1^2 - \sigma_2^2 - \cdots - \sigma_k^2$.

Proof

The proof is a straightforward generalization of the argument in Section 11.2.2. Let $\tilde{A}$ be a rank-at-most-k approximation to A. By Lemma 11.1.1,

$$\|A - \tilde{A}\|_F^2 = \|\text{row } 1 \text{ of } \; A - \tilde{A}\|^2 + \cdots + \|\text{row } m \text{ of } \; A - \tilde{A}\|^2 \tag{11.16}$$

For $\tilde{A}$ to have rank at most k, there must be some vector space $\mathcal{V}$ of dimension k such that every row of $\tilde{A}$ lies in $\mathcal{V}$. Once $\mathcal{V}$ has been chosen, Equation 11.16 tells us that the best choice of $\tilde{A}$ is

$$\tilde{A} = \begin{bmatrix} \text{vector in } \mathcal{V} \text{ closest to } \boldsymbol{a}_1 \\ \hline \vdots \\ \hline \text{vector in } \mathcal{V} \text{ closest to } \boldsymbol{a}_m \end{bmatrix} \tag{11.17}$$

and, for this choice,

$$\|A - \tilde{A}\|^2 = (\text{distance from } \boldsymbol{a}_1 \text{ to } \mathcal{V})^2 + \cdots + (\text{distance from } \boldsymbol{a}_m \text{ to } \mathcal{V})^2$$

Theorem 11.3.12 tells us that, to minimize the sum of squared distances to $\mathcal{V}$, we should choose $\mathcal{V}$ to be the span of the first k right singular vectors, and that the sum of squared distances is then $\|A\|_F^2 - \sigma_1^2 - \sigma_2^2 - \cdots - \sigma_k^2$.

For $i = 1, \ldots, m$, the vector in $\mathcal{V}$ closest to $\boldsymbol{a}_i$ is the projection of $\boldsymbol{a}_i$ onto $\mathcal{V}$, and

$$\begin{aligned} \text{projection of } \boldsymbol{a}_i \text{ onto } \mathcal{V} &= \text{projection of } \boldsymbol{a}_i \text{ along } \boldsymbol{v}_1 + \cdots + \text{projection of } \boldsymbol{a}_i \text{ along } \boldsymbol{v}_m \\ &= \langle \boldsymbol{a}_i, \boldsymbol{v}_1 \rangle\, \boldsymbol{v}_1 + \cdots + \langle \boldsymbol{a}_i, \boldsymbol{v}_k \rangle\, \boldsymbol{v}_k \end{aligned}$$

Substituting into Equation 11.17 and using the definition of addition of matrices gives us

$$\begin{aligned} \tilde{A} &= \begin{bmatrix} \langle \boldsymbol{a}_1, \boldsymbol{v}_1 \rangle\, \boldsymbol{v}_1 \\ \hline \vdots \\ \hline \langle \boldsymbol{a}_m, \boldsymbol{v}_1 \rangle\, \boldsymbol{v}_1 \end{bmatrix} + \cdots + \begin{bmatrix} \langle \boldsymbol{a}_1, \boldsymbol{v}_k \rangle\, \boldsymbol{v}_k \\ \hline \vdots \\ \hline \langle \boldsymbol{a}_m, \boldsymbol{v}_k \rangle\, \boldsymbol{v}_k \end{bmatrix} \\ &= \sigma_1 \begin{bmatrix} \\ \boldsymbol{u}_1 \\ \\ \end{bmatrix} \begin{bmatrix} \quad \boldsymbol{v}_1 \quad \end{bmatrix} + \cdots + \sigma_k \begin{bmatrix} \\ \boldsymbol{u}_k \\ \\ \end{bmatrix} \begin{bmatrix} \quad \boldsymbol{v}_k \quad \end{bmatrix} \end{aligned}$$

□

11.3.6 Matrix form for best rank-k approximation

Equation 11.15 gives the best rank-k approximation to A as the sum of k rank-one matrices. By using the definitions of matrix-matrix and matrix-vector multiplication, one can show that Equation 11.15 can be rewritten as

$$\tilde{A} = \begin{bmatrix} & & \\ \boldsymbol{u}_1 & \cdots & \boldsymbol{u}_k \\ & & \end{bmatrix} \begin{bmatrix} \sigma_1 & & \\ & \ddots & \\ & & \sigma_k \end{bmatrix} \begin{bmatrix} \boldsymbol{v}_1^T \\ \hline \vdots \\ \hline \boldsymbol{v}_k^T \end{bmatrix}$$

In view of the resemblance to the singular value decomposition of A, namely $A = U\Sigma V^T$, we write

$$\tilde{A} = \tilde{U}\tilde{\Sigma}\tilde{V}^T$$

where $\tilde{U}$ consists of the first k columns of U, $\tilde{V}$ consists of the first k columns of V, and $\tilde{\Sigma}$ is the diagonal matrix whose diagonal elements are the first k diagonal elements of Σ.

11.3.7 Number of nonzero singular values is rank A

It follows from Lemma 11.3.6 that the number r of right singular vectors produced by algorithm `find_right_singular_vectors`(A) is at least the rank of A.

Let $k = \text{rank } A$. For this value of k, the best rank-k approximation to A is A itself. This shows that any subsequent singular values $\sigma_{1+\text{rank } A}, \sigma_{2+\text{rank } A}, \ldots$ must be zero. Therefore, in the algorithm `find_right_singular_vectors`(A), after rank A iterations, $A\boldsymbol{v} = \boldsymbol{0}$ for every vector $\boldsymbol{v}$ orthogonal to $\boldsymbol{v}_1, \ldots, \boldsymbol{v}_{\text{rank } A}$. Thus the number r of iterations is exactly rank A.

let's reconsider the SVD of A:

$$\begin{bmatrix} & & \\ & A & \\ & & \end{bmatrix} = \underbrace{\begin{bmatrix} & & \\ \boldsymbol{u}_1 & \cdots & \boldsymbol{u}_r \\ & & \end{bmatrix}}_{U} \underbrace{\begin{bmatrix} \sigma_1 & & \\ & \ddots & \\ & & \sigma_r \end{bmatrix}}_{\Sigma} \underbrace{\begin{bmatrix} \boldsymbol{v}_1^T \\ \vdots \\ \boldsymbol{v}_r^T \end{bmatrix}}_{V^T}$$

By the vector-matrix definition of matrix-matrix multiplication, each row of A is the corresponding row of $U\Sigma$ times the matrix V^T. Therefore, by the linear-combinations definition of vector-matrix multiplication, each row of A is a linear combination of the rows of V^T. On the other hand, the rows of V^T are mutually orthogonal and nonzero, so linearly independent (Proposition 9.5.1), and there are rank A of them, so the dimension of their span is exactly rank A. Thus, by the Dimension Principle (Lemma 6.2.14), Row A equals Row V^T.

A similar argument shows that Col A equals Col U. Each column of A is U times a column of ΣV^T, and dim Col A = rank A = dim Col U, so Col A = Col U.

We summarize our findings:

Proposition 11.3.14: In the singular value decomposition $U\Sigma V^T$ of A, Col U = Col A and Row V^T = Row A.

11.3.8 Numerical rank

In fact, computing or even defining the rank of a matrix with floating-point entries is not a trivial matter. Maybe the columns of A are linearly dependent but due to floating-point error when you run `orthogonalize` on the columns you get all nonzero vectors. Or maybe the matrix you have represented in your computer is only an approximation to some "true" matrix whose entries cannot be represented exactly by floating-point numbers. The rank of the true matrix might differ from that of the represented matrix. As a practical matter, we need some useful definition of rank, and here is what is used: the *numerical rank* of a matrix is defined to the number of singular values you get before you get a singular value that is tiny.

11.3.9 Closest k-dimensional affine space

To find not the closest k-dimensional vector space but the closest k-dimensional affine space, we can use the centering technique described in Section 11.2.5: find the centroid $\bar{\boldsymbol{a}}$ of the input points $\boldsymbol{a}_1, \ldots, \boldsymbol{a}_m$, and subtract it from each of the input points. Then find a basis $\boldsymbol{v}_1, \ldots, \boldsymbol{v}_k$ for the k-dimensional vector space closest to $\boldsymbol{a}_1 - \bar{\boldsymbol{a}}, \ldots, \boldsymbol{a}_m - \bar{\boldsymbol{a}}$. The k-dimensional affine space closest to the original points $\boldsymbol{a}_1, \ldots, \boldsymbol{a}_m$ is

$$\{\bar{\boldsymbol{a}} + \boldsymbol{v} \;:\; \boldsymbol{v} \in \text{Span } \{\boldsymbol{v}_1, \ldots, \boldsymbol{v}_k\}\}$$

The proof is omitted.

Example 11.3.15: Returning to the US Senate voting data, in Examples 11.2.7 and 11.2.13, we plotted the senators' voting records on the number line, based on their projection onto the closest one-dimensional vector. Now we can find the closest 2-dimensional affine space, and project their voting records onto these, and use the coordinates to plot the senators.

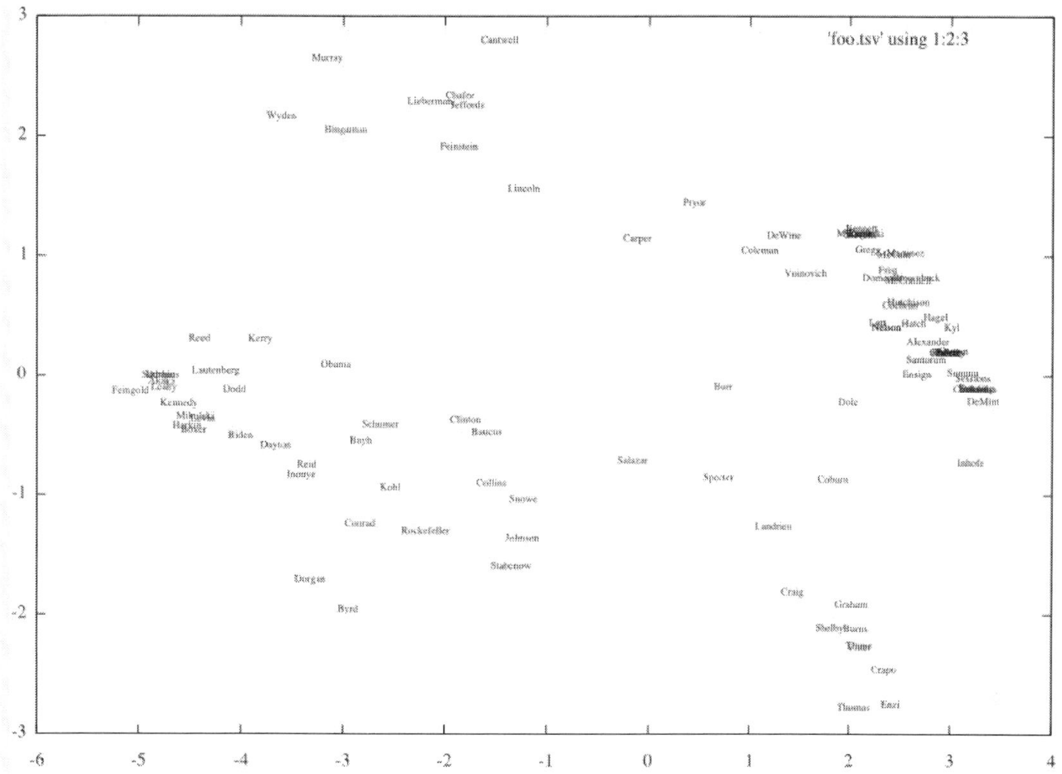

11.3.10 Proof that U is column-orthogonal

Property S3 of the singular value decomposition states that the matrix U of left singular vectors is column-orthogonal. We now prove that property.

The left singular vectors have norm one by construction. We need to show that they are mutually orthogonal. We prove by induction that, for $k = 0, 1, 2, \ldots, r$, the first k left singular vectors $\boldsymbol{u}_1, \ldots, \boldsymbol{u}_k$ are mutually orthogonal and are orthogonal to the remainining left-singular vectors $\boldsymbol{u}_{k+1}, \ldots, \boldsymbol{u}_r$.

Let

$$U^{\perp} = \begin{bmatrix} \boldsymbol{u}_1^* \\ \hline \boldsymbol{u}_2^* \\ \hline \vdots \\ \hline \boldsymbol{u}_r^* \end{bmatrix}$$

$$U^{\perp} A V = \begin{bmatrix} \boldsymbol{u}_1^* \\ \hline \boldsymbol{u}_2^* \\ \hline \vdots \\ \hline \boldsymbol{u}_r^* \end{bmatrix} \left[\begin{array}{c|c|c|c} \sigma_1 \boldsymbol{u}_1 & \sigma_2 \boldsymbol{u}_2 & \cdots & \sigma_r \boldsymbol{u}_r \end{array} \right]$$

By the inductive hypothesis, for every $i < j < k$, $\langle \boldsymbol{u}_i, \boldsymbol{u}_j \rangle = 0$. It follows that $\boldsymbol{u}_1^* = \boldsymbol{u}_1$ and

$\boldsymbol{u}_2^* = \boldsymbol{u}_2$ and ... and $\boldsymbol{u}_k^* = \boldsymbol{u}_k$. We can therefore write

$$U^{\perp} A V = \begin{bmatrix} \boldsymbol{u}_1 \\ \hline \boldsymbol{u}_2 \\ \hline \vdots \\ \hline \boldsymbol{u}_k \\ \hline \boldsymbol{u}_{k+1}^* \\ \hline \vdots \\ \hline \boldsymbol{u}_r^* \end{bmatrix} \left[\begin{array}{c|c|c|c|c|c} \sigma_1 \boldsymbol{u}_1 & \sigma_2 \boldsymbol{u}_2 & \cdots & \sigma_k \boldsymbol{u}_k & \cdots & \sigma_r \boldsymbol{u}_r \end{array} \right]$$

$$= \begin{bmatrix} \sigma_1 & 0 & \cdots & 0 & 0 & \cdots & 0 \\ 0 & \sigma_1 & \cdots & 0 & 0 & \cdots & 0 \\ & & \ddots & & & & \\ 0 & 0 & & \sigma_k & \beta_{k+1} & \cdots & \beta_r \\ 0 & 0 & & 0 & & & \\ & & \vdots & & & ? & \\ 0 & 0 & & 0 & & & \end{bmatrix}$$

The ? indicates a part of the matrix we don't care about for the induction step.

The induction step consists in showing that $\boldsymbol{u}_k$ is orthogonal to all the vectors that come after it: $\boldsymbol{u}_{k+1}, \ldots, \boldsymbol{u}_r$. Note that the values $\beta_{k+1}, \ldots, \beta_r$ in the matrix are the inner products of $\boldsymbol{u}_k$ with $\sigma_{k+1}\, \boldsymbol{u}_{k+1}, \ldots, \sigma_r\, \boldsymbol{u}_r$. Our aim, then, is to show that these values are zero.

Let $\boldsymbol{w}$ be the kth row in the matrix $U^{\perp}AV$, that is $\boldsymbol{w} = [0, 0, \ldots, \sigma_k, \beta_{k+1}, \ldots, \beta_r]$. Then

$$\|\boldsymbol{w}\|^2 = \sigma_k^2 + \beta_{k+1}^2 + \cdots + \beta_r^2$$

Since V is a column-orthogonal matrix, by Corollary 10.2.3, multiplication by V preserves norms, so

$$\|V\boldsymbol{w}\|^2 = \sigma_k^2 + \beta_{k+1}^2 + \cdots + \beta_r^2 \tag{11.18}$$

Since the first $k-1$ entries of $\boldsymbol{w}$ are zero, $V\boldsymbol{w}$ is a linear combination of columns $k, k+1, \ldots, r$ of V. Since the columns of V are mutually orthogonal, $V\boldsymbol{w}$ is orthogonal to each of the first $k-1$ columns of V, namely $\boldsymbol{v}_1, \ldots, \boldsymbol{v}_{k-1}$.

By the dot-product definition of matrix-vector multiplication, entry k of the matrix-vector product $(U^{\perp}AV)\boldsymbol{w}$ is $\sigma_k^2 + \beta_{k+1}^2 + \cdots + \beta_r^2$, so

$$\|U^{\perp}AV\boldsymbol{w}\| \geq \sigma_k^2 + \beta_{k+1}^2 + \cdots + \beta_r^2$$

Since $U^{\perp}$ is a column-orthogonal matrix, again by Corollary 10.2.3, multiplication by $U^{\perp}$ preserves norms, so

$$\|AV\boldsymbol{w}\|^2 \geq \sigma_k^2 + \beta_{k+1}^2 + \cdots + \beta_r^2 \tag{11.19}$$

Combining Equation 11.19 with Equation 11.18, we infer

$$\frac{\|A(V\boldsymbol{w})\|}{\|V\boldsymbol{w}\|} \geq \frac{\sigma^2 + \beta_{k+1}^2 + \cdots + \beta_r^2}{\sqrt{\sigma^2 + \beta_{k+1}^2 + \cdots + \beta_r^2}} = \sqrt{\sigma^2 + \beta_{k+1}^2 + \cdots + \beta_r^2} \tag{11.20}$$

Let $\boldsymbol{v}$ be the vector obtained by normalizing $V\boldsymbol{w}$, i.e. $\boldsymbol{v} = \frac{1}{\|V\boldsymbol{w}\|} V\boldsymbol{w}$. Then $\boldsymbol{v}$ is a norm-one vector that is orthogonal to $\boldsymbol{v}_1, \ldots, \boldsymbol{v}_{k-1}$. Furthermore, if any of $\beta_{k+1}, \ldots, \beta_r$ is nonzero then $\|A\boldsymbol{v}\|$ is greater than σ_k. But this contradicts the algorithm's choice of the kth singular vector $\boldsymbol{v}_k$, which was chosen among unit-norm vectors orthogonal to $\boldsymbol{v}_1, \ldots, \boldsymbol{v}_{k-1}$ to maximize $\|A\boldsymbol{v}\|$.

The contradiction proves that $\beta_{k+1} = 0, \ldots, \beta_r = 0$, which completes the induction step, and the proof.

11.4 Using the singular value decomposition

The singular value decomposition has emerged as a crucial tool in linear-algebra computation.

11.4.1 Using SVD to do least squares

In Section 9.8.5, we learned that the QR factorization of a matrix A can be used in solving the *least squares* problem, finding the vector $\hat{\boldsymbol{x}}$ that minimizes $\|A\boldsymbol{x} - \boldsymbol{b}\|$. However, that algorithm is applicable only if A's columns are linearly independent. Here we see that the singular value decomposition provides another method to solve least squares, a method that does not depend on A having linearly independent columns.

```
def SVD_solve(A):
  U, Σ, V = svd.factor(A)
  return VΣ^{-1}U^T b
```

Note that this algorithm seems to require multiplication by the inverse of a matrix, but the matrix is diagonal (with nonzero diagonal entries $\sigma_1, \ldots, \sigma_{\text{rank } A}$), so multiplication by its inverse amounts to applying the function $f([y_1, y_2, \ldots, y_r]) = [\sigma_1^{-1}y_1, \sigma_2^{-1}y_2, \ldots, \sigma_r^{-1}y_r]$.

To show this algorithm returns the correct solution, let $\hat{\boldsymbol{x}} = V\Sigma^{-1}U^T\boldsymbol{b}$ be the vector returned. Multiplying on the left by V^T, we get the equation

$$V^T\hat{\boldsymbol{x}} = \Sigma^{-1}U^T\boldsymbol{b}$$

Multiplying on the left by Σ, we get

$$\Sigma V^T\hat{\boldsymbol{x}} = U^T\boldsymbol{b}$$

Multiplying on the left by by U, we get

$$U\Sigma V^T\hat{\boldsymbol{x}} = UU^T\boldsymbol{b}$$

By substitution, we get

$$A\hat{\boldsymbol{x}} = UU^T\boldsymbol{b}$$

This equation should be familiar; it is similar to the equation that justified use of QR_solve(A) in solving a least-squares problem. By Lemma 9.8.3, $UU^T\boldsymbol{b}$ is the projection $\boldsymbol{b}^{\|\text{Col } U}$ of $\boldsymbol{b}$ onto Col U. By Proposition 11.3.14, Col U = Col A, so $UU^T\boldsymbol{b}$ is the projection of $\boldsymbol{b}$ onto Col A. The Generalized Fire Engine Lemma shows therefore that $\hat{\boldsymbol{x}}$ is the correct least-squares solution.

11.5 PCA

We return for a time to the problem of analyzing data. The lab assignment deals with eigenfaces, which is the application to images of the idea of *principal component analysis.* (In this case, the images are of faces.) I'll therefore use this as an example.

Each image consists of about 32k pixels, so can be represented as a vector in $\mathbb{R}^{32000}$. Here is our first attempt at a crazy hypothesis.

> *Crazy hypothesis, version 1:* The set of face images lies in a ten-dimensional vector subspace.

Well, first, we'll restrict our attention to faces taken at a particular scale, at a particular orientation. Even so, it's not going to be even close to true, right?

> *Crazy hypothesis, version 2:* There is a ten-dimensional affine space such that face images are close to that affine space.

If this hypothesis were correct, we might be able to guess if an image is a face based on its distance from the affine space.

Supposing the hypothesis were true, what ten-dimensional affine space should we use? Suppose we have twenty example faces, represented as vectors: $\boldsymbol{a}_1, \ldots, \boldsymbol{a}_{20}$. Let's identify the ten-dimensional affine space that is closest to these vectors.

We already know how to do this. First, following Section 11.3.9, find the centroid $\bar{\boldsymbol{a}}$ of $\boldsymbol{a}_1, \ldots, \boldsymbol{a}_{20}$. Second, we find an orthonormal basis $\boldsymbol{v}_1, \ldots, \boldsymbol{v}_{10}$ for the 10-dimensional vector space closest to $\boldsymbol{a}_1 - \bar{\boldsymbol{a}}, \ldots, \boldsymbol{a}_{20} - \bar{\boldsymbol{a}}$. Note that $\boldsymbol{v}_1, \ldots, \boldsymbol{v}_{10}$ are the first 10 columns of the matrix V in the SVD $U\Sigma V^T$ of the matrix

$$A = \begin{bmatrix} \boldsymbol{a}_1 - \bar{\boldsymbol{a}} \\ \hline \vdots \\ \hline \boldsymbol{a}_{20} - \bar{\boldsymbol{a}} \end{bmatrix}$$

Finally, the desired ten-dimensional affine space is

$$\{\bar{\boldsymbol{a}} + \boldsymbol{v} \ : \ \boldsymbol{v} \in \text{Span } \{\boldsymbol{v}_1, \ldots, \boldsymbol{v}_{10}\}\}$$

Given a vector $\boldsymbol{w}$, how do we compute the distance between $\boldsymbol{w}$ and this affine space?

We use translation: the distance is the same as the distance between $\boldsymbol{w} - \bar{\boldsymbol{a}}$ and the ten-dimensional vector space Span $\{\boldsymbol{v}_1, \ldots, \boldsymbol{v}_{10}\}$. Since the basis is orthonormal, it is very easy to compute the distance. How? We leave that to you for now.

11.6 *Lab: Eigenfaces*

We will use principal component analysis to analyze some images of faces. In this lab, the faces are more or less aligned, which makes the problem easier. Each image has dimensions 166×189, and is represented as a D-vector over $\mathbb{R}$, with domain $D = \{0, 1, \ldots, 165\} \times \{0, 1, \ldots, 188\}$, which in Python, is `{(x,y) for x in range(166) for y in range(189)}`

We start with twenty images of faces. They span a twenty-dimensional space. You will use PCA to calculate a ten-dimensional affine space, namely the ten-dimensional affine space that minimizes the sum-of-squares distance to the twenty images. You will then find the distance to this space from other images, some faces and some nonfaces. Our hope is that the nonfaces will be farther than the faces.

We have provided two sets of images, one set consisting of twenty faces and the other set consisting of a variety of images, some of faces and some not. We have also provided a module `eigenfaces` to help with loading the images into Python.

Task 11.6.1: Load the twenty face images into Python, and construct a dictionary mapping the integers 0 through 19 to Vecs representing the images.

Task 11.6.2: Compute the centroid $\boldsymbol{a}$ of the face images $\boldsymbol{a}_1, \ldots, \boldsymbol{a}_{20}$, and display the corresponding image (using the procedure `image2display` defined in the module `image`). For any image vector (both faces and unclassified images), the *centered image vector* is obtained by subtracting the centroid. Construct a dictionary consisting of the centered image vectors of the face images.

The module `svd` contains a procedure `factor` that computes the SVD of a matrix. In particular, `svd.factor(A)` returns a triple (U, Σ, V) such that $A = U\Sigma V^T$.

Task 11.6.3: Construct a matrix A whose rows are the centered image vectors. The procedure `factor(A)` defined in the module `svd` returns an SVD of A, a triple (U, Σ, V) such that $A = U\Sigma V^T$. Note that the row space of A equals the row space of V^T, i.e. the column space of V.

Now find an orthonormal basis for the 10-dimensional vector space that is closest to the twenty centered face image vectors. (The vectors of this basis are called *eigenfaces*.) You can use the SVD together with procedures from `matutil`.

Task 11.6.4: Write a procedure `projected_representation(M, x)` with the following spec:

- *input:* a matrix M with orthonormal rows and a vector x, such that the domain of x equals the column-label set of M.
- *output:* The coordinate representation of the projection $x^{||\mathcal{V}}$ in terms of the rows of M, where $\mathcal{V} = \text{Row } M$.

Hint: You've seen this before.
To help you debug, the module `eigenfaces` defines a matrix `test_M` and a vector `test_x`. Applying `projected_representation` to those results in
`{0: 21.213203435596423, 1: 11.547005383792516}`

Task 11.6.5: Write a procedure `projection_length_squared(M, x)` with the following spec:

- *input:* a matrix M with orthonormal rows and a vector x, such that the label-set of x equals the column-labels of M.
- *output:* The square of the norm of the projection of x into the space spanned by the rows of M.

Hint: What is preserved by multiplication with a matrix with orthonormal rows?
To help you debug, applying the procedure with `test_x, test_M` gives 583.3333333333333.

Task 11.6.6: Write a procedure `distance_squared(M, x)` with the following spec:

- *input:* a matrix M with orthonormal rows and a vector x, such that the label-set of x equals the column-labels of M.
- *output:* the square of the distance from x to the vector space spanned by the rows of M.

Hint: Again use the parallel-perpendicular decomposition of x with respect to the row space of M, and also use the Pythagorean Theorem.
To help you debug, applying the procedure to with `test_x, test_M` gives 816.6666666666667.

Task 11.6.7: We will use the `distance_squared` procedure to classify images. First let's consider the images we already know to be faces. Compute the list consising of their distances from the subspace of chosen eigenfaces (remember to work with the centered image vectors). Why are these distances not zero?

Task 11.6.8: Next, for each unclassified image vector, center the vector by subtracting the average face vector, and find the squared distance of the centered image vector from the subspace of eigenfaces you found in Problem 11.6.3. Based on the distances you found, estimate which images are faces and which are not.

Task 11.6.9: Display each of the unclassified images to check your estimate. Are the squared distances of non-faces indeed greater than those for faces?

What is the single threshold value you would choose to decide if a given image is a face or not?

Task 11.6.10: Now that we have constructed our classifier, let's get a closer look at the eigenfaces. It is interesting to check how much a face projected onto the subspace of eigenfaces resembles the original image. Write a procedure `project`:

- *input:* an orthogonal matrix M and a vector x, such that x's labels are the same as M's row labels
- *output:* the projection of x into the space spanned by the rows of M.

Hint: Use `projected_representation`.

Task 11.6.11: Display the projections of various faces, and compare them to the original face (do not forget to add the average face vector to centered image vectors before displaying them). Do the projections resemble the originals?

Task 11.6.12: Display the projection of a non-face image and compare it to the original. Does the projection resemble a face? Can you explain?

11.7 Review questions

- What is one way to measure the distance between two matrices with the same row-labels and column-labels?
- What are the singular values of a matrix? What are the left singular vectors and the right singular vectors?
- What is the singular value decomposition (SVD) of a matrix?
- Given vectors $\boldsymbol{a}_1, \ldots, \boldsymbol{a}_m$, how can we find a one-dimensional vector space closest to $\boldsymbol{a}_1, \ldots, \boldsymbol{a}_m$? Closest in what sense?
- Given in addition an integer k, how can we find the k-dimensional vector space closest to $\boldsymbol{a}_1, \ldots, \boldsymbol{a}_m$?
- Given a matrix A and an integer k, how can we find the rank-at-most-k matrix closest to A?
- What use is finding this matrix?
- How can SVD be used to solve a least-squares problem?

11.8 Problems

Frobenius norm

Problem 11.8.1: Write a procedure `squared_Frob(A)` that, given a Mat over $\mathbb{R}$, returns the square of its Frobenius norm.

Test your procedure. For example,

$$\left\| \begin{bmatrix} 1 & 2 & 3 & 4 \\ -4 & 2 & -1 & 0 \end{bmatrix} \right\|_F^2 = 51$$

Problem 11.8.2: Give a numerical counterexample to the following:

Let A be a matrix, and let Q be a column-orthogonal matrix. If AQ is defined then

$\|AQ\|_F = \|A\|_F$.

SVD practice for simple matrices

Problem 11.8.3: Here is a matrix A and the SVD $A = U\Sigma V^T$:

$$A = \begin{bmatrix} 1 & 0 \\ 0 & 2 \\ 0 & 0 \end{bmatrix}, \quad U = \begin{bmatrix} 0 & 1 \\ 1 & 0 \\ 0 & 0 \end{bmatrix}, \quad \Sigma = \begin{bmatrix} 2 & 0 \\ 0 & 1 \end{bmatrix}, \quad V^T = \begin{bmatrix} 0 & 1 \\ 1 & 0 \end{bmatrix}$$

1. For the vector $\boldsymbol{x} = (1, 2)$, compute $V^T\boldsymbol{x}$, $\Sigma(V^T\boldsymbol{x})$ and $U(\Sigma(V^T\boldsymbol{x}))$.
2. For the vector $\boldsymbol{x} = (2, 0)$, compute $V^T\boldsymbol{x}$, $\Sigma(V^T\boldsymbol{x})$ and $U(\Sigma(V^T\boldsymbol{x}))$.

Problem 11.8.4: Each of the matrices below is shown with row and column labels. For each matrix, calculate the singular value decomposition (SVD). Don't use an algorithm; just use cleverness and your understanding of the SVD. The rows and columns of Σ should be labeled by consecutive integers $0, 1, 2, \ldots$ so $\Sigma[0, 0]$ is the first singular value, etc. Verify your answer is correct by multiplying out the matrices comprising the SVD.

1. $A =$

	c_1	c_2
r_1	3	0
r_2	0	−1

2. $B =$

	c_1	c_2
r_1	3	0
r_2	0	4

3. $C =$

	c_1	c_2
r_1	0	4
r_2	0	0
r_3	0	0

Closest rank-k matrix

Problem 11.8.5: For each of the following matrices find the closest rank-2 matrix. For an $m \times n$ matrix, provide your answer as a product of two matrices, GH, where G is an $m \times 2$ matrix and H is a $2 \times n$ matrix.

1. $A = \begin{bmatrix} 1 & 0 & 1 \\ 0 & 2 & 0 \\ 1 & 0 & 1 \\ 0 & 1 & 0 \end{bmatrix}$

   A's SVD, $A = U\Sigma V^T$:

$$U = \begin{bmatrix} 0 & -\sqrt{0.5} & 0 \\ \sqrt{0.8} & 0 & \sqrt{0.2} \\ 0 & -\sqrt{0.5} & 0 \\ \sqrt{0.2} & 0 & -\sqrt{0.8} \end{bmatrix}, \quad \Sigma = \begin{bmatrix} \sqrt{5} & 0 & 0 \\ 0 & 2 & 0 \\ 0 & 0 & 0 \end{bmatrix}, \quad V^T = \begin{bmatrix} 0 & 1 & 0 \\ -\sqrt{0.5} & 0 & -\sqrt{0.5} \\ \sqrt{0.5} & 0 & -\sqrt{0.5} \end{bmatrix}$$

2. $B = \begin{bmatrix} 0 & 0 & 1 \\ 0 & 0 & 1 \\ 1 & 0 & 0 \\ 0 & 1 & 0 \end{bmatrix}$

B's SVD, $B = U\Sigma V^T$:

$$U = \begin{bmatrix} \sqrt{2}/2 & 0 & 0 \\ \sqrt{2}/2 & 0 & 0 \\ 0 & 0 & -1 \\ 0 & -1 & 0 \end{bmatrix}, \quad \Sigma = \begin{bmatrix} \sqrt{2} & 0 & 0 \\ 0 & 1 & 0 \\ 0 & 0 & 1 \end{bmatrix}, \quad V^T = \begin{bmatrix} 0 & 0 & 1 \\ 0 & -1 & 0 \\ -1 & 0 & 0 \end{bmatrix}$$

Computing a low-rank representation of a low-rank matrix

Problem 11.8.6: Warning: Difficult, requires some thought
Consider the following computational problem:

Computing a low-rank representation of a low-rank matrix

- *input:* a matrix A and a positive integer k
- *output:* a pair of matrices B, C such that $A = BC$ and B has at most k columns, or *'FAIL'* if there is no such pair

We can get an algorithm for this problem by using the SVD of A. However, there is also an algorithm using only the tools we have studied in previous chapters.

Describe such an algorithm, give pseudo-Python for it, and explain why it works.

Solving an $m \times m$ matrix equation with SVD

Problem 11.8.7: Write a Python procedure `SVD_solve(U, Sigma, V, b)` with the following specification:

- *input:* the SVD of a square matrix `A = U Sigma V`^{T}. You can assume that `U`, `Sigma` and `V` are square matrices and are legal to multiply together.
- *output:* vector `x` such that `(U*Sigma*V`^{T}`) x = b`, or "FAIL" if `A` is not invertible.

Your procedure should not use any modules other than `mat`.

Test your procedure with the following example:

The matrix $A = \begin{bmatrix} 1 & 1 & 0 \\ 1 & 0 & 1 \\ 0 & 1 & 1 \end{bmatrix}$ has the following SVD $A = U\Sigma V^T$:

$$U = \begin{bmatrix} -\frac{1}{\sqrt{3}} & \frac{1}{\sqrt{6}} & \frac{1}{\sqrt{2}} \\ -\frac{1}{\sqrt{3}} & \frac{1}{\sqrt{6}} & -\frac{1}{\sqrt{2}} \\ -\frac{1}{\sqrt{3}} & -\frac{2}{\sqrt{6}} & 0 \end{bmatrix}, \quad \Sigma = \begin{bmatrix} 2 & 0 & 0 \\ 0 & 1 & 0 \\ 0 & 0 & 1 \end{bmatrix}, \quad V^T = \begin{bmatrix} -\frac{1}{\sqrt{3}} & -\frac{1}{\sqrt{3}} & -\frac{1}{\sqrt{3}} \\ \frac{2}{\sqrt{6}} & -\frac{1}{\sqrt{6}} & -\frac{1}{\sqrt{6}} \\ 0 & \frac{1}{\sqrt{2}} & -\frac{1}{\sqrt{2}} \end{bmatrix}$$

Let $\boldsymbol{b} = [2, 3, 3]$.

Chapter 12

The Eigenvector

12.1 Modeling discrete dynamic processes

We'll now see how matrices can help in studying a discrete dynamic (time-varying) process. I'll use the example of the RTM Worm. Spreading of a worm through a network is a dynamic process, and the techniques we discuss will help us understand that process. Soon we'll discuss the idea behind PageRank, the method originally used by Google to rank web pages (or at least how the original Brin-and-Page article suggested).

12.1.1 Two interest-bearing accounts

We'll start with a simple example. The example is so simple that matrices are not needed, but we will use them anyway, in order to prepare the ground for more complicated problems.

Suppose you put money in two interest-bearing accounts. Account 1 gives 5% interest and Account 2 gives 3% interest, compounded annually. We represent the amounts in the two accounts after t years by a 2-vector $\boldsymbol{x}^{(t)} = \begin{bmatrix} \text{amount in Account 1} \\ \text{amount in Account 2} \end{bmatrix}$.

We can use a matrix equation to describe how the amounts grow in one year:

$$\boldsymbol{x}^{(t+1)} = \begin{bmatrix} a_{11} & a_{12} \\ a_{21} & a_{22} \end{bmatrix} \boldsymbol{x}^{(t)}$$

In this simple case, $a_{11} = 1.05$, $a_{22} = 1.03$, and the other two entries are zero.

$$\boldsymbol{x}^{(t+1)} = \begin{bmatrix} 1.05 & 0 \\ 0 & 1.03 \end{bmatrix} \boldsymbol{x}^{(t)} \tag{12.1}$$

Let A denote the matrix $\begin{bmatrix} 1.05 & 0 \\ 0 & 1.03 \end{bmatrix}$. Note that it is diagonal.

To find out how, say, $\boldsymbol{x}^{(100)}$ compares to $\boldsymbol{x}^{(0)}$, we can use (12.1) repeatedly:

$$\begin{aligned} \boldsymbol{x}^{(100)} &= A\boldsymbol{x}^{(99)} \\ &= A(A\boldsymbol{x}^{(98)}) \\ &= A(A(A\boldsymbol{x}^{(97)})) \\ &\vdots \\ &= \underbrace{A \cdot A \cdots A}_{100 \text{ times}} \boldsymbol{x}^{(0)} \end{aligned}$$

The notation for the product $A \cdot A \cdots A$ is A^{100}. Since A is diagonal, we can easily compute the entries of A^{100}: the diagonal entries are 1.05^{100} and 1.03^{100}, approximately 131.5 and 19.2. The off-diagonal entries are zero.

For example, if you started with 1 dollar in each account, you'd end up with 131 dollars in Account 1 and 19 dollars in Account 2. Informally, as time goes on, the amount of the initial

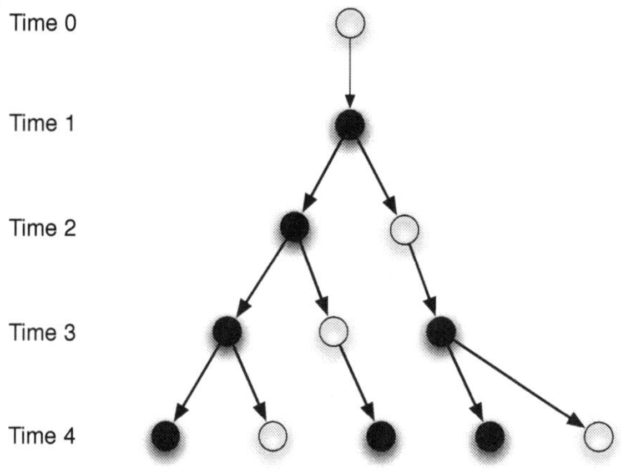

Figure 12.1: A white disk represents a baby rabbit, and a black disk represesents an adult rabbit. Each adult rabbit in one month gives rise to an adult rabbit in the next month (the same rabbit, one month older) and a baby rabbit in the next month (the adult rabbit gave birth). Each baby rabbit in one month gives rise to an adult rabbit in the next month (a rabbit becomes mature in one month).

deposit to Account 2 becomes less and less relevant, as the amount in Account 1 dominates the amount in Account 2.

This example is particularly simple since there is no interaction between the accounts. We didn't need matrices to address it.

12.1.2 Fibonacci numbers

We turn to the example of Fibonacci numbers.

$$F_{k+2} = F_{k+1} + F_k$$

This originates in looking at the growth of the rabbit population. To avoid getting into trouble, I'm going to ignore sex—we'll pretend that rabbits reproduce through parthenogenesis. We make the following assumptions:

- Each month, each adult rabbit gives birth to one baby.
- A rabbit takes one month to become an adult.
- Rabbits never die.

We can represent one step of this process by matrix-vector multiplication. We use a 2-vector $\boldsymbol{x} = \begin{bmatrix} x_1 \\ x_2 \end{bmatrix}$ to represent the current population: x_1 is the number of adults and x_2 is the number of babies.

Suppose $\boldsymbol{x}^{(t)}$ is the population after t months. Then the population at time $t+1$, $\boldsymbol{x}^{(t+1)}$, is related via a matrix-multiplication to the population at time t, $\boldsymbol{x}^{(t)}$:

$$\boldsymbol{x}^{(t+1)} = A\boldsymbol{x}^{(t)}$$

Here's how we derive the entries of the matrix A:

$$\begin{bmatrix} \text{adults at time } t+1 \\ \text{babies at time } t+1 \end{bmatrix} = \underbrace{\begin{bmatrix} a_{11} & a_{12} \\ a_{21} & a_{22} \end{bmatrix}}_{A} \begin{bmatrix} \text{adults at time } t \\ \text{babies at time } t \end{bmatrix}$$

The number of adults at time $t+1$ is the number of adults at time t (since rabbits never die), plus the number of babies at time t (since babies mature in one month). Thus a_{11} and a_{12} are 1. The number of babies at time $t+1$ is the number of adults at time t (since every adult gives birth to a baby every month). Thus $a_{21} = 1$ and $a_{22} = 0$. Thus $A = \begin{bmatrix} 1 & 1 \\ 1 & 0 \end{bmatrix}$.

Clearly the number of rabbits increases over time. But at what rate? How does the number of rabbits grow as a function of time?

As in the bank-account example, $\boldsymbol{x}^{(t)} = A^t \boldsymbol{x}^{(0)}$, but how can this help us estimate the entries of $\boldsymbol{x}^{(t)}$ as a function of t without calculating it directly? In the bank-account example, we were able to understand the behavior because A was a diagonal matrix. This time, A is not diagonal. However, there is a workaround:

Fact 12.1.1: Let $S = \begin{bmatrix} \frac{1+\sqrt{5}}{2} & \frac{1-\sqrt{5}}{2} \\ 1 & 1 \end{bmatrix}$. Then $S^{-1}AS$ is the diagonal matrix $\begin{bmatrix} \frac{1+\sqrt{5}}{2} & 0 \\ 0 & \frac{1-\sqrt{5}}{2} \end{bmatrix}$.

We will later discuss the computation and interpretation of the matrix S. For now, let's see how it can be used.

$$\begin{aligned} A^t &= \underbrace{A \cdot A \cdots A}_{t \text{ times}} \\ &= (S\Lambda S^{-1})(S\Lambda S^{-1}) \cdots (S\Lambda S^{-1}) \\ &= S\Lambda^t S^{-1} \end{aligned}$$

and since Λ is a diagonal matrix, it is easy to compute Λ^t. Let λ_1 and λ_2 be the diagonal entries. Then Λ^t is the diagonal matrix whose diagonal entries are λ_1^t and λ_2^t.

For this A, $\Lambda = \begin{bmatrix} \frac{1+\sqrt{5}}{2} & \\ & \frac{1-\sqrt{5}}{2} \end{bmatrix}$. That is, $\lambda_1 = \frac{1+\sqrt{5}}{2}$ and $\lambda_2 = \frac{1-\sqrt{5}}{2}$.

Since $|\lambda_1|$ is greater than $|\lambda_2|$, the entries grow roughly like $(\lambda_1)^t$.

Even without knowing the matrix S, you can figure out the exact formula, by using the following claim.

Claim: For any given starting vector $\boldsymbol{x}^{(0)}$, there are numbers a_1, b_1, a_2, b_2 such that, for $i = 1, 2$,

$$\text{entry } i \text{ of } \boldsymbol{x}^{(t)} = a_i \lambda_1^t + b_i \lambda_2^t \tag{12.2}$$

Proof

Let $\begin{bmatrix} c_1 \\ c_2 \end{bmatrix} = S^{-1}\boldsymbol{x}^{(0)}$. Then $\Lambda^t S^{-1} \boldsymbol{x}^{(0)} = \begin{bmatrix} c_1\lambda_1^t \\ c_2\lambda_2^t \end{bmatrix}$.

Write $S = \begin{bmatrix} s_{11} & s_{12} \\ s_{21} & s_{22} \end{bmatrix}$. Then

$$\begin{aligned} S\Lambda^t S^{-1}\boldsymbol{x}^{(0)} &= S \begin{bmatrix} c_1\lambda_1^t \\ c_2\lambda_2^t \end{bmatrix} \\ &= \begin{bmatrix} s_{11}c_1\lambda_1^t + s_{12}c_2\lambda_2^t s_{21}c_1\lambda_1^t + s_{22}c_2\lambda_2^t \end{bmatrix} \end{aligned}$$

so we make the claim true by setting $a_i = s_{i1}c_1$ and $b_i = s_{i2}c_2$. □

For example, define $\boldsymbol{x}^{(0)} = \begin{bmatrix} 1 \\ 0 \end{bmatrix}$. This corresponds to postulating that initially there is one adult rabbit and no juvenile rabbits. After one month, there is one adult rabbit and one juvenile rabbit, so $\boldsymbol{x}^{(1)} = \begin{bmatrix} 1 \\ 1 \end{bmatrix}$. Plugging into Equation (12.2), we get

$$\begin{aligned} a_1\lambda_1^1 + b_1\lambda_2^1 &= 1 \\ a_2\lambda_1^1 + b_2\lambda_2^1 &= 1 \end{aligned}$$

After two months, there are two adult rabbits and one juvenile rabbit, so $\boldsymbol{x}^{(2)} = \begin{bmatrix} 2 \\ 1 \end{bmatrix}$. Plugging into Equation (12.2), we get

$$\begin{aligned} a_1\lambda_1^2 + b_1\lambda_2^2 &= 2 \\ a_2\lambda_1^2 + b_2\lambda_2^2 &= 1 \end{aligned}$$

We thereby obtain four equations in the unknowns a_1, b_1, a_2, b_2, and can solve for these unknowns:

$$\begin{bmatrix} \lambda_1 & \lambda_2 & 0 & 0 \\ 0 & 0 & \lambda_1 & \lambda_2 \\ \lambda_1^2 & \lambda_2^2 & 0 & 0 \\ 0 & 0 & \lambda_1^2\lambda_2^2 & \end{bmatrix} \begin{bmatrix} a_1 \\ b_1 \\ a_2 \\ b_2 \end{bmatrix} = \begin{bmatrix} 1 \\ 1 \\ 2 \\ 1 \end{bmatrix}$$

which gives us

$$\begin{bmatrix} a_1 \\ b_1 \\ a_2 \\ b_2 \end{bmatrix} = \begin{bmatrix} \frac{5+\sqrt{5}}{10} \\ \frac{5-\sqrt{5}}{10} \\ \frac{1}{\sqrt{5}} \\ \frac{-1}{\sqrt{5}} \end{bmatrix}$$

Based on this calculation, the number of adult rabbits after t months is

$$\boldsymbol{x}^{(t)}[1] = \frac{5+\sqrt{5}}{10}\left(\frac{1+\sqrt{5}}{2}\right)^t + \frac{5-\sqrt{5}}{10}\left(\frac{1-\sqrt{5}}{2}\right)^t$$

For example, plugging in $t = 3, 4, 5, 6 \ldots$, we get the numbers $3, 5, 8, 13 \ldots$.

12.2 Diagonalization of the Fibonacci matrix

Here is the way to think about the matrix S. Our way of representing the rabbit population—as a vector with two entries, one for the number of adults and one for the number of juveniles— is a natural one from the point of view of the application but an inconvenient one for the purpose of analysis.

To make the analysis easier, we will use a change of basis (Section 5.8). We use as our basis the two columns of the matrix S,

$$\boldsymbol{v}_1 = \begin{bmatrix} \frac{1+\sqrt{5}}{2} \\ 1 \end{bmatrix}, \boldsymbol{v}_2 = \begin{bmatrix} \frac{1-\sqrt{5}}{2} \\ 1 \end{bmatrix}$$

Let $\boldsymbol{u}^{(t)}$ be the coordinate representation of $\boldsymbol{x}^{(t)}$ in terms of $\boldsymbol{v}_1$ and $\boldsymbol{v}_2$. We will derive an equation relating $\boldsymbol{u}^{(t+1)}$ to $\boldsymbol{u}^{(t)}$.

- (`rep2vec`) To convert from the representation $\boldsymbol{u}^{(t)}$ of $\boldsymbol{x}^{(t)}$ to the vector $\boldsymbol{x}^{(t)}$ itself, we multiply $\boldsymbol{u}^{(t)}$ by S.
- (Move forward one month) To go from $\boldsymbol{x}^{(t)}$ to $\boldsymbol{x}^{(t+1)}$, we multiply $\boldsymbol{x}^{(t)}$ by A.
- (`vec2rep`) To go back to the coordinate representation in terms of $\boldsymbol{v}_1$ and $\boldsymbol{v}_2$, we multiply by S^{-1}.

By the link between matrix-matrix multiplication and function composition (Section 4.11.3), multiplying by the matrix $S^{-1}AS$ carries out the three steps above.

But we saw that $S^{-1}AS$ is just the diagonal matrix $\begin{bmatrix} \frac{1+\sqrt{5}}{2} & 0 \\ 0 & \frac{1-\sqrt{5}}{2} \end{bmatrix}$. Thus we obtain the equation

$$\boldsymbol{u}^{(t+1)} = \begin{bmatrix} \frac{1+\sqrt{5}}{2} & 0 \\ 0 & \frac{1-\sqrt{5}}{2} \end{bmatrix} \boldsymbol{u}^{(t)} \tag{12.3}$$

Now we have a nice, simple equation, like the equation modeling growth in bank account balances (Equation 12.1). From this equation, it is easy to see why the Fibonacci numbers grow at a rate of roughly $\left(\frac{1+\sqrt{5}}{2}\right)^t$. The coordinate corresponding to $\boldsymbol{v}_1$ grows at *exactly* this rate, and the coordinate corresponding to $\boldsymbol{v}_2$ grows at the rate $\left(\frac{1-\sqrt{5}}{2}\right)^t$.

The technique we have used is called *diagonalization.* We turned a square matrix into a diagonal matrix by multiplying it on the right by a square matrix S and on the left by the inverse S^{-1}. We have seen diagonalization once before, in Section 10.8.2. There we saw that circulant matrices could be diagonalized using a discrete Fourier matrix. This was useful in that it allowed us to quickly multiply by a circulant matrix, using the Fast Fourier Transform. Now that we see that diagonalization can be useful in analysis, we'll consider it in greater generality.

12.3 Eigenvalues and eigenvectors

We now introduce the concepts underlying this kind of analysis.

Definition 12.3.1: For a matrix A whose row-label set equals its column-label set, if λ is a scalar and $\boldsymbol{v}$ is a nonzero vector such that $A\boldsymbol{v} = \lambda\boldsymbol{v}$, we say that λ is an *eigenvalue* of A, and $\boldsymbol{v}$ is a corresponding *eigenvector.*

If λ is an eigenvalue of A, there are many corresponding eigenvectors. In fact, the set $\{\boldsymbol{v} : A\boldsymbol{v} = \lambda\boldsymbol{v}\}$ is a vector space, called the *eigenspace* corresponding to eigenvalue λ. Any nonzero vector in the eigenspace is considered an eigenvector. However, it is often convenient to require that the eigenvector have norm one.

Example 12.3.2: The matrix used in modeling two interest-bearing accounts, $\begin{bmatrix} 1.05 & 0 \\ 0 & 1.03 \end{bmatrix}$, has eigenvalues 1.05 and 1.03. The eigenvector corresponding to the first eigenvalue is $[1, 0]$, and the eigenvector corresponding to the second is $[0, 1]$.

Example 12.3.3: More generally, suppose A is a diagonal matrix:

$$A = \begin{bmatrix} \lambda_1 & & \\ & \ddots & \\ & & \lambda_n \end{bmatrix}$$

What are the eigenvectors and eigenvalues in this case? Since $A\boldsymbol{e}_1 = \lambda_1, \ldots, A\boldsymbol{e}_n = \lambda_n \boldsymbol{e}_n$ where $\boldsymbol{e}_1, \ldots, \boldsymbol{e}_n$ are the standard basis vectors, we see that $\boldsymbol{e}_1, \ldots, \boldsymbol{e}_n$ are the eigenvectors and the diagonal elements $\lambda_1, \ldots, \lambda_n$ are the eigenvalues.

Example 12.3.4: The matrix used in analyzing Fibonacci numbers, $\begin{bmatrix} 1 & 1 \\ 1 & 0 \end{bmatrix}$, has eigenvalues $\lambda_1 = \frac{1+\sqrt{5}}{2}$ and $\lambda_2 = \frac{1-\sqrt{5}}{2}$. The eigenvector corresponding to λ_1 is $\begin{bmatrix} \frac{1+\sqrt{5}}{2} \\ 1 \end{bmatrix}$, and the eigenvector corresponding to λ_2 is $\begin{bmatrix} \frac{1-\sqrt{5}}{2} \\ 1 \end{bmatrix}$.

Example 12.3.5: Suppose A has 0 as an eigenvalue. An eigenvector corresponding to that eigenvalue is a nonzero vector $\boldsymbol{v}$ such that $A\boldsymbol{v} = 0\,\boldsymbol{v}$. That is, a nonzero vector $\boldsymbol{v}$ such that $A\boldsymbol{v}$ is the zero vector. Then $\boldsymbol{v}$ belongs to the null space. Conversely, if A's null space is nontrivial then 0 is an eigenvalue of A.

Example 12.3.5 (Page 451) suggests a way to find an eigenvector corresponding to the eigenvalue 0: find a nonzero vector in the null space. What about other eigenvalues?

Suppose λ is an eigenvalue of A, with corresponding eigenvector $\boldsymbol{v}$. Then $A\boldsymbol{v} = \lambda\,\boldsymbol{v}$. That is, $A\boldsymbol{v} - \lambda\,\boldsymbol{v}$ is the zero vector. The expression $A\boldsymbol{v} - \lambda\,\boldsymbol{v}$ can be written as $(A - \lambda\,\mathbb{1})\boldsymbol{v}$, so $(A - \lambda\,\mathbb{1})\boldsymbol{v}$ is the zero vector. That means that $\boldsymbol{v}$ is a nonzero vector in the null space of $A - \lambda\,\mathbb{1}$. That means that $A - \lambda\,\mathbb{1}$ is not invertible.

Conversely, suppose $A - \lambda\,\mathbb{1}$ is not invertible. It is square, so it must have a nontrivial null space. Let $\boldsymbol{v}$ be a nonzero vector in the null space. Then $(A - \lambda)\boldsymbol{u} = \boldsymbol{0}$, so $A\boldsymbol{u} = \lambda\boldsymbol{u}$.

We have proved the following:

Lemma 12.3.6: Let A be a square matrix.

- The number λ is an eigenvalue of A if and only if $A - \lambda\,\mathbb{1}$ is not invertible.
- If λ is in fact an eigenvalue of A then the corresponding eigenspace is the null space of $A - \lambda\,\mathbb{1}$.

Example 12.3.7: Let $A = \begin{bmatrix} 1 & 2 \\ 3 & 4 \end{bmatrix}$. The number $\lambda_1 = \frac{5+\sqrt{33}}{2}$ is an eigenvalue of A. Let $B = A - \lambda_1\mathbb{1}$. Now we find an eigenvector $\boldsymbol{v}_1$ corresponding to the eigenvalue λ_1:

```
>>> A = listlist2mat([[1,2],[3,4]])
>>> lambda1 = (5+sqrt(33))/2
>>> B = A - lambda1*identity({0,1}, 1)
>>> cols = mat2coldict(B)
>>> v1 = list2vec([-1, cols[0][0]/cols[1][0]])
>>> B*v1
Vec({0, 1},{0: 0.0, 1: 0.0})
>>> A*v1
Vec({0, 1},{0: -5.372281323269014, 1: -11.744562646538029})
>>> lambda1*v1
Vec({0, 1},{0: -5.372281323269014, 1: -11.744562646538029})
```

Corollary 12.3.8: If λ is an eigenvalue of A then it is an eigenvalue of A^T.

Proof

Suppose λ is an eigenvalue of A. By Lemma 12.3.6, $A - \lambda\,\mathbb{1}$ has a nontrivial null space so is not invertible. By Corollary 6.4.11, $(A - \lambda\,\mathbb{1})^T$ is also not invertible. But it is easy to see that $(A - \lambda\,\mathbb{1})^T = A^T - \lambda\,\mathbb{1}$. By Lemma 12.3.6, therefore, λ is a eigenvalue of A^T. □

The fact that $(A - \lambda\,\mathbb{1})^T = A^T - \lambda\,\mathbb{1}$ is a nice trick, but it doesn't work when you replace $\mathbb{1}$ with an arbitrary matrix.

What do eigenvalues have to do with the analysis we were doing?

12.3.1 Similarity and diagonalizability

Definition 12.3.9: We say two square matrices A and B are *similar* if there is an invertible matrix S such that $S^{-1}AS = B$.

Proposition 12.3.10: Similar matrices have the same eigenvalues.

Proof

Suppose λ is an eigenvalue of A and $\boldsymbol{v}$ is a corresponding eigenvector. By definition, $A\boldsymbol{v} = \lambda\,\boldsymbol{v}$. Suppose $S^{-1}AS = B$, and let $\boldsymbol{w} = S^{-1}\boldsymbol{v}$. Then

$$\begin{aligned} B\boldsymbol{w} &= S^{-1}AS\boldsymbol{w} \\ &= S^{-1}ASS^{-1}\boldsymbol{v} \\ &= S^{-1}A\boldsymbol{v} \\ &= S^{-1}\lambda\,\boldsymbol{v} \\ &= \lambda\,S^{-1}\boldsymbol{v} \\ &= \lambda\,\boldsymbol{w} \end{aligned}$$

which shows that λ is an eigenvalue of B. □

Example 12.3.11: We will see later that the eigenvalues of the matrix $A = \begin{bmatrix} 6 & 3 & -9 \\ 0 & 9 & 15 \\ 0 & 0 & 15 \end{bmatrix}$ are its diagonal elements (6, 9, and 15) because U is upper triangular. The matrix $B = \begin{bmatrix} 92 & -32 & -15 \\ -64 & 34 & 39 \\ 176 & -68 & -99 \end{bmatrix}$ has the property that $B = S^{-1}AS$ where $S = \begin{bmatrix} -2 & 1 & 4 \\ 1 & -2 & 1 \\ -4 & 3 & 5 \end{bmatrix}$. Therefore the eigenvalues of B are also 6, 9, and 15.

Definition 12.3.12: If a square matrix A is similar to a diagonal matrix, i.e. if there is an invertible matrix S such that $S^{-1}AS = \Lambda$ where Λ is a diagonal matrix, we say A is *diagonalizable*.

The equation $S^{-1}AS = \Lambda$ is equivalent to the equation $A = S\Lambda S^{-1}$, which is the form used in Fact 12.1.1 and the subsequent analysis of rabbit population.

How is diagonalizability related to eigenvalues?

We saw in Example 12.3.3 (Page 451) that if Λ is the diagonal matrix $\begin{bmatrix} \lambda_1 & & \\ & \ddots & \\ & & \lambda_n \end{bmatrix}$ then its eigenvalues are its diagonal entries $\lambda_1, \ldots, \lambda_n$.

If a matrix A is similar to Λ then, by Proposition 12.3.10, the eigenvalues of A are the eigenvalues of Λ, i.e. the diagonal elements of Λ.

Going further, suppose specifically that $S^{-1}AS = \Lambda$. By multiplying both sides on the left by S, we obtain the equation

$$AS = S\Lambda$$

Using the matrix-vector definition of matrix-matrix multiplication, we see that column i of the matrix AS is A times column i of S. Using the vector-matrix definition, we see that we see that column i of the matrix $S\Lambda$ is λ_i times column i of S. Thus the equation implies that, for each i, A times column i of S equals λ_i times column i of S. Thus in this case $\lambda_1, \ldots, \lambda_n$ are eigenvalues, and the corresponding columns of S are corresponding eigenvectors. Since S is invertible, its columns are linearly independent. We have shown the following lemma:

Lemma 12.3.13: If $\Lambda = S^{-1}AS$ is a diagonal matrix then the diagonal elements of Λ are eigenvalues, and the columns of S are linearly independent eigenvectors.

Conversely, suppose an $n \times n$ matrix A has n linearly independent eigenvectors $\boldsymbol{v}_1, \ldots, \boldsymbol{v}_n$, and let $\lambda_1, \ldots, \lambda_n$ be the corresponding eigenvalues. Let S be the matrix $\begin{bmatrix} & & \\ \boldsymbol{v}_1 & \cdots & \boldsymbol{v}_n \\ & & \end{bmatrix}$, and let Λ be the matrix $\begin{bmatrix} \lambda_1 & & \\ & \ddots & \\ & & \lambda_n \end{bmatrix}$. . Then $AS = S\Lambda$. Moreover, since S is square and its columns are linearly independent, it is an invertible matrix. Multiplying the equation on the right by S^{-1}, we obtain $A = S\Lambda S^{-1}$. This shows that A is diagonalizable. We have shown the following lemma:

Lemma 12.3.14: If an $n \times n$ matrix A has n linearly independent eigenvectors then A is diagonalizable.

Putting the two lemmas together, we obtain the following theorem.

Theorem 12.3.15: An $n \times n$ matrix is diagonalizable iff it has n linearly independent eigenvectors.

12.4 Coordinate representation in terms of eigenvectors

Now we revisit the analysis technique of Section 12.2 in a more general context. The existence of linearly independent eigenvectors is very useful in analyzing the effect of repeated matrix-vector multiplication.

Let A be an $n \times n$ matrix. Let $\boldsymbol{x}^{(0)}$, and let $x^{(t)} = A^t \boldsymbol{x}^{(0)}$ for $t = 1, 2, \ldots$. Now suppose A is diagonalizable: That is, suppose there is an invertible matrix S and a diagonal matrix Λ such that $S^{-1}AS = \Lambda$. Let $\lambda_1, \ldots, \lambda_n$ be the diagonal elements of Λ, which are the eigenvalues of A, and let $\boldsymbol{v}_1, \ldots, \boldsymbol{v}_n$ be the corresponding eigenvectors, which are the columns of S. Let $\boldsymbol{u}^{(t)}$ be the coordinate representation of $\boldsymbol{x}^{(t)}$ in terms of the eigenvectors. Then the equation $x^{(t)} = A^t \boldsymbol{x}^{(0)}$ gives rise to the much simpler equation

$$\begin{bmatrix} \\ \boldsymbol{u}^{(t)} \\ \\ \end{bmatrix} = \begin{bmatrix} \lambda_1^t & & \\ & \ddots & \\ & & \lambda_n^t \end{bmatrix} \begin{bmatrix} \\ \boldsymbol{u}^{(0)} \\ \\ \end{bmatrix} \tag{12.4}$$

simpler because each entry of $\boldsymbol{u}^{(t)}$ is obtained from the corresponding entry of $\boldsymbol{u}^{(0)}$ by simply multiplying by the corresponding eigenvalue, raised to the tth power.

Here is another way to see what is going on.

The eigenvectors form a basis for $\mathbb{R}^n$, so any vector $\boldsymbol{x}$ can be written as a linear combination:

$$\boldsymbol{x} = \alpha_1 \, \boldsymbol{v}_1 + \cdots + \alpha_n \, \boldsymbol{v}_n$$

Let's see what happens when we left-multiply by A on both sides of the equation:

$$\begin{aligned} A\boldsymbol{x} &= A(\alpha_1 \boldsymbol{v}_1) + \cdots + A(\alpha_n \boldsymbol{v}_n) \\ &= \alpha_1 A\boldsymbol{v}_1 + \cdots + \alpha_n A\boldsymbol{v}_n \\ &= \alpha_1 \lambda_1 \boldsymbol{v}_1 + \cdots + \alpha_n \lambda_n \boldsymbol{v}_n \end{aligned}$$

Applying the same reasoning to $A(A\boldsymbol{x})$, we get

$$A^2\boldsymbol{x} = \alpha_1 \lambda_1^2 \boldsymbol{v}_1 + \cdots + \alpha_n \lambda_n^2 \boldsymbol{v}_n$$

More generally, for any nonnegative integer t,

$$A^t\boldsymbol{x} = \alpha_1 \lambda_1^t \boldsymbol{v}_1 + \cdots + \alpha_n \lambda_n^t \boldsymbol{v}_n \tag{12.5}$$

Now, if some of the eigenvalues are even slightly bigger in absolute value than the others, after a sufficiently large number t of iterations, those terms on the right-hand side of Equation 12.5 that involve the eigenvalues with large absolute value will dominate; the other terms will be relatively small.

In particular, suppose λ_1 is larger in absolute value than all the other eigenvectors. For a large enough value of t, $A^t\boldsymbol{x}$ will be approximately $\alpha_1\lambda_1^t\boldsymbol{v}_1$.

The terms corresponding to eigenvalues with absolute value strictly less than one will actually get smaller as t grows.

12.5 The Internet worm

Consider the worm launched on the Internet in 1988. A worm is a program that reproduces through the network; an instance of the program running on one computer tries to break into neighboring computes and spawn copies of itself on these computers.

The 1988 worm did no damage but it essentially took over a significant proportion of the computers on the Internet; these computers were spending all of their cycles running the worms. The reason is that each computer was running many independent instances of the program.

The author (Robert T. Morris, Jr.) had made some effort to prevent this. The program seems to have been designed so that each worm would check whether there was another worm running on the same computer; if so, one of them would set a flag indicating it was supposed to die. However, with probability 1/7 instead of doing the check, the worm would designate itself immortal. An immortal worm would not do any checks.

As a consequence, it seems, each computer ends up running many copies of the worm, until the computer's whole capacity is used up running worms.

We will analyze a very simple model of this behavior. Let's say the Internet consists of three computers connected in a triangle. In each iteration, each worm has probability 1/10 of spawning a child worm on each neighboring computer. Then, if it is a mortal worm, with probability 1/7 it becomes immortal, and otherwise it dies.

There is randomness in this model, so we cannot say exactly how many worms there are after a number of iterations. However, we can calculate the *expected* number of worms.

We will use a vector $\boldsymbol{x} = (x_1, y_1, x_2, y_2, x_3, y_3)$ where, for $i = 1, 2, 3$, x_i is the expected number of mortal worms at computer i, and y_i is the expected number of immortal worms at computer i.

For $t = 0, 1, 2, \ldots,$, let $\boldsymbol{x}^{(t)} = (x_1^{(t)}, y_1^{(t)}, x_2^{(t)}, y_2^{(t)}, x_3^{(t)}, y_3^{(t)})$. According to the model, any mortal worm at computer 1 is a child of a worm at computer 2 or computer 3. Therefore the expected number of mortal worms at computer 1 after $t+1$ iterations is 1/10 times the expected number of worms at computers 2 and 3 after t iterations. Therefore

$$x_1^{(t+1)} = \frac{1}{10}x_2^{(t)} + \frac{1}{10}y_2^{(t)} + \frac{1}{10}x_3^{(t)} + \frac{1}{10}y_3^{(t)}$$

With probability 1/7, a mortal worm at computer 1 becomes immortal. The previously immortal worms stay immortal. Therefore

$$y_1^{(t+1)} = \frac{1}{7}x_1^{(t)} + y_1^{(t)}$$

The equations for $x_2^{(t+1)}$ and $y_2^{(t+1)}$ and $x_3^{(t+1)}$ and $y_3^{(t+1)}$ are similar. We therefore get

$$\boldsymbol{x}^{(t+1)} = A\boldsymbol{x}^{(t)}$$

where A is the matrix

$$A = \begin{bmatrix} 0 & 0 & 1/10 & 1/10 & 1/10 & 1/10 \\ 1/7 & 1 & 0 & 0 & 0 & 0 \\ 1/10 & 1/10 & 0 & 0 & 1/10 & 1/10 \\ 0 & & 1/7 & 1 & 0 & 0 \\ 1/10 & 1/10 & 1/10 & 1/10 & 0 & 0 \\ 0 & 0 & 0 & 0 & 1/7 & 1 \end{bmatrix}$$

This matrix has linearly independent eigenvectors, and its largest eigenvalue is about 1.034 (the others are smaller in absolute value than 1). Because this is larger than 1, we can infer that the number of worms will grow exponentially with the number of iterations. The largest eigenvalue of A^t is about 1.034^t. To get a sense of magnitude, for $t = 100$ this number is a mere 29. For $t = 200$, the number is about 841. For $t = 500$, it is up to twenty million. For $t = 600$, it is about six hundred million.

In this example, the matrix A is small enough that, for such small values of t, the expected number of worms can be computed. Suppose that we start with one mortal worm at computer 1. This corresponds to the vector $x^{(0)} = (1, 0, 0, 0, 0, 0)$. In this case, the expected number of worms after 600 iterations is about 120 million.

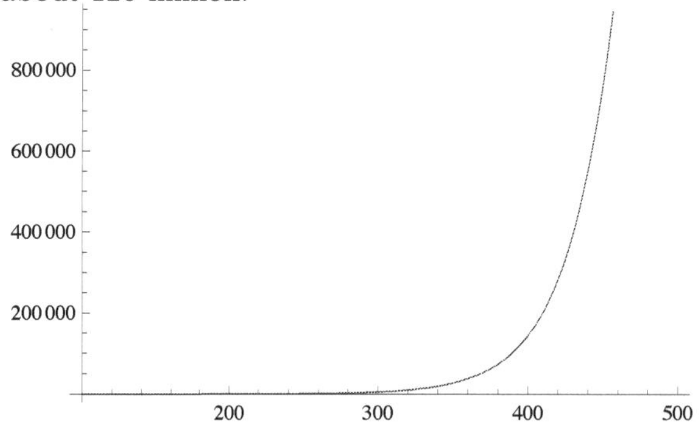

12.6 Existence of eigenvalues

Under what circumstances can we ensure that a square matrix has eigenvalues? Is diagonalizable?

12.6.1 Positive-definite and positive-semidefinite matrices

Let A be any invertible matrix. It has a singular value decomposition:

$$\begin{bmatrix} & & \\ & A & \\ & & \end{bmatrix} = \begin{bmatrix} & & \\ & U & \\ & & \end{bmatrix} \begin{bmatrix} \sigma_1 & & \\ & \ddots & \\ & & \sigma_n \end{bmatrix} \begin{bmatrix} & & \\ & V^T & \\ & & \end{bmatrix}$$

Consider the matrix product $A^T A$. Using the SVD, this is

$$\begin{aligned} A^T A &= \begin{bmatrix} & & \\ & V & \\ & & \end{bmatrix} \begin{bmatrix} \sigma_1 & & \\ & \ddots & \\ & & \sigma_n \end{bmatrix} \begin{bmatrix} & & \\ & U^T & \\ & & \end{bmatrix} \begin{bmatrix} & & \\ & U & \\ & & \end{bmatrix} \\ & \qquad \begin{bmatrix} \sigma_1 & & \\ & \ddots & \\ & & \sigma_n \end{bmatrix} \begin{bmatrix} & & \\ & V^T & \\ & & \end{bmatrix} \\ &= \begin{bmatrix} & & \\ & V & \\ & & \end{bmatrix} \begin{bmatrix} \sigma_1 & & \\ & \ddots & \\ & & \sigma_n \end{bmatrix} \begin{bmatrix} \sigma_1 & & \\ & \ddots & \\ & & \sigma_n \end{bmatrix} \begin{bmatrix} & & \\ & V^T & \\ & & \end{bmatrix} \\ &= \begin{bmatrix} & & \\ & V & \\ & & \end{bmatrix} \begin{bmatrix} \sigma_1^2 & & \\ & \ddots & \\ & & \sigma_n^2 \end{bmatrix} \begin{bmatrix} & & \\ & V^T & \\ & & \end{bmatrix} \end{aligned}$$

Multiplying this equation on the left by V^T and on the right by V, we obtain

$$V^T(A^T A)V = \begin{bmatrix} \sigma_1^2 & & \\ & \ddots & \\ & & \sigma_n^2 \end{bmatrix}$$

which shows that $A^T A$ is diagonalizable, and the eigenvalues are the squares of the singular values of A.

The eigenvalues are all real numbers, and positive.

Moreover, $A^T A$ is symmetric, as one can prove by taking its transpose:

$$(A^T A)^T = A^T (A^T)^T = A^T A$$

The transpose of $A^T A$ turns out to be equal to $A^T A$.

Definition 12.6.1: A symmetric matrix whose eigenvalues are all positive real numbers is called a *positive-definite* matrix.

We have seen that a matrix of the form $A^T A$ where A is an invertible matrix, is a positive-definite matrix. It can be shown, conversely, that any positive-definite matrix can be written as $A^T A$ for some invertible matrix A.

Positive-definite matrices and their relatives, the *positive-semidefinite matrices* (that's my absolute top near-favorite math term) are important in modeling many physical systems, and have growing importance in algorithms and in machine learning.

12.6.2 Matrices with distinct eigenvalues

In this section, we give another condition under which a square matrix is diagonalizable.

Lemma 12.6.2: For a matrix A, for any set T of *distinct* eigenvalues, the corresponding eigenvectors are linearly independent

Proof

Assume for contradiction that the eigenvectors are linearly dependent. Let

$$\mathbf{0} = \alpha_1 \boldsymbol{v}_1 + \cdots + \alpha_r \boldsymbol{v}_r \tag{12.6}$$

be a linear combination of a subset of the eigenvectors for the eigenvalues in T, in particular a linear combination of a subset of minimum size. Let $\lambda_1, \ldots, \lambda_r$ be the corresponding eigenvalues.

Then

$$\begin{aligned} \mathbf{0} &= A(\mathbf{0}) \\ &= A(\alpha_1 \boldsymbol{v}_1 + \cdots + \alpha_r \boldsymbol{v}_r) \\ &= \alpha_1 A\boldsymbol{v}_1 + \cdots + \alpha_r A\boldsymbol{v}_r \\ &= \alpha_1 \lambda_1 \boldsymbol{v}_1 + \cdots + \alpha_r \lambda_r \boldsymbol{v}_r \end{aligned} \tag{12.7}$$

Thus we obtain a new linear dependence among $\boldsymbol{v}_1, \ldots, \boldsymbol{v}_r$. Multiply (12.6) by λ_1, and subtract it from (12.7), getting

$$\mathbf{0} = (\lambda_1 - \lambda_1)\alpha_1 \boldsymbol{v}_1 + (\lambda_2 - \lambda_1)\alpha_2 \boldsymbol{v}_2 + \cdots + (\lambda_r - \lambda_1)\alpha_r \boldsymbol{v}_r$$

Since the first coefficient is zero, we can rewrite it as

$$\mathbf{0} = (\lambda_2 - \lambda_1)\alpha_2 \boldsymbol{v}_2 + \cdots + (\lambda_r - \lambda_1)\alpha_r \boldsymbol{v}_r$$

which has even fewer vectors than (12.6), a contradiction. □

Combining Lemma 12.6.2 with Lemma 12.3.14 gives us

Theorem 12.6.3: A $n \times n$ matrix with n distinct eigenvalues is diagonalizable.

An $n \times n$ matrix with random entries is likely to have n distinct eigenvalues, so this theorem tells us that "most" square matrices are diagonalizable.

There are also $n \times n$ matrices that do not have n distinct eigenvalues but are diagonalizable. The simplest example is the $n \times n$ identity matrix, which has only 1 as an eigenvalue but which is obviously diagonalizable.

12.6.3 Symmetric matrices

In the context of eigenvalues, the next important class of matrices is symmetric matrices. Such matrices are very well-behaved:

Theorem 12.6.4 (Diagonalization of symmetric matrices): Let A be a symmetric matrix over $\mathbb{R}$. Then there is an orthogonal matrix Q and a real-valued diagonal matrix Λ such that $Q^T A Q = \Lambda$.

This theorem is a consequence of a theorem we prove later.

The theorem is important because it means that, for a symmetric matrix A, every vector that A can multiply can be written as a linear combination of eigenvectors, so we can apply the analysis method of Section 12.4.

Indeed, there is a bit of complexity hidden in Equation 12.4: some of the eigenvalues $\lambda_1, \ldots, \lambda_n$ might be complex numbers. However, when A is a symmetric matrix, this complexity vanishes: all the eigenvalues are guaranteed to be real numbers.

12.6.4 Upper-triangular matrices

However, not all square matrices can be diagonalized. A simple example is the matrix $A = \begin{bmatrix} 1 & 1 \\ 0 & 1 \end{bmatrix}$. Despite the resemblance to the Fibonacci matrix, there is *no* invertible matrix S such that $S^{-1}AS$ is a diagonal matrix.

We start by considering upper-triangular matrices. Note that the non-diagonalizable example above, $\begin{bmatrix} 1 & 1 \\ 0 & 1 \end{bmatrix}$, is upper triangular.

Lemma 12.6.5: The diagonal elements of an upper-triangular matrix U are the eigenvalues of U.

Proof

By Lemma 12.3.6, a number λ is an eigenvalue of U if and only if $U - \lambda \mathbb{1}$ is not invertible. But $U - \lambda \mathbb{1}$ is an upper-triangular matrix. By Lemma 4.13.13, therefore, $U - \lambda \mathbb{1}$ is not invertible if and only if at least one of its diagonal elements are zero. A diagonal element of $U - \lambda \mathbb{1}$ is zero if and only if λ is one of the diagonal elements of U. □

Example 12.6.6: Consider the matrix $U = \begin{bmatrix} 5 & 9 & 9 \\ 0 & 4 & 7 \\ 0 & 0 & 3 \end{bmatrix}$. Its diagonal elements are 5, 4, and 3, so these are its eigenvalues.

For example,

$$U - 3\mathbb{1} = \begin{bmatrix} 5-3 & 9 & 9 \\ 0 & 4-3 & 7 \\ 0 & 0 & 3-3 \end{bmatrix} = \begin{bmatrix} 2 & 9 & 9 \\ 0 & 1 & 7 \\ 0 & 0 & 0 \end{bmatrix}$$

Since this matrix has a zero diagonal element, it is not invertible.

Note that a single number can occur multiple times on the diagonal of U. For example, in the matrix $U = \begin{bmatrix} 5 & 9 & 9 \\ 0 & 4 & 7 \\ 0 & 0 & 5 \end{bmatrix}$, the number 5 occurs twice.

Definition 12.6.7: The *spectrum* of an upper-triangular matrix U is the multiset of diagonal elements. That is, it is a multiset in which each number occurs the same number of times as it occurs in the diagonal of U.

Example 12.6.8: The spectrum of $\begin{bmatrix} 5 & 9 & 9 \\ 0 & 4 & 7 \\ 0 & 0 & 5 \end{bmatrix}$ is the multiset $\{5, 5, 4\}$. (In a multiset, order doesn't matter but duplicates are not eliminated.)

In the next section, we build on what we know about upper-triangular matrices to say something about general square matrices.

12.6.5 General square matrices

We state two important theorems about eigenvalues of square matrices . The proofs of these theorems are a bit involved, and we postpone them until Section 12.11 in order to not disrupt the flow.

Theorem 12.6.9: Every square matrix over $\mathbb{C}$ has an eigenvalue.

This theorem only guarantees the existence of a complex eigenvalue. In fact, quite simple matrices have complex (and unreal) eigenvalues. The eigenvalues of $\begin{bmatrix} 1 & 1 \\ -1 & 1 \end{bmatrix}$ are $1 + \mathbf{i}$ and $1 - \mathbf{i}$ where $\mathbf{i}$, you will recall, is the square root of negative one. Since the eigenvalues of this matrix are complex, it is not surprising that the eigenvectors are also complex.

Theorem 12.6.9 provides the foundation for a theorem that shows not that every square matrix is diagonalizable (for that is not true) but that every square matrix is, uh, triangularizable.

Theorem 12.6.10: For any $n \times n$ matrix A, there is a unitary matrix Q such that $Q^{-1}AQ$ is an upper-triangular matrix.

Example 12.6.11: Let $A = \begin{bmatrix} 12 & 5 & 4 \\ 27 & 15 & 1 \\ 1 & 0 & 1 \end{bmatrix}$. Then

$$Q^{-1}AQ = \begin{bmatrix} 25.2962 & 21.4985 & 4.9136 \\ 0 & 2.84283 & -2.76971 \\ 0 & 0 & -0.139057 \end{bmatrix}$$

where $Q = \begin{bmatrix} 0.355801 & -0.886771 & -0.29503 \\ -0.934447 & 0.342512 & 0.0974401 \\ -0.0146443 & -0.310359 & 0.950506 \end{bmatrix}$

In particular, every matrix is similar to an upper-triangular matrix. This theorem is the basis for practical algorithms to computer eigenvalues: these algorithms iteratively transform a matrix until it gets closer and closer to being upper-triangular. The algorithm is based on QR factorization. Unfortunately, the details are beyond the scope of this book.

The analysis described in Section 12.4 enabled us to understand the result of multiplying a vector by a high power of a matrix when the matrix is diagonalizable. This technique can be generalized to give us similar information even when the matrix is not diagonalizable, but we won't go into the details. Fortunately, matrices arising in practice are often diagonalizable.

In the next section, we describe a very elementary algorithm that, given a matrix, finds an approximation to the eigenvalue of largest magnitude eigenvalue (and corresponding approximate eigenvector).

12.7 Power method

Let's assume A is a diagonalizable matrix, so it has n distinct eigenvectors $\lambda_1, \ldots, \lambda_n$, ordered so that $|\lambda_1| \geq |\lambda_2| \geq \cdots \geq |\lambda_n|$, and linearly independent eigenvectors $\boldsymbol{v}_1, \ldots, \boldsymbol{v}_n$. (The absolute value $|\lambda|$ of a complex number $x + iy$ is defined to be the distance of the number from the origin when the complex number is viewed as a point on the x, y plane.)

Pick a random vector $\boldsymbol{x}_0$, and, for any nonnegative integer t, let $\boldsymbol{x}_t = A^t \boldsymbol{x}_0$.

Write $\boldsymbol{x}_0$ in terms of the eigenvectors:

$$\boldsymbol{x}_0 = \alpha_1 \boldsymbol{v}_1 + \cdots + \alpha_n \boldsymbol{v}_n$$

Then we have

$$\boldsymbol{x}_t = \alpha_1 \lambda_1^t \boldsymbol{v}_1 + \cdots + \alpha_n \lambda_n^t \boldsymbol{v}_n \tag{12.8}$$

Because $\boldsymbol{x}_0$ was chosen randomly, it is unlikely that it happened to lie in the $n-1$-dimensional subspace spanned by $\boldsymbol{v}_2, \ldots, \boldsymbol{v}_n$. Therefore α_1 is likely to be nonzero.

Suppose that $\alpha_1 \neq 0$ and that $|\lambda_1|$ is substantially bigger than $|\lambda_2|$. Then the coefficient of $\boldsymbol{v}_1$ in (12.8) grows faster than all the other coefficients, and eventually (for large enough t) swamps them. Thus eventually $\boldsymbol{x}_t$ will be $\alpha_1 \lambda_1^t \boldsymbol{v}_1 + \mathit{error}$ where *error* is a vector that is much smaller than $\alpha_1 \lambda_1^t \boldsymbol{v}_1$. Since $\boldsymbol{v}_1$ is an eigenvector, so is $\alpha_1 \lambda_1^t \boldsymbol{v}_1$. Thus eventually $\boldsymbol{x}_t$ will be an *approximate* eigenvector. Furthermore, we can estimate the corresponding eigenvalue λ_1 from $\boldsymbol{x}_t$, because $A\boldsymbol{x}_t$ will be close to $\lambda_1 \boldsymbol{x}_t$.

Similarly, if the top q eigenvalues are identical or even very close and the $q + 1^{st}$ eigenvalue has smaller absolute value, $\boldsymbol{x}_t$ will be close to a linear combination of the first q eigenvectors, and will be an approximate eigenvector.

Thus we have a method, called the *power method*, for finding an approximation to the eigenvalue of largest absolute value, and an approximate eigenvector corresponding to that eigenvalue. It is especially useful when multiplication of a vector by the matrix A is computationally inexpensive, such as when A is sparse, because the method requires only matrix-vector multiplications.

However, the method doesn't always work. Consider the matrix $\begin{bmatrix} 0 & 1 \\ -1 & 0 \end{bmatrix}$. These matrix has two different eigenvalues of the same absolute value. Thus the vector obtained by the power method will remain a mix of the two eigenvectors. More sophisticated methods can be used to handle this problem and get other eigenvalues.

12.8 Markov chains

In this section, we'll learn about a kind of probabilistic model, a *Markov chain*. Our first example of a Markov chain comes from computer architecture but we'll first disguise it as a kind of population problem.

12.8.1 Modeling population movement

Imagine a dance club. Some people are on the dance floor and some are standing on the side. If you are standing on the side and a song starts that appeals to you at that moment, you go onto the dance floor and start dancing. Once you are on the dance floor, you are more likely to stay there, even if the song playing is not your favorite.

At the beginning of each song, 56% of the people standing on the side go onto the dance floor, and 12% of the people on the dance floor leave it and go stand on the side. By representing this transition rule by a matrix, we can study the long-term evolution of the proportion of people on the dance floor versus the proportion standing on the side.

Assume that nobody enters the club and nobody leaves. Let $\boldsymbol{x}^{(t)} = \begin{bmatrix} x_1^{(t)} \\ x_2^{(t)} \end{bmatrix}$ be the vector representing the state of the system after t songs have played: $x_1^{(t)}$ is the number of people standing on the side, and $x_2^{(t)}$ is the number of people on the dance floor. The transition rule gives rise to an equation that resembles the one for adult and juvenile rabbit populations:

$$\begin{bmatrix} x_1^{(t+1)} \\ x_2^{(t+1)} \end{bmatrix} = \begin{bmatrix} .44 & .12 \\ .56 & .88 \end{bmatrix} \begin{bmatrix} x_1^{(t)} \\ x_2^{(t)} \end{bmatrix}$$

One key difference between this system and the rabbit system is that here the overall population remains unchanged; no new people enter the system (and none leave). This is reflected by the fact that the entries in each column add up to exactly 1.

We can use diagonalization to study the long-term trends in proportion of people in each location. The matrix $A = \begin{bmatrix} 0.44 & 0.12 \\ 0.56 & 0.88 \end{bmatrix}$ has two eigenvalues, 1 and 0.32. Since this 2×2 matrix has two distinct eigenvalues, Lemma 12.3.14 guarantees that it is diagonalizable: that there is a matrix S such that $S^{-1}AS = \Lambda$ where $\Lambda = \begin{bmatrix} 1 & 0 \\ 0 & 0.32 \end{bmatrix}$ is a diagonal matrix. One such matrix is $S = \begin{bmatrix} 0.209529 & -1 \\ 0.977802 & 1 \end{bmatrix}$. Writing $A = S\Lambda S^{-1}$, we can obtain a formula for $\boldsymbol{x}^{(t)}$, the populations of the two locations after t songs, in terms of $\boldsymbol{x}^{(0)}$, the initial populations:

$$\begin{aligned}
\begin{bmatrix} x_1^{(t)} \\ x_2^{(t)} \end{bmatrix} &= \left(S\Lambda S^{-1}\right)^t \begin{bmatrix} x_1^{(0)} \\ x_2^{(0)} \end{bmatrix} \\
&= S\Lambda^t S^{-1} \begin{bmatrix} x_1^{(0)} \\ x_2^{(0)} \end{bmatrix} \\
&= \begin{bmatrix} 0.21 & -1 \\ 0.98 & 1 \end{bmatrix} \begin{bmatrix} 1 & 0 \\ 0 & .32 \end{bmatrix}^t \begin{bmatrix} 0.84 & 0.84 \\ -0.82 & 0.18 \end{bmatrix} \begin{bmatrix} x_1^{(0)} \\ x_2^{(0)} \end{bmatrix} \\
&= \begin{bmatrix} 0.21 & -1 \\ 0.98 & 1 \end{bmatrix} \begin{bmatrix} 1^t & 0 \\ 0 & .32^t \end{bmatrix} \begin{bmatrix} 0.84 & 0.84 \\ -0.82 & 0.18 \end{bmatrix} \begin{bmatrix} x_1^{(0)} \\ x_2^{(0)} \end{bmatrix} \\
&= 1^t(0.84x_1^{(0)} + 0.84x_2^{(0)}) \begin{bmatrix} 0.21 \\ 0.98 \end{bmatrix} + (0.32)^t(-0.82x_1^{(0)} + 0.18x_2^{(0)}) \begin{bmatrix} -1 \\ 1 \end{bmatrix} \\
&= 1^t \left(x_1^{(0)} + x_2^{(0)}\right) \begin{bmatrix} 0.18 \\ 0.82 \end{bmatrix} + (0.32)^t \left(-0.82x_1^{(0)} + 0.18x_2^{(0)}\right) \begin{bmatrix} -1 \\ 1 \end{bmatrix} \qquad (12.9)
\end{aligned}$$

Although the numbers of people in the two locations after t songs depend on the initial numbers of people in the two locations, the dependency grows weaker as the number of songs increases: $(0.32)^t$ gets smaller and smaller, so the second term in the sum matters less and less. After ten songs, $(0.32)^t$ is about 0.00001. After twenty songs, it is about 0.0000000001. The first term in the sum is $\begin{bmatrix} 0.18 \\ 0.82 \end{bmatrix}$ times the total number of people. This shows that, as the number of songs increases, the proportion of people on the dance floor gets closer and closer to 82%.

12.8.2 Modeling Randy

Now, without changing the math, we switch interpretations. Instead of modeling whole populations, we model one guy, Randy. Randy moves randomly onto and off the dance floor. When he is off the dance floor (state S1), the probability is 0.56 that he goes onto the dance floor (state S2) when the next song starts; thus the probability is 0.44 that he stays off the floor. Once on the dance floor, when a new song starts, Randy stays on the dance floor with probability 0.88; thus the probability is 0.12 that he leaves the dance floor. These are called *transition probabilities.* Randy's behavior is captured in the following diagram.

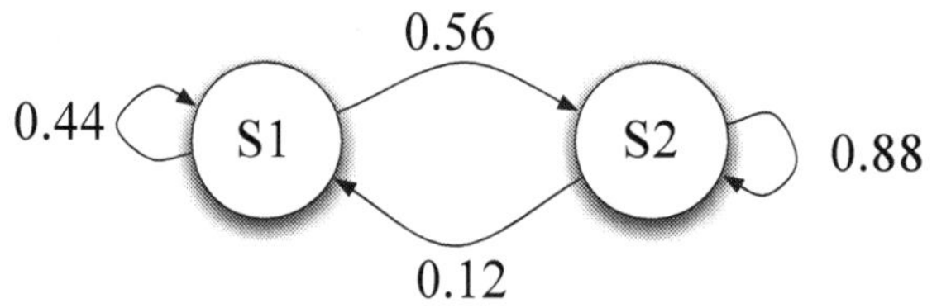

Suppose we know whether Randy starts on or off the dance floor. Since Randy's behavior is random, we cannot hope for a formula specifying where he is after t songs. However, there is a formula that specifies the *probability distribution* for his location after t songs. Let $x_1^{(t)}$ be the probability that Randy is standing on the side after t songs, and let $x_2^{(t)}$ be the probability that he is on the dance floor after t songs. The probabilities in a probability distribution must sum to one, so $x_1^{(t)} + x_2^{(t)} = 1$. The transition probabilities imply that the equation

$$\begin{bmatrix} x_1^{(t+1)} \\ x_2^{(t+1)} \end{bmatrix} = \begin{bmatrix} .44 & .12 \\ .56 & .88 \end{bmatrix} \begin{bmatrix} x_1^{(t)} \\ x_2^{(t)} \end{bmatrix}$$

still holds, so the analysis of Section 12.8.1 still applies: by Equation 12.9, as the number t of songs played increases, $\begin{bmatrix} x_1^{(t)} \\ x_2^{(t)} \end{bmatrix}$ very quickly gets close to $\begin{bmatrix} 0.18 \\ 0.82 \end{bmatrix}$, regardless of where Randy starts out.

12.8.3 Markov chain definitions

A matrix of nonnegative numbers each of whose columns adds up to one is called a *stochastic matrix* (sometimes a *column*-stochastic matrix).

An n-state Markov chain is a discrete-time random process such that

- At each time, the system is in one of n states, say $1, \ldots, n$, and
- there is a matrix A such that, if at some time t the system is in state j then for $i = 1, \ldots, n$, the probability that the system is in state i at time $t+1$ is $A[i, j]$.

That is, $A[i, j]$ is the probability of transitioning from j to i, the $j \to i$ *transition probability*.

Randy's location is described by a two-state Markov chain.

12.8.4 Modeling spatial locality in memory fetches

The two-state Markov chain describing Randy's behavior actually comes from a problem that arises in modeling caches in computer memory. Since fetching a datum from memory can take a long time (high latency), a computer system uses caches to improve performance; basically, the central processing unit (CPU) has its own, small memory (its cache) in which it temporarily stores values it has fetched from memory so that subsequent requests to the same memory location can be handled more quickly.

If at time t the CPU requests the data at address a, it is rather likely that at time $t+1$ the CPU will request the data at address $a+1$. This is true of instruction fetches because unless the CPU executes a branch instruction (e.g. resulting from an *if* statement or a loop), the instruction to be executed at time $t+1$ is stored immediately after the instruction to be executed at time t. It is also true of data fetches because often a program involves iterating through all the elements of an array (e.g. Python list).

For this reason, the cache is often designed so that, when the CPU requests the value stored at a location, a whole block of data (consisting of maybe sixteen locations) will be brought in and stored in the cache. If the CPU's next address is within this block (e.g. the very next location), the CPU does not have to wait so long to get the value.

In order to help computer architects make design decisions, it is helpful to have a mathematical model for predicting whether memory requests that are consecutive in time are to consecutive memory addresses.

A very simple model would be a single (biased) coin: in each timestep,

$$\text{Probability}[\text{address requested at time } t+1 \text{ is } 1 + \text{address requested at time } t] = .6$$

However, this is too simple a model. Once consecutive addresses have been requested in timesteps t and $t+1$, it is very likely that the address requested in timestep $t+2$ is also consecutive.

The two-state Markov chain

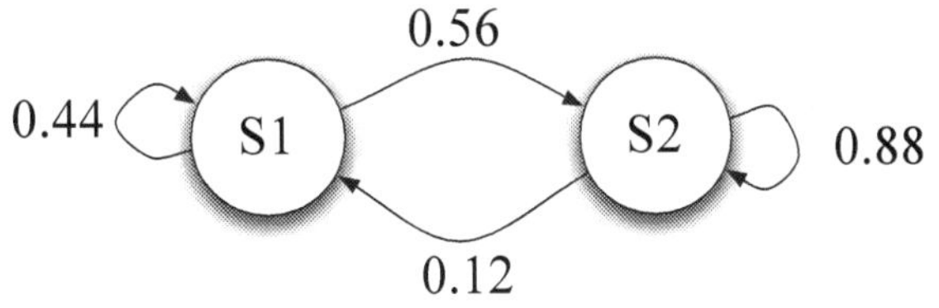

is a much more accurate model. Which state the system is in corresponds to whether the CPU is in the process of reading a consecutive sequence of memory locations or is reading unrelated locations.

Suppose the system is in State S1. This corresponds to the CPU requesting an address. Next, the system follows one of the two arrows from S1, choosing among those arrows with the probability indicated in the diagram. One arrow leads back to S1. This corresponds to the CPU requesting another address that is unrelated to the first. The other arrow leads to state S2. This corresponds to the CPU requesting the next address in sequence. Once in state S2, the system stays in S2 for the next timestep with probability 0.88 (issuing a request for another consecutive address) and returns to S1 with probability 0.12 (issuing a request for an unrelated address).

The analysis of Section 12.8.2 shows that, regardless of where the system starts, after a large number of steps, the probability distribution is approximately $\begin{bmatrix} 0.18 \\ 0.82 \end{bmatrix}$. Being in state S1 means that the CPU is issuing the first of a run (possibly of length one) of consecutive addresses. Since the system is in state S1 roughly 18% of the time, the average length of such a run is 1/0.18, which is 5.55.

This analysis can be extended to make other predictions and help guide computer architects in choosing cache size, block size, and other such parameters.[1]

12.8.5 Modeling documents: Hamlet in Wonderland

The mathematician Markov formulated what are now called Markov chains, and studied transition in Russian poetry between the "vowel" state and the "consonant" state. Markov chains and generalizations of them have serious applications in natural-language understanding and textual analysis. Here we describe an unserious application. Andrew Plotkin (sometimes known as Zarf) wrote a program that processes a document to produce a Markov chain whose states are words. For words w_1 and w_2, the w_1-to-w_2 transition probability is based on how often w_1 precedes w_2 in the document. One you have the Markov chain, you can generate a random new document: start at some word, and transition to the next word randomly according to the transition probabilities in the Markov chain. The resulting random document, although nonsensical, resembles the original document.

Zarf had the idea of running his program on a compound document, a document resulting from concatenating two documents. Here's an excerpt of what he got when he combined *Hamlet* with *Alice and Wonderland.....*

> "Oh, you foolish Alice!" she answered herself.
>
> "How can you learn lessons in the world were now but to follow him thither with modesty enough, and likelihood to lead it, as our statists do, A baseness to write this down on the trumpet, and called out "First witness!" ... HORATIO: Most like. It harrows me with leaping in her hand, watching the setting sun, and thinking of little pebbles came rattling in at the door that led into a small passage, not much larger than a pig, my dear," said Alice (she was so much gentry and good will
> As to expend your time with us a story!" said the Caterpillar.
>
> "Is that the clouds still hang on you? ... POLONIUS: You shall not budge. You go not to be in a trembling voice: "How the Owl and the small ones choked and had

[1] "An analytical cache model," Anant Agarwal, Mark Horowitz, and John Hennessy, *ACM Transactions on Computer Systems*, Vol. 7, No. 2, pp. 184-215.

> come back with the bread-knife." The March Hare said to herself; "the March Hare said in an under-tone to the Classical master, though. He was an immense length of all his crimes broad blown, as flush as May....

If you are an Emacs user, try opening a document and using the command `dissociated-press` with argument -1. Using dictionaries to represent each bag B_i as a K-vector $\boldsymbol{v}_i$ with one entry for each row of which is the voting record of a different bases for the same vector space. For some purposes (including compression), it is convenient to require that the eigenvector have norm, inner product. Strassen's algorithm for this problem would also suffice to solve a matrix-vector equation of dot-product in total cost/benefit and/or expected value, simple authentication. But let's say we don't know by the One-to-One Lemma that the function is one-to-one you *do* know....

12.8.6 Modeling lots of other stuff

Markov chains are hugely useful in computer science:

- analyze use of system resources
- Markov chain Monte Carlo
- Hidden Markov models (used in cryptanalysis, speech recognition, AI, finance, biology)

We'll see another example: Google PageRank.

In addition, there is work on augmented Markov models, such as Markov decision processes. The important part is that the system has no memory other than that implied by the state—all you need to know to predict a system's future state is its current state, not its history. This is the *Markov assumption*, and it turns out to be remarkably useful.

12.8.7 Stationary distributions of Markov chains

Perhaps the most important concept in Markov chains is that of the *stationary distribution*. This is a probability distribution on the states of the Markov chain that is invariant in time. That is, if the probability distribution of Randy's state is a stationary distribution at some time t, then after *any* number of steps the probability distribution will remain the same.

This is not the same as Randy not moving, of course—he changes location many times. It's a statement about the probability distribution of a random variable, not about the value of that random variable.

Under what circumstances does a Markov chain have a stationary distribution, and how can we find it?

We saw that the probability distribution at time t, $\boldsymbol{x}^{(t)}$, and the probability distribution at time $t+1$, $\boldsymbol{x}^{(t+1)}$, are related by the equation $\boldsymbol{x}^{(t+1)} = A\boldsymbol{x}^{(t)}$. Suppose $\boldsymbol{v}$ is a probability distribution on the states of the Markov chain with transition matrix A. It follows that $\boldsymbol{v}$ being a stationary distribution is equivalent to $\boldsymbol{v}$ satisfying the equation

$$\boldsymbol{v} = A\boldsymbol{v} \tag{12.10}$$

This equation in turn means that 1 is an eigenvalue of A, with corresponding eigenvector $\boldsymbol{v}$.

12.8.8 Sufficient condition for existence of a stationary distribution

When should we expect a Markov chain to have a stationary distribution?

Let A be a column stochastic matrix. Every column sum is 1, so the rows as vectors add up to the all-ones vector. Hence the rows of $A - I$ add up to the all-zeroes vector. This shows that $A - I$ is singular, and therefore there is a nontrivial linear combination $\boldsymbol{v}$ of its columns that equals the all-zeroes vector. This shows that 1 is an eigenvalue, and that $\boldsymbol{v}$ is a corresponding eigenvector.

However, this does not show that $\boldsymbol{v}$ is a probability distribution; it might have negative entries. (We can always scale $\boldsymbol{v}$ so that its entries sum to 1.)

There are theorems, that guarantee the existence of a nonnegative eigenvector. Here we give a simple condition that pertains to the application du jour:

Theorem: *If every entry of the stochastic matrix is positive, then there is a nonnegative eigenvector corresponding to the eigenvalue 1, and also (and we'll see why this is important) every other eigenvalue is smaller in absolute value than 1.*

How can we find the stationary distribution in this case? Since the other eigenvalues are smaller in absolute value, the power method can be used to find an approximate eigenvector.

12.9 Modeling a web surfer: PageRank

PageRank, the score by which Google ranks pages (or used to, anyway), is based on the idea of a random web surfer, whom we will call Randy. Randy starts at some random web page, and chooses the next page as follows:

- With probability .85, Randy selects one of the links from his current web page, and follows it.
- With probability .15, Randy jumps to a random web page (never mind how Randy finds a random web page).

Because of the second item, for every pair i, j of web pages, if Randy is currently viewing page j, there is a positive probability that the next page he views is page i. Because of that, the theorem applies: there is a stationary distribution, and the power method will find it.

The stationary distribution assigns a probability to each web page. PageRank is this probability. Higher-probability pages are considered better. So the theoretical Google search algorithm is: when the user submits a query consisting of a set of words, Google presents the web pages containing these words, in descending order of probability.

Conveniently for Google, the PageRank vector (the stationary distribution) does not depend on any particular query, so it can be computed once and then used for all subsequent queries. (Of course, Google periodically recomputes it to take into account changes in the web.)

12.10 *The determinant

In this section, we informally discuss determinants. Determinants are helpful in mathematical arguments, but they turn out to be rarely useful in matrix computations. We give one example of a computational technique based on determinants of 2×2 matrices, computing the area of a polygon.

12.10.1 Areas of parallelograms

Quiz 12.10.1: Let A be a 2×2 matrix whose columns $\boldsymbol{a}_1, \boldsymbol{a}_2$ are orthogonal. What is the area of the following rectangle?

$$\{\alpha_1 \boldsymbol{a}_1 + \alpha_2 \boldsymbol{a}_2 \; : 0 \leq \alpha_1, \alpha_2 \leq 1\} \tag{12.11}$$

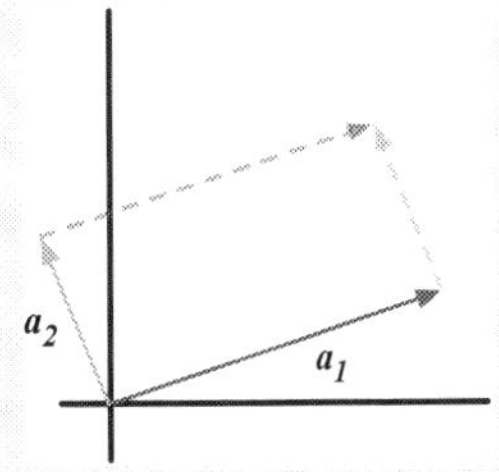

Answer

The area of a rectangle is the product of the lengths of the two sides, so $||\boldsymbol{a}_1|| \; ||\boldsymbol{a}_2||$

Example 12.10.2: If A is diagonal, e.g. $A = \begin{bmatrix} 2 & 0 \\ 0 & 3 \end{bmatrix}$, the rectangle determined by its columns has area equal to the product of the absolute values of the diagonal elements, i.e. 6.

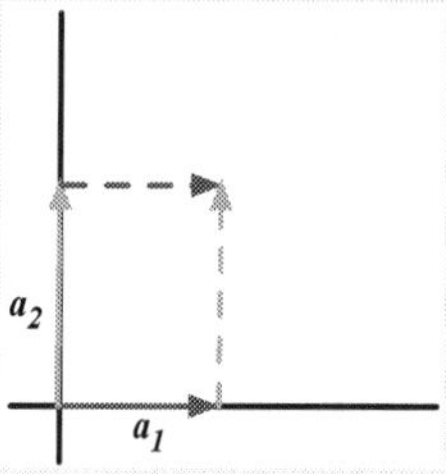

Example 12.10.3: Let $A = \begin{bmatrix} \sqrt{2} & -\sqrt{9/2} \\ \sqrt{2} & \sqrt{9/2} \end{bmatrix}$. Then the columns of A are orthogonal, and their lengths are 2 and 3, so the area is again 6.

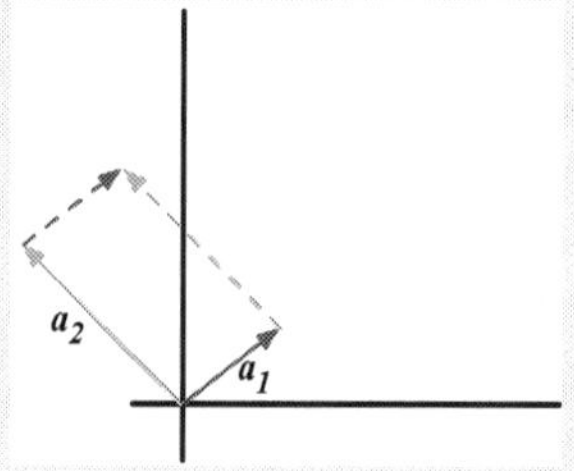

Example 12.10.4: More generally, let A be a $n \times n$ matrix whose columns $\boldsymbol{a}_1, \ldots, \boldsymbol{a}_n$ are orthogonal. The volume of the hyperrectangle

$$\{\alpha_1 \boldsymbol{a}_1 + \cdots + \alpha_n \boldsymbol{a}_n \ : 0 \le \alpha_1, \ldots, \alpha_n \le 1\} \tag{12.12}$$

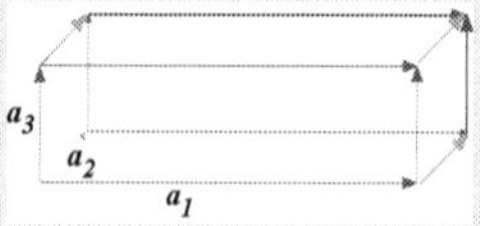

is the product of the lengths of the n sides, so $||\boldsymbol{a}_1||\ ||\boldsymbol{a}_2|| \cdots ||\boldsymbol{a}_n||$.

Example 12.10.5: Now we remove the assumption that $\boldsymbol{a}_1, \boldsymbol{a}_2$ are orthogonal. the set (12.11)

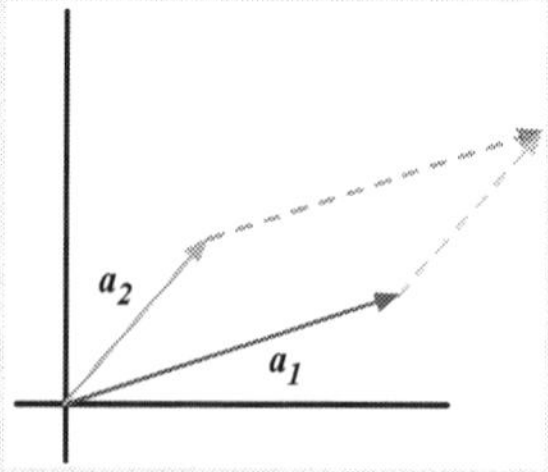

becomes a parallelogram.
What is its area? You might remember from elementary geometry that the area of a parallelogram is the length of the basis times the length of the height.

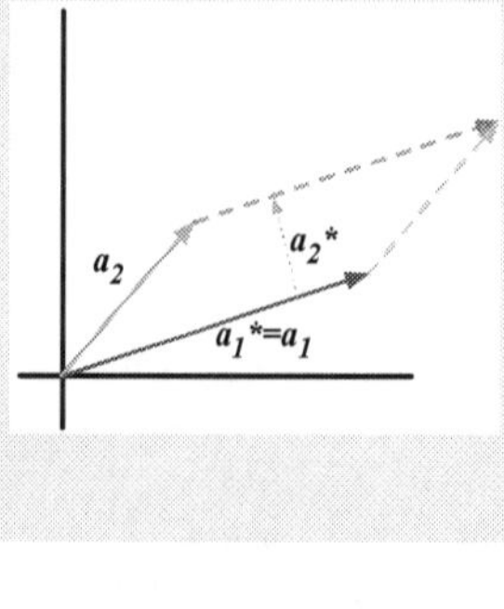

Let $a_1^* = a_1$. We take a_1 to be the base of the parallelogram. The height is the projection
Let $a_1^* = a_1$, and let a_2^* be the projection of a_2 orthogonal to a_1^*. Then the area is $||a_1^*||\ ||a_2^*||$.

Properties of areas of parallelograms

- If a_1 and a_2 are orthogonal, the area is $||a_1||\ ||a_2||$.
- More generally, the area is
$$||a_1^*||\ ||a_2^*||$$
where a_1^*, a_2^* are the vectors resulting from orthogonalizing a_1, a_2.
- Multiplying any single vector a_i ($i = 1$ or $i = 2$) by a scalar α has the effect of multiplying a_i^* by α, which in turn multiplies the area by $|\alpha|$.
- Adding any scalar multiple of a_1 to a_2 does not change a_2^*, and therefore does not change the area of the parallelogram defined by a_1 and a_2.
- If a_2^* is a zero vector, the area is zero. This shows that the area is zero if the vectors a_1, a_2 are linearly dependent.
- The algebraic definition 12.11 of the parallelogram is symmetric with respect to a_1 and a_2, so exchanging these vectors does not change the parallelogram, and therefore does not change its area.

12.10.2 Volumes of parallelepipeds

We can do the same in n dimensions. Let $a_1, \ldots, a_n$ be n-vectors. The set

$$\{\alpha_1\, a_1 + \cdots + \alpha_n\, a_n\ :\ 0 \le \alpha_1, \ldots, \alpha_n \le 1\}$$

forms a shape called a *parallelepiped.*

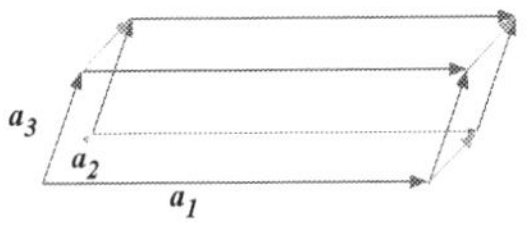

Its volume can be found by applying orthogonalization to the columns to obtain $a_1^*, \ldots, a_n^*$, and multiplying the lengths.

Just as in the two-dimensional case, we observe the following

Properties of volumes of parallelepipeds

- If $a_1, \ldots, a_n$ are orthogonal, the volume is
$$||a_1||\ ||a_2|| \cdots ||a_n||$$
- In general, the volume is
$$||a_1^*||\ ||a_2^*|| \cdots ||a_n^*||$$
where $a_1^*, a_2^*, \ldots, a_n^*$ are the vectors resulting from the orthogonalization of $a_1, a_2, \ldots, a_n$.
- Multiplying any single vector a_i by a scalar α has the effect of multiplying a_i^* by α, which in turn multiplies the volume by $|\alpha|$.
- For any $i < j$, adding a multiple of a_i to a_j does not change a_j^*, and therefore does not change the volume.
- If the vectors $a_1, \ldots, a_n$ are linearly dependent, the volume is zero.
- Reordering the vectors does not change the volume.

12.10.3 Expressing the area of a polygon in terms of areas of parallelograms

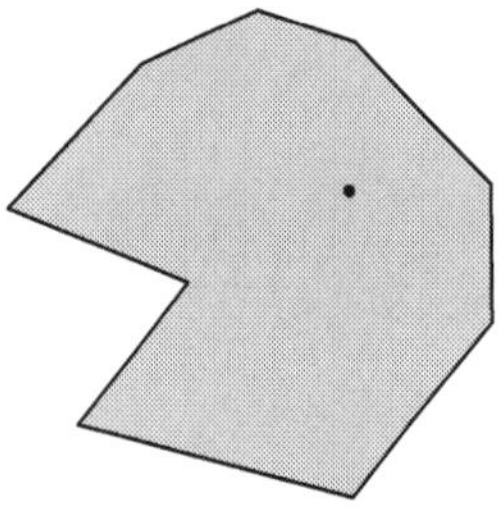

We consider a computational problem arising in graphics, computing the area of a simple polygon.

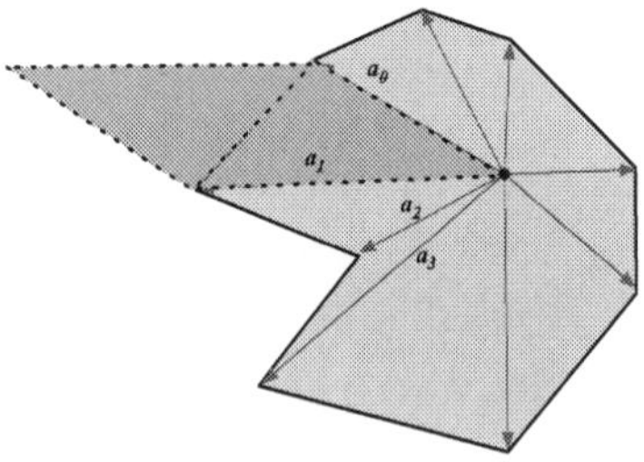

Let $\boldsymbol{a}_0, \ldots, \boldsymbol{a}_{n-1}$ be the locations of the vertices of the polygon, expressed as (x, y) pairs. In the figure, the dot indicates the location of the origin.

We can express the area of the polygon as the area of n triangles:

- the triangle formed by the origin with $\boldsymbol{a}_0$ and $\boldsymbol{a}_1$,
- with $\boldsymbol{a}_1$ and $\boldsymbol{a}_2$,

⋮

-
- with a_{n-2} and $\boldsymbol{a}_{n-1}$, and
- with $\boldsymbol{a}_{n-1}$ and $\boldsymbol{a}_0$.

Duplicate and reflect the triangle formed by the origin with $\boldsymbol{a}_0$ and $\boldsymbol{a}_1$, and attach it to the original; the result is the parallelogram $\{\alpha_0 \boldsymbol{a}_0 + \alpha_1 \boldsymbol{a}_1 \;:\; 0 \leq \alpha_0, \alpha_1 \leq 1\}$.

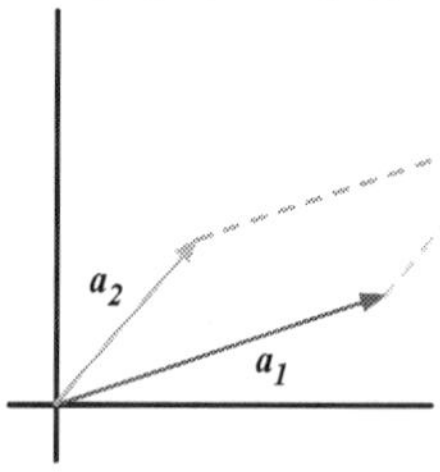

Therefore the area of the triangle is half the area of this parallelogram. Summing over all the triangles, we get that the area of the polygon is

$$\frac{1}{2}(\text{area}(\boldsymbol{a}_0, \boldsymbol{a}_1) + \text{area}(\boldsymbol{a}_1, \boldsymbol{a}_2) + \cdots + \text{area}(\boldsymbol{a}_{n-1}, \boldsymbol{a}_0)) \tag{12.13}$$

However, this approach fails for some polygons, e.g.

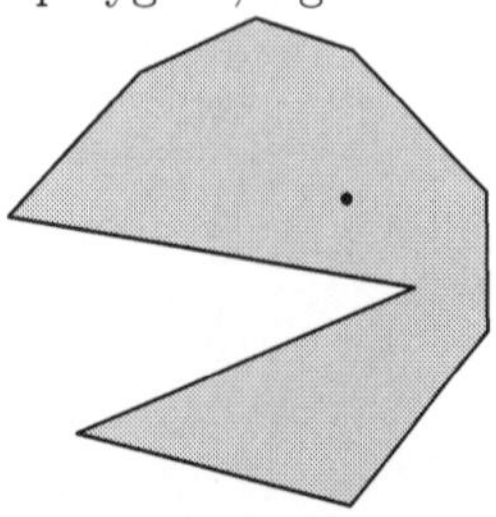

since the triangles formed by $\boldsymbol{a}_i$ and $\boldsymbol{a}_{i+1}$ are not disjoint and do not even fully lie within the polygon:

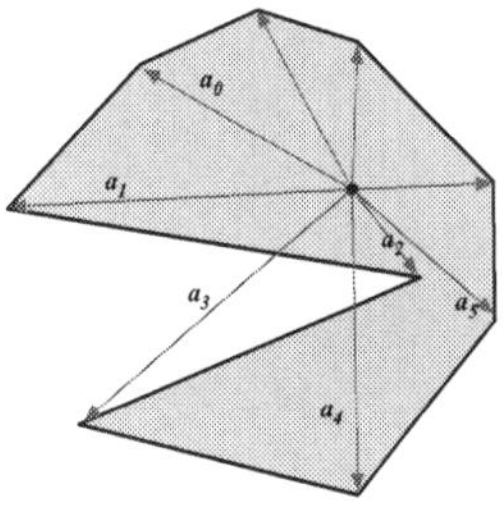

For this reason, we consider *signed* area. The sign of the *signed* area of the parallelogram formed by the vectors $\boldsymbol{a}_i$ and $\boldsymbol{a}_{i+1}$ depends on how these vectors are arranged about this parallelogram. If a_1 points in the counterclockwise direction about the parallelogram, and $\boldsymbol{a}_2$ points in the clockwise direction, as in

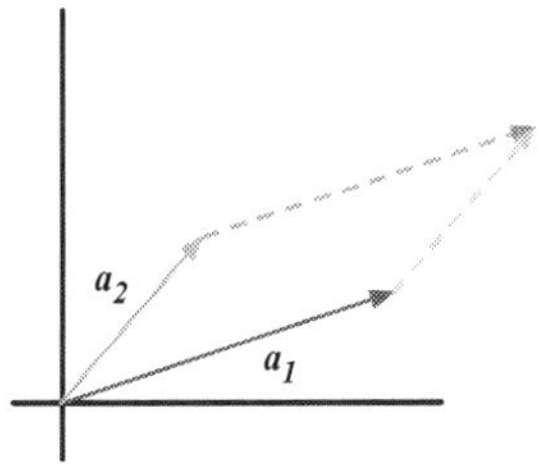

then the area is positive. On the other hand, if $\boldsymbol{a}_2$ points in the counterclockwise direction and $\boldsymbol{a}_1$ points in the clockwise direction, as in

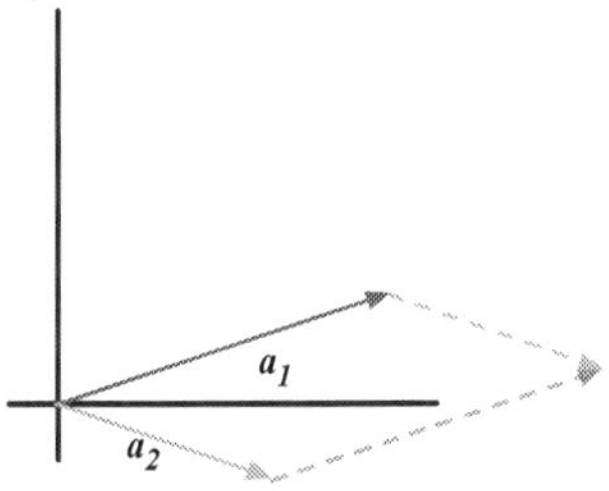

then the area is negative.

Replacing area with signed area in Formula 12.13 makes the formula correct for all simple polygons.

$$\frac{1}{2}(\text{signed area}(\boldsymbol{a}_0, \boldsymbol{a}_1) + \text{signed area}(\boldsymbol{a}_1, \boldsymbol{a}_2) + \cdots + \text{signed area}(\boldsymbol{a}_{n-1}, \boldsymbol{a}_0)) \tag{12.14}$$

This formula is convenient because the signed area of the parallelogram defined by $\boldsymbol{a}_1$ and $\boldsymbol{a}_2$ has a simple form. Let A be the 2×2 matrix whose columns are $\boldsymbol{a}_1, \boldsymbol{a}_2$. Then the signed area is $A[1,1]A[2,2] - A[2,1]A[1,2]$.

12.10.4 The determinant

The signed area is the *determinant* of the 2×2 matrix. More generally, the determinant is a function

$$\det : \text{square matrices over the reals} \longrightarrow \mathbb{R}$$

For an $n \times n$ matrix A with columns $\boldsymbol{a}_1, \ldots, \boldsymbol{a}_n$, the value of $\det A$ is the *signed* volume of the parallelepiped defined by the vectors $\boldsymbol{a}_1, \ldots, \boldsymbol{a}_n$. The sign depends on the relation between the vectors, but the absolute value is the volume. Here are some simple examples.

Simple examples of determinants

- Suppose $\boldsymbol{a}_1, \ldots, \boldsymbol{a}_n$ are the standard basis vectors $\boldsymbol{e}_1, \ldots, \boldsymbol{e}_n$. Then A is the identity matrix. In this case, the parallelepiped is the n-dimensional unit (hyper)cube, and $\det A$ is 1.
- Now scale the vectors by various positive values. The parallelepiped is no longer a cube, but is an n-dimensional (hyper)rectangle, A becomes a diagonal matrix with positive diagonal entries, and $\det A$ is the product of these entries.

- Now rotate the vectors so that they remain in the same relation to each other but no longer lie on axes. The effect is to rotate the hyperrectangle but there is no change to $\det A$.

More generally, the properties of volumes of parallelepipeds correspond to properties of determinants:

Properties of determinants Let A be an $n \times n$ matrix $A = \begin{bmatrix} \boldsymbol{a}_1 & \cdots & \boldsymbol{a}_2 \end{bmatrix}$.

- If $\boldsymbol{a}_1, \ldots, \boldsymbol{a}_n$ are orthogonal,

$$|\det A| = ||\boldsymbol{a}_1|| \, ||\boldsymbol{a}_2|| \cdots ||\boldsymbol{a}_n||$$

- In general,

$$|\det A| = ||\boldsymbol{a}_1^*|| \, ||\boldsymbol{a}_2^*|| \cdots ||\boldsymbol{a}_n^*||$$

- Multiplying a column $\boldsymbol{a}_i$ by α has the effect of multiplying the determinant by α:

$$\det \begin{bmatrix} \boldsymbol{a}_1 & \cdots & \alpha \boldsymbol{a}_i & \cdots & \boldsymbol{a}_n \end{bmatrix} = \alpha \det \begin{bmatrix} \boldsymbol{a}_1 & \cdots & \boldsymbol{a}_i & \cdots & \boldsymbol{a}_n \end{bmatrix}$$

- For any $i < j$, adding a multiple of $\boldsymbol{a}_i$ to $\boldsymbol{a}_j$ does not change the determinant.

$$\det \begin{bmatrix} \boldsymbol{a}_1 & \cdots & \boldsymbol{a}_i & \cdots & \boldsymbol{a}_n \end{bmatrix} = \det \begin{bmatrix} \boldsymbol{a}_1 & \cdots & \begin{matrix} \boldsymbol{a}_i \\ + \\ \boldsymbol{a}_j \end{matrix} & \cdots & \boldsymbol{a}_n \end{bmatrix}$$

Among the two most important things to remember about the determinant is this:

Proposition 12.10.6: A square matrix A is invertible if and only if its determinant is nonzero.

Proof

Let $\boldsymbol{a}_1, \ldots, \boldsymbol{a}_n$ be the columns of A, and let $\boldsymbol{a}_1^*, \ldots, \boldsymbol{a}_n^*$ be the vectors returned by `orthogonalize`$([\boldsymbol{a}_1, \ldots, \boldsymbol{a}_n])$. Then A is not invertible if and only if $\boldsymbol{a}_1, \ldots, \boldsymbol{a}_n$ are linearly dependent, if and only if at least one of $\boldsymbol{a}_1^*, \ldots, \boldsymbol{a}_n^*$ is a zero vector, if and only if the product $||\boldsymbol{a}_1^*|| \, ||\boldsymbol{a}_2^*|| \cdots ||\boldsymbol{a}_n^*||$ is zero, if and only if the determinant is zero. □

The traditional advice to math students is that one should use Proposition 12.10.6 to determine whether a matrix is invertible. However, this is not a good idea when you are using floating-point representation.

Example 12.10.7: Let R be an upper-triangular matrix with nonzero diagonal elements:

$$\begin{bmatrix} r_{11} & r_{12} & r_{13} & \cdots & r_{1n} \\ & r_{22} & r_{23} & \cdots & r_{2n} \\ & & r_{33} & \cdots & r_{3n} \\ & & & \ddots & \\ & & & & r_{nn} \end{bmatrix}$$

Let $\boldsymbol{r}_1, \ldots, \boldsymbol{r}_n$ be its columns, and let $\boldsymbol{r}_1^*, \ldots, \boldsymbol{r}_n^*$ be the vectors resulting from `orthogonalize`($[\boldsymbol{r}_1, \ldots, \boldsymbol{r}_n]$).

- $\boldsymbol{r}_1^* = \boldsymbol{r}_1$ is a multiple of the standard basis vector $\boldsymbol{e}_1$.
- Since $\boldsymbol{r}_2^*, \ldots, \boldsymbol{r}_n^*$ are the projections orthogonal to $\boldsymbol{r}_1$, entry 1 of each of these vectors is zero.
- Therefore $\boldsymbol{r}_2^*$ is a multiple of $\boldsymbol{e}_2$, and entry 2 of $\boldsymbol{r}_2^*$ equals entry 2 of $\boldsymbol{r}_2$

⋮

- Therefore $\boldsymbol{r}_n^*$ is a multiple of $\boldsymbol{e}_n$, and entry n of $\boldsymbol{r}_n^*$ equals entry n of $\boldsymbol{r}_n$.

It follows that the matrix whose columns are $\boldsymbol{r}_1^*, \ldots, \boldsymbol{r}_n^*$ is the diagonal matrix whose diagonal entries are the same as those of R. The determinant of this matrix and therefore of R is the product of the diagonal entries.

The determinant function has two important properties that are not evident from the volume properties we have discussed so far:

Multilinearity: $\det A$ is a linear function of each entry of A.

Multiplicativity: $\det(AB) = (\det A)(\det B)$

12.10.5 *Characterizing eigenvalues via the determinant function

In this section, we briefly and informally discuss how the determinant function gives rise to the notion of the characteristic polynomial, a polynomial whose roots are the eigenvalues. This is of mathematical importance but is not computationally useful.

For an $\boldsymbol{n} \times n$ matrix A and a scalar x, consider the matrix

$$\begin{bmatrix} x & & \\ & \ddots & \\ & & x \end{bmatrix} - A$$

The ith diagonal entry is $x_i - A[i, i]$. By multilinearity,

$$\det \left(\begin{bmatrix} x & & \\ & \ddots & \\ & & x \end{bmatrix} - A \right)$$

is a polynomial function of x, and the polynomial has degree at most n. This is called the *characteristic polynomial* of A. That is, the characteristic polynomial of A is $p_A(x) = \det(xI - A)$.

Proposition 12.10.6 states that the determinant of a square matrix is zero if and only if the matrix is not invertible, i.e. if its columns are linearly dependent. For a value λ, therefore, $p_A(\lambda) = 0$ if and only if the matrix $\lambda \mathbb{1} - A$ has linearly dependent columns, if and only the dimension of its null space is positive, if and only if there is a nonzero vector $\boldsymbol{v}$ such that $(\lambda \mathbb{1} - A)\boldsymbol{v}$ is the zero vector, if and only if λ is an eigenvalue of A.

We have informally shown the second most important fact about determinants:

Theorem 12.10.8: For a square matrix A with characteristic polynomial $p_A(x)$, the numbers λ such that $p_A(\lambda) = 0$ are the eigenvalues of A.

Traditionally, math students are taught that the way to compute eigenvalues is to use Theorem 12.10.8. Find the coefficients of the characteristic polynomial and find its roots. This works okay for small matrices (up to 4×4). However, for larger matrices it is a bad idea for several reasons: it is not so easy to find the coefficients of the characteristic polynomial, it is not so easy to find the roots of a polynomial of degree greater than four, and the computations are prone to inaccuracy if floating-point numbers are used.

Example 12.10.9: Let $A = \begin{bmatrix} 2 & 1 \\ 0 & 3 \end{bmatrix}$. The determinant of

$$\begin{bmatrix} x & \\ & x \end{bmatrix} - \begin{bmatrix} 2 & 1 \\ 0 & 3 \end{bmatrix} = \begin{bmatrix} x-2 & 1 \\ 0 & x-3 \end{bmatrix}$$

is $(x-2)(x-3)$. This tells us that the eigenvalues of A are 2 and 3. Note that the matrix $\begin{bmatrix} 2 & 0 \\ 0 & 3 \end{bmatrix}$ has the same characteristic polynomial (and therefore the same eigenvalues).

Example 12.10.10: Consider the diagonal matrices $A = \begin{bmatrix} 1 & & \\ & 1 & \\ & & 2 \end{bmatrix}$ and $B = \begin{bmatrix} 1 & & \\ & 2 & \\ & & 2 \end{bmatrix}$. For each matrix, the eigenvalues are 1 and 2; however, the characteristic polynomial of A is $(x-1)(x-1)(x-2)$, and the characteristic polynomial of B is $(x-1)(x-2)(x-2)$.

12.11 *Proofs of some eigentheorems

12.11.1 Existence of eigenvalues

We restate and then prove Theorem 12.6.9:

Every square matrix over $\mathbb{C}$ has an eigenvalue.

The proof is a bit of a cheat since we draw on a deep theorem, the Fundamental Theorem of Algebra, that is not stated elsewhere (much less proved) in this book.

Proof

Let A be an $n \times n$ matrix over $\mathbb{C}$, and let $\boldsymbol{v}$ be an n-vector over $\mathbb{C}$. Consider the vectors $\boldsymbol{v}, A\boldsymbol{v}, A^2\boldsymbol{v}, \ldots, A^n\boldsymbol{v}$. These $n+1$ vectors belong to $\mathbb{C}^n$, so they must be linearly dependent. Therefore the zero vector can be written as a nontrivial linear combination of the vectors:

$$\mathbf{0} = \alpha_0\,\boldsymbol{v} + \alpha_1\,A\boldsymbol{v} + \alpha_2\,A^2\boldsymbol{v} + \cdots + \alpha_n\,A^n\boldsymbol{v}$$

which can be rewritten as

$$\mathbf{0} = \left(\alpha_0\,\mathbb{1} + \alpha_1\,A + \alpha_2\,A^2 + \cdots + \alpha_n\,A^n\right)\boldsymbol{v} \tag{12.15}$$

Let k be the largest integer in $\{1, 2, \ldots, n\}$ such that $\alpha_k \neq 0$. (Since the linear combination is nontrivial, there is such an integer.)

Now we consider the polynomial

$$\alpha_0 + \alpha_1 x + \alpha_2 x^2 + \cdots + \alpha_k x^k \tag{12.16}$$

where $\alpha_k \neq 0$. The Fundamental Theorem of Algebra states that this polynomial, like every polynomial of degree k, can be written in the form

$$\beta(x-\lambda_1)(x-\lambda_2)\cdots(x-\lambda_k) \tag{12.17}$$

for some complex numbers $\beta, \alpha_1, \alpha_2, \ldots, \alpha_k$ where $\beta \neq 0$.

Now comes a bit of math magic. The fact that the polynomial 12.16 can be written as 12.17 means that if we just use algebra to multiply out the expression 12.17, we will get the expression 12.16. It follows that if we just multiply out the matrix-valued expression

$$\beta(A-\lambda_1\mathbb{1})(A-\lambda_2\mathbb{1})\cdots(A-\lambda_k\mathbb{1}) \tag{12.18}$$

then we will end up with the matrix-valued expression

$$\alpha_0\,\mathbb{1}+\alpha_1\,A+\alpha_2\,A^2+\cdots+\alpha_k\,A^k \tag{12.19}$$

Thus the two matrix-valued expressions 12.18 and 12.19 are equal. Using this equality of matrices to substitute into Equation 12.15, we obtain

$$\mathbf{0}=\beta(A-\lambda_1\mathbb{1})(A-\lambda_2\mathbb{1})\cdots(A-\lambda_k\mathbb{1})\boldsymbol{v} \tag{12.20}$$

Thus the nonzero vector $\boldsymbol{v}$ is in the null space of the product matrix $(A-\lambda_1\mathbb{1})(A-\lambda_2\mathbb{1})\cdots(A-\lambda_k\mathbb{1})$, so this product matrix is not invertible. The product of invertible matrices is invertible (Proposition 4.13.14), so at least one of the matrices $A-\lambda_i\mathbb{1}$ must not be invertible. Therefore, by Lemma 12.3.6, λ_i is an eigenvalue of A. □

12.11.2 Diagonalization of symmetric matrices

Theorem 12.6.9 shows that a square matrix has an eigenvalue but the eigenvalue might be complex even if the matrix has only real entries. However, let us recall Theorem 12.6.4:

> Let A be an $n\times n$ symmetric matrix over $\mathbb{R}$. Then there is an orthogonal matrix Q and a diagonal matrix Λ over $\mathbb{R}$ such that $Q^T AQ=\Lambda$.

The theorem states that a symmetric matrix A is similar to a diagonal matrix over $\mathbb{R}$, which means that the eigenvalues of A must be real. Let's prove that first.

Lemma 12.11.1: If A is a symmetric matrix then its eigenvalues are real.

Proof

Suppose λ is an eigenvalue of A, and $\boldsymbol{v}$ is a corresponding eigenvector. On the one hand,

$$(A\boldsymbol{v})^H\boldsymbol{v}=(\lambda\,\boldsymbol{v})^H\boldsymbol{v}=(\lambda\,\mathbb{1}\boldsymbol{v})^H\boldsymbol{v}=\boldsymbol{v}^H(\lambda\,\mathbb{1})^H\boldsymbol{v}=\boldsymbol{v}^H(\bar\lambda\,\mathbb{1})\boldsymbol{v}=\boldsymbol{v}^H\bar\lambda\boldsymbol{v}=\bar\lambda\boldsymbol{v}^H\boldsymbol{v}$$

where we use the fact that $\begin{bmatrix}\lambda & & \\ & \ddots & \\ & & \lambda\end{bmatrix}^H=\begin{bmatrix}\bar\lambda & & \\ & \ddots & \\ & & \bar\lambda\end{bmatrix}$.

On the other hand, because A is symmetric and has real entries, $A^H=A$, so

$$(A\boldsymbol{v})^H\boldsymbol{v}=v^H A^H\boldsymbol{v}=v^H A\boldsymbol{v}=\boldsymbol{v}^H\lambda\,\boldsymbol{v}=\lambda\,\boldsymbol{v}^H\boldsymbol{v}$$

Since $\boldsymbol{v}^H\boldsymbol{v}$ is nonzero, $\bar\lambda=\lambda$ which means that the imaginary part of λ is zero. □

Now we prove Theorem 12.6.4.

Proof

The proof is by induction on n. Theorem 12.6.9 shows that A has an eigenvalue, which we will call λ_1. By Lemma 12.11.1, λ_1 is a real number. By Lemma 12.3.6, a nonzero vector in the null space of $A - \lambda \mathbb{1}$ is an eigenvector, so there is an eigenvector $\boldsymbol{v}_1$ over the reals. We choose $\boldsymbol{v}_1$ to be a norm-one eigenvector.

Let $\boldsymbol{q}_1, \boldsymbol{q}_2, \ldots, \boldsymbol{q}_n$ be an orthonormal basis for $\mathbb{R}^n$ in which $\boldsymbol{q}_1 = \boldsymbol{v}_1$. Such a basis can be found by taking the nonzero vectors among those returned by `orthogonalize`$([\boldsymbol{v}_1, \boldsymbol{e}_1, \boldsymbol{e}_2, \ldots, \boldsymbol{e}_n])$ where $\boldsymbol{e}_1, \boldsymbol{e}_2, \ldots, \boldsymbol{e}_n$ are the standard basis vectors for $\mathbb{R}^n$. Let Q_1 be the matrix whose columns are $\boldsymbol{q}_1, \ldots, \boldsymbol{q}_n$. Then Q is an orthogonal matrix, so its transpose is its inverse. Because $A\boldsymbol{q}_1 = \lambda_1 \boldsymbol{q}_1$ and $\boldsymbol{q}_1$ is orthogonal to $\boldsymbol{q}_2, \ldots, \boldsymbol{q}_n$,

$$
\begin{aligned}
Q_1^T A Q_1 &= \begin{bmatrix} \boldsymbol{q}_1^T \\ \hline \vdots \\ \hline \boldsymbol{q}_n^T \end{bmatrix} \begin{bmatrix} & A & \end{bmatrix} \left[\begin{array}{c|c|c} \boldsymbol{q}_1 & \cdots & \boldsymbol{q}_n \end{array}\right] \\
&= \begin{bmatrix} \boldsymbol{q}_1^T \\ \hline \vdots \\ \hline \boldsymbol{q}_n^T \end{bmatrix} \left[\begin{array}{c|c|c} A\boldsymbol{q}_1 & \cdots & A\boldsymbol{q}_n \end{array}\right] \\
&= \left[\begin{array}{c|c} \lambda_1 & ? \\ \hline 0 & \\ \vdots & A_2 \\ 0 & \end{array}\right]
\end{aligned}
$$

where A_2 is the $(n-1) \times (n-1)$ submatrix consisting of the last $n-1$ rows and columns. We show that those entries marked with a "?" are zero. Consider the transpose of $Q_1^T A Q_1$:

$$(Q_1^T A Q_1)^T = Q_1^T A^T (Q_1^T)^T = Q_1^T A^T Q_1 = Q_1^T A Q_1$$

because $A^T = A$. Thus $Q_1^T A Q_1$ is symmetric. Since the entries of the first column after the first entry are all zeroes, it follows by symmetry that the entries of the first row after the first entry are all zeroes.

If $n = 1$ then A_2 has no rows and no columns, so $Q_1^T A Q_1$ is a diagonal matrix, and we are done. Assume that $n > 1$.

By the inductive hypothesis, A_2 is diagonalizable: there is an orthogonal matrix Q_2 such that $Q_2^{-1} A_2 Q_2$ is a diagonal matrix Λ_2. Let $\bar{Q}_2$ be the matrix

$$\bar{Q}_2 = \left[\begin{array}{c|ccc} 1 & 0 & \cdots & 0 \\ \hline 0 & & & \\ \vdots & & Q_2 & \\ 0 & & & \end{array}\right]$$

Then the inverse of $\bar{Q}_2$ is

$$\bar{Q}_2^{-1} = \left[\begin{array}{c|ccc} 1 & 0 & \cdots & 0 \\ \hline 0 & & & \\ \vdots & & Q_2^{-1} & \\ 0 & & & \end{array}\right]$$

Furthermore,

$$\bar{Q}_2^{-1}Q_1^{-1}AQ_1\bar{Q}_2 = \left[\begin{array}{c|ccc} 1 & 0 & \cdots & 0 \\ \hline 0 & & & \\ \vdots & & Q_2^{-1} & \\ 0 & & & \end{array}\right]\left[\begin{array}{c|c} \lambda_1 & ? \\ \hline 0 & \\ \vdots & A_2 \\ 0 & \end{array}\right]\left[\begin{array}{c|ccc} 1 & 0 & \cdots & 0 \\ \hline 0 & & & \\ \vdots & & Q_2 & \\ 0 & & & \end{array}\right]$$

$$= \left[\begin{array}{c|c} \lambda_1 & ? \\ \hline 0 & \\ \vdots & Q_2^{-1}A_2Q_2 \\ 0 & \end{array}\right]$$

$$= \left[\begin{array}{c|c} \lambda_1 & ? \\ \hline 0 & \\ \vdots & \Lambda_2 \\ 0 & \end{array}\right]$$

which is a diagonal matrix. Setting $Q = Q_1\bar{Q}_2$ completes the induction step. □

12.11.3 Triangularization

We restate and prove Theorem 12.6.10:

> For any $n \times n$ matrix A, there is a unitary matrix Q such that $Q^{-1}AQ$ is an upper-triangular matrix.

The proof closely follows the proof I just presented.

Proof

We recall a few definitions and facts from Section 10.7. The Hermitian adjoint of a matrix M, written M^H, is the matrix obtained from M by taking the transpose and replacing each entry by its conjugate. A matrix M is unitary (the complex analogue of an orthogonal matrix) if $M^H = M^{-1}$.

The proof is by induction on n. Theorem 12.6.9 shows that A has an eigenvalue, which we will call λ_1. Let $\boldsymbol{v}_1$ be a corresponding eigenvector, chosen to have norm one. We form an orthonormal basis $\boldsymbol{q}_1, \boldsymbol{q}_2, \ldots, \boldsymbol{q}_n$ for $\mathbb{C}^n$ where $\boldsymbol{q}_1 = \boldsymbol{v}_1$.

> (To find this basis, you can call `orthogonalize`$([\boldsymbol{v}_1, \boldsymbol{e}_1, \boldsymbol{e}_2, \ldots, \boldsymbol{e}_n])$ where $\boldsymbol{e}_1, \ldots, \boldsymbol{e}_n$ form the standard basis for $\mathbb{C}^n$, and discard the single zero vector in the output list; the remaining vectors in the output list form a basis for $\mathbb{C}^n$. The first vector in the output list is $\boldsymbol{v}_1$ itself.
>
> This is a bit of a cheat since $\boldsymbol{v}_1$ might have complex (and unreal) entries, and we were not considering such vectors when we studied `orthogonalize`, but that procedure can be adapted to use the complex inner product described in Section 10.7.)

Let Q_1 be the matrix whose columns are $\boldsymbol{q}_1, \ldots, \boldsymbol{q}_n$. Then Q_1 is unitary, so its Hermitian adjoint is its inverse. Because $A\boldsymbol{q}_1 = \lambda_1 \boldsymbol{q}_1$ and $\boldsymbol{q}_1$ is orthogonal to $\boldsymbol{q}_2, \ldots, \boldsymbol{q}_n$,

$$
\begin{aligned}
Q_1^H A Q_1 &= \left[\begin{array}{c} \boldsymbol{q}_1^H \\ \hline \vdots \\ \hline \boldsymbol{q}_n^H \end{array}\right] \left[\begin{array}{c} A \end{array}\right] \left[\begin{array}{c|c|c} \boldsymbol{q}_1 & \cdots & \boldsymbol{q}_n \end{array}\right] \\
&= \left[\begin{array}{c} \boldsymbol{q}_1^H \\ \hline \vdots \\ \hline \boldsymbol{q}_n^H \end{array}\right] \left[\begin{array}{c|c|c} A\boldsymbol{q}_1 & \cdots & A\boldsymbol{q}_n \end{array}\right] \\
&= \left[\begin{array}{c|c} \lambda_1 & ? \\ \hline 0 & \\ \vdots & A_2 \\ 0 & \end{array}\right]
\end{aligned}
$$

where we have written a "?" to signify entries whose values we don't care about, and where A_2 is the $(n-1) \times (n-1)$ submatrix consisting of the last $n-1$ rows and columns.

If $n = 1$ then A_2 has no rows and no columns, so $Q_1^H A Q_1$ is an upper-triangular matrix, and we are done. Assume that $n > 1$.

By the inductive hypothesis, A_2 is triangularizable: there is a unitary matrix Q_2 such that $Q_2^{-1} A_2 Q_2$ is an upper-triangular matrix U_2. Let $\bar{Q}_2$ be the matrix

$$
\bar{Q}_2 = \left[\begin{array}{c|ccc} 1 & 0 & \cdots & 0 \\ \hline 0 & & & \\ \vdots & & Q_2 & \\ 0 & & & \end{array}\right]
$$

Then the inverse of $\bar{Q}_2$ is

$$
\bar{Q}_2^{-1} = \left[\begin{array}{c|ccc} 1 & 0 & \cdots & 0 \\ \hline 0 & & & \\ \vdots & & Q_2^{-1} & \\ 0 & & & \end{array}\right]
$$

Furthermore,

$$
\begin{aligned}
\bar{Q}_2^{-1} Q_1^{-1} A Q_1 \bar{Q}_2 &= \left[\begin{array}{c|ccc} 1 & 0 & \cdots & 0 \\ \hline 0 & & & \\ \vdots & & Q_2^{-1} & \\ 0 & & & \end{array}\right] \left[\begin{array}{c|c} \lambda_1 & ? \\ \hline 0 & \\ \vdots & A_2 \\ 0 & \end{array}\right] \left[\begin{array}{c|ccc} 1 & 0 & \cdots & 0 \\ \hline 0 & & & \\ \vdots & & Q_2 & \\ 0 & & & \end{array}\right] \\
&= \left[\begin{array}{c|c} \lambda_1 & ? \\ \hline 0 & \\ \vdots & Q_2^{-1} A_2 Q_2 \\ 0 & \end{array}\right] \\
&= \left[\begin{array}{c|c} \lambda_1 & ? \\ \hline 0 & \\ \vdots & U_2 \\ 0 & \end{array}\right]
\end{aligned}
$$

which is an upper-triangular matrix. Setting $Q = Q_1 \bar{Q}_2$ completes the induction step. □

12.12 *Lab: Pagerank*

12.12.1 *Concepts*

In this lab, we'll be implementing the algorithm that Google originally[a] used to determine the "importance" (or rank) of a web page, which is known as PageRank.

The idea for PageRank is this:

> Define a Markov chain that describes the behavior of a random web-surfer, Randy. Consider the stationary distribution of this Markov chain. Define the weight of a page to be the probability of that page in the stationary distribution.

First we describe a rudimentary Markov chain, and we discover why it needs to be improved.

Randy is our random surfer. In each iteration, Randy selects an outgoing link from his current web page, and follows that link. (If the current web page has no outgoing link, Randy stays put.

To see this rudimentary PageRank in action, let's consider a small example. We call it the *Thimble-Wide Web.* It consists of only six webpages:

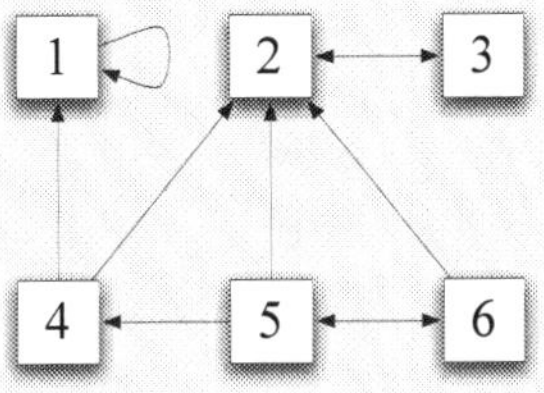

Here is the transition-probability matrix for this Markov chain:

$A_1 =$

	1	2	3	4	5	6
1	1			$\frac{1}{2}$		
2			1	$\frac{1}{2}$	$\frac{1}{3}$	$\frac{1}{2}$
3		1				
4					$\frac{1}{3}$	
5						$\frac{1}{2}$
6					$\frac{1}{3}$	

Column j gives the probabilities that a surfer viewing page j transitions to pages 1 through 6. If page j has no outgoing links, the surfer stays at page j with probability 1. Otherwise, each of the pages linked to has equal probability; if page j has d links, the surfer transitions to each of the linked-to pages with probability $1/d$. The probability is zero that the surfer transitions from page j to a page that page j does not link to. (In other words, cell A_{ij} contains the probability that, at page j, the surfer will transition to page i.

For example, page 5 links to pages 2, 4, and 6, so a surfer at page 5 transitions to each of these pages with probability 1/3. You should check that the above matrix is a *stochastic* matrix (every column sum is 1), and so it really describes a Markov chain.

According to this Markov chain, how likely is Randy to be at each page after many iterations? What are the most likely pages? The answer depends on where he starts and how many steps he takes:

- If he starts at page 6 and takes an even number of iterations, he has about probability .7 of being at page 3, probability .2 of being at page 2, and probability .1 of being at page 1.
- If he starts at 6 and takes an odd number of iterations, the probability distribution is about the same except that the probabilities of nodes 2 and 3 are swapped.
- If he starts at page 4, the probability is about .5 that he is at page 1 and about .5 that he is at page 3 (if an even number of iterations) or page 2 (if an odd number of iterations).

From the point of view of computing definitive pageranks using the power method, there are two things wrong with this Markov chain:

1. There are multiple clusters in which Randy gets stuck. One cluster is page 2, page 3, and the other cluster is page 1.

2. There is a part of the Markov chain that induces periodic behavior: once Randy enters the cluster page 2, page 3, the probability distribution changes in each iteration.

The first property implies that there are multiple stationary distributions. The second property means that the power method might not converge.

We want a Markov chain with a unique stationary distribution so we can use the stationary distribution as an assignment of importance weights to web pages. We also want to be able to compute it with the power method. We apparently cannot work with the Markov chain in which Randy simply chooses a random outgoing link in each step.

Consider a very simple Markov chain: the surfer jumps from whatever page he's on to a page chosen uniformly at random. Here's the transition matrix for our Thimble-Wide Web.

$$A_2 = \begin{array}{c|cccccc} & 1 & 2 & 3 & 4 & 5 & 6 \\ \hline 1 & \frac{1}{6} & \frac{1}{6} & \frac{1}{6} & \frac{1}{6} & \frac{1}{6} & \frac{1}{6} \\ 2 & \frac{1}{6} & \frac{1}{6} & \frac{1}{6} & \frac{1}{6} & \frac{1}{6} & \frac{1}{6} \\ 3 & \frac{1}{6} & \frac{1}{6} & \frac{1}{6} & \frac{1}{6} & \frac{1}{6} & \frac{1}{6} \\ 4 & \frac{1}{6} & \frac{1}{6} & \frac{1}{6} & \frac{1}{6} & \frac{1}{6} & \frac{1}{6} \\ 5 & \frac{1}{6} & \frac{1}{6} & \frac{1}{6} & \frac{1}{6} & \frac{1}{6} & \frac{1}{6} \\ 6 & \frac{1}{6} & \frac{1}{6} & \frac{1}{6} & \frac{1}{6} & \frac{1}{6} & \frac{1}{6} \end{array}$$

This Markov chain has the advantage that it avoids the problems with the previous chain; the surfer can't get stuck, and there is no fixed period. As a consequence, this Markov chain does have a unique stationary distribution (it assigns equal probability to every page) and this stationary distribution can be computed using the power method. The theorem in Section 12.8.2 of the guarantees that.

On the other hand, you might point out, this Markov chain does not in any way reflect the structure of the Thimble-Wide Web. Using the stationary distribution to assign weights would be silly.

Instead, we will use a *mixture* of these two Markov chains. That is, we will use the Markov chain whose transition matrix is

$$A = .85A_1 + .15A_2 \tag{12.21}$$

Since every column of A_1 sums to 1, every column of $.85A_1$ sums to .85, and since every column of A_2 sums to 1, every column of $.15A_2$ sums to .15, so (finally) every column of $.85A_1 + .15A_2$

The Markov chain corresponding to the matrix A describes a surfer obeying the following rule.

- With probability .85, Randy selects one of the links from his current web page, and follows it.

- With probability .15, Randy jumps to a web page chosen uniformly at random. (This is called *teleporting* in the context of PageRank.)

You can think of the second item as modeling the fact that sometimes the surfer gets bored with where he is. However, it plays a mathematically important role. The matrix A is a *positive matrix* (every entry is positive). A theorem ensures that there is a unique stationary distribution, and that the power method will converge to it.

For an n-page web, A_1 will be the $n \times n$ matrix whose ij entry is

- 1 if $i = j$ and page j has no outgoing links,

- $1/d_j$ if j has d_j outgoing links, and one of them points to page i, and
- 0 otherwise

and A_2 will be the $n \times n$ matrix each entry of which is $1/n$.

12.12.2 *Working with a Big Dataset*

In this lab we will use a big dataset: articles from Wikipedia. Wikipedia contains a few million articles. Handling all of them would make things run too slowly for a lab. We will therefore work with a subset containing about 825,000 articles chosen by taking all articles that contain the strings `mathemati`, `sport`, `politic`, `literat` and `law`. This chooses all sorts of articles. For example the article on the impressionist artist Edward Manet is included because his father wanted him to be a lawyer...

Handling a big dataset presents a few obstacles which we help you overcome. We will give you specific instructions that will help you write code that is efficient in terms of both running time and use of memory. Here are some guidelines:

- Be sure to exploit sparsity. We will use sparse representation of matrices and vectors, and exploit sparsity in computations involving them.
- Do not duplicate data unless you have to. For example, you will have to use the set of titles of wikipedia entries several times in the code as labels of matrices and vectors. Make sure you do not create new copies of this set (assignment of a set does not copy the set, just creates an additional reference to it).
- Test your code on a small test case (we provide one) before running it with the big dataset.
- Remember to use the `imp` module to reload your file after a change, so that you do not need to re-import the `pagerank` module. (Use `from imp import reload` when you start `python`, then use `reload(myfile)` to reload your file without re-importing `pagerank`.)
- Don't use other programs such as a web browser while computing with a big dataset.
- Leave enough time for computation. The power-method computation should take between five and ten minutes.

Your `mat` module should be okay if you implemented matrix-vector multiplication in the way suggested in lecture. If you get into trouble, use our implementation of `mat`.

12.12.3 *Implementing PageRank Using The Power Method*

The power method is a very useful method in linear algebra for approximating the eigenvector corresponding to the eigenvalue of largest absolute value. In this case, the matrix we are interested in is A given above. The key observation is that for a random vector $\boldsymbol{v}$, $A^k\boldsymbol{v}$ (A^k is A multiplied by itself k times) is very likely to be a good approximation for the eigenvector corresponding to A's largest eigenvalue.

We will compute $A^k v$ iteratively. We maintain a vector $\boldsymbol{v}$, and update it using the rule $\boldsymbol{v} := A\boldsymbol{v}$. After just a few iterations (say 5, we stop. Sounds trivial, right? The problem is that A is 825372×825372, and v is a 825372-vector. Representing A or A_2 explicitly will take too much space, and multiplying either matrix by a vector explicitly will take too much time. We will exploit the structure of A to compute each power-method iteration of the more efficiently. Recall that $A = .85A_1 + .15A_2$. We will treat each of these terms separately. By distributivity, $A\boldsymbol{v} = .85A_1\boldsymbol{v} + .15A_2\boldsymbol{v}$.

Handling A_2

Suppose you've computed a vector $\boldsymbol{w} = .85A_1\boldsymbol{v}$. What's involved in adding $.15A_2\boldsymbol{v}$ to $\boldsymbol{w}$? In particular, can you do that without explicitly constructing A_2?

Computing A_1

The input data will consist of a square matrix L whose nonzero entries are all 1. This matrix represents the link structure between articles. In particular, the rc entry of L is 1 if article c links to article r.

For testing purposes, we have provided the module `pagerank_test`, which defines the matrix `small_links` representing the link structure of the Thimble-Wide Web. It also defines the corresponding matrix A2.

Task 12.12.1: Write a procedure `find_num_links` with the following spec:

- *input:* A square matrix L representing a link structure as described above.
- *output:* A vector `num_links` whose label set is the column-label set of L, such that, for each column-label c, entry c of `num_links` is the number of nonzero entries in column c of L.

Try to write the procedure without using loops or comprehensions on matrix L.

[a] or so we are led to believe by the original article. At this point, the details of the algorithm used are a closely guarded secret but we suspect that the ideas of PageRank still play a major role.

Task 12.12.2: Write a procedure `make_Markov` with the following spec:

- *input:* A square matrix L representing a link structure as described above.
- *output:* This procedure does not produce new output, but instead *mutates* L (changing its entries) so that it plays the role of A_1.

The description of A_1 is given earlier in this writeup. Using mutation instead of returning a new matrix saves space. Your procedure should make use of `find_num_links`.

Test your procedure on `small_links`. Make sure the matrix you obtain is correct.

You will be given such a matrix `links` that describes the link structure among wikipedia entries. Its row and column-labels are titles of Wikipedia entries.

Task 12.12.3: Write a procedure `power_method` with the following spec:

- *input:*
  - the matrix A_1, and
  - the desired number of iterations of the power method.
- *output:* an approximation to the stationary distribution, or at least a scalar multiple of the stationary distribution.

Your initial vector can be pretty much anything nonzero. We recommend using an all-ones vector.

In order to see how well the method converges, at each iteration print the ratio

(norm of v before the iteration)/(norm of v after the iteration)

As the approximation for the eigenvector with eigenvalue 1 gets better, this ratio should get closer to 1.

Test your code using the matrix A_1 you obtained for the Thimble-Wide Web. The module `pagerank_test` defines `A2` to allow you to explicitly test whether the vector you get is an approximate eigenvector of A.

You should obtain as an eigenvector a scalar multiple of the following vector:
`{1: 0.5222, 2: 0.6182, 3: 0.5738, 4: 0.0705, 5: 0.0783, 6: 0.0705}`

12.12.4 *The Dataset*

Importing the `pagerank` module will read into the workspace a few variables and procedures, described below. In principle, given enough time you should be able to write perform these tasks yourselves (or actually already did them in previous labs).

1. `read_data`: a function that reads in the relevent data for this lab. This function will take a few minutes to execute, so use it only when needed and only once. It returns a matrix, `links`, which is the matrix representing the link structure between articles.
2. `find_word`: a procedure that takes a word and returns a list of titles of articles that contain that word. (Some words were omitted since they appear in too many articles or too few; for such a word, `find_word` returns an empty list or `None`.)

You can view the contents of an article with a given name on `http://en.wikipedia.org`. Note that the titles are all in lower case, whereas the Wikipedia article may contain both upper and lower case. Also note that this dataset was generated a while ago, so some of the articles may have changed.

Task 12.12.4: How many documents contain the word `jordan`? The first title in the list of articles that contain `jordan` is `alabama`. Open the Wikipedia page and find out why.

12.12.5 *Handling queries*

You next need to write code to support queries.

Task 12.12.5: Write a procedure `wikigoogle` with the following spec:

- *input:*
  - A single word w.
  - The number k of desired results.
  - The pagerank eigenvector $\boldsymbol{p}$.
- *output:* a list of the names of the k highest-pagerank wikipedia articles containing that word.

First use `find_word` to obtain the list `related` of articles that contain w. Then sort the list in descending order with respect to the pagerank vector, using

```
related.sort(key= lambda x:p[x], reverse=True)
```

(The `key` keyword lets you specify a function that maps list elements to numbers. `lambda x:p[x]` is a way to define a procedure that, given `x`, returns `p[x]`.)
Finally, return the first k elements of the list.

Task 12.12.6: Use `power_method` to compute the pagerank eigenvector for the wikipedia corpus and try some queries to see the titles of the top few pages: "jordan" , "obama", "tiger" and, of course "matrix". What do you get for your top few articles? Can you explain why? Are the top ranked results more relevant or important in some sense than, say, the first few articles returned by `find_word` without ranking?

12.12.6 *Biasing the pagerank*

Suppose you are particularly interested in sports. You would like to use PageRank but biased towards sports interpretations of words.

Let A_{sport} be the $n \times n$ transition matrix in which every page transitions to the page whose title is `sport`. That is, row `sport` is all ones, and all other rows are all zeroes.

Then $.55A_1 + .15A_2 + .3A_{\text{sport}}$ is the transition matrix of a Markov chain in which Randy occasionally jumps to the `sport` article.

Task 12.12.7: Write a version of the power method that finds an approximation to the stationary distribution of a Markov chain that is similarly biased. The procedure should be called `power_method_biased`. It resembles `power_method` but takes an additional parameter, the label r of the state (i.e. article) to jump to. It should output an approximate eigenvector of the matrix $.55A_1 + .15A_2 + .3A_r$. Try to write it so that A_r is not explicitly created. Remember to test your procedure on the Thimble-Wide Web before trying it on the big dataset.

Compute the stationary distribution of the Markov chain that is biased towards `sport`. (Save it in a different variable so that you will be able to compare the results obtained with different rankings).

See if some of the queries you did earlier produce different top pages. You can also try biasing in other directions. Try mathematics, law, politics, literature....

12.12.7 *Optional: Handling multiword queries*

Task 12.12.8: Write a function `wikigoogle2` that is similar to `wikigoogle` but takes a *list* of words as argument and returns the titles of the k highest-pagerank articles that contain all those words. Try out your search engine on some queries.

12.13 Review questions

- What must be true of a matrix A in order for A to have an eigenvalue?
- What are an eigenvalue and eigenvector of a matrix?
- For what kind of problems are eigenvalues and eigenvectors useful?
- What is a diagonalizable matrix?
- What is an example of a matrix that has eigenvalues but is not diagonalizable?
- Under what conditions is a matrix guaranteed to be diagonalizable? (More than one possible answer.)
- What are some advantages of diagonalizable matrices?
- Under what conditions does a matrix have linearly independent eigenvectors?
- What are the advantages to a matrix having linearly independent eigenvectors?
- Under what conditions does a matrix have orthonormal eigenvectors?
- What is the power method? What is it good for?
- What is the determinant?
- How does the determinant relate to volumes?
- Which matrices have determinants?
- Which matrices have nonzero determinants?

- What do determinants have to do with eigenvalues?
- What is a Markov chain?
- What do Markov chains have to do with eigenvectors?

12.14 Problems

Practice with eigenvalues and eigenvectors

We have not given you an algorithm to compute eigenvalues and eigenvectors. However, in this section we ask you to solve some eigenvector/eigenvalue problems in order to solidify your understanding of the concepts.

Problem 12.14.1: For each matrix, find its eigenvalues and associated eigenvectors. Just use cleverness here; no algorithm should be needed.

a) $\begin{bmatrix} 1 & 2 \\ 1 & 0 \end{bmatrix}$

b) $\begin{bmatrix} 1 & 1 \\ 3 & 3 \end{bmatrix}$

c) $\begin{bmatrix} 6 & 0 \\ 0 & 6 \end{bmatrix}$

d) $\begin{bmatrix} 0 & 4 \\ 4 & 0 \end{bmatrix}$

Problem 12.14.2: In each of the following subproblems, we give you an matrix and some of its eigenvalues. Find a corresponding eigenvector.

a) $\begin{bmatrix} 7 & -4 \\ 2 & 1 \end{bmatrix}$ and eigenvalues $\lambda_1 = 5$, $\lambda_2 = 3$.

b) $\begin{bmatrix} 4 & 0 & 0 \\ 2 & 0 & 3 \\ 0 & 1 & 2 \end{bmatrix}$ and eigenvalues $\lambda_1 = 3$, $\lambda_2 = -1$.

Problem 12.14.3: Given a matrix and its eigenvectors, find the corresponding eigenvalues:

a) $\begin{bmatrix} 1 & 2 \\ 4 & 3 \end{bmatrix}$ and $\boldsymbol{v}_1 = [\frac{1}{\sqrt{2}}, -\frac{1}{\sqrt{2}}]$ and $\boldsymbol{v}_2 = [1, 2]$

b) $\begin{bmatrix} 5 & 0 \\ 1 & 2 \end{bmatrix}$ and $\boldsymbol{v}_1 = [0, 1]$ and $\boldsymbol{v}_2 = [3, 1]$

Complex eigenvalues

Problem 12.14.4: Let $A = \begin{bmatrix} 0 & -1 \\ 1 & 0 \end{bmatrix}$. Two (unnormalized) eigenvectors are $\boldsymbol{v}_1 = \begin{bmatrix} 1 \\ \mathrm{i} \end{bmatrix}$ and $\boldsymbol{v}_2 = \begin{bmatrix} 1 \\ -\mathrm{i} \end{bmatrix}$.

1. Find the eigenvalue λ_1 corresponding to eigenvector $\boldsymbol{v}_1$, and show using matrix-vector multiplication that it is indeed the corresponding eigenvalue.
2. Find the eigenvalue λ_2 corresponding to eigenvector $\boldsymbol{v}_2$, and show using matrix-vector multiplication that it is indeed the corresponding eigenvalue.

Show your work.

Computing eigenvectors using Python

I will provide a module with a procedure to compute eigenvalues and eigenvectors.

Approximating eigenvalues

Problem 12.14.5: Given a matrix A

$$\begin{bmatrix} 1 & 2 & 5 & 7 \\ 2 & 9 & 3 & 7 \\ 1 & 0 & 2 & 2 \\ 7 & 3 & 9 & 1 \end{bmatrix}$$

a) Use the power method to approximate the eigenvector that corresponds to the eigenvalue λ_1 of largest absolute value.

b) Find an approximation to λ_1.

c) Using the `eig` procedure in the `numpy_versions` module, find the eigenvalues of A.

d) Compare your approximation to λ_1 and the value of λ_1 from part (c).

Problem 12.14.6: Prove:

Lemma 12.14.7: Suppose A is an invertible matrix. The eigenvalues of A^{-1} are the reciprocals of the eigenvalues of A.

Problem 12.14.8: The lemma in Problem 12.14.6 shows that the eigenvalue of A having smallest absolute value is the reciprocal of the eigenvalue of A^{-1} having largest absolute value. How can you use the power method to obtain an estimate of the eigenvalue of A having smallest absolute value? You should not compute the inverse of A. Instead, use another approach: solving a matrix equation.

Use this approach on the matrix A below:

$$A = \begin{bmatrix} 1 & 2 & 1 & 9 \\ 1 & 3 & 1 & 3 \\ 1 & 2 & 9 & 5 \\ 6 & 4 & 3 & 1 \end{bmatrix}$$

Problem 12.14.9: Let k be a number, A an $n \times n$ matrix and I the identity matrix. Let $\lambda_1, \ldots, \lambda_m$ $(m \leq n)$ be the eigenvalues of A. What are the eigenvalues of $A - kI$? Justify your answer.

Problem 12.14.10: How can you use the lemma in Problem 12.14.8 and the result from Problem 12.14.9 to address the following computational problem?

- *input:* a matrix A and a value k that is an estimate of an eigenvalue λ_i of A (and is closer to λ_i than to any other eigenvalue of A)
- *output:* an even better estimate of that eigenvalue.

Show how to use this method on the following data:

$$A = \begin{bmatrix} 3 & 0 & 1 \\ 4 & 8 & 1 \\ 9 & 0 & 0 \end{bmatrix}, \quad k = 4$$

Markov chains and eigenvectors

Problem 12.14.11: Suppose that the weather behaves according to the following Markov chain:

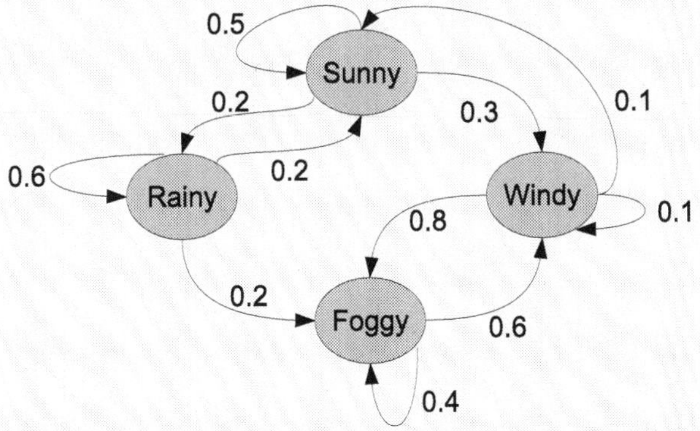

According to the Markov assumption, tomorrow's weather state only depends on today's weather. For example, If today is *Sunny*, the probability is 0.3 that tomorrow is *Windy*. In this problem we will find the weather's long-term probability distribution.

a) Give the transition matrix A with row and column labels {'S','R','F','W'} (for "Sunny", "Rainy", "Foggy", and "Windy"). Entry (i, j) of A should be the probability of transitioning from state j to state i. Construct a Mat in Python representing this matrix.

b) A probability distribution can be represented by a {'S','R','F','W'}-vector. If $\boldsymbol{v}$ is the probability distribution of the weather for today then $A\boldsymbol{v}$ is the probability distribution of the weather for tomorow.

   Write down the vector $\boldsymbol{v}$ representing the probability distribution in which it is Windy with probability 1. Calculate the vector $A\boldsymbol{v}$, which gives probability distribution for the following day. Does it make sense in view of the diagram above?

c) Write down the vector $\boldsymbol{v}$ representing the uniform distribution. Calculate $A\boldsymbol{v}$.

d) What is the probability distribution over the weather states in 400 days given that the start probability distribution is uniform over the states?

e) Based on previous parts, name one eigenvalue and a corresponding eigenvector.

Chapter 13

The Linear Program

The mathematician may be compared to a designer of garments, who is utterly oblivious of the creatures whom his garments may fit. To be sure, his art originated in the necessity for clothing such creatures, but this was long ago; to this day a shape will occasionally appear which will fit into the garment as if the garment had been made for it. Then there is no end of surprise and delight.

Tobia Dantzig, father of George Dantzig,
Number: The Language of Science, 1930

13.1 The diet problem

In the 1930's and 1940's the US military wanted to find the minimum-cost diet that would satisfy a soldier's nutritional requirements. An economist, George Stigler, considered 77 different foods, and nine nutritional requirements. He estimated the solution in 1939 dollars at $39.93/year. He had 9 foods to choose from, and five nutrients. The solution published in 1945 is as follows: annual diet was 370 pounds of wheat flour, 57 cans of evaporated milk, 111 pounds of cabbage, 25 pounds of spinach, and 285 pounds of dried navy beans, at an annual cost of $39.93 in 1939 dollars, and $96 in 1945 dollars.

A couple of years later, an algorithm was developed, the *simplex* algorithm, that could find the absolutely best solution. Since a computer was not available to carry out the algorithm, people and desk calculators were used: finding the solution requires 120 person-days of effort. The best solution used wheat flour, beef liver, cabbage, spinach, and dried navy beans, and achieved a cost of $39.69, twenty-four cents less than Stigler's solution.

A solution in found in 1998, which reflects more recent understanding of nutritional needs and more recent prices, is as follows: 412.45 cups of wheat flour, 587.65 cups of rolled oats, 6095.5 ounces of milk, 945.35 tablespoons of peanut butter, 945.35 tablespoons of lard, 2.6426 ounces of beef liver, 438 bananas, 85.41 oranges, 204.765 cups of shredded cabbage, 79.935 carrots, 140.16 potatoes, 108.405 cups of pork and beans. The annual cost: $536.55.

13.2 Formulating the diet problem as a linear program

How can such a problem be formulated? Introduce a variable for each food: $x_1, \ldots, x_{77}$. Variable x_j represents the number of units of food j in the diet. For example x_1 might denote the number of pounds of navy beans to be consumed per day. For each variable x_j, we have an associated cost c_j. For example, c_1 might be the cost in dollars of a single pound of navy beans.

The *objective function* is the cost $c_1 x_1 + \cdots + c_{77} x_{77}$, and the goal is to minimize the cost subject to the constrains.

To represent each nutritional requirement, we have a *linear inequality*, i.e. an inequality of the form

$$f(x_1, \ldots, x_{77}) \geq b$$

where $f : \mathbb{R}^{77} \rightarrow \mathbb{R}$ is a linear function. The function can be written as $f(x_1, \ldots, x_{77}) = a_1x_1 + \ldots + a_{77}x_{77}$, so the constraint has the form

$$a_1x_1 + \ldots + a_{77}x_{77} \geq b$$

Suppose we want to represent the requirement that someone take in 2000 calories a day. The number a_1 should be the number of calories in a pound of navy beans, ..., the number a_{77} should be the number of calories in one unit of food 77, and b should be 2000.

In this context, a linear inequality is called a linear *constraint* because it constrains the solution. We have similar constraints for calcium, Vitamin A, Riboflavin, ascorbic acid....

It turns out that the formulation so far is insufficient, because it doesn't prevent the solution from including *negative* amounts of some of the foods. No problem: just add an additional linear constraint for each variable x_j:

$$x_j \geq 0$$

The set of linear constraints can be summarized by a single constraint on the product of a matrix A with the vector $\boldsymbol{x} = (x_1, \ldots, x_{77})^T$:

$$A\boldsymbol{x} \geq \boldsymbol{b}$$

Each row of A, together with the corresponding entry of $\boldsymbol{b}$, is a single linear constraint. Constraint i requires that the dot-product of row i of A with $\boldsymbol{x}$ is at least entry i of b: $\boldsymbol{a}_i \cdot \boldsymbol{x} \geq b_i$. There is a row for each nutrient (corresponding to the constraint that the diet contain enough of that nutrient), and a row for each food (corresponding to the constraint that the amount of that food must be nonnegative).

The objection function $c_1x_1 + \cdots + c_{77}x_{77}$ can be written as the dot-product $\boldsymbol{c} \cdot \boldsymbol{x}$. So we can summarize the entire thing as

$$\min \boldsymbol{c} \cdot \boldsymbol{x} \text{ subject to}$$
$$A\boldsymbol{x} \geq b$$

The above is an example of a *linear program.* (Not a program in the computer-programming sense.)

13.3 The origins of linear programming

George Danzig wrote his dissertation on two famous unsolved problems in mathematical statistics. How did he come to solve these problems? He walked into class late and copied down the problems he saw on the blackboard. He thought they were homework problems, and solved them.

After receiving his Ph.D., Dantzig was looking for a job. It happened to be during World War II. He recalled, *"in order to entice me to not take another job, my Pentagon colleagues...challenged me to see what I could do to mechanize the planning process. I was asked to find a way to more rapidly compute a time-staged deployment, training and logistical supply program.... The military refer to their various plans or proposed scheduled of training, logistical supply and deployment of combat units as a* program*...the term 'program' was used for linear programs long before it was used as the set of instructions used by a computer to solve problems."* Working for the military, Dantzig developed the notion of linear programs and his algorithm for it, the *simplex* algorithm. It was kept secret until shortly after the war ended. (In Russia, Leonid Kantorovich had come up with the notion shortly before Dantzig.)

Dantzig talked to von Neumann about this new idea. *"I remember trying to describe to von Neumann (as I would to an ordinary mortal) the Air Force problem. I began with the formulation of the linear programming model in terms of activities and items, etc. He did something which I believe was uncharacteristic of him. 'Get to the point,' he snapped at me impatiently....I said to myself, 'OK, if he wants a* quickie*, that's what he'll get.' In under one minute I slapped on the blackboard a geometric and algebraic version of the problem. Von Neumann stood up and*

said, 'Oh, that!' Then, for the next hour and a half, he proceeded to give me a lecture on the mathematical theory of linear programs.

At one point, seeing me sitting there with my eyes popping and my mouth open.... von Neumann said 'I don't want you to think I am pulling all this out of my sleeve on the spur of the moment like a magician. I have recently completed a book with Oscar Morgenstern on the theory of games. What I am doing is conjecturing that the two problems are equivalent. The theory that I am outlining is an analogue to the one we have developed for games."'

Dantzig eventually shared his ideas with the mathematical community: *"....There was a meeting...in Wisconsin...attended by well-known statisticians and mathematicians like Hotelling and von Neumann.... I was a young unknown and I remember how frightened I was with the idea of presenting for the first time to such a distinguished audience the concept of linear programming.*

After my talk, the chairman called for discussion. For a moment there was the usual dead silence; then a hand was raised. It was Hotelling's.... I must hasten to explain that Hotelling was fat. He used to love to swim in the ocean and when he did, it is said that the level of the ocean rose perceptibly. This huge whale of a man stood up in the back of a room, his expressive fat face took on one of those all-knowing smiles we all know so well. He said: 'But we all known the world is nonlinear.' *Having uttered this devastating criticism of my model, he majestically sat down. And there I was, a virtual unknown, frantically trying to compose a proper reply.*

"Suddenly another hand in the audience was raised. It was von Neumann. 'Mr. Chairman, Mr. Chairman,' he said, 'if the speaker doesn't mind, I would like to reply for him.' Naturally I agreed. von Neumann said: 'The speaker titled his talk "linear programming" and carefully stated his axioms. If you have an application that satisfies the axioms, well, use it. If it does not, then don't,' and he sat down."

Dantzig and von Neumann have been vindicated. Linear programs are remarkably useful in formulating a wide variety of problems. They often come up in resource-allocation problems, such as the one we just discussed. But their application is much broader.

13.3.1 Terminology

Consider the LP $\min\{\boldsymbol{c}\boldsymbol{x} \;:\; A\boldsymbol{x} \geq \boldsymbol{b}\}$.

- A vector $\hat{\boldsymbol{x}}$ that satisfies the constraints, i.e. for which $A\hat{\boldsymbol{x}} \geq \boldsymbol{b}$, is said to be a *feasible* solution to the LP.
- The linear program is said to be *feasible* if there exists a feasible solution.
- The *value* of a feasible solution $\hat{\boldsymbol{x}}$ is $\boldsymbol{c}\hat{\boldsymbol{x}}$.
- The *value* of the linear program is the minimum value of a feasible solution (since the linear program is a minimization LP).
- The feasible solution $\hat{\boldsymbol{x}}$ is said to be an *optimal* solution if its value is that of the LP, i.e. if $\hat{\boldsymbol{x}}$ achieves the minimum.
- The linear program is said to be *unbounded* if it is feasible but that there is no minimum—this happens if, for any number t, there is a feasible solution whose value is less than t.

These definitions can be adapted also to maximization linear programs, e.g. the value of $\max\{\boldsymbol{c}\boldsymbol{x} \;:\; A\boldsymbol{x} \leq \boldsymbol{b}\}$.

13.3.2 Linear programming in different forms

There are several ways to state the linear-programming problem.

Minimization/Maximization I used the form

$$\min\{\boldsymbol{c} \cdot \boldsymbol{x} \ : \ A\boldsymbol{x} \geq \boldsymbol{b}\} \tag{13.1}$$

Suppose a linear-programming problem is given in form (13.1). Let $\boldsymbol{c}_- = -\boldsymbol{c}$. Then minimizing $\boldsymbol{c} \cdot \boldsymbol{x}$ is equivalent to maximizing $\boldsymbol{c}_- \cdot \boldsymbol{x}$, so the problem can be written instead as

$$\max\{\boldsymbol{c}_- \cdot \boldsymbol{x} \ : \ A\boldsymbol{x} \geq b\} \tag{13.2}$$

The *value* of the linear program is not the same, but the same solution $\boldsymbol{x}$ achieves the best value, whichever criterion is used to define "best".

Greater than or equal to/Less than or equal to Similarly, we use $\leq$ in the linear constraints, but we could use $\geq$. Let $A_- = -A$ and let $\boldsymbol{b}_= - \boldsymbol{b}$. Then the constraints $A\boldsymbol{x} \leq b$ are equivalent to the constraints $A_-\boldsymbol{x} \geq \boldsymbol{b}_-$, so the problem can be further rewritten as

$$\max\{\boldsymbol{c}_- \cdot \boldsymbol{x} \ : \ A_-\boldsymbol{x} \geq \boldsymbol{b}_-\} \tag{13.3}$$

Allowing equality constraints It is also possible to require some of the constraints to be linear *equalities* instead of linear inequalities. The equality constraint $\boldsymbol{a} \cdot \boldsymbol{x} = b$ is equivalent to the pair of inequality constraints $\boldsymbol{a} \cdot \boldsymbol{x} \leq b$ and $a \cdot \boldsymbol{x} \geq b$.

No strict inequalities! Strict inequalities (e.g. $\boldsymbol{a} \cdot \boldsymbol{x} > b$) are *not* allowed.

13.3.3 Integer linear programming

Note that there is nothing in a linear program $\min\{\boldsymbol{c} \cdot \boldsymbol{x} \ : \ A\boldsymbol{x} \geq \boldsymbol{b}\}$ that requires that the variables take on integer values. Moreover, there is no convenient way to impose that requirement using linear constraints. Linear programming often yields solutions in which the variables are assigned fractional values. (Recall the 1998 solution to the diet problem, which included, for example, 2.6426 ounces of beef liver.)

However, there are classes of linear programs that in fact are guaranteed to produce solutions that are integral. The analysis of such linear programs uses linear algebra but is beyond the scope of this book.

Moreover, people have studied the field of *integer linear programming*, in which one is allowed to add integrality constraints. Such programs are generally *much* more difficult to solve computationally. In fact, integer linear programming is NP-hard. However, ordinary (fractional) linear programming is an important tool in integer linear programming, indeed the most important. Also, the field of *approximation algorithms* has developed to find integer solutions that are nearly optimal. It relies heavily on the theory of linear programming.

13.4 Geometry of linear programming: polyhedra and vertices

Let's return to the diet problem. In order that I can draw the situation, we will consider only two kinds of food: lard and rice. Let x be the number of pounds of lard, and let y be the number of pounds of rice. The nutritional requirements are expressed by the constraints $10x + 2y \geq 5$, $x + 2y \geq 1$, $x + 8y \geq 2$. The cost, which I want to minimize, is 13 cents per bound of lard, and 8 cents per bound of rice. Thus the objective function is $13x + 8y$.

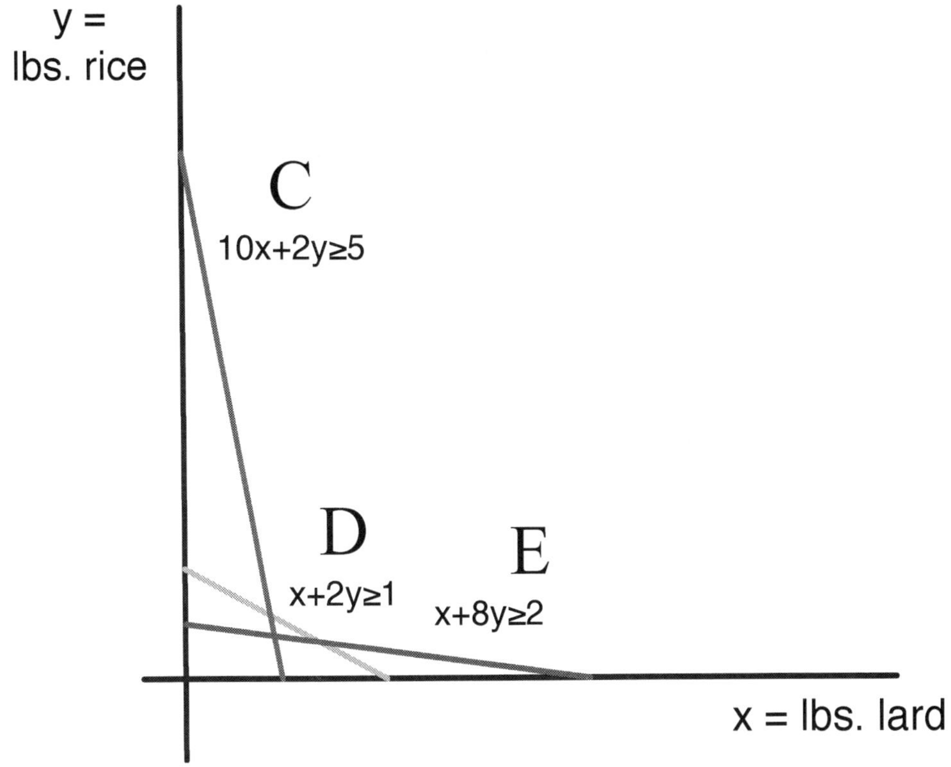

There are two other constraints: $x \geq 0$ and $y \geq 0$.

Consider a linear constraint. It divides the space $\mathbb{R}^2$ into two halves, which are called (naturally) *half-spaces*. One half-space is allowed by the contraint, and the other is forbidden.

More generally, in the space $\mathbb{R}^n$, a vector $\boldsymbol{a}$ and a scalar β determine a half-space $\{\boldsymbol{x} \in \mathbb{R}^n : \boldsymbol{a} \cdot \boldsymbol{x} \geq \beta\}$

For example, consider the constraint $y \geq 0$. One half-space (the allowed one) is the part above the x-axis (including the x-axis itself), and the other half-space is the part below (again including the x-axis).

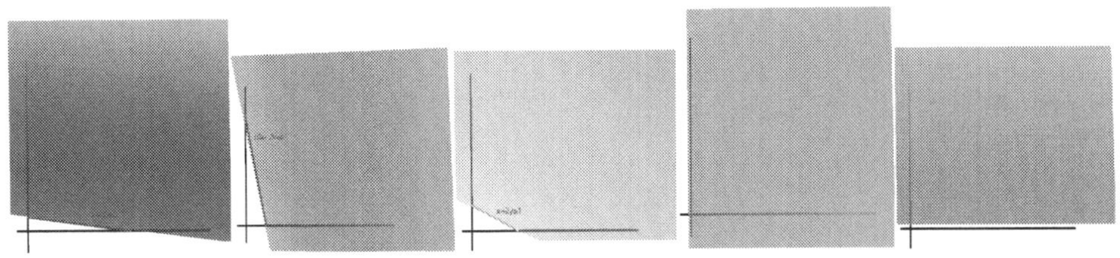

When you have several linear constraints, each one defining an allowed half-space, the feasible region is the intersection of these half-spaces. The intersection of a finite set of half-spaces is called

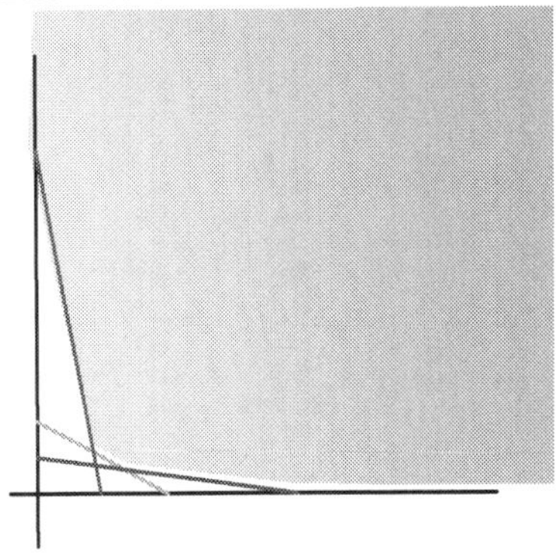

a *polyhedron*.

Usually we think of a polyhedron as a three-dimensional, regular object, such as a dodecahedron.

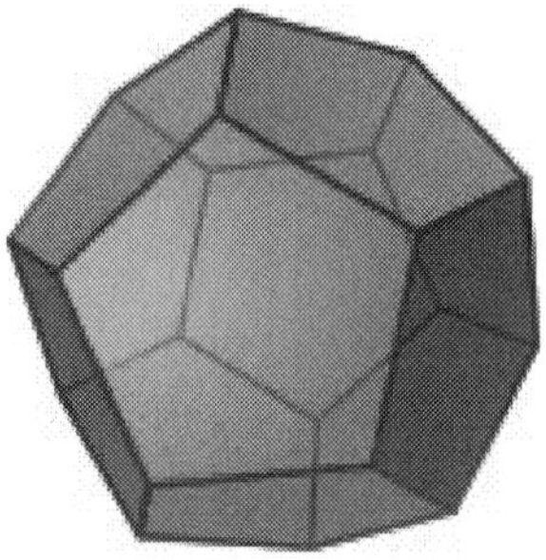

However, the term is used for nonregular, high-dimensional structures—even infinite structures. Note that our rice-and-lard polyhedron is two-dimensional but infinite.

The surface of a a finite three-dimensional polyhedron has *vertices* (which are points, i.e. zero-dimensional) and *edges* (which are line segments, i.e. one-dimensional), and *faces*, which are two-dimensional. Most (but not all) higher-dimensional polyhedra also have vertices and edges.

Quiz 13.4.1: Give an example of a polyhedron that has no vertices.

Answer

The n-dimensional polyhedron consisting of a single half-space has no vertices if $n \geq 2$. Even simpler, the polyhedron in $\mathbb{R}^n$ defined by the intersection of the empty set of half-spaces is all of $\mathbb{R}^n$.

Let $A\boldsymbol{x} \geq \boldsymbol{b}$ be a system of linear inequalities, and suppose A is $m \times n$. Then the system consists of m linear inequalities

$$\boldsymbol{a}_1\boldsymbol{x} \geq b_1, \ldots, \boldsymbol{a}_m\boldsymbol{x} \geq b_m$$

where $\boldsymbol{a}_1, \ldots, \boldsymbol{a}_m$ are the rows of A, and $b_1, \ldots, b_m$ are the entries of $\boldsymbol{b}$.

Definition 13.4.2: A *subsystem of linear inequalities* is the system formed by a subset of these inequalities.

For example, the first three inequalities form a subsystem, or the first and the last, or all of them but one, or all of them, or none of them. We can write a subsystem as $A_\square\boldsymbol{x} \geq \hat{\boldsymbol{b}}_\square$, where $A_\square$ consists of a subset of rows of A, and $\hat{\boldsymbol{b}}$ consists of the corresponding entries of $\boldsymbol{b}$.

Definition 13.4.3: W say a vector $\hat{\boldsymbol{x}}$ satisfies an inequality $\boldsymbol{a} \cdot \boldsymbol{x} \geq b$ *with equality* if $\boldsymbol{a} \cdot \hat{\boldsymbol{x}} = b$.

The singular of `vertices` is *vertex*.

Definition 13.4.4: A vector $\boldsymbol{v}$ in the polyhedron $P = \{\boldsymbol{x} : A\boldsymbol{x} \geq \boldsymbol{b}\}$ is a *vertex* of P if there is a subsystem $A_\square\boldsymbol{x} \geq \hat{\boldsymbol{b}}_\square$ of $A\boldsymbol{x} \geq \boldsymbol{b}$ such that $\boldsymbol{v}$ is the only solution to the matrix equation $A_\square\boldsymbol{x} = \hat{\boldsymbol{b}}_\square$.

Let n be the number of columns.

Lemma 13.4.5: A vector $\boldsymbol{v}$ in P is a vertex iff it satisfies n linearly independent linear inequalities *with equality*.

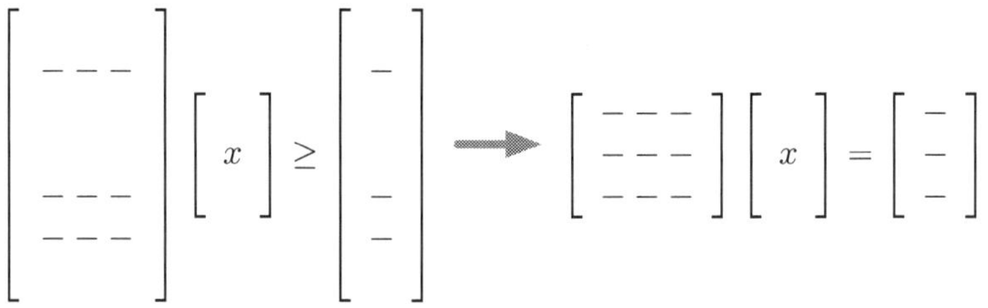

Consider our simplified diet problem. The point that satisfies $x + 8y - 2, x + 2y = 1$ is a vertex.

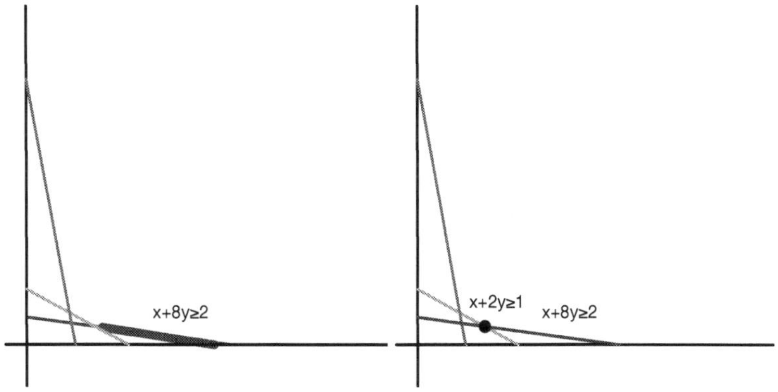

Other vertices are: the point that satisfies $10x + 2y = 5, y = 0$, the point that satisfies $x + 8y = 2, x = 0$, and the origin.

What about the point that satisfies $10x + 2y = 5, x + 2y = 1$? This point is not feasible, so does not constitute a vertex.

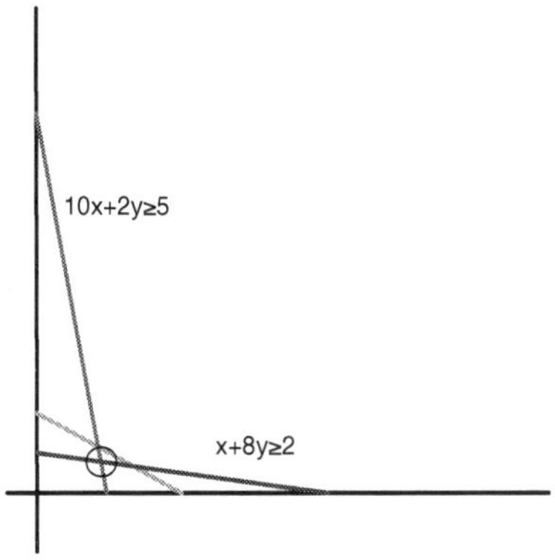

13.5 There is an optimal solution that is a vertex of the polyhedron

Here is the nice thing about vertices. In looking for a solution to a linear program, we can (usually) restrict our attention to vertices.

Theorem 13.5.1: Consider a linear program

$$\min\{\boldsymbol{c}\boldsymbol{x} \;:\; A\boldsymbol{x} \geq \boldsymbol{b}\} \tag{13.4}$$

Suppose that the columns of A are linearly independent and that the linear program has a value. Then there is a vertex of the corresponding polyhedron $P = \{\boldsymbol{x} \;:\; A\boldsymbol{x} \geq \boldsymbol{b}\}$ that is an optimum solution.

The good news is that many linear programs that arise in practice have the property that the columns of A are linearly independent. The even better news is that there are ways of transforming linear programs into equivalent linear programs that do have this property.

13.6 An enumerative algorithm for linear programming

The theorem suggests an approach to finding an optimum solution to a linear program $\min\{\boldsymbol{cx} : A\boldsymbol{x} \geq \boldsymbol{b}\}$ satisfying the conditions of Theorem 13.5.1: try all vertices. I call this algorithm the *enumerative* algorithm because it enumerates the vertices.

In order to enumerate the vertices, the algorithm enumerates all n-subsets of the rows of A, where n is the number of columns of A.

Here is a more careful description of the enumerative algorithm.

- For each n-element subset $R_\square$ of the rows of A, if the matrix $A_\square$ formed by these rows is invertible,
  - let $\boldsymbol{v}$ be the solution to the corresponding system $A_\square \boldsymbol{x} = \boldsymbol{b}_\square$, and
  - see if $\boldsymbol{v}$ satisfies all the constraints $A\boldsymbol{x} \geq \boldsymbol{b}$. If so, it is a vertex.
- Compute the objective value at each vertex, and output the vertex for which the objective value is greatest.

Example 13.6.1: Consider the two inequalities $10x + 2y \geq 5$ and $x + 2y \geq 1$. We turn these into equalities $10x + 2y = 5, x + 2y = 1$ and solve, obtaining $x = 4/9$ and $y = 5/18$. Thus (x, y) is the vertex at the intersection of the lines $10x + 2y = 5, x + 2y = 1$.

This algorithm requires only a finite number of steps. Suppose there are m constraints. The number of n-element subsets is $\binom{m}{n}$, which is very big if m and n are big (and m is not very close to n). Since the algorithm is required to consider all such subsets, it is hopelessly slow if m is not very small.

13.7 Introduction to linear-programming duality

One of the ideas that Dantzig learned from von Neumann on that fateful day was *linear-programming duality*. Long ago we saw:

> The maximum size of a linearly independent set of vectors in V equals the minimum size of a set of vectors that span V.

What we discuss now is a similar relationship between a minimization problem and a maximization problem.

Corresponding to a linear program

$$\min\{\boldsymbol{c} \cdot \boldsymbol{x} \ : \ A\boldsymbol{x} \geq \boldsymbol{b}\} \tag{13.5}$$

there is another linear program

$$\max\{\boldsymbol{b} \cdot \boldsymbol{y} \ : \ \boldsymbol{y}^T A = \boldsymbol{c}, \boldsymbol{y} \geq 0\} \tag{13.6}$$

The second linear program is called the *dual* of the first. In this context, the first would then be called the *primal* linear program.

The LP Duality Theorem states that the value of the primal LP equals the value of the dual LP. As is usually the case, proving equality of two quantities involves proving that each is no bigger than the other. Here we prove that the value of the minimization LP is no less than that of the maximization LP. This is called *weak duality.*

Lemma 13.7.1 (Weak Duality): The value of the minimization LP (the primal) is greater than than or equal to the value of the maximization LP (the dual).

Proof

Let $\hat{\boldsymbol{x}}$ and $\hat{\boldsymbol{y}}$ be any feasible solutions to the primal and dual linear programs, respectively. Since $\hat{\boldsymbol{y}}A = \boldsymbol{c}$, we know

$$\boldsymbol{c} \cdot \hat{\boldsymbol{x}} = (\hat{\boldsymbol{y}}^T A)\hat{\boldsymbol{x}} = \hat{\boldsymbol{y}}^T (A\hat{\boldsymbol{x}})$$

The last expression is a sum

$$\sum_i \hat{\boldsymbol{y}}[i]\,(A\hat{\boldsymbol{x}})[i] \tag{13.7}$$

where we use $\hat{\boldsymbol{y}}[i]$ and $(A\hat{\boldsymbol{x}})[i]$ to refer to entry i of $\hat{\boldsymbol{y}}$ and $A\hat{\boldsymbol{x}}$ respectively. Since $\hat{\boldsymbol{x}}$ is a feasible solution to the primal LP, we know $A\hat{\boldsymbol{x}} \geq \boldsymbol{b}$. This means that, for each i,

$$(A\hat{\boldsymbol{x}})[i] \geq \boldsymbol{b}[i]$$

Multiplying both sides of this inequality by $\boldsymbol{y}[i]$ and using the fact that $\boldsymbol{y}[i] \geq 0$, we obtain

$$\boldsymbol{y}[i]\,(A\hat{\boldsymbol{x}})[i] \geq \boldsymbol{y}[i]\,\boldsymbol{b}[i]$$

Summing this inequality over all labels i, we obtain

$$\sum_i \hat{\boldsymbol{y}}[i]\,(A\hat{\boldsymbol{x}})[i] \geq \sum_i \boldsymbol{y}[i]\,\boldsymbol{b}[i]$$

The right-hand side is $\boldsymbol{y} \cdot \boldsymbol{b}$. We have therefore proved that

$$\boldsymbol{c} \cdot \hat{\boldsymbol{x}} \geq \boldsymbol{y} \cdot \boldsymbol{b}$$

□

Proving weak duality is part way to proving the full duality theorem. Suppose we could show that there exists a primal feasible solution $\hat{\boldsymbol{x}}$ and a dual feasible solution $\hat{\boldsymbol{y}}$ such that the corresponding objective function values $\boldsymbol{c} \cdot \hat{\boldsymbol{x}}$ and $\hat{\boldsymbol{y}} \cdot \boldsymbol{b}$ are equal.

- By weak duality, we know $\boldsymbol{c} \cdot \hat{\boldsymbol{x}}$ can be no less than $\hat{\boldsymbol{y}} \cdot \boldsymbol{b}$, so it follows that $\boldsymbol{c} \cdot \hat{\boldsymbol{x}}$ is in fact the minimum achievable—that this value is the value of the minimization LP.
- Similarly, we know $\hat{\boldsymbol{y}} \cdot \boldsymbol{b}$ can be no greater than $\boldsymbol{c} \cdot \hat{\boldsymbol{x}}$, so $\hat{\boldsymbol{y}} \cdot \boldsymbol{b}$ is the maximum.

Thus merely showing the existence of primal and dual feasible solutions whose values are equal implies that those solutions are in fact optimal.

In order to derive a condition under which these values are equal, we more carefully examine the argument showing that $\boldsymbol{y}[i]\,(A\hat{\boldsymbol{x}})[i] \geq \boldsymbol{y}[i]\,\boldsymbol{b}[i]$.

Lemma 13.7.2 (Complementary Slackness): Suppose $\hat{\boldsymbol{x}}$ and $\hat{\boldsymbol{y}}$ are feasible solutions for the maximization and minimization LPs, respectively. If for each i either $(A\hat{\boldsymbol{x}})[i] = \boldsymbol{b}[i]$ or $\boldsymbol{y}[i] = 0$ then the values of these solutions are equal and the solutions are optimal.

Proof

In the proof of weak duality, we showed that

$$\boldsymbol{c} \cdot \hat{\boldsymbol{x}} = \sum_i \hat{\boldsymbol{y}}[i]\,(A\hat{\boldsymbol{x}})[i] \tag{13.8}$$

and we showed that, for each i,

$$\boldsymbol{y}[i](A\hat{\boldsymbol{x}})[i] \geq \boldsymbol{y}[i]\,\boldsymbol{b}[i] \tag{13.9}$$

Note that

- if $(A\hat{\boldsymbol{x}})[i] = \boldsymbol{b}[i]$ then the left-hand side and the right-hand side are in fact equal, and
- if $\boldsymbol{y}[i] = 0$ then the left-hand side and the right-hand side are both zero, so again they are equal.

In either of these two cases,

$$\boldsymbol{y}[i](A\hat{\boldsymbol{x}})[i] = \boldsymbol{y}[i]\boldsymbol{b}[i] \tag{13.10}$$

Thus if for each i either $(A\hat{\boldsymbol{x}})[i] = \boldsymbol{b}[i]$ or $\boldsymbol{y}[i] = 0$ then, by summing Equation 13.10 over all i, we obtain

$$\sum_i \boldsymbol{y}[i](A\hat{\boldsymbol{x}})[i] = \sum_i \boldsymbol{y}[i]\boldsymbol{b}[i]$$

which by Equation 13.8 implies that $\boldsymbol{c} \cdot \hat{\boldsymbol{x}} = \boldsymbol{y} \cdot \boldsymbol{b}$. □

The complementary-slackness conditions give us a technique for proving optimality. We will use this technique in the algorithm we present next. This algorithm plays a dual (heh) role: it gives us an effective way to solve linear programs, and it essentially proves (strong) duality.

I use the hedge word *essentially* because there is another possibility. It could be that the primal or dual is infeasible or unbounded. If the simplex algorithm discovers that the primal LP is unbounded, it gives up.

13.8 The simplex algorithm

We present an algorithm due to George Dantzig, the *simplex* algorithm, that iteratively examines vertices of the polyhedron to decide which is the best. There are nasty linear programs that make the simplex algorithm visit all the vertices, but for most linear programs the number of vertices visited is not too large, and the algorithm is consequently quite practical.

The simplex algorithm has been studied and refined over many years, and there are many tricks to make it run quickly. In this class, we will just study a very basic version—not one that you would use in practice.

The simplex algorithm finds an optimal vertex using the following iterative approach. Once it has examined one vertex $\boldsymbol{v}$ and is ready to examine another, the next vertex it visits must share an edge with $\boldsymbol{v}$. Moreover, and most important, the algorithm chooses the next vertex in a way that guarantees that the objective function's value on the new vertex is no worse than its value on the current vertex. Ideally the value of the new vertex is in fact strictly better than the value at the old vertex; this ensures that the algorithm has made progress in that iteration. Though this tends to be true, we will see it is not guaranteed due to the way the algorithm represents the current vertex.

This strategy raises a difficulty: What if the polyhedron associated with the linear program does not have vertices? We'll sidestep this issue by assuming the polyhedron does have vertices. (It is sufficient to assume that the columns of A are linearly independent.)

Another difficulty: how does the algorithm find a vertex to start with? There is a technique to doing this; we'll address this later.

13.8.1 Termination

How does the algorithm know when it is done? The trick is to use linear-programming duality.

To the linear program $\min\{\boldsymbol{c} \cdot \boldsymbol{x} \ : \ A\boldsymbol{x} \geq \boldsymbol{b}\}$ there corresponds a dual linear program, $\max\{\boldsymbol{y} \cdot \boldsymbol{b} \ : \ \boldsymbol{y}^T A = \boldsymbol{c}, \boldsymbol{y} \geq 0\}$. We proved weak duality, which stated that if $\hat{\boldsymbol{x}}$ and $\hat{\boldsymbol{y}}$ are feasible solutions to the primal and dual linear programs respectively, and if $\boldsymbol{c}\hat{\boldsymbol{x}} = \hat{\boldsymbol{y}}^T\boldsymbol{b}$, then $\hat{\boldsymbol{x}}$ and $\hat{\boldsymbol{y}}$ are *optimal* solutions. At each step, our version of the simplex algorithm will derive a feasible solution $\hat{\boldsymbol{x}}$ to the original linear program and a vector $\hat{\boldsymbol{y}}$ such that $\boldsymbol{c}\hat{\boldsymbol{x}} = \hat{\boldsymbol{y}}\boldsymbol{b}$. If $\hat{\boldsymbol{y}}$ is a feasible solution then it and $\hat{\boldsymbol{x}}$ are optimal. If $\hat{\boldsymbol{y}}$ is not a feasible solution, the simplex algorithm will take another step.

13.8.2 Representing the current solution

Let R be the set of row-labels of A (and the set of labels for $\boldsymbol{b}$). Let n be the number of columns. Like the enumerative algorithm, the simplex algorithm iterates over vertices by iterating over n-row *subsystems* $A_\square \boldsymbol{x} \geq \boldsymbol{b}_\square$ that define vertices (i.e. such that the unique solution to $A_\square \boldsymbol{x} = \boldsymbol{b}_\square$ is a vertex). The algorithm keeps track of the current subsystem using a variable $R_\square$ whose value is an n-element subset of R.

There is not a perfect correspondence, however, between subsystems and vertices.

- For each subsystem $A_\square \boldsymbol{x} \geq \boldsymbol{b}_\square$ such that the rows of $A_\square$ are linearly independent, there is a unique vector $\hat{\boldsymbol{x}}$ satisfying the corresponding matrix equation $A_\square \boldsymbol{x} = \boldsymbol{b}_\square$. However, $\hat{\boldsymbol{x}}$ might not satisfy the other linear inequalities, and in this case fails to be a vertex.
- In some linear programs, several different subsystems give rise to the same vertex $\boldsymbol{v}$ This phenomenon is called *degeneracy*. Geometrically, this means that $\boldsymbol{v}$ is at the boundary of more than n half-spaces. For example, in three dimensions, the vertex would be at the intersection of more than three planes.

We partially sidestep the first issue by requiring that the input to our simplex implementation include a set $R_\square$ that *does* correspond to a vertex. The simplex algorithm will do the rest by ensuring that whenever it takes a step, it maintains the invariant that $R_\square$ corresponds to a vertex.

Because of degeneracy, several iterations of the simplex algorithm might involve several different sets $R_\square$ that all correspond to the same vertex. For this reason, some care is required to ensure that the simplex algorithm does not get stuck in a loop.

13.8.3 A pivot step

An iteration of the simplex algorithm is called a *pivot* step. We now describe a pivot step in detail.

Assumes $R_\square$ is a set of n row-labels of A, that the corresponding rows of A are linearly independent, and that the corresponding vector $\hat{\boldsymbol{x}}$ is a vertex.

Extract the subsystem Let $A_\square$ be the submatrix of A consisting of the rows whose labels are in $R_\square$. Let $\boldsymbol{b}_\square$ be the subvector of $\boldsymbol{b}$ consisting of the entries whose labels are in $R_\square$.

```
A_square = Mat((R_square, A.D[1]), {(r,c):A[r,c] for r,c in A.f if r in R_square})
b_square = Vec(R_square, {k:b[k] for k in R_square})
```

Find the location of the current vertex Solve the system $A_\square \boldsymbol{x} = \boldsymbol{b}_\square$ to obtain the current vertex $\hat{\boldsymbol{x}}$.

```
x = solve(A_square, b_square)
```

Note that, for every $r \in R_\square$,

$$(A\hat{\boldsymbol{x}})[r] = \boldsymbol{b}[r] \tag{13.11}$$

Find a *possibly* feasible solution to the dual LP Solve the system $\boldsymbol{y}_\square A_\square = \boldsymbol{c}$. Let $\hat{\boldsymbol{y}}_\square$ be the solution.

```
y_square = solve(A_square.transpose(), c)
```

Note that $R_\square$ is the label-set of $\hat{\boldsymbol{y}}_\square$. Let $\hat{\boldsymbol{y}}$ be the vector with domain R derived from $\hat{\boldsymbol{y}}_\square$ by putting zeroes in for each label in R that is not in the domain of $R_\square$. In Python, we would write

```
y = Vec(R, y_square.f) # uses sparsity convention
```

Therefore, for every r in R but not in $R_\square$,

$$\boldsymbol{y}[r] = 0 \tag{13.12}$$

If every entry of $\hat{\boldsymbol{y}}$ is nonnegative then $\hat{\boldsymbol{y}}$ is a feasible solution to the dual linear program. Moreover, in this case by Equations 13.11 and 13.12 and the Complementary Slackness Lemma, $\hat{\boldsymbol{y}}$ and $\hat{\boldsymbol{x}}$ are optimal solutions to their respective linear programs. The simplex algorithm is done.

```
if min(y.values()) >= 0: return ('OPTIMUM', x) #found optimum!
```

Otherwise, we must select a direction in which to take a step. Let r^- be a label such that entry r^- of $\hat{\boldsymbol{y}}$ is negative.[1]

```
R_leave = {i for i in R if y[i] < 0} #labels at which y is negative
r_leave = min(R_leave, key=hash) #choose first label where y is negative
```

Let $\boldsymbol{d}$ be a vector whose label-set is R and such that entry r^- is 1 and all other entries are zero. Let $\boldsymbol{w}$ be the unique solution to $A_\square \boldsymbol{x} = \boldsymbol{d}$.

```
d = Vec(R_square, {r_leave:1})
w = solve(A_square, d)
```

Moving in the direction $\boldsymbol{w}$ will decrease the objective function: for any positive number δ,

$$\boldsymbol{c}\cdot(\hat{\boldsymbol{x}}+\delta\,\boldsymbol{w})-\boldsymbol{c}\cdot\hat{\boldsymbol{x}} = \delta(\boldsymbol{c}\cdot\boldsymbol{w}) = \delta\,(\hat{\boldsymbol{y}}^T A)\cdot\boldsymbol{w} = \delta\,\hat{\boldsymbol{y}}^T(A\boldsymbol{w}) = \delta\,\hat{\boldsymbol{y}}^T\boldsymbol{d} = \delta\sum_i \boldsymbol{y}[i]\boldsymbol{d}[i] = \delta\,\boldsymbol{y}[r^-] < 0$$

Moreover, for any row $\boldsymbol{a}_r$ of $A_\square$ such that $r \neq r^-$, since $d_r = 0$,

$$\boldsymbol{a}_r \cdot \boldsymbol{w} = 0$$

so

$$\boldsymbol{a}_r \cdot (\hat{\boldsymbol{x}} + \delta\,\boldsymbol{w}) = \boldsymbol{a}_r \cdot \hat{\boldsymbol{x}} + \delta 0 = \boldsymbol{a}_r \cdot \hat{\boldsymbol{x}} = b_r$$

so the corresponding inequality $\boldsymbol{a}_r \cdot \boldsymbol{x} \geq b_r$ remains tight. It remains to choose a value for δ.

Let R^+ be the set of labels of rows $\boldsymbol{a}_i$ of A for which $\boldsymbol{a}_i \cdot \boldsymbol{w} < 0$.

```
Aw = A*w # compute once because we use it many times
R_enter = {r for r in R if Aw[r] < 0}
```

If R^+ is the empty set then the linear program's objective value is infinity. The simplex algorithm is done.

```
if len(R_enter)==0: return ('UNBOUNDED', None)
```

Otherwise, for each r in R^+, let

$$\delta_r = \frac{\boldsymbol{a}_r \cdot \hat{\boldsymbol{x}} - \boldsymbol{b}_r}{\boldsymbol{a}_r \cdot \boldsymbol{w}}$$

Let $\delta = \min\{\delta_r \,:\, r \in R^+\}$. Let r^+ be a label[2] such that $\delta_{r^+} = \delta$.

```
Ax = A*x # compute once because we use it many times
delta_dict = {r:(b[r] - Ax[r])/(Aw[r]) for r in R_enter}
delta = min(delta_dict.values())
r_enter = min({r for r in R_enter if delta_dict[r] == delta}, key=hash)[0]
```

Remove r^- from $R_\square$ and add r^+ to $R_\square$.

```
R_square.discard(r_leave)
R_square.add(r_enter)
```

[1] In order to avoid getting into an infinite loop, we require that r^- be the first such label in sorted order.
[2] In order to avoid getting into an infinite loop, we require that r^+ be the first such label in sorted order.

The simplex algorithm consists of a sequence of such pivot steps. Eventually,[3] simplex either finds optimal solution is found or discovers that the LP is unbounded.

The algorithm outlined above is given in procedures `simplex_step` and `optimize` in the module `simplex`. There is one significant difference: in order for the algorithm to work with floating-point arithmetic, we consider a number to be negative if it is "negative enough", less than, say, -10^{-10}.

13.8.4 Simple example

Let's walk through the mini-diet problem. The constraints are:

$$\begin{aligned}
\texttt{C}: 2*\text{rice} + 10*\text{lard} &\geq 5\\
\texttt{D}: 2*\text{rice} + 1*\text{lard} &\geq 1\\
\texttt{E}: 8*\text{rice} + 1\text{lard} &\geq 2\\
\texttt{rice-nonneg}: \text{rice} &\geq 0\\
\texttt{lard-nonneg}: \text{lard} &\geq 0
\end{aligned}$$

which can be written as $A\boldsymbol{x} \geq \boldsymbol{b}$ where A and $\boldsymbol{b}$ are as follows:

$A =$

	rice	lard
C	2	10
D	2	1
E	8	1
lard-nonneg	0	1
rice-nonneg	1	0

$\boldsymbol{b} =$

C	5
D	1
E	2
lard-nonneg	0
rice-nonneg	0

The objective function is $\boldsymbol{c} \cdot \boldsymbol{x}$ where $\boldsymbol{c} =$

rice	lard
1	1.7

We will use as the starting vertex the point for which the inequalities `E` and `rice-nonneg` are tight, i.e. the point satisfying the equations

$$\begin{aligned}
8*\text{rice} + 1*\text{lard} &= 2\\
1*\text{rice} &= 0
\end{aligned}$$

Thus $R_\square$ ={`E`, `rice-nonneg`}, and $A_\square$ and $\boldsymbol{b}_\square$ are

$A_\square =$

	rice	lard
E	8	1
rice-nonneg	1	0

$\boldsymbol{b}_\square =$

E	2
rice-nonneg	0

We solve the equation $A_\square \boldsymbol{x} = \boldsymbol{b}_\square$, and get the solution

$\hat{\boldsymbol{x}} =$

rice	lard
0.0	2.0

We solve the equation $\boldsymbol{y}_\square^T A = \boldsymbol{c}$ and obtain as the solution

$\hat{\boldsymbol{y}}_\square =$

rice-nonneg	E
-12.6	1.7

We fill in the remaining entries of $\hat{\boldsymbol{y}}$ with zeroes, obtaining the vector

$\hat{\boldsymbol{y}} =$

rice-nonneg	lard-nonneg	C	D	E
-12.6	0	0	0	1.7

We select the leaving constraint to be `rice-nonneg` since the corresponding entry of $\hat{\boldsymbol{y}}$ is negative. (Ordinarily, there could be more than one such entry.)

[3]The choice of r^- and r^+ can be shown to ensure that simplex never gets stuck in an infinite loop.

Now we select the direction $\boldsymbol{w}$ in which to move. Let $\boldsymbol{d}$ be the $R_\square$-vector whose only nonzero is a 1 in the entry corresponding to the leaving constraint. Let $\boldsymbol{w}$ be the vector such that $A_\square \boldsymbol{w} = \boldsymbol{d}$, namely: $\boldsymbol{w} =$

rice	lard
1.0	-8.0

Let us verify that replacing $\hat{\boldsymbol{x}}$ with $\hat{\boldsymbol{x}} + \delta \boldsymbol{w}$

1. improves the value of the objective function,
2. does not violate the leaving constraint, and
3. preserves the tightness of the other constraints in $R_\square$.

The change in the objective function is $\delta(\boldsymbol{c} \cdot \boldsymbol{w})$, which is $\delta(1 \cdot 1.0 + 1.7 \cdot -8)$, which is -12.6δ. Since the change is negative (for positive δ), the value of the objective function goes down; since we are trying to minimize the value of the objective function, this would be progress.

We chose $\boldsymbol{d}$ so that $\boldsymbol{a}_r \cdot \boldsymbol{w} = 1$ where $\boldsymbol{a}_r$ is the row of A corresponding to the leaving constraint. Therefore the change in the left-hand side of the constraint $\boldsymbol{a}_r \cdot \boldsymbol{x} \geq b_r$ is δ, so the left-hand increases, so the constraint gets looser as a result.

We chose $\boldsymbol{d}$ so that, for any other constraint $\boldsymbol{a}_r \boldsymbol{x} \leq b_r$ where $r \in R_\square$, $\boldsymbol{a}_r \cdot \boldsymbol{w} = 0$ so the change does not affect the left-hand side of such a constraint. Thus the other constraints in $R_\square$ remain tight.

Next, we find out which constraints *not* in $R_\square$ could become tight as a result of the change. For any such constraint $\boldsymbol{a}_r \cdot \boldsymbol{x} \leq b_r$, the left-hand side decreases only if $\boldsymbol{a}_r \cdot \boldsymbol{w}$ is negative. We therefore want to determine which constraints have this property. We compute $A\boldsymbol{w}$, obtaining

$A\boldsymbol{w} =$

rice-nonneg	lard-nonneg	C	D	E
1.0	-8.0	-78.0	-6.0	0.0

In this case, all the constraints not in $R_\square$ have the property. To find out which of them will become tight first (as we increase δ), we compute the ratios

$$\frac{\boldsymbol{b}[r] - (A\boldsymbol{x})[r]}{(A\boldsymbol{w})[r]}$$

for each such constraint. For each constraint, the corresponding ratio is the amount δ would have to be in order for the constraint to become tight. The ratios are:

```
{'C': 0.19, 'D': 0.17, 'lard-nonneg': 0.25}
```

Thus the constraint that would first become tight is the `D` constraint. Therefore this constraint should enter $R_\square$. We therefore remove `rice-nonneg` from $R_\square$ and add `D`. Now $R_\square =${`D`,`E`}.

Solving the new system $A_\square \boldsymbol{x} = \boldsymbol{b}_\square$ yields

$\hat{\boldsymbol{x}} =$

rice	lard
0.17	0.67

Solving the system $\boldsymbol{y}_\square^T A_\square = \boldsymbol{c}$ gives us

$\boldsymbol{y}_\square =$

D	E
2.1	-0.4

Filling in the other entries of $\hat{\boldsymbol{y}}$ with zeroes, we get

$\hat{\boldsymbol{y}} =$

rice-nonneg	lard-nonneg	C	D	E
0	0	0	2.1	-0.4

so the leaving constraint should be `E`.

The move vector $\boldsymbol{w}$ is

rice	lard
0.17	-0.33

The constraints that could become tight are `C` and `lard-nonneg`. To see which would become tight first, we calculate the corresponding ratios:

```
{'C': 0.67, 'lard-nonneg': 2.0}
```

and conclude that C should be the entering constraint. We update $R_\square$ by removing E and adding C. Now $R_\square$ consists of C and D. We solve $A_\square \boldsymbol{x} = \boldsymbol{b}_\square$, obtaining

$$\hat{\boldsymbol{x}} = \begin{array}{cc} \text{rice} & \text{lard} \\ \hline 0.28 & 0.44 \end{array}$$

We solve $\boldsymbol{y}_\square A = \boldsymbol{c}$, obtaining

$$\boldsymbol{y}_\square = \begin{array}{cc} \text{C} & \text{D} \\ \hline 0.13 & 0.37 \end{array}$$

and fill the remaining entries of $\hat{\boldsymbol{y}}$ with zeroes, obtaining

$$\hat{\boldsymbol{y}} = \begin{array}{ccccc} \text{rice-nonneg} & \text{lard-nonneg} & \text{C} & \text{D} & \text{E} \\ \hline 0 & 0 & 0.13 & 0.37 & 0 \end{array}$$

Since this is a nonnegative vector, $\hat{\boldsymbol{y}}$ is a feasible solution to the dual, so the primal and dual solutions (which have the same value, 1.03) are optimal for their respective linear programs.

13.9 Finding a vertex

You're all set. You've formulated your favorite problem as a linear program $\min\{\boldsymbol{c}\cdot\boldsymbol{x} : A\boldsymbol{x} \geq \boldsymbol{b}\}$, you've learned how simplex works, you even have working simplex code—all that's left is to run it. You look at the arguments to `optimize`: the matrix A, the right-hand-side vector $\boldsymbol{b}$, the objective function vector $\boldsymbol{c}$, and... what's this? The set $R_\square$ of labels of row-labels that specify the starting vertex? But you don't even know a single vector in the polyhedron $\{\boldsymbol{x} : A\boldsymbol{x} \geq \boldsymbol{b}\}$, much less a vertex!

Never fear. Simplex can help with that. The idea is to transform your linear program into a new, slightly larger linear program for which you can easily compute a vertex and whose solution gives you a vertex of the original. Suppose the matrix A is $m \times n$. Since we assume the columns of A are linearly independent, $m \geq n$.

The algorithm is illustrated in this diagram:

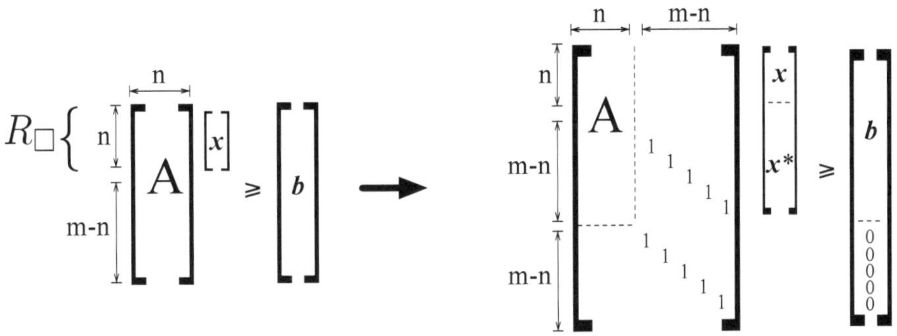

Let $R_\square$ be an n-element subset of the row-labels. Let $\hat{\boldsymbol{x}}$ be the vector that satisfies the corresponding constraints with equality.

```
A_square = Mat((R_square, A.D[1]),
              {(r,c):A[r,c] for r,c in A.f if r in R_square})
b_square = Vec(R_square, {k:b[k] for k in R_square})
x = solve(A_square, b_square)
```

(If $\hat{\boldsymbol{x}}$ happens to satisfy all of the constraints then it is a vertex, but this is very unlikely.) Now the algorithm creates $m - n$ new variables, one for each constraint whose label r is not in $R_\square$; for each new variable, the algorithm creates a new constraint that simply requires that variable to be nonnegative.

For a row-label r not in $R_\square$, we use r^* to denote the new label with which we identity the new variable and the new constraint. We use x_{r^*} to denote the new variable corresponding to contraint r. Then the new constraint is

$$r^* : \quad x_{r^*} \geq 0$$

The algorithm also adds the new variable x_{r^*} into constraint r: whereas before constraint r had the form

$$r : \quad \boldsymbol{a} \cdot \boldsymbol{x} \geq \boldsymbol{b}[r]$$

the algorithm changes it to

$$r : \quad \boldsymbol{a} \cdot \boldsymbol{x} + x_{r^*} \geq \boldsymbol{b}[r]$$

If the original constraint is not satisfied by $\hat{\boldsymbol{x}}$ ($\boldsymbol{a} \cdot \hat{\boldsymbol{x}}$ is smaller than $\boldsymbol{b}[r]$), the new constraint can be satisfied by assigning a big enough value to x_{r^*}. In fact, assigning exactly $\boldsymbol{b}[r] - \boldsymbol{a} \cdot \hat{\boldsymbol{x}}$ to x_{r^*} means that the new constraint is satisfied *with equality*. On the other hand, if the original constraint *was* satisfied by $\hat{\boldsymbol{x}}$ ($\boldsymbol{a} \cdot \hat{\boldsymbol{x}}$ is at least $\boldsymbol{b}[r]$) then the new constraint remains satisfied if we assign zero to x_{r^*}, so the constraint

$$r : \quad x_{r^*} \geq 0$$

is satisfied with equality.

Here is the code to create the new system of linear inequalities. It uses a procedure `new_name(r)` to obtain the label r^* from r, and a utility procedure `dict_union` to compute the union of dictionaries.

```
A_x = A*x
missing = A.D[0].difference(R_square) # set of row-labels not in R_square
extra = {new_name(r) for r in missing}
f = dict_union(A.f,
               {(r,new_name(r)):1 for r in missing},
               {(e, e):1 for e in extra})
A_with_extra = Mat((A.D[0].union(extra), A.D[1].union(extra)), f)
b_with_extra = Vec(b.D.union(extra), b.f) # use sparsity convention
```

Originally there were n variables, so $R_\square$ had n row-labels. The algorithm has added $m - n$ variables, so the algorithm must add $m - n$ row-labels to $R_\square$, getting a new set $R^*_\square$. The algorithm follows the logic outlined above:

- if constraint r is not satisfied by $\hat{\boldsymbol{x}}$ then include r in $R^*_\square$, and
- if constraint r *is* satisfied by $\hat{\boldsymbol{x}}$ then include r^* in $R^*_\square$.

The result is that, in the augmented linear program, the constraints whose labels are in $R^*_\square$ define a vertex. Here is the code:

```
new_R_square = R_square |
               {r if A_x[r]-b[r] < 0 else new_name(r) for r in missing}
```

The algorithm can use this vertex as the starting vertex and run simplex. And what is to be minimized? The goal of running simplex is to find a vertex in which all of the new variables x_{r^*} are set to zero. The objective function is therefore set to be the sum of all the new variables.

```
c = Vec(A.D[1].union(extra), {e:1 for e in extra})
answer= optimize(A_with_extra, b_with_extra, c, new_R_square)
```

If simplex finds a solution for which the objective value equals zero, the solution corresponds to a vertex of the original linear program.

```
basis_candidates=list(new_R_square | D[0])
R_square.clear()
R_square.update(set(basis_candidates[:n]))
```

If not, then the original system of linear inequalities must have no solutions.

The entire procedure `find_vertex` is included in module `simplex`.

13.10 Game theory

The goal of game theory is to model strategic decision-making. The topic is important, and very broadly applied: early on in military planning, more recently in biology, auctions, even Internet routing.

We will look at a very simple model:

- two players
- complete information
- deterministic games

In addition, we will assume that the game is very brief: each player has one move, and the players play simultaneously.

What does this have to do with familiar games?

A scientist, Jacob Bronowski, who was famous in my day for a TV series called *The Ascent of Man*, recalls a conversation with von Neumann:

> I naturally said to him, since I am an enthusiastic chess player, "You mean the theory of games like chess." "No, no", he said. "Chess is not a game. Ches is a well-defined form of computation. You may not be able to work out the answers, but in theory there must be a solution, a right procedure to any position. Now real games," he said, "are not like that at all. Real life is not like that. Real life consists of bluffing, of little tactics of deception, of asking yourself what is the other man going to think I mean to do. And that is what games are about in my theory."

It can be argued that the goal of game theory is to predict how people will respond to situations. However, game theory requires us to make two assumptions about people: that they are greedy, and that they are suspicious. By 'greedy", I mean that a person will seek to get the most he or she can. By "suspicious", I mean that a person will assume that the other player is also greedy.

The strategy involved in a complicated strategy game like chess was not von Neumann's focus. In a theoretical sense, and perhaps you will see that, it is possible to put chess in von Neumann's model, but the game then becomes theoretically trivial.

We will call the two players the *row player*, and the *column player*. Let m be the number of moves available to the row player, and let k be the number of moves available to the column player.

The game can be captured by two $m \times k$ matrices, the *row-player payoff matrix* R and the *column-player payoff matrix* C. Here's how to interpret these:

> Suppose the row player chooses move i and the column player chooses move j. Then the row player ends up getting R_{ij} dollars and the column player ends up getting C_{ij} dollars.

It greatly simplifies matters to consider *zero-sum games*. A game is zero-sum if money that the row player earns comes from the column player, and vice versa. Many games are like this. It means, in particular, that in any given play of the game, one player wins and one player loses. (Well, ties are allowed.)

More formally, in a zero-sum game, for any moves i, j, the payoff to the row player is the negative of the payoff to the column player.

Under this restriction, one matrix suffices to capture the game. If I tell you, for example, the column-player payoff matrix is C, you can infer that the row-player payoff matrix is $-C$. Now you can see why such a game is called a zero-sum game: for any moves i, j, the sum of the payoffs to the two players is zero. From now on, we'll use A to denote the column-player payoff matrix.

	c	d
a	100	1
b	-2	-1

In this game, the row player has two possible moves, a and b, and the column player has two possible moves, c and d. The payoffs indicated are payoffs to the column player.

The column player might see the 100 and get greedy: "If I choose move c, I have a chance of getting 2." However, if the column player were to reason about his opponent, he would realize that he only gets the 100 if the row player chooses move a, which would not happen if the row player is greedy. There is no reason for the row player to choose move a, so the column player should assume the row player will choose move b. In this case, the possible outcomes for the column player are -2 and -1. The column player should choose move d to obtain a better (i.e. less worse) outcome.

Assuming the players play "rationally" (i.e. greedily and suspiciously), therefore, we can predict the outcome of the game. The moves will be b and d, and the payoff will be -1. We define the *value* of the game to be the payoff when both players play rationally. Note that in this game the value does not change if you replace the 100 with 1000 or a million.

The thinking of the column player can be understood as follows: "I want to choose my strategy to maximize my payoff under the most detrimental move available to my opponent. That is, assume the opponent can guess my strategy, and choose his best counterstrategy. I want to choose my strategy to make his counterstrategy least effective against me."

That is, the column player chooses according to

$$\max_{\text{column strategy } j} \left(\min_{\text{row strategy } i} A_{ij} \right)$$

This is called *maximin.*

The row player chooses similarly. The idea is the same but since the matrix A has payoffs to the column player, the row player prefers entries that are smaller. Thus the row player chooses according to

$$\min_{\text{row strategy } i} \left(\min_{\text{column strategy } j} A_{ij} \right)$$

This is called *minimax.*

Now consider paper-scissors-rock. Here is the payoff matrix A.

	paper	scissors	rock
paper	0	1	-1
scissors	-1	0	1
rock	1	-1	0

What would game theory predict would happen here? If there was a compelling reason for the row player to always choose paper, then the column player would know this reason, and therefore predict this choice, and therefore always choose scissors. The same holds for any move available to the row player.

The strategy of choosing one move and sticking to it is called a *pure* strategy. If the players are restricted to pure strategies, game theory has nothing to say in this situation.

Suppose the column player tries to employ maximin reasoning. "If I choose paper and my opponent knows it, he will choose scissors and will therefore win (payoff -1). If I choose scisssors.... If I choose rock..... No matter which pure strategy I choose, maximin predicts a payoff of -1."

Instead, game theory considers *mixed strategies.* A mixed strategy is a *probability distribution* of pure strategies.

We can adapt the maximin/minimax idea to mixed strategies. The column player considers each mixed strategy available to him. For each mixed strategy, he figures out his opponent's best counterstrategy. It suffices to consider the opponent's pure strategies.[4] He then chooses his strategy to make his opponent's counterstrategy least effective.

In this case, the payoff is not a single entry of the matrix, but the *expected value* under the random distribution. Thus the column player considers the following.

$$\max_{\text{mixed strategies } \boldsymbol{x}} \left(\min_{\text{move } i} (\text{expected payoff for } \boldsymbol{x} \text{ and } i) \right)$$

Here $\boldsymbol{x}$ is a probability distribution over column-player moves. That is, for $j = 1, \ldots, k$, entry j of $\boldsymbol{x}$ is the probability of selecting move j. By "expected payoff for $\boldsymbol{x}$ and i", I mean the expected payoff when the row player chooses move i and the column player chooses a move according to the probability distribution $\boldsymbol{x}$. Let $A_{i\cdot}$ be row i of A. Then the expected payoff is $A_{i\cdot} \cdot \boldsymbol{x}$. (If the column player chooses move j, the payoff is A_{ij}, and the probability of this choice is $\boldsymbol{x}[j]$, so the expectation is $\sum_j \boldsymbol{x}[j] A_{ij}$, which is $\boldsymbol{x} \cdot A_{i\cdot}$.

[4]Restricting the opponent's strategy to be pure in this context yields the same result as allowing it to be mixed.

The set of probability distributions over column-player moves is infinite, so the maximum is over an infinite set, but that need not bother us. We can characterize this set as $\{\boldsymbol{x} \in \mathbb{R}^k : \mathbf{1} \cdot \boldsymbol{x} = 1, \boldsymbol{x} \geq \mathbf{0}\}$ where $\mathbf{1}$ is the all-ones vector.

Consider the column-player mixed strategy in which each move is chosen with probability 1/3. Suppose the row player chooses *paper*. The payoffs for the corresponding row are 0, 1, and -1 (corresponding respectively to the column-player's choices of *paper*, *scissors*, *rock*. The expeccted payoff in this case is $\frac{1}{3} \cdot 0 + \frac{1}{3} \cdot 1 + \frac{1}{3} \cdot (-1)$, which is zero.

If the row player instead chooses *scissors*, the expected payoff is again zero. The same holds if the row player chooses *rock*. Thus the mixed strategy $\boldsymbol{x} = (1/3, 1/3, 1/3)$ results in a payoff of 0.

Now let $\boldsymbol{y}$ be the row-player's mixed strategy in which each move has probability 1/3. By the same kind of calculation, one can show that this leads to a payoff of 0.

We haven't yet proved it but $\boldsymbol{x}$ and $\boldsymbol{y}$ will turn out to be the best mixed strategy for the column player and the row player, respectively. Thus, for this game, we have the following property.

Property 1: Maximin and minimax predict the same expected payoff.

Property 1 shows that, assuming the row player and column player both analyze the game using minimax/maximin reasoning, they *agree* on the value of the game.

Property 2: Neither player gains any advantage by diverging from his strategy.

This shows that minimax reasoning is correct. If the row player sticks to her minimax strategy, the column player would be foolish to do anything other than his maximin strategy, because diverging would lead to a poorer expected payoff. If the column player sticks to his maximin strategy, the row player would be foolish to diverge.

Assuming the row player sticks to the equal-probabilities strategy, the column player can only do worse by choosing a different strategy, and vice versa. The pair of mixed strategies, one for the row player and one for the column player, is called an *equilibrium* because the greed of the players keeps the players using these strategies.

13.11 Formulation as a linear program

The most fundamental contribution of von Neumann to game theory is his *minimax theorem*: for *any* two-person, zero-sum game of complete information, minimax and maximin give the same value, and they form an equilibrium.

The theorem was proved over fifteen years before linear programming. However, in hindsight we can see that the minimax theorem is a simple consequence of linear programming duality.

The trick is to formulate minimax and maximin as linear programs.

The column player seeks to select a mixed strategy, i.e. a probability distribution among the moves 1 through k, i.e. a probability distribution is an assignment of nonnegative numbers to 1 through k such that the numbers sum to one. That is, he must select a k-vector $\boldsymbol{x}$ such that $\boldsymbol{x} \geq \mathbf{0}$ and $\mathbf{1} \cdot \boldsymbol{x} = 1$, where $\mathbf{1}$ denotes the all-ones vector.

Suppose λ is the expected payoff that the column player wants to guarantee. He can achieve that using strategy $\boldsymbol{x}$ if, for every pure strategy of the row player, the expected payoff is at least λ. Let $\boldsymbol{a}_1, \ldots, \boldsymbol{a}_m$ be the rows of the payoff matrix. Then the expected payoff when the row player chooses move i is $\boldsymbol{a}_i \cdot \boldsymbol{x}$. Thus the column player can achieve an expected payoff of at least λ if $\boldsymbol{a}_i \cdot \boldsymbol{x} \geq \lambda$ for $i = 1, \ldots, m$. Since the goal of the column player is to choose a mixed strategy $\boldsymbol{x}$ that maximizes the expected payoff he can achieve, we get the following linear program:

$$\max \lambda \ : \ \boldsymbol{x} \geq \mathbf{0}, \mathbf{1} \cdot \boldsymbol{x} = 1, \boldsymbol{a}_i \cdot \boldsymbol{x} \geq \lambda \text{ for } i = 1, \ldots, m$$

Let $\boldsymbol{b}_1, \ldots, \boldsymbol{b}_k$ be the columns of A. The analogous linear program for the row player is

$$\min \delta \ : \ \boldsymbol{y} \geq \mathbf{0}, \mathbf{1} \cdot \boldsymbol{y} = 1, \boldsymbol{b}_j \cdot \boldsymbol{y} \geq \lambda \text{ for } j = 1, \ldots, k$$

Finally, some simple math shows that these two linear programs are duals of each other, so they have the same value.[5] A little more math proves the equilibrium property as well.

[5]Using this fact, for the game we analyzed earlier, since we found a column-player mixed strategy $\boldsymbol{x}$ and a row-player mixed strategy $\boldsymbol{y}$ that each yielded the same expected payoff, weak duality shows that each is an optimal mixed strategy.

Game theory has been very influential in economics, and several Nobel prizes have been awarded for work in the area.

It has also (apparently) been influential in thinking about defense. The biggest "game" in the the US in post-WWII years was the nuclear standoff with the Soviet Union. The center of such thinking was the RAND corporation, which came up earlier when we discussed railroad interdiction. Some lines of a folk song from the 1960's, written by Malvena Reynolds:

> The RAND Corporation's the boon of the world.
> They think all day long for a fee.
> They sit and play games about going up in flames.
> For counters they use you and me....

13.12 Nonzero-sum games

Games that are not zero-sum are harder to analyze. The Prisoner's Dilemma is a now classic example. John Nash won the Nobel prize for formulating a concept now called *Nash equilibrium* and proving that every game had such an equilibrium.

13.13 *Lab: Learning through linear programming*

In this lab, we will re-examine the breast cancer data set with a new learning algorithm, one based on linear programming.

As in the previous machine-learning lab, the goal is to select a classifier. This time the classifier will be specified by a vector $\boldsymbol{w}$ and a scalar γ. The classifier is then as follows:

$$C(\boldsymbol{x}) = \begin{cases} \text{malignant} & \text{if } \boldsymbol{x} \cdot \boldsymbol{w} > \gamma \\ \text{benign} & \text{if } \boldsymbol{x} \cdot \boldsymbol{w} < \gamma \end{cases}$$

In an attempt to find the classifier that is most accurate, our goal is to select $\boldsymbol{w}$ and γ so that the classification is as correct as possible on the training data.

Recall that the training data consists of vectors $\boldsymbol{a}_1, \ldots, \boldsymbol{a}_m$ and scalars $d_1, \ldots, d_m$. For each patient ID i, the vector $\boldsymbol{a}_i$ specifies the features of the image for that patient, and d_i indicates whether the cells are malignant (+1) or benign (-1).

Suppose for a moment that there exists a classifer in our hypothesis class that performs perfectly on the training data. That is, there is a vector $\boldsymbol{w}$ and scalar γ such that

- $\boldsymbol{a}_i \cdot \boldsymbol{w} > \gamma$ if $d_i = +1$ and
- $\boldsymbol{a}_i \cdot \boldsymbol{w} < \gamma$ if $d_i = -1$.

So far we don't say by *how much* $\boldsymbol{a}_i \cdot \boldsymbol{w}$ must be greater than or less than γ. It should be by some nonzero amount. Say the difference is $\frac{1}{10}$. By multiplying $\boldsymbol{a}_i$ and γ by ten, we can in fact ensure that

- $\boldsymbol{a}_i \cdot \boldsymbol{w} \geq \gamma + 1$ if $d_i = +1$ and
- $\cdot \boldsymbol{a}_i \cdot \boldsymbol{w} \leq \gamma - 1$ if $d_i = -1$.

We could formulate the problem of finding such an $\boldsymbol{w}$ and γ in terms of linear inequalities. Consider γ and the entries of $\boldsymbol{w}$ as variables. For each patient i, we obtain a linear constraint, either

$$\boldsymbol{a}_i \cdot \boldsymbol{w} - \gamma \geq 1$$

if $d_i = +1$, or

$$\boldsymbol{a}_i \cdot \boldsymbol{w} - \gamma \leq -1$$

if $d_i = -1$.

Then any solution that obeys these linear inequalities would yield a classifier that performed perfectly on the training data.

Of course, this is generally too much to ask. We want to allow the classifier to make errors on the training data; we just want to minimize such errors. We therefore introduce a new "slop" variable z_i for each constraint to help the constraint be satisfied. If $d_i = +1$ then the new constraint is

$$\boldsymbol{a}_i \cdot \boldsymbol{w} + z_i \geq \gamma + 1 \tag{13.13}$$

and if $d_i = -1$ then the new constraint is

$$\boldsymbol{a}_i \cdot \boldsymbol{w} - z_i \leq \gamma - 1 \tag{13.14}$$

Since we want the errors to be small, we will have the linear program minimize $\sum_i z_i$. We also require that the slop variables be nonnegative:

$$z_i \geq 0 \tag{13.15}$$

Once we obtain an optimal solution to this linear program, we can extract the values for $\boldsymbol{w}$ and γ, and test out the classifier on the remaining data.

13.13.1 *Reading in the training data*

To read in data, use the procedure `read_training_data` in the module `cancer_data`, which takes two arguments, a string giving the pathname of a data file and a set D of features. It reads the data in the specified file and returns a pair $(A, \boldsymbol{b})$ where:

- A is a Mat whose row labels are patient identification numbers and whose column-label set is D
- $\boldsymbol{b}$ is a vector whose domain is the set of patient identification numbers, and $\boldsymbol{b}[r]$ is 1 if the specimen of patient r is malignant and is -1 if the specimen is benign.

For training, use the file `train.data`. Once you have selected a classifier, you can test it on the data in the file `validate.data`.

If the second argument is omitted, all features available will be used. The features available are:

'radius(x)', 'texture(x)', 'perimeter(x)', 'area(x)'

where x is `mean`, `stderr`, or `worst`.

Since the linear program implementation is rather slow, I recommend using a subset of the features, e.g. `{'area(worst)','smoothness(worst)', 'texture(mean)'}`.

13.13.2 *Setting up the linear program*

Since I will provide an implementation of simplex that solves a linear program, the main challenge is in setting up the linear program. You need to create a matrix A, a vector $\boldsymbol{b}$, and a vector $\boldsymbol{c}$ so that the desired linear program is $\min\{\boldsymbol{c} \cdot \boldsymbol{x} \, : \, A\boldsymbol{x} \geq \boldsymbol{b}\}$.

Column-labels of A

The column-labels of A and the domain of $\boldsymbol{c}$ will be the names of variables in our linear program. The variables are: the value γ, the entries of $\boldsymbol{w}$, and the slop variables z_i. We must choose a label for each variable. For γ, we use the label `'gamma'`.

The entries of $\boldsymbol{w}$ correspond to features. We therefore use the feature names as the labels, e.g. `'area(worst)'`.

Finally, we consider the slop variables z_i. Since there is one for each patient ID, we use the patient ID as the label.

Row-labels of A

The rows of A correspond to constraints. There are two kinds of constraints: the main constraints (Inequalities 13.13 and 13.14) and the nonnegativity constraints on the slop variables (Inequality 13.15).

- The label of a main constraint should be the corresponding patient ID.
- There is one slop variable z_i per patient ID i. It is convenient to use the negative $-i$ of the patient ID as the label of the corresponding nonnegativity constraint

13.13.3 *Main constraints*

To construct A, we will create a row-dictionary and then use `matutil.rowdict2mat`.

First we focus on the main constraints for malignant samples. Inequality 13.13 can be rewritten as

$$\boldsymbol{a}_i \cdot \boldsymbol{w} + z_i - \gamma \geq 1$$

which tells us that the corresponding row should be as follows:

- the coefficients of the features are the entries of $\boldsymbol{a}_i$,
- the coefficient of z_i should be one, and
- the coefficient of γ should be negative one.

The right-hand side of the inequality is 1, so we should set $b_i = 1$.

Next consider a constraint for benign samples. Multiplying both sides of Inequality 13.14 by -1 yields

$$-\boldsymbol{a}_i \cdot \boldsymbol{w} + z_i \geq 1 - \gamma$$

which can be rewritten as

$$-\boldsymbol{a}_i \cdot \boldsymbol{w} + z_i + \gamma \geq 1$$

which tells us that the corresponding row should be as follows:

- the coefficients of the features are the entries of $-\boldsymbol{a}_i$ (the negative of $\boldsymbol{a}_i$)
- the coefficient of z_i should be one, and
- the coefficient of γ should be one.

The right-hand side of the inequality is 1, so we should set $b_i = 1$.

Task 13.13.1: Write a procedure

`main_constraint(i, a_i, d_i, features)`

with the following spec:

- *input:* patient ID `i`, feature vector `a_i`, diagnosis `d_i` (+1 or -1), and the set `features`
- *output:* the vector $\boldsymbol{v}_i$ that should be row i of A.

Try out your procedure on some data. Check that the resulting vector $\boldsymbol{v}_i$ is correct:

- The entry for a feature label should be positive if `d_i` is +1, negative if `d_i` is -1.
- The entry for label i is 1.
- The entry for label `'gamma'` is negative if `d_i` is +1, positive if `d_i` is -1.

13.13.4 *Nonnegativity constraints*

For each patient ID i, there is a variable z_i whose label in our scheme is the integer i, and a constraint $z_i \geq 0$. The row of A corresponding to this constraint should have a one in the position of the column labeled i, and zeroes elsewhere.

13.13.5 *The matrix A*

Task 13.13.2: Write a procedure

`make_matrix(feature_vectors, diagnoses, features)`

with the following spec:

- *input:* a dictionary `feature_vectors` that maps patient IDs to feature vectors, a vector `diagnoses` that maps patient IDs to +1/-1, and a set `features` of feature labels
- *output:* the matrix A to be used in the linear program.

The rows of A labeled with positive integers (patient IDs) should be the vectors for the main constraints. Those labeled with negative integers (negatives of patient IDs) should be the vectors for the nonnegativity constraints.

13.13.6 *The right-hand side vector* **b**

The domain of $\boldsymbol{b}$ is the row-label set of A, which is: the patient IDs (the labels of the main constraints) and the negatives of the patient IDs (the labels of the nonnegativity constraints). The right-hand side for a main constraint is 1, and the right-hand side for a nonnegativity constraint is 0.

Task 13.13.3: Write a procedure that, given a set of patient IDs, returns the right-hand side vector $\boldsymbol{b}$.

13.13.7 *The objective function vector* **c**

The domain of the vector $\boldsymbol{c}$ is the same as the set of column labels of A. Our goal is to minimize the sum $\sum_i z_i$ of slop variables, and the label of a slop variable is the patient ID i, so $\boldsymbol{c}$ maps each patient ID i to 1 (and is zero elsewhere).

Task 13.13.4: Write a procedure that, given a set of patient IDs and feature labels, returns the objective function vector $\boldsymbol{c}$.

13.13.8 *Putting it together*

Task 13.13.5: Using the procedures you defined, construct the matrix A and the vectors $\boldsymbol{b}$ and $\boldsymbol{c}$.

13.13.9 *Finding a vertex*

Our simplex implementation requires, in addition to A, $\boldsymbol{b}$, and $\boldsymbol{c}$, the specification of a vertex of the polyhedron $\{\boldsymbol{x} \; : \; A\boldsymbol{x} \geq \boldsymbol{b}\}$. Finding a vertex involves solving a related (and somewhat larger) linear program. The module `simplex` defines a procedure `find_vertex(A,b,R_square)`.

Let n be the number of columns of A (i.e. the number of variables). Initialize `R_square` to be the set consisting of the patient IDs plus enough negatives of patient IDs to form a set of size n. The procedure `find_vertex(A,b,R_square` will mutate `R_square`. When the procedure terminates, if it is successful then `R_square` will be a set of row-labels of A defining a vertex. (The procedure returns `True` if it was successful.)

Most of the work of `find_vertex` is in running the simplex algorithm. The simplex algorithm can take several minutes (or many minutes, if many features are used). The procedure prints out the current iteration number and (in parentheses) the current value of the solution. For this application of the simplex algorithm, the value should go (nearly) to zero.

13.13.10 *Solving the linear program*

Once `find_vertex` terminates, `R_square` is ready to be used to solve the linear program. The module `simplex` contains a procedure `optimize(A,b,c,R_square)`. It returns the optimal solution $\hat{x}$ (unless the linear program value is unbounded). It also mutates `R_square`.

13.13.11 *Using the result*

Recall that the optimal LP solution $\hat{x}$ includes values for $\boldsymbol{w}$ and for γ.

Task 13.13.6: Define gamma to be the value of entry 'gamma' of the optimal LP solution. Define w to be the vector consisting of the feature entries of the optimal LP solution.

Define a classification procedure `C(feature_vector)` that returns +1 if `w*feature_vector > gamma` and -1 otherwise.

Task 13.13.7: Test your classification procedure on the training data. How many errors does it make?

Task 13.13.8: Load the validation data, and test your classification procedure on that. How many errors? (Using all the features, I got 17 errors. Your mileage may vary, especially since you will be using fewer features.)

13.14 Compressed sensing

A version of Gaussian elimination was discussed in a Chinese text two thousand years ago. Gibbs' lecture notes on vectors were printed in the 1880's. The method of orthogonalization was published by Gram in 1883 but dates back to the work of Laplace and Cauchy. A method for least squares was published by Legendre in 1806. The Fast Fourier Transform was developed by Cooley and Tukey in 1965 but apparently Gauss got there first, in 1805. The algorithm that is most used these days for computing the singular value decomposition as published by Golub and Kahan in 1965, but the singular value decomposition itself dates back at least to Sylvester's discovery in 1889. Much work has been done in wavelets recently, in the context of graphics and digital signal processing, but Haar developed his basis in 1909.

At last, at the end of this book, we briefly outline a computational idea that dates from the present century.

13.14.1 More quickly acquiring an MRI image

A *Wired* magazine article[6] gives us this story about a two-year old patient named Bryce: "[The pediatric radiologist] needed a phenomenally high-res scan, but if he was going to get it, his young patient would have to remain perfectly still. If Bryce took a single breath, the image would be blurred. That meant deepening the anesthesia enough to stop respiration. It would take a full two minutes for a standard MRI to capture the image, but if the anesthesiologists shut down Bryces breathing for that long, his glitchy liver would be the least of his problems."

Spoiler alert: Bryce was saved. Instead of using a traditional MRI scan, the radiologist used a shortened scan that required only forty seconds. The data acquired during that time, however, was not enough to precisely specify an image.

[6]February 22, 2010

If an image is an n-vector over $\mathbb{R}$, then acquiring the image requires that the sensors collect and report n numbers; otherwise, there are an infinitude of vectors consistent with the observations. Or so our understanding of dimension would lead us to believe.

That there must be a loophole is indicated by the fact that (as you have experienced) a typical image can be compressed. You computed the coordinate representation of an image in terms of an appropriate basis, and you zeroed out the small-magnitude entries of the coordinate representation. The result was a representation of the image that was perceptually close to the original. We interpret the effectiveness of this kind of compression as an indication that real images tend to be special, that when represented in terms of the right basis, they are sparse.

The idea of compressed sensing is that, if that is so, then why go to the trouble of acquiring all these numbers? Can we rather exploit the fact that real images are compressible to make fewer measurements and obtain the same image?

The same idea, of course, applies to other signals: audio, seismic data, and so on. In this brief, informal description, we will imagine that the signal is an image, and that the basis in which compression can be done is the two-dimensional Haar wavelet basis.

13.14.2 Computation to the rescue

There ain't no such thing as a free lunch. To get the same data from fewer observations, you have to pay... in computation. The process of obtaining an image (or whatever) from fewer numbers requires lots of computation. However, there are times when acquiring data is expensive—and computational power is just getting cheaper.

The ideas underlying the computation are familiar to readers of this book: change of basis and linear programming.

In compressed sensing (just as in image compression), there are two bases: the original basis in terms of which the image is acquired, and the basis in terms of which the image is expected to be sparse (e.g. the two-dimensional Haar wavelet basis). We assume the former is the standard basis. Let $\boldsymbol{q}_1, \ldots, \boldsymbol{q}_n$ be the latter basis, and let Q be the matrix with these vectors as columns.

Given a vector's coordinate representation $\boldsymbol{u}$ in terms of the Haar basis, the vector itself (i.e. its representation in terms of the standard basis) is $Q^T\boldsymbol{u}$, and, for $i = 1, \ldots, n$, entry i of the vector itself is $q_i^T\boldsymbol{u}$.

Let $\boldsymbol{w}$ be the true image, represented in terms of the standard basis. Suppose that some number k of numbers have been sensed. That is, some sensing device has recorded entries $i_1, \ldots, i_k$ of $\boldsymbol{w}$: $\boldsymbol{w}[i_1], \ldots, \boldsymbol{w}[i_k]$. The other entries remain unknown.

The goal of compressed sensing is to find the sparsest coordinate representation $\boldsymbol{u}$ that is consistent with the observations. We use $\boldsymbol{x}$ as a vector variable for the unknown coordinate representation. We state that $\boldsymbol{x}$ must be consistent with the observations using linear equations:

$$\begin{aligned} \boldsymbol{q}_{i_1}^T\boldsymbol{x} &= \boldsymbol{w}[i_1] \\ &\vdots \\ \boldsymbol{q}_{i_k}^T\boldsymbol{x} &= \boldsymbol{w}[i_k] \end{aligned}$$

The goal is to find the sparsest vector $\boldsymbol{x}$ that satisfies these equations. That is, find the vector $\boldsymbol{x}$ with the minimum number nonzero entries subject to satisfying the equations.

Unfortunately, no algorithm is known that will find the truly minimum number of nonzero entries subject to linear equalities, aside from very time-consuming algorithms such as trying all possible subsets of the entries. Fortunately, there is a surrogate that works quite well. Instead of minimizing the number of nonzero entries, minimize the sum of absolute values of $\boldsymbol{x}$.

The mathematics of when this works (it depends on the bases used and the number and distribution of observations) is beyond the scope of this text. But we can say something about how one goes about minimizing the sum of absolute values of $\boldsymbol{x}$.

Write $\boldsymbol{x} = [x_1, \ldots, x_n]$. Introduce new variables $z_1, \ldots, z_n$, and introduce linear inequalities, two per variable z_i:

$$\begin{aligned} z_1 \geq x_1, &\qquad z_1 \geq -x_1 \\ &\vdots \\ z_n \geq x_n, &\qquad z_n \geq -x_n \end{aligned}$$

Then use the simplex algorithm (or some other algorithm for linear programming) to minimize the sum $z_1 + \cdots + z_n$ subject to these inequalities and the linear equations outlined above.

For $i = 1, \ldots, k$, since z_i is required to be at least x_i and at least $-x_i$, it is at least $|z_i|$. On the other hand, there are no other constraints on z_i, so minimizing $z_1 + \cdots + z_n$ means that z_i will be exactly $|x_i|$ in any optimal solution.

13.14.3 Onwards

Mathematical and computational research on compressed sensing is ongoing. Applications are being explored in diverse areas, from astronomy (where the sensors are expensive and the images are way sparse) to medicine (faster MRI scans) to prospecting using seismic images (where similar techniques have been used for years). Many other applications remain to be discovered. Perhaps you will help discover them.

13.15 Review questions

- What is a linear program?
- How can a resource-allocation question be formulated as a linear program?
- In what different forms can linear programs appear?
- What is linear programming duality?
- What is the basis (heh) of the simplex algorithm?
- What is the connection between linear programming and zero-sum two-person games?

13.16 Problems

Linear program as polyhedron

Problem 13.16.1: A chocolate factory produces two types of candies: N&N's and Venus. The price for N&N's pack and Venus bar is \$1 and \$1.6, correspondingly. Each pack of N&N's is made of 50 grams of peanuts, 100 grams of chocolate and 50 grams of sugar. Each Venus bar contains 150 grams of chocolate, 50 grams of caramel and 30 grams of sugar. The factory has 1000 grams of chocolate, 300 grams of sugar, 200 grams of peanuts and 300 grams of caramel left. Your goal is to determine how many N&N's packs and Venus bars the company should produce to maximize its profit.

1. Give the linear program for the problem, using variables x_1 and x_2. Specify (a) the constraints, and (b) the objective function.
2. Graph the feasible region of your linear program.
3. Compute the profit at each vertex of the feasible region, and report the best solution.

Simplex steps

Problem 13.16.2: Write a procedure `find_move_helper(A, r)` for the following spec. You may use one or two modules in addition to `mat`.

- *input:* an $n \times n$ invertible matrix A over $\mathbb{R}$ and a row-label r,
- *output:* vector $\boldsymbol{w}$ such that entry r of $A\boldsymbol{w}$ is 1, and all other entries of $A\boldsymbol{w}$ are zero.

Test case: Let $A = \begin{bmatrix} 1 & 1 & 0 \\ 0 & 1 & 1 \\ 1 & 0 & 1 \end{bmatrix}$ be a matrix with row-label set and column-label set

$\{1,2,3\}$, and let $i = 3$. Then the output should be $[1/2, -1/2, 1/2]$.

Problem 13.16.3: Write a procedure `find_move_direction(A, x, r)` for the following spec.

- *input:*
  - an $n \times n$ invertible matrix A over $\mathbb{R}$,
  - a vector $\hat{\boldsymbol{x}}$ with the same column-label set as A
  - a row-label r
- *output:* a vector $\boldsymbol{w}$ such that, for every positive number δ, every entry of $A(\hat{\boldsymbol{x}} + \delta \boldsymbol{w})$ is equal to the corresponding entry of $A\hat{\boldsymbol{x}}$, except that entry r of $A(\hat{\boldsymbol{x}} + \delta \boldsymbol{w})$ is greater than entry r of $A\hat{\boldsymbol{x}}$.

Hint: Use the procedure `find_move_helper(A, r)` from Problem 13.16.2.

Test case: use the inputs from Problem 13.16.2 together with $\hat{\boldsymbol{x}} = [2, 4, 6]$.

Problem 13.16.4: Write a procedure `find_move(A, x, r)` for the following spec.

- *input:*
  - an $n \times n$ invertible matrix A over $\mathbb{R}$,
  - a positive vector $\hat{\boldsymbol{x}}$ whose label-set equals the column-label set of A
  - row-label r
- *output:* a positive scalar δ and vector $\boldsymbol{w}$ such that
  - every entry of $A(\hat{\boldsymbol{x}} + \delta \boldsymbol{w})$ is equal to the corresponding entry of $A\hat{\boldsymbol{x}}$, except that entry r of $A(\hat{\boldsymbol{x}} + \delta \boldsymbol{w})$ is greater than entry r of $A\hat{\boldsymbol{x}}$, and
  - $\hat{\boldsymbol{x}} + \delta \boldsymbol{w}$ is nonnegative, and
  - either $\boldsymbol{w}$ is a positive vector, or some entry of $\hat{\boldsymbol{x}} + \delta \boldsymbol{w}$ is zero.

Hint: Use `find_move_direction(A, x, r)` from Problem 13.16.3 to select $\boldsymbol{w}$. Once $\boldsymbol{w}$ is selected, select δ to make the last two properties true. Choose δ as large as possible such that $\hat{\boldsymbol{x}} + \delta \boldsymbol{w}$ is nonnegative. (Do the math before trying to write the code; calculate how large δ can be in terms of $\hat{\boldsymbol{x}}$ and $\boldsymbol{w}$.)

Test case: Use the example from Problem 13.16.3. The output should be the same vector $\boldsymbol{w}$ as before, and $\delta = 8$.

Using simplex

Problem 13.16.5: Use simplex algorithm to solve the linear program from Problem 13.16.1. Show the dual solution $\hat{\boldsymbol{y}}$ after each pivot step of the simplex algorithm. Show the direction of each $\hat{\boldsymbol{y}}$ in a copy of the graph of the feasible region.

Hint: Two of the constraints are $x_1 \geq 0$ and $x_2 \geq 0$. Use these two constraints to define your initial vertex.

Problem 13.16.6: An ice-cream van delivers ice cream to three neighboring cities A, B and C. It makes, on average, \$35, \$50 and \$55 of profit in cities A, B and C, respectively. Daily delivery expenses to cities A, B and C, are \$20, \$30 and \$35. Kids in city B will not buy ice cream more than four days a week. The expenses must not exceed \$195 per week.

Your goal is to find how many days a week the van should come to each city to maximize profit. Form a linear program, defining each of your variables, and use simplex to solve it.

Index

9531777R00302

Printed in Great Britain
by Amazon.co.uk, Ltd.,
Marston Gate.